Planting the Seeds of Hope

Purdue University Press Founders Series
Books by Frederick Whitford

Scattering the Seeds of Knowledge: The Words and Works of Indiana's Pioneer County Extension Agents

Enriching the Hoosier Farm Family: A Photo History of Indiana's Early County Extension Agents

The Grand Old Man of Purdue University and Indiana Agriculture: A Biography of William Carroll Latta

For the Good of the Farmer: A Biography of John Harrison Skinner, Dean of Purdue Agriculture

The Queen of American Agriculture: A Biography of Virginia Claypool Meredith

Memories of Life on the Farm Through the Lens of Pioneer Photographer J. C. Allen

Planting the Seeds of Hope

Indiana County Extension Agents During the Great Depression and World War II

Frederick Whitford

Purdue University Press
West Lafayette, Indiana
www.press.purdue.edu

Cataloging-in-Publication Data is on file at the Library of Congress.

978-1-61249-844-7 (hardback)
978-1-61249-845-4 (epdf)

Designer: Dawn L. Minns
Editor: Carolyn A. McGrew
Indexer: Marilyn Augst

All place names mentioned are in Indiana unless otherwise noted.

DEDICATION

Looking back over my thirty years as an Extension specialist makes me more than ever appreciate that I was able to spend a career with the men and women who farm and work in industry. Being able to put together a series of Extension publications along with 6,000 presentations allowed me a chance to interact with farmers, business leaders, applicators, government officials, environmentalists, gardeners, and others. Those accomplishments will fade in time as new educators come to the forefront and new people take over from those I taught during my career.

What I hope will last the test of time are the seven books written about the individuals and events that shaped the early years of Indiana agriculture, Purdue University, and the Cooperative Extension Service in the state. It's true that an author gets all of the credit, but any professional writer will tell you that in the background there are unsung heroes called editors who contribute greatly to how a book takes form from words on paper to a bound book.

I was lucky enough over the course of writing about Extension's history to have Carolyn McGrew in my corner as editor. I say lucky because her editorial skills made me appear on paper to be a better writer than I actually am. A true professional, she took the position that she was a reader and had to understand what was being said. I can honestly say she looked at every line and word, which was evident when I got her edits back on every chapter. Her ability to reduce bulk, find inconsistencies, and select the exact words increased the accuracy of each book and made them more enjoyable to read. Therefore, I would like to dedicate *Planting the Seeds of Hope* to Carolyn McGrew for her many years as my editor and friend.

So God Made a Farmer

And on the eighth day, God looked down on his planned paradise and said, "I need a caretaker."

So God made a farmer.

God said, "I need somebody willing to get up before dawn, milk cows, work all day in the fields, milk cows again, eat supper, then go to town and stay past midnight at a meeting of the school board."

So God made a farmer.

"I need somebody with arms strong enough to wrestle a calf and yet gentle enough to deliver his own grandchild. Somebody to call hogs, tame cantankerous machinery, come home hungry, have to wait for lunch until his wife's done feeding visiting ladies, then tell the ladies to be sure and come back real soon and mean it."

So God made a farmer.

God said, "I need somebody willing to sit up all night with a newborn colt and watch it die, then dry his eyes and say, 'Maybe next year.' I need somebody who can shape an ax handle from a persimmon sprout and shoe a horse with a hunk of car tire. Who can make harness out of hay wire, feed sacks, and shoe scraps. Who, planting time and harvest season, will finish his forty-hour week by Tuesday noon, and then, painin' from tractor back, put in another seventy-two hours."

So God made a farmer.

God had to have somebody willing to ride the ruts at double speed to get the hay in ahead of the rain clouds and yet stop in midfield and race to help when he sees the first smoke from a neighbor's place.

So God made a farmer.

God said, "I need somebody strong enough to clear trees and heave bales, yet gentle enough to yean lambs and wean pigs and tend the pink-combed pullets. Who'll stop his mower for an hour to splint the broken leg of a meadowlark. It had to be somebody who'd plow deep and straight and not cut corners. Somebody to seed, weed, feed, breed and rake and disc and plow and plant and tie the fleece and strain the milk and replenish the self-feeder and finish a hard week's work with a five-mile drive to church. Somebody who'd bale a family together with the soft, strong bonds of sharing. Who would laugh and then sigh and then reply with smiling eyes when his son says that he wants to spend his life doing what dad does."

So God made a farmer.

—PAUL HARVEY

The present extension staff neither deserves nor desires the "lion's share" of credit for whatever has been done in the past, is now being done, or will follow in the future. To all who in any way have assisted in this big program, we wish to express our sincere appreciation, and urge you to continue in the forward movement of Extension work in Indiana, or wherever you may chance to be living in the future. To those who have preceded us in this big extension field, we would like to say "you have built well," and to those with whom we labor today, and those upon whom the tasks of tomorrow will fall, may we say "yours is a most worthy cause—hold high the torch, and carry on."

—*Enos M. Rowe, Extension Agent in Clinton County, Indiana, 1939*

CONTENTS

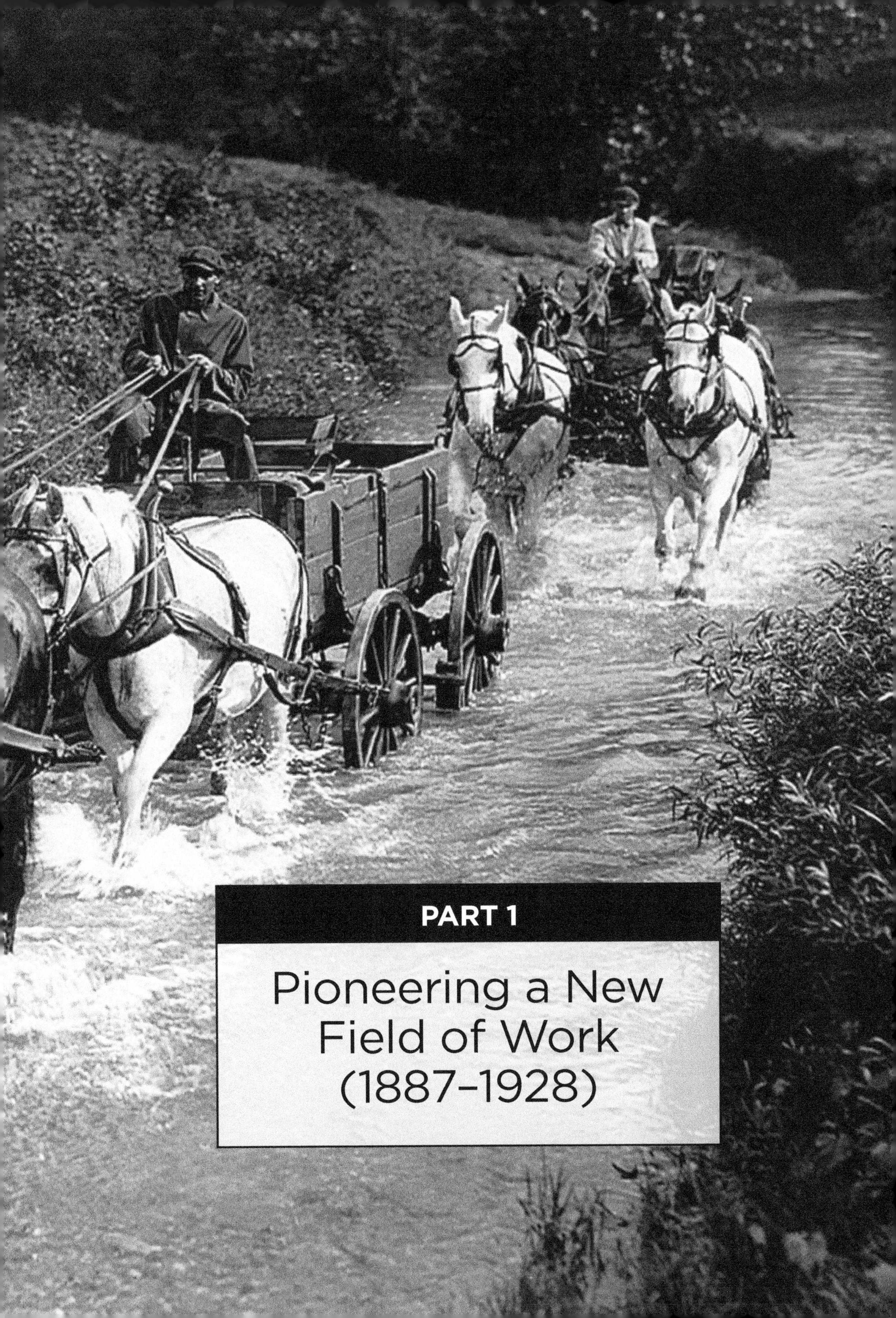

PART 1

Pioneering a New Field of Work (1887–1928)

CHAPTER 1

Neither the Agent nor His Farmer-Constituents Knew Very Much About What to Expect of One Another

In 1887, the United States Congress passed landmark legislation known as the Hatch Act, which supported agricultural research at land-grant universities throughout the country. In Indiana, Purdue University created the new Agricultural Experiment Station for the state in 1889 and set about conducting research in crops, soils, livestock, forestry, and the farm home designed to improve the lives of Hoosier families and communities throughout the state. However, it gradually became apparent that communicating the latest research findings to the people of the state was a perplexing problem.

Experiment station researchers and even the teaching staff at Purdue's School of Agriculture tried many different avenues to share information aimed at helping farmers improve their farms, finances, and home life. In the beginning, the communication came in the form of presentations at Farmers' Institutes, which were meetings for farm families held in counties throughout the state, and in the classrooms of the Winter Short Courses in Agriculture that farmers could attend on the Purdue campus. The scientists also published their findings in research bulletins and press releases to agricultural newspapers and magazines, and they invited the public to attend field days on the campus farms. They even provided information to boys' and girls' agricultural clubs—a precursor to the 4-H program—in hopes that the children would share new production practices with their parents.

While events and activities such as these attracted hundreds of people locally, there were still thousands more farmers who didn't receive information from Purdue. For some, leaving the farm and traveling any distance wasn't an option when there was work to be done, especially at a time when long hours were required to complete tasks by manual labor. For others, their isolation, distance from the program, lack of reliable transportation, and impassable roads during the winter months, when meetings were often held, prevented their participation.[1]

Even if traveling wasn't a problem, researchers still had to overcome the skepticism that farmers generally had for university people and their "book learning." In truth, most farmers were often unwilling to change their practices based on what they heard from campus researchers or read in published research bulletins.

In 1905—nearly two decades after the Indiana Agricultural Experiment Station at Purdue had received its initial funding—the Association of American Agricultural Colleges and Experiment Stations took up the debate about how best to communicate research findings to the agricultural community. Kenyon L. Butterfield, who at the time was president of the Rhode Island College of Agriculture and the Mechanic Arts—today known as the University of Rhode Island—led a committee to explore and report potential options back to association members for their consideration.

Butterfield's committee recommended to the association that "each agricultural college should organize a department of extension teaching in agriculture [to] coordinate with other departments of agricultural work with [a] competent director in charge."[2] By 1906, thirty-nine agricultural colleges across the U.S. would begin disseminating research to growers through extension workers, who would be housed on university campuses or, in some cases, in counties throughout the states they served.[3] The efforts of Butterfield and his committee would eventually be incorporated into the federal Smith-Lever Act of 1914, a seminal piece of legislation relating to agricultural extension work that changed how Purdue and other land-grant universities would interact with farmers, families, and communities throughout the country.

The work by Butterfield's committee was the impetus for creating one of the first extension positions at Purdue. In 1905, the Indiana General Assembly provided $25,000 to fund this position within the experiment station with the responsibility for delivering agricultural research results to growers statewide. George I. Christie was chosen to fill this newly created position. In retrospect, Purdue could not have picked a better person to launch the university's inaugural outreach program because Christie's credentials were impeccable. He grew up on a family farm and earned his undergraduate agriculture degree from Ontario Agricultural College. Then he ventured far from his Canadian roots to work on an advanced agronomy degree with a famous corn breeder named Perry Holden at Iowa State College of Agriculture and Mechanic Arts, which was later renamed as Iowa State University. Christie was a master public speaker who was knowledgeable about farming and who understood the dynamics and importance of county and community relations.

With an in-depth knowledge of corn production, it's not surprising that he focused his initial outreach efforts on this crop, especially since growers depended on it as a cash crop and source of livestock feed. His approach was to use every conceivable method to convey information, including making personal visits to farms, delivering lectures in front of farm groups, conducting corn variety trials across the state, converting technical information into easy-to-understand extension publications, and working with the press to get his views out to a much larger audience.

In December 1905, Christie experienced success with his first major outreach effort when he traveled across Indiana aboard a train popularly known as the Seed Corn Special. During a four-day period, this education train made many stops along its route, where he and other experts gave seminars on corn production, handed out publications, and provided educational displays for attendees. For many, this would be the first time that anyone from Purdue had visited their area. The success of this effort led Christie to launch the two-week Corn Gospel Train the following year.

His second major victory was in 1907, when he offered two- to three-day educational meetings around the state that were topic specific. The annual report for the experiment station indicated that these District Farmers' Short Courses, as they were called, "were organized and given with the idea of carrying to the farmer, information pertaining to the Experiment Station and agriculture in general, in a direct and effective manner. They also aimed to reach a class of people that was unable to attend the courses given at Purdue."[4] While others at Purdue had been involved with bringing the university to the people of the state, Christie had, for all intents and purposes, defined what agricultural extension outreach was and what it would become.

With his access to farmers through the media, at crop and livestock associations, aboard the education trains, and at district short courses, George Christie's name became synonymous with Purdue agriculture. His reputation would propel him to be elected as the secretary of the Indiana Corn Growers' Association before he turned thirty, solidifying his position as a person of importance to Indiana agriculture. With his ever-growing renown and popularity, Christie now had the attention of the movers and shakers of agriculture and, most notably, of the legislators in the Indiana General Assembly and U.S. Congress who were or had been farmers.

As Christie's star continued ascending, the state legislature rewarded his efforts in 1911 by providing additional funding for the continuation and expansion of his outreach program.[5] Christie hit the proverbial jackpot with this legislation, which created a Department of Agricultural Extension within the Agricultural Experiment Station at Purdue. To no one's surprise, Christie was named the first superintendent of the department, placing him in a position commensurate with the dean of the School of Agriculture and the director of the experiment station. Like the other agricultural administrators, he reported directly to the president of the university. His new department would focus strictly on agricultural education and encompass areas such as livestock production, dairying, grain crops, farm home management, and youth instruction. Christie's popularity, support from farmers, and political connections were instrumental in getting the state legislature to pass additional legislation in 1913 that provided funding for county-based agricultural agent positions throughout the state.[6]

The following year, President Woodrow Wilson signed the Smith-Lever Act, which directed the U.S. Department of Agriculture (USDA) to create the federal Cooperative Extension Service within the agency and provided matching funds for states to work with the USDA in creating state Extension programs. This legislation gave Christie's department federal funds to support his ever-growing outreach program for Indiana's farmers and rural communities.

In less than ten years, federal and state laws along with the funding they provided had become essential for shaping Purdue's interaction, intervention, and involvement with people and communities throughout the state. When the dust finally settled, Extension programming became a three-way partnership between the USDA, the state land-grant university, and the counties in the state.[7]

By the time Christie received federal support to place agricultural agents in counties throughout Indiana, he already had some experience with the hiring process. Between 1912 and 1913, he had worked to combine public and private funding to place agents in LaPorte, Montgomery, Parke, and St. Joseph Counties. This helped him understand the importance of finding candidates with the right mixture of talent, experience, and education as well as a good demeanor. He also knew the new hires would need a strong work ethic since they would be expected to work a minimum of sixty hours each week as a salaried employee, including a mandatory half-day in the office on Saturday to be available as farmers stopped by to ask questions.

His preference was to hire graduates from agriculture colleges who had working experience on a farm as a youth or adult.[8] He believed that agents with relevant experience and background knowledge could more easily establish themselves in the county and would be better able to explain to the state's agriculturists, homemakers, and youth how research results were relevant to the farm and home.[9]

Along with Thomas A. Coleman, who was second-in-command in charge of county agent work, Christie tried filling the open positions with well-qualified candidates. Given that these men would be the first in their counties to hold these positions, the success—or even possible failure—of the Extension experiment rested on their shoulders. Any mistakes could jeopardize the program in a county or across the state.[10]

From the onset, the hiring process was restricted by the limited pool of candidates. There were just too few qualified agriculture graduates for the number of positions available, not only in Indiana but across the nation.[11] Attracting someone to the position after graduation, when most had every intention of going back to the farm, required Christie to get them to change their career path. The already small pool of potential applicants shrank even further when the candidates did not meet Christie's and Coleman's standards.

After the hire was made, Christie faced a major obstacle regarding oversight for the position. While Purdue was responsible for identifying the appropriate candidate for the job, it was left to a local county board of education to accept or reject the person presented.[12] Once approved by the board, the new hiree would also have to be approved by the Purdue Board of Trustees, but this was little more than a formality.

The county board of education was made up of the superintendent of schools and a trustee from each of the county's townships. The first contract the county agent signed was for a one-year probation period. Then the board had to determine whether the second and subsequent contracts offered would be for one or two years. Most of the time, the agent was given a two-year contract. In a few cases, though, the agent's contract was reconsidered annually, offering little in the way of job security.[13]

Christie found it troublesome that the state legislature assigned the management of the county agent to the local board of education, while his Department of Agricultural Extension was responsible for providing the agent with training and some semblance of a support structure. Quite possibly, he had little choice in the matter when these positions were funded in 1913. In all likelihood, he needed support from the state's powerful superintendents of schools to get the bill passed, and they wanted the agents to spend part of their work hours providing agricultural instruction in the public schools.

In actuality, Christie and Purdue had little input into whom the local boards retained or removed. Agents were often used as pawns for gain by the political party in power. An agent might be removed simply based on political connections, donations, or leanings.[14] Removing a person not only meant the position would be vacant, but it also meant that the training and instruction Purdue had provided to the previous agent was lost.

It wasn't until 1937 that the Indiana General Assembly would change the law to make county agents employees of Purdue, giving Extension administrators the power to hire, supervise, and retain them.[15] This change came eleven years after Christie had left his position at Purdue to become the president of Ontario Agricultural College.

As the newly hired agents arrived to assume their posts, they stepped off the passenger trains onto the wooden platforms of the railroad depots or drove to the county seat knowing there would be no blueprints to follow, no shoes to fill, no existing programs to build on, and no reports to read. One agent observed that, in the early days, "there was no set precedent by which the agent could work."[16] They would have to be savvy if they were going to build a successful county program.

They would need to generate interest among the clientele and communities they served while setting up the framework for the Cooperative Extension Service within the county where none had existed before.[17] Some put a more positive spin on their challenge, with Greene County agent Arthur Haseman indicating that the first agents were "pioneer[ing] a new field of work. Not only was the work new to the community, but also to the new agent."[18]

Extension agent Chester V. Kimmell accurately summarized what agents were up against in the latter half of the 1910s when he wrote, "In the beginning of extension work in Allen County, the work was new; people did not know what it was all about and they reacted very slowly."[19] And Floyd County agent Charles M. East observed, "[N]either the Agent nor his farmer-constituents knew very much about what to expect of one another. That was the period when most farmers either had a healthy contempt for any young man's ability to help them or regarded the County Agent as one who did or was expected to know everything."[20]

One of the first problems to address was the preconceived notion that agents were nothing more than book farmers who could recite facts and figures but hadn't done the work themselves. Farmers knowingly chuckled that, with all of that knowledge, the agent could make a better living as a farmer than working for Extension. And it wasn't enough that agents were backed by agricultural science originating from Purdue researchers. Farmers as a group questioned that if these university experts knew everything about agriculture and how to farm, why didn't they turn the plows on their own farms?

Farmers' resistance to change was partially rooted in the challenges that were out of their control. Every farmer knew that earning a living from crops and livestock was a risky proposition, even in the best of times. No matter how good their plans were, they had little if any control over environmental and biological variables such as rain (or the lack of it), diseases, weeds, and insects. Acting on advice from someone they barely knew was a chance many were unwilling to take in case it led to poor outcomes. Many producers preferred to settle for what they knew they could produce on their farms rather than risk following an agent's advice down a path that, to the farmer, was unproven.

With widespread and overt negative opinions about the agent and the position, Benton County agent Ellis M. Christen noted, "Perhaps above all, the early county agents had primarily the problem of breaking down the resistance of accepting advice from [a] technically trained individual, and to get the farmer to feel that the purpose of the program was to cooperate with them rather than to dictate to them."[21] But the goal of the agents wasn't to dictate how farmers should farm; rather, they wanted to share improved methods that would increase farmers' production while lowering costs and the amount of effort required.[22]

As agents established the county Extension office, they were left to their own devices to figure out what educational programs were needed locally. To start the process, agents were instructed to meet with the supporters who had raised the required $500 to help fund the position as well as with politically connected individuals, farm leaders, and those who had the ear of the people. These local leaders often included progressive growers and highly respected individuals such as members of the clergy, bankers, and business owners scattered across the townships. For many agents, advice from these people became crucial to developing agricultural outreach by Extension in the county.

Agents broke the ice and got past the cold-shoulder treatment by personally visiting with these individuals, asking them for advice about what to work on and whom could be depended upon to help the agent with activities such as demonstrations. Clay County agent Raymond D. McHargue said, "During the early years of extension work the agents hardly knew what to do, and at this time they were making contacts with people, getting acquainted and attempting to determine some of the major problems."[23] Agents leaned heavily on the influential community members to speak up for them and the Extension program in front of friends and neighbors. These guardians of Extension would, over the decades, become instrumental in keeping the doors to the county office open and the position filled.

But making these contacts alone wasn't enough. The agents knew they also had to get out in the community and meet people where they were. This meant they had to be seen shaking hands at township meetings, churches, Farmers' Institutes, and any other place where people gathered. Being seen during the formative years was key to building some level of comfort with the people.

As time proceeded, Purdue told the agents that they needed to form a local Extension committee to help guide them in regards to educational programs. By choosing the right people, the agent and committee members could work together to develop a strategic plan that addressed the more urgent problems faced by farmers in the county. Often these problems differed across townships, so having influential members from various townships helped the agent customize a plan to meet local needs.

That local input into the process helped convince the people of the county that the agricultural agent was one of them rather than a Purdue person sent to tell them how they should farm. By placing well-respected growers on these advisory committees, agents could lean on them for support. People respected these farm leaders, and that respect began to rub off on the agents, at least to the point that it started to open more doors for them.

As agents looked to get a foothold in the county, one method they relied on was to provide one-on-one interactions. Decatur County agent Matthew M. Hamilton described those efforts, saying, "[I]t appears that the first agents, and they were among the first agents any place, were truly pioneers in their field. Their work was more of a person-to-person contact. That is, he had to locate a farmer who was having some difficulty and then try and help him. This was a very slow, tedious job but the early men here sold the work well enough that it grew quickly."[24]

With many communities more or less isolated, working with key farmers in a community helped spread goodwill faster and helped agents tackle biases held by those who had never met them. Otis Crane, who was named the first agent in Grant County in 1913, took a novel approach to winning the trust of one farmer in his community, as recounted in this 1939 report by Forest E. Condor, the agent in Grant County at the time:

> Mr. Crane tells of one farmer, who had threatened that he would not allow the county agent on his farm. So Mr. Crane called on him, saw that he was fixing a well and asked him for a drink of water, knowing that the farmer would tell him that he'd have to wait until they repaired the well.
>
> During the conversation that followed, the two strangers talked of the farmer's cattle, the county agent took a picture of the herd, they discussed fruit trees and other things of interest to both. By the time the well was repaired, they were fast becoming friends. Soon the farmer asked Mr. Crane his name. "I'm Otis Crane, the county agent," he replied. There was no answer from the farmer, but before Mr. Crane left, the farmer had given him a bushel of pears. The two gentlemen are still very good friends.[25]

Making farm visits was a struggle, considering the deplorable condition of roads around the state. One Crawford County report noted that it was common for the first agent there, Horace Copeland, "to start out in his car to a meeting, hang up in a mudhole, walk the remaining distance to the meeting place, stay all night with some farmer in the community and get his car pulled out and return to his office the following morning."[26]

For years, agents went door to door to meet growers on their home ground. They drove over rutted, so-called roads to reach a remote farmstead and provide assistance. While spending hours traveling over poorly constructed roads was viewed by some as time wasted, the agents understood the importance of these first visits. It gave them a chance to talk directly with the farmer, go out into the field to look at an underperforming or discolored stand of open-pollinated corn, or to examine a sick cow, pig, or sheep. And if the agent dispensed advice that worked, it could result in the farmer spreading the word on how much the agent had helped to remediate the problem, which had the potential to win over even more farmers.

These field visits did more to help the agents than they probably did the individual farmers. Whether the agent was observing field crops or handling an animal, he was gaining valuable, on-the-job training and acquiring needed experience. From these personal conversations, he could assess local attitudes, determine what farmers knew or didn't know, and formulate targeted educational programs.

As the number of farm visits accumulated, the agent was able to transfer that experience from one farm to another. With this newfound experience, prior college training, and updates from Purdue, the agent could start to apply what he knew to new situations. This hands-on learning laboratory made the agent more adaptable, since no two farms were alike. And adding this broad practical knowledge to their résumés helped overcome the bias that agents were only book smart and lacked real-world experience.

While agents faced overwhelming odds against making their positions relevant, there were several underlying factors that would prove pivotal for ensuring that the agents were more likely to succeed than fail. The first was that the agents had proven research readily available. For two decades, the Agricultural Experiment Station at Purdue had conducted field trials on the Purdue campus farm, outlying university farms, and cooperators' farms until the researchers were convinced that the results were repeatable.

Repeatable results meant that farmers who did what Extension asked them to do would see similar results. Agents had data to back up claims they made to their growers on preventing hog cholera; treating oat seed for smut; controlling scab on potatoes; selecting, storing, and testing seed from open-pollinated corn; repairing farm machinery; and learning what it took to grow soybeans.[27] Giving profitable advice meant more money in the grower's pocket, which in the long run would trump the distrust and doubt that farmers had about the validity of an agent's advice.

The second factor leading to the agents' success was that they sought out farmers who were highly regarded in the local community to back the Extension work. In 1939, Gibson County agent Roy Gwaltney explained how important it was to use the reputation of leading farmers across the county to open the doors to the farming community, writing:

> At the inception of County Agent Work, it was more or less like a one-ring circus, with the County Agent, the only performer. The majority of farmers accepted him with their tongues in their cheeks and waited for him to make a fool of himself. There were plenty of farmers who held to the belief that "book learning" was not only unnecessary, but even a hindrance to one on a farm. As a result of this feeling, all the County Agent could do was to find a man, who, with many doubts and misgivings, would allow him to put a demonstration plot on his farm. . . .

> After seeing that the County Agent was right in a majority of the cases, some of the farmers who were more audacious and progressive than others, were prevailed upon to lend their moral support, and finally actual assistance to the Extension program. It was by this process of evolution, that the first local leaders came into existence. From then, until the present time, the part of the local leader has become increasingly important as more and more responsibility is placed upon him. This is true in both the selection of projects and in carrying them through successfully. . . .
>
> As confidence and respect for the County Agent grew in the minds of the farm people, the necessity of him proving everything gradually lessened, and meetings, circular letters and cards were found to be effective in carrying out the programs. Newspapers, which stood aloof for a number of years, gradually opened their columns to the County Agent, and finding that their patrons liked this service, the majority of the papers became good cooperators in the dissemination of information.[28]

By using the goodwill and reputation of respected farmers, agents were able to build trust in themselves and the information they offered throughout the agricultural communities they served.

The third factor for the agents' success was that they used every conceivable method of education to break down the barriers that farmers and the community had placed in front of them. In addition to visiting farms personally in order to get to know area producers, agents also used many of the techniques that had been tried by agricultural researchers at Purdue. Agents spoke at Farmers' Institutes and regional short courses, where they could reach more farmers at once. They took farmers on trips to the Purdue campus farm to see the latest research results in person. And they even taught new agricultural practices to children enrolled in 4-H clubs, who then shared what they had learned with their families. The agents were more successful than the researchers had been with these approaches because the agents had become trusted members of their communities who had worked to develop a bond with people in their area.

One of the most effective educational approaches taken by agents was to conduct a demonstration at an area farm and invite local producers to attend a program there, listen to a presentation, and ask questions about what they saw.[29] These show-and-tell demonstrations covered major topics of the day, such as spraying orchards for pests, inoculating alfalfa with bacteria from soil, controlling hog cholera through sanitation and vaccination, testing soil for acidity and calculating liming needs, and examining different varieties of wheat, soybeans, and corn for quality and yield.[30]

A popular and effective demonstration conducted across the state showed producers how to determine whether a hen was laying enough eggs to justify the expense of feeding her. A group of growers would be shown a technique to identify which hens were laying eggs and which weren't. Then the group would go through a flock of 100 to 200 hens and use that technique to separate the layers from the nonlayers. When the nonlaying hens later produced very few eggs as compared to the laying birds, despite equal treatment and feed, anyone who was initially skeptical would become converts.[31] Demonstrations such as these with clear results went a long way toward building the credibility of the agent.

The last major factor for ensuring the agents' success was the support received from Extension specialists at Purdue. Agents would attend programs at the West Lafayette campus farms, where they would visit field plots and livestock pens to hear about the latest research and participate in hands-on training. As part of this training, they learned how to set up demonstrations and received educational materials, such as publications, pamphlets, posters, and examples of letters that they could send out to farmers.

When the agents returned home to share what they had learned in their communities, Extension specialists would join them there to participate in local demonstrations. Typically, the specialist would perform the first demonstration while the agent watched and learned. Then the agent would conduct the next demonstration, this time with the Purdue staff member watching and providing feedback. Afterward, agents were normally on their own to repeat the demonstration as often as was necessary.

Extension administrators understood that just because the handpicked agents had a degree in agriculture and farm experience did not mean they knew the ins and outs of the research or understood it well enough to pass it along. The agents could easily lose credibility and whatever goodwill they had earned if they tried to convey information when they didn't fully understand it or appreciate the magnitude of the work. Providing training at the campus farms and expert support for demonstrations back in the county increased the chances for the agents to succeed.

Over time, most agents would win over even those who doubted their worth or devalued their advice. In doing so, the successful agents had to give advice as an agricultural scientist, promote the Extension program, and recruit farm leaders to support the program and help with activities such as demonstrations. One agent summarized the effort by saying that the "big job was to sell the educated public that there was a need for a County Agent. . . . The Missionary work that preceded, paved the way for the agents to offer a program, and this program was gradually accepted."[32]

By the time the United States entered World War I in April 1917, the European landscape—including millions of acres of farmland—lay in ruins from nearly three years of fighting, and the loss of agricultural production threatened the possibility of large-scale famine across war-torn areas. The Allied forces looked to the U.S. to provide not only soldiers and equipment, but also extra wheat, meat, and dairy products to help feed the soldiers and people caught up the conflict. To meet the demand, the federal government reached out to the Cooperative Extension Service, calling on county agents and campus specialists to help farmers figure out how to meet USDA quotas for increasing wheat production by 25 percent and brood sows by 20 percent. While "Food Will Win the War" was a slogan used to rally farmers into action, it also represented the reality of the situation.

At the time, George Christie was still in the process of hiring county agents. In fact, the Extension program in Indiana was still so new that only one-third of its ninety-two counties had agents on the job by June 1916, and by June 1917, only forty-seven counties employed an agricultural agent. The situation was much the same across the country, with agents hired for only about half of the approximately 2,800 counties by the time the United States entered the war.[33]

With such an urgent need for Extension personnel, Congress responded by allocating money in 1918 to address the situation. The new funding allowed states to hire emergency agricultural agents to staff county offices, though it would take until the following year before Indiana would be fully staffed. The money was also used to hire on-campus Extension specialists in swine, sheep, dairy, poultry, crop diseases, and insects to work with the county agricultural agents, and it supported hiring fifty-two home demonstration agents to focus on helping families produce and conserve food at home so more would be available for export.[34]

From its inception, the Cooperative Extension Service had worked to convince farmers to adopt more economical means of production. But now the USDA shifted the focus, asking agents to push increased production no matter the cost.[35] As a result, the war would prove to be a baptism by fire for agents, who had only been in their positions for a few years or, worse yet for some, a few months. Would they be able to withstand the demands placed on their time by local, state, and federal bosses? Would they be able to convince farmers to grow enough additional food for the war effort? Those must have been the questions that agents asked themselves over and over again.

With the onset of the war, the agents' work grew exponentially as the federal, state, and local governments placed more responsibility on them. Consequently, their work during this time was often referred to as emergency work.[36] As the assignments piled up, agents found that there were not enough hours in the day to meet all the expectations placed upon them by state and federal agencies. It was a time when many came to realize that they could not personally oversee all aspects of their expanded workload, forcing them to lean heavily on volunteers to help them coordinate and conduct Extension programming and demonstrations.[37]

The scope of the emergency work included efforts such as finding laborers to plant and harvest crops in place of the men who went to war, participating in local draft board efforts to determine who would or wouldn't be exempted from service based on the need at home, and selling war bonds. But helping farmers increase production took up much of the agents' time. And given their short time on the job, the agents saw a lot of success in this area. Their achievements included:

- getting farmers to increase wheat acreage from 1.8 million acres in 1916 to 2.6 million acres in 1917 and emphasizing seed selection, seedbed preparation, the use of fertilizers, the control of specific foliar diseases, and conservation efforts during threshing to increase the wheat harvest from 33.5 million bushels in 1917 to 55 million bushels in 1918;
- promoting the federal government effort to get farmers to increase the number of brood sows, which led to the largest number of hogs in the state's history going to market in 1918;
- leading a campaign to build 10,000 new silos so more feed would be available to support larger livestock herds throughout the year; and
- completing 89,000 office calls, 13,000 meetings, 106,000 letters, 4,000 newspaper articles, 373,000 circular letters, 18,000 farm visits, and with the help of Extension specialists, 1,700 demonstrations and 77 automobile tours to visit area farmers who were using the latest agricultural science to raise crops and livestock.

In addition, the home demonstration agents achieved successes, too. Working alongside the agricultural agents, they had encouraged the creation of 500,000 gardens at vacant properties and in backyards, where crops such as cabbage, corn, potatoes, and tomatoes were grown for families. They also led efforts to preserve what was being grown, presenting more than 200 demonstrations that, according to the 1918 annual report for Extension, covered "canning, drying, salting, storing and other ways of saving vegetables and fruit."[38] In addition, families were shown how to make substitutions for meat, wheat flour, sugar, and fat in order to free up supplies for export.

After the war ended in victory for the United States and the Allied forces, the leaders of numerous nations expressed appreciation for U.S. support of the international effort. From the Extension perspective, agents had shown they were up to the tasks handed to them. If people held any reservations about the need for an agricultural agent and a local county Extension office, those misgivings should have been erased by watching the valuable services agents provided to their communities and country during World War I.

During the war years, American farmers had experienced an economic boom from high prices paid for commodities as people in war-torn areas looked to the United States for food. Extension had pushed farmers to meet or exceed war quotas for wheat, corn, hogs, and cattle, and in return, farmers had made excellent profits from what they grew or raised. They reinvested their profits into their farms, using them to further increase production, expand acreage, and buy more equipment. Unfortunately, these actions would prove to be a financial disaster for many producers in the postwar years.[39]

While farmers had readily taken up the challenge to produce more during the war, after it was over, the oversupply of raw products caused prices to plummet worldwide. As nations in war-torn areas began to restore agricultural production, they reduced their imports of grains and other agricultural products, causing prices to fall drastically as supply far exceeded demand.[40] U.S. agriculture spiraled into an economic depression that, for many farmers, would last throughout much of the 1920s. Many would be forced off the farm when low commodity prices did not cover the loans taken during the war to expand their operations.[41]

The farmers' changing fortunes produced some ill will in the agricultural community toward county agents. Most farmers didn't know anyone personally at the state or federal government, so the agent—the person who, at the behest of the federal government, had promoted increased production at any cost—now endured the brunt of the backlash. In the minds of farmers, the agent had helped create the surplus that had caused markets to crash, so agents took more than their fair share of the blame for the low prices.

The agents had to do an about-face and circle back to the economical production methods they had advocated prior to the war. As prices fell, farmers would need to become more efficient at producing crops and livestock, which meant producing more but at a lower cost.[42]

Clay County agent Raymond McHargue wrote, "After the war, less emphasis was needed upon total production, and . . . more emphasis was placed upon the economical production of quality products. Farmers were encouraged to study their occupation to a greater degree than ever before and make farming a business rather than just a job."[43] But switching directions was easier said than done. Given that a bushel of corn had dropped from $1.50 during the war to 52 cents afterward, farmers couldn't spend money that wasn't there, even if the outcome might be a higher profit.

Possibly the most important Extension outreach during this time centered on convincing farmers to keep detailed production records. Without these records, farmers were, at best, speculating on what they were earning from each of their agricultural activities. Agents continued to promote seed selection, fertilizer, lime, and other ways to improve crops along with the best diets for dairy and beef cattle, hogs, and poultry. But they also began offering programs and demonstrations on production practices, profit margins, and marketing, knowing that the most significant improvements to the farmers' returns would come when producers understood how much it cost to produce a bushel of grain, a pound of meat, a dozen eggs, or a gallon of milk, and how much actual profit their activities returned.

Still, as the agricultural depression wore on, agents continued to be scapegoats for many in the agricultural community. This led one agent to report there was a "hostile attitude on the part of some people toward extension work."[44]

Another part of the hostility directed toward Purdue Extension stemmed from farmers' expectation that the agent was one of them, which on the surface seemed like a good thing. But this led agricultural leaders to wrongly assume that the agents would use their positions to act politically in conjunction with farmers and force elected officials to do something about agricultural prices. However, Extension administrators were not willing to do this overtly. As a result, many farmers who had supported the mission and work of Extension felt irritated and betrayed by the perceived lack of reciprocal support from the organization.[45]

In hindsight, the deck was stacked against the Cooperative Extension Service at Purdue and the county agents to succeed during the formative years of the agency and during the trying times of World War I and the agricultural depression that followed. With so many obstacles facing the new education program, it is somewhat surprising that it not only took hold, but that it also began to flourish across Indiana.

Working through the skepticism and lack of public support of the early days of Extension, the challenges of wartime agricultural efforts, and the economic hardship faced by communities during the 1920s agricultural depression helped agents earn the respect of those they were hired to serve. As Gibson County agent Roy Gwaltney noted in 1939, "The attitude of the farmers gradually changed as the men on the job proved that they were neither impractical nor irresponsible. Farmers gradually learned that a little more knowledge of the mysterious operations taking place in the soil, plants and animals would be to their advantage."[46]

But the turbulent World War I years and the agricultural depression that followed resulted in approximately half of the agents leaving Extension service by 1922. And by 1925, another 40 percent had resigned their positions. The system had survived the trials, tribulations, and rigors of the two events, but it had been costly to those working as agents. What the Extension workers on the Purdue campus and the agents in the counties could not know as the 1920s drew to a close was that history was about to repeat itself. During the next two decades, Extension would be called upon to help the agricultural community weather the Great Depression and rise to the challenges brought about by World War II.

SWP
SWP

PART 2

Outlasting the Great Depression (1929–1939)

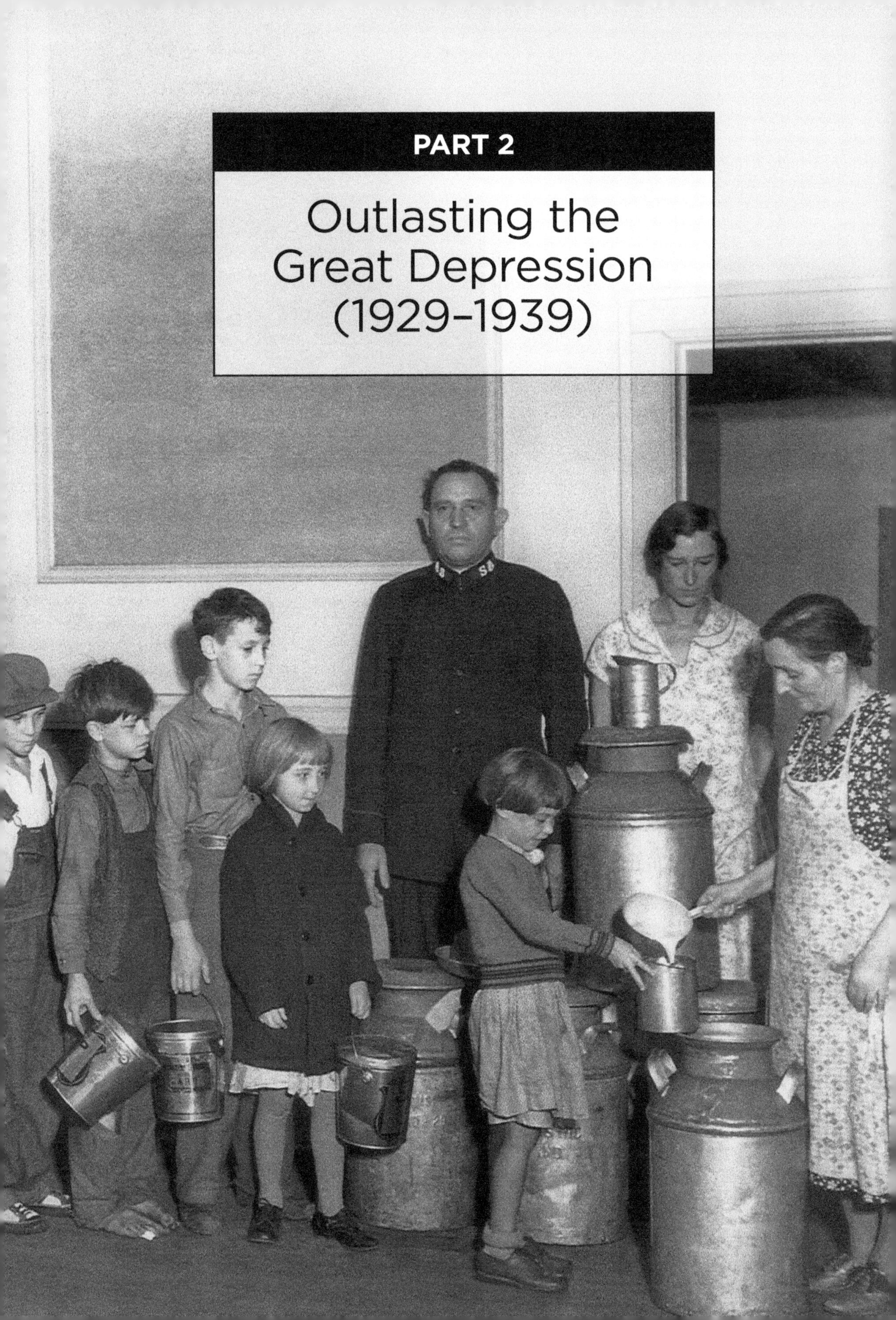

CHAPTER 2

The Shattering, Sledge Hammer Economic Blows of the Depression

During the 1920s, Extension had worked to develop programs to deal with the challenging economic times facing the agricultural community. With market prices barely covering production costs, county Extension agents had focused much attention on getting farmers to adopt efficient production methods, keep production records, and raise as much food for their own families as possible. But when the stock market crashed in 1929, its rippling effect crippled the United States economy, causing what would come to be known as the Great Depression. It brought down some of the wealthiest people in the country and put millions of people out of work.[1]

Unlike the 1920s depression, where farmers and the businesses that catered to agriculture faced tough economic conditions, the national economic meltdown of the Great Depression extended well beyond the farm family. People in towns also suffered from a lack of cash, a shortage of food, and high rates of unemployment, which prompted many men to leave their families in search of work.

As the Depression cast a long and dark shadow over the U.S. economy and people's lives, it became a continuation of the hard times that farmers had already endured during the 1920s. The majority of Indiana farmers who had managed to survive the agricultural depression following World War I were suddenly facing yet another economic catastrophe, when unreasonably high input costs and extremely low commodity prices brought misery to their doorsteps. In writing about that time, Gibson County agent Roy Gwaltney observed, "No one dreamed of a time when farmers could not obtain cost of production for their crops, or when the value of farm products would be at such a disparity with other products."[2]

It was also a hardship on the people of rural Indiana, who depended on a strong agricultural economy to support their local businesses and jobs. And as that economy faltered, so did the tax revenues that supported government spending and debt payments.[3] For everyone, including the county agents, months of suffering—described by one agent as "the shattering, sledge hammer economic blows of the depression"—dragged into years without any relief in sight.[4]

Each bad month was one more in which everyone—farmers, townspeople, businesses, and government entities—would have to tighten their budget belts even further. From the farmers' perspective, expenses had to be greatly curtailed when what they grew had little value in the marketplace.[5] Cash was sometimes in such short supply that farmers had to rely on their horses for field work while their tractors that required oil and gas sat idle in the barn. For farmers and others who did not already own their homes outright, paying their mortgages and seeing to the needs of their families would, out of necessity, take first priority.

Farmers, in general, and Extension, in particular, were better prepared to deal with the issues that this second depression in two decades brought. For one thing, eighty-five of Indiana's ninety-two counties already had mainly experienced agents in place to help guide people through these new economic challenges. Still, agents would find, as Henry County agent William G. Smith noted, that "an abundance of patience, ability for hard work, long hours, tact and persistence are going to be the demand."[6]

One-third of the agents' time would now be spent on government emergency programs, with Depression-era New Deal programs such as the corn-hog and wheat reduction programs significantly adding to their workload.[7] For example, in 1934 agents helped more than 80,000 farmers sign corn-hog reduction contracts with the federal government that paid growers in excess of $20 million.[8] Agents also kept busy helping farmers get crop loans and assisting the rural electric cooperatives with securing enough customers to warrant extending electrical lines into less-populated areas, especially to farms. And as urban and rural families increasingly looked to charities for assistance with each passing year, agents worked with local organizations, such as the American Red Cross and township trustees, to help people regardless of whether they lived in town or on a farm.

From its earliest days, Extension had sought to position itself as a primary source of information and assistance, and the Depression had helped make that a reality.[9] In 1935, the statewide annual report for Extension would record 4,000 county meetings hosting 250,000 farmers, 132,000 office calls to meet with agents, 3,400 news stories written for local papers, and 5,300 farm visits.[10]

Extension workers could not know at the onset of the Depression in 1929 how bad things would get before they got better, but throughout it all, the county agents along with the campus staff would put forth tremendous effort and use as many resources as possible to help people weather the difficult economic climate.

As the economy threw more people out of work and farm commodity prices tumbled downward, it must have occurred to the Extension staff that they had to rise to the occasion of helping so many farm families and others despite having relatively few workers with which to address this challenge. Since Indiana was a farming state, it meant thousands of farmers required assistance with obtaining crop and mortgage loans, enrolling in the government corn-hog and wheat reduction programs, finding employment off the farm, and working with relief organizations.[11] As William Smith in Henry County observed in 1934, it would take an all-out effort to "help both the individual and the groups interpret their own situation in the light of changing conditions, that they might orient themselves as intelligently as possible."[12]

With only so many hours and days in a week and with so much need, Extension staff had to make every second count in order to give their attention to multiple competing demands. This led Dearborn County agent Cory A. Alcorn to observe, "There has been little time for but very little individual work, and in fact not much was attempted. . . ."[13] Indeed, in the early days of Extension, agents had often sought to establish themselves in their communities by visiting area farms in person, but this approach was now considered more of a luxury. Instead, out of necessity, they increasingly began to rely on volunteers to help carry out the mission of Extension.

Without the volunteers, the agents would only have accomplished a miniscule portion of what needed to be done during the dark years of the Depression.[14] For example, in 1933, agent Noah S. Hadley reported that "266 local leaders and permanent committee members have formulated and conducted the extension program of Franklin County. In addition to this number, more than 300 other committee workers have helped with such activities as the 4-H Fair, Farm Bureau Picnic, Husking Contest and Achievement Day."[15] A few years later in Allen County, agent Chester V. Kimmell wrote, "Three hundred thirteen volunteer leaders consisting of 203 men, 70 women, 12 older club boys and 28 older club girls assisted in planning and executing the 1937 agricultural extension program."[16]

Managing an army of volunteers was a full-time job by itself, but some agents found a positive side to getting so many more people involved in Extension work.[17] In 1939, Russell L. Zell, who succeeded Hadley as the Franklin County agent, noted, "Voluntary local leaders have been a major cog in the Extension Program from its beginning. These leaders have become more important in recent years because of the diversity of work guided by the Agent. Their work has been more helpful because, in the main, they are more familiar with special projects with which they assist."[18] Delaware County agent Myron E. Cromer had made a similar observation in 1933, writing, "The Extension Program is built and carried on largely by interested groups within the project" and noted that this approach had resulted in "a more wholesome spirit of co-operation and interest" by the volunteers.[19]

One of the areas where volunteers made significant contributions to support Extension work was through their involvement in demonstrations. During the 1930s, most operations were diversified, growing a variety of cash crops and raising crops to feed different species of livestock.[20] Corn, wheat, hogs, poultry, and dairy dominated agriculture in Indiana at the time, but some farm families also depended upon tomatoes, sweet corn, onions, potatoes, cantaloupes, watermelons, apples, peaches, tobacco, and mint as well as sheep and horses.

This diversity of what was being grown and sold meant individual agents would arrange dozens of demonstrations each year with the help of volunteers so farmers could visit a number of events with useful information for their own operations. For example, in 1934 Extension conducted 9,224 demonstrations statewide that attracted approximately 250,000 people.[21] Some agents credited this realignment with allowing Extension to move from the few to the many, with Fountain County agent Alfred Hesler observing in 1939, "It was not until then that [the] extension service became the motivating factor in country life."[22]

Of course, there was a major caveat associated with the Depression-era advice provided by agents through demonstrations and any other means. Any recommendation that required an outlay of cash would more often than not be discounted or ignored. Improvements needed to be something that farmers could do on their own farms without having to incur much debt in the process, even when agents showed that each dollar spent on a specific project returned more than the initial investment.[23] It just didn't matter what the economics showed if the farmer didn't have the dollars to spend on implementing a change.[24]

Agents looked to the volunteers to help set up the demonstrations and follow through to see that they were being properly managed.[25] For example, project committees assisted with identifying "the limiting factors or the factors that were keeping our farmers from making more money on their farms," as one agent noted, which was important for making sure the demonstrations would focus on strategies of greatest interest to farmers.[26] And local farm bureaus and their members worked side-by-side with the agents to host the demonstrations, whether it was doing varietal trials with corn hybrids or demonstrating how to deworm sheep.[27] Both the agents and the committee members helped to publicize the day's events and worked with the cooperating farmers to showcase the results.[28]

In 1937, a report by Dubois County agent Charles A. Nicholson showed why Extension placed such importance on offering so many demonstrations. He wrote:

> Any extension program to make progress, therefore, must be tested and tried before this group of folks will accept it as sound and true. They won't take for granted that figures and information are sufficient reason to adopt certain practices and projects. They have to see someone else try out any new or unapproved practice; no one wants to start. So this reserved attitude makes a program hard to get under way, but once sound and accepted, they will carry on. . . .
>
> In summarizing the attitude of our people toward extension work, we might make this statement, "It's alright but I'll wait and see how the other fellow gets along first."[29]

Obviously, it was hoped that attendees would repeat on their own farms what they had observed at the demonstrations or share with others what they had learned. As Benton County agent Ellis Christen noted in his 1938 annual report:

> The desired goal in any project is the full completion of a planned program and the distribution of such results to the people in such a manner that they may make use of it. While it is true that results may be carried to people in many ways, results in the county show that the most effective way is for an individual to fully realize the value of the practice to the extent that his interest is sufficient that he will do it himself. Immediately some neighbor observes this practice, even though they have never seen and perhaps never heard of the Extension Project. They start to do the same thing, never associating it with the County Agent, nor realizing the county Extension Program has reached them.[30]

During the 1930s, county Extension agents were required to tackle the Depression by offering anything that might offer recovery and hope. What tied farmers and nonfarmers together was the obvious need for income during those lean years. Townspeople were gainfully employed one day and the next found that their employers had downsized or, worse yet, shuttered their operations. Farmers were in better shape than many in the towns as they had already been through the agricultural depression of the 1920s and clearly understood what they had to do to survive such devasting economic conditions.

While the immediate need was to help people do whatever was feasible to generate more income and reduce expenses, Extension agents saw themselves in a much larger context, which one agent described by saying that "the fundamental of Extension Education is the development of rural people themselves and that the ultimate objectives of Extension Work are to promote the physical, mental, spiritual and social growth of the individual farmer, his wife, and his children."[31] Under the circumstances, though, agents became advisors to all members of their communities—both families and individuals in towns and rural areas alike.

As the demands on their time ramped up, agents were able to keep the important work of the county Extension program going by leaning more heavily on volunteers. In 1939, Greene County agent Arthur Haseman, wrote, "[T]he present, then, finds much of the responsibility shifted onto the shoulders of committees and organizations, with the agents in the key position as advisors and coordinators, instead of working individually in fostering the extension activities. Individuality has been supplanted by group action and self-help by the people themselves."[32] This approach, which arose from an effort to help as many people as possible during a time when agents were pulled in many directions, became an integral part of Extension outreach that still continues today.

CHAPTER 3

Building Support Through Advisory Boards

As county Extension agents took up their new positions, they had to answer to the local boards of education regarding how they spent their time, so it was important to address the priorities and concerns of the communities they served. In order to do this, agents set up county-based Extension advisory boards, which allowed them to get input on what community members wanted while garnering support for the new Extension program locally. Typically, members of these boards were farmers and others in the community who had been handpicked by the agent, with approximately ten to twelve appointed to serve.[1]

While one agent noted that the "policy has been to co-operate with any organizations that were willing to sponsor some of the Agricultural Extension program projects," in truth, the agents tended to choose board members from agricultural organizations such as the county farm bureau, fruit growers association, corn growers association, Jersey cattle club, and poultry association.[2] Members of these organizations were leaders in their respective areas, so they were able to advise the agents on what needed to be done, but more importantly, they were willing to help do the work, such as putting on demonstrations on their farms or on the farms of others they knew the agent could depend on.

However, it soon became clear that the county farm bureaus, which had begun organizing in 1919, had become the main voice directing what programs and projects the Extension agent would pursue in any given year.[3] In 1930, Franklin County agent Ernest F. Brown's comments seemed to express a view common among his colleagues: "Whole hearted support of the active farm bureau [township] units of the county is largely responsible for the progress shown in developing of [the] agricultural extension program in the county. The officers and members of the organization have helped to work out the program and have supplied leaders to carry out plans."[4]

As the farm bureau system was getting established in Indiana during the late 1910s and the 1920s, agents had played a large role—leading the way in recruitment, writing newsletters, and writing the articles of incorporation—and many became heavily involved with these local groups while receiving support from the groups' members in return. Farm bureau members planned, promoted, and participated in the Extension programs, and agents counted on members to volunteer and otherwise help carry out Extension work in the counties.[5] In Fulton County, agent Harry R. Rosenbury reported that the farm bureaus' leadership and members "have considered it their [Extension] program and have provided the leadership in each of the townships who have been largely responsible for the results obtained [at demonstrations put on by Extension]."[6]

During the tough economic times of the Depression, agents also depended on influential farm bureau members to speak up when local Extension programs were facing budget cuts. As county councils around the state sought to save tax dollars by eliminating agents' positions, farm bureau members exerted their political clout and were often successful at providing some level of protection for the agents (see Chapter 4).

In 1934, agent Charles Nicholson in Dubois County illustrated the importance of the local farm bureau in his annual report, writing, "In order for the extension program to firmly establish itself, it must have some local support. The Farm Bureau, who has been the foster-father of all extension activities in this county, is still heading the parade."[7] While others would be included in the "parade," it was clear that the county farm bureaus were out in front leading the procession while other groups lagged far behind.

It's no wonder then that many citizens not involved in farming or farmers not affiliated with the county farm bureaus perceived their county agents as pseudo employees of these groups since the agents took their marching orders primarily from this one organization. This must have been uncomfortable for Purdue administrators and even lawmakers as they realized Extension agents were relying too much on the wants and desires of one group and minimizing the needs of others that were also paying taxes to support the county Extension program.

It seemed that agents were aware of the criticism and tried to address the matter. In Raymond McHargue's annual report for Clay County in 1937, he expressed his concern about working too closely with one particular group, noting, "It has been the general policy in the past to cooperate with all county organizations that have worthwhile motives in mind. However, contact with any one organization has never become so close that criticism might be forthcoming from other organizations or business men."[8] Adams County agent Lawrence Archbold had expressed a similar sentiment in 1930, writing, "Practically the only thing that all the people of the county have in common is the payment of taxes. They are all tax payers. The present county agent's policy is that he is working for the tax payers—that he will do all he can for them, but not to do anything for one group that he cannot do for any other."[9]

Similar language was prevalent in reports from agents around the state, but in reality, it wasn't totally accurate since many agents worked as closely as possible with their county farm bureaus. Instead, it was more likely just rhetoric aimed at appeasing those who were criticizing the agents' closeness to the organization.

Still, some agents reported significant changes in this situation, as Greene County agent Virgil Sexson noted in 1930.

> The most important change in the use of cooperating organizations during the year has been a greater degree of cooperation on the part of the organizations listed outside of the Farm Bureau and less cooperation on the part of the last named organization. This change was due to a rather depressed Farm Bureau morale due to internal troubles and also to a keener appreciation on the part of the other organizations of the value of the extension program and a genuine desire to assist in its development.[10]

In 1937, the Indiana General Assembly passed legislation to transform how Extension would operate. The first major change was to give the director of the Cooperative Extension Service at Purdue complete control for the hiring, training, evaluation, and removal of county Extension agents. This simplified the complex and strained relationship between Purdue and the county boards of education, which had previously shared supervisory responsibilities for the agents.

But a second change required that agents redo their county advisory boards by July 1, 1937, to be inclusive of organizations beyond just farm bureaus and other agricultural associations.[11] In fact, the law mandated that civic and urban groups be represented on these local boards.[12] From a political standpoint, this was a wise move, considering it was often such groups that had been left out of Extension circles and that had tried shutting down Extension programs during the early 1930s.

The resulting advisory boards were typically rather large, ranging from fifteen to twenty-one members.[13] Members of county farm bureaus were still allowed to participate, but they were to be joined by members of organizations such as local rural youth clubs, county fair boards, home economics clubs, granges, Farmers' Institutes, Agricultural Conservation Associations, Dairy Herd Improvement Associations, the Purdue Agricultural Alumni Association, Rotary Clubs, and Kiwanis as well as by individuals such as the superintendent of schools, local bankers, county commissioners, and county council members.[14]

While the number of slots was mostly left to the discretion of the agents, there was one important caveat. Each organization would select the person who would represent it on the board rather than letting the agent choose someone.[15] Agent Cory Alcorn addressed how the 1937 law affected Extension in Dearborn County, writing, "Heretofore our several committees were very largely selected by us, but since July we have asked each organization and group of interested people to select their own member representative to serve on the county committee. This committee has outlined the general phases of extension work and directed the officers to appoint committees for each line of activity."[16]

Soon, the local Extension advisory board began playing a much larger role in how the Extension office was financed by the county.[17] In Bartholomew County, the agent noted, "The entire group [advisory board] made out the budget for the county agent in 1938 and recommended this budget to the county council."[18] It was understood that having broader local support through the advisory board gave the agent a better chance to get the Extension budget approved by the county council.

These advisory boards were also expected to come up with a larger outline of programming for county Extension work.[19] Often, the agent would lead the discussion, becoming the spokesperson for what needed to be done and which projects would have the greatest impact on the local communities. This required the agent to study the local agricultural community and become well versed in how best to support and improve it.[20] It was the agent who collected, summarized, and explained to the advisory board statistical information gleaned from county, state, and federal sources, supplementing the data with his own experiences and recommendations from his closest advisors.[21]

Planning an entire year's program for the local agricultural community became the norm instead of the exception.[22] In some counties, agents held meetings throughout the local townships in order to present the ideas developed by the local advisory board and get input from community members before finalizing programming.[23] In 1938, agent Harold J. Yarling reported that the Hendricks County Extension Advisory Committee "after going over statistical information of the county and much discussion concerning their knowledge of the county's agriculture suggested that committees be appointed for Dairy, Crop Improvement, Beef Cattle, Sheep, and 4-H Club projects. This was done and all committees worked out a program to be followed taking the long-time attitude on their problems." To address crop improvement, Yarling was asked to work on "[a] hybrid corn variety demonstration, Five-Acre corn contest, Soybean variety demonstration, Wheat variety demonstration, Seed Wheat cleaning, Pasture Improvement, 4-H Corn Club, and Corn Fertilizer Demonstration."[24]

Agents benefited from expanding who served on the Extension advisory boards. The larger group of community leaders speaking on behalf of the county Extension program provided important political backing when working with local politicians and community leaders. And agents such as Albert M. Bishea in Daviess County reported in 1937 that this change had started paying dividends almost immediately.

> The coordination of the Agricultural Extension Service, through this advisory committee, with the other organizations of the county has not only helped to develop a more sound program but has brought about a better understanding

> of the work of this office and created a more friendly attitude and willing support. This is evidenced by better financial support of the office and of the various projects carried on, as well as more interest by both farmers and related groups in the extension activities.[25]

While agents took seriously the direction given by their advisory boards, their attention would necessarily be redirected to dealing with individual problems and emergencies as well as addressing the constant stream of people and phone calls coming into their offices. And as the Depression continued to drag on, it added a lot of work to the agents' duties far beyond the plan of work outlined by advisory boards. This fact was well illustrated by Benton County agent Ellis Christen, who in 1939 wrote the following paragraph in his annual report:

> From the standpoint of our Extension Program, these foregoing programs listed under Public Problems are definitely the problems in program planning. The time needed to satisfactorily do the work which comes up during the year is so indefinite at the time the general extension program is planned that proper allowance for them is not made. Therefore, there comes up urgent requests with very short notice. The result is that the planned program is oftentimes neglected in order to care for some current, urgent request in connection with these various projects. We do not know how to remedy it, but will welcome any suggestions.[26]

As the county Extension office had become a local clearinghouse for information, agents such as Christen strived "to keep the program of such size and within those limits where it can be satisfactorily completed, rather than permitting it to become so large and efforts so scattered that there is not a satisfactory completion of the project involved."[27] Agents were left to their own devices on how to best utilize their time, given the demands for attention by those in their communities.

CHAPTER 4

Does the County Agent Do Anybody or Group of Farmers Any Good or Justify His Expense?

As the Great Depression took hold, farmers were more than a little upset that while their crops were bringing less per bushel, the cost of the supplies purchased for the farm remained too high.[1] They were forced to cut purchases for items such as fertilizers, lime, fuel, and other supplies because they didn't have the cash or, for some, could not get bank loans. Dearborn County agent Clyde Madison observed in 1932, "[O]ur people are rather conservative minded and are making an effort to adjust their farm business to present conditions. This is a difficult problem considering high taxes and high prices for machinery and many other things that the farmer must buy."[2]

As the Depression worsened, tax revenues fell off dramatically, so government officials looked at cutting expenditures, primarily by reducing payrolls. They also sought to raise property taxes to replace other lost revenue, which obviously did not sit well with farmers who were already short of cash.[3] For example, in 1931 Benton County raised the average tax per farm from $478 to $560 to make up for other deficits in county funding.[4]

County farm bureaus across Indiana became formidable opponents to any proposals to raise property taxes, but taxes became a single issue that united farmers, whether they were farm bureau members or not. In 1931, Harrison County agent Ronald Johnson reported that local farmers thought their taxes "had risen to an exorbitant rate, and farmers and others have become very much concerned in an effort to discover some means of lowering taxes."[5]

In 1932, Indiana Farm Bureau requested that the Indiana General Assembly call a special session to deal with the tax issue.[6] In meetings with the state tax board, farm bureau leaders maintained that farmland was selling for much less than its valuation. This led the legislators to pass a 20-percent reduction in assessed valuations, which helped lower taxes.[7]

The very next year, the Indiana State Board of Tax Commissioners ordered a 10-percent increase in assessed valuations for real estate outside of incorporated cities and towns—in other words, an increase that would largely be shouldered by farmers. But once again, the farm community and farm bureau leaders pushed back. As Gibson County agent Cyrus Dyer noted, "This meant an increase on farm lands alone and was discriminatory, in the opinion of the Farm Bureau people. . . . The fight was won and the increase was ruled out. This resulted in a saving of about sixteen thousand dollars a year, or a total of sixty-four thousand dollars for the four years to the farmers of the county."[8] Farmers were largely successful in keeping the lid on those wishing to raise taxes by remaining unified in their opposition as tax increases were proposed.

With tax revenues down and county budgets stretched to the limit, farm bureau members demanded that county officials reduce their budgets rather than look toward taxpayers for more money. But while the majority of farmers wanted cuts to all levels of government, most did not want their Extension agent included in those cuts.

With positions hanging in the balance, Extension needed its friends and supporters to cash in their political chips in an attempt to protect the funding for these positions. In 1932, Harrison County agent Ronald Johnson noted the role of the local farm bureau in the funding debate, writing, "Although in the fight for lower taxes some individuals would be willing to abolish the county agent together with some other officials, the county Farm Bureau in its annual meeting voted unanimously to mandate the county council to bring the appropriation for county agent work up in line with other county officials."[9]

Despite the agricultural depression of the 1920s, much of the Extension system had made it through 1929 and 1930 in good shape, even to the point that some counties had increased funding to the agent's office.[10] However, as the Depression worsened, the value of continuing to pay an agent came into question, with one agent summarizing the issue by saying, "Does the County Agent do anybody or group of farmers any good or justify his expense?"[11] Consequently, a small but significant number from the agricultural community began calling for agent positions to be eliminated along with other tax-supported county positions.

In the time since county Extension offices were first established in 1914, agents around the state had helped farmers prevent hog cholera outbreaks, identify and remove cattle infected with tuberculosis from dairy herds, and select, store, and test germination of open-pollinated corn to increase yields, among other accomplishments.[12] But past successes were overlooked by some, who were asking, "What have you done for me lately and at what cost?" And now, when agents needed those they had helped to step forward, a few counties and communities turned their backs on the Extension program, demanding that the office be downsized or abolished.

In 1930, agent Virgil Sexson in Greene County observed:

> The immediate outlook for extension work is of a decidedly mixed complexion. Because of the extreme drought during the past season following two or three years of poor crop yields, farm incomes have been seriously reduced. This in addition to the general depression has caused a financial condition among farmers generally that is rather serious. High taxes have added their share to the general conditions. Under such conditions, farmers are not in a frame of mind conducive to unusual cooperation in any kind of a program. These conditions have rendered it more difficult in some communities and aroused opposition to the county agent's office partly because it represents just another office.
>
> On the other hand, most communities have certain clear minded men and women with vision who realize that a well-directed extension program is needed in times of economic stress more than in periods of prosperity.[13]

A few years later, agent William Smith reported that the "economic stress of 1933 has put the extension service in Henry County on the spot in a way that it has not been for several years. Some have felt that it was a luxury we should do without in these times, others that its purpose was to increase production which we did not need, and still others that if we would just leave the farmers alone, they would get along alright."[14]

Several factors helped explain why some in the agricultural community were questioning the need for an Extension agent. One was that farmers still unfairly blamed depressed prices for farm commodities on Extension's role in increasing production during World War I.[15] Many believed that the push for more wartime production had led to the agricultural depression of the 1920s as the markets adjusted for the postwar situation. Some blamed the agents for the prolonged economic downturn in the ensuing decade because who else could they blame? After all, agents were the official spokespeople for the government policies and quotas at the time. Looking back, Floyd County agent Charles East explained the situation this way in a 1939 report:

> It was during this period, too, that the greatest economic boom and collapse of our generation took place. While neither the boom nor the collapse were the responsibility of the County Agent, both conditions brought changes in the agricultural situation that were his responsibility. When cropping practices that have proved their worth throughout a period of years, are forced to meet an entirely new scale of prices and declining market, something must be done; what is best? In such circumstances the County Agent must offer advice which is based upon the changed outlook. While I wasn't in this county at that time, I was in a similar situation and I assure you that it isn't any easy matter to fulfill one's obligations.[16]

Beyond the agricultural community, those not directly involved in agricultural production were more apt to favor cutting the Extension program because they knew little about it. In 1932, Sexson explained that there "has been a more or less blind resentment against any agency that was supported by public funds. The full fury of this resentment has been directed toward the county agent's office by more or less misinformed and misguided individuals. This condition has made the work difficult."[17] Considering that much of the work helped farmers in the county rather than the general public, it shouldn't have been surprising that wider support for the office was lacking. Why would people agree to pay taxes for an Extension agent or program when they didn't perceive a benefit to themselves or their families?

Across the state, agents' reports reflected the pressure they were under and showed how much individual situations varied. In Adams County, agent Lawrence Archbold knew that the woes of the Depression meant additional work for Extension, but he was fortunate that the county budget included money for his position.[18] In Bartholomew County, agent William A. Douglass wrote in 1931, "Like many of the counties, this county has made no appropriations for extension work next year. Plans are now under way, however, to have the money available."[19] That same year in Cass County, agent James E. Little reported, "If general conditions improve by next September, we will have no cause to worry about the future [of] extension work."[20] However, Little also implied that if the economy failed to make a timely recovery, all bets for funding Extension in the county were off.

In some cases, public officials cut the office stenographer's position rather than the agent's.[21] When the Harrison County Council reduced funding for the local Extension office in 1932, Ronald Johnson canceled some of his planned programs, but he also took the unusual step of using a portion of his salary to hire part-time help.[22]

In Bartholomew, Clark, and Dubois Counties, council members saved a few dollars by holding back funds for the local fairs.[23] This was a new option since the Indiana General Assembly had just removed the mandatory financing requirement for county fairs in 1931.[24]

When Archbold learned in 1932 that the local county council was going to scale back the 4-H program in Adams County, he urged members of the local Extension advisory board to contact council members and persuade them to reconsider. The effort worked, and he later reported that "the 4-H Club Work fund was the only item that was not slashed drastically."[25] The next year when the county council decided to reduce the Extension shows and exhibits budget from $725 to $500, Archbold had members of the advisory board attend the next commissioners' meeting. Sixty men and women filled the meeting room, and Archbold later reported, "[The] commissioners stated that they had made a mistake and that the money would be forthcoming," he said.[26]

However, in Benton County in 1933, agent Ellis Christen bluntly described a more critical situation, writing, "General policies for the year might well be stated by saying that it was one of retaining a county agent. With one man employed the first four months of the year, no one the four next months, and a new man the last four months, no clear general policy can be stated," indicating that this county's program was largely derailed.[27]

In some counties, influential leaders wielded enough power to successfully discontinue the local Extension program.[28] As Daviess County agent John Jensen noted in his 1931 report, "In spite of the fact that several county associations recognize the need for extension work and are working harmoniously and effectively with the county agent in promoting agriculture in the county, the council failed to appropriate for this office for next year. This action was probably taken as the result of urgent demands to cut taxes where ever possible."[29]

In Hendricks County, the Extension program was eliminated on March 15, 1933, after advocates for lowering taxes convinced the local board of education to abolish agent Ellis Hawk's position.[30] Hawk observed in his annual report that this was "a slap at the Farm Bureau. The Farm Bureau committees had been active in demanding the reduction of taxes, and when the Farm Bureau leaders declared for a County Agent, they were accused of being inconsistent and some of the trustees used this means as a retaliatory measure."[31]

At the time this action was taken, Hawk was an eleven-year veteran with Extension, having previously served eight years in Scott County in addition to three years in Hendricks County.[32] When his contract wasn't renewed, he took a position with the Union Central Life Insurance Company in Minneapolis.[33] These types of staff losses were a setback to Extension as veteran agents departed and their experiences and contacts were lost.

The Hendricks County position was left open for four years.[34] With the Depression starting to wane, Harold Yarling took up the position in 1937, reporting that "now there is practically no opposition and the people in general are more than anxious to cooperate in every way possible to make agricultural extension work a success. . . . The County Agent's office also compiled a mailing list of area farmers and wrote a letter to the entire list advising them that the office was again established and requested return of a card listing the subject or subjects of most interest to them."[35] As new agents took on these positions, each would have to rebuild trust and contacts, and reestablish confidence in the local Extension program.

Agents often remarked that at times county leaders might have forgotten or conveniently chose to forget that state law—specifically the Stahl-Yarling Act of 1913—required counties to fund the local Extension position.[36] Under this act, Purdue received state funds to reimburse a county for half of the agent's salary up to $1,000. But the rest of the money had to be provided by the county, and the legislation required the county council to guarantee $1,500 up front toward the agent's salary and office expenses.[37]

There were more than a few counties where officials balked at this arrangement, which ultimately meant they had to cover $500 in office expenditures and any portion of the salary that exceeded $1,000. They thought that this funding structure was unconstitutional since it basically amounted to an unfunded mandate. Local officials argued that the decision to fund an agent's position should have been left to the county, not the state legislature. This would lead supporters of the county Extension position and those siding with local government who wanted to cut the position to challenge each other in court.

In September 1932, members of the Crawford County Council refused to appropriate funds to support the office of Extension agent Willard T. Merrilees. A group of local farmers and Paul Flanigan, editor of the *Crawford County Democrat,* filed a petition in court to force the council to meet its legal obligation under state law. The court granted the council more time to first see if the Indiana General Assembly would eliminate the mandatory county funding for Extension.[38] Merrilees would later report that when legislators failed to change the law, "Judge T. J. Wilson, Corydon, Indiana, ruled [in February 1933] that the minimum appropriation was mandatory and that the council should appropriate that amount at its next regular meeting, which was held in September 1933."[39] Unfortunately, the state attorney general determined that Merrilees was not entitled to be paid retroactively, so he lost a portion of his salary for 1933.[40]

Around the same time, Cyrus Dyer described the mess he was embroiled in as the Gibson County agent. In his 1933 annual report, Dyer recounted a clash with the local county council. At the September 1932 meeting, he had asked council members for $2,950 to fund his Extension office, which was the same as the amount appropriated in 1931. Instead, the council decided not to provide any funds, and after unsuccessful negotiations, the two sides ended up in court.

By the fall of 1933, the matter was still unresolved, so Dyer began asking for 1934 funds as well. As Dyer recalled, at the September 1933 meeting he found a much different council, whose members seemed ready to listen to why they should fund the position. He wrote:

> I went before the council and in a very friendly way talked over the whole situation, assuring them that there were no hard feelings in the matter and recognizing the great pressure brought to bear upon them a year before to reduce taxes, but calling their attention to the fact that due to the newly inaugurated Agricultural Adjustment Act [see Chapter 8] with its attendant responsibilities for the County Agent's office, there had been a decided change in the attitude of the people toward the County Agent and his office. The gentlemen of the council seemed to receive these suggestions very graciously

and later made the minimum appropriation of fifteen hundred dollars for the calendar year of 1934.[41]

But Dyer still wanted money appropriated for 1933 that his office had never received. Unfortunately, the county attorney advised the council members that they could not do so. A trial to resolve the 1933 funding of the Extension office was set to start on Monday, November 27, 1933, but on the Saturday before, both sides came to a compromise where the council would appropriate $1,500 paid in monthly installments of $125 starting in January. Following this verbal agreement, the trial date was canceled.[42]

Unfortunately, with local leaders, state legislators, and the courts wrangling about administrative and legal matters, agents such as Dyer and Merrilees were caught in the cross hairs.

Budgets for Extension in most of the counties seemed to stabilize by 1935 and 1936 as the economy began showing positive signs of recovery.[43] County Extension offices across the state had suffered a number of casualties during the heart of the Depression, but when all was said and done, the Extension program had proved its value in most counties.

Some asserted that funding battles over Extension budgets had made the statewide organization stronger by having the agents answer their critics. By 1935, with the worst of the Depression over, agents such as Enos M. Rowe in Clinton County began writing about the reemergence of Extension. In his annual report that year, he wrote, "Am happy to say that most all of the opposition, which was so active in its destructive criticism of the extension office and program a few years ago, is gone, or practically so, and some of the 'right bowers' of the opposition are now friendly with the office, and what it is trying to get done. We have overcome this opposition largely by the 'silence and hard work method.'"[44] Just four years later, Rowe would report:

> [The county] council was influenced largely by an outside rural organization with questionable leadership. All of these apparent set-backs or handicaps have served to strengthen the program and have since proved to be "blessings in disguise." The organizations which would have undone the extension program and the leaders of that organization have all but been forgotten, while the extension program and the extension office is still functioning and serving the agricultural interests of the fourteen townships.[45]

The situation that agent Ellis Christen described in his 1937 annual report for Benton County represented how the Extension offices in many counties had rebounded from the depressed economic times. He wrote:

> It is seldom, that anyone in the entire county criticizes the expenditure of money for the extension program. An example of this attitude is expressed

> in the local budget for 1938. The budget originally proposed by the local extension representatives was endorsed without change by local groups, recommended by the Board of Commissioners, appropriated by the County Council and passed by the Tax Adjustment Board without any change from the original. So far as is known, this is the first time that a local budget for [the] Extension Program has passed in the same form [as] it was originally submitted.[46]

County Extension offices would go on to survive the ups and downs of the Great Depression, and the agents' work during this gloom-and-doom period helped establish their offices as important county institutions whose value would not soon be challenged again as they had been in the early 1930s.

CHAPTER 5

Live Out of the Garden, the Smoke House, and Cellar

As employers shuttered their businesses and families' cash reserves ran low during the Great Depression, food insecurity became a serious problem each county would have to address. Many people sought out assistance for food from township trustees, charitable organizations, and county Extension agents.[1] In some cases, community relief agencies such as the local American Red Cross partnered with Extension offices, each bringing its unique resources, skills, and experiences to the work of helping the destitute and desperate.[2] But with so many people out of work and seeking significant assistance, aid resources were in increasingly high demand, requiring communities to use whatever means they could to stretch these resources as far as possible.

During World War I and the 1920s agricultural depression, Extension had successfully promoted gardening as a way to increase families' food supplies and promote self-reliance. Lessons learned from those situations were also useful for addressing the widespread food insecurity brought on by the Depression. As a result, using the garden to feed those in need became a signature program for Extension during the 1930s. In a 1931 annual report, Dearborn County agent Clyde Madison observed, "We have been emphasizing the importance of the home garden more this year than ever before. The depression and unemployment make the garden a much more important factor . . . than in ordinary times."[3]

Through their roles as members of local relief committees, many agents were in a good position to advocate for and coordinate gardening programs in their counties.[4] Charles Nicholson explained his role in the Dubois County effort in a 1932 report.

> After the winter woodpile had been exhausted and the burden began to get heavier on the Township Trustee's budget and the Red Cross, it became necessary that the two relief organizations work out some plans by which the expenses might be cut and additional food supplied for the needy. . . . [It] was proposed by the County Agent [Nicholson] that a sum be set aside sufficient to cover the necessary expense of operating a number of individual

> and community gardens in both Huntingburg and Jasper. It was agreed that each unit would receive $200 from the County Red Cross to conduct these gardens. The next stop that was necessary was to list the number of individuals who could and would qualify for the cultivation of a garden. . . . In order that these gardens should be as productive as possible and eliminate disease and insects, the County Agent became the chief advisor.[5]

Relief agencies or the township trustees would often provide the seeds for planting gardens. Agent John Jensen explained that in 1931 the Red Cross in Daviess County provided 500 pounds of seed, which were distributed to those in need.[6] Typically, cabbage, tomatoes, potatoes, and sweet potatoes were provided at no cost for gardeners.[7] In Bartholomew County, gardeners received seed packets that contained "one pound stringless green pod beans, one-half pound Kentucky wonder beans [pole beans], one ounce beets, one-half ounce carrots, one-half ounce lettuce, one-fourth ounce radish, one-fourth ounce parsnips, one-half ounce spinach, one pound sweet corn, one-sixth ounce of Swiss chard and one-half ounce of turnips."[8]

In 1933, agent Ralph Maggart worked with the Carroll County township trustees to create a list of people from each community who could not afford to purchase seeds. The list was submitted to the Governor's Commission on Unemployment Relief, which then paid for the seeds and for seed potatoes.[9] It was estimated that every $100 of garden seed handed out resulted in $1,000 worth of vegetables. This initial investment not only provided needed staples to families, but it also resulted in savings the county could use to meet other financial obligations.[10]

Extension encouraged people to plan their gardens with a variety of vegetables that could be eaten fresh, canned for the winter, or stored for long periods in cellars. The goal was to get families to plant a garden large enough to can a minimum of 60 quarts of produce for each person in the family, with typical crops including beans, cabbage, corn, onions, potatoes, and tomatoes.[11] To help reach that goal, Extension conducted educational events for those who had never gardened before to help them grow the food they needed.[12]

The home garden project was typically spearheaded by the county agricultural agent. Agent Earl Miles conducted a gardening program in Clark County in 1933 and used the phrase "Live out of the garden, the smoke house, and cellar" to encourage participation.[13] In 1932 in Allen County, agent Hugh Heckard put on a series of eight weekly programs on gardening at the YMCA that covered selection of garden plots, preparation of the seedbed and fertilization, varieties of garden vegetables, planting the garden, and controlling weeds, plant diseases, and insects.[14] At events such as these, agents typically distributed Purdue bulletins, such as the garden leaflet *Help for the Home Garden*, to attendees.[15]

With the assistance of the agents, many 4-H clubs also promoted gardening as a project. This got the whole family involved in producing food for the evening dinner and to fill up pantries and cold-storage areas. However, this led one agent to observe in 1931 that many of the local 4-H members had nothing to show for their efforts "since most of their produce was consumed at home," showing the importance of food from the garden to the whole family.[16]

Agents also provided education to the community by writing how-to articles on gardening for their local newspapers.[17] In 1932, Delaware County agent Myron Cromer reported, "Thirty-seven different news articles were run in the newspapers giving timely helps to home garden owners. These articles dealt with such subjects as making hotbeds, making garden plans, amount and variety of garden seeds needed, how to treat seed potatoes, cabbage yellows, how to make and use Bordeaux mixtures, methods of controlling common insects, etc."[18]

It was one thing to grow a garden, but preserving what was grown was equally important. If many had never grown a garden before, it was certain that they would also need to be taught how to preserve what they grew. County home demonstration agents played a significant role in this effort by teaching all kinds of canning and food preservation techniques. Home demonstration agents had first been hired through special federal funding during World War I to help families grow and preserve more food in order to free up supplies for the military and others overseas. These agents had been well received, but after the war, the positions were mostly eliminated when funding dried up. By the start of the Great Depression, there were only eight home demonstration agents remaining in the state. Gradually, the federal government began providing funding once again, so by the time the Depression was ending, Indiana had forty-five home demonstration agents, the majority of which had been hired between 1936 and 1939.[19]

The home demonstration agents along with the eight women on the Extension staff in the Purdue Department of Home Economics targeted many of their outreach efforts to members of home economic clubs across Indiana. With nearly 2,000 active clubs scattered across eighty-five counties, these clubs offered a ready-made network for providing education more efficiently. Often, a Purdue Extension specialist or home demonstration agent would train a few volunteers from each club, who would then relay the information to other club members.

Other Purdue Extension specialists also worked with home economics clubs to provide additional gardening education. Willard B. Ward, an Extension assistant in horticulture from Purdue, developed a set of 20-minute lessons on topics covering nearly every aspect of home gardening, including the proper use of equipment such as a hoe. He then traveled the state to present live demonstrations to groups that included club leaders, garden club members, and others.[20] Part of his efforts included showing participants how rotenone and sulfur dust used in combination or rotenone and pyrethrum would provide control for Mexican bean beetles and cabbage worms, and he even gave local drugstores and hardware dealers a list of recommended pesticide products so they could stock them.[21]

Outreach to clubs helped Extension reach more than 300,000 women in 1936.[22] While home demonstration agents would partner with the county agricultural agent to help others plan and grow their gardens, many of their programs focused on teaching techniques such as how to can vegetables, fruits, and meats; quick-freeze strawberries; and store bushels of vegetables and fruits for longer periods of time.[23] They also provided instruction on other topics relevant for managing a household during the Depression, including making butter, preparing meat, using alternative ingredients when baking, managing home finances, making soap, and repurposing clothing.[24]

One measure of the success of the various efforts to increase food supplies was how much food was grown and saved for the tough times that the winter brought. In 1935, the Fountain County agent reported that 800 gardens tended by those on the trustees' relief rolls yielded more than 30,000 cans of corn, beans, tomatoes, and carrots, with 225 bushels of potatoes stored for later use.[25] In Benton County, a 1937 survey showed that 85 families had followed Extension recommendations, preserving 13,194 quarts of vegetables and 7,852 quarts of fruit. In addition, those families had 790 bushels of vegetables and 174 bushels of fruit stored in their cellars for winter use.[26]

While much of Indiana's population lived in rural and remote communities in the 1930s, Extension also provided support to those in large population centers who needed to raise their own food. In Columbus, Connersville, Fort Wayne, Frankfort, Hartford City, Indianapolis, and elsewhere, Extension was called upon to create community gardens for those who did not have land available for gardening.[27] Purdue was sold on the idea of these relief gardens and assigned H. Earl Young from the Department of Horticulture to head up the project.[28] To promote the program and motivate participants, Young created the slogan "Work will win if we dig in!"[29]

Extension agents tackled this challenge just like they did in World War I and during the 1920s, when vacant lots and other properties were used to create shared community gardens.[30] Businesses such as the International Harvester Company and the Auburn Automobile Company allotted space for families to plant gardens on property the businesses owned.[31] There were also cases where private citizens stepped up to make land available for the Extension-led community gardens. In 1933, Fayette County agent Marion Detrick reported, "Two public-spirited citizens, Anthony Locke and L. A. Frazee, donated the use of tracts of land" for local residents to use in raising their own food.[32]

One of the more successful community garden projects was in Muncie, which an *Indianapolis News* article identified as the "birthplace of relief-work community gardens."[33] The project was supported by the Community Chest and the township trustees. In 1932, Delaware County agent Myron Cromer described the extent of the Muncie garden program, writing, "During these trying times of readjustment, a city with a population of some 46,000 is certain to have a relief problem. . . . A total of 204 acres was donated by different owners to the production of community and home gardens. In addition, there were approximately 2,000 backyard home gardens, and the seed for all of which was furnished by the Community Garden Association. One entire carload of Rural New Yorker seed potatoes were distributed and planted."[34] The following year, Cromer reported working with Purdue horticulture Extension specialist W. Max Austin to teach a spring class where 835 people learned how to cultivate, plant, and harvest the vegetables.[35]

Community gardens were large undertakings, given the severity of the food insecurity faced by people throughout the state. Agents were expected to carry a large share of the responsibility for getting the gardens started and for dealing with problems as the growing season went on. They were also expected to lay out the garden, suggest how much of each vegetable to plant, recommend how to deal with insects and plant diseases, indicate when it was time to harvest, and provide information on the best storage for the different food items. Many people had never grown a garden before, which meant they required more help throughout the year.

Township trustees put much of the responsibility for planting and caring for garden plots on those who had sought aid and requested space to garden.[36] A 1932 report by Cass County agent James E. Little showed the steps taken to make sure recipients of public aid were holding up their end of the bargain.

> This year, through the cooperation of the Eel township trustee; the Red Cross; the county agent; Labon Fisher, Lincoln Junior High School principal; and Mrs. Helen K. Little [county nurse], about five hundred home gardens were planted and cared for. Quite a large percentage of these gardens would not have planted, had it not been for the additional pressure that was placed upon each individual.
>
> During May and the early part of June, calls in the homes of all people receiving aid from the trustee were made voluntarily by Mrs. Little. The purpose of this first call was to learn if the family was planning to have a garden; if so, how large and the general plan for the garden. In cases where there was no plan for [a] garden, they were informed that there would be "no aid" for them.
>
> Many families did not have jars for canning, and in these cases, they were urged to plant larger quantities of the vegetables that could be preserved for winter use without the use of jars. Mr. Harrell, the trustee, was to plow all

> plots for gardens in case the family had no other means and then either he or the Red Cross was to furnish seeds for the plot. The Park Department of the City of Logansport also helped in securing lots and having them plowed for men who were out of work.
>
> During the latter part of June and during July, a second call was made to each of the homes, and the garden at home as well as additional gardens away from home were inspected by Mrs. Little. Large gardens were stressed during this visit, and canning and storing were urged. In cases where a garden was unusually weedy, a report was made to the trustee, and when the call was made for aid, the garden had to be weeded first. This was a very good weapon to hold over their head all of the time, and little trouble was experienced in keeping the gardens clean.
>
> Calls were made in some of these homes later in the fall and it was very gratifying to observe the large quantities of foods that were canned and stored for winter use. This garden work should have lasting effects. No doubt many people canned foods that other wise would have gone to waste; children benefited from the large quantities of vegetables that they had during the summer and that they were put away for winter use. Many families raised and stored vegetables that had not been accustomed to heretofore.[37]

Self-sufficiency and responsibility were key to the Cass County relief garden program, but not every program met with such success. In Dubois County, the agent reported in 1933 that "some of the needy wanted to work, others related that they didn't need to work [and] that they would be fed by the Township Trustee."[38]

In many places, unemployed workers often provided the labor needed to till, plant, and harvest community gardens.[39] One agent observed that these gardens played "a noticeable part in providing occupation for idle hands and in keeping the unemployed off the streets."[40] Another agent reported that the local community gardens "kept a number of the unemployed busy two or three days per week producing their own winter's food supply and, in the end, a vast saving to the tax payers."[41]

In 1932, unemployed workers cared for a 40-acre community farm in Clinton County with financial assistance from the township trustee in the amount of $1,178. Agent Verner J. Mann said there were 310 workers who each contributed time ranging from just a few hours to hundreds of hours for some.[42] The following year, the farm was expanded to include 10 acres of early potatoes and 30 acres of late potatoes; 17 acres of tomatoes; 7 acres each of sorghum and soup beans; 1.5 acres of early cabbage and 3 acres of late cabbage; and 3 acres each of onions, pumpkins, squash, and turnips.[43]

Those who worked the so-called charity or relief gardens received free vegetables for their families, with any surpluses made available for canning.[44] In the case of the Clinton County farm, the harvested products went to a commissary, and workers were given tickets that could then be used to claim produce.[45]

The longer the Depression continued to drag down the American economy, the more important Extension's efforts related to gardening and food preservation became across Indiana. County agents invested precious time setting up individual and community gardens to help those struggling to put food on their tables.[46] Not only did these gardens provide a large amount of food for many people during a time of significant need, but they also lowered the cost of living for families by not requiring them to purchase more expensive food from local stores.[47]

County leaders were well pleased with the work on community gardens that had been done by county agricultural and home demonstration agents. The money that the leaders had provided to support these programs typically paid large dividends since township trustees then had to spend less on assistance provided to those in need.

Gardening programs had the added effect of expanding outreach by Purdue Extension within agricultural and urban communities, showing people around the state that agents were there to serve everyone and not just farm families. These programs had positive outcomes for those living in town and in the country, and those receiving the assistance—many of whom had had little prior interaction with the Extension office—would long remember the help they had received in feeding their families during the rough years of the Depression.

While agents were involved in numerous projects during the Depression, it's clear that helping others raise their own food may have had the greatest impact of all the programs Extension offered during this time. Whether in a private garden or a much larger community garden, tilling the ground, caring for vegetable and fruit crops, and harvesting during the growing season gave people hope that they could weather the hard times.

CHAPTER 6

Cash Is the One Article That Is Scarcest and Hardest to Get

Farmers who had borrowed money to purchase expensive land during World War I—when commodity prices were high—were shocked when land prices plummeted during the years preceding the Great Depression.[1] As prices continued to fall during the Depression, many farmers realized early on that they were not going to make it through these difficult times. From 1913 to 1920, the rate of farm foreclosures was 3.2 per 1,000 farms. By 1926, during the agricultural depression, they had reached 17.4 per 1,000 farms, and by 1933, at the lowest point of the Depression, they were 38.8 per 1,000 farms. From 1930 to 1935, 750,000 farms went bankrupt. And in 1933 alone, more than 200,000 farms were lost for failure to pay their mortgages.[2]

Agent William Douglass did an intensive review of the farm credit situation in Bartholomew County in 1931, which confirmed that falling land prices and low commodity prices were at the heart of the credit problem. That year, he reported:

> A great deal of work has been done concerning the credit situation during the past year. Due to reduced values of property, the percent of indebtedness to prospective values rose to an alarming figure early in the season. This coupled with reduced income made the credit situation very bad. At no time has it been felt by anyone in this region who is posted on the matter that there was any particular shortage of money for loans. The principal banks were carrying cash reserves well above the safety mark, and at the beginning of the year, well after the trouble had commenced there were still ample deposits. The trouble therefore, was entirely or nearly so, due to reduced valuations of property, previous indebtedness and reduced income. In checking over the situation we prepared a list of chattel mortgages [e.g., on farm equipment] outstanding against farmers in the County. There were some 175 farmers on this list. The majority of these mortgages were covering loans under the Indiana Petty Loan Act and carrying interest rates of from 36% to 42% [annually]. These lists were prepared for the benefit of banks and various lenders interested in the credit situation and in bringing help to bear on the situation if anything could be done. When this list was prepared it was thought that a rather serious

> condition was indicated. However further study revealed the fact that the number of mortgages outstanding as well as the number given, was less in 1931 than in 1930, and less in 1930 than in 1929. This tended to emphasize the conclusion that the credit situation was resultant from the reduced values and reduced incomes rather than increased indebtedness.[3]

Early on in the Depression, county Extension agents began reporting that, more than anything, farmers needed short-term credit to get them through spring planting and longer-term loans to meet their mortgage obligations.[4] Farmers were bringing in just enough money from the sale of crops and livestock to cover their obligations, but it took most of their cash. With banks wary of loaning money to farmers with cash-flow problems, farmers often found themselves short of money for planting the spring crops or feeding livestock throughout the winter.[5] Even those who had weathered the first years of the Great Depression were becoming worried that they too were going to run out of money to keep their farms operational and pay their mortgages.

In the 1932 annual report for Dubois County, agent Charles Nicholson explained why some farmers were managing to run a slightly profitable farm in the toughest of times while others were on the verge of bankruptcy or needed an infusion of cash to operate.

> [F]armers as a group have been one of the first classes to adjust their budget and begin to live within their means. Only the necessities of life and farm operations are purchased. A farmer who did not make a purchase of land and equipment some few years ago when prices were abnormally high can still make a living under the present prices and a good many are still making a small profit by watching their excessive purchases and living within their means. On the other hand, the farmer that paid a high price for farm lands and equipment, etc., and went in debt 75% of its original purchase is the fellow that is having a hard time to meet his expenses.[6]

Farmers with collateral who had remained in good standing with their lenders should have been able to count on local community banks to support a loan request.[7] But since nearly 40 percent of U.S. banks had failed or had been forced to merge during the early years of the Depression, the odds were that many farmers had fewer local banks available.[8] When there was money available for new loans, local banks took a more conservative approach.[9]

Farmers falling behind on their mortgage payments created a serious problem for the banks.[10] As land and crop prices fell, cash flow became the sticking point between bankers and farmers. Banks tightened their lending policies and were unwilling to loan money to those carrying large amounts of debt, which included farmers who owed a lot of money on land and equipment.[11] These restrictive loan policies would remain in place as long as the price of land, crops, and livestock remained depressed.

In Bartholomew County, Douglass thought it was his job to intercede with local bankers on behalf of farmers whenever possible. During the Depression, he described petitioning local bank officials to use "conditional sales notes to be given by farmers to secure loans on livestock. After some discussion the bank officials agreed to accept such notes on the recommendation of the County Agent."[12]

On the other hand, farmers who lacked access to local banks had little recourse but to borrow money from private lenders who charged interest rates of as much as 36 to 42 percent.[13] Many of these largely predatory but legal loans would later be repaid using loans with lower interest rates that federal and state governments provided as part of their relief efforts.[14]

The Roosevelt administration and Congress reacted to the financial conditions farmers faced by creating the Farm Credit Administration in 1933 to provide subsidized loans to farmers. This agency consisted of the federal land banks, which focused on providing mortgages to farmers; the Production Credit Association, which made short- and intermediate-term loans available for agricultural production; and the Bank for Cooperatives, which loaned money for rural electric cooperatives, for example.[15]

The federal land banks played a particularly pivotal role in keeping farms solvent during the Depression. In order to help farmers avoid losing their farms to bankruptcy, the government directed these banks to help qualified farmers refinance their existing loans. This was a popular option not only because the federal government offered lower interest rates than local banks but also because loans originating from a federal land bank provided an extended time for repayment.[16] Loans to individual farmers ranged from $3,000 to $30,000.[17]

Indiana Governor Paul V. McNutt directed Lieutenant Governor M. Clifford Townsend to create a program to take advantage of these federal dollars for mortgage refinancing in order to help Hoosier farmers who faced bankruptcy.[18] Townsend outlined the plan in a letter dated October 30, 1933:

> The Indiana Agricultural Advisory Council, appointed recently by Governor McNutt, has adopted a plan which it believes will greatly assist Indiana farmers in securing necessary Federal loans to avoid mortgage foreclosures.
>
> The plan is to set up a conciliation committee in each county of the state. The committee, which will serve without salary, will assist in distressed cases only and will act as a conciliator between the farmer and the holder of the mortgage, and also will assist the farmer in transferring his mortgage to the Federal Land Bank by paying the present mortgage holder with a Federal loan.
>
> The State Council and these county committees will be interested only in distressed cases and it will be necessary to set up the machinery quickly, in order that the greatest benefits may be derived from the Federal Act.

The State Council is asking the County Agricultural Agent and Farm Bureau President of each county to confer immediately and do the following:

1. Nominate a county conciliation committee composed of one banker, one attorney and three farmers, totaling five (5), all of whom should be sympathetic with the movement.
2. Send the nominations immediately to M. Clifford Townsend, Indiana Commissioner of Agriculture, State House, Room 331, Indianapolis. These names, together with their addresses, must be received before November 5th, 1933.
3. Designate a chairman of the county committee before you submit it for approval. Following the setting up on the county committees, complete instructions will be forwarded to the various chairmen and the machinery will be ready to be placed in operation early in November.[19]

Members of the conciliation committees gathered at a state conference in Huntington on December 29, 1933, to receive training on what qualified a farmer for the program, the loans available, and what additional measures were available to help farmers nearing bankruptcy.[20]

It was commonplace for conciliation committees to meet in their local county Extension agent's office to discuss the plight of farmers under review.[21] Each farmer had to submit a signed and completed application along with a financial statement listing outstanding mortgages and banknotes.[22] The only cost to the farmer was a twenty-five-cent notary fee needed to certify that the information being submitted was accurate.[23] If the committee determined that the farmer's balance sheets qualified, the application was forwarded for review by one of the dozen federal land banks, which were located in Columbus, Ohio, or St. Louis, for example.[24]

In Elkhart County, forty-three farmers managed to avoid foreclosure proceedings in 1934 through this program.[25] In 1935, Floyd County agent Charles East noted that the program had resulted in "several farm families being relieved of a crushing burden of debt, and has given their creditors a chance of ultimately receiving a substantial portion, if not all, of the moneys due them."[26] For the banker and the farmer, it was a win-win situation.

Based on the number of applications and the amount of money loaned, there is no doubt that the federal land bank system was a major source of farm credit during the Depression. One agent observed in 1933, "A deluge of applications continues to pour in and the Louisville [federal land] bank now has over twenty-two thousand (22,000) applications [for the low-interest extended loans]."[27] By 1938, there was nearly $900,000 worth of loans in Bartholomew County alone. According to the agent there, all of the loans were in good shape, with the exception of one that was delinquent.[28]

In 1934, Dubois County agent Charles Nicholson painted a rosy picture regarding the dwindling number of foreclosures on farm property, saying, "The adjustment of Farm Mortgages and the more liberal application of the Federal Land Bank has done much to save many a farm. Only one farm foreclosure is on record in this county this year. Records show that many farm mortgages are being released and payments made on others, and few, if any, new ones are being made."[29]

During the Depression, agents across the state reported that farmers were short of cash to make the purchases needed to plant the spring crops, including seed, fertilizer, and equipment. In addition, those raising livestock needed money for feed.[30] With farmers having little cash on hand, Delaware County agent Myron Cromer observed in 1933 that the "continued low price of farm products in comparison with the prices farmers must pay for that which they must buy is rapidly forcing many farmers to look elsewhere for credit, since the banks have ceased extending loans."[31]

The inability by farmers to plant a crop during the Depression was the impetus for the federal government to provide short- and intermediate-term loans to farmers to purchase seed, feed, equipment, poultry, and feeder livestock.[32] Some of these loans were offered through the Farm Credit Administration's Production Credit Association, which required a lien on a future crop or on livestock the farmer owned.[33]

The agency made the final determination regarding which applications were eligible based on the recommendations of the county committees and the documentation submitted along with the loan application.[34] For those who received them, these loans would be just enough to get the fields planted and to purchase feed for livestock in hopes that the resulting crops and livestock would be worth more than the loan amount when marketed.

While it seems that committees did an excellent job selecting farmers who were financially stable enough to pay back the loans through livestock and crop sales when the notes were due, some farmers failed to repay their loans due to extenuating circumstances.[35] For example, agent T. E. Myers reported in 1933 that thirty-two DeKalb County farmers were unable to repay $3,200 in loans because their crops had been destroyed by drought. He also observed, "This loan was beneficial to a majority of the farmers but it is a question as to whether it helped some or just put off the day of judgment."[36]

In addition to borrowing money against a future crop, farmers could also secure a federal loan using an existing crop. In October 1935, U.S. Secretary of Agriculture Henry A. Wallace announced a program offering $150 million in short-term loans to farmers who had stored corn available to use as collateral. This loan program was managed by the Commodity Credit Corporation (CCC), which had been created in 1933 to help provide stability for farm income and prices, and to manage the supply and distribution of agricultural commodities.

In 1935, the Indiana General Assembly passed the Farm Warehouse Act to allow for the creation of county warehouse boards to take advantage of this program.[37] Much like with other federal loans, the first step consisted of a review by a county committee to figure out the details.

Partnering with the CCC, the warehouse board did much of the leg work by issuing warehouse certificates that documented how much corn was being stored and then sealing that corn to prevent its movement. The three-member county warehouse board was selected by Martin L. Lang, the Indiana commissioner of weights and measures. The board used the local Extension office as its base of operation, with the county agent serving as the primary source of information about the federal loan program.[38]

It was typical for the farmer to be guaranteed 45 cents per bushel of stored corn. The farmer was charged 4 percent interest on the amount of the loan by the CCC. In addition, farmers were charged one cent per bushel—not to exceed $20—to cover the cost of the county program. Farmers also had to pay the recording fee on the mortgage and the insurance to cover the loan, which generally had to be paid back within the year.

Each county was asked to create a corn loan board to inspect the quality of the corn and to verify the number of bushels a farmer had stored.[39] Counties hired one or two persons to calculate the amount of corn the farmer wanted to use as collateral and then seal up any openings. All grain under seal was required to be reinspected by the board every ninety days to make sure that it had not been tampered with, was still in storage, and was in good condition.[40]

The warehouse certificate indicated the amount of sealed corn. Inspectors from the Indiana Department of Weights and Measures placed an official seal on each bin, which in essence showed that the corn belonged to the federal government. It was illegal for the farmer or anyone not associated with the county warehouse board to break the seals, and those caught breaking the seal could be fined up to $1,000 and given a two-year prison sentence.[41]

The farmer was allowed to sell the stored corn when the price went above the 45-cents-per-bushel guaranteed price and keep the profit after paying off the loan to the federal government. In some cases, the farmer would let the government take possession of the corn when the market price fell below the amount borrowed per bushel.[42]

Across the state in 1938, there were 1,015 loans that placed 927,945 bushels of corn under seal.[43] These short-term loans provided vital assistance for many farmers to carry them through the rough times of the Depression when cash was so short.

There were still low-income farmers who did not qualify for loans through their local banks, the Production Credit Association, or the Commodity Credit

Corporation. In most cases, farmers were unable to obtain many federal loans if they were too far in arrears or if the percentage of debt they carried was too high.[44] As a last resort, the federal government through the Emergency Crop and Feed Loan Division of the Farm Credit Administration was there to assist even those with little to back up the money they were asking to borrow.[45] The amount that a farmer could apply for under the emergency crop loan was limited to between $25 and $250.

The farmers getting this last-ditch loan had to have been rejected by other lenders. Agent Alfred Hesler in Fountain County noted, "Rejection of this application by the Production Credit Association was considered sufficient evidence that other credit is not available and the applicant then made application to the Emergency Crop Loan Office."[46]

These loans were more general in nature, allowing the farmer to do what was needed to make the farm self-supporting.[47] These loans, with a longer term that allowed up to five years for repayment, would prove to be a last resort for many farmers. In 1937, DeKalb County agent T. E. Myers noted, "[I]f the farmers were not given this assistance, they would be compelled to go on relief and the expenses would be approximately the same. By this method the farmer can be helped to be self-supporting and keep up his morale."[48]

During the Depression, tenants made up a significant proportion of farmers across the United States. For instance, estimates are that by 1930, tenants managed 42 percent of all U.S. farms.[49] In Benton, Jasper, Newton, Tippecanoe, Warren, and White Counties, between 40 percent and 60 percent of the farms were tenant-run.[50]

In sixteen southern Indiana counties, just the opposite was true, with less than 20 percent of farms operated by tenants.[51] In 1937, the Dubois County agent reported far fewer tenants, noting there were "2,209 farms, 1,806 operated by owners, 397 by tenants and six by managers. The farm family does most of the work."[52]

Congress made funds available to farmers who owned their land, but it did not provide equal access to funding for those who farmed someone else's land. In Blackford County in 1933, the agent observed that "many distressed tenants who do not have sufficient collateral were not granted loans," while also saying that "the farmer who had sufficient collateral even though he might be heavy in debt in real estate was granted a loan."[53] It was, in fact, a dual system, with one set of rules for landowners and another for tenants.

Congress tried to even the playing field in 1937 by passing the Bankhead-Jones Farm Tenant Act, which provided federal dollars to help tenants, sharecroppers, and farm laborers purchase their own farms and become landowners.[54] The Farm Security Administration was put in charge of the program, which offered loans with terms of up to 40 years at a 3 percent interest rate.[55] With limited federal money available to support the purchase of land, only one loan out of every hundred was made across the country. By 1941, nearly 21,000 borrowers would be able to use these federal loans to become landowners, with farms costing an average of $5,600 and including an average of 133 acres.[56]

The availability of federal dollars provided the impetus for Indiana to develop a program in 1938 called the Tenant Purchase Plan through which some tenants could purchase farmland under the federal act.[57] With very little money available for the program, participation was limited to a few selected counties each year. Within those counties, only a dozen tenants or fewer were able to get loans to purchase farms, so only the most successful tenant farmers would be chosen.

To facilitate the process, the county agent selected individuals to serve on an advisory committee to review the applications.[58] Agent Alfred Hesler indicated in his 1938 annual report that the Fountain County committee was "looking up farms and cooperating in every way possible to make the project a success. . . . We have insisted on good working men, capable and upright men. Now, we want good farms, where a good man may reasonably expect to meet his obligations on the operation of this farm."[59] The first farm purchased under the Tenant Purchase Plan was in Fayette County in 1938.[60]

In 1938, agent Ellis Christen made the following report about this program:

> Benton County was selected in the summer of 1938 as one of the new counties in the State to be included in the tenant purchase program for 1939. Previous to this, the work of the Farm Security Administration, which was primarily Rural Rehabilitation, was administered from the Monticello Office. With the announcement of the 1939 Counties in Tenant Purchase, a new district comprising Benton and Warren counties was made and an office for the two counties was established in Fowler. . . . The County Agricultural Agent assisted in the recommendation of farmers to secure on the Tenant Purchase Committee. . . .
>
> The County Agricultural Extension Office accepted practically all the applications for the Tenant Purchase of farms from farmers. A total of ninety-one applications were finally received in the county. The committee then selected thirty-five of these applications for further consideration and at the second meeting cut the number to sixteen. They ranked the sixteen in order of preference. On December 1, eight farms had been optioned for the selected tenants and the engineer had made his preliminary investigation. There has been more interest evidenced by the local people than was anticipated.[61]

As Christen indicated, eight farms were available, so the committee ultimately would approve only half of the sixteen finalists.

Agent Warren O'Hara must have been pleasantly surprised when 81 tenant farmers from Hancock County responded to his newspaper article about the program in 1938. The committee had to reduce the higher-than-expected number of applications to a much smaller number for which money was available. From the initial applications, 25 tenant farmers made it through the first two reviews. At the conclusion of the third review, 16 growers were asked to provide intensive personal and financial information to back up their loan applications.

O'Hara then described the remaining process, writing:

> On October 8, the county committee decided to pick out the twelve best looking prospective families [out of the sixteen]. These twelve families were invited to appear before the committee on two nights, Oct. 10 and 14, six families each night. The following ten families were approved and given permission to look for family sized farms, which could be purchased within the allotment limit;—Omer T. Wicker, Fred Ortel, George Simmons, Paul L. Jones, George M. Walker, Oak Snider, Verle Swarts, Raymond Borgman, August C. Mearling, and James Souder.[62]

It was up to the applicant to find the farm of their dreams, which was then inspected and appraised. After that, the tenant had ninety days to complete the purchase.

In the Hancock County program, loans ranged between $5,000 and $7,000, and the farms ranged in size from 53 to 80 acres. O'Hara didn't confirm how many of the ten families completed their purchases, but based on the amount of money the county received, it appears that all did. In 1939, eight Benton County tenants were able to purchase farms of 97 to 160 acres.[63]

As the value of farmland began to rise just prior to World War II, it proved to be an impediment to the tenant purchasing program. Fayette County agent Dudley S. Bishopp summed up the challenge in his 1938 annual report, writing:

> This program is generally recognized as being of considerable help to worthy tenants who otherwise might never become land owners. The greatest obstacle to the program, however, is high prices asked for suitable farms, which not only makes it difficult for the prospective owner to pay for the farm from his income, but it is also difficult to keep the average price for the farm down to the recommendations of the administration without purchasing too small a farm to be practical to operate except under an intensive system of farming.[64]

By 1946, the program had all but ended. While it had helped nearly 44,300 tenants across the United States, hundreds of thousands more were asking for funds to buy property. However, with the country now facing wartime expenditures, federal officials decided that the program's useful life had come to an end.[65]

Farming during the Depression was challenging, but many farmers who faced economic hurdles could turn to county Extension agents for financial advice and to the federal government for loans. And by the mid-1930s, the agents were reporting that incomes and living conditions for those in the agricultural community were improving.

As the upward trend continued, Bartholomew County agent Arvil V. Keesling wrote in 1938, "Banks are making more loans now than has been the case for five years," which showed that most farmers and bankers had survived the worst of what the Depression had thrown at them.[66] In 1939, Fountain County agent Alfred Hesler noted, "With plenty of money in the coffers and eager to lend it, we find that the local banks are easing up and showing some interest in local loans that should rightfully be made to farmers."[67]

Hancock County agent Warren O'Hara reported in 1939 that delivering meaningful help during trying times had left a positive impression of Extension among many in the community—a view echoed by his colleagues around the state. He wrote:

> Following the financial disaster of 1929, farmers were forced to increase production, because of limited liquid assets to meet mounting taxes and fixed prices. The County Extension office worked ardently on sealing down indebtedness, preventing foreclosures and using federal agencies to supplement local money and to such an extent that nearly three dozen farmers personally thanked the office for saving their farms.
>
> This brought respect to the office because the county agent was the only source for helping them plan and execute a workable financial program, securing actual financial help. Some farmers were actually astonished that banks would send them to the county agent office for financial help.[68]

The agents' work in helping farmers secure loans helped win many converts over to the mission of Extension in the county. The agents managed to deliver results at a time when others couldn't keep their promises, and by most accounts, their tireless work during the Great Depression helped solidify the Extension program within the counties. Farmers who, with the help of the agent, were able to get loans to keep their farms, plant the next crop, or feed their livestock would never forget what Extension had done for them.

CHAPTER 7

The Man With the High-Producing Soil, Hen, Cow, and Sow That Kept Operating Expense Down Was Able to Return a Profit

At the beginning of the 1930s, the outlook for agriculture was shrouded in dark clouds and gloomy days. Farmers faced an economic disaster as crop and livestock prices dropped precipitously.[1] In 1929, a bushel of corn had sold for 65 cents. It increased slightly to 71 cents in 1930 before falling to 35 cents in 1931 and to 10–12 cents in 1932.[2] Beef prices per hundredweight dropped from $10.95 in 1930 to $5.42 in 1933. Hog prices per hundredweight fell even more, going from $9.47 in 1930 to $3.94 in 1933. Discouragingly low prices, depleted cash reserves, poor cash flow, and the high cost of production meant many farmers were questioning whether they could continue farming.[3]

In 1933, agent Maurice F. Goode described the serious problems faced by Harrison County farmers, writing, "The attention of the national leaders has at last been focused upon the critical situation with respect to a profitable market for agricultural products. With foreign demand practically gone, and little if any encouragement as to its return, it becomes apparent that agricultural production in this country must be adjusted to effective domestic demand, if parity prices are to be regained again for agricultural products."[4]

Even with the Depression going from bad to worse, better farmers were still making a small return for their work.[5] For example, farm records collected from Henry County farmers in 1931 showed a wide fluctuation of net returns, ranging from one farmer at the top end earning $900 while one at the bottom lost $2,800. The top one-third averaged $576 each while the lower one-third lost an average of $1,300 each.[6]

County Extension agents observed much despair among farmers, leading Floyd County agent Charles East to report in 1932, "[S]addest of all, there is that mental lethargy and discouragement on the part of many people that is perhaps the severest handicap we will have to meet. Because I have an abiding faith in the destiny of this nation, I am confident that we will work out of our present difficulties."[7] The following year he wrote, "I believe that I am properly interpreting the attitude of our farmers, generally, when I describe them as having suffered great losses, having become discouraged, being bitter, being hopeless, and that now the more courageous spirits are becoming to show fighting spirit that hopes for ultimate success, but, even in the event of ultimate failure, prefers to be found still trying."[8]

Those farmers who sought to weather the storm looked to agents such as A. Thomas Marvel in Elkhart County, who noted in 1934 that "as many as 100 farmers came to the office on Saturdays to try and get some information that would be useful to them."[9] Agents and Extension specialists at Purdue were asked to continue providing farmers with educational resources to help them through the tough times, which President Franklin D. Roosevelt endorsed. In 1932, before he was elected as president, Roosevelt expressed his views about agricultural education and research, writing:

> I believe thoroughly in agricultural education. I regard it as one of the most important and essential branches of the whole educational effort that is being carried on in the United States. I am a firm believer also in the value of the cooperative research and experimental work and the extension service which state institutions, in cooperation with the Federal Government, are rendering.
>
> I think it would be nothing short of a disaster if any of this work were seriously curtailed. Particularly in these times when farmers are having such a desperate struggle to maintain themselves, I think it supremely important that they should have the benefit of the expert advice that colleges, experiment stations, and extension services are able to give them, and it is equally important that we should continue to hold out to their children opportunities for an education that will make them something more than field drudges.[10]

Farmers became worried that all of their time spent growing crops and raising livestock was returning so little for their effort. While even the economists could not predict with any certainty how long the Great Depression would last, agents knew the only way out of the economic quagmire was to help farmers become more efficient producers. From an Extension perspective, keeping records was the only way to document whether or not farmers were using their resources in a wise manner.

Starting as early as the 1920s, Extension had pushed and prodded farmers to keep accurate, complete, and up-to-date records of their production costs and returns on expenses.[11] As DeKalb County agent T. E. Myers explained in 1931, "In order to stop the leaks [losses] on the farm, one must first know where they are and the best method for doing this is to keep a Farm Record Book."[12]

While some farmers readily accepted the idea of keeping records, others had a harder time with it, leading one agent to observe, "The natural aversion of most farmers to keeping records seems to make this a rather hard project to get a large spread of practice in."[13] Still, some agents were seeing adoption among farmers, with another agent observing that more farmers were "trying to adapt [their businesses] to changing economic conditions than ever before, largely because they are forced to do so."[14]

Extension used the farm records to help farmers focus on efficiency—not necessarily on increasing acreage or livestock units—as a way to increase farm profits. Agent Virgil Sexson noted this in his 1931 annual report for Greene County, writing, "Lower prices are also forcing farmers to appreciate the absolute necessity for more profitable producing units and proper methods to secure profitable production from these units. The boarder cow, the poor hen, the low-producing acre must be improved or eliminated if satisfactory profits are to result. Interest in marketing that will secure a greater share of a consumer's dollar is also increasing."[15]

Extension agents and specialists continued to push back against those farmers who wanted to rely solely on past experiences to make decisions. Instead, the Extension staff believed that decisions based on experiences could be improved with financial information and future economic projections. As Gibson County agent Roy Gwaltney stated in 1939, "The day has passed when a farmer can be just a farmer. He must, at the same time, be a good farmer and a good business man if he is to succeed. Business methods must be observed on the farm and records, study of general conditions and marketing methods are the three sign posts which guide business. Most farmers keep records of some kind but are not complete enough to permit an analysis of the farm business, as it should be done."[16]

To make the process of keeping records easier for farmers, agents took every opportunity to distribute farm account books developed by Purdue and the federal Agricultural Adjustment Administration.[17] They also used various methods to convince the farmers to keep the records, noting that in addition to using them to evaluate a farm's profits and losses, they could also be used to apply for government crop and livestock reduction programs, complete the state's income tax forms, and obtain credit.[18]

In 1931, Grant County agent Mosher D. Butler cited an example in his annual report to show how record keeping could affect a farmer's bottom line, writing:

> The project has a decided individual advantage too in that it helps the cooperator to wisely make changes in crop selections, it encourages him to market better and pay more attention to seasonal changes in market prices and prices of feed. It makes possible correcting a bad condition in balance in a few seasons and correcting it so it is correct. As an example we have in mind the work of one man operating a small farm. His labor income in 1928 was some $800. In 1929, $1,300, and in 1930, over $1,800. This change took place without the man changing farms, without the addition of a lot of machinery

> and was due entirely to increasing crop yields, changing his proportion of crops to live stock and changing the acreage of some of the crops. My impression is that this may well further increase his labor income by improving the quality of his livestock and by better marketing of his hogs. This particular farmer rated with the lower third during his first year of record keeping, but in 1929 he was high man in a group containing thirty-six other farmers and two of them "Master Farmers."[19]

Extension also enlisted well-known farmers to provide testimonials telling friends and neighbors why keeping records was good business. In his 1935 annual report for Allen County, agent Hugh Heckard directly quoted Marion Township farmer Clem Smith, who had kept records for nine years, as saying, "A farm record balances one's income as it shows in actual figures the profit and loss of each phase of a farmer's major enterprises, namely, poultry, cattle, hogs, and crops. It has also proven to me without a doubt that I had been keeping livestock that were not making me a cent, and the record has enabled me to dispose of those liabilities."[20]

Agents were frustrated that they could not convince more farmers to keep detailed records of their operations. This frustration was clearly revealed in 1938 by Fountain County's Alfred Hesler, who wrote, "With most farmers under mortgage and many farmers already in the red, it's hard to convince them it's worth while to keep records. If more good farmers would keep records, no doubt but that records would show why they are in the red!"[21] And in 1936, Gwaltney in Gibson County observed, "Of all Extension projects, this one meets more 'passive resistance' than any other. Every farmer contacted, agrees they ought to keep records. . . . Possibly, like water dripping on a stone, continued hammering on the subject will win out."[22] The following year, Gwaltney further elaborated, saying, "Farm record keeping seems to be the bug-bear that all farmers are afraid of. Although they agree they should keep them, they will not undertake the job."[23]

Adams County agent Lawrence Archbold wrote in 1930, "The matter of getting farmers to keep and analyze business records is a slow matter at the best and it appeals to the writer that this work is worth while and should go on, but that rapid progress along these lines will not take place until the present club boys and girls grow into manhood and womanhood."[24] The observation about "club boys and girls" was in reference to the fact that most counties had made farm records a project for 4-H members in hopes of getting the younger generation to adopt the practice when they became adults.[25] An example highlighted in 1930 by Floyd County agent Cory Alcorn may have shown the wisdom of this approach since he noted that "the most complete record in the group was kept by an eighth grade boy who had taken farm record work in his school classes. His father has him keep the records of the farm."[26]

Along with county agents, Purdue farm management Extension specialists expended time and resources pursuing the farm record project. This effort had been ongoing since the beginning of the Cooperative Extension Service, but it took on new importance during the Depression. While one goal was to encourage farmers to keep individual farm loss-profit records, another was to aggregate those records into a countywide summary. Some farmers weren't willing to keep records of their own operations, but many were still interested in these summaries and the area-specific information they provided as it allowed them to judge their own operations against those of others.[27]

Agents enlisted farmers who were willing to keep complete records of their operations for the project. Those expressing interest in participating were invited to a half-day meeting that explained the project details.[28] There, a Purdue agricultural economist would discuss how to make an accurate inventory of the farm, what records to keep, how to enter incomes and expenses, and what to expect from Purdue. The agricultural economists would then use the participants' data to prepare a final summary with all personally identifiable information removed.[29]

Those volunteering for the year-long farm record project received one or more visits during the summer from the local agent, a campus Extension specialist, or both. These visits were to verify that the participants were keeping records and registering the information correctly.[30] The visits kept the farmers on track in terms of what Purdue would need in order to make an accurate financial assessment of their farm and of the county.

In 1929, staff from Extension's Division of Farm Management analyzed records from 364 farms around the state. Just one year later, 551 farm records from 23 counties were analyzed.[31] With limited staff available, the number of records analyzed at no cost to the farmer was kept at about 500 each year. As more counties joined the work, individual farmers would have to pay for the analysis. For most years, approximately 200 farmers statewide were willing to pay for having their records reviewed and recommendations provided.

In Allen County, thirteen cooperators agreed to keep records starting in 1929. Otto G. Johanningsmeier, an agricultural economist from Purdue, visited each of the farmers in March 1930 to collect their records after the project ended.[32] In return for their cooperation and complete records, farmers were given an analysis and economic summary of their own data. Johanningsmeier and the local agent also made it a practice to discuss with all participants the strengths and weaknesses of their operations.[33]

The results from the farm record project were used to create charts and figures for each participating county, which individual farmers could then review in order to compare their operations to those included in the project.[34] Data from counties throughout Indiana were combined to provide a statewide summary showing where farmers were making money and which activities cost more than they returned.[35]

In 1935, the Division of Farm Management for Extension at Purdue used the collected data to publish a map of Indiana covering the years 1929 to 1934. For the report, Purdue agricultural economists divided Indiana into eleven areas: Central Grain and Livestock; Corn, Wheat, and Hogs; Dairy, Hay, and Tobacco; Kankakee Grain and Pasture; Northeast Dairy and General Farming; Northeast General Farming; Northwest Dairy; Southern General Farming; Southern Hill Farming; Southwest Corn and Wheat; and Western Cash Grain. The map showed average profits and losses within each of these farming areas by year, starting with a clear drop in incomes from 1929 to 1930. By 1933, the data presented by the economists showed that farmers were starting to make money from their operations.[36]

Of course, the average incomes shown in the Purdue report didn't tell the full story. Closer examination of the data would show that those in the top one-third were actually doing much better than those in the lower one-third.[37]

As the agricultural economists studied the underlying details generated from farm records, they were able to show that cutting expenses was not the sole answer when it came to a farm's profitability. Instead, efficient production was at the heart of profitability, even in the face of lowered prices.[38] As Henry County agent William G. Smith observed in 1932, "It was the man with the high-producing soil, hen, cow, and sow that kept operating expense down that was able to return a profit this year."[39]

The records from across the state showed that small farms tended to diversify while larger farms expanded their acreage.[40] In fact, diversification in both livestock and crops became an important strategy for keeping farms afloat during the Depression.[41] In 1933, agent Walter Rusk reported, "Diversified farming adds much to the stability of Blackford County's agriculture. The average size farm is 98 acres. On these farms you will find swine, sheep, beef and dairy cattle, poultry and good draft horses."[42]

Extension expended a large effort on the farm record project in order to convince farmers to focus attention on productivity. Records highlighted the strengths and weaknesses of a farming operation, which then allowed the farmer to make necessary adjustments to improve bottom-line profits. Reports by agents provided a firsthand perspective of the effects of the project. In 1932, one agent observed, "Even though agricultural prices are depressing and not much income is expected this year, most of the farm record cooperators had a feeling that this record work would give them a more thorough understanding of their farm organization, the better preparing them to follow a balanced farming program in succeeding years."[43] By 1938, another agent reported, "Without records, farmers tend to stress their strong points and slight their weak points which is contrary to good farm management principles."[44]

At the beginning of the Great Depression, Purdue began pushing out marketing and economic education to farmers through the county Extension agents.[45] This was done to provide information farmers could use alongside their operational records to steer their farms through difficult economic times. The information would also help them better manage the profitability of the farm when the economy improved.[46]

Radio and newspapers were used in the effort, but by 1931, the emphasis changed over to providing farmers with live presentations of the economic outlook showing expected trends in farm production, inputs to the farm, and cost of living.[47] Allen County agent Hugh Heckard described how he presented an in-person event in 1933 that also used the radio.

> For several years the Agricultural Outlook has been published by the [Extension Division of] Farm Management of the [Purdue] University but there has been a lack of a detailed program to get the information into the hands of the farmers. This year following the recommended program of the State Office in bringing the Outlook information to farmers of that county, a county meeting was held on February 17. A radio was installed in the assembly room and the Outlook film strip was used to illustrate [Extension agricultural economist] Dr. W[hitney] B. Stout's radio lecture. The farmers were very free to enter into the discussion of Outlook information. As a follow up to this county meeting, a great many individuals who were present at the meeting were personally interviewed to get their reaction on this type of extension work. . . . Many of the men present led discussions in community meetings on the Agricultural Outlook.[48]

These economic meetings, which became known as outlook meetings, were presented at township meetings of county farm bureaus, commodity groups, and livestock associations.[49] The meetings generally took place between December and March, when farmers were planning their upcoming production.[50] Projections by the economists during the winter months allowed farmers enough time to make changes to their operations ahead of the growing season.

Purdue's approach was to enlist volunteers to help the local Extension agent present more county meetings focused on the economy and marketing. To train the agent and volunteers, typically an agricultural economist from Purdue would hold regional conferences to provide participants with that year's economic information, including the current markets and future projections.[51]

Agent Roy Gwaltney in Gibson County reported having at first "to beg farmers to attend the District meeting; now we keep the date a secret, in order to keep our attendance within our quota."[52] In 1932, Cass County agent James Little noted in his annual report that he brought nearly two dozen people to attend the District Indiana Agricultural Outlook Conference at Peru and described the training, writing:

In the morning session, Dr. W[hitney] B. Stout, marketing Specialist from Purdue, spoke on "The Economic Situation." In the afternoon session, Miner Justice, State Statistician, explained the importance of and methods used in the crop and livestock reporting system. Following this were reports by the five county agents present on the Outlook of Corn, Wheat, and Oats, Hogs, Beef Cattle, Dairy Cattle, Sheep, Horses and Mules, Poultry and Poultry products. Following this, Dr. Stout presented "A Program for Adjusting Farming to Present and Prospective Conditions.". . . Each delegate present was asked to help present some information gained at the Outlook meeting at one of the Farm Bureau or other community meetings held in his locality. . . . The Purdue-U.S.D.A. film strip on "Economic Information" was used to present material concerning the cycles and seasonal variation of the various commodities.[53]

The agents worked with the volunteers to help them understand how the information affected farmers in their townships and to answer any questions for which they felt unprepared or were uncomfortable answering.[54] In some cases, the agent or the volunteer did an entire outlook program alone, while at other meetings, they shared presentation duties.[55] Topics often included past, present, and future prices; farm budgets; a review of actual farm records; and information about government subsidy programs.[56]

Chester Kimmell provided this description of how he conducted his outlook meetings in Allen County in the winter of 1938:

At each of the six meetings, the general outlook situation was given by the county agent and the commodity reports were given by volunteer leaders. After the volunteer leaders had given their reports, a general discussion was held in order to bring out the value of the information given and its relationship to the farm problems in the county. These discussions brought out one very important phase in that those present had an opportunity after receiving the economic information to readjust their individual farm layout to meet the situation [in] the coming year.[57]

In Adams County in 1939, agent Lawrence Archbold made sure that leaders of the county home economics clubs were also invited to the economic updates. For the twelve township meetings, he indicated that one-quarter of those attending were women.[58] In his report that year, he observed that "the attendance was much better than had been had for some time, that due to the presence of the women, the gatherings were more attentive and that heckling was not indulged in at any of the meetings."[59]

Agents such as Alfred Hesler in Fountain County devoted significant time to the economic outlook meetings. In 1939, he spent twenty-three days presenting outlook meetings, often delivering tailor-made information to dairy, beef, horse and mule, swine, sheep, and poultry associations. In some counties, a separate outlook was held for tobacco, wheat, or hogs in order to target information to those specific activities.[60] Fayette County agent Marion Detrick held hog outlook meetings in the fall of 1932 and winter of 1933, which gave him an opportunity to warn producers that there were more hogs going to market than expected, indicating that prices would probably be lower.[61]

One agent noted that outlook meetings "started some very heated argument along economics lines which proved that farmers are interested in this sort of information."[62] Another wrote, "The meetings were enlivened with frequent controversial discussions. One panel member held an interesting argument with a member in the audience over 'horses v. Tractors.' Another very interesting sideline discussion between a panel member and the audience was—'did power machinery cause the decline in quality of farm labor?'"[63]

The Extension outlook meetings and other programs focused on increasing incomes through different tactics, but the overall theme was the relationship between production and marketing, as Detrick described in his 1933 report.

> Income can be increased by three means—an increase in the selling price, a decrease in costs, and by efficiency in the use of labor and materials. The outlook figures prominently in locating the markets that offer the most profitable outlet for products and the best time to buy. Production methods must be adapted to the marketing scheme. For example, the prices of dairy products are highest in the fall and winter. Economical fall and winter production require balanced rations. The basis of a practical dairy ration is legume hay. Production and marketing are closely related.[64]

Purdue supplied the agents with charts to illustrate their talks.[65] The USDA also provided charts, pamphlets, and films that supplemented the Purdue information, which also helped agents and volunteers to better communicate the economic information to attendees across Indiana.[66] Agents commonly used these materials to reinforce the message from the outlook meetings throughout the entire agricultural community, with Allen County agent Hugh Heckard noting, "Copies of the Agricultural Outlook reports were placed in the hands of the agricultural leaders of the county and they, along with the local newspapers, were used to broadcast this information to all farmers in the county."[67]

The data presented at county outlook meetings must have been of particular interest to attendees since it was drawn from actual farm records. Detrick reported in 1931 that he used charts with summaries from twenty-four Fayette County farms to show differences in how livestock was managed and the impact on farm income.[68] Using actual records gave him a better chance of getting farmers to notice and understand how to offset less-than-stellar farm incomes.[69]

While outlook meetings were primarily focused on providing information, they also presented Purdue Extension with an opportunity to get feedback about farmers' expectations for the coming year. At the 1936 winter meetings, Kimmell asked Allen County farmers to answer this five-question survey:

1. Are you going to plant more corn in 1937 than you planted in 1936?
2. Will you have the same number of hogs in 1937 that you had in 1936?
3. Are you going to feed your dairy cows during the winter as well as you have fed them in past winters?
4. What are your prospects for hay in 1937?
5. Do you expect to increase your alfalfa in 1937?

He was able to conclude from the survey that "the corn acreage in 1937 would be about the same as it was in 1936; the number of brood sows would remain about the same; dairy cows would be fed about the same as they have been fed in previous winters; roughage and feed would be adequate; those that lost their spring seedlings would plant soybeans in order to have sufficient hay."[70]

It would take quite a few years to get farmers to see how the economic information provided at outlook meetings might be useful to their operations. In 1930, one agent reported being "surprised to notice very little interest in this subject here on the part of Mr. Average Farmer. . . . Either the subject is too technical or makes him think too much."[71] That same year, another wrote, "Presenting economic factors to farmers in a form that they readily can understand is a very difficult matter."[72] But agent Cory Alcorn in Dearborn County eventually observed a change in farmers who attended the meetings, writing, "[A]s the outlook material has been brought to the communities repeatedly for the last few years, an increasing number of leading farmers are coming to understand it and learn how to use it."[73] And by 1935, Extension seemed to be making more inroads as indicated by the fact that 7,500 copies of the publication *Outlook for Indiana Agriculture in 1935* weren't enough to satisfy the demand.[74]

After a slow start, the outlook meetings became one of the more popular programs for both large- and small-scale farmers. Farmers learned to use information from their farms and from those of their neighbors to judge what was happening in local markets. While personal experience was still important, getting farmers to appreciate that local prices were often reflective of national trends became a focus of the education efforts.

Detrick reported progress along those lines in his 1934 annual report for Fayette County, writing:

> A careful observer, however, readily recognizes a growing confidence in statistics and a diminishing confidence in the conclusions drawn from conditions as they exist at any given time. Less and less is being heard of the conclusion drawn regarding future supplies of hogs, for example, from the numbers seen along the road. There is a growing realization of the fact that the small territory that can be observed by any one man is likely to be very deceptive and that statistics alone give the supply and prospect for the entire territory that may influence the supply.[75]

With a growing interest among farmers to examine farm finances more closely, Purdue campus Extension specialists worked alongside county agents to offer farm management schools around the state. These were in-depth educational programs where attendees could study the business side of farming. Some schools lasted one to two days, while others were presented as weekly meetings held during the winter months.[76] In 1934, these schools were presented to approximately 2,400 attendees in twenty-one counties statewide.[77]

As might be expected, specific topics varied considerably from county to county and between regions of the state, but subjects generally included financing, accounting, record keeping, and marketing.[78] Examples of presentations made during the 1930s at farm management schools included:

- The place of pasture in the farm organization
- How I analyze my farm business
- Profitable cropping systems and varieties for the county
- Some farm practices that progressive farmers follow
- The place of the hog enterprise in the farm operation
- Combinations of crops and livestock on two of the more successful farms
- Using records to solve problems on Indiana farms
- The general economic outlook
- Adjusting farming in Indiana to meet the problems
- Producing a high-quality product that is graded[79]

The cornerstone of the farm management schools was the use of farm records from successful farmers.[80] At a 1933 Hancock County event, Otto Johanningsmeier, a Purdue agricultural economist, told attendees, "In times like these it is important that, whether it is an acre, a cow, a hog or whatever we have, that we get the highest efficiency in the unit both in producing and selling."[81] At a 1934 program in Hancock County, he gave farmers the following recommendations:

1. Adjust production of hogs so that they will be ready for market during the time of year when the prices are usually highest.
2. Have pullets ready for production by August or September so they will be laying most eggs during months when prices are high, and cull heavy during spring and early summer.
3. Increase the production of legume crops to build up the land, improve quality of pasture and help balance rations for livestock to reduce cost for protein supplements. Legumes are an excellent crop to use on land taken out of production.
4. Take advantage of the adjustment programs being offered by the [U.S.] Agricultural Adjustment Administration.
5. Keep farm records and study the farm business to help plan the farming program in the future.[82]

At times, individual farmers would also be asked to talk about their operations, sharing their personal records and experiences. Herman Busse from Dearborn County spoke at a 1938 farm management school. According to agent Cory Alcorn, Busse related "how he had used his farm records in analyzing his own farm business, pointing out that for the past few years he had been drifting away from the dairy enterprise into the feeding of beef cattle but that his more recent study of farm management principles through his record analyses and the farm management school had shown him that his farm could better be handled as a dairy farm than as a beef feeding enterprise."[83]

In addition to the farm management schools, Extension also held farm marketing schools around the state that provided more detailed discussions on marketing grain, poultry and eggs, hogs, and sheep. Efforts focused primarily on developing cooperatives for distributing milk, butterfat, eggs, and fruit. Many of these cooperatives were successful because they bought and sold agricultural products using quality standards.[84]

Agent Walter Rusk explained the importance of cooperative marketing for Blackford County farmers in his 1931 annual report, writing:

> Perhaps no other business is due to have so much change and growth in the coming years as the farming industry. A great agricultural revolution is taking place. Farmers have not been slow in adopting improved equipment and new practices. And this efficiency in increasing production creates the problem of organizing to fit production to demand and clearing the channels of marketing through cooperative effort. To do these things, leadership is required; leadership trained in the modern business of farming and experienced in the work of the cooperative organization.[85]

Between February and April 1933, Rusk hosted a ten-part series on cooperative marketing. Most of the lectures focused on marketing quality eggs, poultry, and cream and other milk products.[86]

During the 1930s, Extension would continue to emphasize marketing strategies that helped farmers gain the maximum return for their efforts. Strategies such as keeping milk cooled so it would not spoil, cleaning and grading eggs, and forming marketing cooperatives helped farmers get higher prices from both nearby and out-of-state markets.

As farmers sought to weather the tremendous economic storms of the Great Depression, they became more heavily dependent on Extension to deliver a prognosis of expectations for future markets. Information on national and international marketing trends became important topics of discussion that helped guide farmers as they set their production plans for the upcoming year.

While the larger economic picture was important for planning, county agents and Purdue agricultural economists also repeatedly encouraged farmers to focus on what they could do to better manage their money and to increase their on-farm efficiencies. To that end, agents stressed to farmers the importance of maintaining records to clearly see where their operations were making or losing money. This record keeping, when analyzed alongside expert economic projections, was intended to help farmers plan out a successful path forward for their farming operations in the years that followed.

Agents began observing signs of economic improvements from 1933 to 1935 in many counties, but they were still urging caution. In 1935, Charles East reported that Floyd County farmers had "a more profitable year than any they have experienced since 1929. The fact, however, does not justify the assumption that all Floyd County farmers made money, nor that 1936 should see any mark[ed] departure from policies recommended during the past few years. An improved buying power will permit additional expenditures, but those expenditures should be scrutinized carefully as to immediate or permanent value before being made."[87]

While the 1930s was hard on all farmers, those who kept records of their expenditures and incomes, changed their operations to become more efficient, and learned new marketing techniques were able to better weather the storm of the Great Depression. Extension's role in using the science of agricultural economics to guide them through those changes would forever alter how farmers viewed their operations moving forward.

CHAPTER 8

Farmers Hanging On by a Mere Thread Reached Out for Benefit Payments to Save Their Farms

In 1934, Purdue publicity director Thomas R. "Tommy" Johnston sent out an article written by agricultural writer DeWitt C. Wing that put the Great Depression into perspective for farmers. It noted, "Three 100-pound hogs would buy a 12-inch walking plow, a barrel of flour, a plain, strong chair and a pair of shoes in 1914. During the first eight months of 1933, these four pieces of merchandise would cost a farmer 8 hogs weighing 100 pounds each."[1] At the bottom of the Depression in 1932 and 1933, things looked hopeless for many farmers.[2] As one agent succinctly stated it, "[R]eady cash was hard to find."[3]

Farmers and the agricultural sector had spiraled into chaos as crop and livestock prices fell to record lows and commodities often sold for less than what they cost to produce. In 1932, the price of corn dropped from a range of 50–60 cents per bushel down to 10–12 cents per bushel.[4] A bushel of wheat that had sold for an average of $2 in 1925 was selling for 33 cents by 1931. Even the markets for poultry, which farmers had relied on for a steady supply of cash, saw chickens selling for 8 cents a pound in 1932, which was 42 percent less than they had sold for earlier that same year.[5]

President Franklin Roosevelt had been elected in the fall of 1932 on a platform that he would deal head-on with the Depression facing the country. His New Deal programs were an attempt by his administration and the U.S. Congress to alleviate the pain and suffering that the country was mired in at the time. The magnitude of the programs aimed specifically at the farming community would make U.S. Secretary of Agriculture Henry A. Wallace one of the country's most powerful and influential federal bureaucrats.[6]

The federal government identified two factors leading to depressed agricultural commodity prices: production exceeding national consumption and reduced exports to other countries. Economists from the U.S. Department of Agriculture (USDA) believed that reducing national surpluses would drive up the price of corn, wheat, and pork, which in turn would drive up commodity prices.[7]

The New Deal plans required that farmers voluntarily "do their bit in eliminating national surpluses," according to Clark County agent Earl Miles.[8] If growers agreed to the terms set forth in the government contracts, such as meeting a farm-specific quota, they would be paid predetermined prices for the average yield expected from the acres taken out of production or for the hogs not sent to market. The programs that paid farmers to limit their wheat, corn, and hog production were part of the Agricultural Adjustment Act (AAA) of 1933.[9]

In Indiana, the primary focus was to reduce the acres of wheat and corn along with the total number of hogs, but other federal programs also targeted reductions of commodities such as milk, peanuts, rice, sugar beets, cotton, and tobacco.[10] For the wheat reduction program, the national quota was set at 456.2 million bushels for 1934. Indiana's quota was set at 14.3 million bushels, which was a 15 percent reduction in wheat acres based on the prior three- to five-year state averages.[11]

Corn and hogs were linked together in a separate program because much of the corn raised was fed to hogs. For example, in Indiana, 85 percent of the corn raised was used to feed hogs.[12] Nationally, growers were asked to reduce corn by 20 percent in 1934 based on the previous three-year average and hogs by 25 percent based on the number of litters farrowed and marketed during the previous two years.[13]

It appeared to many outside agriculture that farmers were getting preferential treatment, which Dubois County agent Charles Nicholson explained this way in 1934:

> For a few years the farmer's plight was serious and held the sympathy of everyone, but in some cases today remarks are being made that the farmer has the better of the New Deal. His taxes have been reduced 40%; his farm prices [for crops and livestock] have doubled and in some cases even tripled. He gets a government check on wheat and a government check on corn and hogs. The teacher's salary is mostly paid by the State; his children are transported to school by bus; new county roads are constructed without taxing the farmer a cent. "How can this continue"? is the question asked by the business man and the day laborer.[14]

Paying farmers not to produce while people were in line for food handouts by charities soon became a public relations nightmare for county Extension agents, Purdue, and the USDA. The concern was that better prices for farmers would mean more hardships for consumers who were already short of money to buy basic necessities, including food. Business owners such as local grocers at first opposed the government programs because they thought the consumer would pay more for bread, pork, and lard.[15]

While farmers had little to do with the establishment of the national reduction programs, they still had to deal with this lingering resentment that they were getting special treatment under Roosevelt's policies. Often, the agents were left to explain the benefits to those who were more than a little skeptical about taxpayer money paying for setting aside acres and reducing hog production.

In retrospect, the most controversial aspect was that the corn-hog reduction program required the destruction of large numbers of hogs. At the start of the program, the federal government demanded that pigs weighing from 20 to 100 pounds and pregnant sows be slaughtered to reduce the numbers reaching the market.[16] According to USDA reports, this amounted to approximately 6 million pigs nationally.[17]

In Indiana, farmers provided the resulting meat supplies to the needy through charitable organizations. In some cases, hogs culled from herds were given to the Governor's Commission on Unemployment Relief, Red Cross, and other groups working with the unemployed and transients in the state.[18]

The amount of work necessary to move the government programs forward, especially the much larger corn-hog program, would take every bit of leadership and organizational skill that agents could muster.[19] In 1939, as Gibson County agent Roy Gwaltney reflected on the experience, he wrote, "The County Agent being [the] generally recognized and established Agricultural Agency in the field was drafted and given the responsibility of putting these programs into effect. For four years, the County Agent was so bound up with this work that he had time for little else."[20]

Agents now had to convince the state's farmers that they would be better off financially by signing contracts to grow less of the target crops and raise fewer hogs. In Fountain County, agent Alfred Hesler estimated that 80 percent of the corn acres and 95 percent of the hogs were under contract in 1934, showing the popularity of the program and, in all likelihood, the desperation that farmers were facing.[21]

It would require a superhuman effort and much perseverance by the agents to organize the programs, recruit unpaid volunteers to do the work, educate farmers on the requirements, and distribute checks as soon as possible.[22] By the end of 1934, with one year of Extension work under his belt, Dubois County agent Charles Nicholson indicated that more than 3,000 days of volunteer work had been spent on the government reduction programs, showing the value of volunteers in completing the work. He wrote, "Add to that number the [time spent by the] regular office stenographer and the county agent, and the entire program of this county required the time of 12 years for one individual to complete it thus far. The benefits so far have approached the sum of $172,000 for this county."[23] In Allen County, agent Hugh Heckard also reported investing a huge amount of time on the federal programs in 1934, noting that he provided updates at 134 farmer meetings, wrote 268 news stories, and received more than 4,000 office calls.[24]

Time spent on the commodity reduction programs obviously took time away from other worthwhile projects, causing them to be temporarily suspended or the work assigned to other volunteers.[25] Agent William Smith provided this synopsis of his work with the federal programs in Henry County in 1934:

> Agricultural Adjustment activities required of Extension Agents during the past year have completely over-shadowed customary programs, requiring as they did more than one-third of the time actively and much of the rest in the way of having it continually in mind. Other new endeavors have been shunned and activity in the less important parts of the established order were contracted or put on the shoulders of local committees or individuals. . . . In doing these things one felt much like a tight rope walker juggling three balls and at the same time retaining his poise, equilibrium and public esteem.[26]

Agent after agent reported similar inabilities to continue supporting established county Extension programs because the AAA programs took a majority of their time.[27] In 1933, Enos Rowe postponed his fall poultry work in Clinton County "due to the fact that we were snowed under in our wheat control work, having as high as 91 men in one day in the office, 87 on another day, etc., besides two meetings a day in addition often times."[28]

But those counties that were well organized were able to rely on the project committees to still get much of the work underway. In DeKalb County, agent T. E. Myers credited his volunteers for maintaining a consistent Extension program in 1934, writing, "The extra burden caused by the vast amount of farm relief work undertaken has changed the county extension program very little, except committees have been used to a much larger extent in formulating and carrying on various projects."[29]

The details of the wheat and corn-hog reduction programs were spelled out in law, regulations, and policies administered by the USDA, with both managed in a similar fashion.[30] The first step was determining who was eligible to participate. For the wheat program, policy directives made it mandatory—with some exceptions—that the prior three to five years would be used to determine average wheat yields. This meant that a farmer had to have grown wheat during those previous years in order to qualify for federal payments.[31] For the corn-hog program, farmers had to have at least three brood sows and grow at least ten acres of corn to participate.[32]

Agents received briefings at regional meetings around the state, where Purdue Extension specialists and USDA representatives provided details.[33] With new information, regulations, and policies constantly being added, Extension agents were continually kept in the loop between Purdue and the USDA.

Extension sent growers informational letters about the USDA programs using mailing lists derived from county assessors' offices, then agents hosted meetings to discuss the programs with farmers.[34] Bartholomew County agent Arvil Keesling noted that meeting presentations covered "First: economic necessities for the program, Second: how the plan operates, Third: the effect of the program on the individual farmer."[35] The information provided included details such as who qualified for the program, how the program would be managed, what farmers had to do to get paid for reducing their acres or hog numbers, and the prices paid for their participation.[36] Agents repeatedly emphasized that these were voluntary programs, not a government mandate.[37]

To reach those who didn't attend the meetings, agents placed a detailed explanation about the programs in local newspapers.[38] Farmers were asked to fill out preliminary forms that could be used as the basis for forming contracts and assigning how much each farm could place into the programs.[39] Then the next step was to get the farmers to apply for enrollment.[40]

In a September 1933 letter, Dubois County agent Charles Nicholson sought to reassure and encourage potential participants for the wheat reduction program, writing:

> Do not be excited or alarmed at the number of papers that accompany the application. There are very few that you have to fill out . . . so there are really just two things necessary, the application papers and the map of your farm. Those applications may be filled out in ten minutes if you know your wheat acreage for 1930, 1931 and 1932. . . .
>
> "Will the producers who sign contracts to reduce their wheat acreage fare as well as those who do not? Answer: The Secretary of Agriculture intends that producers who sign contracts shall fare better than those who do not. All the powers granted under the Agricultural Adjustment Act will be used to accomplish this end." With this question confronting us dear farmers, I think it is up to us as wheat farmers in this county to cooperate with our fellow farmers . . . that have signed this agreement. . . .
>
> The signing of this application does not bind you in any respect, but merely gives you the privilege of signing your contract; so it is wise to play safe and go shoulder to shoulder with the big majority of farmers in Dubois County.[41]

Nicholson wrote a follow-up letter dated October 31, 1933, that urged local farmers, tenants, and landowners to sign the application for the wheat program. In it, he said:

> You are asked to do only two things: (a) Reduce your wheat acreage and devote the idle acres to soil improvement. (b) Take the amount of money due for your cooperation for two years' reduction. All of you have shown a fine spirit of cooperation. I am your representative and doing the best I can for you; I have the farmers' interest at heart and I'm sure that when this two-year period is over on our wheat program, you will feel free to speak a word of praise for this office; and if you don't, we will do our best just the same for your interest and good.[42]

Those signing the initial application were asked to elect a permanent committee for their township or community.[43] The chairpersons of the township committees became members of the county-level wheat production control association and corn-hog production control association, which set the policies for how the programs would be managed and administered countywide.[44] The county associations were responsible for encouraging farmers to sign up, distributing the county quotas, ensuring program compliance, preparing the paperwork for the USDA, and distributing checks to the farmers.[45]

Farmers participating in the wheat and corn-hog programs were told the number of acres of each crop they could plant and the number of hogs they could raise.[46] In the 1933 Fountain County annual report, Alfred Hesler summarized how the process worked. His report specifically discussed the wheat program, but the process was similar for the corn-hog program as well. He wrote, "County and state allotments will be made [by the USDA] in Washington computed from the official records on past production and acreage. This allotment for each county will be sent to the county production [control] association committee. It in turn will take this allotment and divide it up among the producers in the county. This list would then be certified to Washington and the adjustment payments made from there."[47]

The township allotment committee was tasked with tallying the production numbers that farmers provided on their applications.[48] To that end, committee members spent months reviewing what farmers had listed. As Allen County agent Hugh Heckard noted, "This was a tremendous task because of the apparent overstatements of producers in the applications. However, the Allotment Committee by their very efficient work were able to find most of the overstatements and successfully brought the total production of hogs in the county under the county's allotment figures."[49]

Once completed, the adjusted summaries were sent to Purdue, where the data was tabulated and given to a state committee for approval.[50] Following that approval, contracts could be issued, with Heckard reporting, "The County's figures were finally accepted by the State Committee in July, and immediately upon their acceptance, eight typists were employed to type the contracts. The sign-up days were completed in the townships on August 18. Following the sign-up meetings it was necessary for the County Allotment Committee to check each contract for accuracy of signatures."[51]

In 1934, Enos Rowe indicated that processing the applications for the corn-hog program created a heavy workload in Clinton County. He reported, "Countless hours of checking and rechecking on the contract and accessory papers followed, with an office force sometimes numbering up to twenty, and day after day, and also many 4-hour extra periods at night, from 6 to 10 P.M. were spent in this work."[52] To a large extent, it was having to track down or verify hog numbers that made processing applications so complex and time consuming.

Agent Walter Rusk talked about the process of getting the final numbers and contracts signed for Blackford County, writing:

> After meeting the quota figures, contracts were then typed in triplicate and were immediately sent to each township, where the township committee-

men held meetings and the farmers signed their contracts. Transmittal sheets were then prepared containing contract numbers, name of contract signers and basic information. The original copy of the transmittal sheet and the white copy of the contract were forwarded to the corn-hog section, Washington, D.C. on June 23 [1934]. The contracts that were delayed for landlords' signatures and corrections were sent in at a later date with transmittal sheets attached.[53]

In 1934, Indiana had 24,280 wheat contracts signed that covered nearly 7 million bushels of reduced production and paid $3.1 million to farmers.[54] The corn-hog program was also popular, with nearly 86,000 farmers signing contracts by midsummer in 1934 and payments reaching $20 million.[55] For many farmers, the money promised through these federal contracts would mean the difference between paying bills such as mortgages and taxes versus going through a foreclosure or declaring bankruptcy.[56]

Under the terms of the federal wheat contracts, participants knew up front that they would receive 27 cents per bushel in 1933, 29 cents per bushel in 1934, and 33 cents per bushel in 1935 for the wheat acreage taken out of production.[57] In 1934, farmers participating in the corn-hog program were promised 30 cents per bushel of corn and $5.00 per head for hogs for reducing their acres and animals.[58] Given that the market price for hogs was $3.49 per head at the time, the promised federal payment was significantly higher.[59]

Farmers were told they would receive their reimbursements spread over two or three payments, depending on the program and year.[60] They would receive their first check soon after the contract was approved by the USDA, with remaining funds provided as soon as the county production control association had certified that a participant was in compliance with the program requirements when the growing season had ended.[61] In general, the contracts sent in to the USDA were paid within three to five weeks if the paperwork was judged to be complete and accurate.[62]

In Clinton County, Rowe indicated that the anticipated wheat program payments had earned farmers a temporary reprieve on paying their taxes.

Taxes, of course, are due in this county, and our County Allotment Committee has made arrangements with our county treasurer, and other officials, to the effect that those who have signed wheat contracts may be granted an extension of time on payment of taxes, by the issuance of an order signed by the County Allotment Committee and the grower, and filed with the county treasurer, to that effect. Something over 250 growers have taken advantage of this movement and deferred payment of their taxes until their wheat adjustment checks come in, without penalty. This has been a big thing in our county, and relieved an otherwise large burden from the shoulders of our wheat growers. We are looking forward to the day when our checks will arrive and folks will have a definite understanding of just what this big movement in wheat is all about, through the actual receiving of the cash payments.[63]

Farmers often received their checks by going to the local Extension office or to other sites predetermined by the county allotment committee.[64] In some cases, farmers were frustrated with how long it took to receive their first payment, but others were more understanding. In the 1933 annual report for Fountain County, agent Alfred Hesler remarked, "Those who have signed [contracts] no doubt have felt that their check would be in their hands soon after signing, but it takes time to check every detail. We will say folks have been mighty patient and we appreciate it."[65] The first wheat checks came in December 1933, helping to make the holidays more joyful for those that received them.[66]

Before farmers could receive their final payment for participating in the federal wheat and corn-hog programs, they first had to prove that they had complied with the program provisions. The county allotment committee was responsible for providing this verification.[67] To determine compliance, each county production control association hired field inspectors to measure each farmer's wheat and corn fields and count the number of hogs on site and sold in order to verify that the grower had not exceeded the established quotas.[68] The government deducted a fee from the farmers' checks to pay administrative costs such as salaries and other program expenses.[69]

In 1934, Warren O'Hara reported that farmers in Hancock County were charged 3 cents per bushel, or roughly $1.50 per farm, to cover the cost of the five field inspectors hired by the county wheat allotment committee.[70] In comparison, the corn-hog program cost local participants 3.5 cents per bushel of corn, or about $11.00 per farm, for the contract, which included the expenses for verifying compliance on both corn and hogs. Trying to deflect local criticism, O'Hara said, "In fairness to the Hancock County Committee, it might be shown that the Indiana State average cost for administration [of corn-hog] expenses was $11.56 per contract, and the state average for every dollar paid out in benefits to the farmers shows 3.854 cents for administration expense."[71]

Those working as inspectors received training on how to inspect a farm and complete the required paperwork to ensure consistency and accuracy across the counties.[72] Compliance inspections for wheat were conducted in late spring, and for corn and hogs, in August and September.[73] Farmers were told when the inspection would take place because they were expected to help measure the fields and answer any questions the inspectors might have.[74] If the participants raised hogs, inspectors would also have to tally up the number of animals on-site and review sales slips to match against the numbers of hogs the farmers had reported. In Fountain County alone, inspectors had to review more than 6,000 sales slips in 1934.[75]

The inspectors determined acreage by walking the fields and pushing a measuring wheel or by using a 100-foot steel chain.[76] Then they compiled the data and submitted the information for each farm along with county summaries to the USDA. Once the government officials approved a county's compliance numbers, those farmers meeting their contract obligations received their last payment.[77]

The results for the wheat program showed that farmers by and large stuck to their planting allotments.[78] For example, in 1934, only 5 percent of those under contract in Clark County exceeded their allotted wheat acres. Those growers were allowed to use the excess wheat as hay.[79] That same year in Bartholomew County, 40 farmers who had planted more wheat acres than allowed were required to destroy the extra wheat.[80]

For the corn-hog program, planted acres had a 10 percent allowance before the farmer was considered to be out of compliance.[81] Those who exceeded this amount were allowed to cut the corn for livestock feed. Participants who exceeded their hog quota would have their payments withheld until the problem was resolved to the satisfaction of the county committee.[82] In 1934 in Dubois County, 936 contract signers were found to be in compliance and eight were in partial compliance, while one contract was not certified because of noncompliance.[83] In Carroll County, 43 farmers were penalized $10.00 per contract for being out of compliance.[84]

Hogs that exceeded contract quotas could be donated to local churches or other groups without the farmer receiving any compensation.[85] When a few Hancock County hog farmers went over their allotted quotas in 1934, O'Hara described the steps taken to resolve the situation, writing:

> As excess pigs above the quotas allowed on corn-hog contracts were scattered throughout the entire county and donation papers signed promised delivery only to [the] nearest railway station, it was decided to arrange a truck to pick up these pigs and deliver them to the organized public relief, Ralph Frank, County Director. Frank arranged with the Commissioners to feed them until slaughter on the County Infirmary farm.
>
> Lawrence Holmes donated his time, and volunteer funds were collected to pay him 5 cents per mile from first stop until delivery. The pigs were collected at a cost of only $11.85 in two full days. The group consisted of 207, of which three were culls and three weighed 250 lbs. each. The average was 38 pounds. It is planned to use the meat locally within the county for poor relief this winter.[86]

In October 1934, with the first cycle of the federal programs coming to an end, farmers were asked to provide input as to whether they wanted to see similar wheat and corn-hog programs for the upcoming season and whether they would support expanding the programs to include additional crops and livestock.[87] Agents held meetings to address these questions, with farmers across Indiana responding with overwhelming support for the existing programs and for expanding them.[88] They also indicated that they would prefer to have one all-inclusive plan rather than to have participation broken down by commodity type.[89]

Agent Hugh Heckard provided the following information on the referendum he held in Allen County in the fall of 1934:

> A schedule of meetings in each township was prepared and at these meetings the corn-hog situation was discussed, and those farmers who cared to were given the opportunity to cast their ballots in a sealed envelope. When the referendum was taken on October 26, a total of 1,402 contract signers cast their ballots. Twelve hundred and ninety of this number voted 'yes.' One hundred ninety-four non-contract signers cast ballots and 164 of this number voted 'yes.' By this referendum farmers gave their whole-hearted approval of the AAA corn-hog program.[90]

While the referendum had indicated farmers wanted to continue the AAA programs, the Roosevelt administration received a major setback toward this goal on Monday, January 6, 1936, when the U.S. Supreme Court ruled that taxing the nation's millers and corn-hog processors to fund these programs was unconstitutional.[91] With the funding deemed unlawful, the programs were discontinued, but the government promised to honor the existing contracts.[92]

Though the federal wheat and corn-hog reduction programs were only in effect for a few years, they would become an important lifeline to growers during economically challenging times. As one agent noted, "In 1933, 1934, and 1935 an emergency existed that most farmers were hanging on by a mere thread and reached out for these benefit payments to save their farm."[93] Indeed, by 1934 and 1935, it seemed that farm life for many in rural Indiana was beginning to take a turn for the better.[94] This was due at least in part to the fact that the federal programs had helped stabilize farm incomes.[95]

From the agents' perspective, both the wheat and corn-hog programs were considered quite a success.[96] It seemed to the agents that the AAA programs had allowed farmers to once again pay their bills and provided some financial stability during times that were anything but stable.[97]

In Clinton County, farmers received more than $700,000 under the adjustment programs.[98] In Allen County, agent Hugh Heckard observed that the $440,000 received locally meant that area farmers were "paying a great amount of debts and purchasing commodities and articles which their farms and homes have needed for years."[99]

The infusion of money into local economies along with an improving outlook for jobs prompted Charles Nicholson in Dubois County to write in 1935, "[The] laboring man hesitates to point his finger at the farmer this year and call him a 'parasite' because he has had little time to do this being occupied in the factory during the day and in many cases two or three hours at night. As long as he has work and is getting a living wage he seems to be satisfied regardless as to what the farmer gets. The right kind of an honest fellow can now go to the bank and get one hundred dollars, whereas a few months ago this condition did not exist."[100] The following year, Thomas Marvel in Elkhart County reported that "better times for 1937 can be anticipated for the rural population of the county, with more interest in the better things of life and more time and money to enjoy a higher standard of living."[101]

Agents could also see that by 1935 farmers were beginning to improve buildings and fences, paint structures, and repair farm equipment—evidence of slowly improving living conditions.[102] Nicholson noted that just having the AAA programs in place in Dubois County resulted in "the spark of enthusiasm having been rekindled in the little farm home; hundreds of farm families are taking a new lease on life."[103]

Nicholson went on to explain that even though the AAA work had been tiring and had tested his patience, "It was the greatest opportunity that the extension program ever had in gaining the confidence of the public, justifying the position of the County Agent and actually delivering the goods."[104] In Allen County, Heckard agreed, saying that the programs were "most beneficial to [the] Extension service in that large numbers of farmers have been brought into contact with the Extension Service who had not previously been identified with it."[105]

Farmers had personally observed the long hours that their county agents had put in to complete the work and get the program participants the money that they so desperately needed. For those who had participated, this solidified the link between the agent and the farmer.

Jefferson County agent Benjamin H. Doddridge drives his Ford Model T Touring Car, ca. 1926. In the early days of the Cooperative Extension Service, agents traveling to rural farms and communities faced rough driving conditions during the winter months, when roads were rutted, and in the spring, when rains flooded low-lying areas. Courtesy of J. C. Allen and Son.

For this farmer, clearing rocks from this field day after day in 1921 must have been a daunting task. In the early twentieth century, farmers often lived a life of subsistence, producing just enough income to get by. The long hours and physical labor left very little time for learning new ways and techniques by attending agricultural programs at Purdue University in West Lafayette. It was the county Extension agent's role and responsibility to bring that education directly to farmers throughout the state. Courtesy of J. C. Allen and Son.

Children stand in line for a turn to receive fresh milk in their buckets in 1932. During the Great Depression, with little money to spare and jobs difficult to come by, families reached out to charities such as the Salvation Army and American Red Cross as well as to county township officials for assistance with obtaining food, clothing, and other support. Courtesy of J. C. Allen and Son.

Agents leaned heavily on well-known and well-respected area farmers to serve on county Extension advisory boards and on committees to conduct demonstrations on area farms. In 1938, Everett Otte (in wagon), who served on the Jackson County Corn Improvement Committee, was a hybrid seed corn grower who had won four gold medals in the local five-acre corn contest. Also pictured is his brother Wilbert Otte, who was the county champion for the 1937 and 1938 events. Courtesy of Purdue University Libraries, Karnes Archives and Special Collections.

A blackboard shows that the topic for this 1944 meeting of the Tippecanoe County Extension advisory board was "A United Effort among Farm Organizations." Up until the mid-1930s, agents had handpicked the individuals who served on the county-level Extension advisory boards. But in 1937, the Indiana General Assembly passed legislation that required the Cooperative Extension Service at Purdue to make those boards more representative of the community. Courtesy of Purdue University Libraries, Karnes Archives and Special Collections.

Members of the Tippecanoe County Extension Advisory Board meet for a planning session in 1939 to establish which topics agent Sheldon Pershing was expected to cover in his Extension program during the upcoming year. The blackboard indicates that officers for the group included chair M. Peake, vice chairs H. K. Bilsland and Mrs. E. Bausman, and secretary D. O. Neidigh. Courtesy of Purdue University Libraries, Karnes Archives and Special Collections.

What Is The Farm Bureau?

What Has The Farm Bureau Done?

Actions speak louder than words—study the following report of activities and you'll have the answer to the above questions. The combined influence of years of active membership has made possible the following achievements in legislation, education, recreation and organization:

Legislation

Organized farm influence supported by Whitley County along with other counties, through the state Farm Bureau, was effective in securing the passage of:

Property tax limitation law (1932) and spreading of the tax base (1933) causing big reductions in tax payments on real and personal property.

Minimum payment of $700 per teacher by state (1937).

Indefinite moratorium on issuance of road bonds (1937).

New county agent law removing the office from politics (1937).

New dog tax law to better protect livestock raisers (1937).

Removal of $2,000,000 General Welfare costs from shoulders of property tax payers (1938).

And others of lesser importance.

Education

The following activities have been supported directly or indirectly:

4-H CLUBS

Helped secure leaders, enrollment and finance.

Enrolled corn and colt club members in State Gold Medal Clubs.

Assisted with 4-H Fair and township achievement meetings.

Presented achievement pins to completing members.

HOME DIVISION

Supported farm ladies in their organization work and their sponsorship of 4-H girls' clubs.

A long list of accomplishments is theirs in their own right.

HOOSIER FARMER

Sent to all members without extra charge to enable them to keep informed on activities of the Indiana Farm Bureau, other county farm bureaus, and affiliated services.

EXTENSION SERVICE

Was represented on County Extension Committee.

Provided community meetings for extension contacts.

Supported short courses and one-day schools.

Promoted corn improvement, farm record keeping, beef and dairy cattle development, alfalfa expansion, soil conservation, sheep dipping, ram exchange, etc.

Social Recreation

To develop talent, enlarge circles of acquaintance, provide good entertainment and promote fellowship.

Gave 36 free community entertainments.

Cooperated in WLS county-wide entertainment.

Conducted Farmer's Picnic—Saturday of the 4-H Fair.

Exhibited "Pioneer" floats in 4-H Fair parade.

Held nine community supper meetings.

Held county get-together meeting in December.

Continued on Back Page

County farm bureaus were instrumental in helping local agents carry out Extension programming. This 1939 promotional flyer listed a number of accomplishments related to Extension work that the Whitley County Farm Bureau viewed as noteworthy, including extensive involvement in a wide range of educational programs and participation on the Extension advisory committee. Courtesy of Purdue University Libraries, Karnes Archives and Special Collections.

While agents were busy administering Depression-era programs, they had to rely on volunteers to keep Extension demonstration work going. In 1934, the Miami County Poultry Committee (pictured) was charged with carrying out the poultry work outlined by the local Extension advisory committee, such as finding cooperators to host demonstrations, monitoring project developments, and keeping the agent apprised of the work. Front row (left to right): Otis Sullivan, Perry Spangler, Mrs. Cecil Riley, Owen Murphy, and Cheston Long. Back row (left to right): A. J. Long, Leon Todd, Frank Musselman, Willis Flora, and Sol Clemens. Todd was a poultry Extension specialist at Purdue. Courtesy of Purdue University Libraries, Karnes Archives and Special Collections.

In 1935, Carroll County Extension held this cattle demonstration at the Robert Bowman Farm near Rockfield, where farmers could ask the agent, campus Extension specialist, and hosting farmer about breeding, feeding, pastures, and marketing. It was important for Extension to continue hosting demonstrations during the Depression to support farmers as they tried to earn a living and use every means possible to keep their farms going. By teaming up with the better farmers in the county for demonstrations, Extension's recommendations were more likely to be accepted and research findings from Purdue shared to benefit farmers throughout the agricultural community. Courtesy of Purdue University Libraries, Karnes Archives and Special Collections.

During the Depression, many families depended on community and home gardens for fresh vegetables for evening meals or produce that could be canned and preserved for winter. Much of the work done in community gardens, such as this one in Tippecanoe County shown in 1932, was done by unemployed men and women who contributed time and effort in exchange for receiving food to put on their tables. Courtesy of Purdue University Libraries, Karnes Archives and Special Collections.

Those managing the Richmond Community Garden depended on volunteer labor from individuals such as these men to harvest the 1932 crop of tomatoes. A local social service bureau was responsible for distributing the produce grown in the garden to those most in need. Courtesy of Purdue University Libraries, Karnes Archives and Special Collections.

At the height of the Depression, many men and women were unemployed but wanted to work. Here, unemployed men from LaPorte harvest 2,600 bushels of potatoes from the county farm in 1933 to help feed those who needed assistance. Courtesy of Purdue University Libraries, Karnes Archives and Special Collections.

A view of an unidentified community garden in 1931 shows women, men, and children removing weeds. Courtesy of J. C. Allen and Son.

A father and son tend to a garden together in 1935. Many children who participated in 4-H made gardening their main project during the Depression, which helped support the family's food supply. Courtesy of J. C. Allen and Son.

Willard B. Ward (foreground), Purdue horticultural Extension assistant, presents a gardening demonstration to the Hoosier Garden Club from Otter Creek Township in Ripley County in 1938. Throughout the Depression, Ward traveled around the state to share his gardening expertise in order to help families who relied on home gardens as a much-needed source of food. Courtesy of Purdue University Libraries, Karnes Archives and Special Collections.

CO-OPERATIVE EXTENSION WORK
IN
AGRICULTURE AND HOME ECONOMICS

Purdue University Agriculture Extension Dept., U.S. Dept. of Agriculture, Decatur County Board of Education, and Decatur County Farm Bureau Co-operating

Extension Service
County Agent Work
"Nutrition Letter"

Dear Madam:

Can The Surplus!!
Newer Methods of
Canning: Fruits, Vegetables
Preserving: Jellies, Jams
Discussed at Leaders Training
Meeting-Held Tuesday June 3
At - Greensburg High School
Starting at 9:30 A.M.

Each leader bring 1quart jar, 1pint jar, and 1 jelly glass - jar rubbers, towel and holder.

Each leader may bring in more fruit, vegetables, or meat she wishes canned. Bring in supplies needed for material to be canned. (Suger, cans, rubbers, etc.) Also Sandwiches and a covered dish.

[signature]
County Agent

In 1930, Decatur County agent Ira Thurston used this flyer to invite home economics club leaders to a program on canning fruits and vegetables. Training programs such as these targeted the leaders, who would then teach what they had learned to the members of their clubs, thereby extending the reach of Extension education efforts. Courtesy of Purdue University Libraries, Karnes Archives and Special Collections.

This 1931 Shelby County home economics program addressed nutrition and cooking topics, which were essential at a time when many families were growing a larger portion of their food supplies and stretching their resources to help make ends meet. Courtesy of Purdue University Libraries, Karnes Archives and Special Collections.

Children watch as their mother prepares garden produce for canning in 1935. Extension taught families how to preserve produce for later use to help them extend their homegrown food supplies. This allowed them to save money on groceries and stretch their budgets during tough economic times. Courtesy of J. C. Allen and Son.

COOPERATIVE EXTENSION WORK
in
AGRICULTURE and HOME ECON.
State of Indiana

Purdue Univ.
U. S. Dept. Agric.
Ripley Co. Board
of Education
Co-operating

Versailles, Ind.
April 12, 1937

Extension Serv.
County Agent
Work-Phone 92

YOU HAVE AN EXTRA FARM HAND ON YOUR FARM THIS YEAR AS PORTRAYED ABOVE. ARE YOU MAKING USE OF HIM?

Farm Records can guide you to financial success if properly used. To be of greatest help in analyzing your farming operations complete records of all transactions should be kept and entered periodically in the farm account book. These records when summarized at the end of the record year indicate which departments of your business have been most profitable, and which have been least profitable or have lost you money. Use your record book as a guide to increased profits from your farming business.

Check on the following for the months of March and April.
Lime and Fertilizer bought or applied.
Cost of Gross feed and seed potatoes.
Baby Chicks bought or hatched with mortality record.
Number pigs farrowed--All feed purchased.

Sincerely yours,
M. J. Huxley
M. J. Huxley
County Agricultural Agent

In a 1937 letter to Ripley County farmers, agent Martin Huxley equated keeping farm records to having an extra worker. During the Depression, Extension encouraged farmers to use records to run more efficient operations and identify where they were making or losing money. Courtesy of Purdue University Libraries, Karnes Archives and Special Collections.

PURDUE UNIVERSITY AGRICULTURAL EXTENSION DEPT. AND U.S. DEPT. OF AGRICULTURE AND STARKE COUNTY BOARD OF EDUCATION COOPERATING

COOPERATIVE EXTENSION WORK
IN
AGRICULTURE AND HOME ECONOMICS
STATE OF INDIANA

EXTENSION SERVICE
COUNTY AGENT WORK

1. ACRES AND YIELDS OF GRAIN AND HAY.
2. BINDER TWINE, MACHINERY REPAIRS AND HIRED LABOR.
3. GRAIN, HAY AND LIVESTOCK SOLD.
4. THRESHING EXPENSES.
5. INSURANCE ON BUILDINGS, LIVESTOCK AND OTHER FARM PROPERTY.
6. TAXES AND INTEREST.

Soon after grain is threshed and hay put up is the best time to enter the acres and yields of such crops. If you are not certain about the yields, use the simple rules printed inside the back cover page of the Purdue Record Books, ask your local committeeman or see me. Time spent in entering these items now will mean twice as much saved next winter. A few minutes will do it now.

Very truly yours,

A. C. Shideler

A. C. Shideler
County Agricultural Agent

In 1935, Starke County agent Arthur Shideler reminded area farmers that recording a few details about their farm operations each day was much easier than waiting until later, when the job might be larger and more difficult. Extension promoted a system known as the Purdue Record Book to help farmers keep their essential information organized. Courtesy of Purdue University Libraries, Karnes Archives and Special Collections.

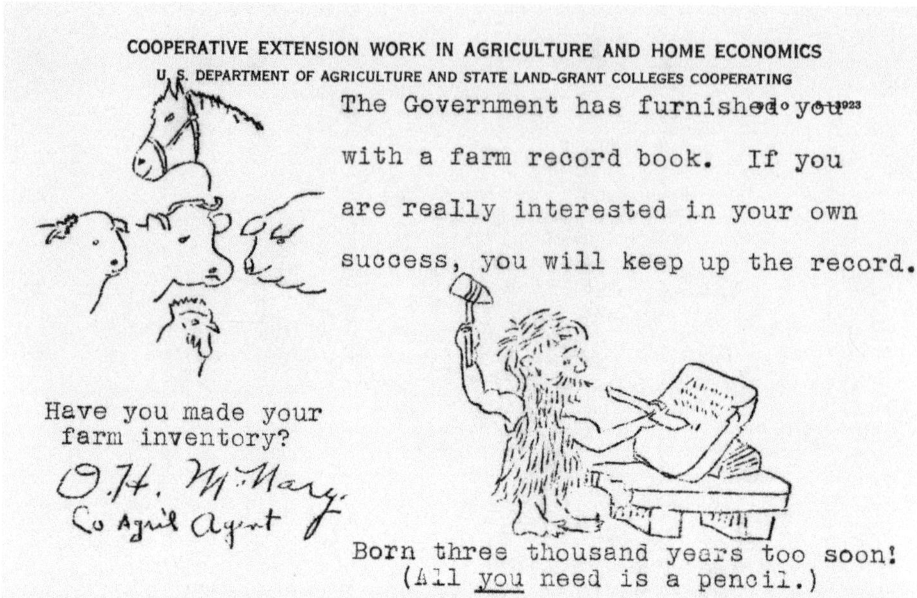

Pulaski County agent Oran McNary reminded those farmers enrolled in the farm record project conducted by Extension's Division of Farm Management at Purdue to stay on top of recording their revenues and expenses in 1935. These year-long records were then aggregated and analyzed by Purdue agricultural economists, who used them to summarize county and regional differences in farm incomes, identify what made farmers money, and report on those practices that resulted in higher incomes. Courtesy of Purdue University Libraries, Karnes Archives and Special Collections.

Purdue agricultural economist Otto Johanningsmeier (right) talks to Earl Maguire of Tippecanoe County about his farm records in 1930. Agricultural economists such as Johanningsmeier would meet individually with participants in the farm record project to discuss the strengths and weaknesses of their farming operations. Courtesy of Purdue University Libraries, Karnes Archives and Special Collections.

Purdue agricultural economist Otto Johanningsmeier (right) discusses farm records with Charles Howell of Rush County in 1935. These one-on-one discussions allowed farmers to use the information they had gathered to see where their operations were making money and which areas needed to be improved or changed. Courtesy of Purdue University Libraries, Karnes Archives and Special Collections.

Extension held township-level meetings, such as this one in Tippecanoe County in 1934, to discuss the details of the contract for the Agricultural Adjustment Administration corn-hog program. Speakers helped farmers better understand what they were signing, what they would be paid, and what they had to do to comply with the requirements of the program. Participation in this voluntary federal benefits program would allow many farmers to weather the economic storm of the Depression. Courtesy of Purdue University Libraries, Karnes Archives and Special Collections.

A man applies lime to a Carroll County farm field to raise the soil pH in 1931. A federal program to provide payments to support restoring the soil's productivity allowed farmers to purchase and apply more lime during the Depression. Courtesy of Purdue University Libraries, Karnes Archives and Special Collections.

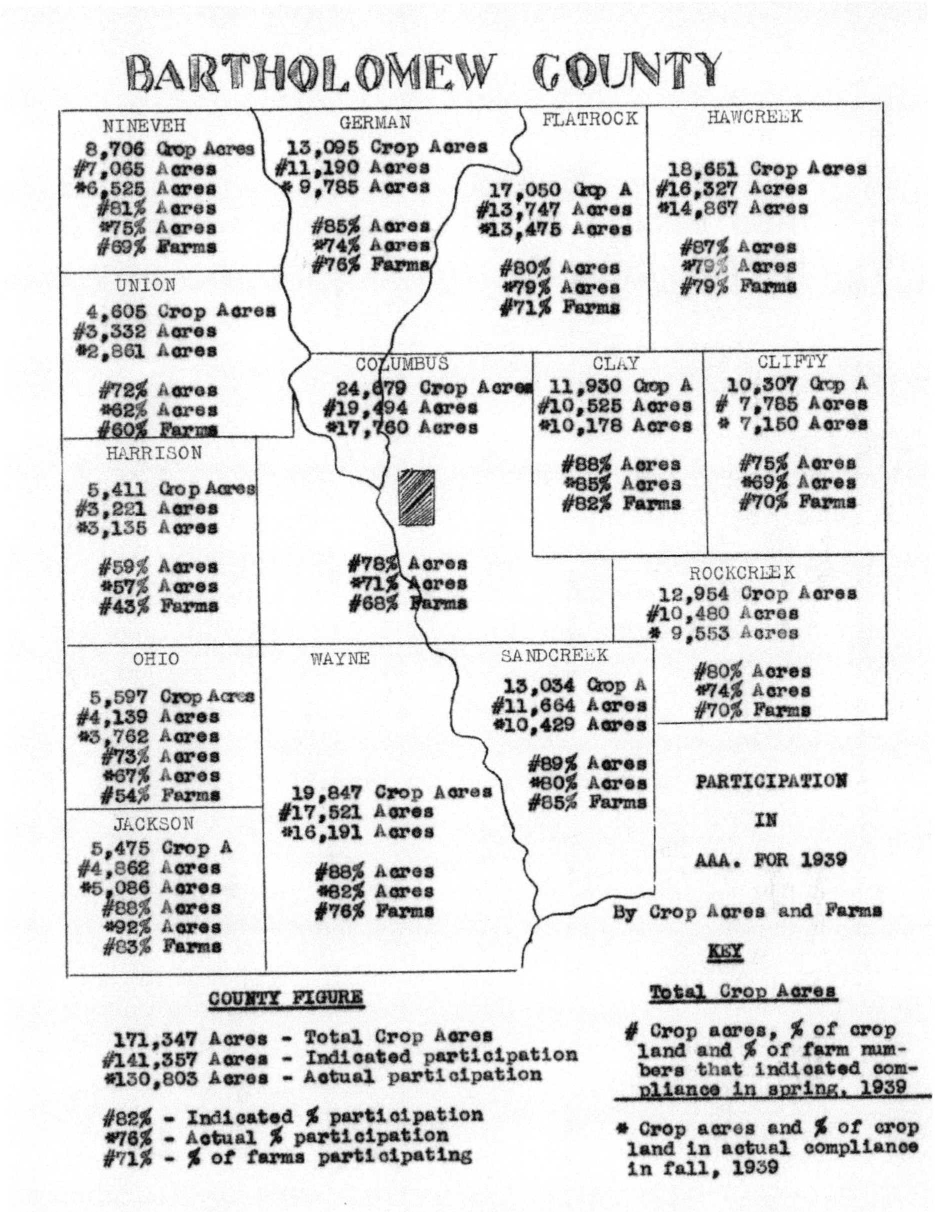

This map shows the information collected in Bartholomew County in 1939 as part of the USDA Agricultural Adjustment Administration program that guaranteed prices for crops grown under government contracts. Courtesy of Purdue University Libraries, Karnes Archives and Special Collections.

CHAPTER 9

Conservation of Soil Is the Solution on Which Will Hang Future Extension Activities

When the U.S. Supreme Court ruled in January 1936 that the Agricultural Adjustment Act was unconstitutional, President Franklin Roosevelt and Congress reacted in a bipartisan way to quickly work around this serious roadblock. On February 29, 1936, Roosevelt signed the Soil Conservation and Domestic Allotment Act (SCDA) as a replacement.[1] While the details of the new law were radically different than its predecessor, the core of both programs was to reduce the nation's surplus of agricultural products while augmenting the price farmers received for their crops and livestock.

The newly minted program focused on reducing corn and wheat acreage while encouraging farmers to adopt practices better suited for improving and conserving the soil. This would be accomplished by paying farmers to reduce erosion, plant permanent pastures, make more lime applications, sow more legumes, and plant trees.[2] In writing about the changes in 1936, Boone County agent Clarence K. Whistler observed, "[C]ounty agents were encouraged that the federal program focused on conserving and improving their [farmers'] lands. While the program worked to refocus attention on soils, it still focused on reducing excess agricultural inventory to drive prices higher. The program is making better farmers and bringing about improved farm and land use practice."[3]

The agents could not have asked for a federal program more favorably aligned with Extension's core values of returning farming to a more profitable status and doing so with practices that improved soils. Whether it was planting cover crops to protect the soil's surface or building up levels of organic matter with manure, Purdue agricultural researchers and Extension specialists pushed the need for farmers to protect, conserve, and improve the soil that their profitability was based on.[4] In fact, as one agent observed, conservation practices would be "the solution on which will hang all other future extension activities."[5]

Agents had noticed with increased frequency that high-yielding crops were slowly depleting farm ground of essential plant nutrients, such as nitrogen, phosphorus, and potassium. In a 1930 annual report, Benton County agent Glenn L.

Reed wrote that the soils lacked much organic matter and the levels of nutrients had been significantly reduced.[6] Another serious issue faced by farmers involved thousands of acres of badly eroded land. This was especially true in many southern counties, where hilly ground that was once forested had been converted to cropland during World War I. Researchers from the Agricultural Experiment Station at Purdue, Extension specialists on campus, and the county agents all were concerned that these less-productive soils would eventually become so degraded that farmers would completely abandon them and farming when the land could no longer be used to make a living.[7]

The SCDA program brought with it a new lexicon, with talk about replacing soil-depleting crops such as corn and discussions about soil-conserving and soil-building practices.[8] As with the previous program, the U.S. Department of Agriculture (USDA) would still pay farmers for not planting corn, but under the SCDA program, those payments would now be tied to implementing specific conservation practices considered to be more beneficial to the soil.

Agents quickly went to work to explain the complicated, convoluted, and ever-changing program to their constituents.[9] At the onset of the SCDA program, one agent noted, "The majority of farmers are of the opinion that the program will not stand up. They say it has been made more complicated [as compared to the Agricultural Adjustment Act program] instead of simplified. There are too many if's and and's, too many conditions and rates of payment for a farmer to know what he can do."[10] But if farmers wanted the guaranteed federal payments, they had little choice but to work out the details, so they relied heavily on agents for advice.

Under the SCDA program, land uses were classified as soil-depleting, neutral, or soil-conserving. Floyd County agent Charles East noted that the USDA defined soil-depleting crops as "field corn, sweet corn, small grains harvest for grain or hay; annual grasses pastured or harvested for hay or grain, seed potatoes, melons, strawberries, sweet potatoes, and other commercial truck and vegetable crops." He went on to report that soil-conserving crops included:

> practically all of the legumes, and the perennial grasses which are grown. If these crops are to be classed as soil-conserving crops when they are grown with a nurse crop, the nurse crop must be clipped green or pastured to prevent grain formation. Soybeans, field beans, [and] cowpeas are soil-conserving if they are not harvested or pastured but are turned under. Forest trees are soil-conserving if planted during the year. Orchards and vineyards, cultivated fallow land, roads, lanes, etc., woodland planted to forest trees and idle cropland of [the] previous year are neutral; they are not included in either base.[11]

In order to participate in the program in 1936, farmers had to agree to replace 15 percent or more of a farm's soil-depleting base with soil-conserving crops.[12] Those who did received a Class 1 payment of $10.00 per acre for planting the soil-conserving crop.[13] In addition, participants could also receive a Class 2 payment of $1.00–$2.50 per acre for implementing new soil-building practices.[14]

As the program continued, program regulations and payment values would change from year to year based on how much federal funding was available. For example, in 1937, the USDA further refined the program by creating two classifications: diversion farms and nondiversion farms. East noted that "diversion includes any farm with a soil-depleting base of 20 acres or more, or with a tobacco base. Any farm which is not diversion is classified as non-diversion."[15] These designations ultimately affected how much money a farmer would receive, so farmers had to recalculate their potential participation earnings annually.

Those who owned diversion farms could enroll up to 15 percent of their soil-depleting acres and qualify for up to $6.00 per acre for planting a soil-conserving crop. In addition, they were given a fixed payment of $10.00 toward soil building, but they had the opportunity to earn a higher amount by implementing practices from the following schedule described by East:

- $1 for each acre in the soil-conserving base,
- $1 for each acre for which diversion payments were made,
- $1.90 for each acre in commercial orchards,
- $1 for each acre of cropland on which only one crop of commercial vegetable was grown in 1936,
- $2 for each acre of cropland on which more than one commercial vegetable crop was grown in 1936,
- $0.50 for each animal unit in excess of 5 which the noncrop plowable pasture will carry during the normal pasture season.[16]

Payments for those with nondiversion farms were fixed at $20.00, with those participants also having the opportunity to earn more by implementing additional soil-building practices.[17]

Farmers in both the diversion and nondiversion categories had the cost of their lime partially offset. They received $1.00 per acre when they drilled a minimum of 500 pounds of limestone into legumes on cropland. The amount increased to $2.00 per acre when limestone was applied to pastures. In Harrison County, 4,500 tons of applied lime became eligible for government reimbursement.[18] Hendricks County agent Harold J. Yarling reported in 1939 that limestone "has been applied that probably would never have been purchased if there had been no adjustment program."[19]

An overarching similarity between the Agricultural Adjustment Act program and the new SCDA program was that both required a large portion of the agents' time at the beginning as they were getting established in the county. Harrison County agent Maurice Goode reported in 1936 that he had devoted seventy days to working on the SCDA program. His time was spent "in advising with the county and community committees, in holding educational meetings in the different communities, in preparing news articles and circular letters, and in training committeemen and farm reporters [field inspectors]."[20] That same year, Crawford County agent James L. Morrow reported spending one-half of his time on the soil conservation program.[21]

Agents and area volunteers were asked to attend district meetings where the details of the SCDA program and any changes for the upcoming year were reviewed and discussed.[22] Armed with the information, each agent and members of the local education committee—typically numbered at six people—would then set up local meetings with farmers to discuss the program for that year.[23] Over time, the members of the education committee took on a much larger role, relieving the agent of this responsibility.[24]

At the local meetings, the farmers were told what they could expect from the federal government in return for participating in this voluntary program and what they would have to do to become eligible for the subsidy payments, which were officially called grants by the USDA.[25] In 1936, Allen County agent Chester Kimmell described the process to educate local farmers about the program, writing, "At each of these meetings, the purposes and objectives of the new soil program were given. The second step was to take individual farm cases where soil-depleting bases had been set up and to show how the local farmer could first become eligible and then just how he could obtain the maximum payment. In each meeting, several such cases were written on the blackboard in order that the program would become more easily understood."[26]

Agents then asked farmers to vote on a slate of candidates to serve on a permanent committee that would be responsible for program activities in that township.[27] Each township elected a chairperson who was then automatically placed as a member of the county board of directors.[28] At the county level, the board of directors elected a chairperson, vice chairperson, and secretary-treasurer, with the county Extension agent serving as an ex officio member.[29] This board then took responsibility for implementing the program and disbursing the federal funds.[30]

As a first step, the board of directors had to hire the people to carry out the many facets of the work. In Fountain County, the program consisted of a paid staff of eight persons in the office and 43 field inspectors.[31] DeKalb County hired 35 field inspectors, while Cass County hired 53 full-time and part-time staff.[32]

The field inspectors visited every farm in a given county to survey the acres of soil-depleting crops.[33] The inspector would then draw a map to indicate the acreage for each crop that the farmer expected to plant. In Allen County, the local board received requests to measure 2,275 farms, with most of the work completed by December 1936.[34] In Cass County, 1,900 of the 2,800 farms to be measured had been completed at the time the agent filed his annual report for 1936.[35]

It took a great deal of time for the field inspectors to walk each farm, so many counties were granted permission to take measurements from aerial photographs instead. This approach was taken in Adams County, with the agent later reporting, "The work to date has been very satisfactory and the comparison with counties measuring by conventional methods shows that the cost for Adams County is approximately fifty per cent less than it would have been by hand measurement."[36] All of the Decatur County farms were measured from photographs in 1936, with the agent noting, "We were very fortunate in getting the Indiana National Guards to do the photographing for us. Our cost for these [photographs] was less than $600 as compared with twice this or more for counties who had private individuals do the photographing."[37]

As part of the survey process, farmers were also required to complete a detailed worksheet, which furnished the necessary information for establishing a soil-depleting base, determining not only the farmer's eligibility but how much the farmer could expect in payments.[38] This required a tremendous amount of work, such as in Allen County, where 5,200 worksheets were completed for the 1936 program.[39]

Each county was given an acreage allotment by Walter Van Nuys, chief statistician of the State Soil Conservation Committee.[40] The allotment was the total number of acres that the county could contract with growers under federal guidelines. James Morrow discussed the Crawford County allotments in his 1937 annual report, writing:

> In March, work with the community committees and community meetings took nearly twelve days. The first work in the month was with the County Committee and consisted of work done in calculating township base acreage quotas after the receipt of the notice of base acreage allotted by the State Committee. The total base acreage for the county was reduced by the State Committee from 31,049 acres in 1936 to 28,977 acres for 1937. Ten days all together were spent by the County Committee and the various Community Committees in the adjusting of bases for the townships and for the individual farms.[41]

In 1936, agent Warren O'Hara reported that Hancock County had 60 percent of its acreage in soil-depleting crops. Once this was determined, he noted, "It was the task of the directors to pro-rate this percentage among the townships in proportion to historical bases established for work sheets for 1935 crops, Assessor's reports and census figures. Immediately after township percentages were approved by the board, the township committees met with the county committees to divide the amount out to each work sheet containing all or a part of a farm."[42]

Farmers participating in the SCDA program were subject to inspections to ensure their compliance. Using ground methods or aerial photographs, each county was required to verify the acres of crops for each farm under contract and the practices the farmers had put in place to earn a government payment.[43]

When all of the farm records were complete, the county would send the information to the state office for payment.[44] In his 1936 report, agent Arvil Keesling explained the status of the verification process in Bartholomew County, writing, "Supervisors were appointed for each township and the outlines of the fields were checked on the pictures. Measurements were made in the county office along with the summary and the application for grant. On Dec. 1st the county had received $10,000 from the first 161 applications. A total of 583 applications were in [West] Lafayette [at the office on the Purdue campus] for a check."[45]

In spite of the complexity of the federal program, the ability of the agents to explain the rules helped increase interest and participation.[46] In 1937, nearly 50 percent of the farm acres in Indiana were enrolled in the conservation program. This covered 57,000 farmers, who were paid nearly $5 million for changing their farming practices.[47] In his 1937 report, agent Everett A. Johnson gave a detailed account of participation by Cass County farmers in the conservation program, noting that he had 1,132 applications that covered almost 200,000 acres. With Johnson's assistance, Cass County farmers were able to obtain $174,000 as payment for participating in the conservation program that year.[48]

While Johnson and other agents played a large role in getting the SCDA program established, their roles soon evolved, with Dearborn County agent Cory Alcorn noting in 1937, "As the [elected] officials of the S.C.D.A. become better able to understand their program and the work involved, we [county Extension agents] have paid less and less attention to the detail involved in these activities, . . ."[49] Instead, Extension staff began focusing their time on providing a wide range of educational programs and one-on-one consultations that farmers needed as they sought to implement the conservation changes on the land taken out of production.[50] With the research on conservation practices already available from Purdue, Extension was ready to supply the information that the farmers needed.

The impact of the SCDA program was real and tangible, with changes in agricultural practices clearly evident in the agents' reports. First and foremost, farmers were paid to modify their farms to be more soil friendly.[51] With so many farmers participating across the United States, the diversion of specific crops had a significant effect. For example, in 1938, corn was reduced from 120 million acres down to 93 million acres.[52] So conservation and crop reduction undertaken through the federal program resulted in price increases for some crops.

As farmers reduced acreage for soil-depleting crops, they began planting more legumes—a practice encouraged by the two federal programs. As a result, counties began reporting huge increases in the number of acres devoted to pastures.[53] Crops such as alfalfa, lespedeza, red clover, and timothy became more common as farmers planted new pastures and renovated old ones.[54]

On one hand, the legumes were important for soil improvement, but they also produced feed for dairy cattle and general livestock.[55] In Bartholomew County, the agent noted that increasing the acres of improved pastures "is definitely increasing the number of beef cattle both on feed and on pasture in the county."[56]

In some counties, alfalfa in particular became more popular because the federal government would pay most of the lime bill under the auspices of the conservation program, which encouraged the use of more lime.[57] In Clay County, it took three to five tons of limestone per acre to grow a successful field of alfalfa, meaning the payments from the government program could be rather sizable for this single practice.[58]

During the 1930s, the federal government provided payments that had helped farmers weather the economic storm, providing much needed financial support for families and communities reeling from the effects of the Depression. Between 1933 and 1938, Cass County farmers received slightly over $1 million from the federal government through the AAA and SCDA programs, which further helped community businesses and the county government.[59] But tying those payments to better conservation practices had also resulted in meaningful changes in the way farmers managed their land. As Fountain County agent Alfred Hesler observed in 1937, "If these programs have done nothing more than to educate the farmers and the public of the great need for conservation of our soil, it has been money well spent."[60]

By the end of the 1930s, agents were reporting how soil conservation had now become part of the psyche of the American farmer. As one agent noted, "Regardless of the economic results of the present [SCDA] program, the facts are evident that farmers, in the majority, are conscious of erosion problems, [and] more are thinking and acting in terms of 'leaving the farm as good or better than I received it.'"[61] And another reported that farmers were "thinking more about the real issues in the Soil Conservation Program and less about the size of their government check for participation."[62] Agents voiced their hope that the soil improvement practices begun under the auspices of the SCDA program would continue even if the government reduced or eliminated federal subsidies for such work.[63]

While the federal program was essential to the economic well-being of the agricultural community, it was key to the reputation and standing of the Cooperative Extension Service at the time as well.[64] As Adams County agent Lawrence Archbold noted in his 1938 report, "Every farmer is now aware that there is an Extension Office in the county and has a better conception of the use that can and should be made of it."[65] With the emphasis on soil conservation and the ability of Extension to furnish access to research to help farmers in this regard, soil conservation would become the mantra of Extension for decades to come.[66]

CHAPTER 10

Erosion Is One of the Major Problems Which Must Be Faced

Shortly after being inaugurated in March 1933, President Franklin Roosevelt signed an executive order to create a program that came to be known as the Civilian Conservation Corps (CCC). The aim of the CCC was to provide jobs for unemployed workers while using their labor to address much-needed environmental and conservation projects across the United States.

The CCC hired unmarried men between the ages of 18 and 25 who had found themselves unemployed during the Great Depression. This federal program would employ more than 3 million men nationwide—including nearly 64,000 men in Indiana—over the course of nine years.[1] These young men had been the first to lose their jobs when factories began cutting positions as the Depression worsened. Those hired agreed to work for the CCC for six months or more and in return received training, food, medical care, and a monthly salary of $30. In addition, 40,000 of the men were also taught how to read and write.

The workers provided labor for a variety of conservation projects on public forest and park lands. As part of their continued employment, each worker sent $20–$25 of their monthly earnings home to help support their families.[2] Agent Charles Nicholson explained that the money CCC workers sent to their families in Dubois County was enough to take many of them off public assistance.[3]

Housing for workers consisted of 1,300 camps across the country that were segregated by race and managed by the U.S. Army.[4] In 1933, the Daviess County agent reported that a local camp included ten buildings, with "one hundred local men being employed for its construction. A total of two hundred men will be assigned to the camp for work in the county."[5]

Workers from the camps often completed projects across multiple counties, sometimes using a substation or auxiliary camps to work in areas away from the main base of operation.[6] For example, workers from a Miami County CCC camp also completed conservation projects in neighboring Cass and Wabash Counties.[7] Camps would be moved to other locations as projects were completed.[8]

Indiana was in line to get thirty-eight CCC camps in 1933, but the actual number would prove to be far fewer during the life of the program.[9] The camps, which were designated according to the task at hand, eventually consisted of approximately 3 camps for work in the national forest (known today as the Hoosier National Forest), 10 camps for work in state forests, 8 camps for drainage work, 6 camps for work in state parks, and one camp for the army.[10] The number of camps varied from year to year, with some closing and others created according to the work projects identified.[11]

CCC projects varied according to the needs of the counties. Examples of the work included:

- redirecting water by digging drainage ditches, building impounding dams for reservoirs, and constructing earth and stone dams;
- preventing erosion by protecting stream banks from washing away, regrading land to reduce gully erosion, and planting grass in waterways;
- improving public lands by planting trees, building roads and hiking trails, constructing fire towers, and promoting wildlife conservation; and
- helping farmers by implementing erosion control practices on individual farms, crushing lime for use on farm fields, constructing field terracing, improving timber stands, and displaying conservation practices on demonstration farms.[12]

Depending on the year, approximately a dozen CCC camps across Indiana were engaged in a variety of erosion and drainage projects. Some work consisted of straightening streams and placing rocks at a river's bend to prevent the moving water from further degrading the bank.[13] Other work involved installing field tiles, building drainage ditches, and dredging ditches that had filled with sediment. This work was important in that it allowed farmers to move excess water away from their fields, which improved crop yields.[14]

It appears that the federal government initially paid for most of the work, but the expectation was that the state or county government, or a farmers' association would provide reimbursement for those expenses. At times, work completed under the auspices of the CCC cost more than what private contractors would have charged. When sediment needed to be removed from the Daily and Walnut Creek Ditches in Blackford County, two county drainage ditch associations got cost estimates from the CCC and private contractors. Agent Walter Rusk later reported, "After many delayed surveys and negotiations it was found these ditch clean outs could be done cheaper by private contracts than with the CCC help."[15]

The drainage projects were often placed under the direction of the county surveyor, who was responsible for maintaining the ditches and knew which properties bordered the drains.[16] In conjunction with the Allen County surveyor, workers from CCC Drainage Camp D-2 in Washington Township completed sixteen ditch projects by 1937, which included cleaning out the ditches using a dragline, tractor, and hand-excavation, and installing field tiles.[17] In Hancock County, agent Warren O'Hara described the scope of the work on the Frank Mingle and Potts Ditches, writing, "A total of 1,483 man days were used in accomplishing the following: 35,065 sq. yards of clearing on 17,421 linear feet of open ditches, 66,329 cu. yds. of excavation on 25,382 lin. feet of ditches, 511 lin. feet of tile ditches were repaired, 3,268 rods of fence, 26 cub. yds. of concrete were used in 7 concrete structures."[18]

Extension specialists from Purdue and county agents sometimes used CCC drainage projects as live laboratories to teach farmers about the project, its intended outcomes, and what was eventually achieved.[19] For example, in September 1936, Extension drainage specialist David H. Harker led a drainage maintenance tour in Allen County along with V. A. Barnett, superintendent of the local CCC camp. At the event, farmers were shown how to repair field tiles and taught different methods for dredging a ditch.[20]

The CCC used demonstration farms to showcase practices that could reduce erosion and improve crop yields. Farms chosen for the demonstrations were normally on badly eroded or highly worn soils where land had been taken out of production. This allowed the CCC to establish newer practices on those farms and show farmers the results of the conservation practices.[21] The work on the demonstration farms was housed in the U.S. Department of Agriculture (USDA), which had assumed control of the Soil Conservation Service—previously called the Soil Erosion Service—in 1935. Today, it is known as the Natural Resources Conservation Service.

To locate those who were willing to participate, the county agent would call a meeting of farmers and camp officials to discuss the requirements. Farmers who were chosen for the program were asked to sign a working agreement with the USDA.[22] Growers had to pay for materials, while the labor was furnished by the CCC.[23] There were many farmers who desperately wanted to enroll their farms but could not afford the cost, especially since participants had to agree to continue the work for three to five years, depending on the project.[24]

These projects aimed to reduce both soil and water loss through the implementation of conservation practices.[25] In his 1938 annual report, Fayette County agent Dudley Bishopp noted that the contracts involved "many different practices such as contour furrows on permanent pastures, terraces, buffer strips, strip cropping; contour cultivation, sod waterways, diversion ditches, and often a change in land use from cultivated crops to pasture and timber. Every farmer [who] visited expressed the same opinion that they did not like the changes at first, but after farming the land under the new arrangement of level cultivation, they like it better than the old way which caused so much loss of soil."[26]

In Blackford County, Charles Gast signed up in 1939, agreeing to participate in the project until 1943. The work done on his 175-acre farm was typical of that done on other farms in the program, which the agent described as preparing soil maps of the farm, developing a new crop rotation system, creating specific recommendations for each field, planting 52 acres of alternating crops in strips, applying 100 tons of limestone, planting 22 acres of pastures, and constructing a diversion ditch, two sod waterways, and a sod buffer strip.[27]

Demonstration farms became a focal point where farmers could see for themselves whether the practices had any usefulness for the way they managed their own farms.[28] Greene County agent Arthur Haseman reported on his association with a local CCC camp, saying, "The CCC camp at Worthington has always cooperated closely with this office in planning their demonstrational farms. Not only are they willing to cooperate in their planning, but their demonstrations are always available for extension projects and demonstrations."[29] This cooperation was important since that camp had signed contracts for demonstrations with 70 farmers across 9,500 acres of farm ground in the county by 1936.[30] The work with the CCC gave Haseman the ability to showcase the latest in soil conservation practices to area farmers.

The program ended in 1942 when Congress diverted the CCC funding toward the ever-growing expenses associated with U.S. involvement in World War II. Young men who were once needed to work in the CCC camps were now needed by the military. The camps and all of the equipment were transferred to the army to be used for the war effort. While the CCC program only lasted a decade, it was one of the most successful conservation programs ever undertaken in the history of the United States.

In addition to conducting demonstrations on individual farms, the Soil Conservation Service also assumed the management of larger tracts of land for what were called demonstrational areas.[31] By working with many farmers within a watershed, the federal agency showed how conservation practices that were incorporated across a larger landscape could improve water quality.

One such area in Benton County, known as the Prairie Soil Conservation Demonstrational Area, focused on controlling erosion by increasing the use of strip farming, grasses, and winter crops.[32] Agent Ellis Christen noted that Soil Conservation Service officials began assessing the area as a potential demonstration site in 1937. After gathering information, the officials invited farmers from within the conservation area to a meeting to discuss what they would be required to do, how long the program would last, and how they would be compensated. Following a positive response to the proposed work from the farmers and from the Benton County Farm Bureau and the Benton County Agricultural Conservation Association, in May 1937 the USDA approved the location for the project.

According to Christen, the demonstration area consisted of "the North East one quarter of Oak Grove Township, all of the South half of Center Township, and the three square miles directly west of Fowler and that section of Parish Grove Township in the South East Corner, extending four miles West and four miles North."[33] The USDA picked up the expenses of the project, paying for the salaries of a project manager and technical staff to deal with farm management and practices to reduce soil erosion on the participating farms. Christen noted that he worked closely with those managing the federal project to get it underway in 1937.[34]

One year later, the Prairie Soil Conservation Demonstrational Area had twenty-one farms enrolled for the required five-year cooperative agreement.[35] Representatives from the Soil Conservation Service and Christen worked together to map out the farms and make recommendations for how each one could reduce erosion.[36]

With soil erosion and reclamation projects receiving attention from several groups—including the federal Soil Conservation Service and Civilian Conservation Corps, the county-based soil conservation districts (see Chapter 43), and county Extension agents—Purdue staff became heavily invested in research and outreach efforts.

One type of outreach consisted of soil conservation schools, which allowed Purdue campus staff and county Extension agents to provide the latest information about conserving soil and methods to reduce erosion on farms. Examples of subjects covered at these schools included:

- Soil conservation and its control through proper land use
- How Indiana farmers are conserving soil
- Why land drainage
- Drainage methods and costs
- Soil erosion and what it means to Indiana farmers
- Practical methods for erosion control
- Crop rotation and better pastures for soil conservation
- The use of limestone and fertilizer in soil conservation
- Balanced fertility
- Small-dam construction
- The value of conserving crops[37]

Key Extension staff from Purdue who were involved in this work included Ruall O. Cole, a soil conservationist specializing in erosion and gully control, and George H. Enfield, an agronomist.[38] After one conservation school program in 1938, Clark County agent Earl Miles described the highlights of Cole's presentation, writing, "The value of conserving soils was brought out by R. O. Cole in his morning discussion when he gave experimental data showing the rapid rate of soil erosion on sloping soils that were left uncovered during winter and through cultivation in the growing season. This was contrasted with practically no loss in [the] case of sod cover. In the afternoon he brought out methods of checking erosion other than sod covers, such as contour furrows, strip farming, contour cultivation, terracing, etc."[39]

That same year, agent Chester Kimmell described a new project in Allen County that involved terracing. In that case, Russell C. Shipman, an Extension agricultural engineer from Purdue, worked with Albert M. Nichter, a field representative of the Soil Conservation Service, to set up three terraces on the farm of Russell Figert. Shipman and Nichter studied the drainage patterns to assess the slope of the fields and determine where terraces should be placed. Figert then constructed the terraces, which would drain the field gradually in order to prevent soil erosion.

In his 1938 report, Kimmell indicated that the terraces would be used as a future demonstration event—another approach Extension used along with the schools to reach farmers.[40] This type of on-farm field demonstration was very effective in showing farmers what could be done to reduce the amount of erosion on their properties. In-person demonstrations such as this allowed agents and others to answer questions on how the practices were installed and explain what benefits the work had shown.

Many field demonstrations on soil conservation incorporated small-dam construction as a method of reducing gully erosion.[41] In his presentations, Shipman often talked about three kinds of dams—sod dams, straw whisker dams, and wire fence dams—that farmers could build to keep water from making a gully deeper and wider. Following a 1938 Cass County demonstration, agent Everett Johnson summarized Shipman's presentation, writing:

> These three kinds of dams should be used by farmers where they will help conserve the surface soil. The sod dams and whisker dams are designed to slow down water near where it falls and thus prevent the amount of surface soil from being washed away. By putting sod in a small ravine, the grass will help sod the ravine quicker than would otherwise be the case and thus prevent the amount of washing. The wire fence dam was constructed at the edge of the field and designed to raise the depth of the washout four feet. Mr. Shipman also discussed the use of sod runways, the use of terraces and contour furrows.[42]

Extension also used tours to highlight soil conservation techniques in use on area farms and their outcomes. Examples of topics covered on these tours included planting different grass species for pastures, using different planting and cultivation techniques to better keep the soil in place, and building permanent dams.[43]

One of the more popular topics was how to use dynamite to open up a ditch.[44] Most farmers liked the idea of using dynamite to make the ditches more efficient in moving excess water away from their fields because it was quicker and cheaper than digging out the ditch by hand or with equipment.[45] For example, agent Warren O'Hara in Hancock County described a dynamite demonstration that lowered a ditch 18 inches along 1,300 feet by using 450 pounds of dynamite.[46]

In a 1938 annual report, agent Chester Kimmell indicated that an Allen County ditch-blasting demonstration had drawn 225 attendees. He reported, "Prior to the meeting the dynamite was set in an open ditch, and just before the dynamite was set off, the entire group was called together in order to explain the value of ditching in this manner. The demonstration proved very successful and the results obtained were much better than expected."[47]

That same year, 200 people attended an event on the Emma Deaver Farm in Hamilton County. Agent James B. Todd reported:

> This demonstration was worked out by the County Surveyor, Dave Harker, State drainage specialist, a dynamite company worker and the county agent. This method of draining an open ditch is very practical on old ditches where there is some water in them. . . . The dynamite used was a straight 50% nitroglycerine dynamite which is more sensitive. The charges were placed 18 inches apart in the ditch bottom and only the first charge capped. The detonation of the first charge sets all the others off so that the explosion appears all at once. Approximately 1,200 feet was loaded in this demonstration by boys from the Frankton [Indiana] C. C. C. Camp. . . . As this was the first demonstration of its kind ever held in the county, it attracted considerable attention.[48]

There was little doubt that the farming practices used to maximize yields across Indiana during World War I had taken a toll on the health of the soil. Continuing many of those practices during the agricultural depression of the 1920s only made erosion problems more visible, even to those who had little understanding of agriculture. Federal, state, and county officials were concerned that if the erosion problems weren't solved sooner rather than later, agricultural production would be diminished.

By the 1930s, farmers only had to look at their fields and property to see that the land was slowly degrading. Whether due to inattention to drainage issues, the wrong cropping system, or planting crops on lands that would have been better suited as pastures or forests, there was much work to be done to improve and preserve the land. Under the auspices of Roosevelt's New Deal, soil conservation became a major focus of federal programs that were designed to alleviate the problem while putting people to work during the Depression. The federal government also provided funding aimed at encouraging farmers to implement practices that would prevent erosion and restore vitality to eroded land.

As federal and local efforts focused more attention on soil conservation, Extension was well positioned to partner with other agencies to offer farmers educational programs that discussed the latest research on rebuilding the soil, reducing erosion, and keeping farmland productive. For all intents and purposes, the 1930s would be a major turning point for how lands were farmed so as to save the important soil resource for future generations.

CHAPTER 11

Land Use Planning Not Altogether a New Idea

In 1938, U.S. Secretary of Agriculture Henry Wallace turned his attention to land management in order to realign the multitude of New Deal agriculture programs into a more cohesive approach that better targeted the needs of individual counties.[1] That July, representatives from the USDA and land-grant universities across the nation met to work out and sign the Mount Weather Agreement, which focused on agricultural planning. According to one agent, this agreement "set up certain objectives and procedures to be used in determining the recommended land use program for the various communities of the county from rural people living in those communities. The work of carrying out the procedures outlined by this agreement is generally spoken of simply as 'Land Use Planning.'"[2]

Wallace and others within the Roosevelt administration wanted U.S. Department of Agriculture (USDA) programs to be better coordinated within each state, and they specifically wanted more input from farmers on programs involved with land management. To help accomplish this, a USDA contact person was assigned in each state to form a land use planning committee along with the state directors of agricultural experiment stations and the Cooperative Extension Service.[3] Purdue assigned agricultural economist J. Carroll Bottum from Extension's Division of Farm Management to lead the county planning campaign for Indiana. Bottum's partner from the USDA was John B. Kohlmeyer, who represented the interest of the federal agency. Both men assisted counties in any way possible with establishing a land use planning committee, developing township and county land use maps, and submitting the required reports.[4]

The key to this effort was to rely on county Extension agents to help develop local land use committees made up of farmers. Agents had experience with federal programs that provided subsidies to farmers for replacing acres planted to soil-depleting crops with soil-building practices. They had screened applications for federal agricultural loan programs and, more often than not, helped farmers complete the required forms. And agents were also in the process of explaining soil conservation districts to farmers. So it seemed logical to once again draw on the agents' experience and contacts to develop these new local committees.

Agents formed the committees with farmers and others they selected to round out the advisory group.[5] Participants were volunteers who would be asked to spend a lot of time with no pay to complete the work.[6] To make sure that the scope of the committee was what farmers wanted, a farmer was elected as chairperson.

The purpose of the county land use planning committee was, as DeKalb County agent T. E. Myers described it, "to determine what kind and amounts of crops and livestock may be most profitably produced in the county over a long time period when the prospective demand for agricultural products and the control of soil fertility and erosion are considered."[7] In the 1939 Fulton County annual report, agent Martin J. Huxley wrote that land use planning was

> not altogether a 'new idea.' Farmers individually have been planning the use of their land from year to year, with the thought uppermost in their minds generally that of securing the greatest returns possible from tilling their soil. Land Use Planning on a community, county, state and national basis is, however, a 'new approach.'. . . Land Use Planning when accompanied by the mapping and classification of all the land within a community according to the present use being made of such land tends to focus attention on the 'long time problems' as well as the immediate problems of that community. Thus through the mapping and classifying of land by people who live on the land, basic facts are obtained. The problem of the various areas mapped are brought more clearly to light, and the foundation laid for more effective planning whether it be for the present or be concerned with a long time program which may take a number of years to bring to fruitful completion.[8]

The idea behind the new initiative was to map each county with the intent of better understanding what the land was best suited for: crops, pastures, forests, or another purpose.[9] Often, township-level committees were appointed to develop local maps, which would eventually be combined into a county land utilization map. In Bartholomew County, five farmers from each of the twelve townships were shown how to make a land utilization map by using a plat map and marking how the land was being used in that township.[10]

After preliminary maps were developed, meetings were held to get input from local farmers on how well the maps represented their townships and whether improvements were needed.[11] In 1939, agent Ellis M. Christen described the process used in Benton County, writing:

> Early in the year a meeting of all Township Committeemen was held with Mr. [J. Carroll] Bottum, who was Extension Specialist in Land Use Planning in Indiana, and Mr. [John] Kohlmeyer, State Representative in this work of the United States Department of Agriculture, explaining the program to the committee, and each township committee made the first land use division on a township map. Minor changes were made on these from time to time and

the second meeting of township committees was held in the fall to make further refinements and to correlate between townships. At the close of the year, [maps of] townships are practically complete and ready for the preliminary [county] map with the exception of possibly Center and Richland Communities. These two townships need some additional work.[12]

The land use planning committees used multiple sources—such as questionnaires and county statistics from Extension's Division of Farm Management at Purdue—to analyze agricultural production in each county.[13] In some counties, the hand-drawn maps developed by the committees were validated by comparing them with aerial photos. Bartholomew County agent Arvil Keesling was surprised at how well the two correlated, showing that the committee members had done their homework and understood how the land was being farmed.[14]

The countywide map formed the basis of area land planning and, over the next decades, zoning. This county map classified the county's land mass into one of two general categories: agriculture and non-agriculture.[15] The non-agriculture sector included land that was unsuitable for growing crops, such as forests and woodlands. In Clark County, the agent noted that it also included "land that in the opinion of the committeemen should be reforested," which represented one-third of the county acreage.[16]

Floyd County subdivided its agricultural lands as red clay, river bottom, truck crop land, and "crawfish land not suited [for agriculture]."[17] According to Keesling, Bartholomew County agricultural land was divided into the following four areas based upon soil type and length of crop rotations:

1. The land unfit for agriculture
2. The land requiring at least a five-year rotation to maintain fertility
3. The land need[ing] a three-year rotation to maintain fertility
4. The river bottom, overflow land on which a two-year rotation might be practiced.[18]

In 1938, Dubois County agent Charles Nicholson described what happened when the countywide land use map was finally presented to the county planning board, noting the involvement of Kohlmeyer and a large contingent of Extension specialists from Purdue.

> Charts showing types of soil, amount of soil erosion, areas of timber land, marginal land, submarginal land and the more productive land in Dubois County, were presented by Mr. Kohlmeyer. The trend of population and the shift in agricultural population were discussed by Mr. Bottum. Mr. [Thomas E.] Shaw [Extension forester] presented the information on forestry. . . . Mr. Mervill O. Pence [Extension agronomist] presented the idea of utilizing a larger amount of the marginal crop land by seeding to grasses and legumes for live stock instead of [to soil-] depleting crops. Such meetings have helped

> a great deal in getting farmers to recognize the problems of correct land utilization. It defines more clearly the need for a Soils Conservation Service District.[19]

In Bartholomew County, Keesling noted that the committee used soil types, crop yields, erosion potential, and other factors to indicate that the county should plant the following acres of crops: corn, 46,000; wheat, 37,000; clover and timothy, 16,000; alfalfa, 8,000; soybeans, 8,000; canning crops, 7,000; oats, 3,000; and annual legume hay, 1,600.[20] The Henry County committee recommended that less corn and oats be planted, with those acres planted to alfalfa, soybeans, and pastures instead.[21] In Fayette County, the committee suggested, among other things, that the dairy cattle be increased by 700 head and the sheep, by 300 head.[22]

In Floyd County, records indicated that the delinquent taxes in certain parts of the farming community were as high as 50 percent, with 25 percent of those farmers receiving public help. Simply put, those farming the less-than-productive acres did not make enough money to pay their bills, including their property taxes. This situation led the committee to suggest that 7,000 acres of these worn-out acres with an assessed value of three dollars per acre be reforested or the land purchased by government.[23]

Those at the USDA and Purdue expected that the land use planning committees would continue to meet as a way of reviewing past work, changing farm practices, and shifting populations.[24] This would allow Extension staff to continually target the evolving needs of agriculture as times changed.

At a 1938 land use planning committee meeting in Greene County, Bottum and Kohlmeyer presented information that focused on the best uses of land within the county.[25] The agent later provided the following details:

> A large area of Eastern Greene is submarginal and should eventually be considered for governmental reforestation. Large areas of strip mine land in the western part of the county should be reforested. The central portion from north to south, including the river bottom land, the marshes and prairie soil should continue to be the principal cash crop area. Better rotational practices and soil treatment should be followed to maintain the natural fertility of this soil. The more level upland bordering on this cash crop area should be devoted to livestock farming, utilizing a maximum of pastures, grasses and legumes, and a minimum of grain crops.[26]

It was one thing to write about submarginal lands, but it was a much different thing when the USDA announced it was releasing federal money to purchase that type of land in the eastern part of Greene County, which prompted "several protest meetings in this portion of the county," according to the agent.[27]

But how would the recommendations issued by the land use planning committees be implemented? In Gibson County, agent Roy Gwaltney indicated there were two possible ways to reach the goals being proposed by the local committee: by legislation, which he thought would be unconstitutional, and/or through educational programs.[28]

Agents generally allowed that the recommendations provided by land use planning committees could be used as part of a county's program of work.[29] But Extension would continue its longstanding tactic to approach the recommendations through education, leaving regulatory solutions to elected officials.[30] And eventually, before questions about implementation could be fully sorted out, matters related to World War II would take precedence, leading counties to put this work on hold until after the war.

In 1942, the USDA land use planning program abruptly ended when Congress diverted the funds supporting it toward the costs of World War II. But even as the federal push waned, local organizers continued to support the work. As one agent noted, the USDA effort had identified "the need of a long time planning program for agriculture and its effect not only on Agriculture but upon the economic stability of the county."[31] So by the time the USDA-led effort ended, most county commissioners had already created a county planning board.[32]

At the start of the war, a few counties had already begun to act on the recommendations of the local land use planning committees, using them to establish a county zoning ordinance with enforcement discretion. By 1942, agent Russell Zell reported that local zoning regulations in Newton County limited the "type of structures to be used for rural residence and limits the subdivision of land when being changed from farm land to residential."[33] In other counties, zoning efforts were still taking shape. In the 1942 Morgan County annual report, agent William J. Record indicated that the local land use planning committee was interested in zoning around cities, "but thus far no active steps have been taken in this connection."[34]

While all but two counties had completed land use maps by 1945, World War II had temporarily taken the focus away from planning efforts for how to best utilize the land.[35] But as the war neared its end, the work of the land use planning committees would form the basis of new committees and regulations to establish zoning for activities that could be conducted in agricultural areas. Based on the county agents' work with the USDA effort, they would be asked to serve on these expanded zoning boards, which would give agriculture a strong voice as decisions were made in the future.[36]

CHAPTER 12

Extension Work Interrupted by Extreme Droughts and Flood

In addition to the economic woes faced by so many Americans during the Great Depression, the decade of the 1930s was also marked by severe droughts from coast to coast, with the midwestern United States being especially hard hit with high temperatures and lack of rainfall.[1] In Indiana, drought conditions began in 1929 in the southern counties.[2] With the challenges brought on by the stock market crash in 1929 followed so closely by droughts that lasted off and on for the next decade, growers wondered if they could survive the one-two punch of the economic downturn and these environmental disasters.[3]

The drought of 1930 was especially widespread, with fifteen states particularly hard hit. Indiana was affected along with Alabama, Florida, Georgia, Illinois, Minnesota, Missouri, Montana, New Mexico, North Carolina, North Dakota, Ohio, Oklahoma, South Carolina, and Virginia.[4] In Indiana, areas further north got enough rain to keep things growing, but thirty-five counties in the southern half of the state would see little to no rain that summer.[5] The hardest-hit counties included Bartholomew, Clark, Crawford, Daviess, Floyd, Harrison, Jackson, Knox, Perry, Pike, Spencer, Sullivan, Warrick, and Washington.[6]

That July, temperatures stayed above 90 degrees Fahrenheit for twenty-five days. The agent in Harrison County reported that high heat and dry soils had destroyed the corn and pastures in the area.[7] In Greene County, the agent indicated it was the driest season on record, with "pastures and meadows burned dry and [a] considerable percent of the corn of the county absolutely ruined by the extreme heat and lack of moisture."[8] These types of reports were common for the whole region.

Winter wheat across the U.S. was especially hard hit by the drought.[9] By August, states east of the Mississippi were seeing drought conditions worsen.[10] The lack of rain also prevented corn from germinating or from growing if it had already reached the surface by the time the ground had become hard. And gardens that people relied upon to feed their families during the Depression were significantly damaged, with many families facing a serious threat to their food security as a result of the loss.[11]

As the drought continued, wells, cisterns, and springs began drying up, and cities were looking at water supplies that were running out. By the summer, towns were asking their citizens to conserve water so there would be enough for personal use and for fire protection.[12]

President Herbert Hoover called a conference on August 14, 1930, to discuss drought relief with the governors of affected states. Hoover wanted on-the-ground information from farmers about how the lack of rain and extreme heat had affected their farming operations and families, so prior to the conference, he directed the U.S. Department of Agriculture to develop a questionnaire for county Extension agents to send out to growers.[13]

Extension agents used the questionnaire to ask members of the farming community for their input and then transmitted the findings to U.S. Secretary of Agriculture Arthur M. Hyde, who was leading the information-gathering effort for Hoover.[14] Agent Alfred Hesler even sent a telegram directly to Hyde indicating that the "crop damage was unprecedented" in Fountain County.[15] Hesler further reported, "Have talked to many farmers of the county and we feel that corn is cut seventy to seventy-five percent now. Young clover is gone. The soybean crop is cut fifty per cent and many will not be harvested for seed. Pastures are gone. Fountain county should take care of herself with roughage for the winter. Grain will be short. We will not have a surplus of feed. As County Agent, we are urging conservation along every line."[16]

To supplement the information gathered through the questionnaires, Dean John H. Skinner of the Purdue School of Agriculture also sent two members of his staff on a fact-finding trip just two days prior to the August 14 meeting in Washington.[17] Thomas Coleman, state leader for county agents, and Leroy E. Hoffman, assistant state leader for county agents, drove through the southern part of the state to talk with local agents and meet with farmers, bankers, and business owners in twenty cities.[18] The information they gathered along with their personal observations would likely have been passed on to Indiana Governor Harry G. Leslie prior to his meeting with the president in Washington.

At the conference, President Hoover and the governors agreed to four key strategies to deal with the worsening drought. These included appointing state-level drought committees, using federally-negotiated discounts to reduce railroad shipping rates for materials delivered to drought-stricken areas, providing government-backed loans to farmers, and distributing 1931 federal road funds early in order to put farmers in the hardest-hit areas to work on road construction projects.[19] Unfortunately, this last strategy would quickly prove to be unsuccessful. By the end of August, the states knew that Congress would not go along with Hoover's request to advance road construction funds to the states in the drought areas.[20]

Upon his return to Indiana, Governor Leslie immediately established the state drought relief committee and named Skinner as the chairman and Coleman as a member.[21] States had also been asked by the federal government to develop county drought relief committees in addition to the state-level committees in order to deal with problems at the local level. The state committees then worked with the county-level committees and coordinated communication to and from a federal committee that President Hoover had established shortly after the initial meeting with governors.[22]

As the drought wore on, the lack of livestock feed became a critical concern for many farmers.[23] By August, there would be 1 million farmers and 12 percent of the country's livestock affected by the drought.[24] Farmers in many drought-stricken areas had to haul water from long distances for their livestock, and as pastures were destroyed by the lingering drought, farmers turned to soybeans as an emergency hay crop.[25] In his 1930 report, agent Ira L. Thurston noted that soybeans were the "only local source of hay for the Decatur County farmers this summer. Although the beans have suffered from the drought, the yield will be approximately a ton per acre."[26]

That same year, William Douglass reported on the condition of the crops in Bartholomew County and how it played out for those raising livestock.

> The drought during the past season was not so severe in this county as in counties further south. The season was extremely dry until August 7th when rain fell on all the county excepting a very small area which has remained dry to the present time. There is a large area in the county of very valuable farm land which is sufficiently sandy to be extremely droughty. On this area almost all crops were practically killed before the rains came. . . . There is practically no area in the county in which there was not a sufficient amount of roughage produced if saved by the individual farmer. However the production of feeding grains was seriously reduced, leaving many farmers in a rather serious situation.[27]

To facilitate drought relief for livestock farmers, Secretary Hyde had asked railroad officials to reduce their freight charges across drought-stricken counties. They agreed to this cut as a service to farmers, who were now importing large amounts of feedstuff, but the companies required proof that the discount would be extended to a farmer in need. The state drought relief committee was able to take advantage of this reduction set in place on August 14 to get half off freight charges when moving hay, feed, and water into the drought areas.[28] Many livestock producers in the southern part of the state would purchase livestock feed from northern Indiana farms, where the drought had not been as serious, while additional supplies were imported from other states.[29]

The proof was left to the county agents, who issued a certification document to verify that the person requesting the lower rate was affected by the drought.[30] Clark County agent Ruall Cole noted that he "acted as a representative of the [U.S.] Department of Agriculture in issuing Freight Reduction permits for the movement of hay, feed, livestock and water," issuing 100 certification permits for local farmers in 1930 so that they could obtain the lower freight charges.[31]

Farmers in Harrison County saved $35 for each of the 102 railcars used, which amounted to a savings of $3,570 in freight charges.[32] In Decatur County, farmers brought in 27 railcars of grain and hay and saved $1,080 in freight rates.[33] In Crawford County, growers ordered 61 railcars—consisting of 38 of oat straw, 8 of mixed feed, 7 of hay, 6 of oats, and 2 of corn—and saved $2,000 when the drought discounts were applied to the deliveries.[34]

Agents tried their best to assist growers by stressing strategies that could help farmers in the short-term. As Floyd County agent Cory Alcorn noted, "The economic conditions require that [farmers] cull their herds and flocks more rigidly than they ever thought would be necessary, because only the best producing cow or hen can pay a profit now. The livestock that is kept must be fed more carefully than ever before. . . ."[35] Even under these dire conditions, agents continued to promote productivity by recommending that farmers eliminate poorly performing livestock and feed the limited supplies of grain and hay to the more profitable animals.

Many hog producers with access to wheat fed that to their animals in lieu of paying a higher price for corn in short supply, but it was common to market underweight hogs and sheep at much lower prices when feed wasn't available or when farmers lacked the cash to purchase it.[36] And as times got tougher, financially strapped farmers began selling off even their productive animals. This led Harrison County agent Ronald Johnson to report that some farmers had "resorted to panicky selling, and large numbers of dairy cattle, poultry, hogs, and sheep were sent to market at very low prices."[37]

Understandably, the shortage of local livestock feed took up a lot of the agents' time in both 1930 and 1931. In Crawford County, Sylvester Scott wrote, "Much of the time was spent assisting farmers with problems arising out of the drought. They were assisted in locating sources of feed and in getting freight rates on this feed."[38] Some agents even worked with trucking companies to import livestock feed and other supplies from other counties.[39]

In the 1930 Fountain County annual report, agent Alfred Hesler described the various ways that farmers had tried to cope with the feed shortages, writing:

1. From one to three hundred shocks of fodder cut for every farm in the county. The cutting of fodder for other than the silo was a practice almost abandoned.
2. Half of our farms shredded fodder.
3. More soybeans put for hay than ever before. Less thrashing for seed.

4. More supplement feeds purchased for hogs and dairy cattle bought than ever before.
5. Greatest trend for better and more economical feeding than ever before practiced.
6. More rye sown for fall and early spring pasture.
7. More straw baled than common.
8. Greater interest in better feeding campaign.[40]

At first, the federal government was hesitant to provide emergency funding to states to help those affected by the drought because it was already dealing with issues related to the Depression. As conditions worsened, though, Congress did provide aid to fund seed loans to farmers whose crops and pastures had been decimated by the drought but were unable to get credit from local banks.[41]

Agent Clyde Madison explained the situation in Dearborn County, noting, "So far our local banks have been able to take care of loans to people who are able to secure loans. There are a number of people who will probably have to have help who cannot secure loans with any kind of mortgages."[42] Many farmers had to depend on these federal loans to survive the drought. Farmers in Harrison County borrowed $20,000 between February 1 and May 1, 1931, with most loans going to purchase seed, but there were more than a few examples where the money was used to purchase food and clothing.[43]

Extension became the conduit for distributing information about the government seed loans, and many agents took the extra step of helping growers fill out the applications. Madison described the process in his annual report for 1931, writing:

> The Federal Seed Loan Office furnished blanks, etc. on which applications were made for such loans. These were rather difficult for the average farmer to fill out properly. So we kept all application blanks, etc. at our office and assisted those wanting loans in properly filling out the blank forms.
>
> This gave us an opportunity to check up on those asking for loans to see if they had local credit. We advised those having local credit not to make application for a loan, as the loan was for only those who had no local credit. We also advised each applicant to keep his loan just as low as possible. The above advice was followed in practically all cases and was responsible for keeping our seed loans to a small figure. A very few resented such advice but they are all glad now that they kept their loans as low as possible. Some wish that they had not gotten a seed loan at all.[44]

Once farmers had the resources to obtain the seed corn, they had to find the supplies. Agents used their contacts to find seed corn to make up for what their counties did not have. With the corn crop destroyed across many counties in 1930, farmers were in desperate need of seed to plant their 1931 acreage regardless of whether it was open-pollinated or hybrid corn. With so much of the supply destroyed in the drought-stricken areas, most of the purchases came from outside these areas.[45]

Working with Indiana Farm Bureau, the state drought relief committee was able to get farmers in north-central Indiana to donate seed corn to those in the southern counties who had little to plant.[46] It was quite a challenge to cover the need, given that 10,000 farmers had insufficient supplies. Each farmer donating corn was asked to provide one or two bushels for distribution to those in need. Each bag had an attached card that described the corn and included the name, address, township, and county of the person who had provided it. Recipients would then tear off a part of the card with their information and send it to the donors.[47] In Harrison County, Ronald Johnson reported, "About two hundred sixty bushels of good seed corn were donated to our Farm Bureau members by brother Farm Bureau members in Shelby and Rush counties."[48]

The 1930 drought was followed by similar conditions in 1934. The experience gained from the earlier drought would be put to use once again in dealing with the miseries posed by the lack of water for humans, livestock, and crops.[49]

It must have seemed during the 1930s that droughts were to be the norm as another one followed in 1936 and Indiana was engulfed in an extreme heat wave. As the *Indianapolis News* reported on July 8, 1936, "The entire state suffered with an average maximum temperature of 103, and the highest temperature in the history of Terre Haute, 106 degrees, was reported."[50] Conditions were so bad that Crawford County agent James Morrow observed it was "the most severe drought that has ever occurred in this County."[51] Agent Arvil Keesling reported, "The drought of 1936 cut the crop of corn in Bartholomew County about 50%."[52]

In Clark County, people were running out of drinking water that summer, leading agent Earl Miles to later report, "The water supply began to get short in July and by August the situation became acute. Springs stopped flowing, wells and cisterns became exhausted and ponds were dried. In many cases farmers with a large number of livestock were able to do nothing but haul water, milk and feed."[53]

A governor's commission on drought relief provided some resources to help get water to those who found themselves without. In a two-month period in Clark County, Miles noted that "water suitable for human consumption was hauled in sterilized tanks by local people owning trucks. 298 orders were filled for 243 different farmers."[54] Unfortunately, he also reported that the "work was discontinued on account of lack of funds; there were 86 orders left unfilled."[55]

People living in Floyd County also saw their wells, springs, and cisterns run out of water that summer.[56] In July, the governor's commission provided five trucks for hauling water provided by the local utility at no cost. Those needing water were asked to contact the Extension office to make the request. By the time the program ended on August 27—when rains began falling—700,000 gallons of water had been hauled to people in the county.[57]

Franklin County received only a few scattered showers from the middle of May to the end of August in 1936.[58] As many as 500 local farmers hauled drinking water for their families or for watering livestock during the summer months. Some had to haul water from sources located fourteen miles away. Agent Russell Zell indicated, "There are sources of water spotted here and there in the county which could be used if properly harnessed and conserved. This problem, water conservation, is an important one."[59] The water shortage prompted officials to begin constructing the county's first dam on the Frank Kirk Farm along the border between Springfield and Whitewater Townships. Zell expressed hope "that this will be only the first of a series of successful water dams, placing an adequate livestock water supply within four or five miles on every farm in the county."[60]

The impact of the drought on crops and pastures caused dairy producers in counties such as Dearborn and Floyd to operate at a loss for the year as they were forced to buy feed on the open market.[61] Obviously, farmers who kept livestock needed feed for their animals not only during the summer months but also in the winter.[62] According to agent Cory Alcorn in Dearborn County, producers "were forced to haul distillers' slop from Lawrenceburg which in many places took the place of both feed and water for their cattle."[63]

Livestock that couldn't be fed were sent to markets prematurely, which meant lost income from selling underweight cattle and hogs. In Crawford County, Morrow reported that the drought "caused heavy selling, at sacrifice prices, of the livestock in the County. It is estimated that over a third of the cattle in the County was sold [in 1936]. Because of the extremely low prices prevailing at that time for any kind of livestock, many farmers sold off their best in order to derive enough income from them to buy enough feed to maintain the ones kept. This has resulted in a lowering of the quality of stock in the County."[64] He further indicated "a great amount of stock consisting of cattle, hogs, and poultry which should have stayed on the farm have been marketed at low prices."[65]

While the 1930s brought repeated droughts to Indiana, counties along the Ohio River experienced a different disaster early in 1937 when the Ohio River flooded. In the first weeks of January, the river began rising along nearly 1,000 miles of its banks. Individuals living alongside the river from western Pennsylvania to southern Illinois would experience a flood that remains today as one of the most destructive in U.S. history.

With the ground already saturated and the river nearing the top of its banks, record rainfalls during a two-week stretch starting January 13 sent the river pouring over its banks.[66] As levees along the river began eroding and failing, the breaks released large volumes of floodwaters that only exacerbated the damage.[67]

People living along the river in portions of Pennsylvania, Ohio, West Virginia, Kentucky, Indiana, and Illinois found themselves inundated with water.[68] Over the course of six weeks, nearly 300 people perished. A staggering 1 million people would find themselves without shelter, food, and fresh water. Estimates were that the flood resulted in $400 million in property damage at the time.[69]

From Jeffersonville to Evansville, towns in southern Indiana that bordered the Ohio River became submerged. In Jeffersonville, the entire town had to be evacuated as the town became uninhabitable.[70] Conditions in Jeffersonville and New Albany were so bad that an eyewitness observed, "[B]oth cities virtually have ceased to exist."[71]

M. Clifford Townsend was the governor of Indiana at the time. Unfortunately, he had only been in office for a few weeks when he was confronted with the disaster. He first responded by sending every unit of the Indiana National Guard to the affected area, directing 5,000 soldiers, engineers, and officers to the flooded towns and cities to help in whatever way would be necessary. He even went so far as to place the flooded area under military control, establishing boundaries across the counties. Only the residents from those areas and those involved in dealing with the aftermath of the crisis were allowed into the area. Under martial law, military courts were set up to deal with civil problems until local authorities could again get back to business as usual.[72]

Legislators from the Indiana General Assembly flew over the flooded areas to see for themselves the extent of the damage. Despite the complete devastation, they realized that the state government could offer only so much relief money since Indiana, like every other state, was already dealing with the economic woes of the Depression.[73]

The most immediate challenge was to care for the needs of the multitude of people who found themselves homeless in the winter or out of work when homes and businesses flooded. At the time, there were approximately 200,000 people living in the flooded areas in Indiana. Of these, 70,000 were now displaced.[74]

As in most disasters, the American Red Cross became the go-to organization to help those in need.[75] Early measures involved establishing temporary shelters in railcars, warehouses, and any other place that could provide protection against the freezing temperatures, sleet, and snow. On January 22, eight inches of snow were recorded in Madison. Three hours west in Mt. Vernon, temperatures of only 13 degrees Fahrenheit complicated relief efforts.[76]

Soup kitchens were established to provide short-term supplies of food and drink.[77] The Red Cross asked health officials to provide vaccinations for diphtheria, smallpox, tetanus, and typhoid in the event of an epidemic.[78] Personal and city wells became contaminated by surface water, so billions of gallons of fresh water had to be trucked in or brought in on railcars.[79]

To help with the recovery effort, J. J. Liddy, superintendent of the Indianapolis Union Railroad, took on the coordination of all railroad activity in and out of the flooded areas. Officials from all of the railroads agreed that transporting food, clothing, medical supplies, and other materials into the affected areas would be done at no cost to the Red Cross.[80]

Getting people to better conditions would require a mass movement of individuals northward. Communities across northern and central Indiana were asked by the Red Cross if they could find room in their towns for those displaced by the floods.[81] For instance, special trains brought 850 refugees to the Indiana State Fairgrounds, where they were examined by physicians and nurses, and provided shelter. Kokomo took in 500 flood victims, while 250 displaced individuals went to Rushville.[82] Churches and other civic-minded organizations would appeal to local residents for food, clothing, and bedding to house those coming to their communities.

Farmers in the flooded areas were especially hard hit. Much of the livestock had drowned as animals were swept away by the fast-moving waters. Farmers who had managed to save some of their livestock now faced the challenge of feeding them through what remained of the winter—a difficult task as hay and other feed supplies had been destroyed.[83] Many farmers also lost their seed corn that they had been saving for spring planting.

Affected counties formed flood relief committees to try to address the losses.[84] Agents in those areas looked outside of the flood zone to find feed and seed for their growers. In Dearborn County, agent Cory Alcorn helped to evaluate damaged corn to see if it could be used as feed for hogs. He later reported being able to "locate a goodly supply of hay and straw at a reasonable price. If credit was needed, the local banks and the Rural Resettlement Administration assisted each one [of affected farmers] at all entitled to consideration."[85]

As with most disasters of this nature, farmers outside of the affected counties helped those in need by providing seed corn for planting the spring crop. For example, Dubois County farmers donated seed corn to farmers in flooded areas, with the local American Legion and Red Cross paying for the collection and distribution expenses.[86] Agent Charles Nicholson provide this description of the volunteer efforts by local farmers:

> Under the direction of the Dubois County Farm Bureau, the Red Cross and with the aid of the extension office, 231 farmers volunteered and came to the rescue of the farmers in the flooded areas. In February the two hundred

> thirty-one farmers donated a bushel of corn to be used for seed by farmers who lost all they had in the flood. The corn was assembled at a store room in Jasper, hand picked, shelled and graded. The result was that 190 bushels of good seed having a germination test of 95% was distributed equally to three points in the flooded area. Posey County, Vanderburgh County and Spencer County farmers received the corn. This corn was handled through the extension offices in those areas.[87]

In response, two agents from the flooded counties—John F. Hull in Vanderburgh County and Oral B. Riggs in Posey County—wrote to Nicholson, saying, "Your farmers are splendid cooperators in a crisis like ours and it is through such acts of courtesy that our farmers were encouraged to keep their heads up and chest out."[88]

While New Albany had been especially hard hit in the flood, Floyd County agent Charles East was able to provide a rather positive update later that year. In his 1937 annual report, he stated that the area had been likened to "the wreck and havoc of a war-torn country. . . . But through Flood Relief measures and Red Cross Charity, homes were rebuilt and businesses re-established."[89]

Each year, Extension agents across the state worked with their local advisory boards to carefully plan which demonstrations and programs they would carry out in order to help local farmers. But in the annual report that agent Sylvester B. Scott filed for 1930, he noted, "The Extension work in Crawford County was greatly interrupted during the year [1930] by the extreme drought. An effective program was outlined last spring by a county committee, but due to adverse conditions this program had to be greatly altered."[90]

Unfortunately, carefully made plans for Extension work would be repeatedly disrupted throughout the 1930s in counties around the state as agents dealt with natural disasters on top of the economic woes of the Great Depression. In a typical year, agents had more than enough work dealing with the demands placed on them by citizens and by officials from their counties, from Purdue, and from the federal government. The repeated droughts and flooding events—along the Ohio River and elsewhere in the state—placed additional burdens on different counties at different times, with the agents in southern Indiana being especially taxed.

Many people in need turned to local Extension agents for whatever advice and assistance they could provide. After all, by the 1930s the agents were well established in their communities and positioned to address the issues at hand. As a result, the agents proved their worth again and again as they quickly refocused their attention on the immediate needs of farmers and the local community. The chaos caused by droughts and flooding along with the challenges of the Great Depression were immense, but Extension remained a consistent presence that people relied on for advice, support, and a helping hand.

CHAPTER 13

There Is Convenience and Satisfaction of Flipping a Switch and Getting Light

By the late 1920s, all signs seemed to indicate that people in agricultural communities throughout Indiana would readily agree to sign contracts with utilities to extend electricity out from the cities and towns and into more rural areas. In 1929, agent Alfred Hesler, who had drawn national attention during the 1920s for his efforts to support electrification in Fountain County, had reported that the county was on track to have complete electrification.[1]

But as the economy collapsed with the onset of the Depression, Hesler noted that "all promotional and educational work [regarding electrification] was abandoned. Instead of new lines being built and additional farms being served with electricity, some 25 or 30 farmers withdrew their contacts."[2] With the fear of how long the Great Depression could last and where the next paycheck would come from, farmers viewed electricity as a luxury that they couldn't afford.[3] And given the common belief that electricity was primarily useful for lighting the home, many people thought that the kerosene lamp put out enough light at little cost. When rural people decided not to sign contracts, utilities that existed at the time decided not to expand electric lines, poles, and infrastructure into less populous townships.

By 1935, with the rural economy seeing some improvements, Hesler was once again reporting positive strides toward electrification, saying, "Instead of farms being withdrawn from lines, additions are being made constantly which shows that farmers are interested in rural electrification and will avail themselves of its use as fast as economic conditions will permit."[4]

While change was slow during the first half of the decade, the farm populace would soon begin to see how much better their lives could be and how much more productive they could become by using electricity to run their farms and homes.

While most urban homes in Indiana had electricity at the start of the 1930s, it was a much different picture for those living in rural areas. By the middle of the decade, 90 percent of homes located outside of urban areas still didn't have electricity.[5] It was likely that the 10 percent that did have it were located on the outskirts of towns, where local utilities could more easily extend their lines to pick up customers.[6]

For farms scattered farther from town, it was too costly to run poles and lines for just a few customers.[7] As Public Service Indiana spokesperson William J. Parvis explained in 1932, "Getting electric service to the farm means more than just the construction of a pole line to support the copper wires carrying the electrical energy. It means that a tremendous investment must also be made in generating equipment, cross-country transmission lines, substations, and other equipment which will assure 24 hour continuous service to the farmer whether he lives near the source of power or hundreds of miles away."[8]

It soon became obvious that if rural America were to get electricity, the federal government would need to invest in the infrastructure instead of leaving it to private utilities.[9] In 1935, President Franklin Roosevelt signed Executive Order 7037, which created the Rural Electrification Administration as a first step toward delivering electricity to rural areas. That same year in Indiana, the state legislature passed the Rural Electric Membership Corporation Act, which cleared the way for counties to seek federal funding to bring electricity to rural areas.[10] The following year, Congress passed the Rural Electrification Act, which provided for the creation of rural electric membership corporations, or REMCs.

Congress allocated the funding for the effort through a loan process whereby a local REMC could borrow money from the federal government to support the construction of the locally controlled and member-owned utility. Those interested in borrowing the money needed to create an REMC and elect a board of directors.[11] The REMC could then apply for a low-interest construction loan for developing and installing the electric infrastructure.

The loan had to be repaid over a period of twenty to thirty years from members' payments to the REMC.[12] Once the loan was paid off, the REMC would own the system, including poles, wires, transformers, and meters.[13]

Agents often organized and promoted meetings where the concept of the REMC and bringing power to farms could be discussed in detail. They also worked the topic of electricity into every conceivable meeting where they had a chance to speak, including at churches, Farmers' Institutes, and local farm bureaus. These events were used to gauge the local residents' interest in having electric service.[14] Many included displays on rural electrification, electrical wiring, and the cost to operate various electric appliances and equipment.

In 1939, James Morrow explained his position regarding the proposed creation of the Crawford County REMC, writing:

> It was not my purpose to urge a man either to take it [electricity] or not take it. I considered that his own personal business. But in making up his mind I

> did want him to have all the accurate information about the proposition that I might be able to give him. Since coming to the county in 1936 I know of no project that has stirred up more interest, comment and discussion among the people in the county than this one has. A great majority of them were interested in getting the information concerning it. Naturally there are differences of opinion among different folks as to whether they can afford it or not.[15]

In his 1936 report, William Smith described the problems he was encountering doing educational programs about electricity, the process, and the cost in Henry County, stating, "In most minds there was little understanding of problems involved, or of the ultimate cost to farmers of the county. At this stage, the extension agent made every endeavor to secure accurate information as to types of construction and possible cost of such a program. Many calls came on the subject, both individually and collectively. Usually when figures were presented, the cost seemed high and enthusiasm would wane for a short time. . . ."[16]

If there was enough interest, community members would elect a board of directors to work as volunteers to move the project forward by making the loan application and signing the contract to borrow money to set the poles, string the wires, contract with a utility to provide electricity, and ultimately deliver power to rural areas.[17] But before they could borrow the money from the Rural Electrification Administration, there were hurdles to overcome, paperwork to complete, and hearings to attend.

While the creation of the REMC went forward, the next step was to see if it was possible to meet the federal government guidelines that required three members per mile on average.[18] The sparse population made this a difficult goal for some counties, with most barely able to reach this number.[19]

Agents helped organize township committees with volunteers who would knock on every door to determine if that household would be interested in paying a five-dollar membership fee for becoming an REMC member and customer.[20] In 1936, the Bartholomew County REMC was able to get five-dollar commitments from 725 members across the 14 townships, while in that same year, the Carroll County REMC obtained 1,280 commitments.[21] And in Henry County, teams of volunteers managed to register 1,300 members by that March.[22]

While many REMCs were typically successful in getting the necessary commitments, some had to continue their membership drives much longer than originally anticipated in order to get enough members.[23] And a few initially encountered a lack of interest on the part of the residents, which meant they could not meet the three-members-per-mile criteria.[24]

A few counties had agents who had decided that their role was limited to providing educational programming and that activities such as membership drives and helping to establish the local REMC weren't in their job description.[25] In Benton County, Ellis Christen set clear boundaries on the work, observing,

"The promotion and organization [of the REMC] is considered outside the extension field. Education of how to use electrical energy is considered within the extension program."[26] This approach wasn't well received by everyone, as he reported in 1938, writing:

> The position of the County Agricultural Agent [Christen] in all the R.E.M.C. activity has been one of cooperation and willingness to give or provide the information desired. At the same time it has been his position that he was not responsible for the organization of a corporation nor the securing of applications for membership into the same. There has been some very definite local criticism to this position. He has been frequently reminded of activities it was claimed other county agents gave in adjoining counties.[27]

Christen concluded by acknowledging a need for education about electricity while noting that Extension also needed to establish a policy that defined the agent's role in securing REMC memberships.

As members were identified, the REMC placed all of their locations on township maps so an engineer could determine the best way to set the poles.[28] Then the REMC had to obtain easement rights from owners of the land on which the poles would be set.[29]

When the membership numbers supported the construction of power lines, the REMC had to submit to the Indiana Public Service Commission—known today as the Indiana Utility Regulatory Commission—an application to operate, proposed maps of where the lines would run, and financial statements.[30] After receiving approval from the commission, the REMC would apply to the Indiana Rural Electric Membership Corporation for final approval for submitting a federal loan application.[31] The state REMC board seemed to serve as a clearinghouse for the federal Rural Electrification Administration by confirming the data, reviewing financial plans, and determining the feasibility of submitted proposals.

Approval by the state REMC board was not guaranteed. In 1936, Bartholomew County representatives were told that their plan was unsatisfactory because it did not meet the three-member-per-mile criteria.[32] In 1937, the Franklin County agent reported that the local REMC wasn't approved by the state, noting, "The necessary mileage was signed, but the various interested communities were separated by so much distance that too much connecting line would be necessary to complete the project."[33]

Once the state REMC board gave its approval, the local REMC could then apply to the Rural Electrification Administration in Washington, D.C., for a loan to construct its electrical system.[34] Upon federal approval, the directors of the REMC, on behalf of the co-op members, signed the contracts moving the money from Washington to the REMC.[35]

With money on hand to move forward, the REMC developed the specifications for erecting the poles, stringing the lines, and setting transformers, and put the construction project out for bid.[36] Most cost estimates used to obtain the construction loans ranged from $1,500 to $2,000 per mile, but actual costs were closer to $1,000 per mile.[37]

In July 1935, Boone County would be the first REMC in Indiana and one of the first REMCs in the nation to be funded.[38] Between that time and January 1939, there would be thirty REMC projects started in the state that would receive almost $11 million in allocations. Twenty of these projects distributed electricity over 6,300 miles of line to 18,000 customers.[39] Boone County used its loan of $567,926 to erect 619 miles of line that brought electricity to 2,106 members.[40]

In Carroll County, work to create an REMC began in September 1935, with 1,000 members signed up by March 1936. The corporation received its initial allocation in September 1936 to build 366 miles of line that would serve 1,300 customers in Carroll County as well as areas of Cass, Clinton, Howard, Tippecanoe, and White Counties.[41] The $365,000 loan covered work done by the Ulen Construction Company that took two years to complete.[42]

Daviess and Martin Counties jointly received $352,000 to cover 354 miles of line that would eventually serve 1,274 members.[43] And Gibson, Pike, and Posey Counties working together as one REMC got $580,000 for the construction of 587 miles of line that would serve 1,800 members in those counties.[44]

Even while the poles were being placed, Extension agents continued to promote electrification and worked to convince more people to become customers.[45] And each REMC continued to add more miles of poles and lines as demand for service increased.[46]

As electric service became available and farmers joined the REMCs, it soon became clear that many farm families would need education about it. In 1936, Henry County agent William Smith offered his perspective on this, writing:

> As the period of construction [by the REMC] came nearer, two apparent educational needs developed. The first was in relation to the prospective customer for current [electricity]. He needed to know something about wiring, current consumption of various appliances, most desirable types of equipment, and possible ways of using electricity economically on the farm. To get started in meeting this need, the services of Seigel Anderson, [Extension] Specialist in Agricultural Engineering, was secured for a series of seven meetings. In these meetings lantern slides were used to show equipment, types of wiring and fixtures were displayed, and the general uses of electricity were discussed. . . .
>
> The other apparent need was for sufficient wiremen to quickly wire the farm premises when construction of power lines began. To meet this need, a committee has been appointed to secure an instructor and receive applications for those desiring to receive this training, organize this group into a school and get it functioning as quickly as possible.[47]

Extension would increasingly be asked dozens of questions, such as how much would it cost for electricity to run a home, how would one go about wiring a house or barn, and whether it was a good idea to invest in home appliances, farm motors, and electric tools.[48] Fortunately, Purdue agricultural researchers and Extension specialists had already begun working on these questions during the previous decade. As early as 1920, Extension had sent out the Home Special Train, which was an education train that traveled around the state to show people all types of electrical appliances, water systems, and lighting fixtures.[49] And since the late 1920s, Extension had hooked up meters to electric brooders, motors, feed grinders, milking machines, water systems, water heaters, stoves, and refrigerators to measure how much energy they used and estimate their operating costs.[50]

In addition to research done on campus, Extension also sought the cooperation of farm families to use equipment and appliances, and record that usage. As Blackford County agent Walter Rusk explained, those who volunteered to help "kept cost studies on the use of different pieces of electrical equipment [and] presented cost facts that proved of great interest. . . ."[51] And agent Harold Yarling noted that this gave Extension "accurate information as to the cost of operating these services with electricity and will give us figures that will be accepted by our people because their neighbors are the cooperators and their conditions are average for Hendricks County."[52]

From this type of testing, Extension was able to provide the following costs for using various types of electrical equipment by the early 1930s:

- 1.4 cents per chick for electric brooding;
- 38 cents for grinding one ton of feed;
- 15 cents to pump water to the farmstead, which had an average of 10 cows, 60 hogs, and 400 chickens;
- 24 cents per ton of milk collected from 15 cows with a milking machine;
- 37 cents to operate a hammer mill to grind up 400 pounds of corn, 300 pounds of oats, and 100 pounds of beans;
- 5.5 cents to clean 30–35 pounds of clothes using a washing machine;
- $4.00 per month for use of refrigerator; and
- $4.60 to operate a range with four burners and an oven for one month.[53]

Providing educational outreach that addressed costs along with potential uses for farm families was important. With money in such short supply, people were very concerned about using electricity when they were unsure of their ability to pay yet another bill. But Extension was trying to get them to view electricity as being "an economic factor rather than a convenience or luxury," as one agent put it.[54]

In 1930, when electric service to farms was still fairly rare, Allen County agent Hugh Heckard had reported, "A survey indicated that a large percent of the farms equipped with electricity were using it only for lighting purposes."[55] That same year, agent Alfred Hesler observed, "It has been demonstrated over and over again that unless a farmer intends to take the full advantage of Electric Power, he cannot afford to install for lights only; however, the farmers in Fountain County are wiring a greater number of their barns and out buildings, which is a positive indication that they expect to add motor-driven appliances as rapidly as possible."[56]

It was Extension's contention that the cheapest rates were obtained by using more power, which flew in the face of what most people believed. For example, in 1936, using electricity cost the members of the Blackford County REMC 6.5 cents for the first 30 kilowatts, 4.5 cents for the next 30 kilowatts, and 2.5 cents for the next 390 kilowatts.[57] REMC billings also required minimum charges, with those minimums averaging between $1.50 and $3.50 per month by the mid-1930s.

Orson G. Lloyd, chief of Extension's Division of Farm Management at Purdue, observed in 1932, "Aside from the comfort and convenience of electricity, those farmers using the most of it obtained a much lower rate. The upper fifth used more than ten times the amount used by the lower fifth of the farmers and it cost them less than one-fourth the rate per kilowatt hour."[58] So Extension wanted to make sure farm families understood this and used their electricity in an efficient manner.[59] Following a 1936 educational event on electricity, Hancock County agent Warren O'Hara wrote, "The purpose of the talk was to show definitely that the more current used made cheaper power; and that all current unused under the minimum was wasted."[60]

With so many farm families signing up for electric service by the mid- to late 1930s, Extension offered a variety of educational events with electricity as the focal point. There were four lines of education that county agents and the campus staff from the Purdue Department of Agricultural Engineering focused on during the early years of the REMCs: (1) explanations of the use of electricity; (2) how to wire a house and barn; (3) studies showing the cost of running appliances in the home and motors on the farm; and (4) membership drives to increase numbers of patrons served. Over the next few years, Extension would put on hundreds of programs statewide, presenting the information with lantern slides, demonstrations, and a U.S. Department of Agriculture film on rural electrification.[61]

These educational events quickly became popular with rural citizens. In 1936, Extension offered 53 events with 2,532 farm operators, homemakers, and others attending. In addition, there were four programs for home economics club leaders and two short courses for vocational agriculture instructors in two counties held that year.[62] The following year, Extension held 93 meetings, with more than 7,300 attendees and 15 conferences that drew another 366 people. Additionally, the Division of Home Economics in Extension presented training programs for 200 home economics club leaders, who then shared the information with club members in their local communities. A one-week short course on wiring held at Purdue drew 76 students, with the state annual Extension report indicating, "The students plan to assist with the wiring of farm buildings during the summer months in areas of the State in which a shortage of wiring contractors is apparent."[63]

Campus Extension staff participated along with the county agents to provide the educational outreach. Truman E. Hienton and Seigel A. Anderson from the Department of Agricultural Engineering at Purdue traveled throughout the state to present programs on the wider use of electricity on the farm.[64] In 1936, Hienton presented a series of twelve lessons to Delaware County farmers that the agent said covered "information on wiring principles, motors, farm water supply, electric equipment for the dairy and poultry farms, feed grinding and household electric equipment."[65] Hienton often used a feed grinder attached to a meter to show how much it cost to grind a specific amount of feed.[66]

Anderson was also a speaker at the 1936 event in Delaware County. Agent Myron Cromer would later observe that Anderson's Wiring the Farmstead presentation on the "principles of wiring and economical uses of electric current on the farm was very timely to those persons who have contracted for electric service and are having the farmstead wired."[67]

Purdue campus specialists from the Extension Division of Home Economics and county agents also offered training to show how electricity could be used to lighten the workload in farm homes. Fountain County agent Alfred Hesler viewed electricity as a life-changing event first and foremost for farm women, writing, "If there is one class of women above the other that needs the relief from drudgery, it is the farm women. The memory of mother with her stooped shoulders and her gnarled hands reflects the need of lightening the burdens by the uses of the electric power washer, the iron, sweeper, the water systems, etc."[68]

Hesler noted that there was a huge disparity in the quality of life for women who had an electric refrigerator to keep food fresh instead of the standard icebox or who used an electric washing machine to clean clothes instead of a washboard and washtub.[69] This is why home demonstration Extension agents such as Bertha Cook in Cass County gave presentations to show women how they could use electrical appliances in their homes. But with only twenty full-time county home demonstration agents listed in the 1936 Extension annual report, Extension also relied on having Purdue campus staff teach lessons on electricity to the leaders of home economics clubs who could then train others in their communities.[70]

Extension also used demonstration farms as showcases where people could see firsthand how electric power was used on the farm and in the farm home. In 1936, Boone County agent Clarence Whistler described one such farm, noting that "electric companies selling electric appliances through local representatives have been invited to install their products on this farm on a co-operative plan. Almost every conceivable electric appliance has been installed from hot pads to milking machines. Periodic open houses will be held on this farm to encourage a wide use of electric energy on the farm."[71]

It would take some time, but eventually, this type of education seemed to work as agents began to report by the end of the decade that more electrical equipment was being purchased and electric usage was on the rise.

Agents commonly noticed that it would take a year or so after a house and barn were wired before rural families began purchasing electrical appliances and equipment. It cost on average seventy-five dollars to wire an average home in 1937, so that might have taken what little cash a family had.[72] In Boone County, Whistler explained it by saying, "It is well to note that after the farmer has paid for wiring his buildings, it is about all the cash outlay he can afford the first year. Farmers and housekeepers want appliances and are applying them as fast as finances will permit. On the first 160 miles energized, the appliances purchased the second year has made a material increase. Also the load has increased in these lines."[73]

Extension tracked the effectiveness of its educational programs by what people were purchasing after receiving information about using electricity on farms. In 1939, agent Ralph Maggart noted that a month-long campaign on electric brooders had resulted in 122 units being sold in Carroll County.[74] He also reported that local farmers purchased 52 ranges, 38 refrigerators, and 73 floor lamps.[75]

That same year in Boone County, the agent reported that farmers had purchased 101 electric brooders, spurred partially by a 33 percent discount from the O. K. Hatchery, which cooperated with the local REMC on the offer.[76] He also cited the results of a survey showing the purchase of the following electrical appliances by county residents:

Type	Quantity	Percent of Households Purchasing
Iron	1,352	91.4
Radio	1,329	89.9
Washer	1,036	70.0
Vacuum cleaner	647	43.6
Toaster	461	31.1
Water pump	360	26.4
Refrigerator	344	23.0
Hot plate	138	9.3 [77]

Surveys like the one taken in Boone County found similar results across the state, with other popular purchases also including electric stoves, churns, mixers, waffle irons, corn poppers, razors, curling irons, and clocks. The high-dollar appliance that made the top of the most-wanted list was the refrigerator.[78]

As far as farm equipment went, electric motors and tools became the items that farmers felt they could not do without. Corn dryers were purchased by those raising commercial seed, and poultry farmers used electric brooders to help increase survival rates of young birds.[79] Dairy farmers found that electric milking machines cut labor by 40 percent over hand-milking. That meant a herd of 15 cows could be milked for the entire month at a cost of 90 cents for the electricity used.[80] Other common purchases were fans, coolers, grinders, cream separators, welders, motors, and water systems.[81]

The historical evidence suggests that helping to bring power to the countryside was one of the greatest contributions by Extension agents and campus specialists during the 1930s.[82] From the agent's perspective, the availability and access to electricity was a game changer for farm families.[83]

Over a relatively short time period, the use of electricity would be more widely adopted and forever alter the daily life of those living on remote farms. While hard economic times slowed the progression during the Depression, it was only a temporary setback. In 1930, one out of every ten Indiana farms had electricity, but by 1940, that number would be one out of every three, showing the rapid adoption of this new technology. And by 1964, most Hoosier farms were connected to electrical lines.[84]

Farmers would soon recognize the value of this energy for running an efficient farm. They realized that the cost of electricity was quite a deal as productivity increased and labor costs dropped. And in the farm home, rural families quickly adapted to timesaving equipment such as washing machines and vacuum cleaners, enjoyed the convenience of having running water indoors, and reduced waste by storing perishable fruits and vegetables in the refrigerator.[85]

In 1938, Greene County agent Arthur Haseman observed, "Electricity has come to take a place alongside the automobile, the telephone, and the radio to make a 'heap of living in the house' called the farm home."[86] Agent Ralph Maggart reported that "electricity in the farm home has done more to elevate the moral[e] of the rural people in Carroll County than any one thing in the past ten years. It has given them a life-long realization to have and to be on an equal [footing] to the city family. There is convenience and satisfaction of flipping a switch and getting light."[87]

The work toward establishing REMCs in the 1930s would be remembered by agents as life-changing for people living in rural areas of the state.[88] Purdue Extension played a major role in that change by helping to deliver power and education to the agricultural community, which ultimately changed how its people worked, played, and lived.[89]

CHAPTER 14

Shall I Sell One Team on a Four-Horse Farm and Buy a Tractor?

By the late 1920s, the horse era appeared to be drawing to a close in Indiana as farmers began replacing their draft horses with tractors on the larger farms. Tractors were considered important equipment and worth the money because they could pull heavier and wider implements such as cultivators, planters, and harvesters. This allowed farmers to work more efficiently and even expand their farms.[1] But many farms in the state were still too small to make the shift, leading Fountain County agent Alfred Hesler to observe in 1936 that "horse production has a very definite place in the livestock program of the county."[2]

As farmers faced the high costs of using mechanical power between the start of the Great Depression and the beginning of World War II, draft horses got a reprieve on Indiana farms. Instead of purchasing fuel, oil, and parts for tractors, farmers simply kept using their horses to plow, plant, and harvest, and fed them with crops from the farm.[3] As Delaware County agent Myron Cromer observed in 1933, "The high cost of tractor power machinery as well as the relative high cost of gas and oil to run [the] same, as compared with the cheapness of farm grains and hay, all combine to popularize the horse for farm work on the average farm."[4]

With the continued interest in using horses for farm work, it became apparent to the agricultural community that there were fewer young horses and mules to replace those that had died or were being retired.[5] In 1936, Hesler reported, "Tractor farming in Fountain County has come along at the same pace as counties of like conditions; therefore, in the past decade, many of the brood mares have passed out of usefulness, but with the sting of the depression, worn out tractors were replaced with many worn out horses."[6] Just a few years later, he observed, "The horse situation in Fountain County is about the same as it is in Indiana and the Nation, that is a gradual decrease in horse numbers. It has been stated that three horses and mules die off while two colts are being born."[7]

The horses put into service were past their prime as they worked the fields.[8] And without younger replacement horses, the older horses were pushed to their limits, leading Elkhart County agent Chris Blosser to report in 1931, "Heat prostrations have exacted a heavy toll on the horse population. One tankage concern alone reported that they had averaged a receipt of ten horses per day for a considerable period and that they took in over two-hundred dead horses during the season."[9]

Low grain prices coupled with a higher demand and better prices paid for working horses led farmers to become interested in raising more colts for the farm and for income.[10] In 1932, Grant County agent Mosher Butler observed, "Horse men are of the opinion that Dobbin [a farm horse] will increase steadily in purchasing power and value so long as corn and oats are cheaper than petroleum products and tractor parts."[11] And people listened closely upon learning that one farmer had sold colts in 1934 for $1,000, which was more money than most were earning from their farms that year.[12] Producers who raised and sold colts from purebred draft horses saw the opportunity to increase their revenue.[13]

In his 1931 annual report for Elkhart County, Blosser also observed, "Owners of draft stable horses report that more mares have been bred during the past season than the previous five years. The owners of stallions report breeding over one-hundred mares each during the season."[14] Those owning purebred stallions made money from fees charged to others who had mares to be bred.[15] Extension put farmers with mares in touch with farmers who had registered stallions, but there were relatively few high-quality stallions available.[16] In 1936, Hamilton County agent James Todd organized a spring stallion show to "(1) provide opportunity for mare owners to compare stallions available for public service; (2) to provide a means of stallion owners advertising stallions for public service; and (3) to advertise these horses for sale."[17]

That same year, Todd also reported that those producing purebred colts found that buyers from around the country and the world were willing to spend good money for these animals. He wrote:

> Due to the large number of pure-bred horses in the county, buyers are coming in every week and taking home a few animals. Many go to other states and Canada. On the other hand breeders here are continually buying horses. Lynnwood farms last week purchased the Grand Champion Stallion of the 1936 Ohio State Fair, Don Again. This week they received 12 foals from France as did also the Conner Prairie Farm. Today there are more pure-bred horses in the county than in any other county of the United States.[18]

For those who could take advantage of the strong interest, the horse industry represented an important sector of the agricultural industry.

To support the interest and need for more horses, Extension had begun providing additional education about horse breeding, feeding, housing, care, and use for farming during the early 1930s.[19] Purdue put on horse clinics—often led by

Roy B. Cooley and Parke T. Brown from the Extension Department of Animal Husbandry—that stressed the need for better breeding stock, the proper feeding of mares and colts, the use of registered stallions for upgrading bloodlines, breaking and training colts, and using multiple horses teamed together to pull larger agricultural implements.[20]

Extension also continued with the popular multiple-hitch demonstrations that had been started during World War I, when there was a shortage of labor on the farm, and also during the agricultural depression of the 1920s, when labor costs could be reduced by combining the strength of multiple horses to pull larger equipment.[21] As one agent noted, Extension focused on these programs "in response to the increasing desire on the part of farmers to accomplish work at tractor speed without the cash outlay for fuel, oil, repairs, etc. required in tractor operation."[22] Another reported that this demonstration was "all the more popular because the price of farm feeds are very cheap, making it advisable to use all the horse power possible."[23]

Community demonstrations were common, and with the help of those experienced with multiple hitches, participants were shown how to arrange the horse teams using eveners and buck ropes.[24] Some attendees also received big-hitch booklets written by the Horse Association of America.[25]

In October 1930 in Clinton County, a crowd of between 3,500 and 5,000 attended a plowing match that used teams of horses hitched together. The *Indiana Prairie Farmer* published a report on this popular event, saying, "Cleveland Combs easily carried off premier honors when he won both the six-horse class and the nine-horse class with horses that he has been working in big hitches for the past two years. Then, to cap the climax, Mr. Combs gave a spectacular demonstration when he hitched 12 horses to a four-bottom plow."[26]

In addition to attending larger events, farmers could get one-on-one assistance as needed from the local county agent to learn how to team up multiple horses. Clinton County agent Verner Mann reported in 1930 that he would "personally assist any farmer who wishes to start a big team. This assistance is given in the shape of making buck ropes and in helping start the hitch the first time. This insures the farmer getting rigged up right in the start."[27] But the following year, Mann reported that the "multiple hitch is now so common that it is not necessary to apply at the county agent's office for assistance. The farmer may easily go to his next neighbor for assistance."[28]

Agents used the Hoosier Gold Medal Colt Club to promote the importance of proper care, feeding, and breeding among those raising draft horses. Sponsored by the Indiana Live Stock Breeders Association, the program awarded a gold medal to colts that reached 650 pounds in 365 days, a silver medal for those reaching

500 pounds, and a bronze medal for those reaching 400 pounds.[29] The competition was under the direction of the agents, who hosted the events, registered participants, and oversaw the judging of the animals.[30]

Delaware County agent Myron Cromer described the competition as being "organized to assist the owner in producing better yearling colts. A colt properly developed to the yearling age has much greater possibilities when matured."[31] But in his 1931 annual report, he also noted that this competition was part of a much larger plan that "attention be given to the production of more and better draft horses, to replace the rapidly increasing numbers of old worn out horses now on farms."[32]

Young colts typically were nominated for the club in July, when the initial weight was recorded.[33] In 1935, agent Maurice Goode described the process used for participants in the Harrison County Hoosier Gold Medal Colt Club, saying, "[O]ne of the stallion owners furnished his truck and a helper to go with the county agent from farm to farm and weigh-up the colts. A large platform scale was taken along and the colts were weighed in this manner."[34] The weights were then recorded in the county Extension office.[35]

One year later, the colts were weighed again.[36] Each animal was brought to a location where a judge could evaluate the horse's conformation and soundness for its breed. If the horse weighed 400 pounds or more, it could be entered in the county, district, and state colt shows.[37] The gold, silver, and bronze medals given by the Indiana Live Stock Breeders Association were presented in each of the different classes of draft horses, including grade mare colts, grade horse colts, and purebred horse colts.[38]

Extension shows dealing with draft horses were popular events that attracted thousands of interested farmers.[39] For example, the Fulton County Horse Show held in April 1935 at the Cole Brothers' Circus Headquarters in Rochester attracted 3,000 attendees.[40]

"Shall I sell one team on a 4-horse farm and buy a tractor?" An agent reported receiving this question from one farmer, but many farmers were wondering the same thing.[41] Even as farmers relied on horses to get the work done during trying times, they continued to weigh their options for using tractors and to calculate the costs and efficiencies of this farm equipment against their horse teams. The Cass County Purdue Agricultural Club held a public debate on the topic at its February 1932 meeting, with the agent reporting:

> Estol Howard and Everett Miller, both of Miami township, gave facts and figures to prove the superiority of the Tractor, while Sam Little of Washington township and W[illiam] Searight of Jefferson township defended the Horses. At the close of the arguments, the meeting was thrown open for discussion

and a lively one ensued. A vote of the audience as to the winners of the debate ended in a tie, while a vote as to which one the audience was going to use this year ended in favor of the tractor.[42]

While work was being done with horses, there was still great interest in tractors and other implements that could be pulled behind them.

By the end of 1939, tractors had almost regained what they had lost during the tougher years of the Depression. As the decade drew to a close, agricultural machines were replacing not only horses and mules but also farmhands. A 1939 headline in the *Palladium-Item* in Richmond that June declared, "Machines Begin Displacing of Hands on American Farms."[43] In the article, journalist Oliver Knight cited a report by the U.S. Department of Agriculture, which stated, "The reduction [in the need for labor] has come about as a result of mechanization in which the combination harvester-thresher, the tractor, motor truck, and larger units of tillage equipment have almost completely displaced the use of horses and the smaller sizes of farm equipment."[44]

That same year, agent James Todd reported that farm power in Hamilton County was largely provided by tractors.[45] A few years before, Todd had invited Hamilton County farmers to witness a tractor plowing contest that showcased plowing power and fuel usage. He had worked with tractor dealers in the county to set the rules for the contest. Operators of the different makes of tractors were told to set the plow to a depth of 7 inches and provide their own plot of land to plow. The fuel in each tractor was drained, then refilled with five gallons of the fuel required by the manufacturer.

The tractors plowed for about an hour, then the fuel was drained to determine the amount used. Fuel costs were calculated for Red Crown gas at 12.1 cents a gallon, Standard Oil low-grade gas at 11.1 cents per gallon, and distillate tractor fuel at 8.9 cents per gallon, with Todd citing the following results:

Make of Tractor	Wheel Equipment	Plowing Time (minutes)	Amount Plowed (acres)	Used (gallons)	Cost (cents)	Cost Per Acre
Case	Rubber	62.5	1.28	0.97	9.2	7.2
Co-op	Rubber	59.0	1.20	2.30	28.0	23.3
Farmall F-20	Rubber	60.5	1.24	1.80	16.7	13.5
Farmall Caterpillar	Steel	59.0	1.30	2.40	22.3	17.2
Allis-Chalmers	Rubber	58.0	1.08	2.70	29.8	27.6[46]

Todd observed in his report, "The results obtained are not to be taken as complete proof of the superiority of any one tractor over another. Mainly it was an initial attempt to give the farmers present a comparison of the tractors under similar working conditions."[47]

The Depression, the scarcity of money, and, to some extent, tradition kept draft horses on Indiana farms for an extended period of time.[48] The use of power equipment piqued the interest of farmers, but even after tractors became widely available, horses and mules continued to power the farm as they had for many prior generations of Hoosier farmers.[49]

The high prices paid for most farm commodities during World War II would eventually provide the funding to allow farmers to expand the size of their farms. This marked the end of the draft horse as the main provider of farm power as farmers replaced their horses with mechanized equipment, completing their work by using tractors to pull plows and combines to harvest grain.[50]

CHAPTER 15

The Average Farmer Has Not Learned the Principles of Economic Uses of Wood Lots

During World War I, U.S. farmers were under pressure to produce more food to support the Allied forces. This led many farmers to clear trees so they could establish new fields to grow more crops. Not only did this increased production help farmers reap the benefits of good prices, but it also helped the country honor its commitment to deliver food supplies abroad to those in need. Unfortunately, the conversion of forests into grainfields did not end well for the land.

In a 1939 report, Crawford County agent James Morrow summarized what had taken place during the preceding decades as forests were cleared to meet the wartime production quotas. He wrote, "Whether or not this putting of land to cultivated crops that was unsuited to cropping purposes was recognized as a serious problem at that time [during World War I] is doubtful, but in the nearly 20 years that have elapsed since, it can be seen now that this extensive clearing of land and heavy cropping was one of the worst things that could have happened to Crawford County soil."[1] Agents around the state could have said the same thing about their counties.

Some of this land was marginal to begin with, which meant that not long after it had been cleared and farmed, the topsoil began eroding and the nutrients were washed away.[2] In 1930, agent Ronald Johnson reported, "Thousands of acres of Harrison County are rough and rolling, most of which has been cut over [had trees removed] at least once, and a number of fields which were cultivated for a time started to wash [erode], and there is much need for more forestry extension work in the county."[3]

Farmers were told that the land could be reclaimed and made productive again, but planting trees, not corn, was the key to that conversion. Agents pushed planting trees as a way to protect the remaining soil, improve the site, and prevent further gully erosion.[4]

The federal government invested resources and manpower to restore and reforest badly eroded properties both on private and public lands. This work was primarily accomplished by the Civilian Conservation Corps (CCC), which in

1933 began hiring unemployed men to plant trees, do erosion control work, build roads and buildings in state parks, and help fight forest fires.[5] (See Chapter 10.)

In 1933, Indiana state forester Ralph Wilcox asked the federal government for 24 CCC camps but received notice that only eight had been approved.[6] The first CCC camp in the state was located in Morgan-Monroe State Forest.[7] In 1934, Indiana would be allotted 24 camps that employed 7,269 men.[8] Each camp consisted of a temporary base that housed up to several hundred workers who lived in tents and completed projects nearby under the direction of army officers. Typically, a trained forester also provided technical assistance to a CCC camp.[9]

Farmers had to register eroded land with the CCC for reclamation. Any work completed was often done at little expense to the landowner.[10] After a county had at least 400 acres of badly eroded land enrolled in the restoration program, the area would be evaluated, and the decision to put a camp in that county would be voted on by U.S. Forest Service and U.S. Army officials.[11] Decisions about where to place camps were based not only on soil erosion problems but also on available camping facilities, water supply, transportation services, and proximity to the work.[12] Once placed, the camps typically remained in an area for a minimum of six months.[13]

CCC workers housed at these camps were instrumental in planting large numbers of trees, building fire towers, and completing many other projects. For example, author David E. Conrad noted that during six years of operation of the camp at the Hoosier National Forest, workers built "14 buildings, 3.5 miles of road, 75 miles of telephone lines. They had planted 1.5 million trees on 1,500 acres, cut 2,000 posts, built three [fire] lookout towers, and completed several hundred acres of erosion control work. They dug six game ponds, fought many forest fires, quarried and crushed 13,000 yards of stone, built a new recreation center, constructed a dam for a small lake, searched for lost children, and assisted in other emergency rescue work."[14] Across the United States, CCC workers planted 1.7 billion trees, built 100,000 miles of trails and roads, and provided 5 million man-days fighting fires, clearly showing their extensive accomplishments.[15] Achievements such as these made the Civilian Conservation Corps one of the more popular Depression-era programs in Indiana and nationwide.

By the 1930s, forestry had long taken a back seat to crops and livestock as an agricultural enterprise on Indiana farms.[16] Many of those who owned forestland had already harvested and sold much of their valuable timber between World War I and the Great Depression, leaving behind nearly worthless stands of unmarketable trees. Often treated as unproductive, this cutover land was grazed by cattle, which further suppressed any natural regrowth and regeneration of the forest since the animals consumed the seedlings.

During this time, county Extension agents were required to list their forestry work efforts in their annual reports. Most entries were single sentences or short paragraphs, such as Benton County agent Ellis Christen wrote in 1937: "The activity in forestry has been confined to the planting of farm windbreaks."[17] In addition to planting trees to serve as wind breaks for protecting homes, agents reported helping farmers reclassify their woodlands with the state to reduce the taxes owed on them and working with growers to establish Christmas tree farms as another source of income.[18]

But these efforts were ancillary to what really needed to be done with privately owned woodlots, which was to manage them in such a way as to increase sustainability and allow for long-term financial returns. Unfortunately, many agents were fighting an uphill battle to generate attention for better timber management among landowners, with Bartholomew County agent Arvil Keesling noting in 1937, "The entire western part of the county should have more forestry work, but as yet interest is not sufficient to secure the desired results."[19] The lack of interest in forestry among landowners could be explained by the immediate need for money that had preceded and continued during the Depression. While Extension talked about the long-term approach to managing the woodlots, farmers knew that selling marketable trees brought in cash needed to pay creditors.

Agents expressed concern about the degradation of the trees in woodlots, especially after the marketable trees had been removed and the land left unmanaged or grazed. In a 1937 annual report, Cory Alcorn painted a rather bleak picture of the remaining forest in Dearborn County, writing:

> We feel that there is a place for more consideration to be given to forestry problems, but folks are slow in showing interest because the returns are so remote. The census shows 8,272 acres of woodland not pastured, 14,378 acres of woodland pastured, which is neither profitable woods nor profitable pasture and should be made either one or the other, and 10,472 acres of idle or fallow land, most of which is probably so badly eroded that it is past profitable cultivation and should be set out in locusts [trees].[20]

Locust trees had the advantage of fixing nitrogen in the soil, being able to regrow more easily than other tree species, and producing wood that could be used for fence posts and firewood.

The year before, Gibson County agent Roy Gwaltney had emphasized the need for taking a long-term approach to improve degraded forestlands, observing:

> Most owners of farm woodlands are prone to consider them as non-profit paying property, and under the plan of management in vogue on most farms, they are right. All good timber has been cut out and all that are left are the worthless species. The land where even these were left was in many cases either too broken or too wet to cultivate. Had it been possible to raise a few

> crops on the land it is doubtful if there would have been one-tenth as much woodland left as there is. . . . These woodlands can be made to pay a profit commensurate to other farm land if given an opportunity. Probably not in the days of the present owner, as it will take them too long to recuperate from the treatment received in the past.[21]

Extension would have to focus on promoting the long-term management of these forestlands on private farm property and keep trying to convince farmers that doing so was in their best interest.

While many farmers around the state hadn't yet adopted better management practices for their forestlands by the mid-1930s, there were some who had begun to make changes. Much of the initial interest started in southern counties, where trees cut from farms supplied the raw material for the mills, and in return, furniture manufacturing businesses provided much-needed jobs. In Floyd County, agent Charles East had reported in 1932, "Forestry as a long time program offers a real possibility of profit, particularly to a community such as this, a large portion of whose wealth is built on wood manufacturing."[22]

Just west of Floyd County in Dubois County, Charles Nicholson pointed out in his 1936 report that jobs were affected by the availability of timber coming off private lands, writing, "Forestry in Dubois County is rapidly being considered as a very much needed and worthwhile extension project. An industry through the operation of fifteen wood working plants located in the county with 2,000 men employed and having a weekly pay roll of $40,000 should command some time and attention of the county, state and national extension forces."[23]

A 1937 census of Dubois County woodlands owned by private landowners found that the majority of the 54,700 acres consisted of four types of timber stands: oak-hickory, beech-maple, pin oak–sweet gum, and black oak–scarlet oak. However, the report estimated that based on the rate of timber harvest at the time, the mills would run out of timber within fifteen years if nothing were done to deal with overharvesting.[24] If the mills shut down, all of the jobs in manufacturing would disappear unless new sources of timber could be found. This was a serious concern since a steady supply of wood created jobs that supported families and generated tax revenues for communities and counties.[25]

With interest growing among landowners, extensive management of existing woodlots and sustainability became key strategies for Extension staff to use to improve private forestlands.[26] Agents looked to Thomas E. Shaw, an Extension forester at Purdue, for help with efforts such as designing and providing forestry demonstrations and holding forestry schools.[27] As Purdue's point person on forestry, Shaw shared his expertise and knowledge on forestry production with the agents, who could then provide instruction for farmers.

With Shaw's help, agents began offering a management plan to woodlot owners to highlight what farmers and other landowners needed to do to enhance the productivity of their timber stands.[28] It consisted of the following six strategies:

• *Protect the forest against fires.* This strategy was paramount for safeguarding the investment of time and resources that went into managing forestlands, so counties in southern Indiana set up fire brigades and, under the supervision of the Indiana Division of Forestry, trained mainly farmers to participate in them.[29] In Harrison County, agent Maurice Goode reported that five fires in 1936 created minimum damage because "of quick work on the part of these volunteer committeemen" and that four fires in 1938 were contained to less than five acres of timber.[30] The state also provided support by building a number of fire towers across the southern counties. According to Charles East in Floyd County, the state even paid for a person to serve full time in a tower in Franklin Township during the spring and fall of 1932.[31]

• *Avoid grazing timber stands.* This single point countered what was actually happening on most farms.[32] A 1930 census in Elkhart County showed 24,000 acres of farm woods, but since only one quarter of these were unpastured, the agent reported that their "productive capacity is being slowly but surely decreased."[33] By 1936, Purdue forestry Extension specialists were reporting some progress in getting farmers to keep their cattle out of their forests. Over the course of the preceding five years, which represented the beginning of forestry as an Extension department at Purdue, the unpastured woodlots increased from 785,000 acres to 972,000 acres. However, the 1936 annual report also stated, "The practice of pasturing woodland presents one of the major forest problems in Indiana," which showed there was still much more work to be done along this line.[34]

• *Properly cut and market timber.* Extension encouraged those who owned forestlands to focus their attention on selling mature trees, removing deformed and undesirable tree species, and working toward getting a stand of white oak and walnut seedlings in the undergrowth.[35] Landowners were advised to cut the largest trees, which would get them the best prices in the market while allowing for the growth of understory trees.[36]

• *Remove inferior trees to make room for good trees.* Many woods had been left to reseed on their own, which allowed undesirable and more aggressively growing species to prevent desirable seedlings, such as those of white oak trees, from establishing themselves in the forest understory. Extension put on many demonstrations, including one 1930 event located six miles east of Corydon on the farm of Herman Meyers, to show landowners how to eliminate undesirable species in favor of promoting growth of desirable ones.[37] Removing the inferior trees was a way to reduce their footprint in the forest understory and open up the canopy. This allowed more light to reach the forest floor, which then promoted growth of oak, hickory, and walnut trees.

• *Use unsaleable wood products.* As Extension staff promoted removal of certain types of growth from forestlands, they also noted that whatever was harvested out of the woodlot had value and that farmers should take advantage of the extra income by not letting anything go to waste. If boards could not be made from the timber, then the wood could be used for fuel, posts, rough lumber, and many other purposes both on and off the farm.

• *Produce the maximum volume of high-quality timber at all times.* When all was said and done, the work on a woodlot was to grow trees that were essential to the market. The message from Extension was that landowners who followed this plan would more than offset their time and cost by being paid top dollar from any tree harvest and that such production practices would promote higher timber values in the future.

In 1939, Fountain County agent Alfred Hesler observed that the "average farmer has not learned the principles of economic uses of wood lots."[38] But Extension sought to change that by offering schools that focused on best-management forestry practices. An event that year in Crawford County covered the following topics:

- Better Timber Crops
- More Cash from Timber Crops
- Soil Erosion and Its Control Through Proper Land Use
- How Indiana Farmers Are Conserving Soil[39]

Many of these forestry schools utilized 16-millimeter educational films provided by a film service at Indiana University with titles such as *ABC of Forestry*, *Big Game and National Forests*, *Good Turns for Our Forests*, *Marking Timber*, and *Nature Study*.[40]

When local media attended the Extension events, they published stories to pass along the key points of the presentations to their audiences. In 1939, Emery Applegate, the farm reporter for the *Fort Wayne News-Sentinel*, wrote an article about demonstrations at an Allen County event, which focused on "planting new trees, cutting out the old trees that have served their usefulness and the advantages of a good woods in order to prevent soil erosion."[41]

Shaw frequently participated in the forestry events put on by Extension around the state to help the local county agent deliver and reinforce the key forestry management points to those in attendance. Following the 1939 Allen County event, agent Chester Kimmell reported, "T. E. Shaw, extension forester, gave a demonstration on measuring trees and also told those present how to tell the growth of a tree. In addition, he gave some very valuable suggestions on forestry plantings and the importance of improving local forests. Many questions were asked relative to cutting timber and pasturing wood lots."[42] That same year, agent Russell Zell noted that Shaw made presentations in Franklin County on "the wise use of timber, including tree stand improvement, proper time of harvesting salable trees, and demonstrated various methods of measuring both standing trees and logs."[43]

Shaw also provided individual consultations as needed. In 1938, he made three trips to Elkhart County to inspect a number of woodlots owned by farmers. During two of the visits, agent J. Howard Brown went along and received one-on-one instruction in the process that would help him advise others in the future.[44]

As Extension continued to focus attention on forestry in the state, an all-out effort was made to supplement schools and demonstrations with other communication methods such as newspaper articles, publications, and consultations. For instance, in 1936, the Extension Division of Forestry at Purdue published a series of circular letters on managing private woodlots that the agents could distribute to farmers. Agents also reported showing a filmstrip titled *Keeping Livestock Out of the Woods in the North Central States* at local meetings.[45] And the 1938 state report for Extension noted that a new Extension publication—No. 221, *The Need for Woodland Improvements in Indiana*—was handed out at many of the 119 woodland management meetings held statewide, which had attracted 5,823 people.[46]

With continued promotion and education efforts by Extension, word got around that farmers should manage their woodlots in addition to managing their crops and livestock. When Extension showed that well-managed woodlots brought significant returns for the landowner, interest in forestry and the proper care of forestlands skyrocketed.[47]

Whether trying to get farmers to replant trees on eroded soils or better manage their woodlots, Extension spent much of the 1930s trying to generate interest in forestry. Farmers who attended a forestry school or an on-farm demonstration offered by Extension became aware that they could improve and manage their own properties if they just followed a few basic practices. The agents were lucky to have Extension forester Thomas Shaw as a resource. He was definitely an asset in making the science of forestry real and practical to those interested in managing their forestland.

As the effects of the Depression started to ease and interest turned to reforestation to prevent erosion and the sustainability of forestlands, Extension found a small but dedicated audience willing to listen to practical advice on how to care for and manage their woodlots to maximize future profits.

CHAPTER 16

Hybrid Corn Is With Us to Stay Until Something Is Found to Take Its Place

During the 1930s, raising corn affected more farmers in Indiana than any other agricultural activity. Simply put, corn was the lifeblood of the state's agricultural industry.[1] As one agent observed, "The business life of the entire County rises or falls with the corn crop."[2]

County Extension agents knew that talking about corn production drew more interest among farmers across the townships than any other subject the agents could discuss.[3] In a 1938 annual report, Decatur County agent Matthew Hamilton noted, "Many farmers are now reached through one of the phases of corn extension work that probably would not be reached through any other county extension work."[4] Corn education got agents inside the door, where they could then offer a full range of education for farmers and their families.

Corn was a cash crop that helped large and small farmers alike pay their bills and provided the money to purchase supplies from local businesses.[5] Its versatility increased its value since farmers fed it to different types of livestock and in various forms—alone; in a blend with wheat, oats, and soybeans; or as silage.[6] And when yields were high or corn was available at lower prices on the market, farmers could increase their numbers of hogs, sheep, and beef and dairy cattle to potentially increase their incomes.[7] Analyses of farm records by Purdue agricultural economists clearly showed that corn fed to livestock brought more money through the sale of meat than what a bushel of corn brought on the market.

Even when the price of corn dropped to dismal lows, Extension continued its educational programs, with agent Mosher Butler in Grant County noting that the Depression "has not [affected] the interest in the problems of corn growing."[8] Most of those programs initially focused on management strategies for open-pollinated varieties, but the 1930s would prove to be pivotal for Indiana corn growers as they saw increased use of commercialized hybrid corn with each passing year. Soon hybrid corn would become the gold standard for row-crop producers and replace the open-pollinated corn that had traditionally been used in the state.

In the early 1930s, open-pollinated corn yielded an average of 25–45 bushels per acre across the state.[9] Agents recognized that these yields were lower than they could be, which was a cause for concern during tough economic times. Lower yields meant the cost of production was higher and profits were lower for each bushel of corn produced.[10]

Based on results from university research, county yield trials, and competitions among growers, there were no logical reasons why corn yields should not be doubled when using sound management based on agricultural research.[11] Even participants in the 4-H corn clubs were achieving these higher yields. In Franklin and Hamilton Counties, 4-H members averaged 77 bushels per acre compared to county averages that were less than half that in the early 1930s.[12]

To increase yields during the first half of the decade, agents encouraged farmers to focus on improving management practices by using adapted open-pollinated varieties, better seed selection, improved winter storage methods, and yearly testing for germination.[13] To that end, agents conducted statewide seed corn campaigns that communicated advice through public lectures, letters to farmers, posters placed in banks and hardware stores, newspaper articles, and radio broadcasts.[14]

During the 1931 campaign, a circular letter titled "Look Ahead: Your Crop Next Year Depends on This Year's Seed Corn" was mailed to every Indiana farmer who raised corn. The overarching theme was that Extension wanted farmers to do a better job in choosing seed for spring planting.[15] Cass County agent James Little reported the following key points of the campaign:

1. Begin seed selection when half or more of the ears are mature, as indicated by ripe husks. The stalks and portions of the leaves should still be green.
2. Select mature ears, slightly drooping, carried on a sound shank of medium size and length.
3. Select ears from strong, erect stalks borne at a convenient height for husking, and surrounded by a full stand of plants.
4. Avoid ears from weak, broken, smutted, or prematurely ripened stalks.
5. Pick two to three times as many seed ears as will be needed for planting. About 12 to 14 medium size ears will plant an acre.
6. Store all field-selected seed the same day as gathered in a dry, airy place, and on racks which keep the ears apart.
7. Seed corn should be thoroughly dried before freezing weather begins or should be stored in a place where it will not freeze.[16]

Following these seven points would give farmers an advantage heading into the spring. Agents provided growers who wanted more information with Purdue Extension Bulletin No. 169, *Selecting and Storing Seed Corn*.[17]

Improving corn yields through better seed selection techniques had been a mainstay of Extension outreach for more than a decade. The message from agents was that farmers could find good seed corn in their own fields if they would just use the techniques promoted by Extension.[18] This approach to improving seed stock had worked for decades, allowing farmers who were willing to do the work to gradually increase their yields by selecting the best ears from their open-pollinated varieties.

Choosing the best seed corn was only one part of the process needed to increase yields. Regardless of how good the seed was, it wouldn't produce a good crop if it wasn't stored correctly. In fact, in some years the seed quality of the open-pollinated corn was so low that one out of three seeds did not properly germinate due largely to poor storage during the winter.

Having even one bad kernel out of every five kernels per ear resulted in significantly lower yields.[19] If there were 640 kernels per ear, then that would represent 128 fewer plants per acre. This is the main reason Extension staff pushed—and, at times, pressured—farmers to give up the practice of selecting the next year's seed crop from the corn bin. Instead, they advised farmers to select seed from the standing crop and then store it properly to protect it from freezing weather.[20]

To see if the corn had been stored properly, agents also advised farmers to complete germination testing on their seed corn in the spring, prior to planting. This testing was accomplished by removing and placing five to seven kernels from each ear of corn onto a muslin cloth called a rag doll that had been wetted and folded, then stored to await germination. After seven to ten days, the rag doll was unrolled, and a person would evaluate how well that set of kernels had germinated and whether the roots showed signs of disease. These readings determined whether the rest of the kernels on the ear should be used for planting or discarded.

Extension typically charged the farmer a penny per ear for conducting the germination test, reading the rag dolls, and separating good and bad ears. This nominal charge paid for the operating expenses for the testing. Any money remaining after paying the bills was often transferred to the local 4-H fund.[21]

Many farmers wanted to test their seed but were unable to spend the penny per ear, which led them to roll the dice and hope their seed would germinate.[22] This prompted Floyd County agent Charles East to observe one year that "farmers planting without testing seed reported their regrets and made firm resolutions to test seed corn hereafter."[23] While it may have been hard to spend the money for germination testing up front, losing income when a crop didn't germinate well was even harder.

Some farmers only tested a fraction of their seed corn supply. In 1932, Bartholomew County agent Arvil Keesling reported, "Two factors were noticeable, one was the lack of money on the part of the farmers. . . . The other factor which played an important role in the testing this year was the fact that a mild winter left the seed corn in good condition. Many farmers brought in small amounts of corn and upon finding their test 95 to 98% good would not bring in the remainder of their seed."[24] With money in short supply, some agents showed farmers who wanted to test their own seed how to make a homemade rag doll and explained how to interpret the results after the seeds had time to germinate.[25]

In Hendricks County, agent Ellis Hawk tried a new approach in 1931 to convince local farmers to test their seed. At a county Farmers' Institute, he distributed envelopes to the attendees, instructing them to collect two seeds each from fifty ears of corn and return the seeds to the Extension office for germination testing at no cost. Hawk later explained, "While this method of mail order testing does not solve the problem of good seed corn, it gives a good idea of seed condition, not only to the farm[er] who tests but [it also] serves as an excellent medium to collect information as to the condition of corn for a series of newspaper articles and thus brings the problem before all of the farmers of the county."[26]

Information from the community corn-testing effort was shared at Farmers' Institutes and farm bureau meetings as well as in newspaper columns.[27] When warranted, the data were used to warn others that the year's seed corn was expected to have a very poor germination rate.

Some agents staged public events to get growers to pay attention to germination test results. In Hancock County, agent Harold Yarling presented a contest in 1930 to convince farmers that they could not tell the difference between good corn that would germinate and bad corn that would not simply by looking at each. He later reported:

> In order to acquaint the Hancock County farmers of the seed condition, a seed corn guessing contest was held during the early part of February. Twenty-five ears of corn [were] placed in two hardware stores and in the Farm bureau Feed store at Greenfield, and the farmers were invited to visit the stores and guess the number of ears in each sample that would germinate strong enough for seed. Part of the ears were crib selected, part early selected and dried in outside racks, and part early selected and heat dried. On a Saturday afternoon rag doll tests of these ears were read before the men who [had] guessed, and the doll reading was explained and winners announced for the prizes offered by the stores. The heat dried corn germinated 71.5%, the rack dried corn 60.5%, and the crib corn 44%. It is interesting to note that all seed corn testers in the County had all the seed they could handle immediately after these results were announced.[28]

In Clinton County, agent Enos Rowe collected seed from the 1935 harvest to do a preemptive sample and germination test. He reported that he had

> secured samples of seed corn from average farmers in each of the fourteen townships of the county and made up a number of rag dolls here in the office. The dolls were germinated [in] ten days here, and then at a big gathering of farmers from all over the county, they were opened and read by [agronomist] M[ervill] O. Pence of Purdue University. Two hundred and fifty farmers

> witnessed the reading of these dolls and the readings proved that less than 50 percent of the corn so assembled would be fit for seed.[29]

Obviously, Rowe hoped that upon hearing that the seed was in such poor shape, farmers would be convinced to test their seed before planting it.

At times, farmers failed to take the agents' warnings seriously, as Everett Johnson from Cass County reported in 1936:

> The first rag-doll tests made in the County Agent's Office indicated that the corn was germinating around 71%, and 57% of the ears tested had one or more dead kernels. A circular letter was sent to all corn growers in the county warning them of the serious condition which our seed corn was in and advising them to germinate every ear before shelling and planting. Although much effort was made to get the corn tested before planting it in the ground, many farmers purchased corn from their neighbors and planted it only to find out that it was low in germination test and [they] were required to replant.[30]

In some cases, such as the one Johnson described in Cass County, poor germination rates required farmers to replant the crop entirely, leading them to unexpected expenditures that far exceeded the cost of germination testing. During the winter of 1935, Purdue published reports stating that

> approximately one-third of the corn from this region [central Indiana] would not grow . . . and in no case was an ear returned for planting where only four out of five kernels germinated. . . . If the corn which was tested had been planted without testing, and the land had been re-planted, the cost of the corn for replanting and the labor involved would have amounted to eighty-three thousand seven hundred dollars. This figures the re-planting and soil processing at three dollars per acre and the seed corn for re-planting at three dollars per bushel.[31]

Across the state, the evidence was clear-cut and consistent, and some agents reported success from their education efforts. In the 1930 annual report for Hancock County, Yarling noted that the preceding years of education on corn production were paying off, writing, "Good farmers from different parts of the county have estimated that 75% of the corn growers have selected corn early, and we feel this estimate is not far off, for we see corn hanging up to dry practically every place we visit and a big part of this number are farmers who have never been in the habit of field selecting."[32] And in Greene County, Arthur Haseman reported, "The campaign has produced satisfactory results. It is a pleasure, upon visiting a farm, to have the farmer proudly invite the agent to see his seed corn selected early and now stored in racks in a well ventilated place."[33]

Farmers grew varieties of open-pollinated corn that many had spent decades perfecting for the soil types on their farms.[34] There were many varieties used around the state, with different varieties popular in different counties. According to agent Everett Parker, in 1930 Hamilton County farmers were selecting "the larger types such as Boone County White and Reid's Yellow Dent for very fertile soil and the smaller, earlier types such as Woodburn's Yellow Dent and Riley's Favorite for the heavier, less fertile soils."[35]

The better farmers continued to select seed by examining their open-pollinated varieties in the field for date of maturity, size of ear, uniformity of stand, height of ear from the ground, number of barren stalks, amount of fodder, standability in high winds, and disease resistance.[36] But during the first half of the 1930s, some agents such as Ellis Christen in Benton County began voicing concerns that the best farmers "have proceeded about as far as possible in the selection of the better varieties of open pollinated strains."[37]

At the time, Indiana hybrid corn was still a relatively new development, so hybrid varieties were just starting to capture more widespread interest among farmers.[38] In 1935, Adams County agent Lawrence Archbold explained one of the reasons that hybrids were drawing attention, writing:

> [I]ndications are that the best yields with open pollinated strains are obtained where a man has followed the recommended practices in selecting the seed in the field before heavy frosts occur, selecting ears having ripe husks from green stalks, drying the corn properly before freezing weather occurs, and then rag doll testing. Even where a man has done this carefully, the indications are that some of the better hybrids will out-yield the open pollinated strains by at least ten per cent.[39]

Researchers from the Indiana Agricultural Experiment Station at Purdue had spent a decade developing lines of hybrid corn for use by the state's growers. Starting in the 1920s and continuing into the 1930s, they had first tested the new lines at campus farms but soon expanded to working with cooperators throughout the state who could conduct on-farm demonstrations designed to evaluate the hybrids using real-world farming practices.[40] Delaware County agent Myron Cromer described one such test conducted with six Purdue agriculture graduates, who agreed in 1928

> to cooperate with Dr. G[eorge] N. Hoffer [a Purdue plant pathologist and one of the pioneers in researching and developing corn hybrids] in testing out on their own farms different corn hybrids. This was done in an effort to check the yielding power of the hybrids in question, against the yielding quality of the corn grown on the particular farm, with the hope of locating a hybrid which might result in a more satisfactory manner, both from the standpoint of feeding quality and maturity for the farmer.[41]

Cooperators investigating the potential of hybrid corn were required to follow specific protocols so that the data collected would have research value to the corn breeders at Purdue.

The university released its varieties, called Hoosier Hybrids, that were categorized by series—400, 600, 800, 900, and 1000—based mainly on maturity dates.[42] The hybrids had to be adapted to local growing conditions, which is why Purdue researchers had field-tested each new hybrid across the state. They anticipated that an individual hybrid might fail in some counties and still produce a high yield in others. They also wanted to measure how well each hybrid performed against the open-pollinated varieties that were usually grown in the area.

To be certain, it wasn't as if the first hybrids from Purdue and other sources were a lock to replace the open-pollinated varieties grown across Indiana. The hybrids had strong competition from the open-pollinated varieties that, by careful selection, had been adapted to specific soils and environments.[43] Even as late as 1939, Carroll County agent Ralph Maggart reported, "Lester Hile of Washington Township had the highest yield of corn, which was 131.0 bushels per acre. . . . It is interesting to note that the highest yield this year was not hybrid corn but was Reid's type of corn which had been bred on the farm. It is a high producing strain of corn that has a lot of show corn in it. This field probably has won more gold medals than any in the state."[44]

The only way the new hybrids were going to dislodge open-pollinated varieties in popularity would be for them to consistently outyield and outperform popular strains, including yellow dents such as Black's, M.A.C., Reid's, Troyer, and Woodburn, and others such as Clement's White Cap, Krug, and Shipley.[45] But figuring out which hybrid varieties to plant on individual farms presented a challenge, with agents fielding questions about this from their constituents.[46] To help local farmers sort out the differences, Extension agents offered demonstrations that showed hybrid varieties being grown under the same conditions as open-pollinated lines. One such trial in Cass County in 1932 tested and compared thirteen varieties of corn, with the agent later reporting, "Krug corn made the best yield of any of these. It was closely followed by Early Reid's Yellow Dent, Purdue Hybrid LTR66 and Plank's XX Hybrid, which made yields of 94 bushels per acre."[47]

Each year these types of trials provided the means by which area farmers could directly compare hybrids and open-pollinated corn to better understand which varieties yielded the best in local growing conditions. After a few years of testing varieties locally, agents generally had more information to use in making recommendations to area farmers. For example, in the 1935 Allen County annual report, agent Hugh Heckard reported, "The Hoosier Hybrid has given the highest yields in the same demonstration during the past three years. Its remarkable yielding ability has caused considerable interest among the corn growers in Allen County and many calls are being received as to its adaptability and where seed sources may be found."[48] And in 1937, Boone County agent Clarence Whistler noted, "The county extension office and Purdue University advised farmers to use hybrid seed of the 600 series. . . . All of the hybrids of the 600 series equaled in production the owner's own open pollinated corn or exceeded it."[49]

For some, even repeated testing did not produce clear results regarding which variety was best for local growers. In 1937, Bartholomew County agent Arvil Keesling reported on the results farmers had obtained from planting the hybrids in larger fields, writing:

> After four years of tests on hybrid corn in variety demonstration plots, an attempt was made this year to have hybrid corn tested on a field basis; thus eliminating the error of small plot activities. Our test plots had indicated that #632 and #845 were the outstanding variet[ies] in our demonstration plots. The county agent assisted in the distribution of 35 bushels of hybrid seed [from Purdue] to 38 growers in the county, pointing out to them that a report would be asked for in the fall as to the results obtained under field conditions. All producers were asked to plant open pollinated corn in or adjoining to the hybrid plot. . . .
>
> The object of this test was not so much to check yields as it was to point out the value of hybrid corn from the standpoint of easy shucking, standing ability, and drought resistance. Every producer has reported that hybrid corn has three distinct values; first, it out yielded open pollinated corn on their own place; second, it withstood even the heavy winds of early fall and is not damaged in the field, and third, it stood up well for wheat sowing, which is important in this county. Most producers report that any one of these three qualities makes hybrid corn the ideal kind to plant. Most of them are going into the market and buying their supplies for 1938 as well as supplies for their neighbors.[50]

Two years later, Keesling offered this update:

> The farmers of Bartholomew County are still trying to find the type of hybrid corn that will best fit their needs. One of the biggest objections is that they cannot find a corn that will repeat itself on their farm as they have seen it on the demonstration plot and on their neighbors' farms. Approximately 40% of the corn acreage [which amounted to 45,000 acres in 1939] was planted in hybrid corn this year, and approximately 15% of those planting hybrid corn were dissatisfied with the results. Every variety of hybrid corn known to be good for Southern Indiana was tried by someone in the county this year. Most of the farmers are sold on the theory of hybrid corn and will continue to use it in their program. The changing from one variety to another is the unfortunate part of the program.[51]

It seemed that in southern Indiana, the hybrids from Purdue and others were not meeting the expectations of the agents. In 1936, agent Earl Miles told local farmers to continue to plant open-pollinated corn until more reliable hybrids were developed for Clark County. He remarked, "Hybrid seed corn has been grown by some few farmers in the county, but as yet, results have not been favorable enough to recommend any hybrid for average southern Indiana conditions, to advocate over the more prolific strains [e.g., Johnson County White] used in this section."[52]

While some farmers around the state still saw better results with open-pollinated varieties, in general, the improved hybrids outyielded the open-pollinated varieties, which obviously meant more money to the bottom line of the farm.[53] In 1934, Franklin County agent Russell Zell reported, "[The] Hoosier Hybrid corn again proved its producing supremacy over various regular corn seed during a demonstration."[54] It would be hard for progressive farmers not to notice when the poorest hybrid in a test plot outyielded the locally grown open-pollinated corn by 11 bushels per acre. And it would have really caught their attention when the more adapted hybrid for the area showed a 26-bushel gain per acre over the same open-pollinated variety.[55]

Depending on the county, planting hybrids in place of open-pollinated varieties increased yields from 2 percent to 25 percent, which resulted in anywhere from 3 to 28 more bushels per acre than the open-pollinated varieties.[56] In 1938, Grant County farmers planted 20 percent of their corn acreage to hybrids, which increased the average yield for the county from 43 bushels per acre to 56 bushels per acre. The agent at the time was Forest Conder, who said, "This increased production has meant an increase in the livestock kept in the county, and as livestock has returned an average of eighty-one cents or thereabouts per bushel of corn fed, the average price per bushel is eighty-one cents rather than forty-six [worth about $487,000]."[57] That same year, Fulton County agent Noah Hadley reported, "The use of Hybrid seed corn came into general use with Fulton County farms in 1938. . . . [A]pproximately 70% of the farmers of this county used some Hybrid seed corn this year. . . . [There was a] 12 to 30 bushel increase per acre by the use of hybrid. . . . [A]pproximately 60% of the total corn acreage of the county this year was grown from Hybrid seed."[58]

In Fayette County, the agent said the adoption of hybrid corn meant 200,000 more bushels of corn in 1938, representing "an increase of $70,000 in farm income even at the present low corn price [of 35 cents a bushel]."[59] In 1936, Fountain County farmers planted one-third of their corn acres with hybrids, resulting in enough additional money to pay the entire tax bill for the county for one year.[60]

While maximum yields were obviously important to corn growers, hybrids could also remain erect during windstorms, resulting in less loss from damaged cornstalks.[61] In 1936, agent Alfred Hesler walked a number of fields after a storm had crossed Fountain County and reported, "It is very evident that the hybrids withstood the storms while the open pollinated corn, field after field, was a menace to the corn picker."[62] In 1938, Clay County agent Raymond McHargue observed, "About the time the corn was tasseling and shooting, a severe wind storm, which approached the proportions of a tornado, hit this field of corn containing the demonstration and blew much of it badly. . . . [The] 829 hybrid [variety] stood practically 100% alongside of Reid's [open-pollinated variety], probably 80% of which was blown over."[63] That same year, Clark County agent Earl Miles stated, "Many farmers feel like the extra price paid for seed is justified for this one characteristic," which underscored the value farmers placed on having corn standing erect until harvest, even following strong storms.[64]

Beyond increased yields and improved standability, farmers also found hybrids better than open-pollinated corn in the following ways:

- higher planting rate for seed,
- better growth when weeds were present,
- better resistance to plant diseases,
- improved uniformity of stands,
- shorter stalks,
- ears lower on the stalk,
- easier to husk,
- fewer barren stalks,
- better drought resistance,
- good livestock gains when fed, and
- earlier maturity, which allowed ears to be harvested prior to late-fall cold snaps.[65]

With such positive comments from those using hybrid corn, one agent wrote in 1936, "The present outlook is for a great increase in the use of hybrid corn the next few years, provided a suitable strain can be developed."[66]

With each passing year, farmers began to see more success with hybrids and planted more of their acres with them. The situation in Boone County showed this steady increase. In 1936, hardly any hybrids were planted, and in 1937, only 8 percent of the corn planted was a hybrid variety. But by 1938, an estimated one-third of the acreage in the county was planted to hybrids.[67] Similar increases elsewhere led one agent to observe, "It can be readily seen that the demand for hybrid seed corn will far exceed the supply."[68]

When Crawford County corn acreage dropped by about 4,000 acres in 1939, the agent still reported that due to "the use of hybrid seed, proper rotation and fertilization, the total corn crop still comes as nearly meeting the needs as the former acreage of ordinary corn."[69] Higher yields on fewer acres meant more profit from each bushel of corn at whatever price it was selling for at the time.

By 1938 and 1939, the corn acres planted to the hybrids varied across the state, ranging from approximately 25 percent to 80 percent, depending on the county.[70] The overall adoption of hybrid corn was slower in some parts of Indiana than many expected, with that hesitation lasting into the 1940s.

There were numerous factors holding back the quicker adoption of corn hybrids. In some cases, the farmer's hesitation came from the local agent's reluctance to completely drop the time-tested, open-pollinated lines of corn from his recommendations. Alfred Hesler knew his progressive farmers had wholeheartedly accepted corn hybrids, but he also stated in his 1936 annual report, "[W]e do feel that the standard varieties still have a place on the average Fountain County farm."[71] In 1937, Hesler would write down the reservations he had against hybrid corn, noting, "We are interested, of course, in the thing hybrid corn will do, but after all, will the pendulum swing too far and cause men to forget the value of open pollinated corn?"[72]

Even Purdue agronomist Mervill Pence, who favored hybrid seed, encouraged farmers to plant only part of their total crop to hybrids as a way of testing them under real growing conditions. Remarks made by Pence were summarized in 1935 by Hancock County agent Warren O'Hara, who wrote, "Much interest was shown in hybrid corn. M. O. Pence of Purdue explained how hybrid corn is developed and that some might be planted locally with profit for comparison; however at present the sack method of gathering mature ears [from open-pollinated varieties] should be depended on for the major crop."[73]

Farmers in many areas of the state could buy hybrid seed corn, but many had problems when they selected seed pushed by the commercial seed industry that had proven to be unadaptable to their locations.[74] When a Hamilton County farmer planted a hybrid produced in Iowa, the agent later reported, "[A] serious objection was that ears were only partially covered with husks and deteriorated greatly due to weather and birds."[75]

Another challenge farmers faced was that some commercial companies lacked quality control on the hybrid seed they sold for premium prices. In 1937, agent Noah Hadley observed that while most Fulton County farmers had reported excellent results with the hybrids they had purchased, a few were not pleased. He explained, "Most of these unfavorable results came from farmers who apparently got inferior seed. A few reported sweet corn kernels showing up in their corn. Others reported a sprinkling of white caps. This indicates that the producer of the seed did not have proper isolation. It is highly important that every effort be made to urge farmers to exercise extreme care in the selection of their hybrid seed corn source."[76]

Of course, there were also those who were simply reluctant to make the switch to hybrids—described by the Gibson County agent as "dyed in the wool hard shells that refuse to accept the superiority of hybrids over open pollinated corn."[77] In a 1939 annual report, Cass County agent Everett Johnson reported, "The larger corn farmers, as a rule, were the first to use hybrid seed for their entire crop. Less informed and bias[ed] farmers have been slower in adopting the practice of paying the premium for hybrid seed."[78]

It was understandable why farmers would not or could not pay more for hybrid seed during the worst years of the Great Depression. When tractors were already parked for lack of money to buy fuel and oil, and farm operations were struggling just to stay afloat, spending more money for seed—money that these households simply didn't have—wasn't an option.

Another short-term setback for hybrid seed adoption was that growers had the mistaken belief that they could buy and plant hybrids one year, then save the seed—as they had long done with their open-pollinated corn—to plant the following year's crop. While saving seed had worked with varieties of open-pollinated corn, with the plants actually improving through careful field selection each year, saving seed from hybrids had just the opposite effect. As one agent noted, farmers learned there was "the expected tendency to lose vitality after the first year" when planting seed collected from a hybrid crop.[79]

Planting seeds saved from hybrids often produced yields that were much less than those of the improved open-pollinated corn strains that farmers had bred over generations. One 1936 study in Greene County showed up to a twelve-bushel loss per acre when using second-generation hybrid seed.[80] In 1937, Grant County agent Forest Conder reported yields of seed saved from hybrid corn placed near the bottom of the list of the corn tested.[81]

In 1935, Johnson commented on Cass County farmers planting seed collected from hybrids in an effort to save money, writing, "At one of the demonstrations, the farmers had occasion to visit a field on the farm that was planted from seed of the second year Hoosier Hybrid Corn. The corn was poor and well demonstrated the fact that Hybrid Corn should not be used for seed the second year. . . . The education work will be continued for the next year as a large amount of hybrid corn is being used for the second and third years and should not be used after the first year."[82] Efforts by farmers to be frugal with seed purchases made them penny wise but pound foolish as they sacrificed the higher yields—and, with them, the higher incomes—that an investment in better seed could bring. In order to get all the advantages from hybrid corn, farmers would have to purchase the premium-priced seed each year, which went against a hundred years of tradition of saving seed from one year to the next.

Some growers grew adaptable Purdue hybrids, then crossed their open-pollinated corn with the hybrids to add new genetic material to their own seed. In 1931, Ralph Maggart reported that a few Carroll County farmers who "have planted seed which was a cross of hybrid and their own are very well pleased with the results. This has been a very desirable way of breeding up their corn. This hybrid has introduced certain qualities on their own corn which they feel is equal to [or] better than a five bushel per acre increase on their whole crop."[83]

The 1930s would be a pivotal decade for the seed corn industry as farmers began to replace their open-pollinated stock with hybrid varieties. Many of those who entered the hybrid seed market early had transformed part of their farming operations into viable seed businesses when they had volunteered to field-test hybrids developed at Purdue and obtained those inbred lines free of charge. Early on, Purdue agricultural researchers had favored volunteers who were agriculture graduates from the university or who had completed one of the eight- to twelve-week Winter Short Courses in Agriculture.[84] But eventually, the university began distributing its hybrid lines to a larger group of growers, as Carroll County agent Ralph Maggart explained in his 1939 annual report, writing:

> For the first 2 or 3 years, R[alph] R. St. John of the Botany Department [of the Agricultural Experiment Station at Purdue] sent to the Purdue Agricultural Alumni the various inbreds to be crossed with their own corn. This seed was to be saved and planted; however, this effort to distribute hybrid seed was impractical, and in 1931 and 1932, economic conditions were not favorable for improved seed. Therefore this plan was given over to the present plan of distribution, of turning the seed over to the commercial producer and farmer growers.[85]

When Purdue started sharing its inbred lines with more growers, it had initiated a commercial seed industry for hybrids that would grow exponentially during the 1930s.

At the week-long Corn School and Stockmen's Convention on the Purdue campus in 1935, experiment station corn researchers called a meeting of farmers interested in producing hybrid seed. As a result of the meeting, the following individuals from Benton County were given Purdue inbred lines to use in producing hybrid seed to sell:

Grower	Acres Planted
Roy C. Atkinson	2 acres
Irwin Bahler	1 acre
Philip Brouillette	1 acre
Melvin Cox	2 acres
Floyd E. Elliott	5 acres
Carl E. Funk	3.5 acres
Harry P. Haynes	3 acres
Glenn L. Kinsell	2 acres
McCabe and Christman	1 acre
Ed. A. Shriner, Jr.	2 acres [86]

This would be the first step in what would become a rapid expansion of commercial production of hybrid seed corn in the county.

Prior to the 1935 growing season, Purdue agronomist Mervill Pence and botanist Ralph St. John met with these farmers at the Extension office in Benton County. As agent Ellis Christen later reported, they would discuss "the isolation necessary, pedigrees, cultural methods, and time and method for detasseling."[87] The following year, Christen described the phenomenal growth he was seeing in the county, writing, "The hybrid corn work was definitely continued with a more than 100% growth in seed production. Where only 30 acres [of] seed were produced in 1935, 287 acres were listed in 1936, 11.5 acres of inbred plots making 1st cross, and two drying plants were constructed."[88] Because of the rapid expansion, he was already suggesting in 1936 that the county be known as the "Hybrid Corn Center of Indiana."[89]

In 1937, Christen provided this update to the farmers' efforts:

> It is probably true, without argument, that Benton County is the leading Hybrid Seed Corn Producing county in Indiana. The first men to become interested in the production of Hybrid seed in Benton County were Purdue graduates, or former students of Purdue University who formerly had undertaken the sponsoring of some local Extension Activities as the Purdue Ag. Alumni of Benton County. The result was, that when these men began the production of hybrid seed, the fact that they had some previous experience, working together, enabled them to formulate a program and carry it out much more rapidly, than had they had no former association.[90]

In other words, by participating in the field tests led by Purdue, these volunteers were getting an education in raising hybrid seed corn that, in just a few years, would lead to a host of them becoming new players in the seed corn industry as it began expanding and flourishing.

Farmers outside of Benton County were also obtaining and growing hybrid lines from Purdue. In 1935, agent William Smith indicated that a number of Henry County farmers had obtained parent seed from Purdue for the production of Hoosier Hybrid 694. When they followed the planting and detasseling instructions provided by university botanists, they were able to get a nice crop of hybrid seed for which others were willing to pay premium prices.[91]

It was important for volunteers to follow the university's protocols and plans when testing hybrid corn. From the time of the first trials, all volunteers were required to file detailed reports back to the Purdue scientists indicating how the hybrids performed under different soil types and management schemes, and how they yielded against the farmer's own open-pollinated corn.[92]

Fountain County farmer Louis N. Samuels of Kingman field-tested Purdue's hybrids in 1937. Agent Alfred Hesler reported on the results by listing the ones that performed better near the top.

Hybrid Number	Qualitative Evaluation
616	Best of all in yield, standing, uniformity of height
608	Not quite so heavy yield
612	Good all around
694	Ears rather low, not so easy to husk
613	Tough shank, smaller ears, not so strong in stalk, tendency to break over after maturity, heavy foliage
800	Best color of any, larger butt ears, uniformity of height of ears not so good
Check	Too many, two on stalk, uneven height of ears on stalks[93]

One of the outstanding traits that farmers quickly picked up on was how well the hybrids performed during stressful growing conditions. Franklin County agent Russell Zell expressed this in his 1938 annual report, saying, "Variations in yield between open pollinates and well adapted hybrids are least noticeable when corn growing conditions are favorable, Mr. M[ervill] O. Pence [Purdue agronomist] pointed out. In other words, it is during the least favorable years when farmers should expect hybrid[s] to show outstanding results above open-pollinated varieties."[94]

Early in the process, as farmers were still learning what worked so they could scale up production, only a handful of farmers in each county were planting a few acres each to produce hybrid seed corn for local farmers.[95] For instance, in 1936 Blackford County had three growers who had used Purdue stock to produce 45, 120, and 130 bushels of seed corn to sell, while in 1937, Hancock County had two farmers who produced 300 bushels of hybrid corn for sale locally.[96] This approach was typical in other counties as well.

With supply falling short of need, a few entrepreneurial farmers began raising more acres of lines provided by Purdue. Agent Arvil Keesling supplied this account for Bartholomew County in 1938:

> There were approximately 18,000 acres of Hybrid corn in the County this year. . . . The intense activity of hybrid seed salesmen this fall indicates that the acreage planted to hybrid next spring will be increased.
>
> There are three producers of hybrid seed in the county at the present time. Ralph Heilman, Hawcreek township is producing 846, 845, 842, and JCW 1. He has the only supply of inbred No. 33 in the State. Mr. Heilman has been active in the seed corn business for twenty years and has a slatted-floor dryer on his farm.
>
> Lloyd Glick [of Clifty Township] is producing three hybrids. He has 632, 845, and 842 this year and has increased his acreage of seed from four acres to sixteen. To care for his seed he recently completed the most up-to-date bin dryer available. This dryer has four bin capacity and should easily handle 3,000 bu[shels]. . . .

> Ira Leatherock of Clifty Township is producing 842 and had recently completed a slatted-floor dryer with much larger capacity than in 1937. His capacity is now approximately 800 bushels.[97]

With only limited amounts available, those growing hybrid corn found a ready market among those willing to pay premium prices to be one of the first to plant the new corn. But for years to come, farmers who wanted to grow hybrids would face a shortage of seed supplies, which kept the price charged by commercial companies at $7–$8 per bushel well into the 1930s. This premium price paid for hybrid corn was substantially higher than the price paid for open-pollinated lines, but the large increases in yields typically more than covered the higher costs.[98]

As more hybrid varieties were developed, it would become more common for agents to introduce ones from additional sources into their local Extension demonstrations. These included varieties developed by other land-grant universities as well as those developed by commercial companies and included in the Extension trials at their request. In some cases, agents would hear about other hybrids because they were used by local farmers or highlighted in the agricultural press. In 1938, Clinton County agent Enos Rowe explained his experience with one of these hybrids, writing:

> [W]e are just beginning in this hybrid seed corn business. In fact in the past few years we have felt that for our particular county Hoosier Hybrid No. 613 and 608 [were] as good as we could find, having led our plot demonstrations each year. . . . On the other hand, however, one of our plots planted to an outside hybrid showed a decided yield over either of these two and leads us to believe that still greater yields will be obtained in the future as soon as there has been time to develop other hybrids to a conventional basis. This one particular hybrid we tried showed a plus of twenty-three bushels over the check plots, as compared to a plus bushels of twelve for 613 and 608, which is indeed quite an item to be remembered.[99]

In Benton County variety tests, hybrids grown in the 1930s came from Purdue as well as from multiple other sources, including Indiana producers such as Ainsworth Seed Company, Kentland; Roy Atkinson, Fowler; Harry Haynes, Talbot; Joe Lamaster, Oxford; and C. E. Troyer, LaFontaine. Out-of-state suppliers included Funk Brothers Seed Company and Woodford County Agricultural Association of Illinois, Henry Fields Seed Company of Iowa, and the Michigan Agricultural College.[100] Other names appearing on bags of hybrid seed corn being tested around Indiana included DeKalb, Farmcraft, Glick, Pfister, Pioneer, and Richard Key—many of which would become giants in the seed corn industry.[101]

Some of these commercial companies put out their own test plots as well to show how their hybrids compared to the open-pollinated lines. Arthur C. Stewart, whose company Stewart Seeds had been offering open-pollinated seed corn for sale since 1918, had added hybrids in 1935.[102] In 1936, he held a tour of his seed corn plots in Decatur County, with agent Matthew Hamilton reporting, "[The] hybrid that Mr. Stewart is producing this year, 819 B, was the highest yielder. This particular hybrid yielded more than 18 bushels per acre than the [open-pollinator] check, a well-adapted variety raised on the place. This year's variety check proved rather conclusively that the 800 series [varieties] are best for this area. A plot was checked here of a second year hybrid [saved seed planted the next year]. As expected, its yield was a little less than that of the check plots."[103] Stewart and others who sold seed corn, such as Pioneer Hi-Bred Corn Company, became adept at continuing their own research, seeking to gain marketplace advantage by constantly developing new and improved corn hybrids.[104]

History would show that the hybrids developed during the 1930s and later by Purdue, other land-grant universities, commercial companies, and others would quickly alter the way corn was grown in the United States and the rest of the world. These hybrids offered so much more in yield potential that farmers began making an investment in new seed every year instead of saving seed from their open-pollinated varieties. After being introduced and performing as promised, hybrids would become widely accepted and still remain as the cornerstone of corn production today.

During the 1930s, Extension used the five-acre corn competitions to promote practices among farmers that resulted in higher corn yields. Purdue and the Indiana Corn Growers' Association had begun holding these competitions in the 1920s, requiring each competitor to raise at least five acres of corn and keep complete production records. Participants who achieved 100 bushels per acre or more would then receive a gold medal from the association.[105]

For Extension, the competition was one of the most effective demonstrations in that it showed what could be done if better practices were employed for growing corn. Most of the competitors were early adopters of new ideas such as applying the right kind and rate of fertilizers, and using more tillage.[106] By the early 1930s, the Henry County agent had even credited the corn contest with the gradual reduction in the planting distance between rows, which went from 42–44 inches down to 38–40 inches.[107]

With the introduction of hybrids into the competition, the 100-bushel mark would soon go by the wayside. Adapted hybrids in the contest generally excelled, producing higher yields relative to any of the open-pollinated lines that had previously won the competition.[108] In 1934, yields in Clark County reached 130 bushels an acre due largely to the introduction of hybrid corn into the competition.[109] And a Hancock County farmer achieved a new state record of 140 bushels per acre by using hybrids in the 1939 contest.[110]

As a result, the competitions also worked to increase farmers' awareness of the role that hybrids could play in their operations, with Adams County agent Lawrence Archbold observing in 1939, "There is no question in the agent's mind but what the 5-Acre Corn Club contest has been the biggest single factor in promoting use of Indiana certified hybrid seed corn and in bringing to the corn grower's attention that high yields of corn are dependent on a high nitrogen level induced by good rotations, good seed bed preparation and proper cultivation. . . ."[111] Of course, as Archbold pointed out, good management practices were also crucial for producing higher yields from hybrids.

In 1939, Hamilton County agent James Todd summed up the impact of the competition on farmers, writing:

> The project has been of great value to the farmers of this county. They have followed the procedure used by those men making the high yields each year. They are today carefully selecting their seed, more are using commercial fertilizer and a large number are rotating their crops in such a way that they can turn under a green manuring crop. They are planting corn at a later date and are moving the corn rows closer together in order to get higher yields. Farmers are now aware that 100 bushel corn yields can be raised and are trying to do it themselves.[112]

While the five-acre contest was, on one level, just a friendly rivalry among those competing to be the best corn grower in the county, it also encouraged farmers to use the latest agricultural research, which ultimately could improve the bottom line of their operations.

Hybrid corn was still a new phenomenon at the beginning of the 1930s, but by the mid-1930s, with Purdue sharing its lines, farmers around the state began producing hybrids for sale to their neighbors and others. While hybrids only accounted for 0.3 percent of corn planted in the state in 1934, just three years later, 29 percent of the crop came from hybrids. Expansion continued at a fast pace, with those totals more than doubling by 1938.[113]

Hybrids had changed Extension education programs on corn, as Ellis Christen's 1938 annual report for Benton County highlighted. He wrote:

> Due to the fact that hybrid corn is generally accepted as a higher yielding corn than any of our open pollinated varieties, most of the acreage in 1938 was planted to hybrid seed. A very rapid transition from open pollinated (principally Krug) to hybrid seed is due first to the yielding ability and secondly to the large acreage and production of hybrid seed in the county [availability]. There are five drying plants in the county, three of them being at Oxford, one at Boswell and one at Fowler. In addition, one of the largest individual producers of hybrid seed in the State of Indiana is Edward J. Funk & Sons, whose land

> and residence is in Benton County but whose drying plant is just across the line in Newton County. . . .
>
> With 85% of the corn in the county hybrid corn, it has been necessary to change the extension program so far as seed corn is concerned because at the present time, very few farmers are saving their own seed. This means that the ordinary publicity and information on the care of seed in the fall is no longer necessary. . . .
>
> It has been said that there is not a farmer in Benton County who does not know about hybrid corn, though approximately 22% of the farmers in the county are either planting none or only a part of their acreage to hybrid varieties. . . .
>
> The program has continually urged the certification of all farm seeds, especially hybrid corn. In the main this goal has been achieved, though there are one or two large producers who feel that they have outgrown the need for further certification. . . . For the individual farmer, the object of the corn project is to continue demonstrations and provide material which will show him continually the need for fertility and good soil building practices. . . . [F]armers generally accept hybrid seed as a reason for increased yield. It is important for them to realize that such increases in yield must come from the fertility in the soil and that superior seed only provides the means of obtaining the yield if the fertility is there.[114]

With the switch away from growing open-pollinated corn, county agents and Extension agronomists at Purdue had to shift their outreach efforts to educate farmers about these new hybrids. Farmers were largely unaware of what it took to produce hybrid seed corn, so much of the initial education dealt with the amount of effort that it took to produce a bag of seed.[115] Farmers learned that the seed cost more because of the care it took to plant inbreds, which included isolation, drying, shelling, and grading.[116]

Agents also had to help farmers learn to grow hybrid corn, which was different than growing open-pollinated corn. As one agent observed, "Granting that the better corn growers must turn to hybrid corn for increased yields, there is still extension work to be done with the average farmer and his production methods before he is ready for hybrid corn."[117] Those who wanted to grow hybrid corn needed to better understand its requirements, including row spacing, variety selection, date of planting, rate of planting, and when to fertilize it.[118]

It would take until the mid-1940s for the commercial seed corn industry to begin producing enough seed corn to supply the needs of farmers and for hybrids to achieve near universal adoption. But as they did, Extension adapted to the changing times and, for decades to come, would continue to show growers how to use hybrids to produce the highest yields from their farm ground.

CHAPTER 17

The Necessity of Knowing the Soil Before a Good Crop Can Be Produced

During the 1930s, Indiana soils were generally deficient in phosphorus, potassium, or both, which prevented farmers from maximizing their crop yields.[1] For example, in 1937 the Dearborn County agent reported that soil testing showed 61 percent of farms had soils deficient in phosphorus, and 46 percent didn't have enough potassium.[2] That same year in Greene County, soil tests confirmed that more than half of the fields needed to be augmented with fertilizers to increase phosphorus and potassium levels.[3]

To compound matters, many farms still lacked the proper soil pH. Tests conducted in Hancock County in 1936 showed that more than half of the fields needed at least one or more tons of limestone added per acre in order to raise the pH to levels that would better support the production of crops such as alfalfa, corn, wheat, cantaloupes, tomatoes, and watermelons.[4] Soils that were too acidic affected the availability of important soil nutrients, such as nitrogen and phosphorus, to plants. After running soil tests in Decatur County in 1935, agent Ira Thurston reported that more than a third registered a pH below 5 and noted, "The 35.1% which tests below pH 5 presents a problem for serious consideration in that below pH 5, very little nitrates are available and the greater part of the phosphates are locked up by the free aluminum and iron. Liming becomes most essential to correct these conditions."[5]

By the early 1930s, fertilizing wheat was more or less an accepted practice (see Chapter 18). However, fertilizing corn was a different matter. With so many fields in need of soil improvements, Extension had begun conducting fertilizer trials for corn. In the 1930 annual report for Clinton County, agent Verner Mann noted, "It has been less than six years that farmers have become interested in commercial fertilizer on this crop. For the past three years, field demonstrations of commercial fertilizer on corn have been conducted."[6] At first, many farmers weren't interested in these trials, but over time, as more nutrients were pulled from the soil and farmers saw a drop in their yields, they became more interested in soil fertility.[7] The push by New Deal programs to prioritize and incentivize soil conservation (see Chapters 9 and 10) would also generate more interest among farmers to replace the nutrients that crops such as corn and wheat removed.

During the Great Depression, farmers had to be convinced to spend what little money they had on fertilizers. The agents brought up the important point that more crops, including corn, needed to be fertilized but that farmers needed to put down the right amount based on the fertilizer product being selected.

Much of the commercial fertilizer used during the 1930s contained little to no nitrogen and instead relied on varying levels of phosphorus and potassium. Then as now, fertilizer manufacturers indicated what percentage of each was included in the formula by the numbers, with the first number representing nitrogen; the second, phosphorus; and the third, potassium. Farmers and county agents often referred to the nitrogen, phosphorus, and potassium in fertilizers as nitrates, phosphates, and potash, respectively.

Common formulations without nitrogen that were used at the time included 0-14-6, 0-14-14, 0-20-0, 0-20-20, 0-21-9, and 0-24-12. Common formulas with nitrogen included were 2-12-6, 4-24-12, 12-24-12, and 13-34-20.[8]

Farmers came to county Extension offices to ask questions about using commercial fertilizers on corn, but the agents understood that there was not a one-size-fits-all response because so much depended on how the land was being farmed.[9] And one agent observed that the "numerous types of soils also help to make the problem a difficult one to answer."[10] Extension's goal was to convince farmers that they needed to appreciate "the necessity of knowing the soil before a good crop can be produced," as one agent put it.[11] Not knowing the nutrient composition of their soils meant farmers were guessing at the amounts of fertilizers to add.

At the time, many farmers who used a corn fertilizer appeared to be locked in to a 2-12-6 formula applied at 125 pounds per acre as their standard. Agents seemed concerned that farmers randomly picked a general product such as this one, regardless of soil sample readings or, worse yet, in lieu of taking any soil samples at all.[12] Agents did their best to show growers that applying formulations based on the actual levels of phosphorus and potassium in their soils would bring a higher yield and a better return on their investment.[13]

In order to address growers' questions with some level of confidence, agents and campus Extension specialists turned to the Hoffer stalk test. Developed by Purdue plant pathologist George Hoffer around 1925, the procedure chemically tested the sap of a cornstalk to determine if the soil was supplying enough nitrogen and potassium to the plant.[14] It was customary for Purdue plant pathologist Charles Gregory to lead groups of farmers at Extension events and point out that the yellow they were seeing along the midrib was an indication that the plant was short of nitrogen. He would then back up his assessment by doing the Hoffer test to prove his point.[15]

To help farmers identify which nutrients might be missing on their farms, Extension offered clinics where farmers could bring in soil samples and cornstalks to be tested by the agent. At one such clinic in Cass County in 1931, agent James Little used the Hoffer test to evaluate the plants for nitrogen and potassium levels, and conducted another test for soil pH. He also noted, "The soils were tested for available phosphate with the Hi-lo-phosphate test and for acidity by the LaMotte-Morgan test. . . . Those who attended expressed themselves as having a better idea of the kind and amount of fertilizer that they could afford to use on their corn and other crops."[16]

Agents also worked with cooperators, or farmers who were willing to set up Extension demonstrations on their farms, in order to conduct various fertilizer demonstrations—sometimes in a single location; other times, scattered across the townships. These demonstrations showed how yields were affected following the application of fertilizers with varying amounts of nitrogen, phosphorus, and potassium.[17]

In 1930, Dubois County agent Charles Nicholson reported that the local growers were well versed on fertilizers due to his previous demonstrations, which he believed had been largely effective.

> From year to year in the study and analysis of the crops grown, such as corn, wheat, etc., our farmers are beginning to determine the needs of the soil and supply these needs and there is also a further tendency to deviate from the cheaper grades of fertilizers, especially those with lower analysis, when double strength may be used and a saving of several dollars per ton may be made. Some demonstration projects have shown clearly that the right kind of fertilizer meeting the needs of a particular plant has shown an increase in dollars and cents above the old scheme of applying and using fertilizer. In the past it was a common opinion of the average farmer that a liberal application of fertilizer would have a tendency to decrease or materially affect the crop production, especially should a short dry season follow such a practice. This fact is more clearly brought out when the statements of one individual farmer are recalled, that is he asked the question "Do you think that by me using 40 pounds of a complete fertilizer to the acre that that is one reason why my corn was fired?" On the other hand his attention was called to the fact that an application of 460 pounds of fertilizer properly balanced applied to the same crop, that of corn, had no appearance of causing the corn to fire or dry up and, if anything, brought out the fact that a heavy application properly balanced would even have the tendency to keep the corn in a much greener stage even though this was an exceptional dry season.[18]

Over time, agents could make recommendations for fertilizing corn based on specific attributes of the farming operation, such as Grant County agent Mosher Butler described in his 1931 report:

> First, on farms east of the river in Grant County, moderate applications of phosphorus and potash will pay as shown by the tests. . . . Secondly, on black land west of the river where wheat isn't fertilized, where the maximum amount of livestock isn't kept and where less clover is grown, the majority of the tests proved the value of phosphorus on corn and proved that nitrogen and potash had little or no value. . . . Thirdly, on farms of high fertility, with maximum livestock feeding and where wheat or small grains are fertilized, neither phosphorous, potash or nitrogen paid.[19]

In addition to conducting local demonstrations and events, Extension also sent out an education train in 1936 to highlight the importance of soil testing. The Baltimore and Ohio Soil Testing Special Train—also known as the Purdue Soil Test Train—allowed farmers to bring representative samples of their soils to the train to have them analyzed for pH, phosphorus, and potassium.[20] The train ran August 3–15, traveling largely in central and southern Indiana. It started at Washington and made twelve stops, completing its run at Rushville.[21] Aboard were Purdue soil specialists Keller Beeson, George Enfield, and Mervill Pence.

The four-car train housed exhibits showing how crop yields were affected by the application of limestone to adjust soil pH and fertilizers to increase nitrogen, phosphorus, and potassium. It included a car where participants could attend lectures, including a morning session on liming and growing alfalfa, and an afternoon session on fertilizing corn, legumes, and wheat on different soil types.[22] It was also equipped to allow the Purdue specialists to analyze a total of 1,604 soil tests for acidity and measure nearly 1,200 soil samples for phosphorus and potassium levels.[23]

It would take the full complement of educational programs and outreach activities by Extension to help farmers understand how to properly use fertilizers to boost their yields and profits.

While the focus early on was to understand how phosphorus and potassium levels influenced corn yields, in the 1930s, Extension demonstrations also looked at the value of adding an adequate amount of nitrogen to boost yields. It was well understood that corn following green manures, cover crops, and animal manure showed increased yields, which were attributed to the addition of nitrogen to the soil.[24] As one agent remarked, "Farmers will always find the more practical use of nitrogen resulting from leguminous crops."[25]

Extension encouraged farmers to plant legumes into a rotation with corn as a method for increasing nitrogen in the soil.[26] In many cases, farmers would plow under a legume sod for the nitrogen benefit, then supplement it with a commercial fertilizer to improve the levels of phosphorus and potassium.[27] In 1932, the

Clinton County agent reported that local farmer George H. Bailey and his son took this approach, noting:

> [They] have averaged better than 90 bushels per acre for the past five years. Another interesting fact is that the piece of ground producing 105 bushels this year has been in corn continuously for the past 30 years. It is a piece of bottom ground and, for the past several years, sweet clover has been sown at the last cultivation. Mr. Bailey is a strong advocate of sweet clover as a soil builder. Two hundred lbs. of 3-18-9 fertilizer was also used this year.[28]

Generally, fertilizers containing nitrogen were not recommended for corn when a legume was part of the rotation, when enough animal manure was applied to the corn crop, or when farmers used the two in combination.[29] Through demonstrations, agents showed more phosphorus and potassium, and less nitrogen often returned more profits to farmers.[30]

However, there were cases where the lack of nitrogen became the limiting factor for attaining higher corn yields.[31] In 1936, Cass County agent Everett Johnson reported that Purdue Extension agronomist George Enfield had shown "that where a small amount of nitrogen was used, the yield was decreased, which is probably because a larger plant is started than there is nitrogen available in the ground to complete the growth. Where nitrogen is needed for corn, a larger amount should be used."[32] The results of a 1937 fertilizer demonstration in Allen County also pointed to a lack of nitrogen as an issue, with agent Chester Kimmell reporting:

> Representative stalks of corn were selected from the Lawrence Blessing fertilizer plots and tested for nitrogen, phosphorus and potash deficiencies. One of the features of this demonstration was the testing of corn being starved from lack of nitrogen. Later in the summer the fertilizer plots were checked and all evidence obtained at the plant tissue meeting led George Enfield to believe that his plant tissue testing was accurate. The fertilized plots did not yield very high although sufficient phosphorous and potash were used.[33]

Farmers were aware that nitrogen was important to crop yields, which is why they adopted practices such as planting cover crops, implementing crop rotations, disking in green manures, and applying commercial fertilizers as a way of maintaining this nutrient in the soil. By the end of the 1930s, some farmers began experimenting with 18-46-0, which supplied higher levels of nitrogen and phosphorus.[34] However, fertilizers containing even more nitrogen would not become more widely available for use by agriculture until the 1940s and 1950s.

Ultimately, agents had to be ready to address the one question that all farmers asked them in their offices, farm fields, farm management programs, and outlook sessions: Did it pay to apply fertilizer to corn?[35] Their answers became a lesson in economic principles for farmers. The reply that many gave was summed up by Daviess County agent Albert Bishea in his 1939 report: "[A] fertilizer containing a balance of phosphate and potash produced more in bushels of corn and in dollars and cents return expended than with fertilizer analysis containing unbalanced phosphate and potash or with nitrogen."[36] Farmers were learning that just because an input such as phosphorus produced higher yields, selling those increased bushels did not necessarily cover the cost of the added material. It always depended on how many extra bushels were produced and what a bushel could be sold for as compared to the cost of the added inputs.

It was clear that applications of fertilizers increased yields, but there was still a reluctance by many farmers to fertilize corn because they didn't think it was an economical practice.[37] With every penny counted in tough economic times, farmers needed to know if the increase in yields would pay for the additional expense of the fertilizer. Obviously, the answer depended on the cost of the fertilizer and the price at which the corn would be sold.[38]

Before the agricultural market collapsed at the onset of the Depression, a farmer in 1930 who spent $2.18 per acre for fertilizer could expect a return of $8.40 per acre when the corn sold for 60 cents a bushel.[39] From 1931 to 1935, the consensus was that it didn't pay to apply fertilizer with the price of corn so low.[40] During that time, it cost more to purchase the fertilizer than what the extra bushels returned as revenue.

In 1931, corn sold for 25 cents a bushel, meaning that the farmers were losing from 8 cents to $1.45 an acre when the cost of the fertilizer was deducted from the selling price. It seems that fertilizer prices had remained high in spite of the economic downturn, leading Gibson County agent Cyrus Dyer to observe in 1932, "Farmers could not see their way clear to pay out money for fertilizer when the prospects for returns from it were unfavorable on account of prevailing grain prices."[41] By 1936, the price of corn had significantly improved, leading agents to tell farmers that for every dollar spent on fertilizer, the expected return was $2–$6 per acre, based on the market price for corn.[42]

During the last half of the 1930s, the increased use of hybrid corn also prompted Extension to further encourage the use of fertilizers. Higher yields derived from hybrid corn logically meant that more nutrients were being removed by the plants. In 1938, Fountain County agent Alfred Hesler wrote that ". . . second year corn, corn following corn, greatly handicaps production. This item is at least one of the new factors that has crept in that most farms are not up well enough in soil fertility to produce two successive corn crops." He went on to say that farmers were "spending a good sum for Hybrid seed corn, therefore they should fertilize to get all returns possible on the investment. It was not so when growing open pollinated corn, farmers did not fertilize generally speaking. We are sure most farmers could have benefited by so doing."[43]

The following year, Hesler bluntly stated, "In increasing corn yields from thirty-six to fifty bushels per acre, it creates a soil fertility problem."[44] Purdue agronomist Mervill Pence reinforced that message at Extension events around the state. After one such presentation in 1938, Franklin County agent Russell Zell reported on Pence's comments, saying that "fertility must be given major attention when hybrid seed is used. He [Pence] pointed out that, with the usual increase in production, soil resources could be drained much faster with hybrid than other varieties of corn. He urged that farmers using hybrid seed follow the very best practices of soil erosion control and fertility building through the use of legumes and fertilizers."[45] At another event that year, Fayette County agent Dudley Bishopp reported that Pence had identified phosphorus and potassium as essential for growing hybrid corn varieties and that farmers could not achieve maximum yields without them.[46]

High-yielding hybrids prompted the agents and campus Extension specialists alike to stress the need to use commercial fertilizers to keep fields productive and to maximize the potential of the new varieties of corn. Agent Raymond McHargue told Clay County farmers whenever he had the chance that a "very remarkable difference in yield was shown by the fertilized corn over the unfertilized corn in all instances."[47] Agents around the state could easily have made these same observations about demonstrations in their counties.

During the 1930s, farmers were gaining a greater understanding of the importance of limestone, nitrogen, phosphorus, and potassium for crop yields. Commercial fertilizers would become even more of a necessity later in the decade as hybrid corn was grown on more Indiana farms. But adding farm inputs increased a farmer's costs, and during the tough financial times of the Depression, agents had to be certain that every dollar invested in commercial fertilizers would bring a higher return that justified the expenditure made by the farmer.

Farmers had a lot to learn about fertilizers, and Extension had to ramp up its educational programs to help them better understand how to use the products; how nitrogen, phosphorus, and potassium affected plant health and growth; and the importance of analyzing soil and stalk samples for nutrient availability and for quantifying deficiencies. Through it all, the county Extension agent bridged the gap between the growers who used the fertilizers and those who sold it. And well into the future, the agent would continue to be an impartial third party whom farmers could count on for sound advice.

CHAPTER 18

Growing Wheat Is One Thing and Growing Quality Wheat Is Another

For Indiana farmers in the early decades of the twentieth century, wheat was an important cash crop that was used as grain to mill into flour or feed to livestock, as straw for animal bedding, and as a cover crop to protect the soil.[1] During the 1920s, wheat had averaged 2 million acres in the state, but it dropped slightly during the next decade to an average of 1.8 million acres. Generally, this was because the annual wheat acreage largely followed the grain markets. When wheat prices were expected to be higher, more was planted, and when prices dropped, so did the wheat acreage.

Early in the 1930s, wheat generally sold for 50–60 cents per bushel. In some cases, this meant that it cost more to produce a bushel of wheat than farmers were receiving in compensation.[2] Other farmers managed to just cover their production cost, basically breaking even. In 1933, wheat prices spiked as high as $1 per bushel for a short period of time, then just as quickly fell back to 74 cents per bushel.[3] In 1935, wheat prices again spiked to $1 per bushel based on a rumor that war in Europe seemed imminent.[4] Those having wheat to sell would have made a rather good profit with speculation driving up the price.

With lower prices paid for wheat during much of the Great Depression, farmers were encouraged "to feed wheat rather than throw it on the already overly loaded market," as one county agent put it.[5] As a general rule, farmers made more money feeding their wheat, corn, and soybeans to livestock rather than selling these crops as bulk, unprocessed grain.

Farmers tried outlasting the underperforming markets at the onset of the Depression by stockpiling wheat in their grain bins, hoping to sell it after prices improved. Unfortunately, the extended time the wheat remained in storage led to increased damage by weevils infesting the grain. Extension advised farmers to check their wheat bins for the presence of weevils and to fumigate with carbon disulfide when the insects were detected. But the insects had been in many bins for a long time, severely damaging the wheat. Under these circumstances, farmers had little choice but to sell the damaged wheat at prices that were discounted far below what they would have gotten by selling earlier.[6]

In a 1938 annual report, Fountain County agent Alfred J. Hesler wrote, "Growing wheat is one thing and growing quality wheat is another."[7] The millers who ground the wheat to make flour for baking had a number of complaints, chiefly that the wrong varieties were being grown or that a field of wheat contained so many varieties that it reduced the quality of their flour.[8]

Several years before that, Gibson County agent Cyrus Dyer had cited the same issue and others, saying, "Many of our growers have given little attention in the past to their seed wheat. This has resulted in mixtures of strains and varieties, which in turn tend to lower the grade at the elevator. In many cases the wheat is mixed with rye to the extent that losses occur from this cause. Another factor becoming worse and worse through the years is the contamination with garlic."[9] Wild garlic, which was a weed, caused an unwanted flavor in the flour and the products produced from it.

As with other agricultural challenges, Extension turned to education to help the farmers improve the grain they produced. The milling industry in Indiana favored, at different times, varieties such as Purdue No. 1, Fultz, Rudy, Trumbull, and Michigan Amber.[10] The Purdue Extension Division of Agronomy promoted Trumbull because of its higher yield and reduced lodging, and because it was less prone to winterkill.[11] When the right variety was selected, both farmers and millers benefited.

To improve what was being sold in the marketplace, agents asked growers to basically start over and purchase certified wheat.[12] Farmers who wanted to grow certified wheat seed to sell to other farmers had to pass a number of field and laboratory inspections to obtain certification from the Indiana Crop Improvement Association. This included the following, according to reports from various agents:

1. Fields must be uniform in appearance and considered to be a good source of wheat seeds.
2. Plant soft red winter wheat varieties recommended by the Agricultural Experiment Station at Purdue.
3. In each field, 98.5 percent of the wheat heads need to be from one variety.
4. There cannot be any bunt or stinking smut. Less than 5 percent of the wheat can show symptoms of loose smut, and there should be very little evidence of scab.
5. The fields must be totally free of wild garlic, wild onion, rye, vetch, and other field crops.
6. The fields must not have more than 32 noxious weeds per acre, including cheat and cockle.
7. The harvested wheat must weigh a minimum of 58.5 pounds per bushel and be free of weed seeds, foreign matter, and damage from pests and the elements.[13]

Farmers who received certification for their seed received premium prices for it, which was $1.35–$1.50 per bushel in 1934.[14]

During the 1930s, wheat yields ranged from a low of 13 bushels per acre to as much as 38 bushels per acre, with the state average being 28 bushels per acre.[15] Some of the variability was due to drought and pest pressures, but much of the lower-yielding fields resulted from farmers not using the latest agricultural research. More often than not, low nutrients in the soil were factors that held back wheat yields, even if the seed was certified.[16]

Up-to-date farmers were convinced that applying fertilizer to 4- to 5-inch wheat in March or early April was worth the additional expenditure.[17] In this process, known as topdressing, the dry fertilizer containing a combination of nitrogen, phosphorus, and potassium was put into a wheat drill and dropped on top of the ground to give the small plants the nutrients necessary to grow quickly.[18] When nitrogen was applied using this method, farmers saw increases in yields ranging from 5 to 15 more bushels per acre.[19]

In 1930, a demonstration in Fulton County showed that yields increased from 36 bushels per acre to 46 bushels per acre when nitrogen was applied to the test plots. Based on the cost of the fertilizer and the selling price for wheat, growers generated a net profit of $3.62 per acre when nitrogen was added.[20] That same year in Floyd County, the agent noted that when 100 pounds of nitrate of soda (16-0-0) was applied to wheat, yields in plots with nitrogen produced five additional bushels per acre, and the plants were a deeper green and showed better growth.[21]

As the United States entered the Depression years of the early 1930s, the cost of fertilizer remained high, and the price for a bushel of wheat lost value. Even as wheat prices began dropping in 1930, one agent noted that some farmers had "declared fertilization was a paying proposition despite the low price of wheat."[22] For example, Cass County farmer James Idle documented his nitrate cost at $4.56 an acre, but this was offset by a 15-bushel-per-acre increase in yield, leading agent James Little to note that Idle had "[a] net gain from the nitrate of $8.00 [per acre], figuring his wheat at 80 cents per bushel and deducting for the cost of Nitrate."[23]

But when faced with the high input costs during the financial challenges of the Depression, many farmers took a much different position on using it. As Fayette County agent Marion Detrick noted, "The low price of wheat has led many farmers to question the profit in using fertilizer on wheat."[24] Agents encouraged farmers to base their decision on the economics of the price paid for the fertilizer and whether any increased yields would bring a return high enough to cover that additional cost when the wheat was sold.[25] Based on anticipated yields and expected market trends, Carroll County agent Ralph Maggart observed in 1931, "Due to the low price of wheat, the Nitrate did not pay for itself and cost of distribution."[26]

Some farmers who still wanted to use fertilizer took a variety of approaches to lower their costs but still get the benefits. Some reduced the amount of fertilizer used on wheat to save money, thinking a little fertilizer was better than none at all. Maggart noted this in his 1931 report, saying that, prior to 1931, 90 percent of the wheat grown in the county received some fertilizer, "but due to the low price and the outlook, the amount of fertilizer was either reduced or cut out entirely. Where farmers had been using 150 to 200 pounds of fertilizer, it was recommended that this be cut in half rather than use no fertilizer at all. It was thought that part of the fertilizer would be available for the clover crop, which generally follows wheat, and in our system of farming is a very important crop. Fertilizer tonnage was probably reduced 50%."[27]

Other farmers purchased less expensive formulations that still contained enough nitrogen and phosphorus to increase yields somewhat but not enough to get the highest possible yields.[28] But there was an economic advantage to buying a formulation that contained higher levels of these two important plant nutrients. For instance, Detrick reported in 1931, "There was a saving of $1.68 per ton in buying a fertilizer carrying twenty percent of plant food compared to a 2-12-2 at prices that prevailed locally this fall. . . . Formulas carrying less than 20 percent of plant food are rapidly disappearing in the fertilizer trade in this county."[29]

In Daviess County, a handful of farmers—thirteen across six townships—managed to get up to 123 pounds of nitrate per acre at no up-front cost at all. The fertilizer was provided by local businesses that wanted to promote their products at a time when sales were declining. In return, participants agreed to allow Extension to host demonstrations on their farms and to give back half of the increased yield to offset the cost of the fertilizer. But the arrangement provided protection for the farmer, with agent John Jensen noting, "In the event that one half of the increased [wheat] yield brought more than the cost of the nitrate, the remainder would be retained by the farmer. On the other hand, if there was no increased yield, no payment would be expected."[30]

As the agricultural economy started to recover and the price of wheat began inching upward, agents once again told farmers that money spent on fertilizers would be profitable. In 1936, one agent reported that topdressing wheat in March with 125 pounds of 2-12-2 returned $2.83 per acre net profit, while another stated that fertilizer applications to wheat on average returned $2.50 for every dollar invested in the product.[31] Toward the close of the decade, as the increased yields more than paid the fertilizer bill, Gibson County agent Roy Gwaltney would write, "The use of commercial fertilizer on wheat is fast becoming a general practice."[32]

One of the more serious weeds in wheat was wild garlic because of the off-flavor it gave to processed flour.[33] Wild garlic in pastures could taint the meat and milk products produced from the beef and dairy cattle that grazed there.

Much of the early successful efforts of controlling garlic in Indiana wheat fields was attributed to Clifford E. Skiver, who was a wheat Extension specialist at Purdue. In a 1932 annual report, Gibson County agent Cyrus Dyer described how Skiver's work simplified the process of controlling this highly destructive weed in wheat.

> Until Skiver became interested in ridding our wheat of garlic, we had the blanket recommendation of plowing in the fall [bringing up the bulbs] and in the spring for three consecutive years to get rid of this pest. Skiver has for two seasons now gone up and down through the garlic infested areas making a close study of the garlic plant and its habits of growth. By digging the plants from week to week all through its growing season, he has learned just when they set on their side bulblets. He has learned too that those side bulblets constitute the principal means of propagation for the plant. He has learned that these bulblets are set on from the last of February until about April first. Consequently, if they are plowed under sometime in February, it is likely that one plowing will almost eradicate the plant. A second plowing the next year seems likely to finish the job in fair shape. This plan so radically simplifies the method of eradication that we believe we will make more progress in the next three years than has been made in the last fifteen.[34]

Those who bought the wheat were willing to pay higher prices to farmers who successfully controlled the wild garlic as well as other damaging diseases and pests.[35]

While farmers were fighting the spread of garlic in their fields, there were many instances where the wheat being planted was contaminated with other weed seeds and other small grains such as rye.[36] Skiver would again play a prominent role in helping farmers solve the problem. He worked to build a truck-mounted cleaner in 1932 that would pull out weed seeds from the harvested wheat. The cost of building the cleaner and for its operation was paid for by the millers and the farmers, who both benefited from this service.[37]

The equipment operator would run the wheat seed through the machine, where weed seeds were removed and discarded. In addition to cleaning wheat seed, many growers also paid to have the weed seeds removed from soybeans and clover.[38] Farmers who used this service in 1932 were charged three cents per bushel, which could be paid in cash, but many of them opted for the alternative of paying in grain. In 1934, the *Extension Service Review* reported that "while times were hard and many toll collections were paid in grain, the machines proved very popular and they were able to reach only a small percentage of the growers requesting the service."[39]

In southwestern Indiana, seven machines were used to remove the weed seeds from wheat, oats, and soybeans during the summer and fall of 1935. In 1936 in

Dubois County, 660 wheat growers pooled together $620 that, in combination with $80 contributed by six millers and the county farm bureau, was used to purchase a movable seed cleaner. This equipment was used to clean and remove weed seeds from a total of 20,000 bushels of wheat at 432 farms that year.[40]

The advantage of these machines was that they were shared and moved from farm to farm to clean and treat the seed, which meant that farmers didn't have to transport the seed to a central location for cleaning. A typical unit was powered by a gas engine and could clean 35 to 50 bushels per hour.[41]

The seed-cleaning enterprise also provided a service for treating wheat against smut by using copper carbonate or mercury.[42] Incidences of buyers severely docking the price paid to growers for smut-infested wheat—including one involving seven carloads of wheat in Hancock County in 1937—would help convince farmers to use this service to treat their wheat before planting.[43]

Over the next few years, the seed-cleaning services continued to expand. In 1937, there were 22 machines operating in Indiana that cleaned 306,000 bushels of mainly wheat for 3,246 farmers. In addition, 80,000 bushels of the cleaned wheat seeds were also treated for stinking smut.[44] By 1938, there were at least 35 portable seed-cleaning machines operating in Indiana.[45]

By 1938, the federal government through the Agricultural Adjustment Act created the Federal Crop Insurance Corporation to provide crop insurance for farmers who wished to participate.[46] Fountain County agent Alfred Hesler described the federal wheat insurance program in 1939, writing, "It is different from any insurance formally available to farmers on growing crops in that it gives protection on production rather than value. It also differs . . . in that the premium is paid in bushels of wheat, or cash which is converted into wheat, which is held as a reserve to pay losses when they occur. Losses are paid in bushels of wheat or cash equivalent at time of loss. The only time the reserve is depleted would be when there are losses."[47]

After a number of private carriers failed to get the program off the ground, the U.S. Department of Agriculture (USDA) was assigned to administer the program.[48] Indiana farmers were required to submit applications to a USDA office in Kansas City, Missouri, but farmers could pay their premiums at the county courthouse.[49]

The new program protected 50–75 percent of a farmer's specified wheat acreage planted in the fall of 1938, with coverage for damage caused by fire, drought, flood, wind, hail, Hessian fly, and winterkill. Claims would be paid based on average yields, which were calculated using a farmer's ten-year average production. For example, a farmer averaging 20 bushels per acre who opted for coverage under the 75-percent policy would be compensated for 15 bushels per acre.[50]

Farmers had to decide how many acres of wheat to put into the insurance program. On average, the premium cost the farmer one to two bushels of wheat per acre, or an average of 56 cents an acre. As Hesler had noted, a farmer could either pay the premium in cash or with stored wheat, with the specific amount determined by USDA officials and based on the price of No. 2 wheat at the time the policy was written.[51]

After the application had been submitted, the federal authorities would provide a quote for the premium cost and the number of acres that had been accepted for enrollment. At this point, any farmer who felt that the cost was too high could decline the USDA offer.[52]

As with any insurance program, the goal was to have more money coming in than what was paid out. Therefore, to make the federal program sustainable, enough wheat growers had to enroll to build up a fund sufficient to cover the claims when the crops were damaged. Agent Warren O'Hara was surprised when 263 Hancock County farmers initially made applications for wheat insurance.[53] In the end, 176 farmers would follow through with enrollment and pay the required premiums.[54] In Delaware County, 190 farmers enrolled 2,200 acres of wheat, which cost them 46 cents per acre on average.[55] Generally, fewer than 200 farmers signed up in most counties.[56]

In 1939, agent Wilbur W. Whitehead reported that Boone County farmer Charles Routh was "the first Indiana farmer to receive a wheat crop insurance check from the Federal Crop Insurance Association as payment for complete loss [of his wheat crop]. Mr. Routh had insured his 9.4 acres and received $124.77. This amount represented coverage of 75% of his lost yield."[57] In Dearborn County, spring floods in 1939 produced serious losses to wheat, with agent Cory Alcorn later reporting, "Those with their crops insured received about $2,000 indemnity while the others stood the loss themselves."[58] And in Bartholomew County, the program paid one in five farmers for some type of loss in 1939.[59]

In the end, farmers had to decide whether it was better to risk a loss or to pay a premium to partially cover unexpected damages, with the latter option winning out in many cases. During the first year, the wheat insurance program proved to be extremely popular among Indiana growers, with slightly more than 11,000 of them paying premiums to enroll.[60]

Wheat was an important commodity that farmers depended upon for cash and on-farm use. Over the course of the 1930s and 1940s, Extension would see the transformation of wheat from a low-producing crop to one with larger yields and better milling properties. Much like with the development of corn hybrids (see Chapter 16), Purdue plant breeders would develop genetically improved varieties of wheat that farmers could secure from those producing certified seedstock.

Extension's role would be pivotal, considering that these newer varieties could only produce the higher yields and better quality grain if farmers incorporated a number of best management practices in their programs. These included following the latest research on planting dates and rates, using certified seed, applying fertilizers containing the right ratio of nutrients in the early spring, removing weed seeds from the harvested wheat, eliminating problem weeds such as wild garlic, and inspecting wheat in storage for the presence of insects. County agents and their Extension partners at Purdue taught the farmers how to use this research, which in turn revolutionized the growing of wheat in Indiana.

Crews representing a Civilian Conservation Corps camp in Bartholomew County in 1936 (top) and another in Madison County (bottom), also from around 1936, worked on environmental projects such as replacing drainage tiles, improving ditches, building trails, and constructing buildings. Under this federal program, unemployed men were hired for six months or more and received a monthly salary of $30 along with training, food, and medical care. Many of these workers would send money home to help their families through the tough times of the Great Depression. Courtesy of Purdue University Libraries, Karnes Archives and Special Collections (top), and J. C. Allen and Son (bottom).

In 1933, workers from the Civilian Conservation Corps camp in Warrick County built a series of brush dams to slow water flow and collect sediment. Agent Robert Endicott was largely responsible for getting the federal government to establish this camp at Boonville. By 1933, the camp had 18,000 acres of eroded and hilly land under contract. With the workers spending $1,000 per month in the local community, the camp provided an important economic stimulus for the county. Courtesy of Purdue University Libraries, Karnes Archives and Special Collections.

Soil erosion was a serious problem that Extension and, later on, county soil conservation districts addressed. Constructing a concrete dam, such as this one in Vermillion County in 1937, slowed the flow of water to help prevent erosion and the formation of gullies. Courtesy of Purdue University Libraries, Karnes Archives and Special Collections.

Running water left unchecked created this gully on a Tippecanoe County farm, pictured here in 1938. Solving a serious erosion problem such as this often required a combination of different techniques that were tailored to individual farms. After the water was diverted by terraces placed in the field on top of the hill, this gully was filled in, covered with sod, and the land no longer farmed. Courtesy of Purdue University Libraries, Karnes Archives and Special Collections.

The top photo shows a closeup of one of eight wire fence dams constructed to prevent gully erosion on the farm of Roscoe Dean in Howard County in 1938, while the bottom photo shows a wider view of the same type of dam built in Warrick County in 1933. Brush, straw, and other debris would be held in place along the fence, then water flowing into it would slow down and deposit soil sediment on the upward side of the fence. In many cases, these dams captured enough soil to fill in the gullies behind the fence. Courtesy of Purdue University Libraries, Karnes Archives and Special Collections.

Pictured is an extreme case of an eroded hillside in Delaware County, which shows how a tree held some of the soil in place while the rest of the hill eroded away. During the 1930s, Extension agents invested a lot of effort in getting farmers to reforest hilly areas that should not have been planted to crops during the previous decades. As this image shows, trees could help stabilize the soil and prevent erosion caused by running water. Courtesy of Purdue University Libraries, Karnes Archives and Special Collections.

Wayne County assistant agent Edward C. Miller (foreground at left) leads a delegation of fifty-two Wayne County farmers on an all-day tour of the Henry County Soil Conservation Demonstration Area in 1938. The test plots showed the effects of planting a ground cover for reducing erosion under different runoff conditions. Courtesy of Purdue University Libraries, Karnes Archives and Special Collections.

Purdue agricultural engineer Russell Shipman (left) and Martin County farmers build a terrace to use as part of an Extension demonstration tour. Courtesy of Purdue University Libraries, Karnes Archives and Special Collections.

This demonstration on the Peterson-Bank Farm in Tippecanoe County was one stop on an Extension erosion tour in 1938. Plowing contour furrows in land being returned to pasture helped reduce the loss of topsoil. Courtesy of Purdue University Libraries, Karnes Archives and Special Collections.

Extension hosted this 1938 dynamite demonstration on the Omer Willits Farm in Howard County. At this type of demonstration, agents would show farmers how to use the dynamite to deepen a ditch and improve water drainage from their fields. Courtesy of Purdue University Libraries, Karnes Archives and Special Collections.

This 1931 photograph shows a fire tower at the Jackson County State Forest—today known as the Jackson-Washington State Forest—located two miles east of Brownstown. During the 1930s, workers from the Civilian Conservation Corps built fire towers across Indiana to protect larger tracts of forestland. Trained individuals used the towers to spot fires and quickly dispatch firefighting crews to the scene. Courtesy of Purdue University Libraries, Karnes Archives and Special Collections.

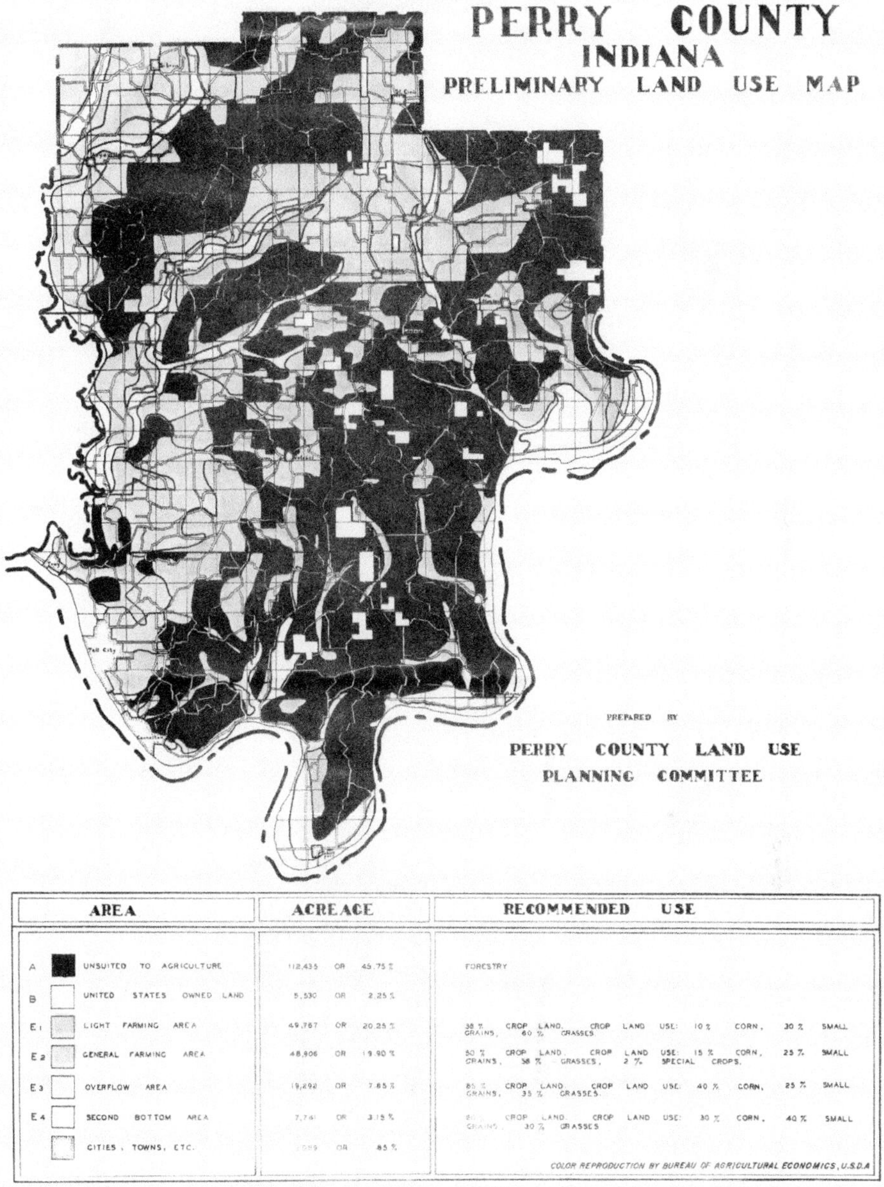

In 1938, U.S. Department of Agriculture (USDA) officials asked counties to create township and county land-use committees to classify land as to its suitability for farming, forestry, and nonagricultural uses. In Perry County, committee members classified land into seven distinct categories. By 1941, the USDA effort stalled when the United States entered World War II. However, following the war, the program would evolve into modern-day zoning boards and ordinances, which regulate how land within a county can be used or developed. Courtesy of Purdue University Libraries, Karnes Archives and Special Collections.

Members of the Center Township Land Use Planning Committee meet in 1941 to discuss land classification efforts in Howard County. The USDA program that classified how a county's lands were used depended on volunteers such as these to gather information and prepare the maps. Pictured are (from left to right) Raymond Dean, Harold Baker, Lee Chambers, and Hershel Jacobs Howard. Courtesy of Purdue University Libraries, Karnes Archives and Special Collections.

Land Use Planning Meetings Next Week

A series of land use planning meetings will be held in Brown County next week. The purpose of these meetings, which will be held in each township, is to classify the land according to the use for which it is best adapted.

Preliminary land use maps have already been made but the land use planning committee does not want to submit these maps to the Washington office until it has been possible to receive the aid and approval of the farmers of the county.

These maps and the information gained at these meetings will be submitted to the Bureau of Agricultural Economics in Washington and future development in Brown County will be based upon this information.

The land use planning committee hopes that a large number of farmers will attend these important meetings and give aid in securing the best information possible. The schedule of meetings are given below:

Nashville—Court House, Monday afternoon, March 11th at 1:00 p. m.

Browning School—Johnson Township, Monday night, March 11th at 7:00 p. m.

Bellsville—Van Buren Township, Tuesday morning, March 12th at 9:00 a. m.

Helmsburg—High School, Tuesday afternoon, March 12th at 1:00 p. m.

Goshen—Hamblen Township, Wednesday morning, March 13th at 9:00 a. m.

In March 1940, a local newspaper article announced that a series of land use planning meetings would be held across the townships of Brown County. Members of land use planning committees held public meetings such as these to solicit input from farmers regarding whether land was best suited for crops, pastures, forests, or another purpose. Courtesy of *Brown County Democrat*.

Here, bags of donated seed corn are loaded into International trucks in Rush County in April 1931 for delivery to farmers affected by drought in Harrison County. Severe droughts plagued many farmers during the 1930s, adding to the misery of the Great Depression. County Extension agents worked with local farm bureaus to secure replacement seed corn for those whose crops had been destroyed. In this effort, 178 Harrison County farmers each received one bag of seed corn that was tagged with the name and address of the Rush County farmer who made the donation. Courtesy of Purdue University Libraries, Karnes Archives and Special Collections.

In 1937, a kerosene lamp provides the light for two women as they read. Using lamps such as this for lighting would become a thing of the past when rural homes were able to hook up to the electricity provided by rural electric cooperatives. Courtesy of J. C. Allen and Son.

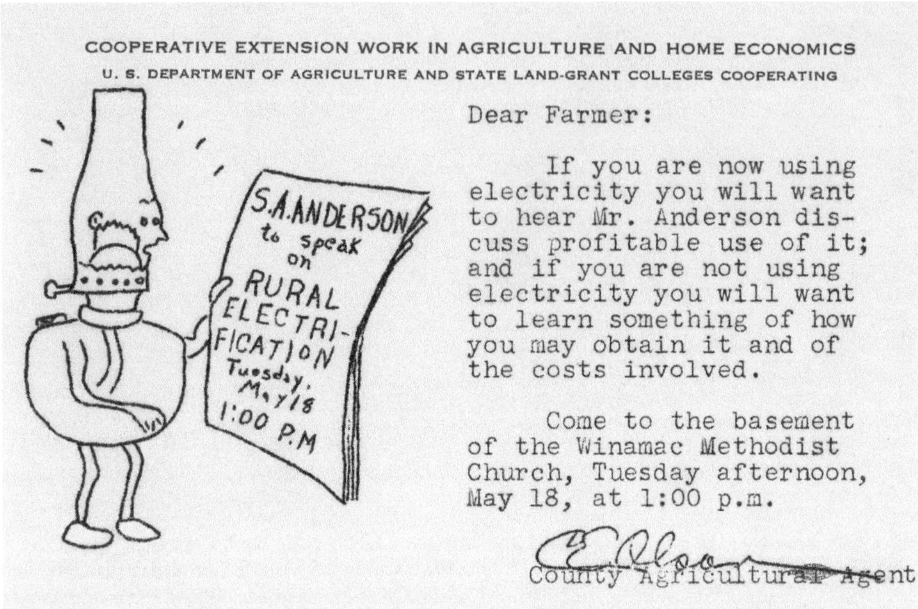
COOPERATIVE EXTENSION WORK IN AGRICULTURE AND HOME ECONOMICS
U. S. DEPARTMENT OF AGRICULTURE AND STATE LAND-GRANT COLLEGES COOPERATING

Dear Farmer:

If you are now using electricity you will want to hear Mr. Anderson discuss profitable use of it; and if you are not using electricity you will want to learn something of how you may obtain it and of the costs involved.

Come to the basement of the Winamac Methodist Church, Tuesday afternoon, May 18, at 1:00 p.m.

County Agricultural Agent

Agent Carl Olson used this postcard to invite Pulaski County farmers to a May 1937 Extension program on electricity in Winamac featuring speaker Seigel Anderson of the Department of Agricultural Engineering at Purdue. Agents and campus Extension specialists often worked together to provide information about how those in rural areas could connect to the power grid as well as the best ways to use electricity for a farm and rural home after the connection was established. Courtesy of Purdue University Libraries, Karnes Archives and Special Collections.

This Union Township float in the 1935 Whitley County 4-H parade conveyed how important electricity was for managing the farm home. Courtesy of Purdue University Libraries, Karnes Archives and Special Collections.

U.S. Representative Finly H. Gray of Indiana (left) throws a shovelful of dirt during the installation of the first electric pole for the Union-Fayette County Rural Electric Membership Corporation in 1938. Once the line was completed in Union County, electricity would reach 250 rural homes there. That year, Union County agent Orlando V. Winks reported, "The breaking of the old oil lamp on the new electric light poles brought a new era to these farm families." Courtesy of Purdue University Libraries, Karnes Archives and Special Collections.

On the Basho Farm in Howard County, this woman uses an electric motor to shell corn in 1933. The electrification of farmsteads by rural electric cooperatives during the Depression allowed farmers to use electric motors to run equipment such as cream separators and emery wheels for the farm as well as lights and appliances such as refrigerators and washing machines for the home. Courtesy of J. C. Allen and Son.

COOPERATIVE EXTENSION WORK
IN
AGRICULTURE AND HOME ECONOMICS
State of Indiana

Purdue University
U.S. Dept. Agriculture
Dearborn Co. Board of
Education Cooperating

County Extension Work
Extension Service
Aurora, Indiana
October 10, 1940

LET THERE BE LIGHT

Isn't it fine to count yourself among the fortunate folk of Dearborn County who now have electricity? I congratulate you. It's remarkable what all a farmer can do with electricity.

We want to make a study of electrical appliances put into use by families since getting current. Will you please help us by checking these two lists and returning this letter to us by next Thursday, October 17? (Use the enclosed envelope.)

Please number 1, 2, 3, etc., the order in which you have already bought any of the following:

_____ Electric iron
_____ Toaster
_____ Floor lamps
_____ Refrigerator
_____ Radio
_____ Vacuum sweeper
_____ Sewing machine
_____ Water pump
_____ Chick brooder
_____ Motor for feed grinding and other uses
_____ Stove
_____ Fan
_____ Mangle

Please number 1, 2, 3, etc., the order in which you intend to buy any of the following:

_____ Electric iron
_____ Toaster
_____ Floor lamps
_____ Refrigerator
_____ Radio
_____ Vacuum sweeper
_____ Sewing machine
_____ Water pump
_____ Chick brooder
_____ Motor for feed grinding and other uses
_____ Stove
_____ Fan
_____ Mangle

What is the one thing you like best about electricity?

Can we help you through the Extension Department in getting greater value or satisfaction from your electricity or any of the appliances? Just write your problems on the other side of this letter.

Very truly yours,
C. A. Alcorn
C. A. Alcorn, County Agrl. Agent

In 1940, Dearborn County agent Cory Alcorn conducted this survey asking those with electricity what electrical equipment, appliances, and tools they had purchased. Alcorn also wanted to know which electrical devices farmers wanted to buy in the future. This allowed Extension to assess the full impact electrical power was having on rural homes and farms. Courtesy of Purdue University Libraries, Karnes Archives and Special Collections.

Farmers near Clarksburg in Decatur County use horses to get gravel from a creek bed in 1921. Before the introduction of tractors, horses provided the pulling power for farm equipment. Draft horses such as those pictured often reached 2,000 pounds and 5–6 feet tall, which meant they were large enough to pull the many types of implements and wagons used on farms. Courtesy of J. C. Allen and Son.

In 1931, Boyd Kotterman plows new ground in Miami County with his two Belgians. Courtesy of Purdue University Libraries, Karnes Archives and Special Collections.

Farmers attend a 1930 Extension demonstration in Clinton County that showed how to hitch together twelve horses to pull larger and heavier farm implements. Those with farms that were too small to warrant the purchase of tractors would continue to rely on horses to pull farm equipment for much of the 1930s. Courtesy of Purdue University Libraries, Karnes Archives and Special Collections.

Farmers with more acres used horses and mules alongside tractors as the transition from animal to mechanical power was underway. This photograph (ca. 1926) shows the different types of equipment used in getting a field ready for planting corn, but it also demonstrates the use of horse-pulled farm implements working with a steel-wheeled tractor. Courtesy of J. C. Allen and Son.

During the 1930s, Extension agents encouraged landowners to manage their woodlots better. This included discouraging farmers from allowing cattle to graze through a woodlot where trees had been harvested, such as these animals are doing. Not only did this practice provide a poor-quality forage, but it also prevented natural reforestation since the cattle consumed young tree seedlings. Courtesy of Purdue University Libraries, Karnes Archives and Special Collections.

Extension set up demonstrations areas where a fence was placed around a woodlot to restrict the area where cattle were allowed to graze (right) and keep them out of the wooded area (left). The visual evidence of natural regeneration could be readily observed in the areas where the cattle had been excluded. Demonstrations such as these helped farmers understand the advantages of Extension recommendations and made them more likely to adopt new practices. Courtesy of Purdue University Libraries, Karnes Archives and Special Collections.

This timber stand remained (top) after trees were selectively cut to build a barn (bottom) for Dr. O. W. Hinshaw in Randolph County in 1939. The oak and beech timber removed from a 17.5-acre tract of the farm's forest provided all of the wood except for the siding and amounted to nearly 30,000 feet of lumber. By using his own resources, Hinshaw not only saved approximately $1,100 versus buying pine lumber, but he also ended up with a stronger barn constructed from hardwoods. Courtesy of Purdue University Libraries, Karnes Archives and Special Collections.

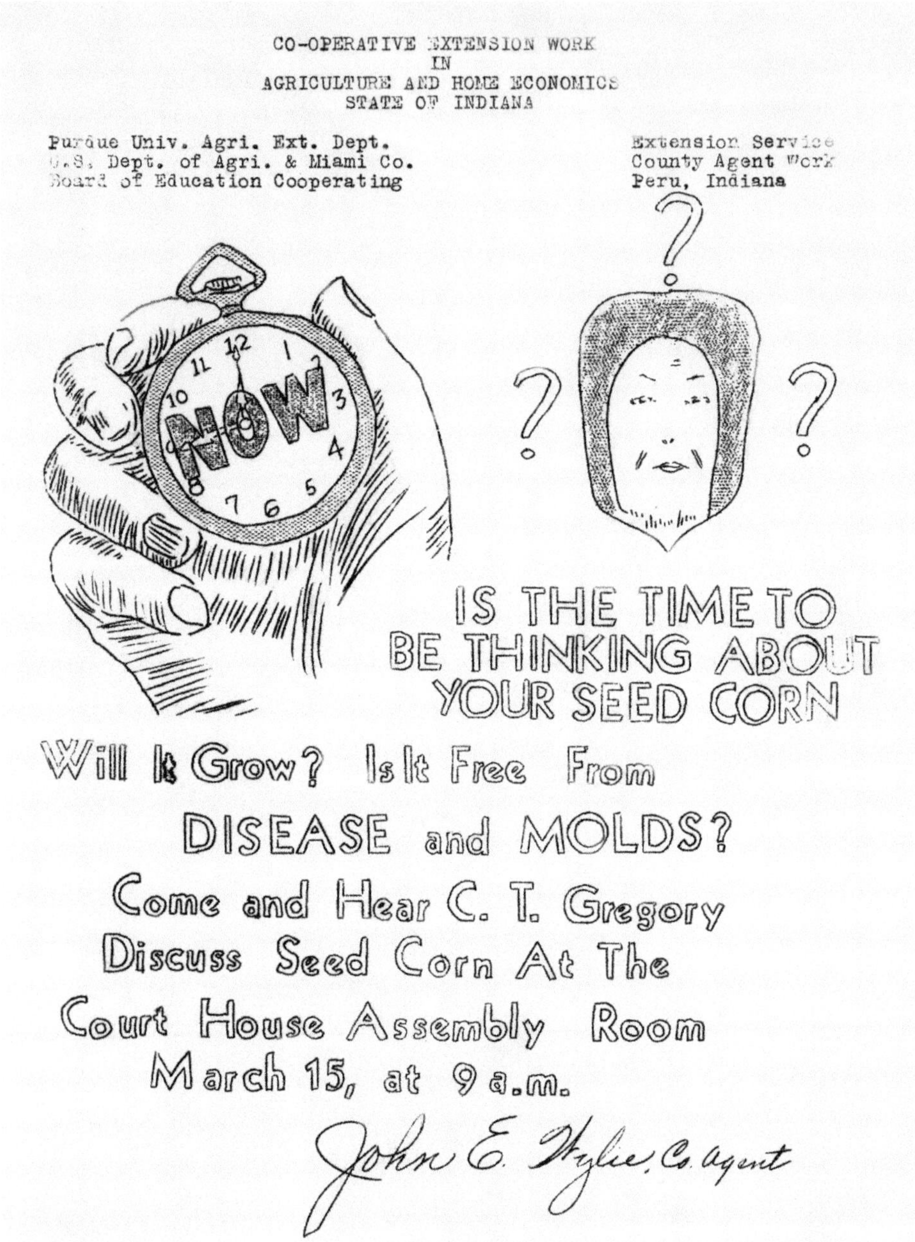
CO-OPERATIVE EXTENSION WORK
IN
AGRICULTURE AND HOME ECONOMICS
STATE OF INDIANA

Purdue Univ. Agri. Ext. Dept.
U.S. Dept. of Agri. & Miami Co.
Board of Education Cooperating

Extension Service
County Agent Work
Peru, Indiana

NOW

IS THE TIME TO
BE THINKING ABOUT
YOUR SEED CORN

Will It Grow? Is It Free From
DISEASE and MOLDS?
Come and Hear C. T. Gregory
Discuss Seed Corn At The
Court House Assembly Room
March 15, at 9 a.m.

John E. Wylie, Co. Agent

In March 1935, Miami County agent John Wylie reminded farmers about the importance of testing their seed corn for germination and diseases. Getting farmers to test their seed before planting was one of the major programs conducted by Extension agents before hybrid seed corn became used more widely. Courtesy of Purdue University Libraries, Karnes Archives and Special Collections.

CO-OPERATIVE EXTENSION WORK
IN AGRICULTURE AND HOME ECONOMICS
STATE OF INDIANA

PURDUE UNIVERSITY AGRICULTURAL EXTENSION
DEPARTMENT, U. S. DEPARTMENT OF AGRICULTURE
AND COUNTY BOARD OF EDUCATION CO-OPERATING

EXTENSION SERVICE
COUNTY EXTENSION WO

WHICH IS YOUR SEED CORN

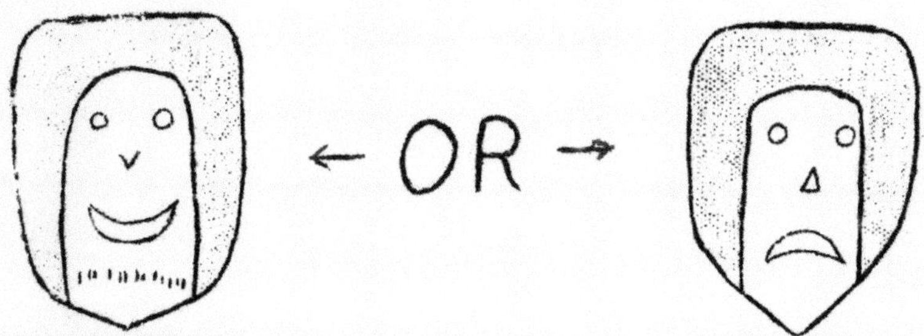

I AM IN DEMAND
BECAUSE
I HAVE

1. Been dried
2. Been TESTED
3. A BRIGHT GERM and FREE of Diseases.

NO ONE WANTS ME
BECAUSE
I HAVE

1. Been FROZEN
2. Haven't been TESTED
3. A weak feeling and sick.

You are invited to attend our Plant Disease and Weed School which will be held February 27, at Assembly Room of the Court House, Peru.

PROGRAM

10:00 A.M.	Methods of preventing weeds, by O. C. Lee
11:00 A.M.	Seedling Rot, by Dr. C. T. Gregory
1:00PP.M.	New Weed Problems, by O. C. Lee
2:00 P.M.	Culling, testing the seed corn for disease control.

Very truly yours,

John E Wylie
County Agent.

JEW/meb

Miami County agent John Wylie invited area farmers to attend a 1936 program featuring two Extension speakers from Purdue: plant pathologist Charles Gregory and weed specialist Oliver Lee. It was common for agents to invite experts from Purdue to back up their advice and recommendations. Courtesy of Purdue University Libraries, Karnes Archives and Special Collections.

Two Marion County students read seed corn tests that evaluated germination rates and the presence of disease. For this testing method, approximately five kernels from each ear were placed in a wet muslin cloth known as a rag doll and allowed to germinate. Those ears with good germination that showed no signs of disease were used for seed corn, then the rest would be used to feed livestock. In 1936, when this photo was taken, Marion County had a capacity to test 130,000 ears, and 550 farmers took advantage of the service. Courtesy of Purdue University Libraries, Karnes Archives and Special Collections.

Purdue agronomists cover these corn plants to prevent self-pollination as they grow hybrid seed corn in 1938. During the 1920s and 1930s, Purdue agricultural researchers were instrumental in developing lines of corn hybrids known as Hoosier Hybrids. These varieties, which were grown in corn breeding plots at the Purdue farm in Lafayette, were distributed to farmers around the state, who then used them to develop their own hybrids. Courtesy of J. C. Allen and Son.

Frank Howat of Pulaski County, pictured in 1938, was one of the first commercial hybrid seed corn producers in Indiana. Here, Howat shakes pollen onto silks, then places a bag over each ear to prevent pollen from other plants from reaching the ear. Courtesy of Purdue University Libraries, Karnes Archives and Special Collections.

In this 1938 flyer, Union County agent Orlando Winks advertises a local hybrid corn variety demonstration. The purpose of on-farm demonstrations was to try and convince farmers to switch from their open-pollinated corn to these newer hybrids that consistently produced higher yields, matured days or weeks prior to first frost, and remained upright during storms. Courtesy of Purdue University Libraries, Karnes Archives and Special Collections.

Fulton County farmer S. R. Wildermuth worked with agent Martin Huxley to plant a corn demonstration trial in 1941. These types of demonstrations were really important because they allowed farmers to observe several varieties grown under the same local conditions and compare maturity dates, standability during inclement weather, tolerance to insects and diseases, and yield potential. Courtesy of Purdue University Libraries, Karnes Archives and Special Collections.

Wayne County farmers take moisture samples from Hoosier Hybrid 600 corn at a field demonstration in 1936. Extension taught farmers how to take these samples so they could gauge their yields more accurately. Courtesy of Purdue University Libraries, Karnes Archives and Special Collections.

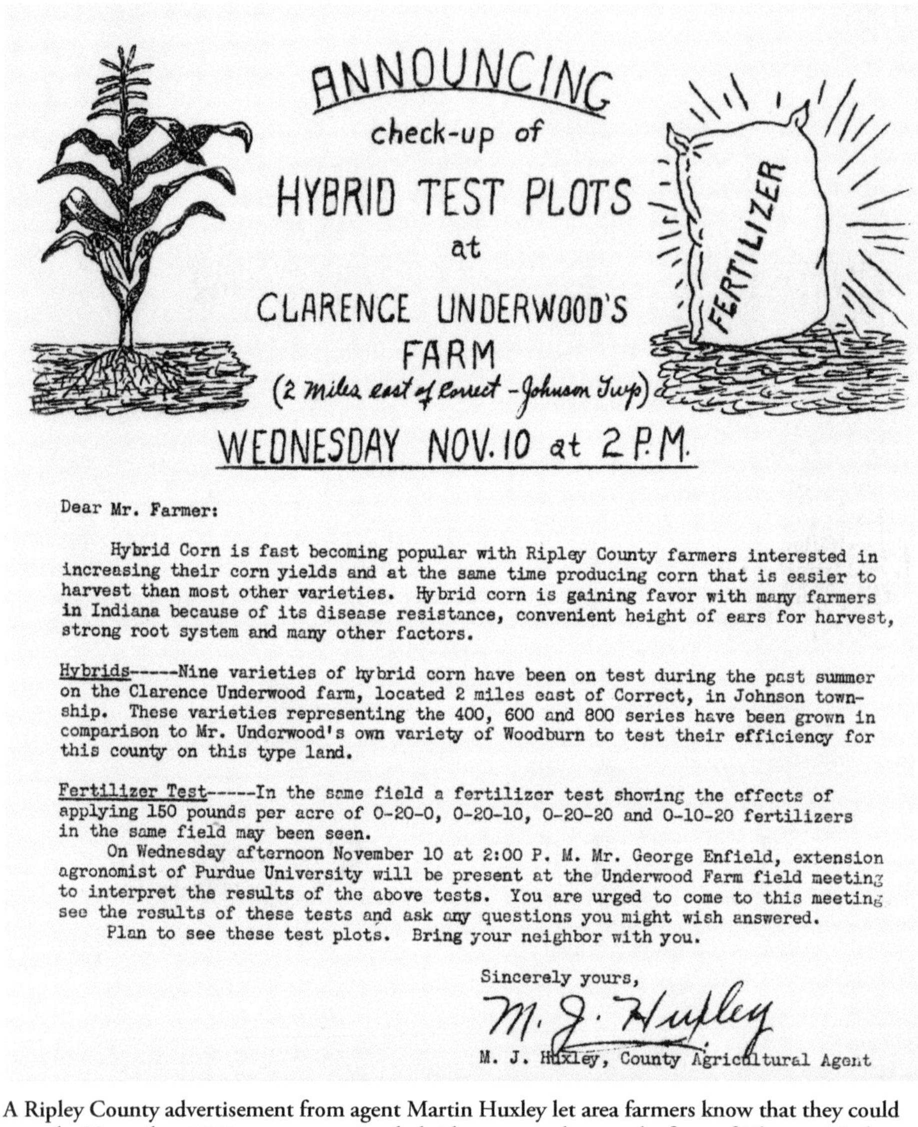
ANNOUNCING
check-up of
HYBRID TEST PLOTS
at
CLARENCE UNDERWOOD'S
FARM
(2 miles east of Correct - Johnson Twp)
FERTILIZER

WEDNESDAY NOV. 10 at 2 P.M.

Dear Mr. Farmer:

Hybrid Corn is fast becoming popular with Ripley County farmers interested in increasing their corn yields and at the same time producing corn that is easier to harvest than most other varieties. Hybrid corn is gaining favor with many farmers in Indiana because of its disease resistance, convenient height of ears for harvest, strong root system and many other factors.

Hybrids-----Nine varieties of hybrid corn have been on test during the past summer on the Clarence Underwood farm, located 2 miles east of Correct, in Johnson township. These varieties representing the 400, 600 and 800 series have been grown in comparison to Mr. Underwood's own variety of Woodburn to test their efficiency for this county on this type land.

Fertilizer Test-----In the same field a fertilizer test showing the effects of applying 150 pounds per acre of 0-20-0, 0-20-10, 0-20-20 and 0-10-20 fertilizers in the same field may been seen.

On Wednesday afternoon November 10 at 2:00 P. M. Mr. George Enfield, extension agronomist of Purdue University will be present at the Underwood Farm field meeting to interpret the results of the above tests. You are urged to come to this meeting see the results of these tests and ask any questions you might wish answered.

Plan to see these test plots. Bring your neighbor with you.

Sincerely yours,

M. J. Huxley

M. J. Huxley, County Agricultural Agent

A Ripley County advertisement from agent Martin Huxley let area farmers know that they could attend a November 1937 program to view hybrid corn test plots on the farm of Clarence Underwood. Firsthand observation of the different varieties at the end of the season allowed growers to compare and contrast the characteristics of the different plants. Courtesy of Purdue University Libraries, Karnes Archives and Special Collections.

Decatur County entrepreneur Arthur C. Stewart from Greensburg, shown here in 1939, was one of the first farmers in Indiana to produce hybrid seed corn on a commercial scale. Stewart's advertising indicates that he was raising Hoosier Hybrids and that his seed was certified by the Indiana Corn Growers' Association. Courtesy of Purdue University Libraries, Karnes Archives and Special Collections.

The hybrid seed industry greatly expanded during the 1930s as farmers adopted this new seed technology in steadily increasing numbers. This would require companies such as the Brodbeck Brothers in Wabash County, pictured here in 1937, to ensure that its hybrid seed was of high quality so that farmers would return as customers. Courtesy of J. C. Allen and Son.

DeKalb County's Karl Pankop (left) was crowned as Indiana's Corn King in 1930 for being the top yielder in the statewide five-acre corn contest. His son, Herman (center), earned the title in 1931, then won it again in 1932 by narrowly beating out his brother, Harold (right). Extension promoted competitions such as these among farmers to highlight what could be achieved when the latest practices based on agricultural science, such as using corn hybrids and recommended tillage and fertilization, were put to the test. Courtesy of Purdue University Libraries, Karnes Archives and Special Collections.

This bronze medal was awarded to Montgomery County's Clarence R. Overstreet, a 1936 Purdue agriculture graduate, for producing 65–85 bushels of corn per acre in the five-acre corn contest held in 1943. The Indiana Corn Growers' Association sponsored the medals given to participants based on their yields. In addition to the bronze medal, a gold medal was awarded to farmers who produced 100 or more bushels per acre, and a silver medal, to those whose yields were 85–100 bushels per acre. Courtesy of Bryan Overstreet.

CO-OPERATIVE EXTENSION WORK
IN AGRICULTURE AND HOME ECONOMICS
STATE OF INDIANA

PURDUE UNIVERSITY AGRICULTURAL EXTENSION
DEPARTMENT, U. S. DEPARTMENT OF AGRICULTURE
AND COUNTY BOARD OF EDUCATION CO-OPERATING

Kokomo, Ind
May 2, 1936

EXTENSION SERVICE
COUNTY EXTENSION WORK

SOIL TESTING CLINIC

WEDNESDAY — MAY 6th

COUNTY AGENT'S OFFICE — 8 A.M TO 4 P.M.

Soil samples will be tested free for potash, phosphate, and lime requirements at my office Wednesday, May 6th. George Enfield, soil testing specialist from Purdue University, will be present to test the soils and discuss their results with you.

IMPORTANT - soil samples to be tested must be air dried. Collect your samples before the meeting and air dry. Do not dry with artificial heat.

Composite samples should be made of the same soil type in the same field by mixing together ten samples from the same type soil. Bring as many composite samples as you care to. Fertilizer formulas for different crops will be recommended by Mr. Enfield after testing your soil.

LET GEO. TAKE CARE OF HIM.

SOLVED — Your fertilizer problems if you bring samples to the meeting.

Very truly yours,

R. C. Stangland
COUNTY AGRICULTURAL AGENT

Howard County agent Roscoe Stangland reminded area farmers that they could attend a May 1936 meeting to have soil samples analyzed for potash, phosphate, and pH at no cost by Purdue soil Extension specialist George Enfield. By using results from soil tests, farmers would no longer have to guess how much fertilizer they needed in order to grow high-yielding crops. Courtesy of Purdue University Libraries, Karnes Archives and Special Collections.

At a 1939 Extension school in Delaware County, Purdue plant pathologist Charles Gregory (holding jar) demonstrated how to do a soil test to florists from seven counties who attended the program at Howell's Greenhouse in Muncie. Courtesy of Purdue University Libraries, Karnes Archives and Special Collections.

Farmers wait to have their soil samples tested aboard the Muck Crops Special, an education train sent out by Purdue Extension in 1939. When an education train stopped in communities around Indiana, farmers could climb aboard to see demonstrations, hear lectures, and receive publications on specific topics. Courtesy of J. C. Allen and Son.

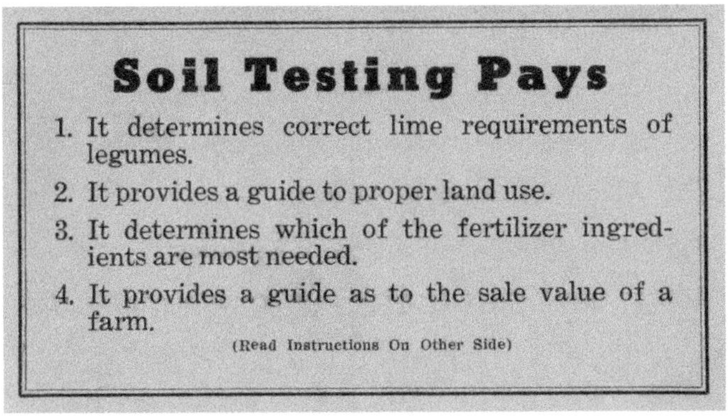

Soil Testing Pays

1. It determines correct lime requirements of legumes.
2. It provides a guide to proper land use.
3. It determines which of the fertilizer ingredients are most needed.
4. It provides a guide as to the sale value of a farm.

(Read Instructions On Other Side)

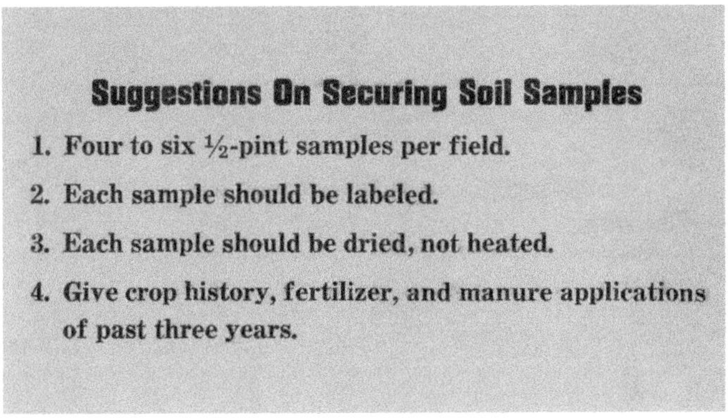

Suggestions On Securing Soil Samples

1. Four to six ½-pint samples per field.
2. Each sample should be labeled.
3. Each sample should be dried, not heated.
4. Give crop history, fertilizer, and manure applications of past three years.

Farmers in Fayette County received this two-sided card in 1939 to remind them about the importance of testing their soil and how to take a proper sample for analysis. In many cases, local Extension agents could complete the testing in their offices for area farmers. Courtesy of Purdue University Libraries, Karnes Archives and Special Collections.

This farmer flattens down a soybean crop in 1932 before plowing it under as a green manure. Today, farmers still use cover crops to improve soil organic matter and texture. Courtesy of J. C. Allen and Son.

In these 1930 photos, Fulton County farmer Boyd Peterson stands in two wheat fields: one that received no nitrogen (left) and one that received an application of nitrogen and produced a much thicker and taller crop (right). While farmers focused on adding phosphorus and potassium to their soils, agents began hosting demonstrations to show the effects of nitrogen on yields and the monetary returns gained from this added crop input. Courtesy of Purdue University Libraries, Karnes Archives and Special Collections.

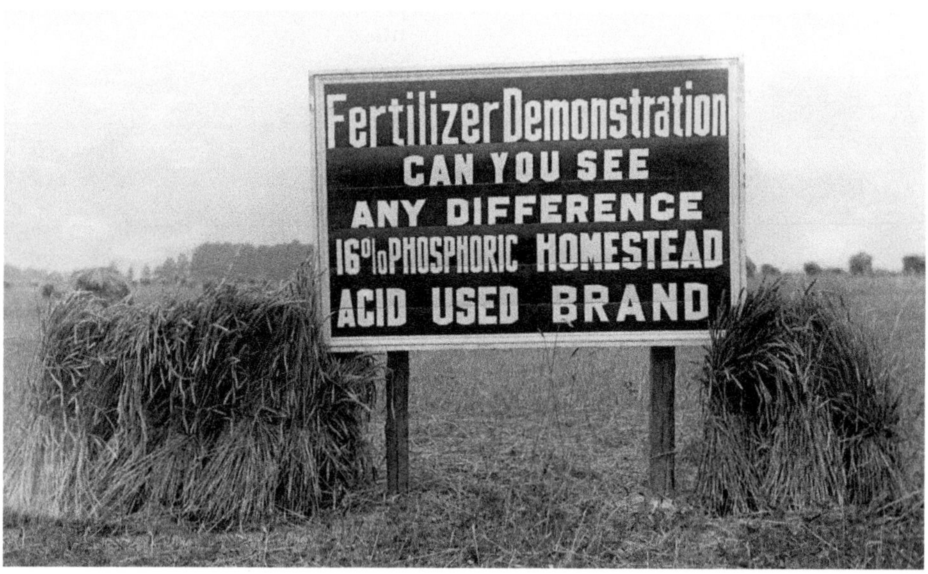

In 1930, Extension made sure to advertise this Pike County demonstration to those passing by. The purpose of this demonstration was to show how the addition of phosphoric acid increased wheat yields. Courtesy of Purdue University Libraries, Karnes Archives and Special Collections.

CHAPTER 19

Farmers on the Lookout for Some New or Different Crop That Offers More Promise for Fair Returns

Farmers knew they were in a financial crisis when corn and wheat prices plummeted. At times, the markets for hogs, poultry, and dairy products weren't much better. Gibson County agent Cyrus Dyer wrote in 1932, "Under conditions prevailing everywhere for the past few years, farmers are continually on the lookout for some new or different crop that offers more promise for fair returns than do such staple crops as corn, wheat or hay."[1] Out of necessity and probably some desperation, farmers reaching the end of their financial ropes had to find new crops to grow on their farms with the hope that these might pay better than what they traditionally produced.[2]

For instance, farmers in the southern counties of Clark, Dearborn, and Floyd diversified by growing tobacco in addition to their usual crops.[3] Like any other commodity, tobacco was not immune to the extreme market swings, taking a price hit in 1932 that resulted in tobacco acres being reduced 30 percent.[4] Still, one agent noted that those who produced "a good quality of tobacco rather than a large quantity" found buyers who were willing to pay higher prices for it.[5]

In Bartholomew County, sweet corn was a significant crop for local growers. The county ranked fifteenth nationally in 1935 for sweet corn production, raising 4,500 acres of varieties such as Country Gentleman, Crosby, Golden Bantam, and Straight Row.[6] Sweet corn yields averaged 2.5 tons an acre, which returned \$12–\$16 per ton in 1931.[7]

In the southwestern counties, cantaloupes and watermelons were important cash crops for farmers. In 1932, Gibson County moved 65 railcars and 85 truckloads of cantaloupes, showing by sheer volume the importance of that crop to area farmers.[8] By 1937, those farmers had formed a melon growers' association and used it to market their crop, ultimately earning them 25 cents more on each of the 30,000 crates of cantaloupes sold by the group.[9] In nearby Daviess County, watermelons had become a major agricultural enterprise in six of the ten townships by 1939.[10]

In Benton County, agent Glenn Reed saw nothing but an upside for local farmers who grew tomatoes and sweet corn for Caroon and Company at Fowler. During the early 1930s, he observed, "These canning crops represented a very distinct aid to the growers this season in view of the unfavorable price level of other farm commodities" and noted that they "are becoming a profitable adjunct to the farm business."[11]

Farmers understood that the market prices they received for the crops they grew were unpredictable—up one day and down the next. They hoped that by diversifying with crops such as cantaloupes, tomatoes, and sweet corn, it would bring a better return because people would be willing to pay higher prices for quality products they consumed directly.

When growing highly perishable but profitable products such as vegetables, producers needed to have buyers in hand long before the crop was harvested. The seventy-five farmers in Elkhart County who grew all sorts of fresh market produce supplied grocery stores and market outlets in and around Elkhart and Goshen with farm-fresh products.[12] In Allen County, local farmers grew 1,200 acres of vegetables and trucked them to Fort Wayne, the largest city in northeastern Indiana, with a population of 115,000.[13]

For farmers living in rural areas without large metropolitan populations nearby, the problem was that the local community could not consume all the fresh products grown in the area. For many growers, the availability of canneries proved to be the answer. Most farmers would specialize in one or two crops that they then sold to processors to turn into canned products. Typical crops included beans, cabbage, cucumbers, onions, peas, pumpkins, sugar beets, sweet corn, and tomatoes. In 1933, Adams County agent Lawrence Archbold wrote, "The outlets for these crops are very important when produced in quantities larger than the local market will absorb; hence, these [canning] outlets must be assured before a campaign for acreage is entered into."[14] Having hundreds of canning plants located around Indiana at the time meant most farmers had one or more close by.[15]

Farmers who sold to processors were required to sign contracts spelling out how many acres of a specific crop they had agreed to plant. Canneries could then use that information to plan the year's production, equipment needs, and number of laborers. In return, farmers were offered guaranteed prices and, at times, bonuses for the quantity and quality of the products delivered.[16] Hamilton County agent Everett Parker wrote in 1930, "Higher quality usually accompanies higher yields, which is the reason the canner is willing to give the bonus of 50 cents to $1.00 [per ton] for all tomatoes yielding over 10 tons per acre. The average yield for those finishing was 12.47 tons per acre. The average yield for the nonmembers was 3.8 tons per acre."[17]

The collective payout for farmers working with the processors could be substantial. In 1926, a single cannery in Bartholomew County paid farmers $625,000 for their crops. It was estimated that two additional packers in the county provided another $75,000 to growers, bringing the total to approximately $700,000.[18] Of course, with the effects of the Depression in full swing a few years later, those farmers received only $140,000 for their tomatoes in 1930.[19]

In addition to individual farmers selling directly to consumers or canning companies, there was renewed emphasis on forming cooperatives, where farmers worked together to sell larger quantities of their products to markets in Indiana and other regions of the United States. In 1932, DeKalb County farmers produced spinach and tomatoes for a canning factory in Ashley, but they also grew other vegetables. When they produced a surplus of carrots and cabbages that exceeded local demand, they organized into a marketing association to send their vegetables to buyers elsewhere who wanted both quality and quantity.[20]

Across Indiana, the most important crop farmers added to their farms during the Great Depression was tomatoes. As Carroll County agent Ralph Maggart succinctly put it in his 1937 annual report, "Tomatoes. A new cash crop!"[21] His counterpart in Henry County indicated that potatoes and tomatoes were "both popular cash crops in the county, the potatoes being consumed locally and the tomatoes canned for shipping."[22]

Blackford County farmers who wanted to sell tomatoes had their choice of canneries to contract with, including Clamme Canning Company and Rynerson and Son Canning Company in Hartford City and Arthur Canneries in Montpelier. They could also work with canneries in nearby counties, such as Eaton Canning Company in Delaware County and Marshall Canning Company and Upland Canning Company in Grant County.[23] It was similar for Henry County growers, who had seven canneries to sell their tomatoes to.[24] In 1937, Carroll County growers had canneries from Frankfort, Terre Haute, and Chicago vying for their crop.[25] With intense competition among those needing tomatoes to run their plants, there were 64,000 acres under contract to canneries in the state in 1931. By 1935, this had increased to 97,300 acres.[26]

Extension staff knew that tomatoes were the perfect crop to help farmers make more money during the Depression.[27] In 1930, Dubois County growers were making $50–$175 per acre on their tomato crop, which agent Charles Nicholson reported was "by far more profitable production per acre than any crop which they might grow on that type of land."[28]

Alfred Hesler provided one specific example of the production cost and net return of one Fountain County farmer. This individual, who was from Troy Township, had 16 acres planted to tomatoes. Averaging 20 tons per acre, he sold the 320 tons of tomatoes for $1,364. He had $600 in the crop—including plants, $100; land rental, $85; cultivation, $65; trucking, $150; and picking, $200—so his net profit was $764. Hesler concluded that the profits from the tomato crop would pay the grower's taxes for almost three years.[29]

Early on, prices paid by canners were based solely on the tonnage, but during the 1930s, the processors transitioned to offering prices that also took into account the quality of the crop as indicated by U.S. Department of Agriculture (USDA) grading standards.[30] Often, the agents had to explain to the growers how to determine grades when, as one agent put it, "they had not been sold on this basis before."[31]

There was a rather substantial difference in the prices paid for the various grades. In 1931, growers were paid $16 per ton for USDA No. 1 tomatoes, while the price for USDA No. 2 tomatoes was only $10 per ton.[32] In 1938, Carroll County farmers received from the Campbell Soup Company $18 per ton for No. 1 grade tomatoes and $9 per ton for No. 2 grade tomatoes. The agronomist for the company indicated that the expected yields in 1938 were six tons of No. 1 tomatoes per acre, which led agent Ralph Maggart to observe, "This gives a cash income of about double corn."[33]

Canneries began paying premium prices when growers hit targeted yields and filled their hampers with No. 1 grade tomatoes. Blackford County agent Walter Rusk explained the pricing change in his 1935 report, saying that there was "a new premium plan the [Clamme Canning] Company is offering for their growers who produce quality tomatoes, which is 20 cents per ton for six ton of number 1 tomatoes, 30 cents per ton for eight ton of number 1 quality, and 50 cents per ton for 10 ton or more of the number 1 tomatoes. A grower must produce three acres or more to be eligible for the premiums."[34]

At times, specific processors recognized growers by awarding them with medals of distinction for their achievements. In Henry County, 120 growers attended an achievement day hosted by the Mt. Summit Products Company. There were eight growers who qualified for gold medals by delivering a crop consisting of 75 percent USDA No. 1 tomatoes with 3 percent or fewer culls. Twenty-two growers earned silver medals for producing a crop of 65 percent USDA No. 1 tomatoes with fewer than 4 percent culls.[35]

Roscoe Fraser was a well-known Extension specialist from Purdue whom Indiana tomato growers respected and trusted for his advice. If there was any sort of gathering of tomato growers, the agents knew that he would gladly attend if at all possible. He walked fields, worked with individual growers, and conducted his popular tomato schools statewide.[36] According to one agent, Fraser basically "stressed the importance of a well prepared seed bed on special selected fields, [and] using good, thrifty, healthy plants, along with plenty of fertilizer. He also gave picking hints and urged picking of all No. 1's whenever possible."[37] Adams County agent Lawrence Archbold reported in 1938 that Fraser "was given the floor and in his inimitable manner, with the help of slides and motion pictures, held the audience for one and one-half hours."[38]

In March 1935, Fraser conducted a spring tomato school in Blackford County at the old high-school gymnasium. Agent Walter Rusk recalled one of the funnier moments from the event in his annual report, writing:

> A discussion on the various plant foods necessary for successful plant growth was held and plans for combatting both disease and worm enemies of the tomato plants were discussed. After his talk Mr. Fraser invited his listeners to ask questions. This they did and found new ways of combatting the enemies of the tomato plants. A humorous discussion of the proper way to eradicate the tomato worm was heard with several of the visitors giving such means as "siccing" the dog on them, stabbing them with pins, cutting them with scissors, teaching pet crows to eat them and many other quaint ideas.[39]

Rusk would later go on to produce filmstrips to show at a tomato school in 1940, using local tomato growers as actors to explain topics such as seeding methods, seedbed preparation, fertilizers, diseases, harvesting, and grading.[40]

In 1936, Fraser would return to Blackford County to conduct what would become known as picking schools. Three events drew 240 attendees, who learned about differentiating grades while handpicking tomatoes and what constituted a cull. The main theme for the schools that year was "Pick Red Ripe Tomatoes and Secure an Extra Profit."[41]

In the 1930s, nearly all potatoes consumed in Indiana were imported, given that seventy out of the ninety-two counties did not produce enough to fulfill local demand.[42] For instance, in southwestern Indiana, Gibson County and seven counties surrounding it formed a "potato pocket," as one agent termed it, where 500,000 bushels of potatoes were imported each year.[43]

With such a large percentage of potatoes produced elsewhere, the state's growers had an opportunity to raise this staple crop for a guaranteed market.[44] In 1934, Grade 1 potatoes sold for 57 cents per bushel, while those of lower quality traded for less.[45] Since farmers invested just 18 cents per bushel in the crop that year, it meant that potatoes were making them money even during the Depression.[46]

Extension put forth a lot of effort to introduce the latest research on what it took to raise potatoes profitably. From the Purdue perspective, two of the key players were Extension horticulturist Roscoe Fraser and Willard Ward, an Extension assistant in horticulture. Both made presentations to farmers at programs around the state to discuss how to prepare the seed bed, apply seed treatments and fertilizers, and spray for insects.[47]

Extension advised farmers who were growing potatoes for the market to plant certified potatoes, which were verified to be disease-free prior to planting, in place of seed pieces derived from their homegrown potatoes. In 1933, Adams County agent Lawrence Archbold observed, "One of our farmers, Otto D. Bieberich, put out two test plots this year. In one plot he used home grown potato seed, and in the other, Northern grown certified seed. The first plot yielded at a rate of fifty bushels per acre, while the latter yielded at the rate of three hundred bushels per acre."[48]

While Allen County produced 450,000 bushels of potatoes in 1930, this still amounted to a deficit of 33–50 percent of what was consumed in the county.[49] Across Indiana, growers only produced 37 percent of the 11.3 million bushels being consumed, which showed there was a potential to make money by growing potatoes.[50]

In order to increase the supply to address the shortage and help growers maximize the income produced from potatoes, Allen County agent Hugh Heckard arranged several demonstrations using certified potatoes from Michigan in 1933. He provided this description of the effort:

> A part of the extension program this year was to demonstrate the profit in certified seed. The certified seed growers of Iron County, Michigan sent five bushels of seed to this county for use in the demonstration. This seed was given to five farmers in various parts of the county. Each farmer planted the seed in the same plot with seed of his own selection which had been produced locally. On the potatoes within the same plot there was given identical treatment, as to soil management, fertilizing, spraying and cultivation.[51]

Heckard went on to explain that the result was that yields from the certified potatoes were dramatically higher, producing 101 bushels per acre as compared to the 24 bushels per acre for the common potatoes.

Extension had found value in using competitions such as the five-acre corn contest (see Chapter 16) to educate growers about the best practices for increasing yields, so in 1930, it instituted a similar contest for potatoes called the 400-Bushel Potato Club. The goal of this competition was similar to others in that it "demonstrate[d] both the possibility of the potato crop and practices that make it successful," as one agent put it.[52]

In 1930, contest participant William Lampe in Allen County produced a high yield of 410 bushels per acre, earning a gold medal from the Indiana Vegetable Growers' Association.[53] In 1934, the top statewide honors for club members went to Elmer Mitchell from Rochester. The Fulton County agent reported that Mitchell "raised a gross yield of 700.2 bushels per acre. This is the highest yield of potatoes ever grown in Indiana and the second highest yield ever grown in the United States under non-irrigated conditions."[54] With the average potato yield for the state coming in around 110–125 bushels per acre at the time, these contest results showed that most farmers were not raising anywhere close to what could be produced from certified seedstock if properly managed.[55]

Mired in the Great Depression, farmers were desperate to find anything or any way to reduce expenses and bring in cash to keep the farm operating during the worst of times. Many corn and wheat farmers became novice growers of tomatoes, potatoes, and other specialty crops as a way to diversify their operations and generate new revenue.

Farmers who chose to raise additional crops during these turbulent times could receive guaranteed prices up front from one or more nearby processors for crops such as tomatoes and sweet corn. This helped reduce their risk at a time when they could ill afford to take a loss. But farmers who took the plunge also faced a sharp learning curve on what it took to successfully grow these horticultural crops, making advice from the county agent, campus Extension specialist, and cannery agronomist more essential.

While the Depression was a nightmare for farmers, the lessons learned about raising additional crops during that time would be essential as the 1940s would soon bring on new agricultural challenges related to World War II.

CHAPTER 20

Not More Cows but Fewer and Better Dairy Cows Is the Imperative Need

In 1930, approximately 10 percent of all Indiana farms were classified as dairy farms. However, many farms without that classification still kept dairy cattle. For example, 2,500 out of 2,650 Hamilton County farms had dairy cattle, but only about 20 percent were classified as dairy farms based on herd size.[1] With approximately 700,000–800,000 head of dairy cattle in the state between 1930 and 1935, dairying remained a mainstay for many Indiana farmers during the Great Depression.[2] When prices for whole milk or butterfat went up or down in value, the farmers' financial statements followed suit.

In some counties, dairying was the largest sector of the agricultural industry.[3] This included Elkhart County, where dairy farmers milked 18,000 Ayrshire, Brown Swiss, Guernsey, Holstein, and Jersey cows in 1936. The agent there noted that, collectively, producers supplied "milk and cream for two cooperative creameries, two private creameries, two condenseries, [and] many milk distributors in the county," which in turn had a distribution system that reached South Bend, Chicago, and even Cleveland.[4]

In the dairy counties, a large portion of farm income was derived from the production and sale of dairy products.[5] By 1932, the price dairies received for milk and cream products was so low that it discouraged many producers from continuing in the business.[6] In 1931, butterfat had sold for 59 cents per pound, but a year later, it had dropped to 40 cents per pound.[7] Whole milk, which had sold for 10.1 cents per quart in 1931, dropped to 8.8 cents per quart a year later.

Harrison County agent Ronald Johnson observed in his annual report for 1932, "In spite of the depression, the dairy breeders of the county have maintained their morale remarkably well."[8] But that same year in Floyd County, agent Charles East wrote, "Dissatisfaction with market methods, prices and requirements has not tended to sweeten the dairyman's temperament."[9] But he also noted that feeding good cows and producing clean milk would prove profitable.[10] Even during tough times, some of the more efficient dairies were maintaining their profits as high as possible in spite of the Depression-era prices, while others were forced to sell off their herds and close their milking barns.

By 1935, prices for dairy stock and products had improved slightly, prompting a renewed interest in dairying.[11] Those dairies that had held on through the economic downturn saw better prices by 1937. Butterfat once again brought 59 cents per pound, while whole milk returned 11 cents per quart.[12]

Tuberculosis, also known as TB, was a major disease among humans during the early twentieth century, and there were legitimate concerns that TB bacteria were being passed from dairy cows to milk and dairy products consumed by children and adults. As a greater number of TB-cases were reported, the American public demanded action from local, state, and federal governments.

From 1918 to 1940, the U.S. Department of Agriculture (USDA) worked with states to coordinate a program to detect TB in dairy cattle. A veterinarian would administer a shot at the base of the animal's tail, then come back two days later to see if the area had reacted. Animals with a positive reaction were destroyed, and the farmer received a government subsidy for the loss. In the early years, the program started off as voluntary, with individual dairies receiving TB-free certifications, but it soon evolved into one requiring all dairies in a county to participate if they expected to sell their animals, milk, or butterfat.

A county and its dairies were certified as being TB-free when less than 0.5 percent of the cattle tested positive for the bacterium.[13] In 1930, Dearborn County achieved the distinction of being the first Indiana county to be labeled as a modified TB-free area.[14]

In order for a county to maintain its certification, its dairies had to be retested every three years. The process involved testing all previously TB-positive herds from three years prior as well as enough additional animals to represent 20 percent of all dairy cattle in the county.[15] Passing this round of testing meant the county would be certified as TB-free for another three years.[16] Conversely, not completing the testing or completing it and identifying cattle infected with bacteria above the established threshold meant losing the hard-earned federal accreditation, which in turn limited the movement of dairy products and cattle from that county.[17]

The dairies relied upon county governments to foot the bill for retesting herds in order to maintain accreditation. It typically cost a county $500–$3,500 to have all of its dairy herds tested.[18] Agent Everett Parker reported in 1930 that local officials had furnished $1,500 to pay for the testing in Hamilton County.[19]

Fayette County had passed its two previous testing rounds in 1927 and 1930, but the accreditation was scheduled to end in May 1933 unless its herds were retested. Agent Marion Detrick provided the following update in his 1933 annual report:

> It was learned that the county would be penalized in the markets if the status was lost. At the time the subject of retest was under consideration, some cattle shipped from the county for slaughter were found tuberculous. The herds

> from which the cattle were shipped and another herd into which a cow from the herd had been sold were tested by a representative of the U.S. Bureau of Animal Industry, and a total of fifteen tuberculous cattle were found. These facts were presented to the local authorities; and arrangements were made for a retest. . . . The county's status has been reestablished for another three years.[20]

It was clear that the potential harm from not seeking accreditation—both economically and from a health standpoint—far outweighed the few thousand dollars that retesting county herds would cost.

When Fountain County started its retesting process in 1930 in cooperation with the Indiana state veterinarian, the U.S. Bureau of Animal Industry, and local veterinarians, agent Alfred Hesler sent a letter to every dairy operation explaining the process and expectations.

> 1. The T. B. work is under State and Federal supervision.
> 2. You will be notified [of] the date of your test.
> 3. You will be expected to be present and assist.
> 4. You are advised to build a chute.
> 5. Everybody tests; the test is free.
> 6. Dr. C. T. Howard is in charge of the County test and all questions will be referred to him.[21]

Hesler later noted that the participating veterinarians traveled 13,000 miles over the course of eleven weeks.[22] During this period, a total of 13,415 head of cattle in 1,753 herds were given a shot at the base of the tail, with a veterinarian later checking the injection site for a reaction. Individual animals that tested positive for TB were destroyed, then the affected herd was effectively quarantined and required to pass three additional tests before being certified as TB-free.[23]

Only sixteen animals reacted positively for the bacterium that year, so the accreditation for Fountain County was extended until 1933. The whole process cost about 25 cents per animal.[24]

The impact of the government TB testing program for cattle was significant for reducing these bacteria in dairy products. When the program started in 1918, 10 percent of dairy cattle tested positive for TB. When the government program ended in 1940, every county in the U.S. had detections of less than 0.5 percent. Over the duration of the program, it would take more than 5,000 veterinarians to test 25 million cattle for bovine tuberculosis, and nearly 4 million animals would be slaughtered.[25]

While dealing with tuberculosis was a serious concern for Indiana's dairy industry in the 1930s, some agents believed that Bang's disease—which is known more commonly today as brucellosis—was more destructive.[26] Caused by *Brucella* bacteria, the disease could be transmitted from infected to healthy animals and cause a cow to abort a fetus, produce weak calves, or become infertile. Infected animals also had decreased milk production and weight loss. The bacteria could also infect people through the consumption of contaminated products from infected animals, including unpasteurized milk. The disease in humans—often known as undulant fever until the 1940s but commonly called brucellosis today—produced symptoms such as fatigue, headaches, fever, and joint pain.

In 1930, agent William Douglass reported on local efforts to address the disease in Bartholomew County, writing:

> Abortion control work in the past has consisted in encouraging farmers to blood test their cows for abortion and to weed out reactors. Very little cooperation has been secured from the veterinary profession. Early in the past year we called a meeting of the veterinarians practicing in the county which was addressed by Dr. Frank Brown, State Veterinarian of Indiana, on the subject of abortion control. Dr. Brown advocated the methods which have been previously advocated from this office in this county, with the result that the local veterinarians have given considerably more cooperation to the program. One circular letter outlining the abortion control methods was sent to the entire county. The constant agitation of control work has resulted in most farmers becoming familiar with the nature of the disease and methods of control. A large amount of vaccination is still practiced and advocated by veterinarians but most farmers who seek to control the disease are doing so by [a] weeding out process.[27]

Early on, the state would test herds for brucellosis on a voluntary basis.[28] But by 1931, only thirty herds tested free of the disease for the entire state.[29] Behind the scenes, dairy producers were worried about what would happen to their herds if the disease were detected.

In 1932, agent Lawrence Archbold provided this assessment of the situation in Adams County and reported that farmers were starting to take more notice as their losses had increased:

> One of the outstanding pieces of work undertaken in the [dairy herd improvement] association this year was the survey conducted to find Brood Cows. A tabulation of the production of all dams and their daughters in the association was made with the hope of finding some dams whose daughters would show a consistent increase in production over their dams. These cows would then be considered as brood cows and would be looked to as the mothers of desirable herd sires. The findings disclose the fact that true brood cows are very scarce, but the tabulation did show in a very graphic manner the very heavy losses that Bang's Disease was causing in the herds.
>
> When the dairymen realized not only the financial losses they were encountering due to Bang's Disease but also its threat of destruction to their breeding program, they came to realize that some very definite steps must be taken to

> control or eradicate this disease. In cooperation with Dr. C[harles] C. Dobson of the State Veterinarian's office, seventeen herds were signed up on the abortion control plan under the Indiana Livestock Sanitary Board. . . . The testing will begin as soon as there is a slight upward adjustment in the economic conditions.[30]

The push to eliminate brucellosis ramped up in 1934, when a cooperative campaign was initiated between federal and state agencies.[31] Agents helped publicize the effort by discussing testing at farm bureau meetings and by writing newspaper articles describing the process.[32] At first, testing was done at no charge to the herd owner, with the USDA Bureau of Animal Industry picking up the veterinarian's salary, supplies, and transportation as well as the laboratory costs.[33]

Agent Everett Johnson provided this detailed account of the Cass County program:

> In July 1935 the first contracts were signed with the federal government for the eradication of this disease. When the owner signed the contract, he agreed to have the cattle on his farm tested at the expense of the government and agreed to sell for slaughter any positive to the agglutination test for Bang's disease. The cattle branded as reactors were to be appraised by the federal government representative, and if the animal sent to market did not bring the appraised value, they would pay indemnity to the herd owner up to twenty-five dollars if a grade animal (or fifty dollars if a registered animal).
>
> At ninety-day intervals the veterinarian returned to the herd and drew another blood sample for test. When the herd received four tests without a reactor or suspect, the herd was considered clean and a clean herd certificate was issued the herd.
>
> This certification is good for one year, at which time it is necessary for the owner to hire a licensed veterinarian to draw blood samples from his herd for a retest. If the cattle all pass the test, the owner is issued another clean herd certificate. If the cattle do not pass the test, the herd is again put under federal supervision and it must again pass four clean tests.[34]

At first, many dairy owners were more or less skeptical of the government's efforts to eradicate the disease, but later, producers felt that their dairy herds doubled in value after obtaining a brucellosis-free certificate.[35]

Typical infection rates for the disease in cattle ranged from a low of 5 percent to a high of 20 percent.[36] While beef cattle were included in those statistics, most of the infection numbers were tied back to dairy cattle.[37]

Initially, tests were conducted voluntarily on a farm-by-farm basis, but in 1938, there were efforts to make the test mandatory within a community or county. That year, Johnson offered this update for Cass County:

> The latest plan in progress is being worked for the first time in Indiana here in Cass County. Petitions are being circulated by farmers with which they hope to get the signatures of the owners of seventy percent of the cattle stating that they favor having all the cattle tested in the community. If the men are successful in obtaining a favorable reaction to this plan, a crew of veterinarians will be sent to the county for the testing, which would be a big step towards making our county or some communities a Bang's free area.[38]

However, soon after—on July 1, 1939—the testing program ended when the Indiana General Assembly failed to provide the necessary funds to match the federal dollars that covered the indemnification to growers for infected animals.[39] In his 1939 annual report for Cass County, Johnson stated, "Since the indemnity is not being paid, local cattlemen have not been keeping up their accreditation on their herds to the place they were before."[40] And in the Daviess County report that year, Albert Bishea observed, "Private tests are still being made but cannot take the place of a definitely organized program."[41]

While a lot of work was done in the 1930s to detect and eradicate brucellosis in dairy herds, the United States didn't make a concerted effort to eradicate it until 1954. Today, all fifty states are classified as brucellosis-free.

By the early 1930s, hundreds of Indiana dairy herds were enrolled in county-based dairy herd improvement associations (DHIAs) through Extension. These groups, which had first been formed in Indiana as cow testing associations in 1913, focused on measuring the economic return of each animal by recording the pounds of milk and percent butterfat produced along with the amount of feed consumed.[42]

An association was established when a group of dairy producers—typically twenty-six, though enrollment numbers sometimes varied—agreed to pay the salary and expenses, including mileage, of a tester hired by Purdue. A typical tester was paid a salary of $2.00–$2.75 per day.[43] The membership agreement lasted for one year, though some dairies renewed theirs annually.

Testers collected important data for each animal—inputs such as cattle feed and outputs such as butterfat and milk production—to determine the overall efficiency of an operation. They also shared new ideas from dairy Extension specialists at Purdue as well as management practices learned from other dairies.[44] In short, they became personal consultants to the dairies they served as well as honest appraisers of the practices used by each producer.[45]

Edward A. Gannon, a well-known dairy Extension specialist from Purdue, put the dairy herd improvement association work in perspective, saying:

> When rations are balanced and fed to cows on the basis of production, much more profit will be made and much grain will be saved because

> the low-producing cow is usually overfed while the good profitable cow is usually underfed. Cows should be fed a balanced ration, based on each cow's individual production. The dairyman should know his cows by keeping daily or weekly milk records of production. This will provide a sufficient guide in feeding which, if properly followed, will result in satisfactory returns.[46]

Bartholomew County agent William Douglass showed an example of those satisfactory returns in his 1931 annual report, writing that:

> 110 cows, which averaged 361 pounds butterfat, produced an average of $115.13 above feed cost per cow. While the 116 cows, which average 247.5 lbs. butterfat per cow, averaged only $51.15 above feed cost per cow.
>
> It will be noted that the high producing cows consumed more feed per cow but that an increase of 7.4% of feed cost, resulted in an increase of 55% value of product; and 27% return per $1 invested in feed.[47]

By analyzing 1931 records from the Hendricks County DHIA, agent Ellis Hawk was able to describe what it would take for a dairy to make an annual income of $2,000 over the cost of feed:

> The low cow produced 225 pounds of butter fat, had an average feed cost of $63 and returned an income over feed cost of $50. The average cow in the association produced 334 pounds of butterfat with a feed cost of $70 and an income over feed cost of $87. The average of the high cow selected from each herd produced 461 pounds of butterfat with a feed cost of $76 per cow and an income over feed cost of $132. Using these cows as an index, it would be necessary to milk 15 of the high producing cows, 23 of the average cows, or 40 of the low producing cows to make an income of $2,000 over feed cost.[48]

The numbers cited by Hawk clearly showed that it was important for producers to remain efficient and manage their herds wisely. The emphasis on efficient production was paramount during the Depression because demand for dairy products had fallen. Delaware County agent Myron Cromer summarized the situation in 1933, writing, "The demand for [dairy products] is not great due to under consumption and restricted buying ability."[49]

The reduced demand and continued high production levels led to a surplus of milk and butterfat production that, in turn, fueled low prices. This led Cromer to also write, "[N]ot more cows but fewer and better cows is the imperative need. . . . Today as never before the importance of economic production cannot be over emphasized. Low producers not only are expensive to keep—both from the standpoint of feed cost and labor—but they also contribute their portion to the large quantity [market surplus] of dairy products."[50] In Henry County, three dozen cows were sold to a butcher in 1930 when DHIA records showed them to be unprofitable.[51]

As participants used the collected data to improve their own operations, data from across the county allowed agents to provide useful information about local averages. In Decatur County in 1931, the records of DHIA members showed an average production per cow of 351 pounds of butterfat, which led to a profit of $2.79 for every $1.00 spent for feed, which was a typical return for dairies at the time.[52] Based on the same records, it cost an average of 17 cents to produce one pound of butterfat and 78 cents to produce 100 pounds of milk for those dairies that year.[53]

The collected totals also allowed members to judge their own herds against the low-, average-, or high-production herds in an association. For example, in 1938, the Clark County agent reported DHIA results, writing, "The high herd in butter fat production produced an average of 394 pounds of [butter]fat per cow and the low herd, 156 pounds, while the highest average profit per cow was $111.43 and the lowest, $28.68."[54] Knowing information like this might reassure producers whose numbers matched the higher totals while encouraging those who matched the lower numbers to make improvements.

From year to year, the results achieved by DHIA herds—namely, more pounds of milk and butterfat—remained rather consistent across the state. For example, Adams County agent Lawrence Archbold reported in 1933 that when the association was started in 1925, the average butterfat produced per cow at participating dairies was 248 pounds, while just a few years later in 1932, the average had risen to 365 pounds.[55] With these types of increases commonplace for member dairies, the value added by participating in an association was clear. Indeed, just a few years later, Archbold would report, "While you have shown that there are herds in the county that belong to the testing association that are producing an average of 400 to 500 pounds of butterfat annually, the county average per cow is only 190 pounds of butterfat."[56] And Carroll County agent Ralph Maggart reported in 1933 that dairy farmer Ralph Rinehart had said that "the money spent in this work has paid him [Rinehart] a larger return than any other investment on his farm and he often says, 'I wish I knew as much about my other businesses as I know about dairying.'"[57]

But even though the documentation showed that following the production practices recommended by Extension led to higher profits, some producers still faced hard times. Clay County agent Harry L. Royce said the low prices of 1931 "made many of our people wonder whether they should stay in [the dairy business] or reduce or get clear out. With this state of mind existing, it was impossible for us to secure sufficient interest to continue our dairy herd improvement association."[58] Others who participated in DHIAs, such as dairy producers in Hendricks County, reduced their costs by testing their milk production every other month instead of monthly.[59] However, even throughout the dark years of the Great Depression, the value provided through participation in DHIAs meant that most would continue to operate.

In 1917, the Indiana State Dairy Association had started using testing records to give awards through its newly formed 300-Pound Cow Club. To gain entry to the club, producers had to have ten cows that each produced 300 pounds of butterfat, which earned a bronze medal.[60] When eight cows out of the ten each recorded 400 pounds of butterfat, the dairy received a silver medal. A dairy received a gold medal when six cows each produced 500 pounds of butterfat and the remaining four each produced 300 pounds or more. These goals had to be accomplished during a 365-day period.

Records that were used to confirm the butterfat production of individual cows also confirmed their genetic link to a specific bull, and in 1929, the association also began recognizing bulls that produced high-yielding offspring with Proven Dairy Sire awards.[61] Bulls were given gold, silver, or bronze medals based on criteria related to their daughters' production records.[62] In January 1929, medals were awarded to twenty-one proven sires at the association's annual convention held at the Purdue campus in West Lafayette.[63] These awards not only added prestige to the dairy that received them but also added much value to the bull and his offspring.[64]

In 1930, agent Lawrence Archbold reported:

> The first proven sire for Adams County was checked by Elvin T. Wallace, Extension Dairyman [from Purdue], in November this year. The animal is a Holstein named Cloverleaf Duke Ormsby owned by D[an] D. Schwartz, Monroe Township, Berne, Indiana. This bull will be awarded a bronze medal on the performance of eight of his daughters who produced over 300 pounds of butterfat each [during a period of 365 consecutive days]. Four of these daughters produced over 400 pounds of butterfat and one over 500 pounds of butterfat. Their average age was 2.6 years.[65]

In 1932, Carroll County agent Ralph Maggart offered his views on the importance of having a bull with the genetic potential for producing offspring with high milk and butterfat production, writing:

> The fact that many daughters of pure bred sires are now on test will give to these Guernsey men the real value of the herd sire. It also points out the ability of certain cows to transmit the production to their offspring. . . .
>
> The records of the bull's daughters are the only true measure of his real value as a transmitting sire. The use of such bulls means dairy herd improvement at a more rapid rate. Dairymen are being encouraged to use sons of these Meritorious Sires out of good cows.[66]

It soon became clear that one of the top counties in the United States for high-producing genetics was Indiana's own Adams County. Archbold's 1935 description of the Adams County Dairy Herd Improvement Association provides some insight into its success in producing high-quality dairy animals. He wrote:

> The first reason that a man usually gives for joining a testing association is that of culling out his low producers; next is to learn how to feed a balanced ration according to the body weight and production of the remaining animals. The Adams County association has been going long enough that these are now secondary reasons for the existence of the work. The work has progressed to the point now that foundation brood cows can be found by studying daughter and dam comparisons. At the same time, by using these figures, it is possible to learn that bulls are doing an effective job of transmitting increased production to their off-spring. This is evidenced by the fact that each year one or more meritorious bulls are found in Adams County herds.[67]

In 1936, R. R. Graves and M. H. Fohrman of the USDA Bureau of Dairy Industry wrote that of the "twenty-six Indiana herds mentioned in the superior germ plasm survey reported in the 1936 USDA *Yearbook of Agriculture*, five of the herds are from the Adams County Dairy Herd Improvement Association, namely; D. F. Mazelin, Sol[omon] Mosser, D[an] D. Schwartz, Jacob J. Schwartz, and Peter D. Schwartz."[68]

Better dairy producers sought to improve their herds by finding foundation brood cows and sires with proven production records, either from among their own animals or by purchasing them based on production records.[69] Purebred bulls were especially sought after, and producers who wanted one with a proven track record were going to pay top dollar for them.[70] One bull could influence the productivity of an entire herd, which justified the investment.[71] In Allen County, eight sires of known production were added to the county's herds in 1930.[72]

Purchases of these expensive bulls dropped during the first half of the 1930s, but they began to rebound starting in 1934. That year in the Carroll County annual report, Maggart shared this anecdote about a local farmer:

> With a butterfat production of 377 pounds for twenty-four cows, a problem that is confronting [dairy producer Ralph Rinehart] is his sires that will hold his production level. This fall, in looking for a sire, he drove more than five hundred miles and spent at least a week's time. For eight years he has been driving a Model Ford car. He made this statement: "I could buy a new car and finance it and my neighbors would not think anything about it, but if I go out and buy a $200 or $300 bull and would have to borrow the money, my neighbors would laugh all day at me."[73]

It would become more common with each passing year for dairy producers to rely on production records in order to identify animals with the best potential for increasing production. And in the process, they also realized that a single bull of proven pedigree backed by production records from its offspring could do more to increase a herd's performance than almost anything else they could do.

Prior to 1931, the state of Indiana did not have a well-coordinated milk regulatory program. From 1906 to 1921, infrequent dairy inspections were conducted by food inspectors from the Indiana State Board of Health. In 1921, the agency hired a dairy bacteriologist to use a mobile laboratory to test the quality of the milk when requested by cities. That year, an agency report showed that "most Indiana cities have a very inferior milk supply."[74] As a result of the testing, cities such as Greenfield, Liberty, Marion, Muncie, Warsaw, and Winona Lake passed local ordinances requiring milk to be pasteurized or to meet certain local standards.[75] Despite the fact that testing was conducted by the state agency, milk quality continued to be regulated primarily by local ordinances.

It wasn't until 1927 that Indiana hired its first full-time dairy inspector, with two more inspectors added by 1929. In 1931, the Indiana General Assembly allocated funds to the Indiana State Board of Health to add three more milk inspectors and one dairy bacteriologist to its staff, in part because it had little choice.[76] The previous year, surrounding states had begun rejecting Indiana milk and products such as sweet cream because they did not meet the more stringent regulations established by those states. Out-of-state milk inspectors were making routine visits to Indiana dairies, where they were appalled by the unsanitary conditions for collecting, storing, and processing milk. High bacteriological counts in the milk meant that it and other dairy products were rejected by these inspectors.[77]

The loss of milk markets by 1930—both within Indiana and across state lines—had raised concerns within the dairy industry. John Taylor, director of the Bureau of Dairy Products at the Indiana State Board of Health, reported:

> This sudden interest [in increased regulation at Indiana dairies] was brought about largely by the closing to Indiana of many out-of-state markets and also to lowered consumption, to a great surplus of milk and milk products, and to lowered prices. The closing of out-of-state markets has meant the loss of hundreds of thousands of dollars to Indiana milk producers. Realizing the seriousness of the situation, several progressive dairy products dealers requested the cooperation of the entire industry and of this Bureau in a program of improving farm production conditions and milk quality.[78]

With the political support of the Indiana dairy industry, the Bureau of Dairy Products would pass a series of sweeping regulations in April 1931 that dealt with every aspect of dairy production.[79] The rules addressed a bacteriological standard for milk; a wide-ranging set of building requirements for dairy barns and milk houses; how milk was handled, bottled, and capped; the need for concrete or metal tanks for cooling and storing milk; and guidelines for drivers who picked up milk from dairies to reject it if it was warm or smelled bad. While times were tough during the Depression, Frank C. Wilson, director of dairy products inspection for the state, observed, "Compliance with the regulations will result in increased quality of the product offered to the consumer. With this increase in quality will doubtless come an increase in consumption which leads to greater profits."[80]

The dairy regulations had an almost instant impact on the industry, with more than 7,000 milk houses constructed from July 1931 to June 1932.[81] In his 1931 report, Clark County agent Lester E. Moore noted, "Practically all milk shippers have remodeled their barns and built milk houses."[82] That same year, agent William Smith stated, "The building of milk houses and remodeling of dairy barns to conform with the sanitary regulations of the State Board of Health has been one of the chief concerns of Henry County Dairymen. The demand for plans and information concerning these things has been heavy."[83] To support the construction of the new milk houses around the state, Purdue agricultural engineers and dairy Extension specialists had produced building plans and published an Extension leaflet titled *Farm Milk House* in 1931, which agents then shared with their constituents.[84]

The agent became the go-to person for answering questions from producers and distributors about the regulations. In 1935, Arvil Keesling in Bartholomew County indicated that his office "received practically all of the questions regarding milk license and have passed onto those involved the necessary information, so that they can comply. . . ."[85]

Grant County agent Mosher Butler described a problem that confronted local dairies in 1931, when bacteria counts often reached 100,000 or more and prevented the fresh milk from being sold into the Marion County market:

> The low quality of the milk was due entirely to sanitary measures or the lack of them and to the fact that the local board of health has no system adequate to help the individual dairyman, and at that time the local board of health further more was unpopular with these men. In addition the men look toward each other more as competitors than as friends, and two large distributors of pasteurized milk were doing all they could to get control of the market, thus depriving the consumer of fresh milk and injuring the producer of this commodity or finally putting him out of business. . . .
>
> [T]he writer [Butler] consulted the Purdue Dairy department and by correspondence consulted the department at Washington, D.C., at Cornell University and the Boards of health at Indianapolis, La Porte, Indiana, and New York City. A meeting was then held at which all the large producers of fresh milk were present. The grades recognized by La Porte were described and the system of establishing these grades explained. La Porte was selected because, from the standpoint of sanitation in fresh milk, they had the highest standards.
>
> At this meeting it was decided that the best way to meet these standards was to have a school which, if successful, would result in a higher grade of milk bacteriologically and a better feeling between the individual men and the

> local health body. Arranging such a school was a new procedure to the writer and the men too. First, though, we enrolled the men. Then we decided that the school should continue at least six weeks. During the school the writer agreed to call meetings as often as seemed advisable and to make frequent unannounced inspections at the farms, inspecting the complete *[illegible]* of equipment and methods. The members of course agreed to follow the suggestions of the writer after each inspection. To check with the results I found on the farms, we arranged with Purdue and the State Board of health to have plate counts made of milk samples. . . .
>
> Several of them started with high counts, in fact two of them being above 100,000. However after the first week's inspection and scoring, the counts dropped and continued to drop and the men's score on appearance, cleanliness, cooling methods and handling improved. . . .
>
> Before the six weeks were finished, all of the milk samples were running below 20,000 and several regularly below 10,000. These samples were taken at the farm at the time in the history of the milk the producer lost control of the bottle. To check with this, some samples were taken from door steps, in grocery stores and at the farms at different ages and at different cooling methods and temperatures. . . .
>
> During the meetings it developed that the local consumer was not well informed about the merits of good fresh milk and it also developed that local pasteurizers were exaggerating the merits of their product.[86]

Butler's remarks and description of his involvement in the process of getting the bacteria counts down showed what agents did to help their farmers comply with the regulations and, in the process, maintain a market for their products.

The 1930s began with a major challenge to the Indiana dairies, which was to produce clean and safe dairy products demanded not only by Indiana consumers but also by those in other states where milk and products derived from them were exported for sale. During the decade, dairy producers made major advances in how their herds were managed. Science was no longer a novelty, and instead, information was used to make the dairies more efficient and profitable. This meant using records on individual animals to identify the most productive cows and cull those that underperformed, introducing better feeding rations, improving the genetics of the herd through better sires, and working toward eliminating tuberculosis and brucellosis from their livestock and dairy products.

The shift from county-based regulations to state-based ones—supported to a large extent by the producers themselves—helped improve the dairy industry's output of quality products, which in turn enabled efficient producers to survive the challenges of the Great Depression. By the end of the decade, dairy production as a sideline was a thing of the past, which meant only the better-managed dairies remained.[87]

These immense changes, which revolutionized the dairy industry in the state, would better position Indiana dairy farmers to meet the challenge when the federal government called on them to produce more milk products for the soldiers fighting in World War II.

CHAPTER 21

Sheep Have a Place on Most Every Farm

In the early 1930s, sheep were more or less a passing interest for most Indiana farmers. County Extension agents reported that only a small percentage of farms had a flock. One agent noted that the farmers who have sheep kept them "on farms of a scrubby and inferior quality."[1]

In 1928, a Clay County survey indicated that 4 percent of the 2,500 farms had sheep. When agent Harry L. Royce saw these results, he questioned local farmers as to why they didn't raise sheep and received a variety of answers. Some of the problems cited included dogs killing sheep, the need to treat for external and intestinal parasites, and lower market values for lambs that weren't docked or castrated.[2]

Compared to grain prices, which had dropped precipitously at the start of the Great Depression, sheep prices had remained generally stable, so agents understood that some of the issues cited by Royce were worth addressing through education.[3] In 1930, agent Ralph Maggart wrote, "Even though farm prices of sheep and wool have declined, good ewes have returned the farmers of Carroll County a profit. Perhaps there is nothing on the farm yet that will return the profit for money invested as sheep."[4]

Generally, Indiana farms at that time had the capacity to handle a flock of 25 to 50 ewes.[5] In reality, a more common flock size was 15, meaning most farmers had too few sheep to generate an economic benefit.[6] As one agent put it, farmers "consider sheep as scavengers, keep them for the woods they will clean up, and consequently fail to reap desirable returns from the dollars invested in sheep."[7]

But the view of sheep as scavengers would slowly start to fade as the 1930s progressed, the agricultural economy showed signs of recovering, and new markets started to emerge, including in Indianapolis. By 1935, agent Walter Rusk supported the assertion that sheep and wool were profitable endeavors, writing, "Blackford County farmers find the production of sheep as profitable a livestock enterprise as any one engaged in. The county has a well established reputation with Eastern wool buyers with a production of quality wool."[8] He also noted that his county had more Corriedale, Hampshire, Merino, and Shropshire sheep per square mile than other Indiana counties, which he largely attributed to bluegrass pastures.

In 1935, DeKalb County agent T. E. Myers reported, "Although sheep are used on a majority of the farms as a side line, nevertheless practically a third of the income on the farm is derived from the sale of sheep or wool; therefore it is evident that the majority of the farmers should be interested in better management of sheep."[9] To encourage better management, Extension offered a variety of educational opportunities for Indiana sheep farmers during the 1930s.

Agents were fortunate to have sheep Extension specialist Claude Harper as a resource at Purdue. Not only was he a sought-after speaker for the events they hosted, but he taught them about improved methods for feeding, housing, and caring for sheep, which allowed the agents to better help local growers.[10] Harper also wrote educational letters on specific management topics, which agents could then send to farmers who raised sheep.[11] He wrote a 1933 series called "Make Indiana Lambs and Wool Excel" that included the following letters:

- "The Breeding Season Is the Beginning of a Sheepman's Year," on selecting the breeding ram and ewes and how to feed them during the breeding season;
- "During the Gestation Period (146 days)," on housing, feeding, and general management;
- "When the Lambing Season Arrives," on dry and warm housing, providing attention to the flock around the clock, making sure that lambs are nursing, and feeding the ewe after lambing;
- "During the Suckling Period," on docking and castrating young lambs, feeding lambs in protective pens, keeping milk fat on lambs, and saving the best ewe lambs as replacement animals in the flock;
- "Before and After Shearing," on producing quality wool, and packing and storing fleece; and
- "Sanitation and Management," on keeping feed troughs and water containers clean, treating sheep to eliminate disease-causing pests, and controlling stomach worms and tapeworms.[12]

One management strategy promoted by Extension was for farmers to improve their flocks by buying the best rams that they could afford. A high-quality ram could improve the genetics of a whole flock in the same way that a good bull could improve a dairy herd.[13] (See Chapter 20.) In some cases, agents were directly involved in helping farmers locate rams for purchase.[14] In 1933, Harper conducted a ram-selection demonstration in Franklin County, with the agent later reporting, "[Harper] pointed out the value of a short legged, deep bodied, wide ribbed ram as the sire of market lambs. He said that the short blocky ram will sire lambs that will fatten, while lambs from a rangy ram will not put on sufficient fat at an early age."[15]

In addition to demonstrations, Extension hosted other sheep management events, such as tours of successful sheep farms.[16] In 1936, a tour of Hancock County farms owned by Raymond W. Lantz and L. H. Collins featured Purdue sheep Extension specialist Henry Mayo, who spoke on the importance of a good ram, the use of high-quality hay for winter feeding and pastures for grazing, and raising 70- to 90-pound lambs for the preferred May and June markets.[17] In 1938, a tour organized by Extension in Allen County highlighted better management techniques, with agent Chester Kimmell providing this summary of the event: "At each stop, good sheep management was studied and in each case the farmers had used a purebred registered ram, had fed good legume hay and a little grain during the winter and were creep feeding their lambs. Those attending the tour asked many questions pertaining to selection of rams, winter feeding and management, dipping, creep feeding, castrating and docking, and summer pastures."[18]

Another popular Extension event was the sheep school, which was typically an all-day event that immersed participants in the latest agricultural research from Purdue.[19] In Allen County, the first sheep school was held in 1937 at the St. Joseph Consolidated School and drew sixty-seven attendees. The first session in the morning was a discussion on breeding the flock with desirable rams, a need that was deemed even more critical considering that the Extension sheep committee who helped plan the event "had a hard time finding a good class of breeding rams," in the county, according to Kimmell.[20] Other activities during the day included a presentation by two veterinarians on controlling parasites in sheep, an outlook on the 1937 sheep and wool markets by Paul Mitchell of Extension's Division of Farm Management at Purdue, a sheep-shearing demonstration by an employee of the Chicago Flexible Shaft Company, and a lecture by Harper on pastures and best management practices.[21]

Harper was also a presenter at a 1935 sheep school in Franklin County, where he spent a considerable amount of time showing farmers how to grade sheep according to the standards that the buyers used to determine prices. He stood inside a pen and pointed out the good and bad qualities of each animal as farmers watched from outside. As he talked, he marked the sheep with different colors to indicate the four grades of sheep: blue for choice, red for good, green for medium, and yellow for common. He based his grading process on the lamb's form, weight, and finish, or meat and fat at the time of sale. The use of different colors allowed the farmers to visually identify the grades and better understand the characteristics that separated one class from another.[22]

By providing this type of in-person instruction, Extension gave attendees the tools they needed to improve their management practices and the returns for their operations.

During the 1930s, agents knew that the average flock of sheep in Indiana was plagued with internal and external parasites that negated any improvements the farmers made.[23] The best genetics could not overcome the harm that these parasites caused to the health of sheep.

In 1930, Clay County agent Harry Royce observed a problem with internal parasites in local sheep flocks, saying, "The breeding ewes on most of our farms were at that time heavily infested with intestinal parasites which of course resulted in a rather unthrifty condition of the ewes, inferior suckling qualities and a resultant lack of thrift and profit on their lambs."[24] To help educate farmers about this type of problem, often a local veterinarian would do a postmortem examination to show that stomach worms had caused the problems with the unthrifty lambs.[25]

To control the stomach worms, sheep received an internal dose of copper sulfate, which was known as a drench.[26] Breeding flocks required at least two stomach drenches that had to be supplemented by rotating the pastures that sheep grazed from year to year.[27]

Sheep were treated for external parasites such as ticks, lice, and mites by immersing them in a trough filled with a solution of sulfur arsenic, which was a process known as dipping.[28] While the dipping process was simple enough, most farmers didn't have the equipment to complete the treatment.[29] As a result, many counties used funds provided by the local farm bureau or even the federal government to build portable dipping operations.[30]

In 1939, agent Raymond McHargue described the use of a portable dipping tank in Clay County, which he indicated was one of the first being used cooperatively in the United States at the time. He wrote:

> The portable sheep dipping tank started in 1934 and has been operated by Fred Roeschlein every year since it was built. The tank was originally built by the Clay County Farm Bureau which turned it over to the County Sheep Committee in 1938 after they received all of the original investment in the tank. This was one of the first cooperative sheep dipping outfits in the United States and has proved to be quite popular. The original blueprints made by Purdue University were taken from this tank, although they have been altered considerably at the present time.[31]

In the spring of 1936, Adams County had to establish a portable dipping program for sheep when the Indiana Livestock Sanitary Board informed agent Lawrence Archbold that scab, which was caused by a parasitic mite, was prevalent in the county. Archbold described the situation in his annual report that year:

> A total of 14 farm flocks were then or had been under quarantine. Sales of sheep, through community sale rings, had been stopped unless a veterinarian's certificate was provided, indicating the animals were free of scab. This representative [of the livestock board] further stated that he was hampered in

> treating the quarantine flock by reason of the lack of adequate dipping facilities. He also stated that due to lack of dipping, the sheep were badly infested with ticks. He asked that the Extension Service do something to help provide a portable dipping outfit and help educate our sheep farmers in the need and benefit of dipping.
>
> This office, knowing that the Resettlement Department [of the federal government] was empowered to make loans for projects of community value, immediately contacted Charles Poe, [of the] local Resettlement office, and suggested that a portable sheep dipping outfit would be a project of real community value. He agreed and induced a client of his office, Virgil Draper, to undertake the construction and operation of an outfit. When he [Draper] was ready to take to the road, he had invested $165 in a trailer-mounted dipping tank, with a self-contained heating unit. In the meantime the extension office informed the farmers of the extent of sheep scab and tick infestation in the county through news stories. Meetings were called at the farms where quarantine flocks were dipped under the supervision of the Indiana Livestock Sanitary Board and those present were shown the scabby sheep and they watched the dipping operations.
>
> From July 1, 1936, to September 2, 1936, Mr. Draper dipped 2,766 sheep at ten cents a head. This figure is practically double the number of animals involved because 95% of the flocks were treated twice at intervals of fourteen days. Mr. Draper is quite well pleased with his venture, as he figures his labor income to be 50% of his gross receipts. The flock owners interviewed know the treatment is good for the flocks and estimated the additional gains on the lambs and increased wool clip to be worth $1.50 per head. This would mean [an] approximately $2,000 increase [in] returns for the flocks in the four townships in which the flocks have been dipped so far.[32]

Farmers interested in having a portable tank at their farm were advised by letter that the sheep needed to be penned in the barn or in a lot on the scheduled day.[33] Farmers were normally charged 8–10 cents per head for the service.[34] In 1939, Cass County hired Mark Jones from Harrison Township to handle the dipping duties, compensating him 6 cents per sheep out of the farmer's 10-cent fee and using the balance to pay for supplies and other expenses.[35] In some cases, the rate charged varied based on the number of sheep dipped, with Clay County setting a rate in 1939 of 10 cents per head for flocks of 35 animals or fewer, 9 cents each for flocks of 35 to 50 sheep, and 8 cents each for flocks of more than 50 animals.[36]

These portable dipping units were very popular with farmers, especially when the economy started to improve. By the mid- to late 1930s, it was common for a county to dip 2,000 to 18,000 sheep each year.[37]

When farmers spent the money to control internal and external parasites in their sheep, the benefits were clear.[38] In Adams County, not only did farmers see at least $2,000 in increased returns following the 1936 dipping effort, but Archbold also reported continued improvements the following year, noting that "the fleeces were cleaner and sheared about one pound heavier; that lambs gained more rapidly and they were able to market them at an earlier age."[39] And Russell Zell told Franklin County farmers that spending money to have the sheep dipped paid dividends back to them by a 3-to-1 margin, explaining that "every animal dipped would produce one pound additional wool or its equivalent in mutton. Wool has averaged pretty close to 30 cents a pound this year. By this method of calculation, farmers would net $927 for an expenditure of $309. In other words, for every ten cents spent, thirty cents should be expected [in] income."[40]

Most lambs were sold in the fall and winter, which meant that farmers received reduced prices during these times since the supply was greater. Extension wanted the lambs to hit the markets in June instead. Getting the lambs to market in the summer would bring farmers higher prices due to a shortage of lambs.[41] One agent explained this strategy by writing, "Our sheep growers are being encouraged to breed their ewes for January, February and March lambing and we are encouraging them to remove the rams from the flocks about the first to the middle of November, and to sell or cull out all ewes which do not prove to be with lamb at that time."[42]

Following lambing season, the farmers had to address docking, or shortening, the tails of their lambs and castrating the male sheep. The markets returned higher prices for these animals, so farmers who didn't follow Extension's advice regarding these management practices could expect a lower profit for their animals. Docking a lamb's tail prevented fecal matter from accumulating around the tail area, which better protected their wool and flesh from being damaged by flies, while castrating the male lambs made them easier to handle as they matured.

In a 1930 report, one agent stated that lambs often went to market "long tailed and 'bucky' [describing a male lamb that wasn't castrated]."[43] In 1936, Benton County agent Ellis Christen indicated that "only 50 per cent of the lambs produced in the county each year are docked, and less than 20 percent of the Ram Lambs are castrated."[44]

Male lambs that were not castrated, or bucks, were priced one dollar less than a castrated male, or wether.[45] In 1930, ewe and castrated lambs brought $14.50–$15.00 per head, while uncastrated lambs brought $13.50–$14.00 per head.[46]

Docking and castration were typically done at the same time, before lambs reached a month old, so Extension would have to convince sheep farmers that doing both procedures at the same time would not negatively affect how well the lambs grew or put on weight.[47] This was accomplished through demonstrations, which, as one agent reported, "show a very satisfactory method of docking and castrating. This has had much to do with improving the market quality of our buck lambs and has eliminated no small loss often encountered from summer fly infection."[48] The aim of the demonstration was to allow a farmer to do the work under supervision and instruction so that, as Allen County agent Chester Kimmell put it, "the sheep men would then be able to go home and castrate and dock their own lambs."[49]

To get the lambs filled out in time for the higher-priced markets, Extension promoted feeding them a properly balanced ration, just like other livestock on the farm.[50] With the Indianapolis Producers Commission Association being a major buyer of lambs, purchasing large numbers of lambs to raise to marketable weights offered a ready market.[51]

Established in 1922, the association was a cooperative marketing organization of the Indiana Farm Bureau, Illinois Producers Association, and National Live Stock Producers Association. Farmers delivered sheep to a central location, then the association sold them in larger lots and charged the farmers a commission. Selling in larger lots brought participants better prices for their livestock.[52]

In 1937, Boone County farmers imported 65,000 lambs to finish feeding to market weights and then sell through the Indianapolis marketing association.[53] And in 1939, agent Alfred Hesler provided the following details about Fountain County farmers who worked with the association to finish lambs:

> The feeding of 10,000 western lambs yearly in Fountain County is creating a county wide interest. Lambs may be bought in the early fall, August or early September, and clean up the farm [by eating green vegetation]. About half of these lambs are fed on Contract under the Supervision of the Indianapolis Producers Commission Association. . . . [I]t has made Indianapolis the leading fat lamb market of the United States. . . . The remarkable fact is that fat lambs are going to market from this area in less than a 90-day feeding period.[54]

With available pastures, water, and grain, raising market lambs became a profitable business for growers. Extension advised farmers to feed lambs grain mixed with chopped hay and to supplement this mixture with soybean oil meal.[55] Extension also recommended putting lambs on a good pasture with a variety of plants, including alfalfa, alsike, bluegrass, and sweet clover.[56]

To encourage sheep farmers to adopt best management practices for raising market lambs, the Indiana Livestock Breeders Association and Extension sponsored the Gold Medal Sheep Club.[57] The goal of the competitive club was to send lambs to market at the heaviest weights, which required farmers to use the up-to-date management practices. Farmers raising sheep could see that club winners were feeding their lambs better, castrating and docking them, treating them for internal and external parasites, and using better rams to improve flock genetics.[58]

Those wanting to enter a flock of sheep had to have the male lambs castrated and all lambs docked. Farmers had to enter at least 25 weighed lambs to represent the entire flock. Weights were taken after the first lamb was 150 days old.[59] Medals were awarded based on the average weight of the flock, with gold medals given to flocks that averaged 75 pounds per animal; silver, for 70 pounds per animal; and bronze medals, for 65 pounds per animal.

In 1937, Blackford County had two contestants that met the requirements to receive medals. Lloyd Rice earned a gold medal for raising 34 lambs that averaged 82 pounds per animal, while Frances Johnson received the bronze medal for having 43 lambs averaging 67 pounds per animal.[60] That year in Wayne County, Harvey Whitton raised 43 lambs that averaged 85 pounds per animal, earning him gold medal status.[61]

Extension put forth tremendous effort in getting farmers to better understand what it took to raise lambs so that they could maximize their profits. Sheep experts such as Claude Harper and the agents teamed up to deliver the message that good money could be made if the sheep were properly handled and cared for, while the Gold Medal Sheep Club demonstrated the possible outcomes when farmers implemented the best management practices.

In 1936, the lambs from an average ewe returned a gross profit of $14.55, and in some years, selling the wool brought in additional cash.[62] But farmers who wanted to market the fleece had to learn how to shear their own sheep because there was an acute shortage of commercial shearers.[63]

Extension hosted events where sheep specialist Henry Mayo from Purdue would demonstrate how to handle a ewe and shear it quickly and efficiently. Growers were also shown how to tie the fleece into a bundle.[64] Electric shears made the process much easier, but not all farmers had electricity available on their farms.[65]

Extension loved competitions, and sheep-shearing contests were no exception.[66] At these events, each contestant had to shear one sheep, with scores assigned based on the following criteria: speed, 30 percent; freedom from injury to sheep, 20 percent; skill and handling, 25 percent; and completeness of shearing, 25 percent.[67]

The first state sheep-shearing contest was held on March 15, 1933, in Hartford City.[68] There, winners from county-level events competed for the title of grand champion, with Roy Keller from Jay County taking top honors by hand-shearing his sheep in 10.4 minutes.[69]

Five years later, the first national sheep shearing contest was held at the International Live Stock Exposition in Chicago, where eight contestants who had won their state events competed. William Willman from Hartford City represented Indiana at the competition, while Clarence Jeffers from Princeton, Illinois, took home the $100 first prize.[70]

With large crowds in attendance, events such as these offered yet another way to reach audience members with the recommendations made by Extension agents and university specialists.

As farmers entered the early 1930s under the cloud of the Depression, some turned to raising sheep as another source of income to help offset lower crop prices.[71] Extension offered encouragement and support by teaching farmers how to manage their flocks and maximize their returns as they marketed their sheep. But the focus on sheep as a viable venture that began in the 1930s would only last until the United States entered World War II. With the price of agricultural products skyrocketing, sheep once again became a sideline business on most Indiana farms as farmers pursued other agricultural avenues to earn their incomes.

CHAPTER 22

Runts and Diseased Pigs Seldom Lift the Mortgage

By 1930, Extension had spent a dozen or so years educating farmers about raising hogs profitably. This included what constituted a proper diet for pigs and how to properly house the sow and her pigs, control internal parasites with medications, and prevent hog cholera—known today as swine fever. In Indiana, county Extension agents had begun to attack the hog cholera problem head on as soon as they took up their positions starting in 1912.

In fact, the effort to vaccinate against the disease was one of the earliest successes by Extension. As some hog producers vaccinated their herds as recommended by the agents and prevented the losses that those with unvaccinated herds suffered, it helped to establish the credibility of the agents.[1] Recounting the work of his predecessor, Grant County agent Forest Conder wrote in 1937, "If Otis Crane had done nothing else during his term as county agent, his work in treating hogs for cholera was pay enough for what little he cost us."[2] Those lessons learned during prior decades became that much more important during the Depression for those trying to profit from raising market hogs.

Despite up-and-down prices during the Depression, hogs had remained as a staple on Indiana farms even as federal regulations and price supports became an important part of hog production both in the state and nationwide.[3] (See Chapter 8.) While some farmers only kept a few sows, others made a living selling hogs for the markets up to the limits allowed by the federal Agricultural Adjustment Administration.[4] In 1933, the federal agency had contracted to pay farmers to decrease their inventory of hogs in order to reduce market surpluses and increase prices. Prior to that, in 1930, Indiana had 3.3 million hogs in inventory worth about $34 million, with the United States reporting 55 million hogs nationwide. By 1935, federal government contracts had reduced the Indiana herd to 2.5 million, while the U.S. hog inventory fell to 37 million.[5]

Many counties in Indiana depended on hogs for their largest share of farm income from livestock. By 1934, the largest producer of hogs in Indiana was Rush County, with 80,000 head. This was followed by counties with more than 50,000 hogs on farms, including Boone, Clinton, Grant, Henry, Madison, and

Randolph.[6] In a 1936 annual report, agent Myron Cromer observed how reductions had affected Delaware County farmers, writing, "Notwithstanding the fact that the swine population of the county has decreased materially in the last six years due to the drought and the Agricultural Adjustment Programs, the returns from swine still continue to be the largest source of income."[7]

The top tier of hog producers had made major advances in improving their management practices during the 1920s. But agents still had a lot of work to do with other producers, whose hogs "were not being fed balanced rations, were not being raised under sanitary conditions and, as a consequence, were not being marketed at the peak of the market," as one agent described it.[8] Part of the larger problem was that using the same ground to raise successive litters promoted intestinal parasites such as roundworms. As Gibson County agent Roy Gwaltney observed in 1939, "Pigs are allowed to be farrowed in any old lot, any time. There are hog lots in the county that have been used for that purpose for 20 years or more and are still used."[9]

In a 1932 annual report, Fulton County agent Harry Rosenbury shared his thoughts about the many problems associated with young pigs in particular, writing, "Most hog troubles are little pig troubles. Little pig troubles are due chiefly to inadequate equipment at farrowing time, failure to get the pigs on the ground quickly enough, infestation of parasites from running in foul hog lots, and failure to get the pigs on the proper feed quick enough."[10]

With fewer hogs being raised on Indiana farms, those continuing to do so now needed more than ever to produce economical gains quickly and to focus their attention on what it would take to still make money during the Depression. To help them succeed, Extension continued to address issues related to inefficient hog production by continuing an education effort it had begun in 1925.[11] Called the Thrifty Pig Project, this Extension program was based on four strategies developed through Purdue agricultural research and aimed at quickly moving hogs to market:

- farrowing spring pigs by May 1;
- raising pigs under sanitary conditions to encourage rapid and economical weight gain;
- feeding a balanced ration to promote growth; and
- selling 200-pound hogs six to eight weeks before most were typically sold and prices declined.[12]

Agent James Little provided this advice to Cass County producers in 1932:

1. Begin before breeding time to feed both the sow and boar suitable rations and provide plenty of exercise for them. Have brood sow gaining 1 to 1.5 lb. per day at breeding time.

2. Provide sanitary conditions at farrowing time by (1) scrubbing out farrowing house with hot water and lye; (2) wash off sow's udder and sides; (3) either have sow and litter farrowed on clean ground or move them there from central farrowing house within the first week or ten days.
3. Start young pigs to eating grain from a creep. A mixture of 70# ground corn, 30# ground wheat, and 10# good tankage or skim milk, fed in a self-feeder has given very satisfactory results. Pigs started by this method do not lose their baby fat and never show any ill effects of weaning. A self-feeder helps prevent robbing [one animal eating another's share of feed], keeps feed cleaner and gains are generally faster and more economical than trough feeding.
4. Dry feeding suckling pigs and sows rather than slop feeding will prevent most cases of scours.
5. Castrate pigs six weeks or under.
6. Vaccinate young pigs at 8 to 10 weeks after they have lost their natural immunity.
7. Full feeding pigs pays best.
8. Plan to have spring pigs marketed in August and September and fall pigs, the last of February or March.[13]

Feeding hogs a healthy and balanced diet was a key issue promoted by Extension during the Depression, and it was central to the Thrifty Pig Campaign. Simply put, it was expensive to feed hogs, but while cutting back on feed saved money in the short term, it cost the producer much more in the long run.

In 1932, Myron Cromer had a very clear message for Delaware County hog producers, telling them, "One of the limiting factors to maximum returns from hogs is the failure to push them along on a full feed. Too many farmers by not full feeding fail to finish their hogs when the market shows the higher [price] trend."[14] Agents such as Cromer knew the only way to get hogs to market earlier, when prices were higher, was to feed them better.

Purdue hog Extension specialist John Schwab recommended pigs be fed a 70-20-10 ration consisting of 70 pounds of cracked corn, 20 pounds of cracked wheat, and 10 pounds of tankage. He also suggested that producers use a self-feeder to allow animals access whenever they wanted to eat. More importantly, this mixture of grain allowed the young pigs to rapidly reach 100 pounds before their diet was switched to a ration of corn and tankage.[15]

During the 1930s, amid the depressed grain markets, growers learned it was more cost-effective to feed homegrown grains such as corn and soybeans to hogs, which brought a better return in pounds of meat versus bushels of grain sold.[16] To promote even more growth, Extension recommended that producers use a practice known as hogging off, which consisted of letting hogs feed in fenced-in fields of corn, soybeans, wheat, and alfalfa.[17] However, producers were warned that the fields being hogged off needed to be rotated every two years to reduce the prevalence of roundworms.[18]

Fountain County agent Alfred Hesler indicated that farmers would benefit from following at least some of the management practices promoted by Extension, noting, "[I]n most cases even a modified plan is better than the raising of pigs around an old barn in old hog lots."[19] In 1930, agent Everett Parker demonstrated how implementing a single practice had affected production and selling prices in Hamilton County.

Thrifty Pig Project Practice	Average Number Pigs per Litter	Average Date of Marketing	Average Price
Houses scrubbed and pigs grown on clean ground	7.3	September 15	$10.65
Houses not scrubbed and pigs kept in old lots	6.1	November 25	$8.50[20]

Extension repeatedly emphasized that the timing of delivering hogs to market made all of the difference in the prices received. Producers were told over and over again that better feeding allowed spring and fall hogs to reach the high-priced markets in March, April, August, and September, when demand outstripped supply.[21] In 1931, hogs sold for $7.67 per hundredweight in the early markets, but by November, the price had dropped to $5.00 per hundredweight when more hogs were being sold.[22] That same year, a hog producer from Hendricks County sold 65 hogs at five months and ten days for an average price of $15.00 per head. A group of hogs sold just two months later, when most of the hogs were reaching the market, only brought $9.00–$11.00 per head.[23]

By the end of the decade, swine operations run by farmers who followed Extension recommendations were showing the positive effects. In 1938, Gibson County producers had 90 percent of their spring litters reaching market at six months and weighing 180–210 pounds.[24] That same year, agent Warren O'Hara used his annual report to summarize how far the Hancock County hog producers had progressed, writing, "A check in November on 65 farms showed 558 sows farrowing on the average date of March 24 weaning 3,234 thrifty pigs. This averaged six weaned pigs per sow. . . . The average date of marketing the spring pigs was Sept. 15 to 18 with the average of 3 weeks earlier marketing than five years ago. The average weight at marketing the first 2,628 (81.2% weaned) was 210 pounds."[25] Just a year later, he noted, "Extension work in hog production has changed the one litter per year to the two litter per year system and has resulted in marketing of hogs at earlier ages and from six weeks to two months earlier in the fall."[26]

Agents such as Ira Thurston used statistics to demonstrate that the Thrifty Pig Project made growers money. He reported:

> A recent survey by the county swine committee revealed approximately 13,000 brood sows in the county. . . . [and] approximately 39,000 additional pigs from the same number of sows could be raised. Records throughout the state kept by practical hog men show an average of 1.5 additional pigs raised per litter by following the practical program discussed at the hog school. . . . At market prices of $4.00 per hundredweight [in 1933], this would give the Decatur County farmers $312,000 with very little additional cost, if any, by following a good practical program of care and proper feeding of the brood sow and pig.[27]

The Thrifty Pig Project promoted practices designed to reduce losses of young pigs and send more hogs to market, which seemed diametrically opposed to the federal government mandate to reduce supplies. Agents must have been asked about that numerous times, leading Walter Rusk in Blackford County to cite a Purdue agricultural economist who had found the opposite to be true: "Dr. W[hitney] B. Stout discussed the outlook for the swine industry and emphasized the fact that the growing of thrifty pigs was not in opposition to the A.A.A. program but that farmers who produced lower priced pork required fewer hogs to make a total income and thus were not responsible for putting surplus hogs on the market."[28]

Purdue Extension would push the Thrifty Pig Project—or at least parts of the program—through a variety of outreach efforts. One strategy involved sending farmers timely Grow Healthy Pig letters that were prepared by Purdue hog Extension specialist John Schwab and mailed periodically to area farmers by their local agent.[29] Arvil Keesling, the Bartholomew County agent, noted that hog producers liked receiving the letters, writing, "[T]hey would much rather receive timely information on swine production instead of having the full information in bulletin form, because it not only gave them information but served as a reminder, which might be forgotten if all the information was presented at one time."[30] Those receiving their first letters in 1930 were also sent the Purdue bulletin *How to Handle the Brood Sow and Litter* to help answer any questions that the letters brought up.[31]

In 1931, Schwab prepared the following letters:

Letter 1. The Life Cycle of the Round Worm
Letter 2. More on Round Worm
Letter 3. Brood Sow Management
Letter 4. To Keep Pigs Healthy, Start Them on Feed Early
Letter 5. As the Pig Is Bred, Fed and Managed, So Will Be the Hog

Letter 6. Feeding Spring Pigs for Market
Letter 7. A Few Farm Facts Are Convincing
Letter 8. A Plan to Grow Thrifty Fall Pigs
Letter 9. Start Fall Pigs on Feed Early
Letter 10. Grow Thrifty Pigs
Letter 11. Feeding Fall Pigs for Market[32]

In 1934, there were 12,000 farmers from 67 counties who received the updates, with the monthly letters proving to be extremely popular with farmers. According to the annual statewide Extension report for that year, 98 percent reported reading the letters, 76 percent saved them as a reference, and 82 percent made changes based on the recommendations contained in them. Farmers also believed that the information in the letters added an average of around $114 to each of their operations annually.[33] Allen County agent Hugh Heckard observed that a local veterinarian kept "all the letters on his desk for reference to any farmer seeking information on swine management."[34]

A second strategy used by Extension was to offer regional hog schools, which also proved popular with producers and were well attended. In 1934, Extension conducted 37 hog schools around the state that averaged 117 attendees per event.[35] As was the case with most Extension programs dealing with hogs, Schwab would often provide instruction on specific topics. Many of Schwab's lectures and demonstrations covered the same topics he addressed in his letters, but by attending an event, participants could ask questions in person. Some titles of presentations by Schwab and others included:

- Improvement of the Breeding Herd
- The Management of the Brood Sow
- Care from Brooding to Weaning
- Homemade Rations
- Starting Young Pigs for Health and Ability to Gain Rapidly for Market
- Better Breeding Stock
- Hog Outlook for Farmers
- Facts Learned from Experimental Feeding Trials about Farm Feeds, Protein Feeds, Minerals, and Pastures
- Factors Affecting the Marketing of Hogs
- A Successful Farm Hog Program
- What Buyers Look for When Buying Hogs
- Control of Parasitic Troubles[36]

Controlling parasitic worms was one of the more popular lessons taught at the hog schools. Most of the time, a veterinarian would dissect the runt of a litter to clearly show its abdomen full of roundworms.[37] This type of demonstration reinforced how proper sanitation practices promoted by Schwab and the agents could reduce these types of infections.[38] According to Decatur County agent Ira Thurston, "Producing pork free of internal parasites and on a balanced home grown feed has been the source of more incoming dollars than any change of system on the other types of livestock."[39] Fayette County's Marion Detrick summed up this work by saying, "Runts and diseased pigs seldom lift the mortgage, and neither do hogs marketed in November, December, May, or June return maximum profits for their owners."[40]

During the 1930s, it was common for the regional events to include agricultural economists from Purdue who discussed current and future prices and marketing trends. With these added market discussions, the programs covered nearly every aspect of breeding, raising, feeding, and marketing hogs in Indiana.[41]

A third strategy involved a statewide event on hog production. Each fall, the Purdue Department of Animal Husbandry held a Swine Day program at its research facilities in West Lafayette. Agents and interested growers from around the state drove to campus in a caravan or took the train to attend. The day-long school provided an opportunity to highlight the latest findings by Agricultural Experiment Station researchers.[42]

The hope was that the several hundred growers in attendance would implement the suggestions that were based on scientific research methods.[43] After the 1938 Swine Day, Bartholomew County agent Arvil Keesling reported, "Ten farmers have already reported that they are feeding a protein mixture mixed like the one that gave the best results at the University Farm."[44] The agents who attended also gained valuable information to share at local meetings and during other conversations with farmers.[45]

A fourth strategy offered a statewide contest known as the Hoosier Ton Litter Club, which was overseen by Schwab and sponsored by the Indiana Livestock Breeders' Association. Started in 1922, the club promoted competition among farmers to see who could produce a litter of pigs weighing 2,000 pounds at six months old.[46] Essentially, the club was a way for Extension to showcase the accomplishments of top hog producers as a way to teach other farmers what could be achieved by following the best management practices. To reach the 2,000-pound milestone, eight to thirteen pigs in a litter had to survive and be raised under management practices that were recommended by Extension. The Indiana Livestock Breeders' Association recognized farmers for their efforts by offering a gold medal to litters weighing at least 2,000 pounds, a silver medal for litters reaching 1,800 pounds, and a bronze medal for litters surpassing 1,600 pounds.

By relying on a variety of outreach strategies, Extension continued an aggressive education campaign throughout the 1930s to bring producers the latest information about raising hogs for the markets. This effort to reach as many farmers as possible showed how important the hog economy was to Indiana agriculture.

Like corn and dairy cattle, hogs were an integral part of the state's rural economy. Throughout its early years, Extension devoted many efforts to distributing research findings that would move the industry forward and improve how hogs were bred, raised, fed, and marketed. But it would take the Great Depression to push farmers raising hogs to do a better job in all aspects of their programs. Times were tough for hog producers during the 1930s, but those that excelled still made enough profits to keep their operations afloat during turbulent times. On the other hand, those who were slow to adapt to economical production methods would be forced to exit the market.

As it turned out, Purdue's engagement with hog producers up to and during the 1930s would reap much-needed dividends during World War II, which would bring major demands for increased production to meet wartime demands.

CHAPTER 23

A Bushel Basket of Eggs Brings In as Much Money as 100 Bushels of Corn

During the Depression, most farms had a flock of chickens that for many became the one agricultural activity that would save them from foreclosure. This was very similar to what had happened during the 1920s agricultural depression, when poultry had been called a "mortgage lifter."[1] This showed just how important this stream of income was for farmers.

In the 1930s, most farmers kept poultry. For example, a 1937 survey in Dearborn County showed that 1,840 farms out of 2,000 had poultry.[2] In 1931, agent Ralph Maggart categorized the flocks in Carroll County according to size, reporting, "one flock numbering above 1,000 birds; 15 or 20 flocks above 500 birds; 150 flocks from 300 to 500 birds; about 500 flocks from 100–200 birds; about 700 with flocks of less than 100 birds; and 300 farm flocks with less than 50 birds."[3] These numbers indicated that some farmers raised poultry as a business within their farming operations.

In 1930, Dearborn County agent Clyde Madison reported, "Practically all of our farmers keep poultry, the average size flock being about two hundred. This is a profitable industry. In fact, the prices for poultry produce have stayed up better than the prices for most any other farm produce."[4] The question was, Could Extension help farmers raise poultry more profitably through better management so they could take advantage of the markets?

During the 1920s, Extension had emphasized good management practices for poultry, including feeding better rations, providing better housing, reducing disease, culling flocks of unproductive birds, and improving record keeping. This actually set the stage for continued advancement during the next decade. In 1930, agent Earl E. Stinson in DeKalb County indicated, "The income [from poultry] can be increased by one third without adding materially to the cost of production if certain well established practices were followed on these farms."[5] Being able to increase income without spending a lot of money was an important selling point for farmers during the tough economic times. However, getting farmers to use established and proven practices was one of two hurdles for Extension. The second, equally difficult hurdle was getting farmers to treat poultry with the same respect that was given to corn, hogs, and dairy.

This was not a new problem, but it was one that would persist until Depression-era egg prices caught the attention of farmers.[6] In 1932, Gibson County agent Cyrus Dyer recalled how poultry flocks had not been considered farm work at one time by most men but how that less-than-positive attitude had changed as the agricultural economy weakened. He noted:

> This county has been too busy with corn, wheat, hay, and hogs to spend much time fooling around with such things as poultry. Raising chickens has been considered a woman's work. Our farmers were interested in bigger things. During the last few years however some peculiar turns have been taken by affairs on our farms. This season our farmers have seen a bushel basket of eggs bring in as much money as 100 bushels of corn. They have figured up the business on a large farm and discovered that they operated at a loss. In checking over the wife's business they have seen that she made some money on chickens. These things have tended to increase the respect of the average farmer for chickens as a part of the farm business. As a result there is a growing interest in the farm flock. As a matter of fact I think the county as a whole has been passing up about a million dollars a year that they might have been getting had more attention been given to the farm flock.[7]

During a time when every penny counted, poultry represented an opportunity for farmers to make more income by better managing the resources they already had.

While Extension had offered much education and instruction during the 1920s to improve poultry management in Indiana, there was still more work to be done. This was the case in Hendricks County, which by 1931 ranked sixth in the state for the number of poultry but only twentieth in the money made from those flocks. Agent Ellis Hawk wrote, "Possibly part of this is due [to] the fact that poultry is rather a sideline with most farms, yet the improvement of farm practice in the line of feeding, breeding and management would make a very important addition to the income of most Hendricks County farms."[8]

Poultry management included the same concepts as used for other livestock: feed balanced rations, cull unproductive animals, keep animals healthy, and improve sanitation practices.[9] However, Extension chose to focus on campaigns aimed at keeping chicks alive by reducing internal parasites, paralysis, and colds.[10] Keeping more chicks alive to become layers and broilers was key to the continued success of a poultry operation.

Across Indiana, county agents and poultry Extension specialists at Purdue engaged with producers through several statewide efforts, including the Grow

Healthy Chick Campaign, Grow Healthy Pullets Campaign, and Better Baby Chick Campaign.[11] These educational outreach programs were aptly named, considering most poultry producers lost 20–40 percent of their chicks before they could return any money to the grower.[12]

At the county level, the Grow Healthy Chick Campaign often started with the agent surveying the practices used by poultry producers to raise chicks.[13] Agent James Little reported the findings from a 1930 Cass County survey, writing:

> Out of 494 cards sent out, we received reports from 125 farms which reported on 43,444 chicks hatched last year. . . . Of the reports which we could tell something about, there were 33,646 chicks which had a mortality of 7,585 chicks or 22.6 per cent mortality. In the number of deaths reported, the most reported were due to Coccidiosis [a parasitic disease] and crowding. An average of 9 chicks more to the 100 chicks started were raised from blood tested parent stock over that of flocks in which the parent stock had not been tested. The mortality rate was about the same for all breeds of chickens.[14]

Purdue poultry Extension specialists William Kohlmeyer, Leon Todd, and Stephen M. Walford made presentations to farmers about the program's six management principles designed to keep more baby chicks alive.[15] In his 1930 annual report, Little summarized one of the presentations made by Todd to a small group of Cass County poultry producers:

1. CLEAN CHICKS: Get chicks or eggs from flocks bloodtested for Bacillary White Diarrhea or known to be free from B.W.D. and hatched in clean incubators.
2. CLEAN BROODER HOUSE: Scrape and sweep all equipment; scrub with boiling lye water, 1 lb. to 20 gal. Use good disinfectants; keep floors tight and smooth.
3. CLEAN GROUND: Move house to ground that has not had poultry or poultry manure on it for one year. Keep hens off this range. If not movable, use concrete or ½ inch mesh wire floor at least 6 to 8 feet deep and full length of brooder house and high enough off ground to clean easily.
4. CLEAN LITTER: Use cut straw, planer shavings, bright clean chaff from haymow, or corn cobs run thru hammer mill with ¼ inch screen. Clean house twice each week; remove all old litter.
5. CLEAN FEED: Feed clean and well balanced all-mash ration in hoppers or troughs which are made so as to prevent droppings from getting into the feed. Have the feed and water before the chicks when they are placed in brooder. Keep mash and water available at all times.
6. CLEAN MANAGEMENT: Do not place more than 350 chicks in 10 x 12 ft. brooder house. Remove cockerels at ten to twelve weeks of age. Heat house several days until stove can be easily regulated before putting in the chicks. Watch chicks to determine necessary warmth. Chicks should

> be active and feel satisfied at the edge of hover [canopy to funnel heat]. House should be free from drafts and dampness.[16]

As was the case for many of the Extension-led campaigns, enrollment cards were sent out to farmers with an explanation letter that discussed the advantages of participating. Some agents also placed posters in window displays at hardware stores to promote the benefits of the program.[17] Hatcheries, feed stores, and elevators helped promote the campaign to their customers as well.

Enos Rowe was a new agent in Clinton County when he used a novel approach to draw attention to raising more baby chicks in 1933. He explained that he used ten baby chicks, "which I water colored to attract more attention and placed them in the east window of my office in the court house, where all could watch their growth and development." Over the next eight weeks, he weighed them and displayed the results on cards in the window. He later reported that, as a result of his display, "many over the county have adopted our system of raising baby chicks."[18]

Farmers who enrolled their flocks in the Extension program were provided with information on all aspects of raising profitable poultry. They were given a form for recording the number of chicks that had died.[19] And at appropriate times during the growing cycle, agents mailed letters written by poultry Extension specialists to participants.[20] Agent Arvil Keesling sent the following circular letter to Bartholomew County poultry producers in 1935:

> Dear Poultry Producer:
>
> "I housed 140 good pullets from 300 chicks started last year." Can you truthfully make that statement? Do you want to? As a matter of fact, many Bartholomew County People have done so.
>
> After all, the real satisfaction lies, not in talking about the results, but in banking the profits from such a success. You know, when a chick dies, you are not only losing the cost of the chick, but also the food that it has eaten, and its proportionate share of the time, the coal, the equipment and housing cost which must be divided amongst the chicks that survive. And along with the fact that every chick that dies is one less to go on the market or into the laying house, we must remember that pullets from a high-mortality brood of chicks seldom, if ever, make profitable layers—and that's what we're raising them for. . . .[21]

The letter continued by providing detailed management information designed to promote a healthier environment for keeping the chicks.

Most baby chicks were obtained from commercial hatcheries, so Extension emphasized that farmers first needed to make sure any young birds they had purchased had been tested for diseases such as bacillary white diarrhea (pullorum) and coccidiosis.[22] Blood-testing to ensure disease-free chicks was an important first step because mortality at twelve weeks of age was significantly higher in untested chicks.[23] Farmers who provided eggs for hatcheries received a premium price when the eggs had been obtained from tested hens.[24]

In another letter Keesling sent to local poultry producers, he discussed the dangers of one particular disease that could devastate their flocks, writing:

> You've all seen chicks that get droopy, don't seem to lengthen out, stand around under the hover all day, occasionally give a painful "peep," often give off whitish, sticky droppings, and eventually become all pasted up behind.
>
> Perhaps you have even tried to raise them, and perhaps you tried to get eggs from the pullets that survived. Just about heartbreaking, wasn't it?
>
> The chicks were no doubt suffering from Pullorum Disease (formerly miscalled Bacillary White Diarrhea). As you might guess, most of them died. But some pullets lived through, carrying the Pullorum germ in their blood streams and, later on, in their ovaries. These infected pullets, if allowed to remain in the breeding flock, would lay infected eggs, and the eggs that hatched would give chicks loaded down with disease germs. These chicks would pass the germs to others in the same incubator (no matter what type of incubator it happened to be) through infected chick down and dust: and in the brooder house, the originally infected chicks (and those infected in the incubator) would infect still other chicks through germ-laden droppings, and so the disease runs its cycle.
>
> The only effective means of controlling this disease is to eliminate from the breeding hens, birds which are carriers of the disease. This can only be done satisfactorily by the recognized agglutination method when administered by competent conscientious testers, and when these reactors are removed and the eggs from tested stock are hatched in a separate incubator.
>
> There is no pill or remedy for chicks infected with Pullorum Disease, although many so-called cures lull many poultry folk into a feeling of false security.[25]

Some growers seemed to think that the testing process had the potential to spread the disease within their flock, a concern that Dubois County agent Charles Nicholson addressed in a 1930 report, writing, "Last year [1929] approximately around 5,000 birds were blood tested and an accurate check on the baby chicks hatched from these flocks showed that there was no serious outbreak in any of the baby chicks obtained from these flocks having been tested by the agglutination test; and it has been definitely stated that the people are beginning to demand baby chicks from tested flocks chiefly due to the results which are noticeable the past three years."[26]

To help educate producers about the various poultry diseases and how to reduce them, county Extension offices featured display racks of publications on many individual diseases, and farmers could request free copies of publications for more information.[27] Agents also encouraged producers to avoid keeping animals in cramped quarters since crowded conditions made it easier for a disease to spread quickly within a flock.[28]

Growers who put into practice some or all of the six steps of the program had significantly lower mortality within their flocks, which led to more layers being added to the flock or more broilers sold to the meat market.[29] Agents reported that following the main points outlined in the campaign reduced flock mortality rates by 10–30 percent.[30]

Producers who followed all six of the management steps in the program fared better than those that didn't adopt as many, with results declining as farmers adopted fewer of the practices. For example, Harrison County had 90 flocks enrolled in the Grow Healthy Chick Campaign in 1930. This included a total of 48,077 chicks, of which 9,661—or 20 percent—died overall. But when the losses were broken down according to how many of the six steps were followed, the results were clear, with agent Ronald Johnson reporting, "[T]he nineteen flocks following all six of the recommended practices lost only 8.2%; the 32 flocks following five points lost 11.6%; the twenty-four [flocks] following four points lost 19.6%; the eight [flocks] following three points lost 20.6%; the five [flocks] following two points lost 37.8%; and the two [flocks] following only one point lost 79%."[31]

Other counties reported similar results that year and the next. In 1931, the average flock mortality rate in Daviess County was 26 percent, but it went as high as 45 percent when none of the six practices was used.[32] In Clay County, 13 percent of the chicks died when producers followed all six points, but one-third died when fewer practices were used.[33] Cass County reported a flock mortality rate of 7 percent when all of the points were followed but up to 28 percent when fewer were adopted.[34]

The statistics proved that the program worked, so agents used them as a selling point to show that growers who took the recommendations seriously could raise more replacement chicks and pullets for their flocks. In 1935, Keesling shared the following information with Bartholomew County producers as a way to convince them to improve their management practices:

> Here are the results of the Grow Healthy Pullet Campaign, which I completed recently and the results of the survey made in March. Compare your own results this year with the average of the Table 1 as well as with each group as I have separated them.
>
> Everyone wants better than average results. If you were above the average, fine,—you have something to "live up to"! If they were below, I am sure the campaign can help you another year in raising a fine lot of healthy, well-developed pullets at low cost.

Table 1

	Number of Flocks in Group	Number of Chicks in Group	Total Loss in Group	Percentage of Loss in Group
All [six] points followed	32	13,563	964	7.0
All but clean chicks	7	1,817	308	17.0
All but clean ground	60	25,060	2,864	11.4
All but clean chicks and clean ground	15	5,291	1,307	24.7
All others	21	7,244	980	13.5[35]

In addition to agents sharing information directly with growers, the findings on how to reduce chick mortality were published in most newspapers, and the collected information was also presented at Extension meetings.

Farmers were shown that using the latest management research in conjunction with keeping records on all aspects of raising poultry brought them higher profits. If there was any doubt about the value of research and record keeping, the on-farm demonstrations where the farmers gave their testimonials about the value of each would have sealed the deal for those on the fence.

Buyers paid farmers higher prices for eggs when they met higher quality standards. But to meet those standards, producers had "to recognize eggs are a perishable product," as one agent noted, and take more care in handling them prior to the sale.[36]

To promote better egg handling by farmers before the sale, Alfred Hesler shared the following advice with Fountain County farmers in 1934:

> Poultry raisers who sell their eggs on a grade or quality basis will find it profitable to give particular attention to the temperatures under which their eggs are kept, states Leon Todd, Purdue university poultry specialist. Fertile eggs start to germinate at 69 degrees, and the higher the temperature, the greater the deterioration. . . .
>
> Ordinarily eggs are of top quality when first laid, and Todd recommends the following practices for avoiding loss from summer heat:
>
> 1. Each evening confine all broody hens in a small coop, so they will not sit on the newly laid eggs.
> 2. Collect the eggs at least twice each day. This is a good practice to follow the year round.
> 3. Keep the eggs in a cool place, preferably below 60 degrees. If available, a basement free from odors, and with some moisture is desirable.

4. Do not case today's eggs until tomorrow, thus allowing the animal heat to get out of them before they are cased.
5. Avoid exposing the case to sun and wind when taking eggs to market.[37]

It was customary for farmers to store eggs for a week in order to collect enough to sell, so it was imperative that the eggs be stored at cooler temperatures, especially during the hot summer months, so they would be fresher when sold.[38]

Other quality standards involved cleaning the eggs and packaging them by size.[39] Extension staff were repeatedly asked whether it paid to spend the time to grade eggs according to the standards set by the U.S. Department of Agriculture.[40] Carroll County agent Ralph Maggart shared the following information in 1931 to show how the farmer profited by producing quality eggs:

> Since the opening of the Poultry Cooperative Plant at Monticello, there is real need for the production of quality products. First, the consumer demands quality products; the organization and working of the cooperative plant insure the producer of better prices for better products. This extra price when figured over a period of time will more than pay for additional labor and certain adjustments necessary. At the present time the grading plan has a differential of 7 cents per dozen between Henry White Eggs and the first grade. This is almost ½ cent per egg, and if a bird lays 100 eggs per year, it would figure 50 cents per bird. With a flock of 100 birds, this would make an additional income of $50, which is rather high returns for such minor changes necessary.[41]

The following year, Maggart reported that cleaned and graded eggs produced by a flock of 100 birds brought $8 above local prices, while another farm with 150 hens made $10 more than the average price for ungraded eggs.[42]

In 1938, agent Lawrence Archbold reported that a 1937 auction of graded eggs in Versailles, Ohio, brought premium prices, noting, "[I] saw 234 cases of graded and candled eggs sold in approximately 35 minutes. The prices were substantially greater than those prevailing in Adams County for like grades of eggs, and it is entirely possible that some eggs from Adams County can be moved into this auction."[43]

One effect of producing graded eggs meant that the eastern states and other locations where prices were higher received the best eggs, leaving local markets with lower-quality eggs.[44] With larger, fresher, and cleaner eggs, the higher-priced markets in Indianapolis, Detroit, New York, and elsewhere got their fair share of eggs from Indiana. But these buyers wanted larger quantities, which meant farmers would have to pool their eggs to create larger shipments.[45]

In Hamilton County, the local farm bureau got involved in trucking eggs to buyers in Indianapolis. Larger egg producers in the county would bring crated eggs to the farm bureau store, and from there they were driven to Indianapolis.

After paying the transportation cost, in 1933 producers netted three to four cents more per dozen eggs than what they would have received locally.[46]

In 1932, Greene County agent Virgil Sexson described the cooperative-selling approach used by local farmers, saying:

> [C]rates of eggs will be delivered to the car at Linton, where the local manager will receive them and load them in the car. Each shipper will pay at the car an amount sufficient to cover transportation and local handling costs. Each crate will carry the shipping tag of the producer's own commission firm. The car will move to New York City consigned to a trucking concern that will deliver the crates according to the shipping tags for an agreed handling charge. Settlement will then be made direct from the commission firm to the individual shippers.[47]

In some counties, poultry cooperatives also marketed broilers in addition to eggs, with one plant in Monticello returning a premium price of six cents more per pound than the market price to members.[48] The higher price provided an incentive for farmers to raise broilers to sell out of state while ensuring that the buyers had access to larger quantities at one time.

In 1936, Benton County agent Ellis Christen reported that one of the most concentrated production centers of winter broilers in the United States was located in and around Boswell and Oxford.[49] A few years later, his annual report included additional details about sales of the broilers:

> One of the big problems in broiler production has been the marketing of the broilers. More than 90% of the broilers produced are sold to dealers from the Chicago Area. Common practice has been for these dealers to come into the area and contact local buyers as to who have broilers ready for market. These buyers would then go to the farm and buy the chickens. The buyer then would pay the local dealer one half cent per pound for the information he gave as to their location. Most producers felt some local organizations could provide this same information to buyers [so] that producers could receive the half cent additional in selling price which had been going to some local buyer who never owned nor handled the birds. . . . [An] association was formed. . . . Mr. Dexter served as supervisor during the season last spring and buyers coming in the community would contact him and he would give them a list of the number of birds ready for sale and no charge was made to the buyer.[50]

With poultry paying good prices, many farmers took advantage of growing broilers instead of eggs. While at the time it was a niche market, over time these farms prospered by focusing their attention on products that consumers wanted.

In order for farmers to justify keeping a flock of chickens, it had to produce enough eggs to cover expenses and return a reasonable profit. To help egg producers determine their profits, the Department of Poultry Husbandry at Purdue provided a calendar for recording how many eggs were produced each day over the course of a year.[51] Dividing the total number of eggs by the average number of hens allowed the producer to calculate how many eggs each hen laid and then figure out production costs.

Putting the data on a per-hen basis removed the size of the flock from the equation and allowed farmers to make a hen-to-hen comparison with other farms. In 1932 in Benton County, the agent used the data from 161 egg calendars to calculate what an average flock produced and the number of eggs the top flocks were laying. While the comparison to average production was useful, it was the highest number of eggs produced by other area growers that interested farmers the most. One of the top producers in Benton County was Mrs. George Steele from Oxford, whose flock produced 197 eggs per hen that year.[52] In 1937 in Adams County, assistant state leader of county agents Louis M. Busche reported that some "flock owners have hens that will produce 200 eggs each, [and] many flocks will average 175 eggs each; the average hen for the county produces only 90 eggs."[53]

Extension often used demonstration farms to showcase the best farmers who were using Extension recommendations, and poultry was no different. Agents eagerly took poultry producers to these area farms to hear directly from the cooperating farmers and ask questions about what they saw. The agents also used information collected at these farms to make presentations and write columns in local newspapers.

Farmers whose operations were poultry demonstration farms recorded every aspect of their work, including the money spent on the flock and the money earned from selling eggs. As Allen County agent Hugh S. Heckard recounted, "An extension program illustrating better practices in poultry production must necessarily have some local information upon which to base these recommendations. Therefore, the Poultry Demonstration Farm project has been selected as a means of collecting some information on poultry production and also, at the same time, demonstrating results which can be obtained from the use of approved practices."[54]

One of the approved practices was making sure the flock received a diet with enough protein. While farmers were feeding whole yellow corn as the sole diet, Purdue researchers had developed a poultry feed ration and advocated supplementing corn with ground wheat and oats, meat scraps—and later, soybeans—and alfalfa leaf meal.[55] The ration consisted of mixing 500 pounds of corn with 200 pounds of wheat, 200 pounds of ground oats, and 100 pounds of meat scraps. Another option was to mix 500 pounds of corn with 350 pounds of wheat, 90 pounds of meat scraps, and 60 pounds of alfalfa leaf meal.[56]

Adding protein increased feed costs, so Purdue poultry Extension specialist Leon Todd used the following table at a 1932 Cass County program to show the results obtained by spending more on feed:

	Corn Diet	Purdue Blended Diet
Eggs per bird	60	179
Average value per dozen	28.2	28.2
Egg receipts per bird	$1.40	$4.16
Pounds of feed used	62	78
Average price per cwt.	$1.76	$1.94
Feed cost per bird	$1.09	$1.51
Total return per bird	$0.31	$2.65 [57]

Attendees must have been shocked by the discrepancy in earnings between the corn diet and the Purdue blended diet. Extension knew that any discussions that linked practices to profits would generate interest among those taking a serious approach to raising poultry. Even when egg prices were low during the early 1930s, farmers were still profiting.[58]

In 1933, one Benton County flock of 123 birds brought a profit of $150, while another area flock of 213 only earned $60, showing how management—or the lack thereof—affected the grower's bottom line.[59] That same year in Delaware County, one farm with 150 birds averaged 156 eggs per bird and netted a profit of $101, while another producer earned $275 from 283 birds, each laying on average 186 eggs.[60]

Keeping detailed records allowed producers to calculate their own profit margins to determine if the income generated by the flock was acceptable and make adjustments as needed to increase returns.

While poultry products weren't typically the top source of income for farmers during the 1930s, in many counties they had become a second or third source, providing just enough cash to help farmers survive the lean years of the Depression.[61] Because poultry was one of the bright spots in agriculture, Extension hosted a number of poultry schools where producers could obtain specific information on the latest research and best management practices for increasing the value of the farm flock.[62] Most discussions focused on using the information collected from the poultry demonstration farms to highlight what was being done, what it cost, and the revenue returned.

Information reported by agent Earl Miles following a Clark County poultry demonstration farm in 1938 provided details on profits, but what the grower did to attain these numbers was more important to Extension. He wrote:

> The Simonton flock (Utica twp.) has a complete record of operation for 5 years. During this period it has never returned less than $1.00 per bird for labor and has an average of $1.80 per bird. This variation has been due largely to cost of feed and price of eggs since the average number of eggs per bird has not varied more than 3 any one year. . . . The mortality of the laying flock has never exceeded 8%.
>
> An area of less than 10 acres has been used to brood on and house the laying birds. Each year the chicks are started in February and March on clean alfalfa pasture, cockerels removed at 6 to 8 weeks of age and the pullets moved to new ground from one to two times during the summer. Eggs are sold to a hatchery and the entire flock is tested for Pullorum disease each fall.[63]

Poultry schools commonly attracted anywhere from 100 to 400 people.[64] Adams County agent Lawrence Archbold had developed a reputation as a poultry expert, so other agents often invited him to assist with their poultry schools.[65]

The topics varied from year to year, but many fell along similar lines that included:

- Feeding the laying flock for increased egg production
- Laying rations that have proved successful
- Increasing the egg income by improving quality
- Poultry house construction and ventilation
- Brooding chicks
- Feeding and rearing baby chicks
- Chick sanitation
- Care of pullets on the range
- More profit through hatching eggs and broilers
- Preventing theft loss by tattooing
- Controlling diseases through sanitation
- Practical poultry breeding
- Marketing poultry products[66]

The Question and Answer Poultry School held in Corydon offered a first-of-its-kind approach to attendees in 1931.[67] Harrison County agent Ronald Johnson described the event in his annual report that year, saying:

> [It] came nearer answering the everyday questions arising in the poultry yard than any poultry school in recent years. As requested in letters and newspapers, poultrymen sent in questions for discussion. The poultry school committee sorted them by subjects, and ten questions on each of the five subjects selected by the committee were printed in the county papers as a part of the publicity. . . . The question and answer school consumed a morning and an afternoon session, and this was preceded by a night meeting on Outlook and Marketing, two timely subjects.[68]

With so much excitement about poultry and what it meant to farm income, poultry schools allowed attendees to hear Purdue agricultural researchers and Extension specialists talk about the latest findings for improving flocks. This information was often supplemented by data collected from county poultry producers, which made the information even more relevant.

In 1935, Allen County agent Hugh Heckard shared a testimonial from Mrs. Ralph Etzler of Monroe Township to show the influence that poultry schools had on those in attendance. She had stated:

> At the time of the Poultry School our flock of chickens were diseased and I was at the point that I did not know what to do next. I even took several sick chickens in a crate along to the Poultry school in the hopes that I could learn what the trouble was. As a result of the poultry school, my flock was tested for Tuberculosis (however, finding only two reactors), then I followed the advice given me and sold every chicken on the farm. This was hard to do because I missed the egg money and the eggs for cooking. We cleaned everything thoroughly and let our large Purdue type chicken house be empty all summer. We bought blood-tested chicks from a hatchery recommended by one of the speakers at the Poultry School. We put a new floor in the brooder house and moved it to a new location. We had better luck with these chickens than we ever had before. I had been so discouraged with chickens that I thought it not worth while to bother with them, but I hated to leave a modern $500 chicken house vacant. My pullets are laying now at less than seven months old.[69]

Raising poultry for the egg and broiler markets would see a renaissance during the Great Depression that would continue into the 1940s as more farm families came to rely on poultry as a source of income. Receipts from egg and broiler sales paid for farmers' living expenses and helped keep their mortgages current. Along the way Purdue agricultural researchers, Extension specialists, and county agents contributed important work to transform farmers' more or less indifferent attitudes toward poultry into ones that were focused on managing their flocks well.

CHAPTER 24

The Life of an Extension Worker Is an Honorable Occupation and an Interesting One

The year 1939 represented a significant milestone for the Cooperative Extension Service because it marked twenty-five years since the agency had been established by the 1914 Smith-Lever Act. For the benefit of posterity, Clyde W. Warburton, director of the Federal Extension Service, asked all county agents to write in their annual activity reports a brief synopsis of how the work had evolved since the initiation of the national education program, what Extension meant to rural people, and what it personally meant to work as an agricultural Extension agent.[1] Those writing these anniversary-year reports relied heavily on prior annual reports, discussions with leading growers, and in a few cases, interviews with former agents who had previously occupied their positions.[2]

One of the links that connected Extension workers in 1914 to those in 1939 was that agents dealt with problems facing farmers and the community each and every day—it was just part of the job. As Allen County agent Chester Kimmell noted:

> Extension work is very interesting considering all of the problems that existed 25 years ago and are still with us. Extension work does not become monotonous because of the diversified interests, questions, problems, meetings and demonstrations. A good agent can render a wonderful service to his people and his life's work can be very enjoyable. The work is purely educational and can best be accomplished by spending more of his time in organizing various projects and then directing leaders to execute the job. This is the satisfaction of extension work.[3]

The other similarity that created a kinship between the agents of the past and those writing the reports in 1939 was that, through the years, the recommendations given by Extension were continually based on new research in combination with the agent's experience gained from walking fields or handling livestock. For instance, Bartholomew County agent Arvil Keesling observed, "Agriculture has kept pace with the times. . . . The varieties of soybeans now being planted by the farmers of the county are entirely different. Corn varieties have changed, with hybrid varieties of corn being the latest improvement. In 1914 Burkhart corn was considered a standard variety and today it is almost impossible to find seed to stock for Burkhart except in canning factories. The vaccination of hogs was new in 1914 and is now a standard practice among farmers."[4]

Fayette County agent Dudley Bishopp made similar observations, noting the improvements that had been achieved by farmers as new science and ideas replaced those that had preceded them. He wrote:

> Swine producers now are producing 6 to 8 pigs per litter to market weight in five to seven months. These pigs are almost universally free from worms . . ., and most farmers regularly vaccinate their pigs for hog cholera and self-feed a balanced ration to their pigs from the time they are old enough to eat. It is a rare individual who does not plant hybrid seed corn today, and most farmers are using nothing but hybrid seed. And instead of using just any kind and quality of fertilizer, more and more farmers are bringing soil samples to the extension office for testing to determine the amount and analysis of fertilizer to use.[5]

Clearly Extension's ability to disseminate research to the farmers was having the desired effect. In the earlier and more formative years, the agents successfully taught farmers what they needed to do to prevent hog cholera, treat oats for smut and potatoes for scab, and select higher-yielding open-pollinated corn.[6] These problems were largely replaced with new challenges as tractors and power machinery displaced horses, electricity was used instead of candles and kerosene lamps, soybeans went from an experimental crop to part of the farm's usual rotation, and hybrid corn became more common than open-pollinated varieties.[7]

Agents writing about the early years of Extension painted a picture that was remarkably different than the one that existed in 1939. Charles Nicholson, who had been an agent in Dubois County since 1927, indicated, "The early extension projects in this county were like the tombstones of the early pioneers—few with limited remarks. As time has gone on, the extension projects have taken foothold and have been firmly established and have gained County and State recognition."[8] The roads that the first Extension agents had paved had largely been replaced with new paths and directions, and as the agents at the time of the anniversary asserted, were more appreciated in 1939 than they had been previously.

Keesling summed up a major difference between those holding the agent's job in the 1910s versus the 1930s, noting, "When the work began, there was no set precedent by which the agent could work."[9] In the Benton County report, agent Ellis Christen elaborated on the idea that the first agents were truly pioneers who had been sent to the counties to build a program without the benefit of any blueprints or footsteps to follow, writing:

> At the beginning of Extension work in the County, the County Agent made many of his contacts by individual calls and demonstrations. Practically all of the meetings held were in connection with Public Schools or farmers groups organized for a definite but specific purpose. . . . The subject matter presented at any group was given by the County Agent. In March 1919, the Benton

> County Farm Bureau was brought into existence largely through a series of meetings held over the county in which the County Agent assisted. . . . It was largely the leadership developed by this organization which became the voluntary local leaders of the extension program. With the crash of 1929 and the resulting demoralization of not only farm but all prices, this farm organization in Benton County lost fully half of its active membership and in 1939 has not recovered from that set back. It was, however, the leadership developed and the interest created during the period of active support which has carried on the Extension Program in the County during the last half of the twenty-five years of Extension Work in Benton County.[10]

Agents often mentioned in their historical accounts that the number of farm visits they personally made became fewer with the demands placed on their time. This made perfect sense as World War I, the agricultural depression of the 1920s, and the Great Depression of the 1930s required the agent to depend on subject-matter committees to help them conduct the customary Extension programs.[11] Work handled by committees freed up the agent's time to work on programs that government officials asked them to manage under state and national initiatives.

However, those first agents had little choice but to personally make those farm visits in order to meet the farmers in the county, which led Christen to observe that Extension work had undergone three educational phases during its first twenty-five years:

> First, and early in the history was the method of individual contact and individual demonstrations which gradually grew into a community interest or project. During this period, the already assisting groups such as Public Schools, Local community groups, and Farmers' Institutes were used as the means of contact. The second step was the growth from these local interests to the emphasis being placed on selected subjects or projects, covering not only the community but a County wide area. Lastly was the County wide program covering the entire field, consequently increasing the demands on the local Agent's time and thereby making it necessary that the emphasis for any given year be placed on one or two major subjects in which there was wide spread interest. During all this period of development in teaching methods, individual contacts and many demonstrations were used to show a farmer how.[12]

Often, farmers did not connect the work they saw on neighboring farms with the agent's efforts. As a particular practice was copied from one farm to the next, credit was often given to the farmer who had made the initial change. The agent who had trained that farmer was not directly seen in the exchange of information between farmers. Christen addressed this by saying:

> In the whole history of the Extension Organization in the County, there has been a continual teaching which has enabled many farmers to solve some of their problems without any direct contact with local Extension Representatives. This has been largely due to the development of local leadership and

> also, in part, to a better fundamental education. Thus, when a profitable practice is demonstrated on a farm in a community, many neighbors who observe that it is profitable, are able to rather accurately copy the practice and many never associate the local County Agent or the Extension Program with what he is doing.[13]

A number of agents noted in their reports that, early on, opposition to the county Extension agent system was twofold, with objections based on using local funds to support the position and the perception that many of the agents were book smart but not experienced.[14] Christen's account stated, "Perhaps above all, the early county agents had primarily the problem of breaking down the resistance of accepting advice from [a] technically trained individual and to get the farmer to feel that the purpose of the program was to cooperate with them [the farmers] rather than to dictate to them."[15]

By 1937, Gibson County agent Roy Gwaltney wrote that this opinion had changed over the course of time, noting, "People in general are relying more and more upon the Extension Office as a source of reliable and timely information. It would seem that old antagonism for anyone who falls in the class of 'book farmer' is dying out and as the service continues to prove its reliability and worthiness, and as the older generation passes the reins to the younger one, we will find the value of Extension Service has grown, in the estimation of the people."[16]

In the formative years, it was the responsibility of the agent to bring local farmers to demonstration plots organized by Extension.[17] But Clay County agent Raymond McHargue stated that, over time, it became understood that the work was a cooperative effort between farmers and the agent. He wrote, "In my opinion the greatest differences in extension work now and 20 years ago, is the fact that farm people are used to leaders in their community to a much greater degree now than several years ago and are made to feel that the extension program is a large part their program and not necessarily a county agent's program."[18] As agents relied more on committees to accomplish the needed work, farmers were taking more ownership of the projects, which would pay tremendous dividends during times of war and economic depression.

Those helping Extension with all aspects of the work were not run-of-the-mill farmers, of course.[19] They were men and women who were respected in their communities and who used their influence to give their neighbors the confidence to adopt the latest agricultural research.

From an Extension perspective, those volunteering to do the work had to follow the guidelines set up by the agent, see the project through to the completion, and be available to talk about the findings. Blackford County agent Walter Rusk wrote the following in his 1939 annual report:

> As farmers became more versatile in organization work and began to ask for more assistance from the county agent office, the project committee method of conducting extension methods of teaching was adopted and, as a result, reached a greater number of people. Leaders were trained, [and]

> farm and home demonstrations were conducted with a result that a greater number of people were influenced in changing their attitude and adjusting their farm business to the rapid change in economic and social conditions. With farm people working together in various organizations, they soon learned that by helping others, they were helping themselves. When Jim Smith or John Jones produced a yield of 100 bushels of corn per acre or a litter of hogs that went to market weighing one ton at 200 days of age or a flock of poultry that netted two dollars per bird or a dairy cow that made a record of 500 pounds of butterfat per year, their influence as community leaders spread in influencing other farmers to make even greater records.[20]

This meant agents had to be choosy about who they asked to serve on their Extension advisory boards and project committees because many of the decisions about what to work on would be left to those groups. As Adams County agent Lawrence Archbold wrote in 1939, "During the 25 years of the life of extension work, a great many people have given and are giving of their time as local volunteer leaders. Extension workers, themselves, are learning how to properly train these leaders; and together, the extension workers and volunteer leaders are learning how to . . . evaluate their problems."[21] This collaborative effort made Extension and, specifically, the agent more effective during times of war and adversity.

Most agents reflecting back on past activities and comparing them to what they knew in 1939 would have agreed wholeheartedly with the statements put forward by Fayette County agent Dudley Bishopp, who said, "[E]xtension work of today is more fully reaching that goal which all extension members are striving for: to work out a program that strikes at the problems that are holding back the progress of agriculture and affecting the well being and happiness of the people of the county."[22]

It was no small matter when Bishopp wrote in his report that by 1939 the farmer had realized "the importance of scientific research into his many problems and looks to the local extension representative to keep him informed on the new better method of crop and livestock management."[23] Each new advancement created its own set of questions. For example, did hybrid corn need more fertilizer? What were the seeding rates? When should it be planted? Was one variety better suited than another? So Extension specialists, agents, and agricultural researchers had to address whichever questions arose as agriculture progressed. Thus, the agent had to be motivated with a desire for lifelong learning. Keeping up with advantages meant, as Bishopp noted, "that information for the betterment of rural people can be disseminated. As conditions are constantly changing and scientific research is developing new and better practices, there promises to be a continued need for Extension workers."[24]

Agents collectively mentioned that the Depression-era federal Agricultural Adjustment Act played a major part in solidifying their position in the county. Many of the rank-and-file farmers who had never directly consulted Extension before turned to the agents for help with explaining the programs administered by the Agricultural Adjustment Administration (AAA) and obtaining the federal government payouts. This led Clay County agent Raymond McHargue to observe:

> During the 1930s, prices became low, over production was apparent, and many farmers were unable to see the need for extension work. However, since the beginning of the farm board [Federal Farm Board] which was followed by the new deal legislation, the duties of the county agricultural agent have been greatly increased and probably doubled. However, it is apparent that the Triple A program has been of much help rather than a hindrance to extension work in Clay County. Through the Triple A program and Agricultural Conservation program, many farmers have been thrown into contact with extension work who would otherwise not have been contacted by the county agent's office. Through the educational programs, farmers have been made to feel that there are problems which are of vital importance that were given very little thought a few years ago.[25]

Agents offered mixed views of their positions. On one hand, many felt that being a county Extension agent was an honorable and worthwhile endeavor. As Earl Miles in Clark County observed, "The county agent has adjusted himself to the changes, growing with the programs, working with no regard as to hours working for improvement in social and economic conditions in his particular county, to a place where it would be a source of pride of those concerned."[26] And Fulton County agent Martin Huxley observed, "County Agent work is an interesting, healthful and inspiring vocation. It is an exacting profession, demanding a larger proportion of the agent's time than many other positions of similar standing. Among other things, it requires tact, a working knowledge based on both education and experience, a large measure of executive ability, a sincere desire to render practical service to agriculture and an unlimited capacity for work to be an efficient and effective county agent."[27]

Working around the clock for the betterment of the county in which the agents served was a noble aspiration, but comments made by those who were employed in 1939 also showed that it was physically taxing and mentally stressful. Agents were routinely asked to sacrifice their personal lives for the good of others, leading one agent to write, "One wonders sometimes how many years the human machine can hold up at this pace and then asks, 'Who will want me as Agent when I'm 50 or 60?'"[28]

In fact, the day-in, day-out grind of the job made it difficult for many agents to see themselves making a career of it.[29] Consequently, the tenure for most agents was short and the rate of burnout was high, leading many to seek careers elsewhere.[30]

Gibson County agent Roy Gwaltney expressed his worries about the stressful situation that the agents found themselves in, writing:

> Extension work is a profession where a man who likes to be of service to his fellow man can get a great deal of satisfaction. It is a profession, however, which will take the best years of a man's life and leave him without any source of income. As the salary in most cases is not sufficient to allow him to make any savings, and after a man prepares himself and works as a County Agent for ten, fifteen, or twenty years, it is going to be hard for him to find other employment. There should be some kind of a retirement fund set up in order that these [individuals] might be guaranteed security after their period of usefulness is over. Ample compensation in case of injury while on duty is another safeguard that should be available to one whose duty demands that he travel the highways both day and night.[31]

As Gwaltney pointed out and many other agents would also bring up in their 1939 reports, low pay and retirement benefits—or the lack thereof—were two major issues agents confronted. Apparently, the Indiana agents were not paid enough from federal funds to make them eligible for retirement benefits under the federal government. Huxley addressed this in his report, saying:

> It is the opinion of the writer of this report that Indiana county agricultural agents should receive a greater proportion of their salary from federal funds and should be made eligible for final retirement on pension. In view of the great amount of work of a governmental nature that county agents are performing, this would seem only fair. County Agents of other states receive a much larger proportion of their salary from federal funds which automatically gives them preference in federal retirement. Given reasonable assurance of retirement with adequate income after spending one's best years in extension teaching work, I believe County Agent's work as a profession would be made more secure, greater confidence in one's own future would be obtained and County Agent work made more attractive as a life work.[32]

Ellis Christen in Benton County echoed many of the same thoughts expressed by his colleagues and also covered other points that he believed needed to be improved if agents were to remain for any extended period of time:

> Education is a slow process and it is the rate of improvement that tends to discourage the Extension Worker. No extension worker can be very successful or enthusiastic, unless she or he derives a great deal of personal gratification for helping someone else. It is such gratification that serves as a part of the Extension Worker's compensation. Selfishness and jealousy have no place in the Field Agent's life. It is apparent that if a personnel of ability, and a tenure of permanence is to be maintained, the Federal and State Departments must work together with field workers and taxpayers on a stable and higher plane of compensation. Either the plane of [the] Extension Field must be rated to the level of commercial enterprises or commercial firms will continue to take many of the better personnel from Agricultural Extension. If the experienced and more successful extension workers continually leave, then the Extension

> work over the entire county must be satisfied to remain at its present level or to improve itself slowly.
>
> At the same time, the Extension Worker must continue to improve himself professionally; practically he must have an opportunity to increase his or her knowledge in those fields in which there has been previously a minimum of technical training. Most Extension workers, I believe, recognize continually that there are specific subjects on which they need further information in order to best do the job in their county.
>
> With the tendency to increase the number of extension workers in any county from one to perhaps, two or three or four workers in a county, there must be more detailed instructions on relationships and closer supervision by state and federal authorities. County Extension workers cannot be considered as infants but as qualified individuals who do know their local problems. Programs cannot be dictated successfully from the top down. Programs must be planned by the farmer and home maker themselves. It becomes the job of the extension worker, regardless of position, to help complete that program. While the county worker is at the end of the line, he knows more about the local problem than does the specialist or administrator. The County Agent needs assistance on how to do the job. If the extension worker, regardless of his or her position, attempts to dictate the program to the people, then our farmers will not in general accept it as their responsibility. The acceptance of the program by the people as a part of their duty will make for the most rapid progress.[33]

At the time, Christen and other agents were upset that there was a disconnect between the counties and the administrators at the state agricultural colleges and the U.S. Department of Agriculture. Allen County agent Chester Kimmell was very direct in his criticism in this regard, writing, "Probably one of the biggest mistakes in extension work is brought about by the fact that our leaders in the colleges and United States Department of Agriculture do not contact the extension agents and see their problems as they see them. The least that the Director of the Purdue Extension Service and the Director of the Purdue [Agricultural] Experiment Station can do is to attend all of the district county agents' conferences once a year."[34]

Raymond McHargue in Clay County believed that the administrators also needed to build more stability into the position. He explained:

> I feel that the life of an extension worker is an honorable occupation and an interesting one. There is, perhaps, one problem which should receive the attention of the State Agricultural College and the U.S. Department of Agriculture. Extension work always had been on a somewhat uncertain basis and those engaged in the work have not been able to make plans which they were sure could be carried out several years hence. Although extension work is on a more stable basis now than three years ago, an agent working on the

> minimum salary would be very poorly paid, and my one suggestion would be to remove still more of the uncertainty from extension work.[35]

The "more stable basis" McHargue mentioned likely referred to the change in the chain of command for the agents. In July 1937, the Indiana General Assembly had changed the management structure for Extension in the state, making Purdue administrators fully responsible for hiring, training, evaluating, and retaining agents.[36] (See Chapter 1.) This single move by the state legislators gave agents more job security since they would now be evaluated based on performance.[37]

In time, the county Extension agent position would move from a semi-political one to that of a highly regarded professional educator and community resource. With the transition to fall wholly under the Purdue administration along with the inclusion of retirement benefits, pay raises, and lifelong learning opportunities, the position ultimately became a career job.

The twenty-fifth year of Extension was an important milestone given that in 1914 no one knew if this type of educational approach would work. This brand-new system of outreach carried out by rookie and inexperienced county agents was attempting a large-scale experiment to reach every farmer in the country. The goal was simply to get research in front of the farmers in hopes that the changes in practices would lead to more profitable farming and an improved standard of living.

While the acceptance of the agents' advice and the adoption by farmers of scientific research from Purdue occurred more slowly than most had hoped, nevertheless the evidence is clear that, by 1939, both were part of the agricultural fabric. By all accounts, the educational experiment had succeeded.

In 1931, workers use equipment and plants provided by a canning company to set tomato plants on the farm of Joseph Holscher of Fowler in Benton County. During the Great Depression, Extension promoted tomatoes as an important cash crop. Growers often contracted with a canning company in advance, as Holscher had done for his 20 acres of tomatoes. Learning how to produce this crop profitably during the Depression would become important during World War II, when tomatoes were identified as an essential war crop by the federal government. Courtesy of J. C. Allen and Son.

Growers learned how to grade tomatoes according to federal standards at this Hancock County event in 1936. Farmers were paid more by the canneries for products that were graded No. 1, so understanding and meeting these standards brought a higher return on the crop. Courtesy of Purdue University Libraries, Karnes Archives and Special Collections.

William Lampe harvests potatoes in Allen County in 1930. During the Depression, Indiana imported the majority of its potatoes, so some farmers found it was cost-effective to grow them for their local markets as a new cash crop with a guaranteed demand. Courtesy of Purdue University Libraries, Karnes Archives and Special Collections.

By 1939, Extension was telling farmers that they could raise 400 bushels of potatoes per acre with the right management practices. Reaching these levels, as was done by Allen County's Charles Hutson (middle) in 1932, provided significant income to farmers, helping them to pay mortgages during the hard times of the Depression. Courtesy of Purdue University Libraries, Karnes Archives and Special Collections.

A tobacco crop dries at the Brown County farm of Charles Seitz in 1937. Some farmers in southern Indiana, especially in the counties bordering Kentucky, grew tobacco as an alternative crop during the Depression. Courtesy of Purdue University Libraries, Karnes Archives and Special Collections.

Ralph Conder weighs milk at the J. C. Edgerton Dairy in 1938 as part of his work for the Grant County Dairy Herd Improvement Association (DHIA). The Depression forced dairy owners to closely examine the amount of milk and butterfat each animal produced relative to the cost of feeding the animal. Fees to participate in the local DHIA covered the tester's salary and expenses, and in return, members received detailed production records that helped them manage their herds more efficiently. Courtesy of Purdue University Libraries, Karnes Archives and Special Collections.

In 1938, a worker caring for dairy cattle in Tippecanoe County provides an amount of feed linked to the production of each animal. Cows that produced more milk and butterfat were fed more, and their offspring were used as herd replacements. Courtesy of J. C. Allen and Son.

CO-OPERATIVE EXTENSION WORK
in
AGRICULTURE AND HOME ECONOMICS
STATE OF INDIANA

Purdue Univ. Agri'l
Extension Department
U.S.Dept. of Agriculture
Co. Bd. of Education
Co. Farm Bureau
Co-operating

EXTENSION SERVICE
COUNTY AGENT WORK

Winamac, Indiana

THE FELLOW WHO GETS ACROSS

E. T. Wallace of the Purdue Dairy Department will hold three important meetings with the dairymen of Pulaski Co.:

Francesville	- Library	- Mon. Eve.	- March 24
Winamac	- Court Room	- Tues.Eve.	- March 25
Monterey	- School Bldg.	- Wed. Eve.	- March 26

TIME - 8 o'clock p.m. O. H. McNARY, COUNTY AGENT.

In 1930, Purdue dairy Extension specialist Elvin T. Wallace was the featured speaker at a series of meetings held in Pulaski County that focused on the dairy industry. To help dairies become more profitable, Extension encouraged management strategies such as using production records to determine which cows to keep in the herd, providing better feed for high-producing cows, setting a goal of 300 pounds of milk per cow annually, and using a proven sire to improve the genetics of the herd. Courtesy of Purdue University Libraries, Karnes Archives and Special Collections.

Farmers from seven Indiana counties and Ohio attend a 1933 Extension event on the T. T. Wallace Farm in Adams County. At the time, Adams County was noted as having some of the most productive dairies in the United States. Courtesy of Purdue University Libraries, Karnes Archives and Special Collections.

A 1932 event held on the Dan D. Schwartz Farm in Adams County included a discussion about the production records of five cows linked back to the genetics of one bull. Tracing increased productivity and profit to an animal's lineage became a popular topic at Extension programs for dairies during the 1930s. Courtesy of Purdue University Libraries, Karnes Archives and Special Collections.

In 1939, Charles Grandlienard exhibited Queen Esther's Fairy (top) at the Wells County Fair. The accompanying sign (bottom) allowed fairgoers to better understand this cow's production records. Extension used this type of exhibit to highlight important information for its constituents. Courtesy of Purdue University Libraries, Karnes Archives and Special Collections.

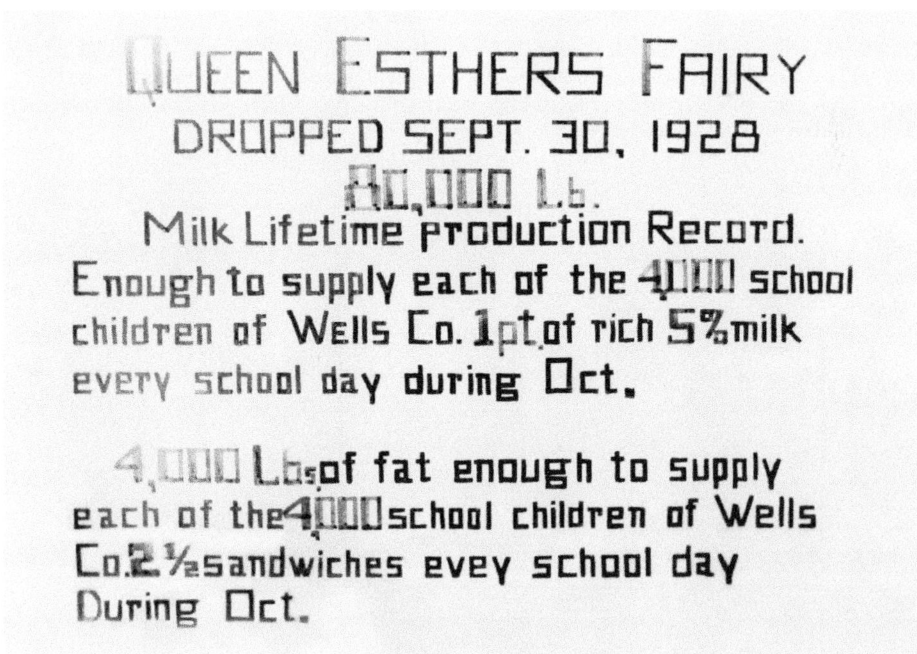

Ira Eby from Wakarusa in Elkhart County holds his Ayrshire bull as another man looks on in 1930. Extension encouraged dairy owners to improve their herds by breeding them with proven sires. These were bulls that had produced high-yielding offspring, which could be verified by the production records of their daughters. Courtesy of J. C. Allen and Son.

In 1932, the St. Joseph County dairy tour hosted by Extension stopped at the Reliable Dairy. This gave tour participants a chance to see how other dairies operated and the use of new technologies and management practices. Courtesy of Purdue University Libraries, Karnes Archives and Special Collections.

This 1933 illustration from Vanderburgh County focused attention on the importance of producing milk in a sanitary manner for consumers, especially children. Early in the 1930s, new regulations passed by cities, counties, and the state focused on ensuring that milk contained fewer bacteria. This helped protect the health of the consumers while expanding market opportunities for Indiana dairy producers. Courtesy of Purdue University Libraries, Karnes Archives and Special Collections.

Farmers such as this one in Union County in 1939 had little choice but to upgrade their dairy processing facilities if they expected to sell milk under new and stricter regulations. Courtesy of Purdue University Libraries, Karnes Archives and Special Collections.

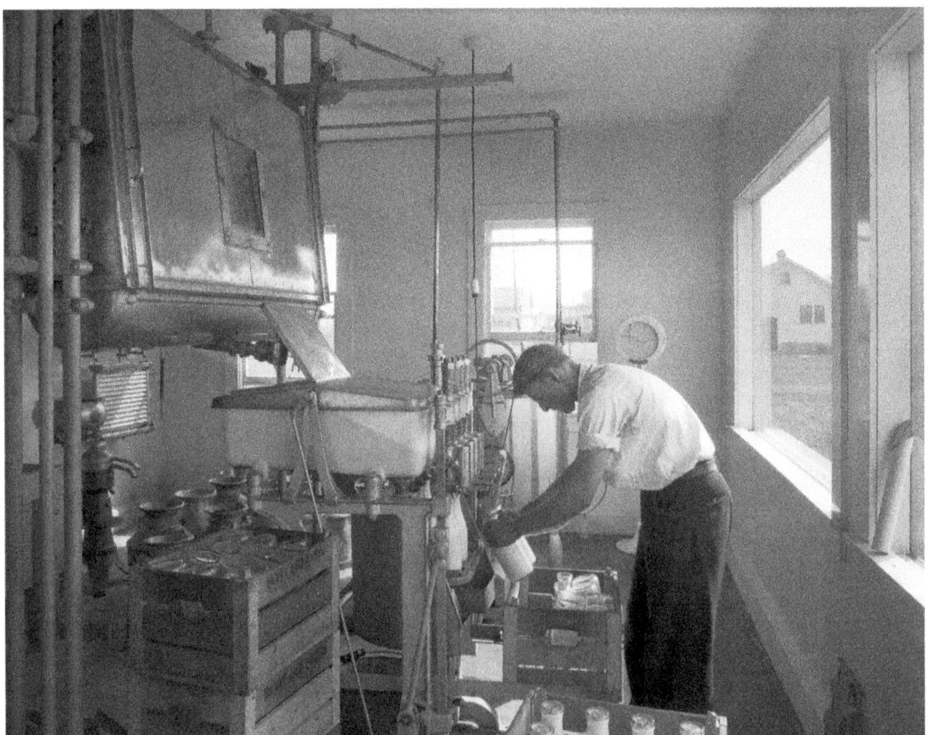

Fred Schornstein, shown bottling milk in 1932, owned the Betty Lou Dairy Farm in Kokomo in Howard County. The inside view of this facility was typical of those dairies that had upgraded their operations to meet the continually changing dairy regulations at that time. Courtesy of J. C. Allen and Son.

In 1934, a man loads a delivery truck in Huntington County at the Big Elm Dairy owned by Chancy Theobald of Andrews. This dairy wanted consumers to know that its herd was blood-tested, indicating it was free of tuberculosis. This helped build consumer confidence in the safety of its products. Courtesy of J. C. Allen and Son.

A flock of sheep waits for a portable dipping machine to arrive at the Fountain County farm of M. M. Young in 1938. Many county farm bureaus built portable dipping equipment to remove ticks and lice from sheep at a cost of 8–10 cents per animal. Extension calculated that this process, which resulted in a larger fleece and a higher gain per animal, returned three times the original cost to the owner. Courtesy of Purdue University Libraries, Karnes Archives and Special Collections.

CO-OPERATIVE EXTENSION WORK
IN AGRICULTURE AND HOME ECONOMICS
STATE OF INDIANA

PURDUE UNIVERSITY AGRICULTURAL EXTENSION
DEPARTMENT, U. S. DEPARTMENT OF AGRICULTURE
AND COUNTY BOARD OF EDUCATION CO-OPERATING

EXTENSION SERVICE
COUNTY EXTENSION WORK
RICHMOND, IND.

JUNE 15, 1936

Are your lambs market toppers?

WHICH TYPE OF RAM ARE YOU USING?

THE RAM IS ALWAYS COUNTED 60% OF THE FLOCK. A GOOD RAM STAMPS HIS LAMBS WITH DESIRABLE MUTTON QUALITIES AND IMPROVES THE FLEECE OF EWES SAVED FOR BREEDING PURPOSES FROM THE AVERAGE FARM FLOCK.

NOW IS THE TIME TO MAKE YOUR SELECTION OF A CHOICE RAM TO SIRE YOUR NEXT YEAR'S LAMB CROP. RAMS OF THIS TYPE ARE NOW AVAILABLE FROM WAYNE COUNTY BREEDERS AT REASONABLE PRICES. FLOCK OWNERS OF A NEIGHBORING OHIO COUNTY PURCHASED 18 SUCH RAMS IN WAYNE COUNTY A YEAR AGO AND EXPECT TO PURCHASE ADDITIONAL RAMS HERE THIS MONTH. OUR OWN FLOCKS NEED THE SERVICES OF THESE RAMS NEXT FALL. A SALE LIST IS BEING MADE UP TO ASSIST YOU IN LOCATING RAMS FOR SALE. WE WILL GLADLY HELP YOU MAKE YOUR SELECTION NOW.

IF WE MAY AID YOU IN SORTING YOUR LAMBS FOR MARKET, COME IN AND WE'LL SET A DATE. THE NEIGHBORING FLOCK OWNERS MIGHT BE INVITED IN AT THE TIME WE GRADE YOUR LAMBS.

YOURS FOR CHOICE MARKET TOPPING LAMBS.

VERY TRULY YOURS,
S W Milligan
S. W. MILLIGAN
COUNTY AGRICULTURAL AGENT

Wayne County agent Scott Milligan reminded those raising sheep in 1936 about the importance of selecting a quality ram when breeding ewes. At the beginning of the Depression, Extension encouraged Indiana farmers to raise more sheep because the market prices for sheep were stable at a time when prices for other commodities had fallen. Courtesy of Purdue University Libraries, Karnes Archives and Special Collections.

The Clay County Farm Bureau managed this mobile sheep dipping operation (top and bottom), charging Charles Romas of Brazil 10 cents per head to treat his flock for external parasites in 1935. Sheep would be pushed up the ramp toward a vat containing a chemical mixture that would control ticks, lice, and mites. Courtesy of J. C. Allen and Son.

Delaware County farmer Arthur W. Shideler feeds his lambs in 1938. During the Depression, many farmers purchased young lambs from western states to pasture and grow to market weights. This was a profitable endeavor at a time when farmers were looking for new sources of income. Courtesy of J. C. Allen and Son.

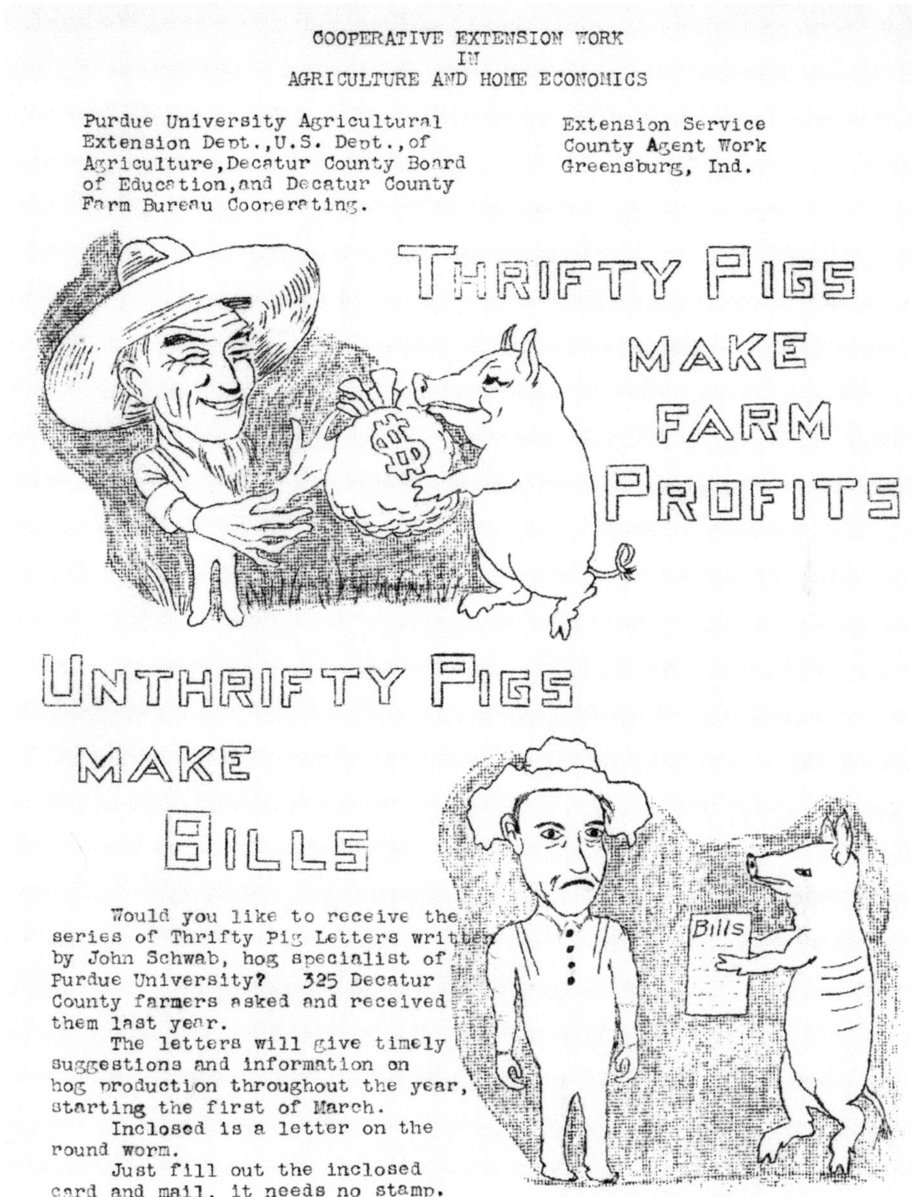
COOPERATIVE EXTENSION WORK
IN
AGRICULTURE AND HOME ECONOMICS

Purdue University Agricultural
Extension Dept., U.S. Dept., of
Agriculture, Decatur County Board
of Education, and Decatur County
Farm Bureau Cooperating.

Extension Service
County Agent Work
Greensburg, Ind.

Would you like to receive the series of Thrifty Pig Letters written by John Schwab, hog specialist of Purdue University? 325 Decatur County farmers asked and received them last year.

The letters will give timely suggestions and information on hog production throughout the year, starting the first of March.

Inclosed is a letter on the round worm.

Just fill out the inclosed card and mail, it needs no stamp.

Yours for thrifty pigs,

I. L. Thurston

County Agent

In 1931, agent Ira Thurston sent Decatur County farmers this circular letter encouraging them to sign up for a series of letters written by Purdue hog Extension specialist John Schwab. With hog numbers reduced during the Depression, the correspondence focused on sharing timely information aimed at helping producers run more efficient swine operations that would finish hogs for market sooner. Courtesy of Purdue University Libraries, Karnes Archives and Special Collections.

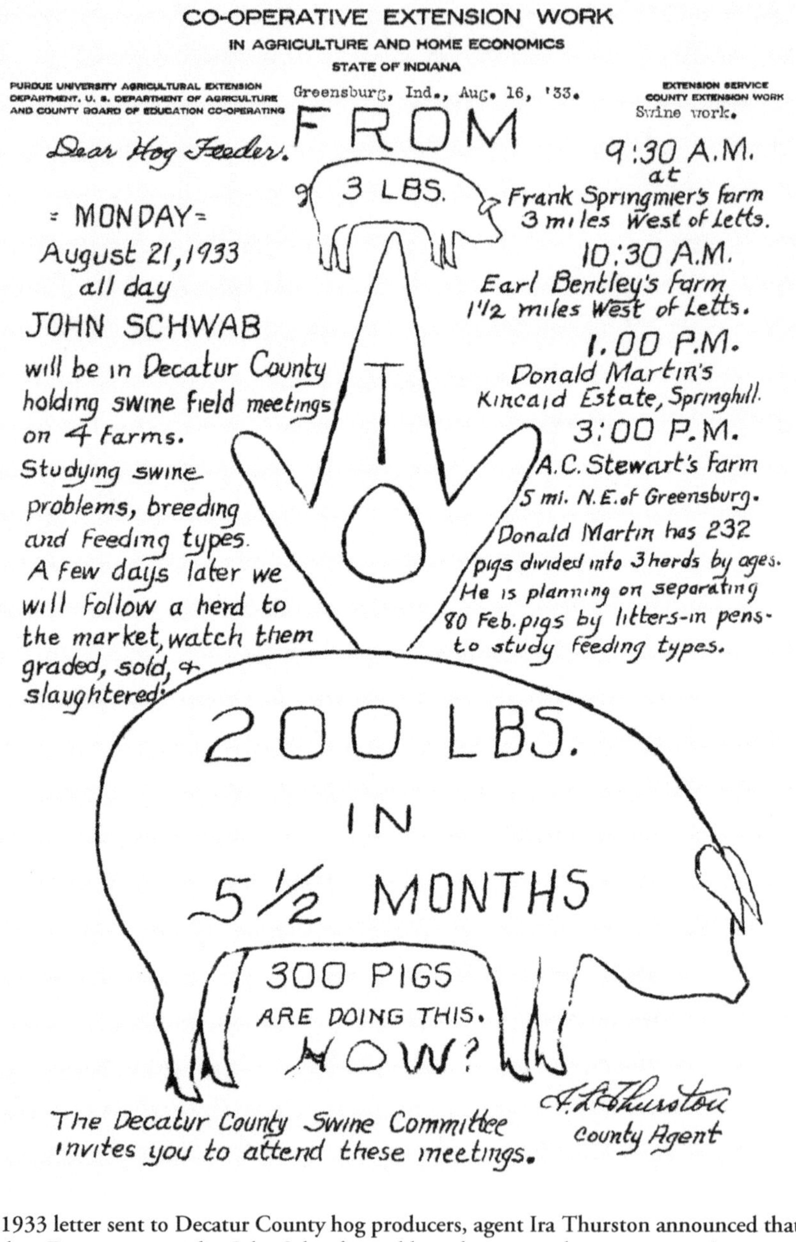

CO-OPERATIVE EXTENSION WORK
IN AGRICULTURE AND HOME ECONOMICS
STATE OF INDIANA

PURDUE UNIVERSITY AGRICULTURAL EXTENSION
DEPARTMENT, U. S. DEPARTMENT OF AGRICULTURE
AND COUNTY BOARD OF EDUCATION CO-OPERATING

Greensburg, Ind., Aug. 16, '33.

EXTENSION SERVICE
COUNTY EXTENSION WORK
Swine work.

Dear Hog Feeder.

= MONDAY=
August 21, 1933
all day
JOHN SCHWAB
will be in Decatur County
holding swine field meetings
on 4 farms.
Studying swine problems, breeding and feeding types.
A few days later we will follow a herd to the market, watch them graded, sold, + slaughtered.

FROM 3 LBS. TO 200 LBS. IN 5 1/2 MONTHS

300 PIGS
ARE DOING THIS.
HOW?

9:30 A.M.
at
Frank Springmier's farm
3 miles West of Letts.

10:30 A.M.
Earl Bentley's farm
1 1/2 miles West of Letts.

1.00 P.M.
Donald Martin's
Kincaid Estate, Springhill.

3:00 P.M.
A. C. Stewart's farm
5 mi. N. E. of Greensburg.

Donald Martin has 232 pigs divided into 3 herds by ages. He is planning on separating 80 Feb. pigs by litters-in pens-to study feeding types.

The Decatur County Swine Committee
invites you to attend these meetings.

I. L. Thurston
County Agent

In this 1933 letter sent to Decatur County hog producers, agent Ira Thurston announced that Purdue hog Extension specialist John Schwab would conduct a one-day program in August at four area farms. Schwab was a much-sought-after speaker who spent a large portion of his time making presentations at demonstration programs, short courses, judging events, and marketing tours held around the state. Based on Schwab's research, agents such as Thurston told producers that getting a three-pound baby pig to 200 pounds in five to six months would return the highest profit. This opened the door for Extension to promote better diets and better management for hogs to accomplish this weight gain. Courtesy of Purdue University Libraries, Karnes Archives and Special Collections.

In 1934, this Extension meeting at the James Stryker Farm in Parke County reviewed the steps required to make hogs ready for market in 180 days. In order to reach the higher-priced September hog markets, farmers needed sows to deliver pigs in February, rather than in May or June. Then, young pigs would be placed on self-feeders as soon as possible to improve feed consumption and nutrition. Courtesy of Purdue University Libraries, Karnes Archives and Special Collections.

In 1939, this Grant County exhibit at the Indiana State Fair showcased the importance of Purdue's Thrifty Pig Project to the financial success of hog producers. The list on the right side of the display shows the variety of methods that Extension used to promote the project and better management techniques among hog producers. Courtesy of Purdue University Libraries, Karnes Archives and Special Collections.

In 1929, Purdue's ninth annual Swine Day drew approximately 700 attendees to a university farm in West Lafayette. By the start of the Great Depression, Purdue agricultural scientists had already conducted decades of research, so they were ready to help hog producers navigate the tough economic times ahead using proven management strategies that would maximize profits. Courtesy of J. C. Allen and Son.

At the 1936 Shelby County Hoosier Ton Litter Show, Arthur Wisker exhibited his litter of hogs in front of a large crowd. This competition was one of several outreach methods that Extension used to encourage producers to follow recommended management practices to achieve the best results with their livestock. In many cases, participants would be interviewed as to what they had done to raise the showcased litters, which then encouraged other producers to adopt the same strategies. Courtesy of Purdue University Libraries, Karnes Archives and Special Collections.

In 1934, Hazel Miller (right) of Lagro in Wabash County gathers eggs with her daughter. Poultry, which had been a major financial lifeline for farmers during the 1920s agricultural depression, continued to provide a stable income during the Great Depression that followed. Those who had put into place proven management practices increased their net profits from selling eggs. One strategy was to gather eggs in a wire basket to allow them to cool quickly, which helped maintain egg quality. Courtesy of J. C. Allen and Son.

Miami County agent John Wylie advertised that an Extension poultry school would be held in January 1934 in Denver, Indiana. Since so many families relied on the income that their flocks provided during the Depression, events such as this were popular, typically drawing 100–400 people. Courtesy of Purdue University Libraries, Karnes Archives and Special Collections.

Extension conducted poultry programs around the state, such as this Lawrence County event held in 1934 at the Moses Fell Annex, a Purdue demonstration farm in Bedford. By attending, producers gained access to poultry experts and the latest research results from Purdue. Courtesy of J. C. Allen and Son.

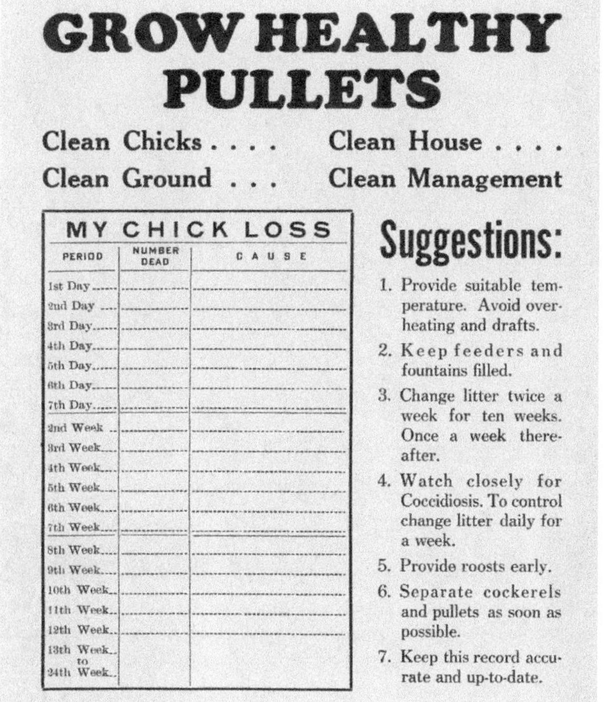

GROW HEALTHY PULLETS

Clean Chicks . . . . Clean House . . . .
Clean Ground . . . Clean Management

MY CHICK LOSS

PERIOD	NUMBER DEAD	CAUSE
1st Day		
2nd Day		
3rd Day		
4th Day		
5th Day		
6th Day		
7th Day		
2nd Week		
3rd Week		
4th Week		
5th Week		
6th Week		
7th Week		
8th Week		
9th Week		
10th Week		
11th Week		
12th Week		
13th Week to 24th Week		

Suggestions:

1. Provide suitable temperature. Avoid overheating and drafts.
2. Keep feeders and fountains filled.
3. Change litter twice a week for ten weeks. Once a week thereafter.
4. Watch closely for Coccidiosis. To control change litter daily for a week.
5. Provide roosts early.
6. Separate cockerels and pullets as soon as possible.
7. Keep this record accurate and up-to-date.

Chicks Started........ Date Started.................... Breed....................

This chick mortality record was distributed to Bartholomew County poultry producers in 1931. Extension promoted the use of this form, which included suggestions and important reminders, to help poultry farmers raise as many chicks as possible. Courtesy of Purdue University Libraries, Karnes Archives and Special Collections.

In 1934, agent Ralph Maggart (third from right) teaches an audience of Carroll County farmers how to cull unproductive hens from their flocks as a child watches from the roof. Removing hens that weren't laying eggs was critically important during the Depression since it reduced farmers' expenses and raised their profits. Courtesy of J. C. Allen and Son.

In White County in 1934, these individuals identify underperforming hens and conduct blood tests on the flock of Harry Hagerty in Brookston. Culling unproductive birds and removing poultry that carried diseases were two strategies that improved egg production and reduced mortality rates in a flock. Courtesy of J. C. Allen and Son.

In 1930, a man and woman use a 1928 Chrysler Roadster to transport eggs from a poultry farm in Tippecanoe County. Courtesy of J. C. Allen and Son.

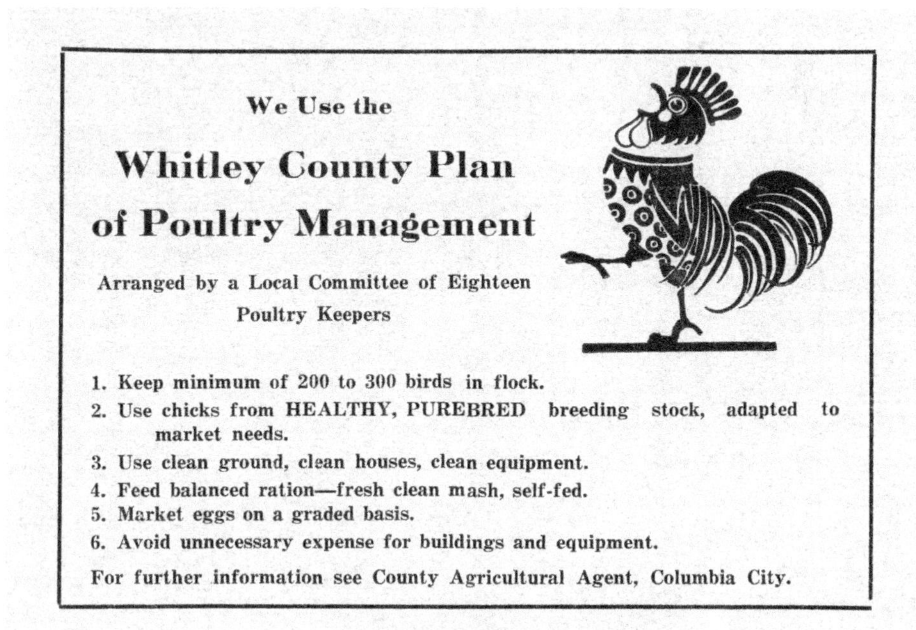

We Use the

Whitley County Plan of Poultry Management

Arranged by a Local Committee of Eighteen Poultry Keepers

1. Keep minimum of 200 to 300 birds in flock.
2. Use chicks from HEALTHY, PUREBRED breeding stock, adapted to market needs.
3. Use clean ground, clean houses, clean equipment.
4. Feed balanced ration—fresh clean mash, self-fed.
5. Market eggs on a graded basis.
6. Avoid unnecessary expense for buildings and equipment.

For further information see County Agricultural Agent, Columbia City.

In 1933, the Whitley County agent distributed this card that poultry producers could tack up to the hen house door to remind them of a six-point management plan for maximizing profits from their flocks. Courtesy of Purdue University Libraries, Karnes Archives and Special Collections.

GET YOUR BUS READY AND
TAKE THE

ON THE CASS-MIAMI COUNTIES
POULTRY TOUR

SCHEDULE

9:30 - 10:15 - Lee Long's - 1 Mile East & 1 Mile North of D-nver
10:30 - 11:00 - Chas. Long's - 1 Mile East & ½ South of Deedsville
11:15 - 11:45 - Lloyd Carson - 1 Mile South of Macy Road 31
12:00 - 1:30 - Picnic Lunch at Nyona Lake
1:45 - 2:15 - Mrs. L. Hill - 1 Mile E. & ¾ Mile No. of T. Mile
2:30 - 3:00 - Omer Maus - 1 Mile E, ¼ Mile N. of Twelve Mile

Speaker: Leon Todd, Poultry Extension Specialist.

Topics: Quality Eggs, Culling Pullets, Housing & Care of Pullets.

E. A. Johnson
Cass County Agri. Agent.

John E. Wylie
Miami County Agri. Agent.

DATE - AUGUST 21

Poultry tours were a popular way to give farmers a first-hand look at successful management strategies. (Top) An August 1934 tour across Cass and Miami Counties featured Purdue poultry Extension specialist Leon Todd and included stops at five poultry farms. (Bottom) This 1931 poultry tour drew a large crowd to the Fred Mangus Farm in LaPorte County. Courtesy of Purdue University Libraries, Karnes Archives and Special Collections.

In 1934, Mrs. Perry Spangler of Miami County prepares to store eggs in a root cellar until they are shipped. Storing eggs at cooler temperatures kept them fresh longer and brought higher prices at market. Courtesy of J. C. Allen and Son.

(Top) A worker prepares eggs for market in 1941. Cleaning debris off of eggs, grading them for size before packing, and maintaining freshness improved the price paid for them in the marketplace. Courtesy of J. C. Allen and Son.

(Bottom) This 1933 photo shows an example of crated eggs from Miami County farmers prior to shipping by rail. Farmers had to work cooperatively to produce enough eggs to attract the interest of wholesalers in New York, Boston, and other large metropolitan areas. Courtesy of Purdue University Libraries, Karnes Archives and Special Collections.

Farmers deliver eggs to a refrigerated railcar at Denver for cooperative shipping in 1934. That year, Miami County poultry producers capitalized on higher prices by shipping large quantities of eggs to the New York market. However, the eggs that were shipped by the Denver Egg Car Shipping Association had to meet quality standards for size, freshness, and cleanliness in order to capture those higher prices. Courtesy of J. C. Allen and Son.

Workers grade and pack eggs inside the Swift and Company plant at Logansport in 1934. Courtesy of J. C. Allen and Son.

MASSEY HARRIS

PART 3

Soldiers of the Soil During World War II (1940–1945)

CHAPTER 25

Fitting the Extension Program to Wartime Conditions Has Required Some "Give and Take"

In 1940, many countries had already been or would soon be drawn into the war that had begun when Germany invaded Poland in 1939, but the American public had little interest in sending soldiers into direct military action overseas. Americans had not forgotten the extreme sacrifices made by their sons as soldiers and daughters as field nurses in World War I, even though twenty years had passed since the hostilities had ended. Still, even before entering the war, the United States provided behind-the-scenes support for its foreign allies through the Lend-Lease Act, which President Franklin Roosevelt signed into law in March 1941.[1]

The Lend-Lease Act gave the president the legal authority to loan to Great Britain—and, in time, to fifty other countries—military aid that would be paid in kind at a later time. This aid came in the form of cash, food, fuel, and equipment to countries whose efforts in the war were considered to be protecting the national security of the U.S.[2] In this manner, the U.S. could provide much-needed supplies to its allies without putting military boots on the ground, which, at the time, seemed to appease the public, Congress, and even some of the president's most trusted military advisors.[3]

The importance of the food shipments, in particular, was well articulated by Robert H. Brand, a British diplomat in Washington, D.C., who said:

> It [donated food supply from the U.S.] is enough to keep Britain in the war. It furnishes more than one-fourth of Britain's total supply of the rich protein foods we so badly need. . . . When Denmark, the Netherlands, and Belgium were occupied, we lost overnight about three-fourths of the bacon and eggs and dried and condensed milk that we normally imported. . . . When your first Lend-Lease ship arrived in May, 1941, it brought four million eggs, sixty tons of cheese, and one thousand tons of flour. But that first ship also brought something else. It brought a million tons of hope. It let us breathe again.[4]

While it was true that the U.S. was sending significant military aid and humanitarian supplies across the Atlantic Ocean, Roosevelt did not have public support nor congressional backing for sending soldiers to fight in faraway places. This political posturing—where the U.S. had one foot in and one foot out of World War II—was the best that Roosevelt could do at the time to help political friends of the U.S. overseas without declaring outright war against the aggressors.

Through the aid program, Roosevelt had also directed American naval warships to protect convoys in the Atlantic Ocean under a policy of neutrality. But Americans woke up on October 31, 1941, to read that a German U-boat had torpedoed the USS *Reuben James* near Iceland, killing more than a hundred sailors aboard.[5] It would be one of the first provocations that slowly changed Americans' attitudes about withholding troops from the war.

Whatever neutrality the U.S. had tried to maintain ended on December 7, 1941, when the Imperial Japanese Navy Air Service attacked the U.S. Naval Station Pearl Harbor in Honolulu, Hawaii. Early that Sunday morning, wave after wave of fighter planes dropped bombs while submarines fired torpedoes on the military base, killing 2,403 soldiers and civilians, and damaging or destroying eight battleships and 300 airplanes.

The Japanese government may have hoped that its air raids on docked battleships and parked fighter planes at Pearl Harbor would send a clear signal to the United States to stay out of the global conflict in Southeast Asia, but it proved to have just the opposite effect. The assault turned a pacifist public into one now ready to defend the country against foreign aggressors here and abroad, and created a national go-to-war conversation.

Following the attack, Roosevelt would no longer have to wage war from the sidelines through the Lend-Lease program.[6] With bombed ships still smoldering in the harbor and men fighting for their lives in hospitals, on December 8 he delivered his now-famous speech: "Yesterday, December 7, 1941, a date which will live in infamy, the United States of America was suddenly and deliberately attacked by naval and air forces of the Empire of Japan. . . ." In addressing a joint session of Congress, he asked the senators and representatives to declare war on Japan.[7] Congress voted to declare war on Japan that day and, days later, also declared war on Germany and Italy, making the United States an active combatant in a war that would consume the first half of the decade.[8]

As the war effort ramped up, so did the need to increase food supplies to support that effort. By 1942, the U.S. was sending a minimum of 400 million pounds of food overseas each month as part of the Lend-Lease program. An average freighter carried in its cargo quantities of products that came from 3,824 farms, which journalist Frederick Simpich summarized in a 1943 *National Geographic Magazine* article titled "Farmers Keep Them Eating" as:

- 6,000 barrels of dried [powdered] eggs, equal to a year's work for 229,137 hens.
- 6,000 barrels of dried milk, a year's work for 2,783 cows.
- 16,522 cases of evaporated milk, a year's work for 304 cows.
- 20,000 boxes of cheese, a year's work for 3,037 cows.
- 14,500 big cans of pork, the meat from 5,021 hogs.
- 16,800 boxes of lard, the fat of 27,632 hogs.
- 6,061 sacks of flour, the wheat from 838 acres.
- 26,111 cases of canned vegetables, equal to 40 acres of tomatoes, 100 acres of snap beans, and 102 acres of peas.[9]

The focus on wartime needs meant, as one agent later termed it, "a shifting from those projects not food producing to those which are classed as strictly food producing."[10] LaGrange County agent Riley L. Case wrote in 1942, "War work is exacting and must come first. Production and more production must be the cry."[11]

Extension programming took a 180-degree turn from what had been pushed during the Great Depression, when the focus had been on controlling cost and increasing efficiency. When the U.S. entered the war, Extension quickly returned to the script used during World War I, which was to push programs that promoted maximum production at all cost.[12]

In his article, Simpich framed the importance of the farmer to the war effort by saying:

> *In the fields.* That's where American farmers, including women, girls, and school children are fighting now—fighting frost, heat, dust, drought, mud, flood, and insect pests, growing our biggest crops in history. Food is as much a munition as TNT. Farm tractors and milk wagons, like tanks and cannon, are war machines. Farmers don't get killed and wounded on battlefields, get decorated with medals, or have to sleep in mud and snow. Yet without this "soldier of the soil" all armies would soon have to quit, for it is still true that an army travels on its stomach.[13]

Porter County agent Stewart Leaming acknowledged Extension's role in this effort in his 1943 annual report, writing, "The country is fully aware of the importance of food as a war weapon. It is determined that nothing shall be left undone to see that ample supplies of food are produced. The extension service, because of its long experience in working with farm people, is looked upon for leadership in the food production activities."[14] Just as they had during World War I, county Extension agents would again become the intermediary to connect federal and state governments with the agricultural community.

During that war, agents had been successful in helping farmers significantly increase agricultural production. But in the postwar years, farmers had experienced a severe agricultural depression as countries began producing their own food once again, resulting in overproduction by American farmers. Naturally, this made some farmers hesitant about the economic effect of increasing production to meet World War II demands, as Orange County agent Ross Atkinson explained in his 1942 annual report:

> Farmers as a group are more aware of the danger of inflation than most other people, however, they are sensitive to inflationary government spending as a cause for after the war depreciation as much as inflationary farm prices should be. Therefore they are conservative about increasing livestock populations as a result of these fears. Because there is a justification for these fears on the part of farm people, it sometimes becomes a duty of [the] extension service to help educate national leaders to existing conditions and educate farm people as to their duties as food producers for the world.[15]

The agents' ability to understand both the government's demands and the concerns of the agricultural community they served would help them navigate this situation.

The connections established by agents within their local communities would also serve them well in the role of a go-between, as St. Joseph County agent Eugene Bird described in 1942:

> Opportunity for maximum service [by Extension during World War II] seemed to be open because of two fundamental obligations. First, because of a long period of service (1919–present), the county extension agent [Bird] has assumed certain obligations of leadership and citizenship in the community. Because of wide acquaintanceship with rural and urban organizations and leaders, the community has apparently expected him to function as advisor and leader to many programs affecting rural people. After all, there are few people who know the problem of all communities and many individuals and organizations through the whole county quite as well as he does. This is due, of course, to the fact that he has worked with groups and individuals over a long period of time and perhaps has earned a measure of confidence and faith among those served. Many civil leaders, therefore, have turned to the Extension Service for advice and suggestions on personnel and procedure for their particular programs.[16]

While Bird had been an agent for more than twenty years by the beginning of World War II, other agents were just starting their Extension careers. Charles D. Chambers, who was hired as the Perry County agent in 1941, quickly learned that his new job was "a big undertaking."[17] And Edward S. Bogue in Pike County observed in 1942, "The work in the county this year has been exceedingly difficult for the new agent, due to the conditions brought about by the international situation."[18] These inexperienced agents lacked the necessary community connections at first. In order to survive the rigors of the position at this moment in American history, they would have to become quick studies as the demands placed on them increased substantially by the day and month.

Purdue Extension administrators made it clear that the ninety-two county agricultural agents had little choice but to get the work done or their services would no longer be needed.[19] Ross Atkinson, who was an experienced agent in Orange County in 1942, offered the following metaphor to describe what his boss at Purdue was telling him: "War time in a county agent's office is like a ten year old boy loading hay in the old-fashioned way with two pitchers on both sides. However, Mr. T[homas] A. Coleman [associate director of Extension] once told a group of agents that, "If you can't stay on top of the job, you had just as well get off." Therefore we claim to have gone up with the load, and hope that we can ride it on in, up to the reasonable expectation of those who depend on us for cooperation."[20]

In order to get done what the government and Extension administrators were asking them to do, agents began consolidating and even eliminating programs long identified as being essential to the county.[21] Though there was little choice in the matter, some agents and members of the Extension committee expressed concern as wartime demands crowded out traditional Extension work, with one agent later observing, "We must not get too far off the beam that we lose sight of the long time program for the county."[22]

The reality was that many important Extension programs would be dropped. As Steuben County agent Thomas M. Grabill explained, "Fitting the extension program to wartime conditions has required some 'give and take.' Take time from one activity to give to another."[23] And agent Lloyd E. Cutler noted in his 1942 annual report that "the extension program in Lake County became a part of the national defense program. As the year progressed, an increasing amount of attention was paid to those parts of the extension program which were most important to our war effort, and those parts least important were eliminated or given proportionately less attention."[24]

Simply put, agents would need to spend most of their time addressing myriad demands that arose from the war, especially regarding the production and conservation of food needed at home and abroad.[25] There have been plenty of well-deserved accolades for the American people during World War II—for the bravery and heroism shown by the soldiers and nurses, the sacrifices made by the public, and the role that farmers played in winning the war by making sure that people at home and abroad were fed. However, what's been underreported is the critical role that county Extension agents played in mobilizing the agricultural community during the war years and managing the complex issues of this turbulent time. The following chapters offer a glimpse into the work of the agents during this critical point in American history.

CHAPTER 26

The County Agent Is Expected to Be a Walking Encyclopedia on Government Programs

Prior to the December 1941 attack on Pearl Harbor, the United States federal government was already promoting increased production of grains and meat under the auspices of the Lend-Lease program. To support that work at the local level, state and county agriculture defense boards were created in July 1941 to implement related federal policies.[1] These local boards were intended to speak on behalf of rural people and, as one agent later noted, "to see that farmers get all of the material that they will need to produce the added items being requested of them to produce."[2]

Following the U.S. declaration of war, these defense boards were renamed as war boards on January 7, 1942.[3] While the boards would still promote and support increased agricultural production, the name change reflected the nation's move from a defensive posture to a war footing.

Members of the county war board represented various U.S. Department of Agriculture (USDA) agencies. The county agricultural Extension agent represented the Cooperative Extension Service and served as the board secretary.[4] Other board members came from the Agricultural Adjustment Administration, Farm Security Administration, Forest Service, Production Credit Association, and the State Soil and Water Conservation Committee.[5]

In February 1942, agents and other board members were called to district meetings around the state to discuss the role that the boards would play in the local war effort. For better or worse, the county war board would soon be one that all farmers would come to know. Stewart Leaming observed in the 1942 Porter County annual report, "While it was pointed out early in the development of the organization [war board] that it was to be advisory in nature and designed to coordinate the activities of the various agricultural agencies, it soon became evident that this Board was to have administrative functions also. . . ."[6] And Owen County agent Fred I. Hoover noted, "Following December 7th the character of the Extension Work took on a new form that accelerated all projects dealing with needed production and in addition a whole myriad of new duties sprang up under the all-embracive USDA War Board."[7]

During the early years of the war, board members' duties evolved and expanded to eventually include the following:

- assigned county-level food production goals;
- appointed a committee to handle meat rationing, licensing of meat dealers, and licensing of those slaughtering livestock;
- promoted the federal milk subsidy program among the dairies;
- worked with the Selective Service draft boards to determine an applicant's eligibility for deferment;
- granted temporary releases to farmers to work in the war industries during the winter months;
- located laborers for farm work;
- granted applications for electrical extension connections in rural areas;
- allocated limited supplies of copper wire for electrical extensions;
- coordinated needs surveys of farm-related supplies such as machinery, tires, etc.;
- allocated rationed farm machinery and trucks through a farm machinery rationing committee;
- offered farm machinery schools and clinics to teach farmers how to better care for equipment;
- approved applications for converting steel wheels to rubber tires on farm implements;
- created a county farm transportation subcommittee to assist farmers in completing federal paperwork for obtaining fuel for farm production;
- distributed the county lumber allotment to retailers;
- approved building permits when construction of farm buildings exceeded more than $1,000;
- approved the repairs to residential farm homes if costs exceeded $200;
- approved applications for Regional Agricultural Credit Corporation loans;
- organized rural scrap drives;
- distributed sodium chlorate for weed control; and
- approved applications to purchase pressure cookers, milk cans, metal pipes, and stoves.[8]

Given the diverse responsibilities and nature of the assigned duties, agents would spend a considerable amount of their time on war board projects.[9] By 1943, one agent noted that the board was meeting fifty to sixty times over the course of the year.[10] The work of the board "touches the life of practically every one," as one agent put it.[11] Parke County agent Noah Hadley indicated that the "biggest job of the [USDA War] Board has been to guide farmers through a maze of governmental red tape, confusion, and obstruction."[12]

From the agent's perspective, serving on the board as an intermediary between the government and farmers was a delicate balance, with Hadley observing, "It has been the policy of this board to do everything possible to keep farmers supplied with necessary production farm materials. A sincere effort has been made to do this and, at the same time, abide by the letter of regulations. However, in many cases, the spirit rather than the letter has had to be the rule. Some houses had to be wired to retain necessary labor, and some corn had to be purchased above ceiling price to keep hogs going."[13]

The extent to which board members participated in the projects required of them varied across the counties. What would develop as the war continued was that the agent was expected to do the brunt of the work.[14] As Roscoe C. Stangland reported for Howard County in 1942, "The county agent had the choice of either 'doing the work' or letting the activity fail and taking the blame along with the other members."[15]

As war efforts ramped up, people would immediately begin to feel the government's control of everyday life, but perhaps none would feel it as much as farmers.[16] As Posey County agent Oral Riggs explained in 1943:

> Farmers have adjusted tremendously to comply with the various war restrictions. Because of the nature of farming, with his automobile, his truck, tractor, gasoline pumping engine, kerosene stove, and other equipment, he finds himself more greatly handicapped by government regulations than the average citizen. The government naturally organizes its various divisions according to the most convenient type of organization. Since this is done, it makes it necessary for the farmer to apply one place for his automobile gasoline, another place for his truck, and make different applications for his tractor gasoline with certain other supplementary forms used for his stoves and incubators. By the time the farmer gets his tires, his repair permits for his buildings, and sundry other things needed, he begins to get considerably discouraged.[17]

At the beginning of the war, multiple government-mandated organizations had been set up in each county, each with its own directives and regulations. Agents quickly saw a serious duplication of effort among the organizations, which frustrated farmers. It was often left to the agent to bring agencies together to discuss who was doing what and get them to coordinate their activities within the agricultural community.[18] As Montgomery County agent Gordon A. Sowers explained:

> Immediately following the Pearl Harbor incident, the various groups and social organizations of the county, together with all of the county's governmental agencies, were individually working on various phases of the war effort, resulting in much overlapping of activities and adding to the general state of

> confusion. One agency was intensifying its effort on one project, and another agency was doing similar work, and there was soon a realization that there would have to be a unified program in the county in order to handle the emergency efficiently and effectively. . . . As a result, it was decided that all war assignments pertaining to rural people must be decided on by an advisory committee and must clear through the county agent's office. This put a definite responsibility on the county agent for the successful carrying out of wartime activities among the county's farmers.[19]

People turned to the agents for accurate information as rumors became rampant and spurious claims quickly spread across the county.[20] In 1942, Spencer County agent Earl F. Downen reported, "[O]ne of our first aims has been to 'spike the rumors,' and maintain a well informed and educated public."[21]

Many questions posed to agents had little to do with Extension programming. Instead, inquiries were often related to the agents' role on their county war boards. Because of this position, the local agent became the go-to person for explaining how the various government programs worked. In his 1942 annual report, agent Roscoe Stangland observed:

> "See Your County Agent" seems to be the "by word" of a large number of organizations and others in Howard County when asked any question which they cannot answer. This situation has developed with the confusion from the vast number of new and little understood ration, salvage, and production programs in connection with the war. This situation, while congratulatory to the Extension Office, has taken up a large part of the agent's time and permitted less time for regular projects.[22]

That same year, Sullivan County agent Jean C. Sliger echoed Stangland's assessment, writing, "[The] county agent is expected to be a walking encyclopedia on government programs, the rationing system, war drives, and whether the newspaper has given the correct information concerning Selective Service."[23] And LaGrange County agent Riley Case reported, "The agent's office is becoming a place of general information for wandering citizens seeking more gasoline, machinery, tires or sugar. Men and women want help to fill out their many rationing blanks."[24] Fortunately, the agent would be there to offer local citizens assistance on many fronts for the duration of the war.

The agents were in a perfect position to answer these wide-ranging questions from the public. In addition to serving on the local war board, they were also board members or advisors to groups such as the Agricultural Adjustment Administration, Civilian Defense Council, Crop Goal Committee, Defense Transportation Committee, Employment Service, Farm Security Administration, Price Administration Committee, Rural Electrification Administration, Soil Conservation Service, Surplus Marketing Administration, and Victory Garden Committee as well as initiatives such as war bond sales and Red Cross and salvage drives.[25] By virtue of being well connected to other local leaders, the agent was one of a few individuals within the county with comprehensive knowledge of the varied programs and the policies, rules, and regulations being pushed down to the counties by state and federal government officials.[26] And if the agents didn't already know the answers, they knew who to ask.

Working between the various agencies and programs might have left little time for the agents' typical Extension work, but Riggs in Posey County noted the importance of shifting his attention, saying:

> The tremendous amount of war work that is necessary to be done is taxing the capacity of the county extension office to do what is requested of them. About everything that has anything to do with agriculture seems to land in the county agent's office. At first it seemed impossible to manage so many different things in this office, but after we began to get organized, we realized that it was much better to have something to do with the harmonizing of all programs in the county than to be running into and conflicting with these programs in the activities that we were trying to carry on. Consequently, we have tried to cooperate in every endeavor that has come our way.[27]

Around the state, agents may have expressed concerns about handling the increased wartime demands in addition to traditional extension programming, but ultimately, they rose to the challenge because farmers and other community members were depending upon them for assistance.

To address the domestic challenges brought on by World War II, the U.S. Department of Agriculture mobilized communities across the nation through local war boards. It was an effort aimed at encouraging farmers to meet the increased need for agricultural production while helping them work through the government rationing programs and wartime labor shortages. County Extension agents were tapped to play a major role in supporting this work.

As the United States entered the war, the government quickly stepped in with sweeping regulations and policies administered by departments that had not previously existed. These regulations directly affected agriculture, especially regarding food production goals and how fuel, machinery, and labor would be obtained as farmers faced a shortage of these items. Farmers looked to the county agents for help in working through the new requirements, paperwork, and permit requests. Involvement in such critical activities would require the agents to alter their regular plan of work in order to spend more time addressing the needs of the farmers and, indeed, of the country. This placed the county agent at the center of local war efforts and made the agent one of the more influential people in the community.

CHAPTER 27

Production Goals That Looked Impossible Were Reached

Even before the United States entered World War II, it had committed food support for the war effort through the Lend-Lease program. After joining the Allied forces in December 1941, the U.S. would be one of the few countries involved where farmland remained unspoiled by the conflict. As a result, American farmers were told that it was up to them to produce enough pork, eggs, milk, and vegetables to supply not only the needs of the nation but also those of the Allied combatants and the people living in war zones.[1] They would also be called upon to provide more grain, fiber, and timber to address wartime needs.

The U.S. Department of Agriculture (USDA) was assigned the task of mobilizing farmers to produce what was needed.[2] It was a huge task, but it wouldn't take much encouragement to get farmers and their families to put in the extra work. As Vermillion County agent Sumner Brown reported in 1942:

> The farm leaders at the beginning of 1942 were those men who had either farmed during the first world war or had served their country during that war. They were mindful of the fact that great quantities of foods would be needed to feed our soldiers at the front, and since the passing of the Lease-Lend [Lend-Lease] Act, there would be an unlimited demand on the American farmer for this food. They also knew that the midwestern Farming Area would by necessity furnish a large part of this food. And it was to this conclusion that they did harness their best efforts toward "All Out Production."[3]

The phrase "All Out Production" referenced by Brown was one of several motivational messages used by the government to unify Americans, especially farmers, behind this massive effort. "Food will win the war and write the peace" became one mantra that would be repeated throughout the war.[4] Others included "Food for Freedom," "Food for Defense," and "Food for Victory"—all of which would emphasize the important role that food would play in winning the war.

Indeed, farmers would now be pushed to send more agricultural products to the markets and the military. But they would have to do so while faced with growing shortages of typical inputs, which over time would include labor, machinery and trucks, tires, fuel, fencing and building supplies, limestone, and fertilizers.[5] The dichotomy of producing more food and fiber in the face of such handicaps should have been untenable under most circumstances, but as one agent noted, "[F]armers were determined to produce their utmost, using what was available."[6]

With so much production being shipped overseas, the United States quickly went from national surpluses to national scarcities, leading Steuben County agent Thomas Grabill to observe, "Probably no year has seen such change in attitudes, objectives, evaluations, and plans as has occurred in 1942. The outbreak of the war on December 7 plunged us into a wartime economy, and almost overnight American agricultural policy changed from a program of controlled surplus to one of producing abundantly."[7]

With food as part of the national strategy to win the war, USDA officials estimated and set national production goals, or quotas, for milk, pork, poultry, eggs, corn, soybeans, wheat, tomatoes, and other commodities.[8] With the country already committed to the Lend-Lease program, in September 1941 the USDA had announced food production goals for twelve states in the Midwest for 1942. According to press releases during this time, these goals were "designed to achieve the largest output of food in history so that ample supplies may be available for Great Britain and this country."[9]

The 1942 production goals for Indiana were:

Commodity	USDA Quota	State Production (estimate)
Milk (pounds)	3,612,000,000	3,465,000,000
Eggs (dozen)	147,000,000	128,167,000
Hogs (pounds)	1,545,292,000	1,403,645,000
Beef and Veal (pounds)	616,831,000	536,375,000
Lamb and Mutton (pounds)	48,255,000	47,630,000
Soybeans (acres)	1,090,000	814,000
Commercial Vegetables (acres)	228,500	227,200
Potatoes (acres)	50,000	48,000
Farm Gardens (numbers)	181,400	163,114
Oats (acres)	1,300,000	1,385,000[10]

These quantities were divided among the counties using past production records and livestock inventories.[11]

Extension surveyed farmers to determine whether the production goals set for individual counties were in line with what producers thought could reasonably be achieved.[12] Madison County agent Walter C. Haynes reported that a spring survey in 1942 had predicted "an increase of 15% in milk production, 20% in eggs, 28% in hogs and 12% in beef."[13]

In a few cases, agents reported that the surveys had found the crop and livestock goals to be too low or too high, so those agents had to submit adjusted values to the USDA Agricultural Adjustment Administration in Indianapolis.[14] Warren County sent the following revised estimates for 1942 back to the federal office:

Commodity	State Allocated Increases (%)	Revised Goals (%)	Percent Change
Dairy	4	21	+17
Milk	4	17	+13
Eggs	10	30	+20
Hogs	12	36	+24
Beef Cattle	15	11	-4
Soybeans	31	22	-9[15]

With revisions such as these, state officials could better predict whether Indiana could meet the federal quotas. In time it would become clear that farmers in Indiana and across the country would do all they could to ramp up production, but as the war continued to destroy more farmland and affect more people overseas, the federal government would ask for even more. As the state annual report for Extension would note in 1943, "Even the all-time record production of American agriculture failed to meet the demands of armed forces, lend-lease and civilians for food."[16]

Indeed, just one year into World War II, federal officials asked everyone involved in agriculture to meet the ever-increasing demand by the Allies for meat, grain, milk, and eggs by raising the quotas. Indiana's share of the national food production strategy for 1943 was:

Commodity	State Production
Wheat (acres planted)	1,134,000
Corn (acres planted)	4,275,000
Oats (acres planted)	1,280,000
Barley (acres planted)	120,000
Soybeans for Beans (acres harvested)	1,325,000
Sugar Beets (acres planted)	10,000
Potatoes (acres planted)	55,000
Sweet Potatoes (acres planted)	3,000
Cattle and Calves	1,073,000
Sheep and Lambs	684,000
Hogs to Farrow: Spring Pigs	681,000
Hogs to Farrow: Fall Pigs	662,000
Milk Cows	814,000
Chickens	33,367,000
Eggs (dozen)	149,158,000
Turkeys	412,000[17]

With one year of war work behind them, agents expressed trepidation about how they could possibly get farmers to meet the increased county goals on crops and livestock when many of the needed inputs were unavailable or in such short supply that obtaining them was dubious at best. Madison County agent Walter C. Haynes noted that "1943 looks like one of uncertainty. There promises to be more and more government regulations, yet a call to produce more agricultural products than ever before with less labor, equipment, fertilizer, some kinds of feeds and less travel allowance than last year."[18]

This situation and reports such as this would continue for the duration of the war. In 1943, Marion County agent Horace E. Abbott would write, "With approximately one-third of the nations' pork products moving across the seven seas to feed the fighting men of the United States and our Allies . . . and with an even greater proportion of American foodstuffs scheduled for export during 1943, the production problem of American farms for the coming year becomes easier to understand."[19] That year, Miami County agent John Wylie would also put the growing need into perspective, observing, "With the biggest production of crops and livestock ever recorded in the United States, we are facing a shortage of food. This is no reflection on the farmers of the State. This shortage is caused by the abnormal drain on our food supply by the War, and these demands will increase as the Allies liberate territories that have been over run and bled of all their food."[20]

Meeting the increased goals in order to fulfill U.S. wartime obligations presented farmers with an enormous task. This meant that the agents had to help farmers achieve the maximum output possible despite numerous limitations. Howard County agent Roscoe Stangland observed, "The year 1943 will require the county agent to head the agricultural educational work and to get maximum agricultural production from every farm, every farmer and every machine."[21] Marshall County agent Lyman Butler echoed the same ideas in his 1943 report, writing, "The goal set in this effort is to make every acre yield the largest possible amount and increase on every possible farm the production of pork, poultry, and dairy products."[22]

Throughout the war years, farmers would face increased demands with each new growing season, and Extension would continue to push for more production. In his annual report for Marion County, Abbott quantified what meeting the goals in 1943 meant for the Allied soldiers, writing:

> These figures converted to food reveal the fact that the increase in soybeans will provide a sufficient amount of this type of food for the diets of 200,828 soldiers for one year. The increase in tomato production will supply the tomato requirements in the diets of 778,080 soldiers for one year. The increase in the production of canning peas and corn was sufficient to supply this requirement for 16,123 soldiers for one year. Fresh vegetables produced in 1943 in excess of our 1942 production were sufficient to supply 144,080 persons for

> one year. The increased pork production will supply meat for 8,000 soldiers for one year. The increased poultry meat will supply the meat in the diets of 3,200 persons for one year and the increase in the production of eggs will supply the annual requirements of 43,636 very active persons for a period of one year. The increased bushels of corn in 1943 as compared to 1942 is sufficient to produce the extra 400 tons of pork produced during the current year and also to supply the cereal requirements of 22,196 soldiers for [a] one year period. These figures were determined according to the estimated requirements for very active men 20 years or over as listed on page 337 in the book titled "Food and Life" published in 1939 by the United States Department of Agriculture.[23]

While the wartime challenges faced by farmers and Extension would be many and constant, statistics such as these showed the impact of their work.

With farmers being so instrumental to the war effort, national speakers began traveling around the country to address agricultural communities and encourage them to do everything possible to meet the government's demand for production. In 1942, John Birky, a former member of the Canadian Air Force, spoke to a Porter County audience, telling them that as a soldier he had been fed "liberally on a carefully selected diet featuring plenty of cereals, cream, milk, eggs and other protein and vitamin rich foods" and asking, "Will Porter County farmers meet their crop quota this year?"[24] This showed the pressure placed on farmers to achieve or exceed what some thought were unattainable government goals.

To motivate farmers to rise to the challenge, the government held an event at the beginning of 1943 aimed at coalescing all actions toward food production at all cost.[25] Agents were told to bring farm families together on January 12 for a farm mobilization rally.[26] In fact, rallies would be held locally in communities across the nation so attendees could attend speeches by local dignitaries before listening to a national radio broadcast from Washington, D.C.[27]

In his 1943 annual report, agent Edward Bogue provided details about the Pike County agricultural mobilization rally held at the Lyric Theater in Petersburg, writing:

> Three hundred and twenty-five farmers and businessmen attended a meeting which included a musical program of patriotic songs, [and] a talk by V. B. Phillips, Chairman for the County War Board, who gave the county food goals and the general agricultural situation in the county. The most appealing part of the program was given by Frank Engleman, a local farmer and father of Pike County's first boy to make the supreme sacrifice for his country. Mr. Engleman urged every farmer to make every effort to produce food and fiber and to cooperate with neighbors to end the war so other boys could come home to a prosperous and free life.

> The principal address was given by Vic Hebbler of Huntingburg. He stressed neighborhood cooperation to utilize labor and machinery and encouraged more efficient practices in both field crops and livestock as a means of getting increased production.[28]

Following the local presentations, rally attendees in Pike County as well as in counties throughout Indiana tuned in to a national radio broadcast urging farmers to redouble their efforts to increase production and conveying the serious message that more food was needed.[29] Speakers for the broadcast included President Franklin Roosevelt, agriculture secretary Claude Wickard, General Dwight D. Eisenhower, a soldier, a mother of a son who was fighting in North Africa, and an Indiana farmer.[30] Roosevelt summed up the situation when he declared, "[F]ood is the life line of the forces that fight for freedom."[31]

During World War II, farmers experienced economic boom years like they had never seen before. Jay County agent Chester A. Langston reported in 1943 that "a greater demand for agricultural products has meant a higher income level for all farmers. . . . Good production years, accompanied by war-time prices, have placed farmers in the best position they have been in for some time."[32] Nationally, income for the agricultural sector had increased by 85 percent in just three years—going from $8.7 billion in 1939 to $16.1 billion in 1942—while the cost of farm inputs had only risen by 25 percent during the same time period. This meant that farming was once again a money-making venture.[33]

Farmers were well aware that the higher market prices would not last forever. Those who had farmed during World War I had seen postwar prices collapse from the wartime highs to more normal values. In 1944, agent Herman Madaus told Warren County farmers, "It may be safely stated that present prices of farm products cannot be maintained after the war is over. For this reason farmers should be advised to practice conservation in negotiating their investments for any long period of time in liquidating debts."[34]

With the extra income, farmers began reducing debts they had accumulated during the Great Depression and, in some cases, even before.[35] Many were eager to pay down their mortgage notes, with some paying them off rather quickly.[36] Some farmers put money in the bank or bought bonds as a way of hedging their bets against what might happen after the war ended.[37] As one agent put it, "[T]he farmers have prepared themselves for any let-down in prices, should that occur."[38] The Ohio County agent reported in 1942 that there were no farmers delinquent in their taxes—another indication that farm families were doing better financially.[39]

Land prices were rising as well. In Putnam County, they rose on average about 60 percent above prewar prices, with acreage for some of the better farms valued at twice as much as it was prior to the beginning of World War II.[40]

In 1943, Marion County agent Horace Abbott observed, "Many mortgages have been materially reduced and property valuations definitely increased, which have greatly widened the margins of equity to create a much sounder economic condition than has existed for many years past."[41] And agent Oral Riggs described the economic situation for Posey County farmers in 1944, saying, "Although it has been twenty-five years since World War I, farmers have not fully recovered from the debts that were contracted just previous to the flop after the last war. Last year and this year, however, farmers have been paying off their debts rapidly, both the farmers who were recovering from the last war and those who have contracted debts at more favorable periods."[42]

Some farm families wanted to invest a portion of their recent earnings in new appliances and equipment as they sought to enjoy some of the things that a higher standard of living brought them.[43] Later in the war, agents would also note in their reports that more farmers were remodeling existing homes and farm buildings or constructing new ones.[44]

By 1945, the prices of farm inputs were beginning to rise, with one agent predicting, "The rise in costs of what the farmer buys will catch up with the rise in farm commodity prices so that there will be less spread."[45] Agents kept apprised of the economic trends in order to better advise their growers.[46] The agents wanted farmers to think twice about buying land at inflated prices to avoid getting too far into debt.[47]

The following year, Lake County agent Lloyd Cutler reported, "Reasonably good production of farm crops plus unusually high prices has made 1946 a better than average year for Lake County farmers. Most of them say that they think farm prices are entirely too high and they are planning for the future rather conservatively so that they will not get caught like many of them did in the last depression."[48]

The increased production achieved by Indiana farmers during the war years cannot be overstated. Ramping up production to meet government quotas certainly included some positive outcomes for the agricultural community, including the higher commodity prices and better incomes. However, the increased production also had the following negative repercussions within the processing and distribution systems:

- *Feed shortage due to more livestock and poultry.* With the USDA asking for increases in pork, poultry, beef, and mutton, producers substantially increased their livestock numbers. In 1942, Tippecanoe County reported a 45 percent increase in hog litters farrowed over those of the previous year. But that increase meant they would need tens of thousands more bushels of corn to get the hogs finished for market.[49] With other counties experiencing similar increases, a feed shortage developed across the state as farmers ramped up the numbers of livestock raised on their farms. The situation would persist throughout the war, causing ongoing challenges for farmers who had responded to the federal government's request to supply more protein for the war effort. (See Chapter 28.)

• *Processing and distribution facilities at capacity.* Agent Ross Atkinson was pleased to report significant increases in livestock, eggs, and milk produced in 1942 as Orange County farmers expanded the size of their operations and focused on producing what the USDA considered essential. He noted, "Never before has so much finished livestock gone to market, never before have the creamery and cheese factories been flooded with milk, never have they shipped so much milk to the Louisville milk shed, and never have so many hatching eggs gone to the hatcheries and market eggs to market."[50] While that is exactly what everyone wanted, moving all of that extra farm production through the processing and distribution channels posed a challenge. It didn't take long for the livestock markets in Indianapolis and Cincinnati to become clogged as the numbers reached levels never seen before.[51] As the terminal markets became glutted, farmers were asked to not ship their livestock until the backlog could be resolved. But this meant the livestock had to be fed and cared for longer, which increased costs and reduced profits for the grower. (See Chapter 28.) Those handling milk and other dairy products, eggs, and crops such as tomatoes faced similar pressures in processing and distributing the extra production that the government wanted.

• *Less acreage for nonessential war crops.* From 1942 to 1943, production of the more important grain crops such as corn and soybeans increased by double-digit percentages.[52] For example, in order to meet 1943 goals, Jay County farmers put in an extra 8,000 acres of corn, wheat, soybeans, and tomatoes in fields that had previously been planted to oats, hay, and pastures.[53] Every acre like these grown to meet the government's needs meant fewer acres planted to the crops they replaced. Through its production goals program, the federal government changed what was being grown on Indiana farms.

• *Lack of storage for increased bushels of grain.* In Warren County, the agent reported in 1942, "At the present time, local grain elevators are filled with corn and soybeans which cannot readily be moved, thereby creating a congestion of the marketing channels of these commodities."[54] Part of the problem was that the military had locked up many of the available railroad cars to move its supplies, making it difficult for the elevators to get enough cars to move the grain to terminal markets. Further down the production line, processing plants were working at full capacity to process grain into products needed by the public and the military, but they could only handle so much grain each day and had limited capacity to store the overflow.[55] The increased grain production quickly filled and exceeded what on-farm storage bins and local elevators could hold.[56] This lack of storage space affected how much some farmers grew the following year as they assumed that the situation wouldn't change.[57] (See Chapter 29.)

• *Lack of modern machinery.* Farmers who had increased crop acreage found it difficult to purchase farm machinery even though they had the money. This was because the production of new equipment had been cut back so materials used to manufacture it could be diverted to military equipment. With the demand for farm equipment outstripping supplies, the government put a rationing program in place. As expected, less equipment coupled with less labor meant some fields would not be harvested. There were scattered reports from around the state that the lack of combines resulted in beans remaining in the fields longer than expected or remaining unharvested.[58] (See Chapter 36.)

This list of challenges stemming from the increased production created monumental downstream obstacles that the federal government had not anticipated and county Extension agents had to deal with in order to move those increases into the channels of trade.

As the United States shipped men and military supplies overseas, the government turned to the American farmer to increase farm production and outputs to levels never seen in this country's history. The farmer would have to work under conditions where the government was taking a significant share of fuel, oil, steel, food, and labor for the war, but this created shortages at the expense of the farming community.[59] With so many obstacles facing agriculture, it seemed certain that many of the production goals would go unmet or, over time, be unsustainable. However, Scott County agent John McCulloch reported just the opposite in 1942, noting, "Production goals that looked impossible were reached. Projects that looked too big were completed. Planning that farmers said could not be done, was done. Neighbors that never cooperated, made perfect cooperation."[60]

In 1944, Posey County agent Oral Riggs expressed his pride for contributing to the war effort, writing, "We are proud to have had an important part during a most critical period in the history of the United States in producing the biggest corn crop in history, the biggest wheat crop in history, and sending the most tons of meat to the market in history. All of this was done under serious handicaps of help, machinery, and weather."[61]

The power of Purdue agricultural research pushed by the Extension specialists and county agents helped move the production needle forward even during trying times.[62] Agent Sheldon Pershing pointed that out in his 1942 annual report for Tippecanoe County, saying, "Hog production exceeded all records,—eight years of major emphasis on sanitation and careful management had established the essential practice on many farms, and quickly were adopted on many other farms under the pressure of war-time needs."[63]

World War II had changed the way science was used and accepted by the farming community. It was no longer a foreign concept to farmers, and the attitude that book learning had no place in agriculture had slowly evaporated. As Riggs had observed in 1943, "No doubt as time goes on, we in extension work and those who are listening to our recommendations will realize more and more the value of science and book learning in agriculture."[64] Notably, Riggs's framing of book learning in a positive light was just opposite of the views that had been expressed by farmers for many decades prior.

It's true that many farmers had increased their acreage and livestock herds to take advantage of the higher commodity prices, but this required a tremendous effort by the entire farm family, who had to work longer hours to get all of the work done.[65] Agent Willis Stall provided this excellent narrative of the difficulties families faced as they strived to meet their production goals:

> Jackson county farm people have also had to readjust many of their plans in conformity to the National "Food for Victory" program. During the present year they have done a splendid job of producing more than their quota of most of the products asked for. This they have done in spite of less machinery and with less help, due largely to the fact that the War industries in our own county have hired practically all available labor, at prices much higher than farmers can afford to pay. Along with this has been the drafting of farm boys, without any regard to agricultural considerations or the agricultural necessity, so far as we can see. It was not until November [1942] that our Draft Board gave any consideration to agricultural needs in classifying young men for service, although the National Director had recommended this during the month of February.[66]

What kept the American agricultural community giving its all were the constant reminders through government slogans such as "Food will bring the boys home earlier." Farmers understood that meeting their responsibilities could mean that their sons and daughters or those of their neighbors would return home sooner.[67] One agent summed it up by saying, "Loyalty to their sons and daughters in service and contempt for the enemy moves them to give their utmost in 'winning the war to this cause.'"[68] As it turned out, giving their utmost meant that even with the government-imposed hardships placed on agriculture, Indiana farmers would largely meet production quotas assigned to their counties and to the state as a whole.[69]

CHAPTER 28

Higher Hog, Dairy, and Poultry Prices Created an Interest Like Never Before

As the United States entered World War II, President Franklin Roosevelt understood clearly that the destruction across Europe and elsewhere was going to lead to massive famine and the collapse of Allied nations if the U.S. did not provide food. It would be no small task for American farmers to produce the quantities of agricultural products that the war would demand. Meat, primarily in the form of pork and poultry, along with powdered milk and eggs, and certain canned fruits and vegetables such as tomatoes became classified as war foods, indicating how important they were for feeding civilian populations and soldiers.[1]

During 1943 and 1944, the U.S. population would receive 67 percent of the meat supply produced domestically, while the remainder would be used to supply the U.S. military, Allied forces, friendly nations, and liberated areas, and placed in an emergency reserve.[2] In order to produce enough protein for the war effort as well as for the American public, Roosevelt and the U.S. Department of Agriculture (USDA) had set up a national quota system by the fall of 1941, which was then divided among the states and passed down to the counties (see Chapter 27). At the local level, the county Extension agents became the voice of the government, encouraging area farmers to produce what they were being asked for in order to meet the allotted quotas. It was the perseverance of the agents pushing for the increased production along with the commitment of the farmers to reach their war goals that would allow the U.S. to raise historic amounts of food.

Producers increased their hog numbers almost immediately after the U.S. government called for more hogs as part of the Food for Freedom program.[3] Purdue agricultural researchers and Extension specialists had spent the previous thirty years revolutionizing how hogs were bred, fed, and marketed, so they were ready to help farmers navigate the new production demands.[4]

Farmers who normally kept five sows increased their herds to fifteen or twenty.[5] With significant increases on individual farms, counties soon reported major increases in breeding sows, with some doubling their numbers. For instance, Starke, Montgomery, Pulaski, and Warrick Counties increased breeding sows from the spring of 1942 to the spring of 1943 by 29 percent, 33 percent, 50 percent, and 95 percent, respectively.[6] Raising more breeding sows on the farm was a first step, but getting more pigs from each litter to the market would be critical for meeting the government quota.

To encourage farmers in Huntington County to achieve maximum production, agent Walter Rusk sent out a flyer in 1941 that read, "[H]og producers are handling each year nearly 7,000 sows. These sows farrow an average of 6 pigs per litter which require eight and a half months to be finished for market. Yet, according to reliable information, a sow, to be a profitable unit, should produce at least 7 pigs per litter which should be finished and on the market in six months or less. Does this indicate a problem on your farm?"[7] In this example, one extra pig per litter would mean 7,000 additional hogs going to market.

The increase in 1942 was nothing short of spectacular, with one agent describing it as "10 million more head of hogs being produced for market this year than was ever produced in the U.S. previously."[8] For example, prior to the war, Perry County typically marketed 8,000 hogs each year, but in 1942, that number more than doubled to 15,000.[9] Putnam County went from marketing 40,000 hogs in 1940 to 80,000 in 1943.[10] But the elation of making good money from selling more hogs soured quickly when the corn used to feed them became scarce and expensive.[11] Randolph County agent Lewis E. Husted reported in 1943 that "with the stimulus of war prices, livestock numbers were greatly increased with the result at the beginning of the year [that] feed would be one of our major problems."[12]

Some counties produced enough corn to feed the livestock raised there. Tippecanoe County was one of those counties, but by 1944, it would reach maximum livestock numbers based on the acreage of corn.[13] Other counties had to rely on outside supplies to feed their livestock, with Parke and Scott Counties each importing about half of the corn needed to feed out hogs.[14] It helped that much of the state's corn acreage was grown to feed livestock, but it still wouldn't be enough to cover the increased livestock numbers.[15]

As livestock numbers increased and feed supplies dwindled, many farmers were warned by their agents to keep only the livestock that they could feed from their own corn acres.[16] But without adequate feed, livestock producers would not be able to meet the government quotas, and many would begin liquidating their swine herds by sending underweight hogs to market.[17] Sending animals to the markets unfinished and reducing livestock herds based on the anticipation that there would not be enough corn to feed livestock was not what the government wanted.

Federal officials had tried alleviating the feed shortage by selling government-owned wheat as a substitute for corn as early as 1942.[18] Knowing that this wheat would soon be available, agents had hosted meetings in 1942 to urge farmers to save their corn and buy the less expensive government wheat.[19] Farmers paid a relatively low price for government wheat that year, ranging from 84 cents to $1.08 per bushel.[20]

Individual farmers placed their wheat orders with the USDA Agricultural Adjustment Administration, then the grain was shipped by rail, with each railcar holding 1,500–1,800 bushels of wheat.[21] In 1942 in Morgan County, 120 farmers bought nearly 24,000 bushels of government wheat, which filled 14 carloads.[22] That same year, Marion County and seven surrounding counties purchased 174,300 bushels of government wheat that filled roughly 115 railcars.[23]

Based on the low price, it was more economical to feed wheat instead of corn, so raising livestock for the government could remain a profitable farm venture.[24] To describe the use of wheat as part of the ration, agents handed out two Purdue Extension leaflets—No. 230, *Feeding Wheat to Livestock*, and No. 236, *Properly Fed Sows Farrow Large, Vigorous Pigs*—both of which were published in October 1942.[25] Agents distributed these to farmers and fielded their questions, repeatedly telling them to use more ground wheat in feeding livestock and to use as much pasture as possible.[26]

In addition to using wheat as a substitute for corn, agents told farmers that putting hogs on pastures could save 10 percent of the corn normally fed. Pastures with clover and alfalfa were preferred, but under emergency conditions, rye, sudangrass, and rapeseed pastures could be used.[27]

In the 1942 annual report for Morgan County, agent William Record documented how one of his producers had made money by feeding wheat and using pastures, writing:

> One farmer, Mr. R. D. Sellars, of Brown township, made the following report on a bunch of hogs fed with this wheat [from government supply] on alfalfa pasture, with a commercial supplement.
>
> - August 29, started 52 head weighing 5,270 pounds
> - October 2, sold 52 head weighing 8,500 pounds
> - Gain: 3,230 pounds
>
> This was a gain of better than 1.8 lbs. per day per head. The hogs consumed 153 bushels and 30 pounds of the feed wheat at a cost of $125.87 and 1,500 pounds of the commercial supplement at a cost of $56.25, or a total feed cost of $182.12. This made the feed cost per pound of gain 5.6 cents. The hogs were sold at 15 cents a pound.[28]

With the introduction of wheat as a livestock feed, government planners thought that the livestock needed for war purposes could be achieved. Marion County agent Horace Abbott reported in 1942 that the "extensive use of Government wheat in the feeding program of the county indicates that livestock production will continue to be on a high level for several months, limited only by a possible shortage of labor available to handle this livestock."[29]

But by 1943, the shortage of feed posed such a serious threat to the war effort that counties formed feed conservation councils to deal with the situation, which many anticipated would worsen before it got better. In April 1943, Pike County sent ten agricultural leaders to a district feed meeting in Washington, Indiana, to listen to spokespersons from the Indiana Feed Conservation Council, Purdue, and the U.S. Department of Agriculture.[30] As one agent later reported, the presentations largely focused on "the voluntary program for stretching protein feeds. Suggested ration changes, substitute feeds, pasture suggestions, and crop changes were the topics used for discussion."[31] Those in attendance, including agents and Extension specialists from Purdue, quickly relayed to producers what they had learned from these district meetings by leading local discussions on why there were shortages of feed and what producers could do to change their feeding rations.[32]

By 1944, with the feed shortage persisting, Miami County agent John Wylie observed that farmers had "increased livestock numbers beyond available feed supplies and face the problem of adjusting their farm business to this situation. . . ."[33] In Wells County, agent Mervin Smith reported that low inventories of corn resulted in the "liquidation of a considerable portion of the livestock through sales of herds at auction and selling on the markets."[34]

Tippecanoe County agent Sheldon Pershing described the situation around Indiana and elsewhere in 1944, writing, "The nation's need for corn has become critical in early spring. At the same time the government corn 'freeze' order [temporarily restricting purchases to protect the supply for essential industries] caused a near panic at the stockyards as feeders dumped their 'near' finished livestock to conserve corn. . . . This is probably in keeping with the over-all adjustment of livestock numbers to the amount of feed in prospect."[35]

Simply put, without adequate feed supplies, farmers were making the difficult choice to reduce their production levels. But this also meant that they risked not meeting government quotas for this important part of the wartime food supply.

It's not surprising that the large influx of hogs meant the stockyards didn't have the facilities to deal with the historic production numbers.[36] By the start of 1943 and especially in 1944, hog producers lost confidence in the market when the stockyards were backed up and the guaranteed prices set by the government were being openly ignored.[37] Farmers ready to market hogs were told to wait their turn in line, but each day's delay increased feed costs and reduced profits.

In August 1943, agent John McCulloch sent a letter to 650 hog producers in Scott County advising them to market their hogs at lighter weights to mitigate "the severe marketing situation which has begun to take place at the present time."[38] In 1944, Whitley County agent Blanchard Widney described how problems in the market had affected those raising hogs for the government, writing:

> The collapse of the hog market early in the year had a depressing effect on interest in swine production. . . . Markets would not receive hogs as rapidly as offered and altogether refused some weights. Many strange and unusual practices were reported which made producers accept abnormally heavy shrink, sharp grading and unjustifiably low prices. Government floor price regulations were frequently flaunted by buyers with one scheme and another. There was a general feeling among hog raisers that hog dealers and packers enjoyed a "hey-day" of profit at farmers' expense such as never before occurred in the livestock markets of the United States. This experience combined with a poor corn crop caused a heavy liquidation of breeding stock and a low ebb of interest in swine production.[39]

The guaranteed prices set by the government had induced farmers to expand their operations, but Widney's account indicated that these had not been honored. In the 1943 Steuben County annual report, agent Thomas Grabill had also noted that buyers ignored these mandated prices."[40]

In his 1943 annual report, agent Joseph Clark included a quote by a Kosciusko County farmer that seemed to blame Clark for being part of the problem. Clark wrote, "The hog production in Kosciusko County was in good hands this year as now indicated by the feed lots. Hogs—Hogs—and more hogs. As one farmer said, 'Our County Agent said we need more hogs to win the war, and he said I should use a lantern to save pigs and I did; and now I can't sell them.'"[41] Agents such as Clark who had succeeded in getting farmers to raise more hogs were now the ones who had to face those same farmers when government promises of guaranteed prices failed to materialize.

By the end of 1944, hog producers had cut back to more normal hog numbers, with Clark noting that the "hog marketing 'glut' decreased the number of fall litters."[42] The lack of feed supplies contributed to farmers sending fewer animals to market. Even with wheat provided by the government, there just wasn't enough grain to feed livestock herds by 1944. Pike County agent Edward Bogue observed, "Part-time farmers who had increased hog production and who were purchasing feed are practically 'out of business.'"[43]

The wartime feed shortage was challenging enough, but when drought reduced corn yields in 1944, it wreaked further havoc on livestock feed supplies.[44] Agent Lewis Husted summarized the feed situation among Randolph County livestock producers by saying that:

> the feed situation became extremely acute and necessitated a considerable reduction in the amount of livestock kept. Farmers continued to import wheat for feed as long and in as large amounts as it was available. Corn was brought in from farther west until transportation conditions made this impractical. As a result of the feed conditions, the county is closing the year with a much reduced livestock inventory and probably insufficient feed to carry it until next year's crop.[45]

The combined challenges facing livestock producers had a negative impact on attendance at the once popular hog schools offered by Extension. Agent Elmo A. Chatham wrote in 1944 about low attendance at the Perry County hog school, saying, "[It] can doubtlessly be said that this is an example of 'supply and demand' affecting even attendance. There seems to be too many people disappointed in hog prices for this to be a very attractive subject at this time."[46]

Hog producers had significantly increased production at the start of the war and showed that they were more than willing to do what the government had asked. However, with inadequate grain supplies and the inability of the markets to absorb the extra animals, growers had little choice but to adjust to these conditions and reduce the number of hogs produced.

The USDA classified milk as another war crop, raising the minimum price and setting national quotas. Once again, prices drove up production, with Orange County agent Ross Atkinson observing in 1942, "After Pearl Harbor national and state leaders were worrying about increasing dairy production. When they doubled the average price, they much more than doubled the production of the average cow."[47]

Purdue agricultural researchers had worked with dairy farmers for decades to improve their operations, so those who had followed the university's recommendations were now positioned to answer the call for increased production and reap the benefits. In Whitley County, Blanchard Widney reported, "A big production increase of more than 25 per cent followed the improved milk prices of the national war program. The increase was so great that handlers found their equipment inadequate to process the supply. . . . The price influence and patriotic appeal caused their owners to put into practice the good management which they already knew."[48]

In Ohio County, farmers milked approximately 3,200 cows in 1941. The county goal in 1942 was for each cow to produce 12,918 pounds of milk, which on average amounted to an additional 11 pounds per cow per day. For 1943, these farmers were asked to increase daily milk production by one pound per day, which would create an additional 1 million pounds of milk for the war effort.[49]

Statewide, dairy producers would receive an additional 30–70 cents per hundredweight, or per one hundred pounds, for milk and 4–11 cents more per pound for butterfat as a subsidy from the government.[50] This subsidy program was so popular that approximately 80 percent of Noble County farmers were participating in 1944, according to the agent.[51] And that same year, the Porter County agent reported that the dairy business "is the most profitable enterprise at present due in part to government subsidies."[52]

Better prices and more cows led to greater volumes of milk being shipped to processors. In March 1942, farmers sold an average of 56 pounds of milk each to the Kraft Cheese Company in Ridgeville, which was a 20-pound increase over the previous March.[53] At face value, increased production was a good thing, but the processors were soon overwhelmed by the huge influx of milk.[54] In Orange County, the agent reported in 1942 that the Paoli Cheese Factory "had to cut whole milk patrons from their route and stop some routes all together in order to limit their supply to the amount that their skim milk dryer could process, running twenty-four hours per day."[55]

By 1943, milk containing 3.5 percent butterfat brought $3.20 per hundredweight, which was the highest price that the dairies had received in nearly twenty years.[56] But the cost of feed had also significantly increased in 1943, raising concerns that milk production might be reduced.[57] With corn and other feed for dairy cows in short supply, farmers were asked to rely more on crops that were grown on the farm, such as a legume hay or pastures.[58]

In an attempt to keep farmers thinking about milk production rather than the escalating price of feed, the War Food Administration put a dairy feed payment program into place in October 1943.[59] As was the case for most government programs targeting farmers, it was left to the agent to explain how the program worked and what the government was offering in return for cooperation.[60]

The Miami County agent told local producers that if they wanted to participate in the government feed program, they needed to keep the sales slips for the milk and butterfat sold to buyers.[61] With records to back up their claims, farmers would then receive a feed subsidy of 25–50 cents per hundredweight of milk. For cream, the feed subsidy was 3–6 cents per pound of butterfat.[62] The dairy feed subsidy proved to be a popular federal program with Indiana dairies.

Farmers incorporated a variety of methods to increase dairy production during the war years. Of course, agents reminded them that one way to increase production quickly was to minimize any spoilage of the resulting dairy products. In 1942, Miami County agent John Wylie wrote, "'Bugs,'—the kind that grow in milk, won't bite you but they sure can ruin a can of milk. . . . Two—just two—simple remedies will stop this headache. (1) Scrub, and we mean scrub, the milking utensils thoroughly. (2) Cool the milk immediately, and keep it cool until collected. Scrub and cool."[63] In a later report, Wylie offered this perspective on the loss of milk due to quality issues:

> Approximately 45 million pounds of milk are rejected each year because it becomes sour. There are 3.5 ten-gallon cans of milk lost per year, per farm in Indiana, at a present cash value of $10.54. This would pay for 88 gallons

> of tractor gasoline, which would plow 29 acres of land, or would cultivate 176 acres of corn. This would also supply one soldier with one pint of milk daily, for ten months. Cool milk, promptly, to 60 degrees F, or below. This will save a tremendous loss in milk, feed, labor and trucking.[64]

But the higher dairy prices had also created an interest like never before on how to increase production through better feed and care for cows and through improving the quality of the herd. For years, agents had shared information gathered from Dairy Herd Improvement Associations to show how those who followed better management practices could increase the production of their herds. Many farmers who had previously ignored this information and the dairy programs put on by Extension became more interested in light of the government quotas and the better prices. In 1942, Montgomery County agent Gordon Sowers reported that one of the more requested Extension bulletins addressed how to balance feed rations to maximize milk yields and increase per-cow profits.[65] That same year, LaPorte County agent Everett L. Hartman indicated that local dairy farmers were now asking questions about "improved breeding stock, adequate pasture, legume hays, more common use of protein supplements, and more common practice of feeding balanced rations according to production."[66]

Interest also surged for purchasing or sharing breeding bulls that had proven production records (see Chapter 20) as another avenue for increasing dairy production and meeting government quotas.[67] In 1942, the American Jersey Cattle Club established the Victory Bull Campaign, which allowed dairy owners to apply for the opportunity to breed their Jersey herds with a donated bull with high-yielding genetics.[68] In discussing the program in Ohio County in 1942, agent George I. John wrote:

> The long term goal of the County dairy program is to increase the milk production efficiently. The dairy "Food for Freedom" object is the production of the maximum of milk possible from the dairy resources available. The goal as set up for Ohio County is an increase of 4% or more, or 12,918,000 lbs. or more in 1942. . . .
>
> B. H. Scranton, Rising Sun, Indiana, is sharing some of his greatest Jersey dairy herd inheritance with neighbors. He has had one bull registered under Selective Registration and selected by the American Jersey Cattle Club, New York City, for the nation-wide Victory Bull Campaign. Distribution will be made in August [1942].
>
> More than 1,000 Victory Bulls are to be given as an out-right War-Time Gift to America. Of these, 35 are coming from Indiana. There have been more than 25,000 applications received from dairymen in all states. A total of 673 have been received from farmers in Indiana.
>
> The Scranton calf is from a two-year-old that has a D.H.I.A. [Dairy Herd Improvement Association] record of 419 lbs. [butter]fat. Her mother has 536 lbs.

> [butter]fat as a 3-year-old, and her grand dam has 509 lbs. [butter]fat as a 6-year-old. All these records were made under supervision of Purdue University.
>
> The significance of the production carried by this Victory bull becomes readily apparent when we recognize that the average dairy cow in America produces slightly over 175 lbs. [butter]fat a year. . . . Inheritance carried by these promising youngsters should go a long way toward increasing average milk production in the herds they go into, and in building a living demonstration of the value of production bred registered Jersey Sires.[69]

Another aspect of herd improvement that also drew tremendous interest around this same time was the new development of the technology known as artificial insemination.[70] Under the auspices of the American Dairy Scientific Institute in Elgin, Illinois, dairies in Indiana were asked to organize county-level artificial insemination associations.[71] One of the early associations for dairy cattle was established in Washington County in 1941.[72]

In writing about the initiative in 1942, Marion County agent Horace Abbott stated:

> This program has enabled the average herd owner to avail himself of the service of these bulls of high known productivity and, therefore, is enabling these herd owners to build future herds with a probably higher degree of production than those that might have been built without the artificial insemination program. Other advantages as expressed by farmers using the association's bulls are the elimination of the overhead involved in keeping a bull as a part of the individual farm herd and also the elimination of the risk and liability involved.[73]

One challenge some dairy producers faced in meeting government quotas was that the large dairies required a lot of manual labor at a time when much of the labor pool had left for military service or better-paying industry jobs.[74] One agent described the situation by observing, "The fluid milk is produced on larger farms where they depend more on hired help and it takes more labor to produce the fluid milk."[75] In some cases, the lack of farmhands resulted in dairies downsizing herd numbers or going out of business.[76]

However, the use of electric milking machines during the war allowed those with access to this technology to operate with less labor. Milking machines were in use prior to the war, but many dairies did not have the financial means to purchase them. With the shortage of labor and high dairy prices, the milking machine would become universally adopted as a must-have piece of equipment for dairy producers.

Agents began hosting what became known as fast milking demonstrations, where farmers would visit dairies that used electric milking machines.[77] Starke County agent Arthur C. Shideler offered this perspective of such an event in 1944:

"The purpose of the demonstration was to show that fast milking was not only a labor saver to the busy dairyman but also resulted in more milk per cow, less udder troubles and less danger of the spread of garget [inflammation of the udder] and mastitis in the herd from infested to healthy animals."[78] Additionally, the milk obtained using the machines didn't have as much bacteria and foreign debris.[79]

After agent Horace Abbott offered fast milking demonstrations at the Marion County fair, he estimated that 1,000 dairy producers had stopped by the exhibit over the course of five days. Milking a cow in three minutes must have caught the attention of all dairy producers.[80] The more that this equipment was shown in use, the more farmers wanted to buy this technology, but the limited supply produced under wartime rationing measures could not keep up with the strong demand.[81]

The more advanced dairy producers had long ago joined county Dairy Herd Improvement Associations (DHIA) so that a tester could evaluate the monthly production of their herds (see Chapter 20). DHIAs required each tester to receive training for one week at Purdue with dairy husbandry Extension associate Gerald A. Williams and other staff members in the Department of Dairy Husbandry.[82] The tester evaluated each cow in the herd in order to calculate her milk volume, butterfat content, and feed costs so owners could cull low-producing cattle and adjust feed rations based on output.

During World War II, the problem was that no sooner than testers had been hired, trained, and sent to the counties to work, they were served their draft notices.[83] The position became a revolving door as one tester replaced another.[84] In other cases, the person would turn down the position in favor of working in the manufacturing plants, which were paying significantly higher wages.[85]

Members of most DHIAs felt lucky to retain a tester once they were trained. The Warrick County association had been fortunate to have employed Clint Hessler as a tester for seventeen years by 1942. In fact, the agent noted that at the age of eighty-seven, Hessler "was the oldest tester and point of service and age of any tester in the United States."[86]

As the military draft and industry claimed available workers, Morgan County agent William Record reported "a general shortage of qualified testers was showing up in Indiana and a number of associations were experiencing difficulty in keeping a tester on the job."[87] Consequently, many DHIAs had to discontinue their programs for the duration of the war.[88]

In an effort to increase supplies of protein, the U.S. government asked poultry farmers to increase egg production in 1942 by approximately 10 percent over the previous year. This meant that Indiana egg producers would need to supply

147,000,000 dozen eggs to reach their portion of the nationwide quota.[89] In fact, more than 50 billion, or 4.2 billion dozen eggs, would be produced across the United States in 1942.[90] The quota included an allotment of 14 percent of that year's egg production for the military and for other countries as promised under the Lend-Lease program.[91] (See Chapter 25.) Powdered or dried eggs became a major food item both domestically and abroad.

Poultry meat was also included in the government quotas. In 1942, U.S. Secretary of Agriculture Claude Wickard asked U.S. poultry producers to market 200 million winter broilers for 1943.[92] Miami County agent John Wylie explained the request, writing, "On account of the large demand for meat by the armed forces and our Allies; we are facing a shortage of about 600 million pounds of meat. This meat can be produced this winter by raising winter chickens. The plan as set up by the USDA is to get 1 million farmers to raise 200 or more additional winter chickens to 3 pounds or more in weight. Indiana's quota is 20 million pounds."[93]

Whether eggs or meat, poultry products were needed to help meet the nutritional needs of the military and other populations during the war. Women would play an especially important role in supplying these products since they were often responsible for managing the poultry operations on family farms.[94]

Eggs made money for farmers, just as pork and milk did.[95] Depending on the poultry operation, farmers could make anywhere from 81 cents to $4.76 per bird on egg production during the war.[96] And it didn't hurt that most farmers could take advantage of the better market prices since, as one agent put it, "[p]oultry is a very definite part of the farm business on practically all of our farms."[97]

Those who participated in programs to sell better-quality eggs that were graded according to government standards found they could make more money.[98] In addition to paying attention to the color, size, and cleanliness of the eggs, farmers were advised to keep them cool in order to preserve higher quality.

Farmers also used egg shipping associations to improve their marketing advantages. One of the older ones was the Denver Egg Car Shipping Association in Miami County, which had started in 1930. Through this group, egg producers had been able to ship 427,000 cases of eggs from 1930 to 1945 to places such as New York, where they earned approximately $560,000 more than what their eggs could have sold for locally.[99]

As with other wartime agricultural products, the increase in production for eggs presented challenges to the distribution system. The Ripley County agent reported that farmers who were supplying eggs to the Cincinnati market in 1943 were "asked not to ship any more for the entire summer" as there were transportation problems since the military required so many railcars and there was little storage space to keep them.[100] The following year, a glutted market caused price drops, which upset producers.[101]

Even though the government guaranteed a minimum price a producer would get for a dozen eggs, some producers didn't receive this amount. In 1944, eggs were selling for 18–22 cents per dozen instead of the 27 cents per dozen the government had guaranteed.[102] In his 1944 annual report, Pike County agent Edward Bogue described the situation faced by many poultry farmers that year, writing:

> A very serious egg marketing problem was worked out in cooperation with the War Food Administration of Jasper on May 25. Produce companies of Petersburg who had been issued [a] license to buy eggs for the government were not participating in the program and were paying the farmers of Pike County only 18 cents per dozen for eggs. A new marketing center was established in Petersburg on the advice of the county extension office and on the morning of May 24th, the marketing center was paying the support price of 27 cents and all retail establishments who purchased eggs were contacted so that they could pay the support price and in turn sell their eggs to the marketing center.[103]

Still, many flocks were sold off as farmers who didn't get the best returns for their efforts looked for better opportunities elsewhere.[104]

During World War II, Indiana farmers responded positively to the demand by the federal government for more protein to feed Allied troops as part of the U.S. Food for Freedom program. The government backed up its request by paying higher prices for meat and eggs, and for dairy products that included products ranging from whole milk to dry skim milk as well as evaporated milk, butter, and cheese.[105] The need for protein combined with the government's willingness to pay premium prices meant that, for the most part, those raising livestock during World War II earned a good profit from their efforts.[106] And due largely to Purdue agricultural research and Extension outreach to distribute that information to farmers—work that had been going on for decades prior to the war—producers were in a great position to meet the government demands for more protein in the form of pork, milk, chicken, and eggs.

CHAPTER 29

For Patriotic Reasons as well as for Profit, Acreage Has Been Expanded

During the Great Depression, the United States government paid farmers to reduce the acres planted to corn due to what was called soil-depleting characteristics.[1] But when the demands of World War II required farmers to plant more corn in order to feed more livestock, the need for more protein for soldiers trumped the concerns about corn depleting nutrients in the soil.[2]

For decades, farmers had received educational programs by Extension on growing corn, so yields of 100 bushels per acre were no longer an aberration. This was largely due to the availability of hybrid corn and using management practices that included adequate amounts of fertilizer and crop rotations.[3] Whitley County agent Blanchard Widney, who had by 1942 spent more than twenty years in Extension, summarized the state of affairs for corn production that year, stating, "The rank and file of growers are much more 'corn conscious' now than they were a decade ago."[4]

World War II brought about permanent changes to how corn was grown in Indiana. Farmers were able to produce more corn per acre than anyone could have imagined just a decade before through three significant developments, which included the following:

- *The use of hybrid corn largely replaced open-pollinated corn.* During World War II, planting hybrid corn increased yields as much as 25 percent over the commonly grown, open-pollinated lines that farmers had traditionally relied on.[5] In Jackson County, the agent noted in his 1942 report that adoption of hybrid seed corn had increased yields by an average of 22 bushels per acre, which resulted in an extra 440,000 bushels harvested. This was a significant increase at a time when farmers were being challenged to meet government quotas for this essential crop.[6] Yield gains like these across the state provided more corn for use within the county or for export to other counties to feed livestock. With corn prices ranging from 80 cents per bushel in 1942 to as high as $1.20 per bushel in 1943, these extra gains in yield also boosted farm income.[7]

Planting hybrid corn was no longer considered an oddity, with at least three-quarters of the state's corn growers having replaced their open-pollinated lines that they had grown for generations by the early 1940s.[8] It had taken years of repeated demonstrations and educational programs to overcome the hesitation that farmers had about making this change.[9] (See Chapter 16.) As Marion County agent Horace Abbott would report in 1944, "Changing with more rapidity than any other single farm enterprise, hybrid corn has completely replaced open pollinated varieties within the last ten years . . . [producing an] average increase of approximately 20 bushels per acre over average yields produced in the days of open pollinated varieties."[10]

The use of hybrid seed corn became so widespread during World War II that open-pollinated corn was now no longer worth incorporating into county variety trials.[11] Still, there was a minority of farmers who continued to plant the open-pollinated corn, believing this grain made better livestock feed.[12]

By 1944, Jay County agent Chester Langston observed, "It took a long time to sell farmers hybrid seed corn, yet now they are anxious and are on the constant look-out for new varieties that are the result of the [university] experiment stations."[13] In addition to capturing farmers' shifting views on using hybrids, Langston's comments also highlight the fact that farmers now understood there were differences in hybrid varieties and that they could look to the Agricultural Experiment Station at Purdue and those at other land-grant universities as well as to commercial seed companies for new releases and informational updates.

A couple of years before that, the Knox County agent had reported that area farmers "have stopped asking the question, 'Which is more profitable, hybrid or open pollinated corn?', but they are asking the question, 'Which is the best variety of hybrid for Knox County?'"[14] But this introduced a new challenge for Extension agents such as Ross Atkinson in Orange County, who observed:

> The fact that some hybrid corn varieties are accepted in every part of the county as a valuable, unquestioned means of economically increasing production creates another problem in new proportions. The growing question is which varieties? During the time that the [Agricultural] Experiment Station [at Purdue] was able to carry on numerous field tests in numerous fields over the state, there was proven evidence of the desirability of a relatively small number of hybrids yet of sufficient variety to satisfy all the local conditions. . . . To be sure, there is still plenty of factual information to satisfy the open minded student of the problems at hand. Yet, on the other hand, unproven varieties are being sold in the county which give satisfaction. When a farmer has success with a new thing, he very naturally recommends it to his neighbors, even though he does not have the information as to how the new corn would compare with others which could possibly be even better.[15]

To answer the question of the adaptability of new hybrids, Extension ramped up its demonstration plots to test the varieties of hybrid seed corn being released by Midwestern land-grant universities and commercial companies. County demonstrations featuring these allowed farmers to make side-by-side comparisons of yields, standability, and disease resistance to validate or refute the claims being made by individual farmers and the seed corn industry.[16] Agent Willis Stall noted in 1942 that these demonstrations allowed Jackson County farmers "to keep abreast of new and important developments as they occur."[17]

• *Fertilizer applications became standard practice.* While corn hybrids legitimately earned much of the credit for increasing corn yields, agents admitted that the wider use of fertilizers was equally important in achieving this.[18] It didn't hurt for agents to remind farmers in 1942 that the cost to apply 0-20-20 fertilizer was $2.27 per acre while the net return from increased yields was $7.35 per acre.[19]

In 1943, Warren County agent Herman Madaus weighed in on why he thought local farmers were accepting the expenditure on fertilizers, writing:

> The general use of hybrid corn has convinced farmers that the better and more active root system of hybrid plants make it necessary to supply more fertility in the soil in order that maximum yields can be produced. It has been found that hybrid corn gives abundant response to fertilization. A few years ago, the fertilization of corn was practice[d] by only a comparatively few farmers in the county but the more general use of hybrid corn has created an interest in supplying more commercial fertilizer and better soil practices.[20]

Using fertilizers also allowed corn to mature up to ten days earlier, which was critical for getting corn to full maturity in the northern counties before the first hard freeze.[21] Extension recommended that farmers apply fertilizers containing phosphorus and potassium with formulas such as 0-10-10, 0-12-12, 0-14-7, and 0-20-20.[22]

Purdue agricultural researchers had shown that applications of nitrogen increased yields as well. Applying 500 pounds of 8-8-8 fertilizer gave farmers a net increase of $5.30–$8.67 per acre.[23] However, it was unfortunate for farmers that the supply of nitrates was being diverted to the military and converted into explosives. In his 1942 annual report for Jackson County, Stall wrote about this conversion as "a situation that has had to be met with a good deal of tact and explanation, because it is so closely related to the War program."[24] Some farmers used barnyard manure to overcome the shortage of nitrogenous fertilizers.[25]

By 1945, it was typical for three-quarters of Indiana farmers to use 75–150 pounds of fertilizer per acre.[26] Of those farmers who applied fertilizer, 74 percent used ones with phosphorus and potassium in combination; 25 percent used ones with nitrogen, phosphorus, and potassium; and 1 percent applied ones with either phosphorus or potassium.[27] By the time the war had ended, applying fertilizer to hybrid seed corn was standard practice.

• *Better corn management principles were used more consistently.* For decades, the five-acre corn contests sponsored by the Indiana Corn Growers' Association had showcased how to incorporate best-management practices to improve corn yields.[28] Farmers were usually eager to see what other growers were doing to drive up yields and profit, so Extension used the participants' variety selection, fertilizer rates and application methods, seedbed preparation, and crop rotations as teaching tools for those who didn't participate.

Many of the practices used by the best farmers in the competition were derived from Purdue agricultural research. When to plant was a recurring question that Extension could answer based on this research. Farmers began preparing the soil by plowing in March and April, and, by and large, followed Extension recommendations to plant corn between May 21 and May 30 to get the highest yields.[29] By 1942, farmers were also more inclined to plant their corn in 34- to 40-inch rows based on the science that showed this was a best practice.[30]

Extension promoted the proven practices gleaned from the county competitions and quantified the increases in yields and profits.[31] In 1942, agent Walter Rusk explained that he compiled information collected from the Huntington County Five-Acre Corn Club into "a printed circular that is distributed to all farmers of the county and many farmers accept it as a guide toward improving their corn production practices."[32]

It was the culmination of decades of work with the corn crop along with the introduction of hybrids that allowed growers to produce the quantity of grain needed to help feed the expanded number of hogs. While there were still major shortages of corn at times during the war, this problem would have been greatly exacerbated if it were not for the yields the improved corn crop provided.

During the 1920s, soybeans had started gaining in popularity among Indiana farmers as a soil builder, a cash crop, and a hay crop.[33] By 1939, farmers across the state planted 9.6 million acres of soybeans, but wartime demands would push that number to 13.1 million acres by 1945.

Repeated droughts in the 1930s that had killed off pastures and alfalfa had convinced dairy producers to rely more on soybeans as a hay crop that would provide dairy cattle with an important source of protein.[34] Soybeans were often grown for hay because they did better than alfalfa on acidic soils.[35] With so much interest in soybeans, Extension had conducted educational programs during the 1920s and 1930s to help farmers understand what was involved in raising this crop successfully.[36] Agents used field demonstrations to show the differences between varieties based on whether the crop would be used for livestock feed, seed, or oil.[37]

By the end of 1941, the U.S. Department of Agriculture (USDA) had declared soybeans as a war crop and set federal quotas for the 1942 growing season.[38] To meet the demand, farmers in Indiana and other midwestern states, such as Illinois, Iowa, Michigan, and Ohio, were asked to increase their acreage by 20 percent. But due to poor planting conditions for corn, they actually ended up increasing soybean acreage by 140 percent for that year.[39]

As with other crops, the government had backed its requisition for more soybean production with federal price guarantees.[40] In 1942, the government guaranteed that farmers would receive $1.60 per bushel.[41] The price shifted to $2.10 per bushel in 1944 and to $2.05 per bushel in 1945 as the demand for products made from soybeans escalated.[42] With the price supports in place, Whitley County agent Blanchard Widney reported in 1943, "Soybeans have assumed a new place of prominence as a cash crop because of the war. For patriotic reasons as well as for profit, acreage has been expanded at the expense of wheat and oats."[43]

Those owning combines prior to the war or lucky enough to get one of the federally rationed machines during the war (see Chapter 36) were able to take advantage of the government-subsidized prices by planting more acres.[44] In 1943, a cooperative of 400 soybean farmers from Indiana and Kentucky processed nearly 400,000 bushels of soybeans, which produced 3 million pounds of oil and 21,000 pounds of meal.[45]

Farmers wanting to cash in on soybean prices had turned to Extension for recommendations on variety types, planting dates, seeding rates, tillage, crop rotations, and other management questions. By World War II, plenty of data had been generated across the state on what it took to produce the highest yield of soybeans from an acre of ground.[46] Agents used Extension trainings and field days to share best practices with farmers. Decades of research by Purdue agricultural scientists meant that Extension specialists and county agents were ready with their recommendations, which was critical as the demand for meal, margarine, and cooking and industrial oil rose because of the war.[47]

In 1943, agent William Record recommended the following management practices to Morgan County farmers:

1. Growing earlier maturing varieties of soybeans with high oil yielding potential;
2. Sowing soybeans in rows to keep down weeds and permit earlier harvesting;
3. Fertilizing; and
4. Using a winter cover crop after the wheat was harvested.[48]

The release by USDA plant breeders of newer, high-yielding varieties such as Lincoln in 1943 helped the war effort even more.[49] With proper management, Lincoln had the potential to yield 50 bushels per acre—an increase of 18–30 bushels more than other varieties.[50] This variety also offered an earlier-maturing bean, met the needs of the market, and fit well into a planting rotation that included wheat.[51]

Much like with corn hybrids, the demand for better soybean seed exceeded the supply, so agents assisted farmers in finding seed supplies, often conducting germination tests to verify the quality.[52] In describing his efforts to locate soybean seed for Morgan County growers, Record wrote in his 1943 report:

> Early in the year it was apparent from a survey made through meetings and through the mail that farmers would respond whole-heartedly to the call for an increased acreage of soybeans for harvest, but that there was a serious shortage of good, germinable seed beans. This office and vocational departments of the county tried to cooperate by testing samples of soybeans brought in and to locate farmers with ample supplies of germinable beans. The bulk of the soybean seed needed, however, was brought in from outside sources through grain dealers and truckmen. . . . [T]he acreage of soybeans grown in the county this year was twice as large as last year, with a much greater proportion of it being harvested for grain.[53]

In the years after World War II, farmers would continue to expand their soybean acreage, eventually increasing it to make soybeans the second leading crop in the state after corn.

With the federal government pushing farmers to plant more corn and soybeans to meet quotas during World War II, the resulting increase in grain quickly overwhelmed the available storage space. With so many railcars being used by the military, elevators faced a shortage of railcars to move the grain out of their facilities, and processing facilities were limited on how much grain they could accept or store at a time.

The inability to move surplus grain out of the counties meant that finding additional storage space became a priority for farmers and agents. In 1942, agent Scott W. Milligan reported that the elevator storage in Wayne County was completely filled, but storage was still needed for more than 20,000 bushels.[54] The following year, Marion County agent Horace Abbott would report, "The greatest demand has been for granaries and cribs for the storage of farm-produced grain, especially soybeans which could not be absorbed by regular market channels as rapidly as they were being produced and harvested."[55]

Agents reported that storage space was so limited that farmers turned stalls, corncribs, chicken houses, spare bedrooms, and empty houses into bins to store the excess grain.[56] How to store 350,000 bushels of soybeans produced in Parke County in 1942 became a popular topic of discussion, with the agent indicating they were stored in "brooder houses, hog houses, corn cribs, and every place that could be made available."[57]

The federal Agricultural Adjustment Administration responded in 1942 by building thousands of storage bins. The wooden structures were made in sections, bolted together, then placed on concrete blocks and roofed with wooden shingles. The bins were painted white to improve their appearance as they dotted the rural landscape.[58]

There were 114 storage bins built in Warren County alone. With each holding either 1,600 or 2,800 bushels of soybeans, the total storage capacity was approximately a quarter million bushels of soybeans in this one county.[59] When Knox County farmers harvested 40,000 acres of soybeans in 1942, they depended on the government bins for storing the 100,000-bushel overflow.[60]

But as the harvested grain poured out of midwestern farms, the government-financed construction lagged behind due to a shortage of lumber and labor.[61] This clearly showed how pushing for more yield created cascading challenges as the grain was moved from the farm to the distribution points. Federal and state agencies—and, to some extent, growers—would have to deal with the domino effect when yields of corn and soybeans outpaced the ability to store, move, and process them.

The collective work by Purdue agricultural researchers and Extension staff supported the rapid adoption of new practices by farmers that helped them increase their yields of field crops during the early 1940s. With the government's quotas in place and prices paid for corn and soybeans being much higher than normal, farmers had both a patriotic desire to help with the war effort and a financial incentive to grow more of the crops that the government wanted. However, those increases led to negative consequences that affected the ability to handle the large surpluses and use the additional output for the war effort. It was important to solve these problems because corn and soybeans formed much of what was fed to livestock, which in turn became a protein source for soldiers and those overseas who were affected by the war.

CHAPTER 30

Tomatoes Have Become a Major Crop

With food for soldiers and civilians in high demand during World War II, farmers were encouraged to grow crops that could be canned because of the advantages they offered. Canned products had a longer shelf life than fresh vegetables, could be transported in larger ships going overseas, and offered a nutritious meal to any soldier with a can opener. Tomatoes were one of the most sought-after crops in this category along with others such as beans, beets, cabbages, peas, pumpkins, spinach, and sweet corn.[1]

Indiana became an important state for canned products because of the concentration of canneries that had existed there before the war. Marion County was the epicenter of the canning crops industry. As agent Horace Abbott noted in his 1942 annual report, it was "home of one of the largest canning factories in the world, and Marion County is also one of the largest tomato-producing counties in the United States."[2] That year, 800 area farmers had dedicated one-third of the county's acreage to growing crops that could be canned.[3] At the time, Indianapolis had the Columbia Conserve Company, Morgan Packing Company, and Stokely–Van Camp Canning Company, which were collectively large enough that they could take an unlimited amount of tomatoes and other canning vegetables from growers.[4]

As the canning industry prospered in Marion County, Abbott spent more time dealing with issues more or less unique to canneries and to truck farmers. These were the farmers who grew many different types of specialty crops and delivered them to market themselves in their trucks and trailers. In 1943, Abbott reported, "The Extension Service will cooperate with the canners in helping secure acreage, holding canning crops schools and other grower-canner problems. The Extension Service will help recruit labor, not only for the harvesting of canning crops, but also labor for the planting and culture of these crops."[5]

In the spring of 1942, the U.S. Department of Agriculture (USDA) had released production quotas dealing specifically with canning crops, putting out a call nationally for 40 million cases, or approximately 240 million gallons.[6] That year, Indiana's growers were asked to supply 5 million cases, which was 67 percent more than what was canned during the previous year.[7]

The government-assigned quotas were distributed among the counties based on past production. For example, Pike County growers were asked to plant 600 acres of tomatoes; Orange County growers, 250 acres; and Perry County growers, 125 acres.[8] Vigo County growers typically planted 240 acres of tomatoes each year, but their 1942 quota was three times their normal acreage.[9]

While each county received a specific quota, the county war boards did not assign a specific target to growers who customarily grew tomatoes. Since the canneries typically contracted with farmers to secure produce in advance, company representatives would work with the agents to locate and contract with growers. In 1942 in Marion County, Abbott reported, "The Extension Service helped the canning factories get more acreage by appealing to the farmers to plant canning crops."[10]

Extension wanted tomato growers to make an official commitment to the processors by signing advance contracts, which kept the canneries operating and helped to ensure that the county would meet its quota. However, the growers could choose to market their crop in one of three ways: contract a certain number of acres to one processor; split the acreage between two or more processors; or grow tomatoes without contracts to possibly get better prices from processors later.[11] For those who were willing to risk it, the last option could potentially bring a higher return, but depending on the growing season, the gamble didn't always pay off. In 1943, around one-fifth of Marion County tomato acres—or approximately 1,200 acres out of 5,891 total acres—were grown without contracts.[12]

The number of canning factories in and around a county often determined which of the three options a farmer took. For example, Ripley County only had Nass Corporation, leaving farmers there with limited options. By comparison, there were nineteen canning factories in Marion County and seven more in neighboring Johnson County, which meant there was more competition for the produce among the area companies.[13]

Tomatoes became an essential crop early on in World War II. The push to plant more acres to tomatoes meant more farmers were getting into tomato production for the first time. In 1942, Jefferson County agent Leo P. Geyer reported, "Tomatoes came in for considerable attention in our county this year, due to demands for a bigger pack to aid in the War."[14]

The increased price being paid for tomatoes meant those who did grow them would be paid well for their crop. In Pike County, agent Edward Bogue even noted the interest among small-scale farmers, writing, "Tomatoes have become a major crop in the northern section of the county. With small farms and with a high percentage of farmers falling in the low income group, it is important that cash crops be grown as much as possible. The only developed market for a cash crop in this territory is tomatoes, and with family labor generally available, this crop has proved to be very satisfactory as a cash crop."[15]

Extension staff conducted and participated in many meetings to support growers and emphasize the importance of tomatoes under the Food for Freedom program.[16] One popular and renowned speaker at many of these events was Purdue tomato Extension specialist Roscoe Fraser, who also frequently presented at educational meetings offered by the canning companies.[17] He would address subjects such as disease and insect control, proper fertilization rates, direct seeding, and USDA grades as well as "the importance of securing good color and delivering red ripe tomatoes to the canning factories," as one agent reported.[18]

As the war continued from one year to the next, tomato schools offered by Extension focused on the patriotic duty of growing tomatoes. Bogue offered this account of the 1944 Pike County tomato school:

> One hundred and twenty-seven county tomato growers attended the Tomato School on March 28th. In addition to Mr. [Roscoe] Fraser, specialist conducting the school, Pfc. [Private First Class] Richard Leaf of the Billings General Hospital of Indianapolis explained the importance of food to the fighting men. Pfc. Leaf had been wounded in the battle of Guadalcanal. Al Turmail of the Quartermaster Depot and buyer for canned food for the military department in Indiana spoke on the importance of canning crops to the armed service and told of the percentage of food canned by the commercial canneries, which would be taken by the government this year.[19]

At events such as these, Extension specialists and agents covered planting varieties suited for canning, some of which were developed by the Agricultural Experiment Station at Purdue.[20] They also addressed fertilization of the crop, which was an important topic since at least one county cited the increased use of fertilizer as being "responsible for the county easily surpassing its quota of production."[21]

Extension staff also taught growers how to grade tomatoes using USDA grades. At the start of the war, packing houses had begun grading tomatoes and paying more for better grades as a way to encourage growers to deliver higher-quality produce.[22] This change in the way tomatoes were priced prompted Extension to offer grading schools to teach farmers and farm workers the best picking strategies.[23]

Prices paid for USDA No. 1 and No. 2 grade tomatoes were approximately $20 per ton and $13 per ton, respectively, in 1942.[24] The following year, canning factories were paying $26 per ton for No. 1 tomatoes and $16 per ton for No. 2.[25] For 1944, as demand continued to exceed the previous year's supply, the government once again raised its subsidy for tomatoes. That year, processors paid $32 per ton for No. 1 tomatoes and $20 per ton for No. 2.[26]

In 1942, the average farmer produced 5–7 tons of harvestable tomatoes per acre.[27] In 1943 in Marion County, where tomatoes were a primary crop, growers averaged 6.4 tons per acre. But just a few years later in 1946, Howard County producers averaged 18–20 tons per acre, showing the great advancement that farmers had made in increasing tomato yields during the war years.[28]

Some of those advancements stemmed from clubs and contests that recognized the better farmers with medals or other awards for their ability to produce higher yields or a higher percentage of top-grade tomatoes. Extension often encouraged farmers to participate because the experience promoted the use of best-management practices based on scientific research and brought higher returns to participants. Canneries also encouraged participation since the improved quality that resulted gave them a better product for processing. In 1942, Frank Backer and I. D. Meyer of Tippecanoe County won a statewide tomato competition by producing the record yield of 14.82 tons per acre, with 93 percent of their crop graded as No. 1. Their harvest went to the Ladoga Canning Company for processing.[29]

Expectations were that higher government-guaranteed prices would induce farmers to take full advantage of planting more acres to tomatoes during the war years. But there were serious issues with raising tomatoes, including problems that started at planting and continued through processing.

From the grower's perspective, those problems included a shortage of field laborers to plant and harvest the crop and a lack of trucks to transport tomatoes from the farm to the canneries. When farmers did deliver their produce to a cannery, they often faced long wait times to unload due to the increased production. At times, it took thirty-six hours to offload a farmer's truck, with delays causing spoilage of many tomatoes.[30]

In light of these issues, farmers became more hesitant to increase their tomato acres after 1942. By 1943, their hesitation was such a serious issue that the chairman of the Jackson County War Board, Will K. Isaacs, sent a letter to local growers on May 28 pleading with them to grow tomatoes. He wrote:

> We are in a critical situation. Something must be done AT ONCE. It becomes more apparent each day that we are not going to plant as much tomato acreage in this county as the Government has requested our canners to pack. . . . The situation is critical because tomatoes have to be raised where the factories are. They can't be raised 200 miles away. We, in this county, live where the packers are. Our Government needs tomatoes for our armies and others. This is purely a matter of helping Win the War quickly.
>
> We very strongly urge: (1) Those of you who have already contracted to increase your present acreage of tomatoes if at all possible. (2) Those of you

> who have not contracted to see your canner and do this at once. Any amount, regardless of how large or small, is vitally needed and must be produced. We hope you and everyone in this county will use your influence to bring about the production of at least another 1,000 acres of tomatoes in Jackson County, THIS YEAR. We must act quickly if this is done. . . .
>
> We have carefully avoided the assignment of individual farm goals in tomato production. It is the belief, however, of this Board, that we are not expecting too much when we believe that every farm should help. In particular, we believe that every farm on which there is a farmer classified in 2c or 3c [for Selective Service] should produce tomatoes this year. Doing this is fighting the battle AT HOME. Please help us get this job done.[31]

That same year in Jay County, agent Chester Langston reported that farmers had reached the point where they were "busy enough without planting any additional crops."[32] In some cases, agents attributed the reduced acres of tomatoes to a preference among farmers to plant an alternative crop instead. This included soybeans, which were easier to grow, required less labor, and could be mechanically harvested with combines.[33]

At the same time, the processors faced challenges of their own. In 1943, Tipton County agent Walter M. Clary reported the following concerns that had been expressed by canners during a meeting with local officials:

> 1. If U.S. Employment Service labor is similar to [what] it was in 1942, their service is not needed. The canners were not satisfied with the department's service.
> 2. Lumber cannot be secured to build shelter houses for pickers and factory workers.
> 3. Tomato factory labor "freeze" is causing year around labor to leave for defense factories. This essential labor will not be available during the tomato packing season.
> 4. Tomato canners believed there would be sufficient tomato pickers if this transit labor [migrant workers] secured gasoline and tires to make the trip.
> 5. Tomato canners believed the labor shortage will be the most critical in the factory. The best source of labor (peelers) is in the community around the factory. These housewives have the experience, etc.; however, [they] do not want to work because the extra money is not needed at the present time. . . .[34]

With the labor shortage being a primary concern for both growers and processors, agents invested a lot of time to find the needed laborers to plant, harvest, and process the crop. In Marion County, Horace Abbott reported in 1943, "Last spring, labor was the chief drawback on growing tomatoes. After the Extension Service and the canners assured the farmers that labor would be available for harvesting the crop, the farmers were more free to contract for tomatoes."[35] Migrant laborers from southern states, Jamaica, and elsewhere as well as prisoners of war from Germany and Italy were used to supplement the local labor pool (see Chapter 34).

In some cases, farmers worked on their farms during the day and in the canning factories at night, joined during the evening shift by office workers, homemakers, and others.[36] A report by agent Leo Geyer acknowledged how people from all walks of life were working side-by-side in a peeling line at a Jefferson County canning factory, with him concluding, "In other words people were . . . doing the important job just to be patriots."[37]

While alternative workers such as these helped ease the labor shortage, many were inexperienced in what it took to quickly and efficiently peel tomatoes, which posed another problem for canners. Two Tipton County processors—Jaqua Canning Company and Ray Brothers and Noble—solved that issue by training their inexperienced workers using a black-and-white film titled *The Proper Method of Peeling Tomatoes*, which had been produced by the Purdue Department of Horticulture.[38]

Unfortunately, most of the challenges experienced by growers and processors were hard to solve in light of wartime constraints, so they continued to persist throughout the war years. While some counties would meet their quotas, the difficulties associated with raising tomatoes meant Indiana growers would not plant enough acres to reach their assigned quotas.[39]

CHAPTER 31

The Total Increase in Home Production and Consumption Would Be a Staggering Amount of Food

Throughout World War II, the government primarily focused on increasing and supporting agricultural production efforts. But while U.S. farmers rose to the challenges put to them, in retrospect, those years are often remembered for the sacrifices made by women as they dealt with widespread rationing, embraced victory gardens, and undertook extensive food preservation efforts that would provide for their own families as well as for the military. Less known, perhaps, is that the leadership to support those efforts in Indiana came from Extension through specialists at Purdue and the county agricultural agents, who were joined in their efforts by home demonstration agents and part-time war food assistants.

Just like during World War I, the federal government once again funded these latter two positions to help deliver the educational programs aimed at food production from the home garden and poultry flock. These professionals would find willing partners in the thousands of county- and township-level home economic clubs throughout Indiana.

In 1928, the federal government had provided some limited funding to make home demonstration agents permanent members of the Extension staff in counties, but by 1935, only thirteen Indiana counties had a full-time home demonstration agent on staff. Two years later, that number had grown by just nine more agents. During the 1930s, it was typical for organizations such as area home economics clubs to provide funds for the travel and office expenses of these agents. This generally amounted to around $450–$600 annually.[1] With expenses guaranteed upfront, the county council then petitioned Purdue to appoint a home demonstration agent, whose salary would be paid from federal and state funds.[2]

As World War II erupted, it seems that federal dollars to support the home demonstration agents increased, which allowed the hiring of additional agents. In Indiana, there were 45 of these agents in 1939, 48 in 1940, and 54 in 1944. As the war put tremendous pressure on the food supply in the United States, the federal War Food Administration provided enough funding to hire 58 war food assistants starting February 1, 1944, to work collaboratively with the home demonstration and agricultural agents across Indiana.[3] These special hires were expected to make individual house calls to rural and urban homes, host garden meetings, and conduct food preservation demonstrations.

As home demonstration agents joined county Extension staffs across the state, they had an immediate impact on the community through their work to increase the number of home economics clubs, reinvigorate the girls' 4-H clubs, and build the social life for rural women through tours and activities.[4] It would be largely through the county home economic clubs that the advancement of rural women would be seen and felt across Indiana as the war placed demands on the home.[5]

By 1946, the emergency war food assistant positions were no longer mentioned in the Department of Agricultural Extension annual reports, likely indicating that their funding was gone and the positions eliminated. Women leaders must have worried that the number of full-time home demonstration agents would also be slashed following the war. However, that did not happen. During the post–World War II boom, these positions continued to increase, and by 1950, there were seventy-five across the state. Home demonstration Extension agents were now considered as essential to their communities and counties as the agricultural Extension agents.[6]

Early in 1942, women had to start adjusting how they managed their homes and fed and clothed their families. Soon, widespread rationing affected the supply of basic food items such as meat, cheese, coffee, sugar, and juices as well as other items such as clothing, shoes, and nylons.[7] All families, regardless of income, were affected by the shortages as ration books became the currency of the day.

The rationing of meat was particularly challenging for most American families. By March 1943, the U.S. military and the Allied forces were consuming one-fourth of the meat produced domestically, meaning that supplies had to be rationed to ensure access by everyone.[8] County war boards, which operated under the U.S. Department of Agriculture (USDA), established meat committees to oversee the supply chain. These committees were responsible for issuing permits to butchers, farmers, and other slaughtering operations in an effort to maintain control.[9] The USDA also set up regulations to limit the black market on meat.[10] The meat committees would operate until meat was removed from the rationing list in 1944.[11]

As the war progressed and additional food items were rationed or in short supply on store shelves, a course titled Efficient Management in Wartime Meal Preparation became popular among the home economics clubs. Offered by Purdue home economists, this particular training was subdivided into two parts: (1) Meat and Other Food Alternatives, and (2) Time and Labor Savers in Wartime Meal Preparation.[12] Topic titles included Yeast Bread and Rolls, Less Time from Garden to Table, and Quick Preparation of Vegetables.[13]

The Purdue School of Home Economics along with the Department of Agricultural Extension in the School of Agriculture provided education like this on many subjects to support women as they cared for their homes and families during the war. In addition to offering food-related topics, such as gardening, preserving food, and improving family nutrition, Extension specialists also covered areas such as understanding farm and home finances, reusing and mending clothing, and dealing with diseases and medical emergencies.

The Purdue experts often used a train-the-trainer approach to offer these educational opportunities. County home economic clubs would vote on which topics were of interest, then each club would send two women to attend the training sessions with Purdue Extension specialists over a period of several weeks or months. The club representatives would later share what they had learned with members of their clubs.[14]

This train-the-trainer approach had long been used by Extension and the School of Home Economics in order to expand the outreach provided by a limited number of experts to a large audience statewide. Through the years, this approach had often been used for programs of interest to women, especially during the Great Depression, when the family budget needed to be stretched to conserve scarce resources. Now the lessons taught during the dark days of the preceding decade would be revisited and updated to address the new challenges brought on by World War II.[15]

Through the home economics clubs, Purdue Extension specialists and home demonstration agents had a willing and enthusiastic partnership with thousands of rural women across Indiana. In 1939, there were 1,424 clubs with 41,492 members. By 1945, there were 1,645 clubs in Indiana with a total of 45,666 members.[16] It was this surge in the number of clubs and participants that allowed Purdue home economists and Extension specialists to help rural women make significant contributions to the war effort while dealing with the many sacrifices that would be asked of them.

One unique wartime program that came out of the kitchen was the saving of waste fats from cooking for use in making munitions. In a 2014 article for *The Atlantic* titled "Turning Bacon into Bombs: The American Fat Salvage Committee," author Adee Braun provided the following description of this national program:

> During World War II, the U.S. government urged Americans to save excess fat rendered from cooking and donate it to the army to produce explosives. . . . [F]ats are used to make glycerin, and glycerin is used to make things blow up. One pound of fat supposedly contained enough glycerin to make about a pound of explosives. . . . The premise was simple: Engage women in the war effort right from their kitchens. "A skillet of bacon grease is a little munitions

> factory," announces a booming voice in the Disney propaganda cartoon. "Every year 2 billion pounds of waste kitchen fats are thrown away—enough glycerin for 10 billion rapid-fire cannon shells.". . . Housewives were directed to strain their leftover fats (no bacon bits in bombs, please), and store them in a "wide-mouth can." Once a pound or more was collected, the fat was to be handed over to any one of 250,000 participating butchers and retail meat dealers or 4,000 frozen food plants who would turn the fat over to the army. The donor got four cents a pound for the fat, and in December 1943 when lard and butter began to be rationed, the government started offering two ration points per pound as well.[17]

Based on annual reports by agents, Hoosier women were active participants in this project.[18] In many cases, the county agricultural agent was responsible for getting grocers and others to receive the saved cooking fat from their customers.[19] The collected fat would be sold to brokers, who then supplied it to the military.[20]

In Huntington County, metal barrels were placed in all of the butcher shops and large grocery stores, which facilitated the collection of more than 8 tons of fat in 1942.[21] In LaGrange County, butchers paid 5 cents per pound for waste fats largely from cooking meats and collected 1,122 pounds of fat.[22] In Newton County, the money collected from selling fats was used in 1943 to offset the costs of the 4-H camp.[23]

Early on during the war, there were legitimate concerns raised about the possibility of food being in limited supply during the war. The reports from the agents were clear that the public would feel the full effects of the global war on the supply of food they were accustomed to purchasing. Some of the agents' observations included:

- The Lend-Lease program would consume a large share of what was produced in this country each year.
- The military would have first priority on meeting its food requirements.
- The expectation was that the United States would have fewer imports, thus restricting some food items commonly purchased in stores.
- Many food items being diverted for the war effort would cost more to purchase when available.
- Some food items would be in such limited supply that the government would impose rationing.
- The large number of railcars needed for military use would disrupt the distribution of food items.
- Without fresh produce from gardens either fed in season or preserved for later use, people—especially children—would not receive enough calories and vitamins to sustain a healthy lifestyle.

- A shortage of labor on farms and at the processors might result in limited supplies of fruits and vegetables that households normally relied upon.[24]

Agents quickly began encouraging a greater reliance on homegrown food, including meat, fruit, and vegetables. Any of these items not purchased at local groceries would leave more available for other uses, particularly for the military and foreign aid.[25]

The push was on for every American with access to ground to use it to stretch the nation's food supply.[26] For example, one effort intended to help stretch the limited supplies of meat and protein saw the federal government encouraging everyone who had even limited available spaces to raise poultry.[27] The idea behind the project was for people to raise enough "backyard" chickens to meet their needs for eggs and meat, which allowed the production from the larger farms to be used by the military. To support this campaign, Purdue Extension specialists and agents distributed leaflets that explained what it took to raise and feed chickens.[28]

In Lake County, the agent reported on the effort locally in 1943, writing, "Meat rationing caused many city dwellers who had never raised chickens before to raise them or try to raise them under all sorts of conditions. Several people raised chickens in garages, one man raised them under his front porch, [and in] a few instances people were reported to have been raising them in rooms of their homes."[29] That year in Marion County, agent Horace Abbott reported, "It is not uncommon to find on the lawn of city residents in the most exclusive residential areas a small building housing a few hens that are furnishing the family's egg supply."[30] Abbott also indicated how popular the program had become, saying:

> A large increase in the production of poultry and eggs will result from the establishment of small victory flocks among the urban areas of the county. . . . In fact, poultry has become so numerous within the limits of the city of Indianapolis that the attention of the City Council has been called to the fact that poultry was prohibited within certain areas but the county took no action to prevent these flocks being kept due to the urgent need for supplementary food production under war conditions.[31]

But it was the Victory Garden Program that quickly became the main focus for homegrown food production. During World War I, an effort to get people to plant victory gardens, or war gardens, had helped increase the national food supply, so a similar effort was quickly set up during World War II. The Victory Garden Program was outlined by the National Defense Gardening Conference in Washington, D.C., in December 1941. In Indiana, the rollout for the program took place in March 1942, with one agent observing by the following year that the "subject was certainly one of the most widely discussed of any project in the history of extension work, and was closely tied up with Food Preservation and

Storage."[32] Much of the effort in Indiana focused on increasing the number of gardens, which meant that campus Extension specialists and county staff, including agricultural and home demonstration agents, needed to deliver gardening information to both established and new audiences.

The plan for implementing the Victory Garden Program in Indiana was jointly developed by the Purdue School of Home Economics and the Department of Agricultural Extension in the School of Agriculture. The Civilian Defense Council—a group with responsibility for organizing local protection and preparedness efforts—established county- and township-level committees to lead the program locally and coordinate with the county Extension office. Committee members included representatives from the defense council and the local farm bureau as well as individuals with expertise in home economics, including local schoolteachers.[33]

The overarching strategy by Extension was to encourage every family to plant a garden, whether on their own property or on ground provided by local authorities or businesses.[34] To help with this massive gardening project, the Purdue Department of Agricultural Extension furnished pass-through federal funds to hire part-time emergency war food assistants and provide secretarial support. The role of the new assistants, as described by Jackson County agent Willis Stall, was to work in conjunction with home demonstration agents "producing and preserving the food supply of vegetables, fruits and meats for the coming winter."[35] To that end, these assistants along with home demonstration agents would conduct live demonstrations showing how to properly preserve food when gardeners were ready to harvest their produce.[36]

It seemed an easy program to sell, with Stall reporting, "Practically everyone insisted on helping in the Victory Garden Program."[37] In addition to working with the Civilian Defense Council, agricultural and home demonstration agents worked to generate interest among nearly every imaginable community group, including farm bureaus, churches, garden clubs, home economic clubs, neighborhood leaders, businesses, labor unions, parent-teacher associations, and 4-H clubs. Of all these groups, none was more important than local home economics clubs.[38]

In his annual report for Marion County, Abbott described how much Extension had relied on the home economics clubs to get the Victory Garden Program off the ground in the spring of 1942, writing:

> During past years the Marion County Homemakers' clubs were the nucleus around which the garden program was carried on. There are thirty-nine clubs in the county with a membership of nearly one-thousand, most of them being rural women. On February 16th the club presidents of all the Homemakers' Clubs were called into a meeting to outline the 1942 garden program. Each president was requested to appoint a Victory Garden chairman for her club and bring this chairman to this meeting with her. This year the Victory Garden Program was a major project with every Homemakers' Club in the county.

> Throughout the year the Extension Office provided the Garden Chairman of each club with bulletins, mimeographed material, and information letters on gardening. At each monthly meeting of the clubs, some time was devoted to gardening. Each member of a Homemakers' Club was asked to contact four other families in regard to the Victory Garden Program.[39]

Campus Extension staff also provided support for the effort. During the spring of 1942, home garden Extension specialists Willard B. Ward and Carl Smith traveled throughout Indiana promoting victory gardens. Ward, who was long noted for his work with home gardening programs, explained how the new program would work and its expected outcomes. At local events, he and county agricultural agents handed out enrollment cards to generate interest, while others received cards in the mail from Extension or other sponsoring organizations.[40]

In order to educate the garden leaders in the community, the Purdue Department of Horticulture conducted Victory Garden Leader Training Schools in 1942. Agent Walter Rusk offered the following description of an event that year, writing, "Sixty-eight leaders representing the 15 Home Economics clubs and 42 other women's organizations in Huntington county attended the county Victory Garden Leader Training School held at the Huntington YMCA when Mr. Carl Smith, Purdue Garden Specialist, presented instructional material which proved to be of great help in assisting these leaders to carry the information back to their respective clubs."[41]

Each person who signed an initial enrollment card to plant a victory garden would later be asked to sign a victory garden pledge. Those signing the pledge to plant a garden were given an official Hoosier Victory Garden window sticker sponsored by the Indiana Home Economics Association.[42]

Gardens were planted on land owned by the gardener or on vacant lots made available for no or low cost by industry and the government.[43] In Huntington County, the mayors' offices in Huntington, Andrews, and Warren worked together to create a map showing where garden plots could be secured on vacant lots.[44] And in Madison County, the Guide Lamp Corporation freed up 12 acres at its Anderson facility for 166 garden spaces, each measuring 35 feet by 100 feet.[45]

Sponsoring organizations in many counties would plow the plots located on vacant lots since most of the gardeners didn't have access to a tractor and plow.[46] In 1943, Lawrence County agent John Armstrong described how a local business offered plowing as well as additional incentives for participants, writing, "The best example of a successful industrial group in a cooperative garden was the Bedford Dairy. The manager, Chick Wood, rented a five-acre plot, which was plowed and prepared by tractor and divided into 11 garden plots. These were drawn for by numbers. The seed, fertilizer and insecticides were purchased cooperatively on advice of the agent. Prizes in war bonds and stamps were offered by the Creamery for best garden products of various kinds."[47]

As more gardens were planned or actually planted, the agents received substantially more requests for information on how to grow crops, deal with insects, and store the produce.[48] Agents relied heavily on Purdue Extension leaflet No. 228, *Home Garden Pest Control*, which would become an important source of gardening information since insects and plant diseases were two of the more limiting factors in growing a successful garden.[49] Tippecanoe County agent Sheldon Pershing noted that other popular leaflets addressed "planning, varieties, soil preparation, insect control, canning and storage."[50] The garden publications were often financed by the civilian defense organizations and seed dealers, and distributed through the mail and by neighborhood leaders.[51] (See Chapter 39.)

Purdue Extension put on gardening clinics, schools, and demonstrations across the state to provide further education for participants who had never planted a garden before.[52] Presenters delivered information through lectures and by showing motion pictures and lantern slides.[53] In one of the more unique teaching approaches, Purdue Extension entomologist Glen E. Lehker used colored chalk to illustrate his presentations.[54] Some events also included hands-on training, with one popular activity being how to apply dusts to control insects.[55]

Soon, agents were seeing positive results from the educational efforts, with Owen County agent Fred Hoover reporting in 1944, "Better gardens are now the rule rather than the exception."[56] And Delphine Smith, the Orange County emergency war food assistant, identified the following improvements in the gardens planted in 1945:

- Buying disease-resistant varieties of vegetables;
- Selecting good seeds and plants;
- Early and consistent use of insecticides;
- Shallow cultivation;
- More and better varieties of vegetables;
- Better garden planning;
- Closer spacing of rows in garden;
- Using a food budget;
- Proper use of fertilizer;
- Proper depth of seed planting;
- Rotation to different areas of garden;
- Proper preparation of seed bed;
- Planting of vegetables together that mature at the same time;
- Arranging rows north and south, if possible;
- Planting tomatoes, beans, etc., near taller crops to prevent damage from the sun; and
- A well-planned fall garden.[57]

Preservation through canning and storage were two additional components of the national gardening program.[58] The Extension team—including the home demonstration agents, the emergency war food assistants, and staff from the Purdue School of Home Economics and Department of Agricultural Extension in the School of Agriculture—conducted hands-on demonstrations and workshops to show how to correctly preserve fruits, vegetables, and meats for winter use for the family and for shipping surplus canned goods directly overseas to soldiers.[59] Extension staff advised gardeners on what produce could be canned, what could be frozen or dried, and how to best use preserved foods in recipes.[60]

Since a large number of people were new to canning, many of them lacked the experience needed to make the process successful. Some spent hours canning, only to find later that the contents had spoiled due to improper handling. Russell Zell noted in his 1943 annual report for Newton County that getting vegetables from the garden to the can in one hour helped reduce spoilage.[61] Instructors emphasized washing the produce before cutting and processing.[62] There were also other problems in the canning process to address, as Pike County agent Edward Bogue noted in his 1944 annual report: "Many farm women were found to be using very poor methods in canning, seemingly expecting a high percentage of spoilage. In many cases, their efforts to avoid spoilage was destructive to the vitamin contents because of over-cooking and much was accounted for in their attempts of canning too large a quantity at one time. The improper use of different types of lids contributed to much of the spoilage."[63]

Huntington County emergency war food assistant Frances Smith conducted seven food preservation demonstrations in 1944, which included the following topics:

- Sterilizing the cans by turning them upside down in a pan of boiling water and letting them steam for fifteen minutes.
- Preparation of the food, washing, shelling, peeling, steaming, etc.
- Precooking of starchy vegetables.
- Filling the cans, emphasizing a loose pack with the product swinging in the liquid.
- Sealing, explaining carefully the use of each of the four main types of closures on the market this year.[64]

Additional tips such as these mentioned by Evelyn Manley, the Jefferson County emergency war food assistant, helped ensure a better outcome for home canners:

1. Use only fruits and vegetables of high quality.
2. Can products as soon as possible after gathering.
3. Use only jars and lids which give an air tight seal.
4. Read carefully directions accompanying new equipment.
5. Get reliable canning instructions and follow them carefully.
6. Purchase canning supplies and equipment early in season.
7. Have adequate storage space for canned fruits and vegetables.[65]

Those who canned at least 60 quarts of fruit and 50 quarts of vegetables for each person in the family in 1942 received for their achievement a Hoosier Garden Club pin from the Indiana Home Economics Association.[66] By 1943, individuals received a medal from the association for canning 130 quarts per person.[67] While reaching the number of quarts required to medal seemed a difficult mark to achieve, Pulaski County agent Carl Olson offered this perspective on the numbers: "[T]he garden reports for 1942 showed an average of 116 quarts of food canned, dried, and frozen per family member. This amount would provide less than a pound a day throughout the year. In view of the fact that persons needed from 2.5 to 5 pounds of food per day, families were urged to increase their home production and storage of food this year."[68]

People who were willing to can as much as possible from their gardens encountered other problems. Pressure cookers were in such short supply that they had become rationed by the federal government.[69] Anyone who wanted to purchase a pressure cooker had to make an application to a county committee under the direction of a county farm machinery rationing board.[70] (See Chapter 36.) While pressure cookers were removed from the federal rationing list in 1944, they still remained in short supply for many years after that.[71]

When sugar was rationed in April 1942, there was great concern that this would be a serious impediment to home canners.[72] Sugar was an important ingredient in the canning process because it improved the texture, taste, appearance, and color of the preserved foods.[73] Counties took immediate action to address the shortage, with one agent noting, "The reason for haste was that the strawberries were ripe, but the sugar was lacking."[74]

Agent Guy T. Harris indicated that those canning in Ripley County would have to register to get the sugar they needed, writing:

> During the last part of May [1942], registration for canning sugar was held at each of the nine school sites with the high school principals in charge. Each family was rationed one pound of sugar for each four quarts of fruit to be canned. In addition one pound of sugar per individual was allowed for preserving purposes. One-half of the year's supply of canning sugar was issued at this first registration. At this time they were informed that all that was necessary to receive additional sugar for canning was to write to the rationing office and a certificate would be mailed to them.[75]

In Jackson County, agent Willis Stall described the following approach:

> As early as April [1942], it was apparent to us that an inconsistency had developed in the sugar rationing program. After having spent considerable time on the importance of growing and preserving fruits and vegetables in the Food for Victory program, we were informed by our local Rationing Board that 5 pounds of sugar for canning purposes per family was all that could be secured.

> We investigated this matter and discovered early in August that the matter of fixing this amount was done by our local Rationing Board, who had authority to change this amount if they wished to do so. After discussing this with a newly appointed member of the Board, we found them quite willing to cooperate by bringing this matter up for attention at a regular meeting of the Sugar Rationing Panel on August 12, and on that same day additional sugar was made available to those households or persons who had used their family quota and who desired to do additional canning of fruits and fruit juices.
>
> The Rationing Board further cooperated in this matter when it was brought to their attention that this county was one of less than half a dozen counties in Indiana where a definite maximum was set up. It was felt that this accomplishment, to a considerable extent, relieved a rather serious situation and again brought about a degree of consistency between the Sugar Rationing program and the Food for Victory program.[76]

Not every garden yielded produce that needed to or could be canned, so additional Extension demonstrations highlighted the benefits of storing vegetables and fruits in cellars, outbuildings, and large clay drainage tiles.[77] A 20-by-24-inch tile held 3.5 bushels of produce, a 24-by-24-inch tile held 6 bushels, and a 24-by-30-inch tile held 8 bushels.[78]

Horticulture Extension specialist Willard Ward traveled around the state to demonstrate the use of drainage tiles for storing food. He recommended placing tiles in a location easily accessible from the kitchen.[79] Then, as one agent described, the tile was "placed on end on bricks in a hole just large enough to accommodate the tile. Hardware cloth, some straw and an inverted old metal wheelbarrow bed were used for a cover."[80]

Simply put, Extension specialists and agents spent so much time and effort on teaching people to properly preserve and store their homegrown foods because minimizing food loss meant more food would be available for domestic consumption and stretch the wartime supplies further.

Freezing offered another method to preserve foods for later use, so county home economics clubs became outlets for providing education on this topic.[81] Often, the local emergency war food assistant provided classes for club members on what it took to properly freeze fruits, vegetables, and meat. In 1944, the Newton County assistant presented ten freezing demonstrations, with an additional six freezing demonstrations provided by club leaders to their fellow club members.[82]

While some farm families had purchased freezers to store meats, fruits, and vegetables at home, many others lacked this appliance or did not have electricity to run one.[83] Those without home freezers could rent a frozen food locker, if one was available in their community.[84] In 1939, agent Arvil Keesling offered this account of a new Bartholomew County food locker:

> On February 25 the cold storage lockers which had been erected at the end of Fourth Street were opening to the public. Four hundred lockers have been installed for rent by those desiring to use the service. The total capacity of the plant is 1,000, but a portion is now being used for butter storage. The lockers were built on a cooperative plan through the Farm Bureau service cooperative.
>
> The usage of lockers was relatively slow with about ten per month being rented for the first seven months. In October an ambitious program of usage was announced by the cooperative in which they offered the use of a locker free for three months provided the user would pay a small key charge and cutting fees. As a result of that program, over one hundred fifty lockers have been taken, making a total in use of about two hundred twenty-five. Since the starting of this special campaign, four butchers have been employed to care for the meat and are still very busy. The reasoning back of the offer is that farmers can become acquainted with locker service more quickly by using the service than by being sold a locker. Apparently this reasoning is correct.[85]

Much of the interest in providing frozen food lockers and working through the federal government for funding was handled by the agent's office.[86] Depending on the year, the person who wanted government-backed funding had to have commitments from at least 150 individuals who had agreed to pay rental fees.[87]

In September 1943, the government released enough building materials to construct approximately 350 frozen food locker plants nationwide. Based on interest, the Union County War Board decided to sponsor one locally. Harold Woodruff from Liberty expressed his desire to own and operate this facility as a commercial enterprise.[88] Agent James L. Guthrie reported that the county war board "helped Mr. Woodruff draw up the necessary papers, assisted with the advertising and the locker rentals. The plant will have a capacity of 560 lockers and at present 370 advance rentals have been made. The project is before the WPB [War Production Board] in Washington [D.C.] awaiting final action at the present time."[89] The plant would receive approval and open in 1944.[90]

In 1942, Warren County had two frozen food locker facilities, while in nearby Fountain County, a facility in Attica had 250 individual lockers.[91] By 1946, there were four refrigerated locker plants in Wells County, and Spencer County had 800 individual locker boxes, with one location at the Hammond Frozen Food Locker Plant.[92]

The Victory Garden Program was highly successful in Indiana, with thousands of gardens planted across the state by many people who had never even thought of growing their own food before.[93] According to agent Leo Geyer in 1943, "There were more gardens in Jefferson County than at any time since '17–'18 [during World War I]."[94]

That same year, the Perry County agent noted that 95 percent of rural residents and 50 percent of townspeople had participated.[95] Estimates from LaPorte County indicated that residents there had planted 8,000 gardens in 1943, and Wayne County reported 7,500 gardens planted in towns and rural areas in 1944.[96]

In the 1943 annual report for Orange County, the agent stated that people "literally 'grew up' out of a home garden as a principal source of family food supply."[97] As Marion County agent Horace Abbott observed in his 1942 annual report, raising food for home consumption was one way that a family could contribute toward their personal well-being and gave everyone the feeling that they were helping the war effort.[98] He also explained, "The additional food produced in Marion County through these enterprises has exceeded the goal assigned to the County and has made extensive contributions to the nation's food supply as well as to decrease the normal demands by Marion County residents on [the] commercial food supply."[99]

Horticulture Extension specialist Willard Ward pointed out that "in 1943 victory gardeners had produced at home the equivalent of some 260,000 car loads of produce in the United States."[100] The quantities canned at home further proved that the Victory Garden Program had been successful.[101] Using information provided by those reporting on their home canning operations allowed counties and the state as a whole to quantify the amount of food preserved and number of bushels stored.[102] For example, in 1943 the LaGrange County agent noted that 310 families reported canning "27,568 quarts of fruit, or 89 quarts per family; 28,568 quarts of vegetables, or 92 quarts per family, 8,987 quarts of meat, or 29 quarts per family."[103] Those same families were able to freeze 573 pounds of fruit, 115 pounds of vegetables, and 4,050 pounds of meat. They also home-cured 20,594 pounds of meat, or 66 pounds per family, and stored 849 bushels of fruit and 3,655 bushels of vegetables, which amounted to 3 bushels and 13 bushels per family, respectively.[104] In Pulaski County, the value of the food canned and stored in 1942 was placed at $100,000.[105]

Preservation and storage of homegrown food reached a new high in 1943.[106] That year, agents across the state indicated that 6.5 million quarts of fruits and vegetables had been canned at home.[107] By 1945, Purdue Extension reported that the value of the fruits and vegetables grown and preserved from home gardens was worth $26 million.[108]

People in Indiana—primarily women—accomplished much during World War II by growing victory gardens and preserving what they grew. They were up to the task of adapting to wartime conditions, which required them to work around the shortage of food supplies and the rationing of commonly used items. As a generation, they became more reliant on what they grew in the garden, canning enough food to last through the winter and ensuring that their families were properly fed. As one agent observed, "It is difficult to measure the results secured, but on a county-wide basis the total increase in home production and consumption would be a staggering amount of food."[109]

But as early as 1944, interest in the Victory Garden Program was waning as the strain of the war took its toll.[110] That year in Marion County, Abbott noted, "The public was justified in questioning the need for a Victory Garden this year."[111] Abbott might have based his observation on the following:

- Gardening was not as easy as most people thought it would be.
- The anticipated food shortage was not as bad as many had anticipated.
- The gardening, canning, and storage of fruits and vegetables had been done so well in 1943 that the cupboards and pantries were well stocked for the following year.
- When many of the commonly used products were removed from federal rationing lists, people returned to the grocery stores to purchase what they needed.
- As more women entered the workforce, they found they had less time to care for a home garden.
- Many lacked the transportation to get to gardens that were located away from their homes.[112]

The federal government did its best to reinvent the program in 1946, even after the Allied forces had won the war. The hope was that transitioning from the Victory Garden Program to a famine emergency relief program would reinvigorate interest, with the aim being to prevent waste at home as one way of helping people who were still experiencing postwar food shortages.[113] But while a lot of excellent work was done after the war, the interest generated in home gardening was not repeated to the same degree as it had been during the war.

CHAPTER 32

The Armed Forces Have Taken 1,500 Men, Including Farmers. Why Wouldn't It Create Many Problems?

Even though the United States had not yet committed its military to the Allied efforts overseas in 1940, the government began preparing for the possibility. On September 16, 1940, Congress passed the Selective Training and Service Act, which required men to register for a national draft. Following the Japanese attack on Pearl Harbor on December 7, 1941, the number of U.S. soldiers increased rapidly through voluntary enlistment in addition to involuntary conscription, and the federal government quickly embarked on a multibillion-dollar spending spree to furnish the military with tanks, guns, ships, planes, munitions, and clothing.

County agricultural agents soon found themselves in the center of wartime decisions when the government asked them to determine the effect the draft would have on agricultural production. Basically, the agent became part of the local county system that advised draft boards about whether a farmer or agricultural worker could be sent overseas to fight or should instead be kept on the farm to help meet the critical agricultural quotas. In this role, agents were forced to make gut-wrenching decisions, even when pressured to keep farmers and their sons out of the military.

With the draft apparatus in place and registration underway, the federal Selective Service boards—more commonly known as draft boards—began calling up tens of thousands of men from all walks of life to serve their country on battlefields in Europe, North Africa, and East Asia. The number of draft boards varied by county. For example, St. Joseph County had six, Lake County had ten, and Marion County had fifteen.[1] Members of the draft boards, which consisted of three to five volunteers, were told not to release any information regarding their work, including about government-established quotas.[2] With a national goal of having 9 million men in uniform by the end of 1943, the nation's 6,400 draft boards would eventually need to draft 200,000 to 300,000 men per month—a rate that would continue through the end of the war.[3] From late 1940 to March 1947, when the draft ended, 10 million men would be inducted.[4]

At first, the legislation required men between the ages of 21 and 35 to register with their draft boards on October 16, 1940.[5] After the U.S. officially declared war, the draft was expanded to include men from 18 to 45, and then eventually up to age 64.[6]

The first lottery took place on October 29, 1940, with the first men entering military service the following month.[7] At first, the military allowed the men to choose which branch of the army to enter—the air corps, quartermaster corps, or signal corps—with many choosing the air corps.[8] But by January 1942, one Indiana newspaper reported, "The local draft board yesterday received notice that men in the Army no longer have the privilege of choosing their branch of service, but must serve wherever they are ordered."[9]

Many draft boards had drafted all eligible men under the age of 28 by the start of 1942, so they began drafting men ages 28 to 36 to fill military quotas for February and March.[10] One problem encountered by Indiana draft boards was the poor health of those being drafted. Out of a hundred men drafted, only forty-seven passed the physical examination administered by local draft board doctors or military physicians. Some conditions that prevented men from passing included insufficient and defective teeth, skeletal defects, vision impairments, heart troubles, mental disorders, hernias, venereal diseases, and height and weight problems.[11] But it didn't take long for the government to relax its standards, meaning those who had received deferments, such as for poor vision or defective teeth, would be reevaluated and possibly drafted.

Members of county draft boards were told how many men they needed to draft, with names selected by lottery from those who had previously registered.[12] With Pearl Harbor freshly seared upon the American psyche, draft boards chose any and all capable men to serve, regardless of their occupations at the time of the military call-up. According to Posey County agent Oral Riggs, "When the war first started and the selective service system was set up, the local selective service board took the attitude brought about by pressure from the citizens from the county, that all boys were alike and that they should all go."[13] But the draft boards soon came under fire as farmers questioned why so many young men from rural Indiana were being drafted without regard to the negative impact it would have on agricultural production.[14]

With the first year of American military involvement in the war drawing to a close, St. Joseph County agent Eugene Bird wrote about the drafting of farm workers, noting that "contrary to national assurances, the food production program is being ruthlessly disregarded by the local board in St. Joseph County as of November 1942."[15] In 1943, agent Mervin Smith also addressed the shortage of agricultural workers, observing that the "greater portion of the approximately 1,200 Wells County young men in the armed services, have been drawn from the rural portions of the county. This, coupled with the large number who have gone into industrial plants at Fort Wayne and other cities, has drawn heavily on our farm labor supply."[16] In the 1943 annual report for Posey County, Riggs offered his take on this issue, saying, "[T]he Selective Service took the attitude that agriculture had huge surpluses piled up and that they were not worrying about farm production. Of course, we all know that the picture changed."[17]

As additional men were pulled from the agricultural community, farmers began complaining to agents that the shortage of workers would hamper their ability to produce the commodities that the government needed. In 1942, Marshall County agent Lyman Butler wrote, "[The] Selective Service System has taken several hundred additional young men from the farms for the Armed Forces. Many of these men could, no doubt, serve America better as producers of food than as fighters," which was a belief shared by agents across Indiana.[18] Butler indicated that more than 1,500 men from his county were now in the military, leading him to ask, "With this drain on the population, why wouldn't it create many problems?"[19]

As more men were drafted or left the farm to work in industry, the government was forced to face the possibility that the lack of available agricultural workers could seriously jeopardize the ability to produce the food and fiber critical for the war effort. This put pressure on those making decisions at the national level to solve the labor problem on farms while still supplying the needs of the military and industry.

In 1942, U.S. Senator Millard E. Tydings from Maryland tried solving the labor problem for agriculture by adding an amendment to the Selective Training and Service Act that provided a deferment to any agricultural worker whom the draft board considered essential to farming.[20] Tydings explained his position, noting this was "not to upset your basic military requirements but which will at the same time safeguard for the farms sufficient labor as will keep the food coming to our soldiers and sailors and our civilian population and our allies."[21] It was a recognition from the federal government that tilling the ground and milking cows were as important to the nation's interests as carrying a rifle and driving a tank.

However, those who were able to obtain agricultural deferments were at first considered to be unpatriotic and cowards. In Kosciusko County, agent Joseph Clark observed, "In the last war, one who used more sugar than he was supposed to was called yellow [coward]. This time only farm boys who are producing food and ask for deferments seem to be yellow, a lack of actual understandings [by those who complained]."[22] In Posey County, Riggs reported a similar situation in 1943 among young men seeking agricultural deferments, writing, "Such deferments received brought much ill will upon the one receiving them, and one boy, after receiving deferment for essential farm work, finally joined the army. This boy was badly needed on the farm. He joined because of the pressure and the talk in his neighborhood. As a result of all this, the farm boys were rapidly depleted and help was becoming very scarce."[23]

In 1942, agent Leo Geyer explained that the chairman of the Jefferson County draft board had said "that from now on every draft would pull more and more full time farm boys into the army and he wanted a yard stick which he might use in measuring the boys for deferment."[24] During the spring of 1943, orders came down from L. Marshall Vogler, director of the Indiana State War Board, and Colonel Robinson Hitchcock, director of the Indiana State Selective Service, that county affiliates for their agencies were to work together in providing deferments to workers who were considered crucial to a farm's operation.[25] This placed the draft boards in the untenable position of needing to supply men to the military while not drafting those who were productive farmers or farmhands.

The process was that the draft board would submit to the U.S. Department of Agriculture war board in the county the names of those being considered for the draft. Then the war board—and, in particular, the county Extension agent as the secretary—would provide a recommendation as to each individual's importance to agricultural production. In theory, the draft board would take the recommendations of the war board before assigning each person under review their all-important classification: Class 1, available for military service; Class 2, deferred because of agricultural occupation; Class 3, deferred because of dependents; and Class 4, unacceptable for military service due to criteria such as health, vision, and dental issues.

In his 1943 annual report, Jay County agent Chester Langston described his work relating to farm deferments, writing:

> It has been the duty of the local extension office to see to it that the Selective Service Board at all times had sufficient data on hand for all individuals of draft age in order that they could make the proper decision relative to an inductee to determine his draft status. This fine working relationship has allowed those boys who are actually performing a service here at home in producing the necessary foods, to remain. Those living in rural areas who did not have a sufficient volume of business on their farms to justify deferment were placed in 1-A [available for military service] and sent to the army.[26]

Agents such as Langston used government guidelines called war production units to help separate those who needed to remain on the farm from those who could be drafted.[27] On January 5, 1943, an *Indianapolis News* article explained how war production units would be calculated:

> Beginning soon after January 12, every farmer in Indiana will receive "maximum war production" work sheets to serve as guides for selective service boards in determining deferment of Hoosier farm workers. . . . L. M[arshall] Vogler, chairman of the Indiana United States department of agriculture war board, announced the plan to distribute forms worked out on a nation-wide basis with the approval of Claude R. Wickard, [U.S.] Secretary of Agriculture. Purpose of the forms will be to determine the number of farm laborers required

> for maximum production in various types of farm work. The findings will be translated into war production units, with one additional employee being permitted for each sixteen units. . . . Work sheets will contain an inventory of the farmer's crops, livestock, machinery and labor . . . [and] production of sixteen war units of essential farm products may properly be considered a person necessary to the war effort. This will not be an iron-clad rule. Under the plan, if a farmer had thirty-two production units, he could qualify for one laborer [in addition to himself].[28]

The number of war production units varied among counties and changed over the course of the war, but the evaluation process remained basically unchanged.[29] Each county war board was given tables showing how a unit was calculated. This partial listing drawn from various newspaper reports shows the calculation of a single war production unit based on the crop acreage, type of livestock, and whether the products being produced were classified as essential or nonessential:

Essential Farm Products	**Number of Animals or Acres That Equal One War Production Unit**
A. Beef Cattle	
Farm herds	12 animals
Feedlot	20 animals
B. Chickens	
Broilers	600 animals
Egg production	75 (per 100 birds)
C. Hogs	20 animals
D. Milk and Related Products	1 cow
E. Barley, Wheat, Sorghum, Hay, etc.	15 acres
F. Corn	5 acres
G. Fruit Trees	1 acre
H. Truck and Canning Crops	1 acre
broccoli, Brussels sprouts, cabbage, carrots, onions, snap beans, tomatoes, spinach, etc.	

Nonessential Farm Products (no war unit credits)

A. Specialty crops, including cantaloupes, hops, popcorn, and watermelons
B. Vegetables, including artichokes, cucumbers, eggplant, iceberg lettuce, pumpkins, radishes, squash, etc.
C. Other crops not listed[30]

It was clear from the tables that the government was attempting to shift production from nonessential commodities to those listed as essential and put pressure on farmers to increase their production of the essential ones. Those with smaller crop acreages or smaller livestock herds had to expand their operations in order to qualify for the deferment. For example, 5 acres of corn would have to be enlarged to 80 acres or a herd of 12 beef cattle would need to be expanded to 192 head in order to qualify for 16 war production units. Having one unit assigned per dairy cow showed that the government placed a high value on dairy products.

Most farmers used a combination of different crops and livestock numbers to reach the 16-unit threshold, but sometimes that posed challenges. In his 1943 report for Orange County, agent Ross Atkinson cited this example:

> A young married man with [a] small family, with [a] life time [of] experience on farm. He had bought and paid for a farm a few years back. Apparently the operations were normal, successful and all was as it should be. Selective Service called for a report, which indicated unsufficient units to justify deferment. The young man could think of nothing to be done about it. With further consideration, he agreed as well as the neighbors and township leaders that he had plenty of farm, plenty of barn, and plenty of feed, for the support of three or four milk cows, which would more than push him across the line. Yes, he would like to have them, and could milk them, but he did not have the money with which to pay for them. There he was, a successful young farmer with several hundred dollars worth of assets and not a dime of indebtedness, and had no idea how money could or should be borrowed and used.[31]

It's unclear whether Atkinson's involvement had been able to keep the farmer behind the plow instead of marching off to war.

Farmers with operations that were too small to qualify for a deferment based on their existing war production units had some flexibility to hire out to farmers with larger operations, which in turn helped the smaller-scale farmers reach deferment status.

All of the work required to verify each farm worker's status took a lot of the Extension agents' time.[32] The number of applicants the county war boards had to review and make recommendations on regarding deferments was, at times, overwhelming.

For example, in 1943 Pike County had 312 men whose draft status had to be reevaluated.[33] In 1944, Ripley County had 700 registrants for the draft board to review.[34] That same year, Parke County reclassified 400 farmers who had previously been assigned as 2-C (deferred because of agricultural occupation) and 3-C (deferred because of both dependents and agricultural occupation) to 1-A. From this revised list, 45 of the reassigned men were ordered to go take a physical examination that, if they passed, meant they would be called up for active duty.[35] In Jasper County, the Extension office was asked to supply farm worksheets on all

2-C, 3-C, and 4-F (rejected for military service) registrants working in agriculture in order to determine if their status was still valid.[36]

To request a deferment, a registrant first had to fill out a farm worksheet, which provided an account of the farming operation.[37] Typically, the agent helped individuals complete this form, which would eventually be used by the county war board to determine whether that person would be considered as a worker who was essential to agricultural production in the county.[38]

The county war board was expected to verify that the applicant's information was, in fact, truthful and accurate. In many counties, the board members left this task to the agent, which led Eugene Bird in St. Joseph County to write in 1944, "The agricultural agent inherited this task because he is secretary of the War Board, and the other members insisted that he handle Selective Service. . . . It has been the most unpleasant and nerve-racking job that we have ever had in the Extension Service."[39] Sullivan County agent Jean C. Sliger made a similar observation in 1943, saying, "The worst feature of War Board work is that of being an advisor to the Selective Service Board. No one likes to judge whether or not a man shall go into the army or shall stay at home, but of course it must be done, and in this particular county the board has good men who try to look at every angle before making a decision."[40]

Given the gravity of the decisions, counties soon developed a system that relied on anonymous reviewers to help with the process, as Posey County agent Oral Riggs described in 1944:

> Since the responsibility was falling heavily upon this office on this proposition, it was realized that some good system would have to be worked out to prevent the County Agent from dictating who went to the Army and who did not. This system of the selective [service] furnishing the information on special forms and looked over by confidential committeemen from all parts of the county was the program worked out which enabled us to process more than 1,500 within a period of about thirty days.[41]

The names of those helping the agents were not disclosed to the public. Their reviews, comments, and suggestions were submitted to the war board without their names attached as part of the documentation accompanying the registrant's petition.[42] By keeping the reviewers anonymous, it was thought that each applicant could be fairly evaluated.[43] It also meant that family members could not use their positions in the community or personal friendships to place undue pressure on those making the decisions.

In Posey County, the anonymous review process seemed to work as intended, with Riggs admitting to being surprised that the local reviewers "never reported a single case of clear-cut padding the report; in fact, we are all amazed at the accuracy."[44] But in St. Joseph County, Bird was perturbed when the draft board leaked information about the process to the draftees in 1945. In his report that year, he explained:

> [Members of the draft board] informed many of the registrants that they were being sent to the Army because of action of the St. Joseph War Board. We believe that this was a distinct breach of confidence and there was a question in our minds as to whether or not we ought to respond to any future requests from this Board until we could have some assurance that they would handle this matter confidentially. In one instance at least, the list was shown to one of the registrants, and in several others, they were told that they were placed in an unfavorable position by the War Board.[45]

Using the war production units as the basis for the decision, the war board members classified the person under review as essential or nonessential for the farm operation, then made their recommendations to the draft board.[46] If the draft board agreed with these recommendations, the deferments were assigned for six months, at which time they would be reviewed again.[47] It became evident to all involved that as more men were needed by the military, an initial deferment would not necessarily be renewed at the subsequent review.

The draft boards generally agreed with the recommendations of the local war board. However, agents also reported times when their recommendations were totally ignored by the draft boards, and men identified as critically important to agriculture were drafted in spite of the recommendations of the war board.[48] There was nothing the agents could do when draft boards chose to go in a different direction. Bird shared his thoughts on the situation in the 1944 annual report for St. Joseph County, observing:

> No project has taken as much time and energy and involved as much grief as the discussion of the personal problems of farm boys and their parents in connection with Selective Service. Most of them seemed to come to us for information and advice, and there was no alternative but to hear them patiently and courteously. Beyond giving them information on how to proceed, there has been little that could be done in most cases.
>
> The problem became so involved because the rural board for this county has seldom had more than one farmer on it and often none. The four or five city business men and laborers, in spite of their apparent effort to deal wisely with the various cases, were simply incompetent to pass judgment on the merits of relative cases involving farm operations. Up to date no headway has been made in correcting this situation. . . . Because so many were treated discourteously by employees of Board 2, they came here seeking facts and a courteous hearing, which they always got.[49]

While drafting decisions should have been based solely on facts, it appeared that in some cases personalities and politics played a more important role, which made it seemed rigged in favor of some prominent farm families. In 1944, agent Willis Stall recounted such a situation in his annual report for Jackson County, writing:

> The Agricultural War Board of this county has definite knowledge and information which shows that there are certain farm families in this county having 2, 3 and 4 male members of service age, with none of them in the service. Within the same county, and in some cases within the same neighborhood, there are similarly situated farm families with two or more (in several cases 3 and 4) in the service. In certain other families, there are members whose agricultural necessity cannot be justified who have never been placed in any but deferred classifications.[50]

The agents had nothing to do with such discrepancies, but they had to deal with the fallout as parents came to them to express concern about the unfairness of a draft that favored one family over another. Switzerland County agent Oran McNary pointed out, "We were bothered by parents trying to have us get boys out of the army in several cases which we felt were unjustified."[51]

Some people thought the local agent could use his contacts to help get a family member discharged from the military. In the early years of the war, federal rules precluded the agent from helping to get someone released from military service. But by late 1943, agents were allowed to advise those seeking to have someone discharged from the service in order to work on a farm about the procedures they would need to follow.[52] In 1943, John Armstrong explained how he had become the focus of public attention in Lawrence County, writing, "Due to publicity, Army and Navy announcements, or something, some 300 mothers, fathers, wives or sweethearts undertook to enlist the County Agent's assistance in getting him [individuals] released from the Army or Navy to come home and farm. It was the toughest and hottest problem to handle we had in 1943."[53]

At the time, Pulaski County agent Carl Olson wrote that he now had another duty, which was to help parents fill out the paperwork needed to obtain an early release for their sons. He explained:

> Information was broadcast over the radio that essential farm workers who had already been drafted and were now in the Army might obtain a release if they were needed on the farm during 1943. The report stated that the County Agent could give assistance in obtaining these releases. The County Agent and War Board received numerous requests to give help in obtaining these releases. The assistance given was that of presenting facts concerning the farming operations of the parents asking for their sons' release.[54]

The agents had let everyone know that the discharge letters would be based only on the need for farm help and that requests not associated with farm work were not eligible for consideration for early releases. Arthur Wissman included this account in his 1945 annual report for Jennings County: "During August, eight letters were written for farm families stating farm programs and requesting discharge of bona-fide farm boys that will actually come back to the farm. Some of these are badly needed to complete this year's farm operations. Later in August, notice was sent that discharge request letters should be sent only in case of extreme emergency, so no further letters were sent."[55] Since only extreme cases were considered by authorities by 1945, that sealed shut the potential for early releases for most already serving in the military.

Jefferson County agent Leo Geyer noted in 1944 that when the draft boards "were instructed to take those under 26 and to weigh agricultural deferments carefully, the local board took [it] to mean to take all farm boys regardless."[56] Nationally, 625,000 men under the age of twenty-six had received a farm deferment as of May 1944, so giving draft boards access to these men provided a major source of potential draftees.[57]

With so few eligible men left to draft, the draft boards had little choice in the matter unless they drafted older men. In January 1944, the director of the U.S. Selective Service System, General Lewis B. Hershey, stated, "We must have food, but all farmers do not produce food or other products in quantities substantially more than they consume. Those must go if physically fit."[58] Each registrant with a deferment status would now have to prove that he was irreplaceable for agricultural production.[59]

By April 1944, nine out of ten men drafted came from farms, with one Indiana newspaper reporting that the draft boards were calling "for able-bodied men from now on, regardless of the effect on industry and agriculture."[60] But agents continued to do whatever they could to preserve labor for agricultural production, with Geyer filing this report for Jefferson County that same year:

> A compromise was then worked out that was agreeable with the local selective service that took off the pressure of heavy drafts made of mostly farmers. This compromise was that each farmer under 26 would be asked to fill out a War Board Questionnaire and then we of the Agri[cultural] War Board would place them as to their relative importance and that the Selective Service would start with those of least importance and fill their quota by coming up on the list and, in that manner, always leaving the more important at home. This made a tremendous lot of work for the Extension Service to work out the man hours of production on each one. The rating of importance took about five meetings of the Agri[cultural] War Board but once completed did not cause any trouble except where investigation by the draft board, they found that the applicant made a supposedly false statement. . . .
>
> This rating has to be continuously rearranged each month and it did seem rather a heavy burden on the Extension Agent's office, but it probably was worth the effort considering how splendidly our farmers got along with their work. And one will have to admit [agricultural] production was important and that labor was necessary to get it.[61]

In most counties, the review process was initiated when the Extension office received the names of the men who had previously been assigned 2-C or 3-C classifications. The agent would then send a letter asking each of the men to come in and complete a questionnaire dealing with their farming operation, leading one agent observe, "The task of securing the information on II-C and III-C registrants was one of the most difficult jobs given the Extension Office."[62] After the questionnaire was completed, the Extension office prepared another form, which the registrant had to get notarized and take to the draft board.

Members of the war board then met to discuss the information submitted by each registrant and used the following criteria to determine where to rank each on the draft list:

> 1. The registrant and other able-bodied men on the farm should have at least 20 man power units or more. Plus consideration of the following:
>   a. The productivity of the soil. (A good 100-acre farm should produce more food properly farmed than a much larger farm of low productivity poorly managed.)
>   b. The registrant's ability as a farmer and producer of livestock.
>   c. The equipment on the farm. (Is the equipment on the farm being used to full capacity or is the registrant producing only enough to warrant a deferment?)
>   d. The age, number and physical condition of other farm labor on this farm.
>   e. Would the production of this farm be materially reduced if this registrant be inducted in Service?[63]

Each registrant was placed in a descriptive category, which Jay County agent Harold K. Rippey explained in his 1945 annual report, writing:

> Group 1. To include those individuals that were absolutely necessary to continue production on the farm.
> Group 2. To include those individuals that were smaller operators or who were needed on the farm for dependency of their parents or others.
> Group 3. To include those individuals that could be spared from the farm.[64]

Using the above criteria, 213 Jay County men who were previously given a 2-C or 3-C classification were reevaluated in 1945. Of those, 74 remained in Group 1, 51 were assigned to Group 2, and 45 lost their farm deferment and were expected to be drafted. The agent indicated that the remainder couldn't be classified until he received additional information.[65]

As the draft continued, this evaluation process would be used by many counties in one form or another to continually review the 2-C and 3-C registrants.[66] And as the war continued, an increasing number of them would be inducted into the military.

Registrants who were reassigned from a deferred category to a 1-A classification could take their cases to a district appeal board. County war boards were also given the authority to appeal to the district board on behalf of those identified as important to farming. This was especially true in cases where a war board had recommended that a person be deferred. The district boards usually approved the recommendations that favored the original decision by the county war board, leading one agent to observe, "After several cases were reviewed by the District Appeal Board, the local War Board's recommendations carried more weight with the local Selective Service Board."[67]

With each reevaluation of those who had previously been deferred, more men working on the farm found themselves being inducted.[68] By March 1945, Indiana was facing a July 1 deadline to send an additional 18,000 men to the armed forces. At the time, the *Lafayette Journal and Courier* reported that Colonel Robinson Hitchcock, the state Selective Service director, said, "For every young farmer we don't take, we will have to take a man up to 38 years old from the ranks of essential war workers or a man with a family."[69] The Lawrence County draft boards in Mitchell and Bedford would later report in 1945, "It has been our policy to send all farmers to [the] army if not essential on farms of [the] county. Furthermore, we send boys if and when they quit being essential. This will continue to be our policy."[70]

As the casualties mounted, the draft boards had no choice but to draft even workers who were considered essential in order to replenish the depleted military ranks. In his 1944 annual report, Jackson County agent Willis Stall noted, "Manpower became further depleted as additional men were sent to or joined the armed forces. Draft officials reclassified and inducted a large number of farm men and boys into the services during the months of April and May, well after the time 1944 farm plans were made."[71]

Ross Atkinson explained how the process went off track in Orange County in 1944 when the draft boards were asked to take all men from 18 to 26, paying special attention to those who were farming:

> We were informed that a change had been made and [war production] units were not to be considered, therefore there would be no need for any further reporting from the War Board. For several days things drifted along without comment, and a number of low-producing farm boys were taken into service, quite properly. However, some of the most essential farm boys began to get their 1-A classifications. The War Board held an extra long session. The various members of the board reported on what they had heard of how things were doing in other counties. These reports indicated that Orange County had less than half as many men deferred for agricultural production than most all other neighboring counties. There[fore] we reasoned that we should not openly seek an opportunity to keep men from military service, but, on the other hand, if an employer and/or employee on an essential operation

> were interested to the extent of . . . appealing the case and would make sworn statements of the farm operations, and if these operations averaged as much as twenty units per man on the farm, that they (the War Board) would make investigation, and if it verified the original claim, that they would go definitely on record with selective service. . . .[72]

During 1945, Whitley County agent Blanchard Widney reported that the local draft board "was unsympathetic with the rural labor problem to the extent that in February they had left only 70 of the boys, [ages] 18 through 25, on the farms. By June, it was estimated, more than half of these had been inducted."[73] In Starke County, Arthur Shideler said casualties from the Battle of the Bulge had put more pressure on the draft boards to find men to replace those who had been badly injured or killed. In his 1945 annual report, he wrote, "In December [1944], following the military reverses on the Belgium Bulge in Europe, the Selective Service Board was required to re-examine all Class II registrants for essentiality to Agriculture."[74]

From the very beginning of World War II, draft boards across the United States were confronted with the need to determine which men should carry guns and which should plow the ground. As the war progressed, more men were taken from agricultural work to fight overseas, leaving a shortage of labor on farms.

During World War II, Indiana did its part by sending 338,000 men and 118,000 women to serve the United States, with 13,370 men making the ultimate sacrifice as they defended their country.[75] The responsibilities of working with the draft boards put extreme pressure on the county Extension agents, both from the agricultural community they served and from members of the draft boards. Being placed in the middle of weighty decisions about who would or wouldn't be sent to foreign battlefields was an intolerable position. The agents were well aware that their decisions had life-or-death consequences, but they were charged to fulfill this role by the federal government, so they did their best to meet their obligations.

CHAPTER 33

All Agricultural Workers Seeking Employment in Industrial Factories Must Have a Statement of Transfer From the County Agent

Prior to World War II, hired farmhands provided the manual labor that still predominated field and livestock work at the time. But just as farmers were being asked to ramp up their production, the once plentiful laborers were being sent overseas by the thousands as the Selective Service draft turned plow boys into soldiers and sailors.[1]

At the same time, industries being converted to produce military hardware and supplies needed labor as well, so men and women were drawn from the farm sector to take jobs paying much higher wages than farmers could match.[2] As members of the county war boards, Extension agents would find themselves dealing with this shortage of agricultural labor from 1942 to 1946.[3]

After the Pearl Harbor attack, the U.S. government took control of the industrial manufacturing facilities under the auspices of the War Production Board. Manufacturers were told to retool their plants to make the hardware and other supplies that the army and navy needed as the military ramped up in personnel and equipment. Household names such as Bendix, Firestone, B. F. Goodrich, Goodyear, Oliver, Singer, Studebaker, and U.S. Rubber Company turned over all or part of their facilities to produce what the government needed.[4]

Manufacturing plants running at full capacity put tremendous pressure on industry to fill open positions, so it was natural for workers from rural communities to be drawn to the wages that businesses offered as inducements.[5] In 1943, Steuben County agent Thomas Grabill recalled, "Skilled farm labor rapidly moved into the defense plants at wages of such high rates that farmers cannot possibly expect to compete in retaining that labor. The sad part about the situation is that skilled farm labor cannot be replaced on the farm, and in many cases the defense job in which the farm laborer is finding employment can be done by persons without any special skills."[6]

Lawrence County had long been known for its stone quarries, which commonly provided part-time employment to local farmers.[7] At the start of the war, many of these quarries were converted over to factories making aluminum, airplane engines, tanks, and landing craft.[8] By March 1943, six of the quarries needed 1,600 men and 800 women to meet their production needs. With such a huge draw on the local workforce, farmers found labor for their operations to be nonexistent.[9]

Skilled laborers such as blacksmiths, carpenters, mechanics, and welders—once plentiful in rural communities—were also taking their tools and specialized skills to the higher-paying jobs in industry.[10] This concerned Posey County agent Oral Riggs, who wrote, "There was one thing that was very clear, and that was that men who were qualified to work in farm repair shops were finding better jobs in working in the war plants. As a result, approximately two-thirds of the men who were working in these repair shops last year were now working in the war plants, leaving the shops extremely short on help."[11] Workers with special skills could not be replaced with inexperienced workers, which put a secondary hardship on farmers who needed these services to keep their farm equipment running.

The more manufacturing plants retooled to produce goods for the military, the more desperate the labor situation became for agriculture.[12] Government facilities—such as the training station for bombers at Columbus in Bartholomew County; the ammunition plant at Charlestown and military warehouse depot at Jeffersonville, both in Clark County; and the ammunition testing facility at Madison in Jefferson County—had taken many workers from the state's farms.[13] What were the farmers supposed to do in counties like Rush County, where half of the 2,100 farms required part- or full-time labor?[14]

Some farms needed many workers to run them.[15] For example, dairy farms were labor-intensive operations, and without the necessary workers, the only option was to reduce the herd size. The government did not want the lack of access to labor to lead to decreased farm production at a time when more per acre and per animal was needed. But as St. Joseph County agent Eugene Bird noted in 1942, "The heavy demand for factory workers and the inability for farmers to meet (with present food prices) the competition of wages set by the government in munition and in airplane factories is a first cause in limiting food production."[16]

With the higher wages being paid by the government-supported manufacturing industries, it's no wonder that so many men and women left the farm and other jobs to take these positions. Agent Horace Abbott reported in 1942, "It is not uncommon for a laborer to secure wages in one day for his services in a defense plant equal to one week's wages formerly earned on a Marion County farm."[17] Men and women from Lawrence County took government jobs paying $8–$24 per day in surrounding areas, including in Burns City, Charlestown, and Madison, and at Camp Atterbury. With workers getting wages they never expected to earn in their lifetimes, more than 10,000 people from in and around Lawrence County eagerly worked for industry.[18]

In Pulaski County, women were being hired in such large numbers that a Pulaski County home economics club disbanded when members no longer attended meetings.[19] And in some cases, factories hired workers as young as fourteen years old to fill their open positions.[20]

While farmers needed part- and full-time workers to provide the manual labor, they were unable to compete with the wage scales paid by the war industries. That's not to say that farmers didn't try to be competitive, but the odds were stacked against them. In Madison County, farm wages went up 40 percent in 1943, but even that was much lower than what could be earned in a factory.[21]

For many leaving the farm, money was an important incentive, but the factories also offered set hours, shorter work days, and less manual work.[22] With better pay and working conditions, many no longer found farm work appealing.

At the start of the war, a full-time, married farm worker earned $75–$100 per month.[23] It was customary for a farmer to also provide a married man with a house for his family to live in, a place to plant a home garden, milk, poultry, and other amenities, which amounted to an additional $10 per week.[24] On the other hand, single men working full time received $45–$60 per month, with a few securing room and board as part of their compensation.[25]

Those workers not living on the farm were paid $2 a day for routine work, which increased to $3 per day when the work involved harvesting the crop.[26] At these relatively lower wages and with the long and physically demanding work days, workers didn't think twice about going to the manufacturing plants to earn a living.[27]

Seasonal helpers with experience were paid $4–$7 a day for operating a tractor, while corn shuckers received 10–12 cents per bushel.[28] Picking tomatoes earned workers 8–12 cents per hamper.[29] But seasonal workers knew that as soon as the fall crop was harvested, they would no longer be employed, making the year-round factory job more appealing.[30]

In the summer, school-aged students represented an important source of labor for agriculture.[31] Boys and girls between the ages of 14 and 16 received 40–60 cents an hour for work such as harvesting vegetables, picking fruit, and detasseling corn.[32]

During World War II, the government required counties to follow minimum wage scales to pay those working the fields. Horace Abbott reported the following pay structure for Marion County in 1944:

Task Involved	Pay Rate
Apples from ground	10 cents per bushel, $1.50–$4.00 per day
Apples from trees	15–20 cents per bushel, $1.50–$4.00 per day
Beans	2 cents per pound, 80 cents–$4.92 per day
Hay	50 cents per hour, $4.00 per day
Spinach	40 cents per hour, $3.20 per day
Tomatoes	10–12 cents per hamper, 80 cents–$6.00 per day
Turnips	12–15 cents per bushel, $1.10–$7.20 per day[33]

Not only were the hired hands leaving the farm behind, but growers were also heading to the factories in record numbers because they could earn more money there than they did by farming.[34] More than half of the Point Township farmers in Posey County were working in Evansville at the war plants, and some of the best farmers and farm leaders in Scott County were also heading off to the defense plants to cash in on higher paying jobs.[35]

In the 1942 annual report for Marshall County, agent Lyman Butler told a story of Marshall County family that provided insight into the times, stating, "The writer personally knows of some farms where both the farmer and his wife are working at Kingsbury and on different shifts. On one farm of a little over 300 acres the wife works in a South Bend factory and, aside from harvesting hay, soybeans, and corn, this young farmer, a renter, by the way, is doing all the work."[36]

Agents understood that the labor situation would remain critical as long as industry continued to pay workers well.[37] As a result, they had to use every conceivable resource to help farmers find the labor needed for the farms (see Chapter 34).

In the spring of 1943, the federal government took the extreme action of telling farmers and farm laborers they could not leave the farm to work in the manufacturing plants without first getting permission from the county war board. It was the only way to control the flow of workers from agriculture to industry. As rumors had spread ahead of the government's change in policy, many quickly made the transition to the manufacturing sector to avoid being kept on the farm.[38]

According to Madison County agent Walter Haynes, the new government policy, issued through the War Manpower Commission, was "that all agricultural workers seeking employment in industrial factories must have a statement of transfer from the county agent," which put the agent in a rather precarious position.[39] That decision included those classified under Selective Service registration as 2-C (deferred because of agricultural occupation) and 3-C (deferred because of dependents and agricultural occupation).[40]

It was common for farmers, full- and part-time farm workers, and tenants to ask the agent for a transfer to industry during the winter months, when work on the farm was slower.[41] Farmhands were required to get a release from the farmer who employed them that indicated they could be spared for the winter.[42] Workers who were released received a referral card from the agent that was to be shown to the appropriate officials hiring them for an industry position.

In 1944, Madison County agent Walter Haynes wrote about the transfer process, saying, "This line of work had practically no rules to go by, no one seemed to be in authority, and most releases were given in accordance with each individual case."[43] That same year, Wells County agent Mervin Smith explained the process he followed for granting transfers.

> During the winter months a considerable number of these young men apply for temporary transfers permitting them to work in other industry until they are needed for spring work on the farm. It has been our policy to have these registrants first apply to the War board, and if the application is approved there, it is sent to the Selective Service Board for the granting of a temporary release. If this is granted, the application is then sent to the County Agent's Office, where the transfer is issued. We have endeavored to limit these transfers to work in related agricultural industry, such as soybean processing, elevators, feed mills, creameries, etc., unless the applicant possesses special qualifications that would enable him to render greater service in the War effort in some other occupation.
>
> In the case of these temporary transfers granted to II-C and III-C registrants, the limitation placed on their employment in the other industry has been either to some specified date, or stated "Until needed in agriculture."[44]

In Knox County, 92 men were released from the farm during the winter of 1945, and 71 were refused transfer requests because they were needed on a farm.[45]

While work releases were common for those who had not been drafted, Selective Service boards took a harder stance on allowing transfers for those with military deferments or at an age when they were subject to the draft.[46] In 1944, the Steuben County agent reported, "The policy of the local Selective Services Board was to grant no permission to farmers between the ages of eighteen and twenty-six for this part time industrial employment," while the Jefferson County agent wrote, "It was also the policy of the committee to not release any farmer who had made application for draft deferment."[47]

Most releases from farm work started November 1, following harvest, and lasted until March 1, which coincided with preparations for spring planting.[48] In a few cases, the period could be extended to April 15 since potatoes, onions, and other specialty crops were planted later in the spring.[49] But agents adjusted dates and release conditions depending on the circumstances. The Jefferson County Agricultural War Board set the policy that "no farmer was given any release after June 1st except for limited time during slack season after harvest of crops. This was thought to be necessary by the committee in order to insure maximum farm production."[50] And when an early frost was predicted in 1944, Switzerland County agent Oran McNary refused releases until the tobacco crop had been harvested.[51]

Those released from farm work were expected to return to farming at the allotted time, but enforcing their return was lax at best, with Pulaski County agent Carl Olson citing it as a principal problem in 1944.[52] Occasionally, individuals were allowed to remain at a factory job indefinitely if leaving it would cause financial hardship.[53]

The agents tried their best to balance the needs of agriculture and industry as they made decisions about transfers, but Steuben County agent Thomas Grabill described some of the challenges he faced in 1943 as he weighed the decision, writing:

> The disturbing element in the requests is that regardless of what decision is made on the part of the county extension agent, someone is disappointed. The hired hand who wishes to seek employment in a defense plant, if refused, is disgusted, and if on the other hand the request was granted and the hired hand left the farmer's employment, then the farm operator would be inclined to voice his criticism. One other condition which is difficult is the refusal of the farm operator who actually has sufficient work to be done on his farm and home, but who would like to reduce his debt load by securing some of the high wages being paid in industrial plants.[54]

The next year found Grabill still trying to balance competing interests, as he reported:

> There are cases where farmers' sons request the transfer and if refused on the basis of unfilled demands for agricultural help, it would not change the situation any. The individual refused a transfer under those conditions would remain at home disgruntled and disappointed. The only alternative in that case is to grant the transfer, as the individual requesting the transfer should be employed rather than remaining at home doing nothing even though some farmers might need the help badly.[55]

The government's need for labor to support the war industries in addition to the ongoing military draft would lead to significant issues for labor needs related to agricultural production. The following are examples of some of those issues, which agents would continue to deal with throughout World War II:

- There was a serious shortage of skilled labor for farms and ag-related businesses.[56] In short supply were those who could drive a tractor, repair equipment, and take proper care of livestock.[57]

- Farmers were unlikely to expand their operations much further to meet increasing government quotas in light of the shrinking labor pool. In Jackson County, Stall noted, "Our farm people became resigned to this condition early in the year [1943], and planned accordingly by undertaking only a little more than they were going to be able to accomplish with their own family help, and through trading help within the neighborhood."[58]
- Farmers refused to grow government-prioritized crops when labor was not guaranteed. Tomatoes were a high-priority commodity that the government desperately needed, but without labor, farmers could not or would not expand their acreage beyond what they could reasonably manage on their own (see Chapter 30). Edward Bogue indicated in the 1943 Pike County annual report that the "primary problem at the time was that farmers wanted a guaranteed supply of tomato pickers before they would sign contracts with the canning companies. According to all indications at that time, the quota of [a] 25% increase in the acreage of tomatoes would not be reached."[59]
- Operators of small farms were being drafted, which resulted in the farm ceasing all production.[60] In other cases, farmers had to reduce the size of their operations when their sons volunteered for or were drafted into the military.[61]
- Farmers sold a part of the farm when labor could not be found. This led Grabill in Steuben County to report in 1942, "Farm labor problems are extremely critical and are becoming more so as reflected in the large increase in the number of farm sales where farmers have decided that adjustments must be made if they are going to be able to operate without additional labor."[62]
- Farmers were leaving their ground unworked as they headed to the industrial defense plants. Jackson County agent Willis Stall reported in 1942, "Most of the residents of these sections are gainfully employed away from home—during the present year largely in defense work."[63]
- The livestock-based operations such as dairy, hogs, and poultry, which were heavily dependent on manual labor, found themselves in dire straits. These operations were forced to adapt to the lack of labor, reduce their size, or look for mechanical equipment to make up for the lack of labor.[64] Some farmers scaled back dairy herd sizes to match available labor, with one agent noting that this had been done at a time when the government was wanting more milk powder for the troops.[65]

- Dairy Herd Improvement Associations were largely dissolved due to the lack of available milk testers.[66] In Steuben County, the agent noted in 1942 that the dissolution of these associations was "due to a lack of testers rather than [a lack of] interest on the part of [dairy] farmers."[67] Dairy testers were typically young men of draft age, so while Purdue still trained them, their numbers were continually depleted as the military continued to draft them.[68]
- Sheep went undipped for external parasites when there were no commercial operators available to do the work.[69]
- Crops were left in the fields to rot or not planted at all in the spring. In Knox County, several thousand bushels of apples were left on the trees when apple pickers were unavailable, even though the orchard owners were willing to pay sixty cents an hour for the work.[70]
- Those wanting to volunteer for farm work but who were unfamiliar with what the job entailed required training. For instance, picking tomatoes was a manual task that anyone in good health could do. However, the picker needed to develop an eye for picking the right tomatoes based on quality standards.
- Processing plants could not unload the trucks with their customary speed due to the lack of workers. Agent Stall observed that on two occasions "more than half of the farm trucks in Jackson County were tied up with loads of sweet corn and pumpkins where they waited at one of our factories from 60 to 70 hours to unload."[71]
- The longer hours of farm work proved to be unhealthy for older men. The Miami County agent observed that older men were "overworked on account of labor shortage and have broken their health."[72]

County Extension agents were forced to make tough decisions related to the acute labor shortage for agriculture during the years of World War II. The federal government tried to stem the migration of workers from farms to industry in order to maintain production levels. Unfortunately, with industry and the military taking so many people, there were only so many workers available. This left the agricultural community holding the bag as men were sent off for military service and the remaining workers gravitated toward the higher pay offered by industry. All the agents could do was to try to secure another source of agricultural labor to pinch-hit during a time when labor was nearly impossible to find.

CHAPTER 34

Farm Women and Children Will Ride the Machinery to the Desired 10 Percent Increase in Production of Meat, Milk, and Eggs

As the shortage of agricultural labor during World War II threatened the production levels the United States so badly needed, the Farm Security Administration of the U.S. Department of Agriculture (USDA) along with the U.S. Department of Labor were charged with recruiting workers for American farms.[1] For instance, in 1943 Tippecanoe County got twenty-five full-time farm workers from southern Indiana who were recruited by the Farm Security Administration. Unfortunately, this was just the tip of the iceberg for the number that were needed there and in most Indiana counties.[2]

Jay County agent Chester Langston described problems he encountered with the assistance that federal agencies were providing farmers, writing:

> [T]he Farm Security Administration was securing year round hired men in the county. Through their efforts five men were secured. Their program consisted of recruiting in southern Indiana, in labor surplus areas, bringing them to Purdue University for a short training period and then sending them throughout the state where needed.
>
> Of these five sent to Jay County, none stayed the full year. Many were dissatisfied soon after their arrival. The dissatisfaction was not due to lack of wages, living conditions or housing, but was just the fact that they were away from their regular homes, [and] friends and neighbors were not the same. Wives became dissatisfied, and it was not long before they all left. They seemed to work well for a while and were capable, but upon earning enough money to return home, they did so.[3]

With federal agencies having limited success in recruiting workers and farmers clamoring for more help, on April 29, 1943, Congress "dumped into the laps of the Extension Department," as one agent bluntly stated it, the task of finding farm labor.[4] Extension would now have to find agricultural workers, transport them, and provide training to those without farm labor experience, with the federal government providing funding to help the states and agents accomplish these tasks.[5] Purdue was only one of a handful of land-grant colleges in the Midwest that kept the program instead of passing the federal funds to other agencies.[6]

With the changeover in program responsibilities from federal bureaucrats to the states, agents in Indiana were appointed as representatives of the U.S. Employment Service operating under the auspices of the Emergency Farm Labor program.[7] John B. Kohlmeyer of the USDA Bureau of Agricultural Economics was named the state supervisor of the program and given the responsibility for dispersing the federal funds that would pay for part-time labor assistants and secretarial support in the counties.[8]

In May 1943, Kohlmeyer met with ten agents at the Purdue Memorial Union in West Lafayette to work out a labor program for Indiana, which obviously would depend heavily on local contacts the agents had with farmers and businesses.[9] Harry J. Reed, the director of Extension in Indiana, instructed agents statewide to begin setting up permanent county labor committees to help guide them in their farm labor recruiting efforts.[10]

In his annual report that year, Jasper County agent Frank E. DeLaCroix explained the seriousness of this responsibility, stating, "A large portion of the County Agent's time was spent with the labor problems of the county as it was felt that this was a very critical project and would determine the amount of agricultural production within the county."[11] Agent John Armstrong expressed a similar sentiment in his report for Lawrence County, writing that the farm labor program "was the biggest war job. Obviously an intimate relationship exists between increased food production and farm labor."[12]

It's not surprising that the agents seemed overwhelmed with this newest project added to their already crowded schedule. It was probably a job that none of the agents really wanted to do, but they had little choice in the matter, given the demands from their bosses at the USDA and Purdue.[13]

Agent Willis Stall provided these details about the Jackson County labor program in 1943:

> During the month of August, which is the month of the greatest farm activity, we managed to live from one day to the next, trusting that no one would need help and that not too many people would want to work at farm jobs on any particular date. During this month we had on file not more than three or four requests for help but very little help available. We reached the conclusion during this month that our farm people had taken fairly good care of the situation in most of our communities by working out plans of helping each other that seemed to do the work. A few requests that came for larger groups for work in bean picking, tomato picking, orchard work and the like were handled by asking these people to apply through Bartholomew County, where Italian War prisoners were available, and where their services could be secured. . . .

> We will advise them [local farmers] to do as most of our people have apparently done this year, to be certain where help is coming from before the farm program is undertaken. In other words, to undertake no more farm work during the coming year than can be handled by the family and what other help is certain, will be our advice.[14]

Surveys were commonly used to assess the labor problems in each county.[15] Often, neighborhood leaders (see Chapter 39) conducted these surveys, which addressed labor needs, available supply, and farmers' crop production intentions.[16] Stall had described one such effort in 1943:

> The two commonest comments that were heard early in the year from farm people were: "Where are we going to get help?" and "What are we going to do about farm machinery?" Believing that the situation was serious, we determined to do something about it. On February 5 we wrote a letter to all our men neighborhood leaders and asked them to canvas the situation in their neighborhood and find out whether anyone in that neighborhood needed full-time help. At the same time, we asked them to locate anyone in the neighborhood not engaged in farming who was not employed, or only partially employed.[17]

Agent Sylvester B. Scott in Johnson County broke down area labor needs according to specific tasks, with the list showing a need for 3,605 workers while only 1,720 were potentially available:

Farm Category	Numbers Needed	Numbers Available
Setting tomatoes	672	226
Picking tomatoes	995	272
Wheat harvest	415	186
Sweet corn pulling	395	153
Silo filling	348	n/a
Orchards	20	140
Canning factory	n/a	743[18]

The results of a 1944 Lawrence County survey could have represented the state of agricultural labor throughout Indiana. It found that two-thirds of rural laborers there were working off the farm, largely at manufacturing plants. This amounted to 2,400 workers that were no longer available to the 600 area farmers. In addition, 1,800 men from the rural parts of the county served in the armed forces.[19] These conditions only reaffirmed what every agent was saying: that the labor shortage was real and that two primary causes were workers who had moved to defense industry jobs and those who were serving in the military.

Most of the larger farms employed at least one full-time worker, while smaller farms could only afford part-time seasonal help, so it was common for surveys to report a higher need for part-time workers during the growing season.[20] In

1943, one out of every 20 Jay County farms needed a full-time employee, while seven out of 10 needed help to get crops planted or harvested.[21] That same year, Marshall County needed about 1,000 workers to work for a matter of days or weeks but not year-round.[22] And the Putnam County agent reported that "272 additional workers would be needed during hay harvest, 141 for harvesting small grain, 200 for corn picking, 84 during tomato picking, 40 for weeding and harvesting potatoes, and 123 for silo filling."[23]

As a general rule, the farmer typically provided half of the farm's labor, with his wife and children accounting for an additional third.[24] But there was a limit to the amount of work that the farmer and his family could do without hiring additional help.

The shortage of labor also affected more than just the farm. The number of short-term workers needed by processors for canning field beans, peas, sweet corn, and tomatoes seemed insurmountable.[25] With many of the available workers preferring the higher pay, regular hours, and better working conditions offered by the defense plants, agents would have to look outside of their counties and at nontraditional sources of labor in order to find the agricultural workers they needed.

Working in cooperation with county labor assistants, Extension set up recruitment and placement programs to connect those who were available for agricultural work with those who needed workers. Labor assistants were often employed full-time in the county between May and September.[26] Some counties used other sources of funding to keep these assistants in place through the fall of the year as well.[27]

The labor assistants worked with local business owners and, as one agent noted, had police officers "watching for floaters and any person who looked like they were looking for a job. These would be sent to the county agent's office and were soon placed on a farm."[28] They also interviewed high-school students who might be interested in working during the summer and peak harvest seasons.[29] They had to have potential workers identified in advance since, as Pulaski County agent Carl Olson noted, some farmers "waited until the morning they needed the labor, then called in and expected the office to have labor out there in the next few minutes," which of course did not happen in most cases.[30]

To help facilitate work arrangements, agents established placement centers around the county where farmers could register their needs and those looking for farm work could sign up.[31] These centers were strategically located where people could get to them with less driving to offset tire and gasoline rationing (see Chapters 37 and 38), so it was common to find them at grain elevators, grocery and general stores, food banks, telephone exchanges, Extension offices, furniture stores, lumberyards, and offices of town clerks.[32]

Information collected at each center was added to a master list at the Extension office.[33] Agents often reported that they were spending a fair share of their time working on labor problems in addition to the time spent on the project by the labor assistants.[34]

The results were mixed. In 1943, the Porter County agent reported that he had filled the full-time farm positions with married men and their families, who lived on the farm, but he was hardly able to find any single men over the age of seventeen to take farm jobs.[35] That same year, the Howard County agent noted, "All requests for farm workers, with the exception of two or three, have been filled . . . so that most people are beginning to realize that the Extension Service is now responsible for the farm labor program."[36] The following year, Howard County reported 1,131 farm placements, with the largest number provided for tomato production.[37]

Agents evaluated their programs using the value of their time and that of the labor assistant along with the number of positions filled to determine how much it cost to supply the workers.[38] In 1943, it cost Lake County 68 cents each to place 330 farm workers, which the agent reported was lower than the cost in adjoining counties.[39] The Sullivan County agent indicated it cost an average of $3.51 each to place 293 workers.[40] Eighty workers in Union County cost $7.87 per placement.[41] Agents considered the program to be expensive, "but it did fill the need," as one acknowledged.[42]

While agents had little to no choice but to help with the labor shortage, some were critical of farmers needing labor and workers unwilling to work. In the 1944 annual report for Lawrence County, John Armstrong lodged his complaint, writing, "The biggest lemon in the Farm Labor set-up is that we are asked to furnish help to farmers that never are able to keep men under the best of conditions. They pay too little and are most difficult to please."[43]

Farmers would cope with the challenges posed by an acute labor shortage during World War II through many means. Not unexpectedly, they relied on men who were too old for the draft and, to some extent, on returning servicemen. In a few cases, agents mentioned that Japanese Americans from the West Coast came to Indiana to work as farm laborers, but these individuals, who were likely from incarceration centers, were few in number.[44]

With such a high demand for agricultural production and the amount of effort it required, a large portion of the needed labor would have to come from individuals other than the hired farmhands that farmers usually employed, which meant they would now rely on nontypical sources of labor.[45] This included the following:

• *Labor from the Farm Family.* The lack of hired labor meant, first and foremost, that farmers and their families had to work longer hours to meet the daily workload of the farm.[46] In 1942, St. Joseph County agent Eugene Bird reported, "We have never seen so many women helping with farm work and believe that already the maximum has about been reached in this field," and a few years later he stated, "Most of the two-man farms which are generally operated by the owner and the hired man were, this year, operated without the hired man. . . . Many older men, women and boys have found it necessary to operate farm equipment and work in the fields."[47] Kosciusko County agent Joseph Clark echoed that by observing that young boys and girls "were taking the place of a man on the farm, operating tractors and tractor-drawn equipment, helping on hay [balers], combines and etc."[48]

Faced with a shrinking labor pool, farm families simply had to work longer hours to get the crops planted and harvested. In a 1943 observation about their efforts, Huntington County agent Walter Rusk explained how the work of the farm family was critical to whatever successes were attributed to Indiana farmers, writing, "A rainy period during the last of May and first of June delayed corn planting and bean seeding to where farmers began to be discouraged and thought it would be impossible to get the crops in. But it was definitely demonstrated that it was the farm woman, the farm girl, and the farm boy who got on the tractors and drove them day and night to get the crops planted."[49] That same year, agent Ross Atkinson observed, "[W]e and Orange County agricultural leaders feel safe in predicting that the farm women and children will ride the machinery to the desired 10% increase in production of meat, milk and eggs in 1943."[50]

Of course, the farm family could only do so much work, so for many families, the lack of hired labor meant their only option was to limit the number of dairy cows or hogs, or to plant fewer acres.[51] Agent Bryce T. Bond specifically mentioned in 1944 that Warrick County farmers were reducing the number of hogs they carried due to the unavailability of labor.[52] And that same year, Wayne County agent Scott Milligan reported that some "herds have been sold, [and] others reduced to numbers that can be handled by the operator and his immediate family."[53] Cutting back on milk and pork production posed a serious threat to the government's need for these two specific commodities, but without access to more workers, some farmers had little choice.

• *Labor and Machinery from Other Farmers.* In 1942, Orange County agent Ross Atkinson wrote, "The exchange of farm work is an old, old custom."[54] While this custom had dropped off during years with plenty of available labor, farmers quickly revived the practice of swapping out their own labor and machinery for time or equipment from others.[55] By cooperating with each other, many farmers could solve some or all of their labor issues.[56]

Since machinery was in short supply, neighborhood leaders surveyed farmers to determine the location of combines, corn pickers, and other equipment in the county so these resources could be shared as needed.[57] (See Chapter 36.) Agent

Fred Hoover noted, "Since many Owen County farms are relatively small compared to some areas, the owners cannot afford to own a complete line of harvesting equipment. So one farmer may own a combine and another a corn picker or a pick-up [baler], and by exchanging they all get their crops harvested by their own labor and equipment."[58]

Agents made significant efforts to identify "relatively scarce machinery, such as combines and pickers, so that the greatest possible use of such machines might be had," as Marshall County agent Lyman Butler explained in his 1943 annual report.[59] To make up for the shortage of new equipment, Knox County agent Henry Benson noted that the agents' goal was "keeping the tractor going day and night."[60]

Typically, agents kept a list of farmers who owned equipment and were willing to do work for hire, or custom work.[61] However, at least one agent reported that some farmers were "reluctant to go out and do custom work for neighbors as long as their own work at home was pressing," which obviously made sense.[62]

• *Labor from Migrant Workers.* Nationwide, 230,000 workers from outside the United States worked to stem the shortage of agricultural labor throughout World War II.[63] Indiana depended heavily on migrant workers coming from states such as Arkansas, Kentucky, Mississippi, Missouri, Tennessee, and Texas.[64] When labor was needed in Huntington County to harvest the 1943 tomato crop, 125 small-scale farmers from Arkansas and Missouri came to work in the county.[65]

In addition to hiring workers from other parts of the U.S., Indiana farmers employed foreign migrants. In most cases, the War Food Administration Labor Division brought these workers to Indiana after their work in other regions of the country had been completed. Migrants came from the Bahamas, Jamaica, and Mexico.[66] In 1945, the Morgan-Adams Canning Company in Vermillion County brought in 24 Mexican migrants to harvest 900 acres of sweet corn and 90 acres of green beans, employing them from August 24 to October 24.[67]

There were indications that agricultural companies such as the Heinz Company could bypass the county Extension agent and directly contact John Kohlmeyer, the state supervisor of the Emergency Farm Labor Office, to bring migrants to this country through their own recruiting efforts abroad.[68] But typically those who wanted to hire migrant workers went through the agent. In turn, the agent called Kohlmeyer, who then made an official request to the War Food Administration Labor Division in Indianapolis.[69] Such a request required officials from that agency to make a site visit to inspect the housing provided to the migrants in order to ensure that they had adequate living arrangements.[70]

In 1942 in Huntington County, 135 Mexican families were housed in empty school buildings that a tomato cannery had rented.[71] In 1945 in Wells County, transient laborers were housed in the former barracks of the Civilian Conservation Corps.[72] In other cases, migrant workers stayed in vacant houses on farms where specialty crops were grown.[73]

When the Caar Canning Company in Redkey and the Nass Corporation in Portland wanted to bring migrants to Indiana, they turned to Jay County agent Harold Rippey for help. Working with the U.S. Army Quartermaster Corps at Camp Atterbury, Rippey was able to borrow beds, mattresses, tents, blankets, cooking equipment, and other items needed for housing and feeding the workers.[74] It was Rippey's responsibility to return the borrowed items to Camp Atterbury.[75] When local sources such as military bases were unavailable, equipment for housing and feeding migrant workers could be rented from the Emergency Farm Labor Office at the Purdue campus in West Lafayette.[76]

In many cases, agents also helped transport migrant workers between farms. In 1944, Jasper County agent Frank DeLaCroix indicated that he and the county labor assistant helped transport sixty-nine Jamaicans from Winamac to the William Gehring Farm in Gillam Township. DeLaCroix wrote, "These workers were secured on a subcontract basis from canning companies in southern Indiana. The fine cooperation and speedy actions of the State Farm Labor Organization made it possible to recruit these workers and have them at Mr. Gehring's farm within three days after the order was placed."[77] That same year, the Steuben County agent and two area farmers transported ten Jamaicans from the Waterloo train station to their local work assignments.[78]

Wages for migrants were not set arbitrarily by local farmers nor by the state Extension office. As Pulaski County agent Carl Olson explained, "When you brought in foreign workers, it was necessary to hold a labor hearing to determine [the] prevailing wage rate."[79] District labor meetings required by the county wage board and open to the public were called where prevailing wages were discussed and agreed upon.[80] Many of these meetings were held at the local Extension office, and the agent often served on the board and participated in making decisions about what was considered fair compensation.[81]

It is impossible not to appreciate the amount of work migrants did in order to help growers and support each county in the efforts to meet wartime production quotas. Whether from another state or another country, these workers were essential for harvesting many types of crops such as tomatoes, apples, and peaches, and for helping to keep the canneries running at full capacity.[82]

• *Labor from Prisoners of War.* The first prisoners of war (POWs) were brought to the United States in April 1942 because the European prison camps were filled to capacity and because the country needed labor to meet the demands of war.[83] By the end of 1942, estimates were that as many as 30,000 POWs were arriving in the U.S. each month.[84] Eventually there would be approximately 372,000 POWs from Germany, 50,000 from Italy, and 4,000 from Japan.[85] By the conclusion of the war, half of the POWs in the U.S. had provided important agricultural labor.[86]

There were 175 main camps and 511 smaller area camps for POWs scattered across 46 states. Indiana had three of the larger camps, which were:

Camp Atterbury at Edinburgh	Housed 3,500 Italians starting in 1943 and later 10,000 Germans; branch camps at Austin, Eaton, Morristown, Vincennes, and Windfall
Camp Thomas A. Scott at Fort Wayne	Housed 600 Germans from the Afrika Korps captured in North Africa in 1943
Fort Benjamin Harrison at Indianapolis	Housed 250 Italians and 300 Germans from the Afrika Korps in 1944[87]

In order to establish a branch camp, there had to be enough work to support at least 300 POWs. In 1945, Lloyd Cutler indicated that Lake County was unable to get a branch camp when a "survey indicated, however, that there was not this much farm work available for this number of prisoners, and labor unions in this part of the county would not agree to their use by industries that were short of help."[88]

Base camps would transfer POWs to smaller branch camps as needed to become laborers for farmers and canneries, which agents considered as they calculated the availability of workers. In 1945, Pike County agent Barnett Jewell had encouraged farmers to increase their acreage of tomatoes and other crops in the spring because he anticipated that war prisoners would be available to help with the harvest.[89]

At the time, the United States abided by the Convention relative to the Treatment of Prisoners of War that was signed at Geneva, Switzerland, on July 27, 1929. This spelled out principles regarding the humane treatment of POWs that the U.S. and other signers agreed to. Those held as captives could neither be forced to work against their wishes nor work in dangerous jobs.[90] Prisoners were asked if they wanted to volunteer for assignments such as harvesting crops or working at a cannery.[91] As one agent noted, "[O]nly prisoners who were willing were used and only a few proved unacceptable to the employers."[92]

Agents, farmers, and processors had to complete forms and jump through many hoops before the military would agree to send POWs out to work. For instance, in 1945, Captain Frank Bodenhorn, who commanded Camp Scott, attended a meeting of representatives of the local labor committee in Huntington County to discuss the regulations for securing POWs for agricultural work. Bodenhorn told the committee members that the canning companies were expected to work through the county Extension office to first get a certification of need.[93]

Farms and companies such as canneries that were close enough to an established POW camp could put in a request for the much-needed workers. At first, government policies only allowed POWs to work within a 25-mile radius of their camp, but with wartime labor in such short supply, the radius was eventually expanded later in the war, allowing them to work further away.[94] In Lake County, Cutler reported in 1945 that the "Farm Bureau plant near Schererville got some prisoners from the prisoner-of-war camp at Thornton, Illinois. Five truck gardeners [who grew specialty crops for delivery to nearby markets] used from 10 to 20 prisoners each during the busier part of the truck [crop] growing season. At potato digging time Henry Ewen, who had about 300 acres of potatoes east of Gary, was unable to get help from ordinary sources but was able to get 30 prisoners-of-war to save his crop."[95]

In some cases, labor from POWs wasn't used because of the red tape involved.[96] And as with most governmental processes during the war, the paperwork only got more complicated over time, which frustrated Johnson County agent Sylvester Scott. In 1943, he wrote:

> The Italian Prisoners of War located in Camp Atterbury [were] the thing that saved the situation [shortage of labor]. These prisoners did a good piece of work and we had a very satisfactory arrangement with the camp in making these workers available until the [federal War] Manpower Commission stepped into the picture and then the "Red Tape" makes it impossible to get these workers. Before the Manpower Commission entered the picture, these Prisoners could be gotten on a day or two notice. It now takes three weeks. This situation must be corrected before next spring or our tomato crop cannot be planted.[97]

In 1944, Madison County agent Walter Haynes described the complicated process he navigated to get POWs to help with local labor needs, writing:

> Much time and effort [were] expended in trying to reduce the red tape regarding the German war prisoners located at Windfall, part of whom were to be used for farm labor, picking tomatoes, snapping sweet corn, etc. After a conference at Tipton, a trip to Windfall and many long distance phone calls, a county wide contract was signed by the Madison County Farm Bureau to act as the contracting agency for war prisoners. About six different canning plants also signed contracts for war prisoners on farms. It was agreed to require a deposit in advance, with the order for workers. Due to the distance to Windfall and also to many restrictions and regulations regarding the use of war prisoners on farms, there were few requests. However, their work was quite satisfactory. A total of 21 orders were handled through this office. These orders were made by eight different farmers. They involved a total of 1,625 man hours of work. The total amount of money involved was $812.50 of which $82.88 was refunded for transportation.[98]

In addition to working in Madison County, the POWs from the Windfall camp also provided labor in other area counties, including in Howard County, where they worked in forty food processing plants.[99] For 1945, agent Everett Johnson reported that the German prisoners put in 5,600 hours of labor in Howard County, showing the reliance by some counties on work from POWs.[100]

Prisoners largely helped with perishable crops, such as harvesting tomatoes in the field or working in the plants that processed them.[101] In 1943 in Morgan County, POWs set, hoed, and picked tomatoes.[102] That same year in Madison County, 800 German prisoners housed in a camp near Elwood worked in the canning factories with American soldiers standing guard over them.[103] Howard County employed some German prisoners of war from the Windfall camp to pick tomatoes but hired more to work as tomato packers.[104] In Shelby County, which had a branch camp located at Morristown, 400 POWs worked at seventeen canning plants.[105]

In Knox County, 300 German POWs from a camp established in Kimmell Park at Vincennes harvested most of the 1945 tomato crop processed by the Vincennes Packing Corporation. When not harvesting tomatoes, the prisoners worked in area orchards, picked corn, and dug up sweet potatoes. They also provided labor on drainage projects, such as cleaning ditch banks.[106]

Farmers and industry leaders liked having the same prisoners return to the same farms and processors. Morgan County farmers who had used labor by POWs told agent William Record that "they had the good fortune to have practically the same individuals each day and that the prisoners proved to be some of the most satisfactory orchard help they had ever had."[107]

The POWs did not work for free. According to Article 24 of the 1929 Geneva Convention, prisoners of war were to be paid based on prevailing wages. In Marion County, local farmers held a hearing to set the prevailing wage for POWs.[108] Prisoners from Camp Scott at Fort Wayne were paid 60 cents an hour for detasseling corn and picking apples and peaches, and 50 cents an hour for other types of jobs, which was the same as the wage paid to other farm laborers doing this work.

The wages earned by the POWs were put into savings accounts, then given to each man at the end of the war.[109] While in the camps, the POWs received scrip, rather than cash, from their earnings to spend on items at the camp.[110] At the end of the war, POWs were repatriated to their home countries, with all war prisoners held on U.S. soil returned by 1946.[111]

• *Labor from Nonfarm Elementary to High-School Students.* Students between the ages of eleven and seventeen who did not live on farms were recruited and paid to do farm work during the summer and seasonally during the school year.[112] Most often, it was boys who were hired, but in some instances, girls were hired, too. While the younger students provided some of the labor, teenagers made up the majority of this workforce. The students' help was needed in the early spring, when crops were being planted; in the summer, when hay had to be stacked and crops such as beans, strawberries, tomatoes, apples, and peaches were ready to pick; and in the fall, to harvest various other crops.[113]

Some of the students hired were Victory Farm Volunteers (VFV). This group was organized under the federal Crop Corps, a program designed to supplement the shortage of agricultural workers. Nationally, 2.5 million youths would make significant contributions to the war effort through this program.[114] Each VFV participant needed a parent or guardian's permission and had to pass a physical.[115] For signing up, participants were provided a membership card and a letter of recognition, and in some counties, they also received at least twelve hours of farm training.[116] Those working at least 25 hours received an embroidered Victory Farm Volunteer insignia to wear on their clothing.[117] For contributing at least 100 hours of service, participants were awarded a U.S. Crop Corps certificate, which included the volunteer's name, township, and number of hours of work performed.[118]

All students who provided agricultural labor while schools were in session had to work out a class schedule to accommodate their release.[119] In 1943, Porter County agent Stewart Leaming noted that while school principals were "in sympathy with the labor situation and will do anything within reason to adapt the school management to the emergency," they were also concerned that "excusing boys from school for farm work too frequently may cause them to lose interest in their classes."[120]

Students who missed school days to work on farms were allowed to make up their assignments without penalty or losing attendance credits.[121] In Montgomery County, agent Gordon Sowers reported in 1944, "Only students whose scholastic standing was high enough to justify absence from school were selected and then not to exceed two days each week."[122]

In a 1943 annual report, agent Arthur Shideler detailed how students were assigned to jobs and how the school kept track of them, writing:

> The Starke County schools cooperated this fall in a plan whereby students may be excused from school without penalty to assist farmers. The farmer telephoned his request to the High School principal or Extension Agent. This information was recorded on an "approval" blank which indicated how many students were wanted, the number of days needed, when work was to start and the job to be done. The boys who wish to work were then assigned to this job. At the end of the employment period, the farmer reported the days worked by each boy on a form provided him. This report is filed with the principal so that the student will not be penalized for his absence.[123]

Schools took different approaches when releasing students for farm work, even within the same county. In 1942 in Lake County, agent Lloyd Cutler reported that some students left school for two hours in the morning to pick asparagus but that "older school boys too young to enlist in the armed services or work in the mills" could be dismissed from school for one week between April 15 and May 15 to help with farm work.[124] In 1943 in Ripley County, Sunman delayed school for three weeks so students could help harvest tomatoes, while in Milan, the school went to half days until the tomato harvest had been completed.[125]

Student labor gave assistance to those requesting farm help for short periods of time, especially to harvest crops such as tomatoes, which had to be hand-picked.[126] In 1944 in Montgomery County, six to eight boys were grouped together to pick tomatoes for ten cents a hamper. It was customary for the grower to transport the student workers to and from the farm daily until the crop was harvested. There was little choice in the matter since farmers could not house a large number of students overnight.[127] Some farmers required the students to carry their lunches, while others furnished the noon meal.[128]

Farmers growing commercial potatoes in Warrick County depended heavily on student labor to pick up the potatoes lying on the ground. Agent Bryce Bond explained that the boys were paid "on a sliding scale according to their ages and each boy was assigned a certain area according to the age to pick up the potatoes and in this way the younger boys had a smaller territory to cover and it proved to be a very satisfactory way to work these smaller boys."[129] Children from eleven to thirteen years old were also paid thirty cents an hour for removing weeds from mint in Marshall County in 1943.[130]

While students from the city schools provided meaningful labor, they were so inexperienced that some training was required.[131] Extension often provided a minimum amount of training to these boys and girls to get them better oriented regarding the work.[132] But in 1943, Marion County agent Horace Abbott reported on concerns that growers had shared with him about these workers:

> Early in the season the growers were resentful about using untrained high school boys and some of the other unskilled labor which was available. They were justified in taking this attitude for this type of labor was much more expensive even on a unit pay basis than was the skilled labor. An example of this was one grower who was paying 1.5 cents a pound for picking green beans. With 30 pounds in a bushel basket, this should have figured out [to] 45 cents a bushel. Due to the quality of work done by these inexperienced high school boys, the actual cost per bushel for harvesting the green beans after they were sorted over and repackaged into a salable product was 85 cents per bushel. In past years with experienced help, this grower said he did not have to grade and repackage his green beans when they came in from the field.[133]

Ultimately, farmers had little choice but to hire these inexperienced students from the city since many of those with experience on farms were already assisting their parents or working for friends and neighbors.[134]

Some farmers seemed to have a better experience with the student workers and reported being pleased with their work. In Jay County, agent Chester Langston indicated in 1944 that the students "have proved willing to do their share and are doing a good job compared to their age and experience."[135] Orange County agent Ross Atkinson wrote that high-school students "became farm laborers for the first time in their lives and even though corn planting, soy bean sowing, and alfalfa harvest came at the same time, a one hundred percent job was done."[136]

In more than a few cases, the student workers provided much-needed labor in dire situations. When Knox County orchard owner J. W. Kimmell needed additional labor to harvest a heavy crop of peaches in 1943, the county labor assistant sent out a plea for help to his colleagues in the surrounding counties. One hundred fifty students from Dubois, Lawrence, and Vigo Counties were sent to the orchard for ten days. They were housed in a camp that was approved by the Indiana Board of Health and equipped with cots, bedding, and mess equipment borrowed from Camp Atterbury. One camp supervisor and one field supervisor were provided for every twenty-five boys.[137]

That same year, the Pike County agent said that students working at the W. C. Reed and Sons Orchard near Vincennes picked 60,000 bushels of apples that would not have been harvested otherwise.[138] And in Montgomery County, a tomato farmer told the agent in 1944 that "had it not been for the assistance of Extension personnel and the school boys, his crop would not have been all saved in proper time."[139]

One of the biggest demands for student labor was in the production of hybrid seed corn. Commercial businesses turned to Extension to help find enough workers. The majority of the companies signed an agreement with John Kohlmeyer at the state Emergency Farm Labor Office at Purdue, who put the burden on agents to find detasselers, bus drivers, and field supervisors for the seed corn industry. To be hired, the students needed to be at least fourteen years old with a height of at least five foot three inches and a weight of 120 pounds or more.[140] Many seed operations hired both boys and girls but placed them in separate fields.[141]

In 1944, Montgomery County was asked to provide seventy-five detasselers, three bus drivers, and three supervisors.[142] Agent Gordon Sowers gave this detailed account of the complexity involved in finding workers for detasseling assignments:

> School bus drivers were selected with the thought in mind of them being qualified and willing field supervisors as well as capable drivers and reputable citizens. These haulers were personally visited by [the] Emergency War Food Assistant and all the details of their work explained; necessary forms for township trustee permission filled out; gasoline stamp requirements met and sent through proper channels; details of obtaining necessary insurance coverage

> worked out; and an understanding reached to be ready on call, the first of which was on July 15. . . . Each bus was loaded with boys and girls in groups of proper number to be most effectively supervised in the field.
>
> Early in June the actual registration of detasseler applicants was begun in the Office of County Extension Agent and each application was examined for proper qualification such as age, weight, etc., and divided in bus loads by sexes. . . . As calls came for detasselers, each recruit was notified by phone or mail where bus was to be met and at what time. Name of driver was given each recruit and instructions how to dress and to bring along noon lunch was also included in notice.
>
> A copy of list of recruits notified to appear was also given [to the] driver and supervisor. . . .
>
> The detasselers were paid 50 cents an hour and free transportation furnished. A Liaison Camp supervisor was on the job with his car and rendered first aid when, as, and if needed and also returned those becoming too ill to continue work to their home or to a physician for proper care.
>
> Haulers were given $10 per day for their bus plus $7 per day for their own services as Field Supervisors. All this was paid by Seed Corn Growers.
>
> Liaison Camp Supervisor was given 5 cents per mile for his car and $7.50 per day for his services. This was paid by Emergency Farm Labor Office [at Purdue].[143]

Camps to house those doing detasseling work for seed corn companies were located in counties where the companies needed help. Some of the ones that hired students included Corn Belt Hybrid Seed Company, Boswell, and Farmcraft Seed Company, Oxford (Benton County); Woodward's Hybrid Corn, McCordsville (Hancock County); Edward J. Funk and Sons, Kentland (Newton County); George F. Chester and Sons Seed Company, Valparaiso (Porter County); DeKalb Seed Corn Company, Lafayette (Tippecanoe County); Pioneer Hi-Bred Corn Company, Tipton (Tipton County); and Crow's Hybrid Corn Company, Milford, Illinois (Iroquois County).[144]

Funk and Sons had one of the largest detasseling camps in the state, which was located at the Newton County Fairgrounds in Kentland. In all, the company hired 700 students from twenty-one different counties.[145] The local agent, Russell Zell, had a detailed working knowledge of the company's camp and provided the following description of its operation:

> Physical make-up of the camp was provided by Edward J. Funk and Sons. Eight large tents capable of housing 500 boys were rented; a screened-in kitchen was constructed between two dining halls secured for the program from two local churches. Hot-water showers were provided. Recreation and entertainment facilities were arranged by Labor Assistant [Alvin C.] Cast

> assisted by 16 supervisors furnished by ten State Labor Offices. . . . Transportation and field supervision was furnished by Edward J. Funk and Sons.[146]

At night, Zell showed motion pictures, helped to conduct church service programs, and even wrote a newsletter for the students.[147]

It was common for other agents to occasionally visit students from their counties at the Newton County camp to make sure that everything was working as planned.[148] In 1945, agent Barnett Jewell provided this report after visiting eighty-five students from Pike County:

> They [those inspecting the camp] found the camp well organized and the boys well satisfied with the sleeping and eating facilities at the camp. During their stay, they ate three meals with the boys and the following menus were served: Breakfast—plates consisted of scrambled eggs, sweet rolls, and two kinds of cereal served cafeteria style with tomato juice, jelly, fruit, butter, cream and sugar, coffee and all the milk they could drink on the tables; Lunch boxes served in the fields consisted of a variety of three sandwiches, orange or banana, and milk; Evening meal—two large frankfurters, diced cream potatoes, green beans, peas, beets, slaw, apricots, butter, jam, catsup, bread, coffee and milk. Second helpings were available at the morning and evening meals.[149]

Extension agents were pleased when students largely remained on the job until the work was completed.[150] During a two-week period at the camp, Pike County was among three counties whose workers received a rating of 100 percent by Funk and Sons based on their performance and attendance. The company recognized individuals from the selected counties with a chicken dinner.[151]

• *Labor from 4-H Youth.* In addition to the paid student workers, individuals who were 4-H club members volunteered time toward the war effort and receive credit for their work through the 4-H Victory Club.[152] In 1942, assistant county agent Charles J. Murphy wrote to Marion County youth to encourage their involvement in the war effort, saying:

> With the shortage of farm labor that faces the farmer today, the older members of our rural organization will not only be milking their own cows, feeding their own hens, growing their own pigs and producing their own lambs and corn, but will be helping Dad and Mother to do a better job of growing and producing all agriculture commodities necessary to accomplish the thing that our government has pledged itself to do.
>
> To accomplish this goal, it may be necessary to discontinue some of the activities we all enjoy doing. We may not be able to play as much or to go swimming as often as we would like to; we may have to postpone camping trips or give up fishing even though we feel sure the fish will bite. But I am

sure that every American boy and girl will be glad to sacrifice these privileges voluntarily for the interests of their older brothers and neighbors who today are enlisting in the various branches of our government's military services. We want them, as well as our allied neighbors, to have all the milk, butter, eggs, hams, steaks, and vegetables they will need during these troublesome years of fighting.[153]

In 1943, 242 Rush County 4-H members reported engaging in nearly 201,000 hours of work in the program.[154] That same year, 143 Parke County 4-H members reported working 80,854 hours. With one day equal to 10 hours, agent Noah Hadley in Parke County noted that these volunteers had contributed roughly 8,085 days of work through the 4-H Victory Club. His annual report listed the number of days logged for each type of activity:

Field of Service	Total Days Worked
Farm machinery repair	29
Beef cattle	148
Corn	1,039
Dairy	1,239
Farm records	27
Garden	639
Poultry	885
Soybeans	213
Swine	926
Safety and first aid	8
Health	16
Recreation	43
Salvage (iron, rubber, tin, and fat)	30
Sales of war bonds and stamps	2
Sheep	80
Potatoes	134
Baling hay	33
Janitor service in school, church	71
Other farm and homework	2,368
Other fields of service	145 [155]

In 1944, the Warrick County agent reported that 4-H club members provided nearly 65,000 hours of labor, wrote more than 11,000 letters to those in the armed forces, helped collect 45,000 pounds of scrap metal, and sent 2,500 bags of milkweed pods to Michigan.[156] (See Chapters 39 and 40.)

Vigo County agent Charles L. Brown described in his 1943 report how 4-H Victory Club members were recognized for their contributions. Those who volunteered 100 hours received a letter V made of green felt. At 300 hours of work, members received the letter I. After that, it would take 200 additional work hours to earn each of the remaining letters needed to spell out the word *victory*. Brown also reported that six Vigo County participants completed more than 2,000 hours that year, including Ruth Ann Dugger, Teddy Ray Figg, Wilma Frederick, Marilyn Havens, George Puller, and Josephine Rockwood. Statewide, the top 4-H volunteer for 1943 was Nancy Marcella Ross, who logged 2,609 hours, which was quite an accomplishment for this student.[157]

• *Labor from Nonfarm Women.* Through the years, much has been written about the role of women in the workforce during World War II. As so many men left for military service, it's commonly understood that women stepped in to fill manufacturing jobs that might otherwise have been done by men—jobs that provided military equipment and supplies critical to the war effort. But what isn't as well known is that women also made a significant contribution in the agricultural sector. A program known as the Women's Land Army (WLA) was launched in April 1943—also under the auspices of the federal Crop Corps program, like the Victory Farm Volunteers program—and by the end of the war, these 1.5 million women represented 22 percent of U.S. agricultural workers. Prior to the war, in 1940, women only represented 8 percent of agricultural workers.[158]

From 1943 to 1947, the U.S. Congress allocated funds as part of the Emergency Farm Labor Service to help recruit and train women as farm laborers for the WLA. Experience was not a requirement to join the WLA, but participants needed to be physically fit to do farm work. They earned the wages of an unskilled worker, which was 25–40 cents an hour.[159]

Women in the WLA and those from communities around the state were especially appreciated in tomato production, where they picked the crop and worked in processing plants. In Huntington County alone, hundreds of tons of tomatoes would have spoiled in the yards of the canneries without the aid of the women working in the evenings.[160]

In Marion County, Horace Abbott reported in 1942 that several area farmers had employed women to harvest tomatoes, noting, "Some of these women could pick a hundred hampers of tomatoes a day, which is a good day's work for any man. It was not an uncommon sight to see a woman take small children to the tomato field with her while she picked tomatoes."[161]

• *Labor from Local Community and Business Volunteers.* In the evenings and weekends during harvest, when many hands were needed to get the work done, employees from local businesses provided temporary labor to help farmers.[162] It was common for the agent or county labor assistant to contact the local American Legion, labor unions, and service clubs to recruit workers for short-term assignments.[163] While any labor was appreciated, most of these workers came from the urban areas, which meant they required training.[164]

While the war industries were legitimately blamed by farmers for shrinking the agricultural labor pool, there were instances when the factories agreed to supply temporary workers for the farm.[165] Many of these industry workers had prior farm experience, so their availability was greatly valued in terms of the equipment they could operate and the care they could provide to livestock. In 1943, volunteers such as these from industries in Knox County were asked to work four to six hours per day when the need for agricultural labor was high.[166] That same year, Newton County agent Russell Zell noted that "the planting and harvesting of hundreds of acres of crops were directly attributed to volunteer business men."[167] And in Wabash County, agent William Delaplane reported the following swap of labor:

> When it became apparent that farmers would have difficulty in getting their corn planted, this man [county labor assistant] promoted a drive in the factories for men with farm experience who were willing to work a few hours each day in helping with the planting. . . . Since we have had a very favorable fall for harvesting and most of the farm crops have come in in good shape, the War Board, Selective Service, and the Extension Office are trying now to reciprocate by releasing farmers who have some extra time for work in factories until March 1st.[168]

• *Labor from Conscientious Objectors.* In a few cases, farmers had access to hire conscientious objectors, or the men who, when drafted, would not carry a gun or fight based on their moral or religious principles.[169] There were two main classes of conscientious objectors: Class 1-A-O, which included the individuals who were willing to provide noncombatant service for the military, and Class 4-E, which included those opposed to any military service.[170]

Nationwide, 72,000 men registered as conscientious objectors, but only 52,000 were granted this special status. Around the country, 12,000 of them would do work deemed to be of national importance in Civilian Public Service (CPS) camps.[171] In Indiana, there were 863 conscientious objectors working in these camps, which was one of the largest number of objectors in the United States.[172]

After a CPS camp was established in Sullivan County, agent Jean Sliger reported in 1942 that farmers nearby felt so strongly "against this camp because they [those at the camp] are conscientious objectors" that many farmers refused the service being offered.[173] Still, the men from the camp completed some work in the area, including helping install and remove fencing; establishing terraces, terrace outlets, and gully control; clearing and cleaning channels and levees; excavating channels, canals, and ditches; planting trees and improving forest stands; and working on fire suppression, surveys, and research data for state and regional Soil Conservation Service offices.[174] But the camp was moved in August 1943, with Sliger reporting that the camp "was never popular with the mass of people."[175]

Another camp, CPS Camp 28-01, opened in Pulaski County in April 1942 and continued operations there until May 1946.[176] It was located at the Jasper-Pulaski Game Preserve just north of Medaryville. Men from the camp planted trees in abandoned strip mines and completed various jobs for area farmers. To get help from the workers at the camp, farmers submitted a request to the agent, who then certified the need. In 1944, agent Carl Olson certified that farmers needed 112 workers in the fall and noted, "These men [from the camp] assisted in the harvesting of soybeans and corn and with other work such as ditching, carpentering, and other general farm work." He also noted that these men were only used for labor because "this is practically the only source of available labor in the county [and] farmers are taking full advantage of it."[177]

In 1945, following the end of World War II, Howard County agent Everett Johnson summed up the work to find agricultural labor for the war effort by writing that the "emergency farm labor program was one of the biggest responsibilities of the Extension Agent."[178] In just the last twelve months of the war, county Extension agents had—with the assistance of 81 county labor assistants—reported placing 1,826 full-time workers, 36,946 seasonal workers, and 13,489 prisoners of war in agricultural positions across the state.[179]

When it came to finding labor to run farming operations and other agricultural industries during the war, agents had been put in a difficult position. With the military recruiting or drafting thousands of men, the supply of labor was greatly diminished. In addition, high-paying positions in the war industries largely took the men and women that were left. As a result, Extension faced the constant problem of where to find workers that the farmers needed for planting and harvesting crops, feeding and milking the dairy cows, and picking crops such as apples and tomatoes.

Farm families rose to the challenge, with every family member pitching in to work longer days and take on more responsibilities. And neighbors helped one another by swapping labor and equipment with one another in order to get everyone's fields planted and harvested, and livestock grown for the markets, with agents helping to coordinate the effort.

As more labor was still required, though, agents would be forced to look at every possible avenue to fill the void, including from volunteers, children, women, prisoners of war, and migrants from the United States and other countries. This challenge would tax the Extension system to the fullest, but in the end, the agents and labor assistants came through by finding help from whatever sources they could, providing aid to the agricultural community and the country during a time of critical need.

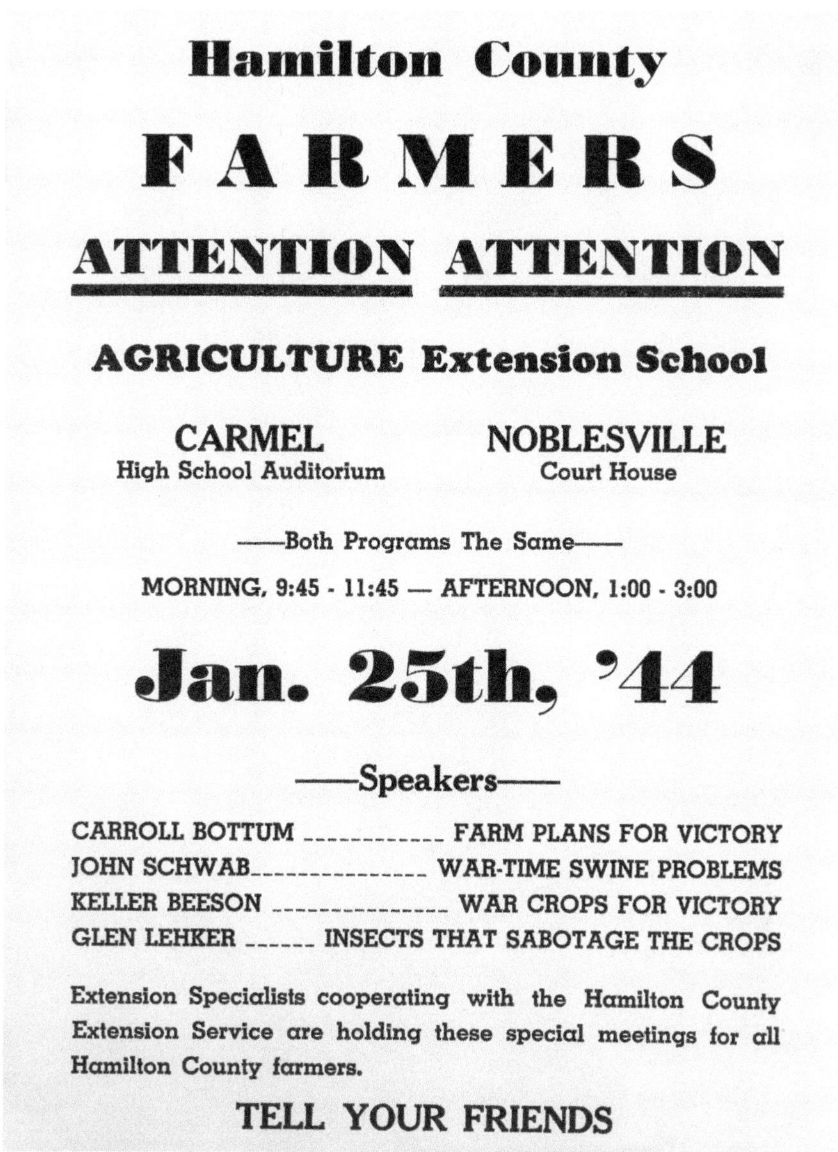

Hamilton County

FARMERS

ATTENTION ATTENTION

AGRICULTURE Extension School

CARMEL	NOBLESVILLE
High School Auditorium	Court House

—Both Programs The Same—

MORNING, 9:45 - 11:45 — AFTERNOON, 1:00 - 3:00

Jan. 25th, '44

—Speakers—

CARROLL BOTTUM ____________ FARM PLANS FOR VICTORY
JOHN SCHWAB_______________ WAR-TIME SWINE PROBLEMS
KELLER BEESON _______________ WAR CROPS FOR VICTORY
GLEN LEHKER ______ INSECTS THAT SABOTAGE THE CROPS

Extension Specialists cooperating with the Hamilton County Extension Service are holding these special meetings for all Hamilton County farmers.

TELL YOUR FRIENDS

Extension schools, such as this one held in Hamilton County in 1944, allowed topics related to World War II to be discussed in greater detail. Such workshops were important in helping Indiana farmers meet their federal agricultural quotas. Courtesy of Purdue University Libraries, Karnes Archives and Special Collections.

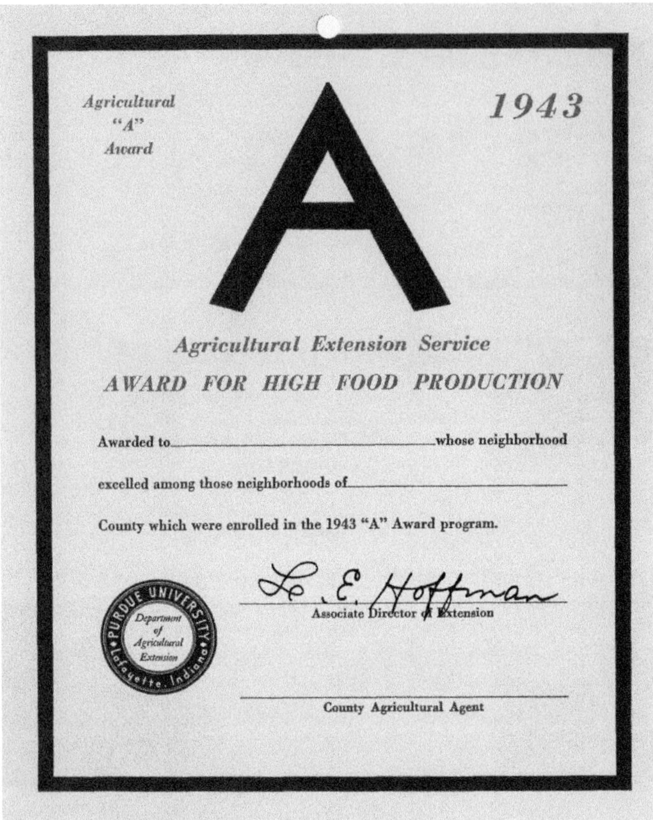

Agricultural
"A"
Award

1943

A

Agricultural Extension Service

AWARD FOR HIGH FOOD PRODUCTION

Awarded to ________________ whose neighborhood

excelled among those neighborhoods of ________________

County which were enrolled in the 1943 "A" Award program.

L. E. Hoffman

Associate Director of Extension

Purdue University • Department of Agricultural Extension • Lafayette, Indiana

County Agricultural Agent

The Purdue Department of Agricultural Extension recognized farm families for their efforts to increase production and help meet war demands by giving them the Agricultural "A" Award. Winners of this prestigious award were provided with a certificate (top) that could be posted on a barn or fence for others to see (bottom). Courtesy of Purdue University Libraries, Karnes Archives and Special Collections.

CO-OPERATIVE EXTENSION WORK
IN AGRICULTURE AND HOME ECONOMICS
STATE OF INDIANA

PURDUE UNIVERSITY AGRICULTURAL EXTENSION
DEPARTMENT, U. S. DEPARTMENT OF AGRICULTURE
COUNTY BOARD OF EDUCATION AND COUNTY
EXTENSION COMMITTEES CO-OPERATING

December 2, 1942

EXTENSION SERVICE
COUNTY EXTENSION WORK

IT TAKES GOOD MATERIALS TO TURN OUT
A GOOD FINISHED PRODUCT

SWINE SCHOOLS
DEC. 9-1942

SPEAKERS

John Schwab- Extension Swine Specialist-Purdue University

Subject-Winter feeding bred sows for vigorous pigs at farrowing time, how to save pigs from over laying, start pigs on feed for good health and economical gains.

J. Carroll Bottum- Extension Economist-Purdue

Subject-Economic Situation as it pertains to hog production.

These meetings will last two hours, each.

PLACES

Morning Session
9:30 A.M.- Mexico School

Afternoon Session
1:30 P.M.-Clay Township
Voc.Agriculture Building

COME AND BRING YOUR NEIGHBORS!

Very truly yours,

John E. Wylie

John E. Wylie
County Agricultural Agent
Miami County

JEW:md

This flyer advertised two Purdue Extension staff members—John Schwab, hog specialist, and J. Carroll Bottum, agricultural economist—as the speakers for December 1942 swine schools held in Miami County. With World War II requiring an increase in hog production, both feeding and economics would be crucial to helping farmers accomplish that. Courtesy of Purdue University Libraries, Karnes Archives and Special Collections.

John Schwab (right foreground), a well-known and sought-after hog expert from Purdue, addresses attendees at a February 1942 swine school in Montgomery County, providing information about feeding and caring for pigs. Attendees included 85 farmers, six vocational agriculture teachers, and 50 students from the vocational agriculture classes. Courtesy of Purdue University Libraries, Karnes Archives and Special Collections.

In 1945, William Wichman in Daviess County used Extension recommendations such as feeding balanced rations and providing plenty of pasture to raise these hogs more efficiently. Extension agents promoted the achievements of producers such as Wichman at local meetings and in newspaper articles in order to encourage other producers to improve their management practices as well. Courtesy of Purdue University Libraries, Karnes Archives and Special Collections.

In this 1943 photo, Blackford County hog producer C. R. Smith stands near his rather famous Duroc sow. Smith received national recognition that year for raising 70 pigs that had been produced by this sow over seven consecutive litters. During World War II, farmers were asked to do everything in their power to send more hogs to market to help supply the wartime demand for protein. Courtesy of Purdue University Libraries, Karnes Archives and Special Collections.

The J. R. Miller Modern Dairy of Akron in Fulton County, pictured in 1940, was one of hundreds of Indiana dairies that would be asked to ramp up their production of milk and dairy products during World War II. Much of the increased supply was turned into powdered milk, evaporated milk, cheese, and other products for the soldiers of the Allied forces. Courtesy of J. C. Allen and Son.

CO-OPERATIVE EXTENSION WORK
IN AGRICULTURE AND HOME ECONOMICS
STATE OF INDIANA

April 11, 1942

PURDUE UNIVERSITY AGRICULTURAL EXTENSION
DEPARTMENT, U. S. DEPARTMENT OF AGRICULTURE
COUNTY BOARD OF EDUCATION AND COUNTY
EXTENSION COMMITTEES CO-OPERATING

EXTENSION SERVICE
COUNTY EXTENSION WORK

OBJECTIVE:

"The objective of the Dairy Program this year must be the production of the maximum amount of milk possible from the dairy resources available."

Dear Dairyman:

THE FOLLOWING SUGGESTIONS WILL HELP YOU IN

PRODUCING YOUR SHARE OF MILK FOR THE FOOD FOR FREEDOM PROGRAM

No. 1 <u>If</u>- just a two-letter word, but, can it spell grief! Next summer, for instance, when pastures are short, will you be saying "If I had only seeded sudan grass for pasture, I would not be short now." (Rate of seeding - 25 to 30 pounds, seeded after May 15.)

No. 2 A salad makes the meal taste better. Silage is the salad on the old cow's menu. Every dairyman milking 10 or more cows <u>should</u> be thinking about silage for next winter. (Plans for temporary silos will be available.)

No. 3 You know, feed could be higher than taxes next fall. Although you couldn't finish hogs on legume hay and silage, a cow won't do bad on that diet, with just a little grain. Soybeans, planted for hay, this spring, will help keep that milk "pouring out" next fall. Two tons of good hay per cow will be good property - <u>next fall</u>, <u>but</u> we have to do our planning and planting this spring.

With the wartime requirements for more dairy production under the U.S. Food for Freedom program, Extension agents such as John Wylie in Miami County reminded dairy producers in April 1942 to plan ahead to meet the increased need for cattle feed. Courtesy of Purdue University Libraries, Karnes Archives and Special Collections.

CO-OPERATIVE EXTENSION WORK
IN AGRICULTURE AND HOME ECONOMICS
STATE OF INDIANA

PURDUE UNIVERSITY AGRICULTURAL EXTENSION
DEPARTMENT, U. S. DEPARTMENT OF AGRICULTURE
COUNTY BOARD OF EDUCATION AND COUNTY
EXTENSION COMMITTEES CO-OPERATING

EXTENSION SERVICE
COUNTY EXTENSION WORK

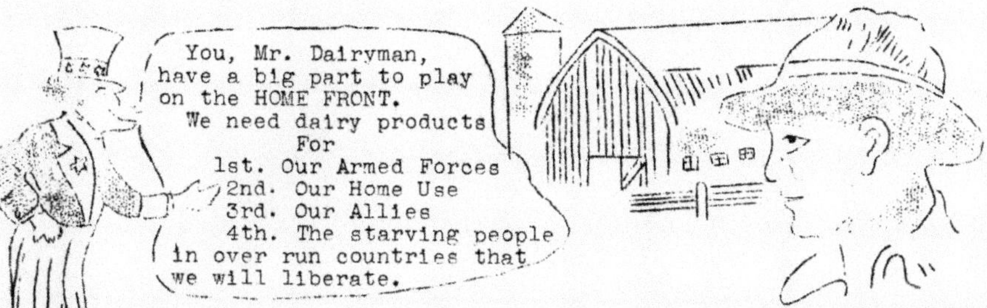

Dear Dairyman:

Uncle Sam has given us a job to do and we can't let him down.

The men in the Service are giving their all for us, so let's do a good job better this year.

Our Dairy School, which will be held in the Peru High School January 18, 1944 will help us with our 1944 program.

<u>Program</u>

1:30 P.M.	Dairy Outlook and Needs for 1944 E. Z. Gannon, Extension Dairyman of Purdue
2:30 P.M.	Bang's Disease and Calfhood Vacination Dr. C.R. Donham, Head of the Veterinary Dept. Purdue
7:30 P.M.	Milk Production Under War Conditions - Mr. Gannon
8:30 P.M.	Mastitis in Dairy Cattle - Dr. Donham

Come and bring a carload of your neighbors.

Very truly yours,

John E. Wylie

John E. Wylie
County Agent

JEW:bl

This letter inviting farmers to a January 1944 dairy school in Miami County presented the increase in production as a patriotic duty, noting that additional supplies would be for soldiers, domestic use, allies, and people in war-torn countries. The agenda included presentations on mastitis and Bang's disease, or brucellosis, which were two serious diseases of dairy cattle that threatened to reduce production. Courtesy of Purdue University Libraries, Karnes Archives and Special Collections.

During the war, it was common for dairies to place patriotic messages on their bottles. Pictured are examples from Covalt's Dairy (left) and Whitney's Dairy (right), both of which were located in Muncie in Delaware County. Courtesy of John Cleland.

CO-OPERATIVE EXTENSION WORK
IN AGRICULTURE AND HOME ECONOMICS
STATE OF INDIANA

PURDUE UNIVERSITY AGRICULTURAL EXTENSION
DEPARTMENT, U. S. DEPARTMENT OF AGRICULTURE
COUNTY BOARD OF EDUCATION AND COUNTY
EXTENSION COMMITTEES CO-OPERATING

Peru, Indiana
December 31, 1941

EXTENSION SERVICE
COUNTY EXTENSION WORK

MIAMI COUNTY POULTRYMEN ARE ASKED TO PRODUCE 12 % MORE EGGS IN 1942

Dear Poultrymen:

LET'S DO OUR PART. UNCLE SAM WANTS MORE EGGS AND WE CAN PRODUCE THEM.

Our Poultry School, which will be held Friday, Jan. 9th at the Denver Community Building, will help you obtain this increase.

PROGRAM

A.M.

10:00-10:45	Farm Practices Affecting Market Value of Eggs E.H.Matzen-Marketing Specialist
10:45-11:15	How I Manage and Feed my Flock-Local Poultryman
11:15-12:00	Exploded Ideas in Poultry Raising Joe W.Sicer-Poultry Specialist
12:00- 1:15	PITCH-IN DINNER

P.M.

1:15- 1:25	Report of Denver Egg Car- Cheston Long
1:25- 2:20	War and the Indiana Poultryman- E.H.Matzen
2:20- 3:10	Securing Economical Laying Rations-J.W.Sicer

COME AND BRING YOUR NEIGHBORS!

Very truly yours,

John E. Wylie
County Agricultural Agent
Miami County

Federal officials requested that Indiana produce an additional 147,000,000 dozen eggs in 1942 as part of a nationwide increase to meet the wartime demand. This January 1942 poultry school was aimed at helping Miami County producers raise their egg production by 12 percent as their contribution to the statewide goal. Courtesy of Purdue University Libraries, Karnes Archives and Special Collections.

Frank Blacker of Romney in Tippecanoe County pours fresh water into a fountain for his three-week-old White Plymouth Rock chicks in 1944. In addition to being asked to increase egg production, producers were assigned federal quotas on poultry meat as well, with much of that increased production going to support the war effort overseas. Courtesy of J. C. Allen and Son.

Farm Management School

TUESDAY ALL DAY FEB. 15, 1944

GRAND THEATRE IN AURORA

9:30 a. m. — 3:00 p. m.

A FINE HELPFUL PROGRAM

$25 - WAR BOND - $25

8 — OTHER VALUABLE PRIZES — 8

ALL FREE to Dearborn County Farm Families.

Today you stand face to face with this question: "What is the Best Program for MY Farm in 1944?"

SOME QUESTIONS THAT WILL BE COVERED IN THIS FARM MANAGEMENT SCHOOL:

1.—How long can I expect farming to continue profitable?
2.—Shall I reduce or increase my hogs, dairy, poultry, or sheep?
3.—What's the fertilizer, labor, and machinery situation?
4.—How can I keep my business soundly financed?
5.—What plans have some of my neighbors?
6.—What are some of the problems that will confront farmers in the post-war?

This program is built to help you make every acre, every machine, and every animal on your farm, do its dead level best in 1944.

This is YOUR invitation. — Plan NOW to come.

SPONSORED BY—

Dearborn County Agricultural Extension Committee
American State Bank of Lawrenceburg
Peoples National Bank of Lawrenceburg
Dillsboro State Bank of Dillsboro
Moores Hill State Bank of Moores Hill
First National Bank of Aurora
Aurora State Bank of Aurora

BUT............ READ ALL ABOUT THIS FINE PROGRAM AND THOSE VALUABLE PRIZES on the next page..........

—AND—

———BE ON TIME———

A farm management school held in Dearborn County outlined the agenda topics important to agriculture in early 1944. Throughout World War II, farmers would express concern about the potential for a postwar agricultural depression like the one that had followed World War I. Consequently, agricultural economists from Purdue remained in high demand at Extension events held during the war to discuss current and potential economic conditions. Courtesy of Purdue University Libraries, Karnes Archives and Special Collections.

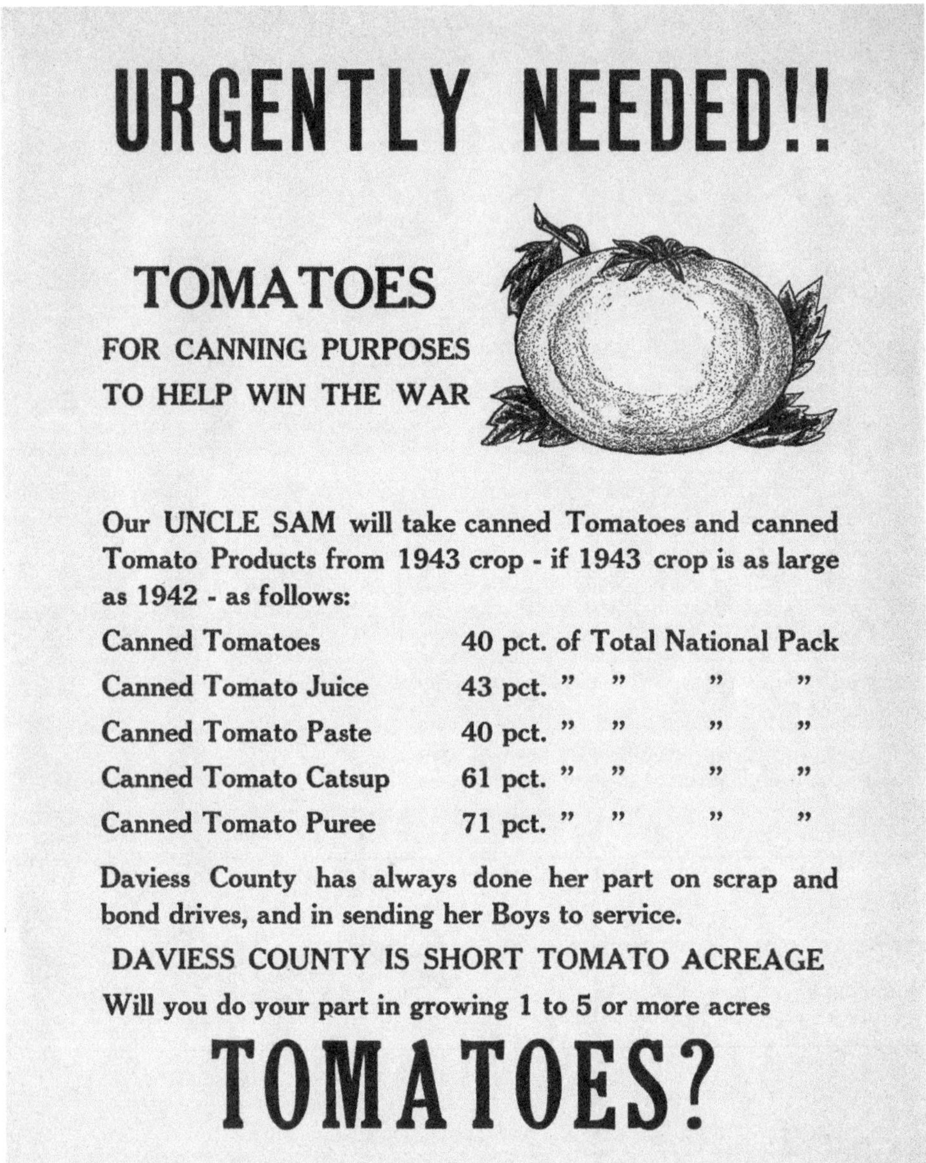

URGENTLY NEEDED!!

TOMATOES

FOR CANNING PURPOSES
TO HELP WIN THE WAR

Our UNCLE SAM will take canned Tomatoes and canned Tomato Products from 1943 crop - if 1943 crop is as large as 1942 - as follows:

Canned Tomatoes	40 pct. of Total National Pack
Canned Tomato Juice	43 pct. " " " "
Canned Tomato Paste	40 pct. " " " "
Canned Tomato Catsup	61 pct. " " " "
Canned Tomato Puree	71 pct. " " " "

Daviess County has always done her part on scrap and bond drives, and in sending her Boys to service.

DAVIESS COUNTY IS SHORT TOMATO ACREAGE

Will you do your part in growing 1 to 5 or more acres

TOMATOES?

Daviess County Extension agent John F. McKee distributed this flyer early in 1943 to encourage area farmers to grow more tomatoes to support the war effort. Tomatoes were a much-sought-after commodity by the federal government because they could be canned and shipped overseas. Courtesy of Purdue University Libraries, Karnes Archives and Special Collections.

A group of boys sets tomato plants in Clinton County in 1944. The work required to plant and harvest tomatoes meant growers had to use nontraditional labor sources to increase production levels for this essential war crop. Courtesy of J. C. Allen and Son.

Young men and women living at the Vigo County Home for Dependent Children—also known as the Glenn Home—raise tomatoes for the war effort in 1945. Courtesy of Purdue University Libraries, Karnes Archives and Special Collections.

Tomato processors were an important cog in the government's effort to secure the quantities of food required during World War II. Receiving, sorting, and processing tomatoes at a cannery should have been easy, but the lack of labor at times curtailed production. In order to keep the lines moving, companies looked toward women from nearby towns as a source of labor. Courtesy of J. C. Allen and Son.

CO-OPERATIVE EXTENSION WORK
IN AGRICULTURE AND HOME ECONOMICS
STATE OF INDIANA

PURDUE UNIVERSITY AGRICULTURAL EXTENSION
DEPARTMENT, U. S. DEPARTMENT OF AGRICULTURE
COUNTY BOARD OF EDUCATION AND COUNTY
EXTENSION COMMITTEES CO-OPERATING

March 24, 1942

EXTENSION SERVICE
COUNTY EXTENSION WORK

LET'S EAT
HOMEGROWN FOOD

FOOD FOR FREEDOM

Dear Friend:

We are calling a meeting of the committee on War Gardens, Thursday afternoon - March 26th, at 1:30 P.M., in the Assembly Room of the Court House.

At this meeting we will discuss and formulate plans for this needed work in our county.

We are selecting on this committee representatives of Home Economics Clubs, Farm Bureau, A.A.A., and local farmers and farm women.

We feel this meeting is highly important and if you cannot attend, will you please have some one there to represent you.

Very truly yours,

John E. Wylie

John E. Wylie
County Agricultural Agent
Miami County

JEW
md

In March 1942, Miami County agent John Wylie invited members of the county's home economics clubs, township farm bureaus, and other interested parties to attend a meeting about organizing war gardens, which were also known as victory gardens. During the war, growing more food in home and community gardens supported the nation's need to increase domestic food production. Courtesy of Purdue University Libraries, Karnes Archives and Special Collections.

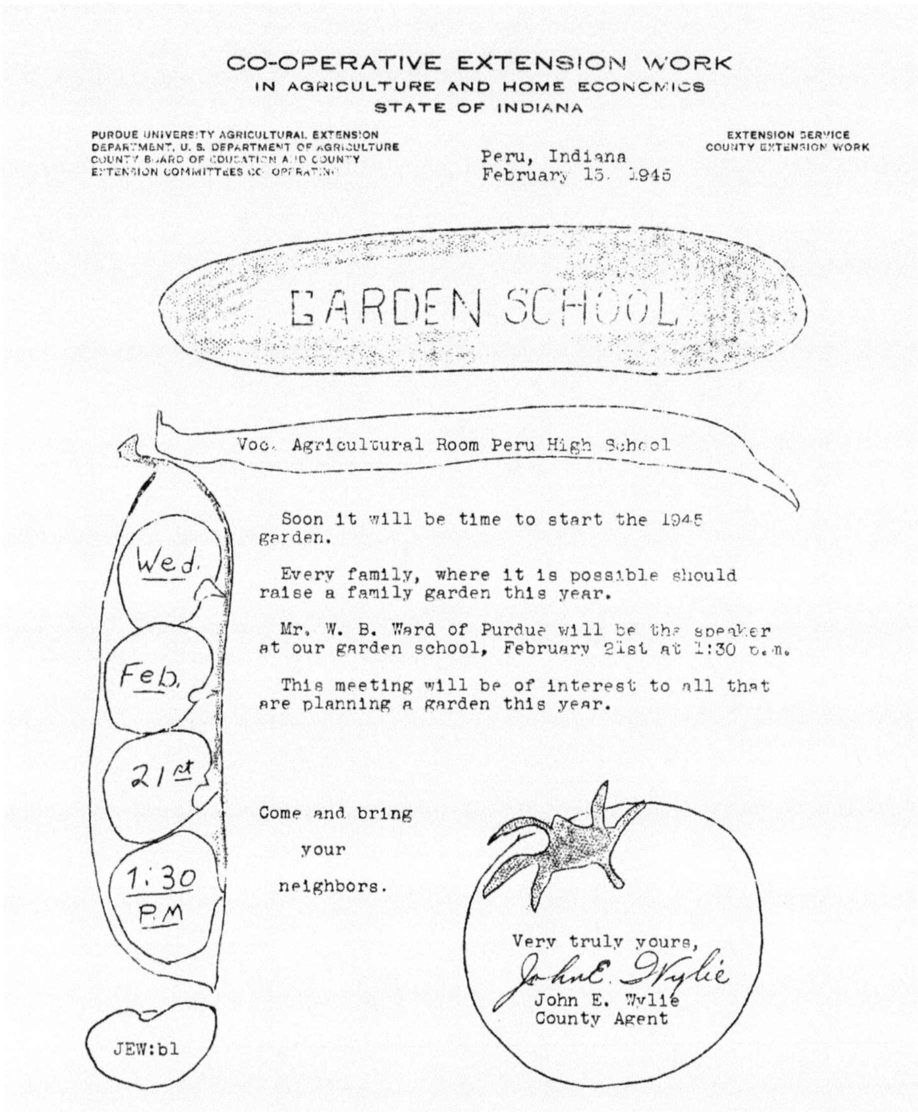

CO-OPERATIVE EXTENSION WORK
IN AGRICULTURE AND HOME ECONOMICS
STATE OF INDIANA

PURDUE UNIVERSITY AGRICULTURAL EXTENSION
DEPARTMENT, U. S. DEPARTMENT OF AGRICULTURE
COUNTY BOARD OF EDUCATION AND COUNTY
EXTENSION COMMITTEES CO-OPERATING

EXTENSION SERVICE
COUNTY EXTENSION WORK

Peru, Indiana
February 15. 1945

GARDEN SCHOOL

Voc. Agricultural Room Peru High School

Soon it will be time to start the 1945 garden.

Every family, where it is possible should raise a family garden this year.

Mr. W. B. Ward of Purdue will be the speaker at our garden school, February 21st at 1:30 p.m.

This meeting will be of interest to all that are planning a garden this year.

Wed. Feb. 21st 1:30 P.M.

Come and bring
your
neighbors.

Very truly yours,

John E. Wylie
John E. Wylie
County Agent

JEW:bl

Throughout the war years, Purdue horticultural Extension assistant Willard Ward was a frequent speaker at county gardening schools held around the state. Courtesy of Purdue University Libraries, Karnes Archives and Special Collections.

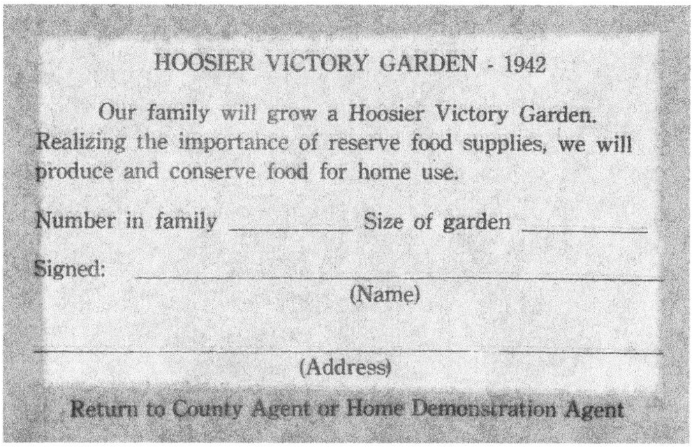

HOOSIER VICTORY GARDEN - 1942

Our family will grow a Hoosier Victory Garden. Realizing the importance of reserve food supplies, we will produce and conserve food for home use.

Number in family __________ Size of garden __________

Signed: ______________________________
(Name)

____
(Address)

Return to County Agent or Home Demonstration Agent

Individuals who planned to plant a Hoosier Victory Garden were asked to sign a pledge card (top) indicating the number of family members and the size of the garden they intended to plant. Each person who planted a victory garden received a window sticker (bottom) to show that they were helping the war effort by growing some of their own food supply. Courtesy of Purdue University Libraries, Karnes Archives and Special Collections.

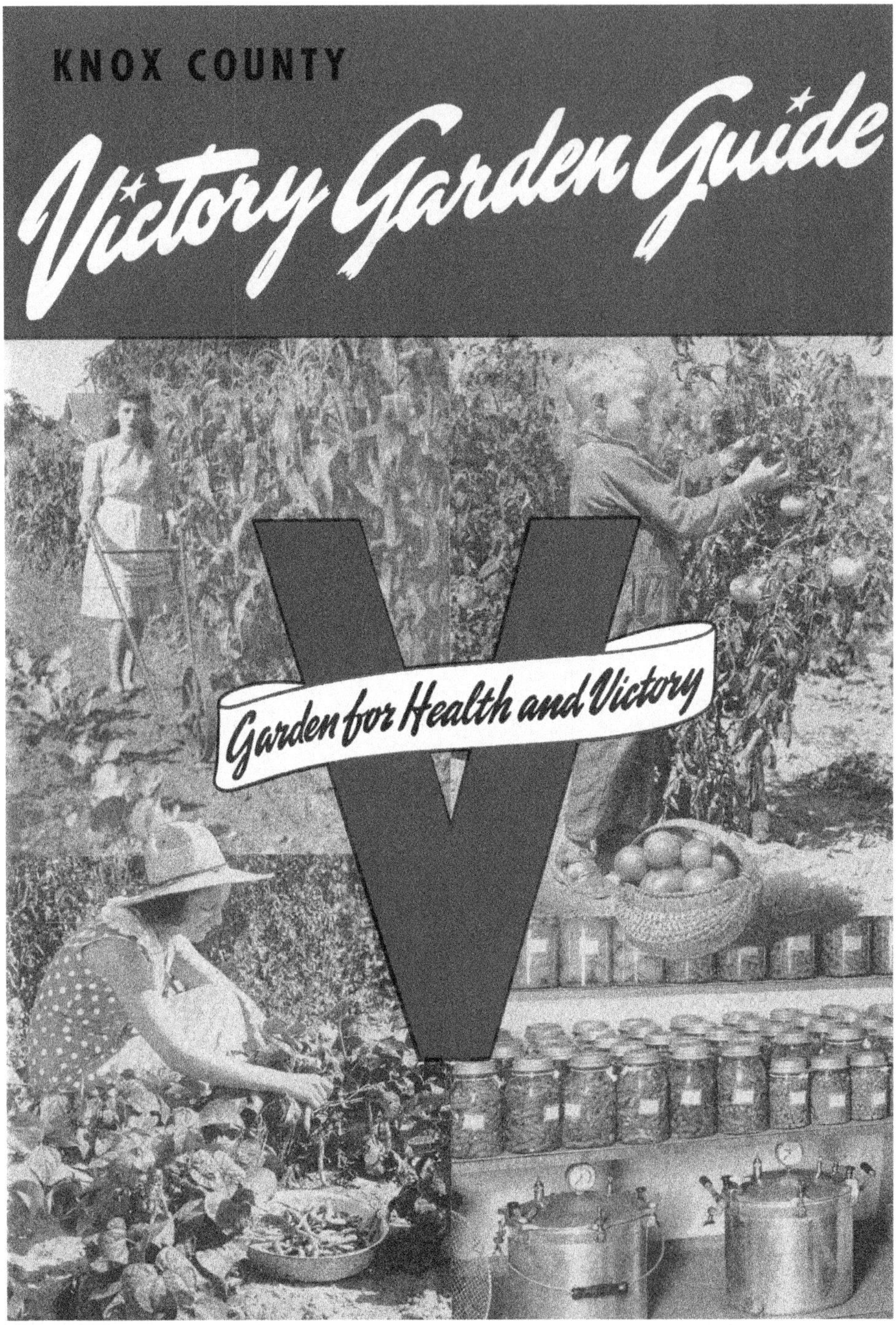

Those who planted victory gardens received a guide that explained how to lay out the garden, which seeds or plants to select, when to fertilize, and how to deal with insects, weeds, and diseases. This guide was important because for many participants, this would be their first garden. Courtesy of Purdue University Libraries, Karnes Archives and Special Collections.

Planting a victory garden, such as these people from Tippecanoe County were doing in this 1944 photo, was considered a patriotic endeavor during World War II. Often, the entire family was involved in planting and harvesting the crops, which provided fresh food for the evening meal and canned fruits and vegetables for winter use. Courtesy of J. C. Allen and Son.

A home economics Extension specialist from the Purdue campus (foreground) demonstrates canning techniques to leaders of Vigo County home economics clubs in 1943. These leaders would in turn teach what they had learned to members of the clubs they represented, thereby ensuring more people could benefit from the information provided. Courtesy of Purdue University Libraries, Karnes Archives and Special Collections.

Two members of the Otter Creek 4-H Club—Jack Miller (left) and Jack Ruszler, both from North Terre Haute in Vigo County—demonstrate how to preserve vegetables in 1944. Miller is showing the use of a food storage tile that Extension agents promoted during World War II. Courtesy of Purdue University Libraries, Karnes Archives and Special Collections.

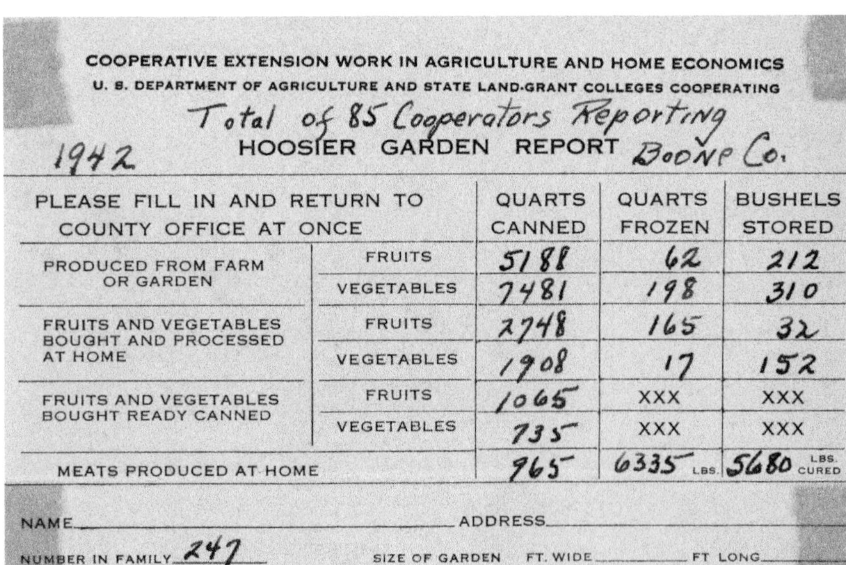

COOPERATIVE EXTENSION WORK IN AGRICULTURE AND HOME ECONOMICS

U. S. DEPARTMENT OF AGRICULTURE AND STATE LAND-GRANT COLLEGES COOPERATING

Total of 85 Cooperators Reporting

1942 HOOSIER GARDEN REPORT Boone Co.

PLEASE FILL IN AND RETURN TO COUNTY OFFICE AT ONCE		QUARTS CANNED	QUARTS FROZEN	BUSHELS STORED
PRODUCED FROM FARM OR GARDEN	FRUITS	5188	62	212
	VEGETABLES	7481	198	310
FRUITS AND VEGETABLES BOUGHT AND PROCESSED AT HOME	FRUITS	2748	165	32
	VEGETABLES	1908	17	152
FRUITS AND VEGETABLES BOUGHT READY CANNED	FRUITS	1065	XXX	XXX
	VEGETABLES	735	XXX	XXX
MEATS PRODUCED AT HOME		965	6335 LBS.	5680 LBS. CURED

NAME ______ ADDRESS ______

NUMBER IN FAMILY 247 SIZE OF GARDEN FT. WIDE ______ FT LONG ______

In 1942, Boone County agent Wilbur Whitehead prepared this garden report to show the amount of canned, frozen, or stored food items that were processed by 85 local cooperators. Extension agents collected this information in order to quantify for county and state records how much food had been raised by gardeners. Courtesy of Purdue University Libraries, Karnes Archives and Special Collections.

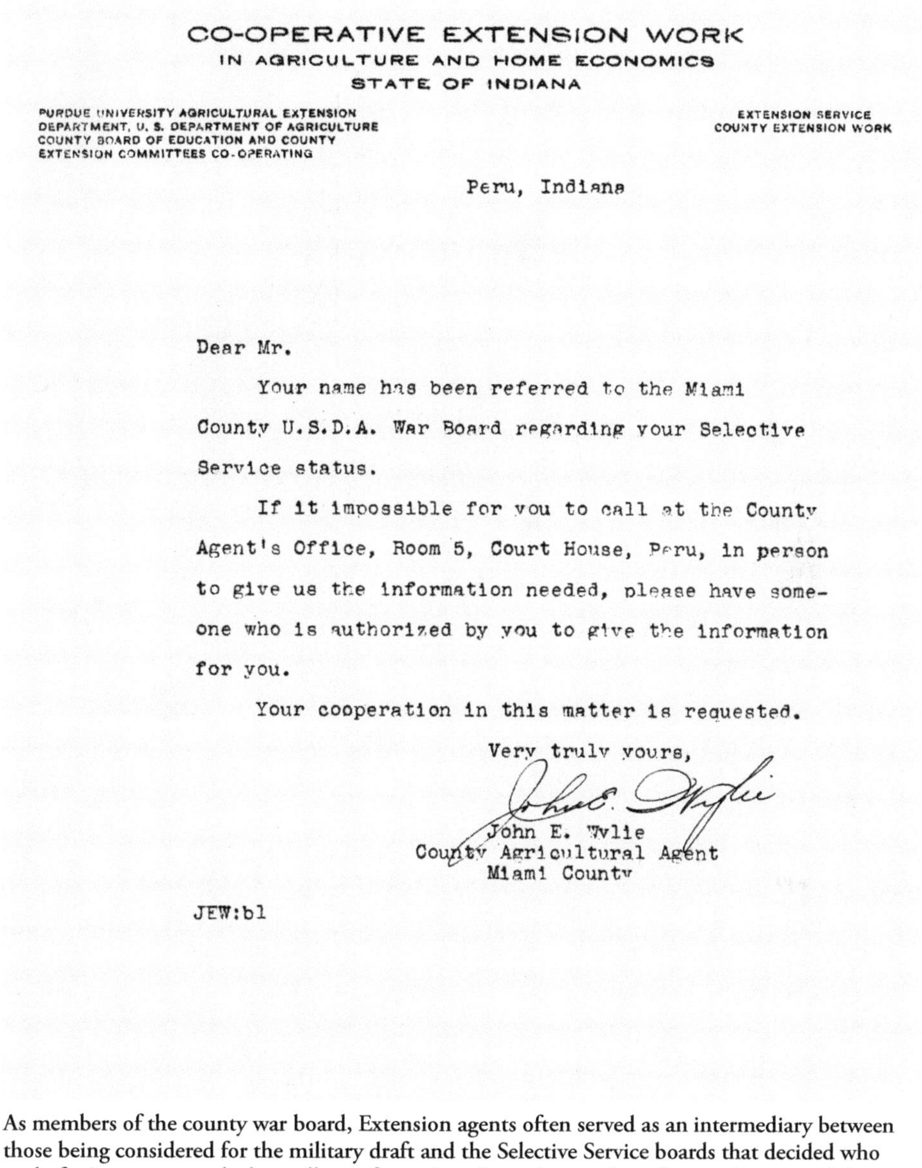

CO-OPERATIVE EXTENSION WORK
IN AGRICULTURE AND HOME ECONOMICS
STATE OF INDIANA

PURDUE UNIVERSITY AGRICULTURAL EXTENSION
DEPARTMENT, U. S. DEPARTMENT OF AGRICULTURE
COUNTY BOARD OF EDUCATION AND COUNTY
EXTENSION COMMITTEES CO-OPERATING

EXTENSION SERVICE
COUNTY EXTENSION WORK

Peru, Indiana

Dear Mr.

Your name has been referred to the Miami County U.S.D.A. War Board regarding your Selective Service status.

If it impossible for you to call at the County Agent's Office, Room 5, Court House, Peru, in person to give us the information needed, please have someone who is authorized by you to give the information for you.

Your cooperation in this matter is requested.

Very truly yours,

John E. Wylie
County Agricultural Agent
Miami County

JEW:bl

As members of the county war board, Extension agents often served as an intermediary between those being considered for the military draft and the Selective Service boards that decided who to draft. Agents were asked to collect information about the number of acres of crops and the number and type of livestock on farms where draft-eligible individuals worked. Since increased agricultural production was essential to the war effort, the main question agents had to determine was whether those individuals were more important to agriculture than they would be to the military. Courtesy of Purdue University Libraries, Karnes Archives and Special Collections.

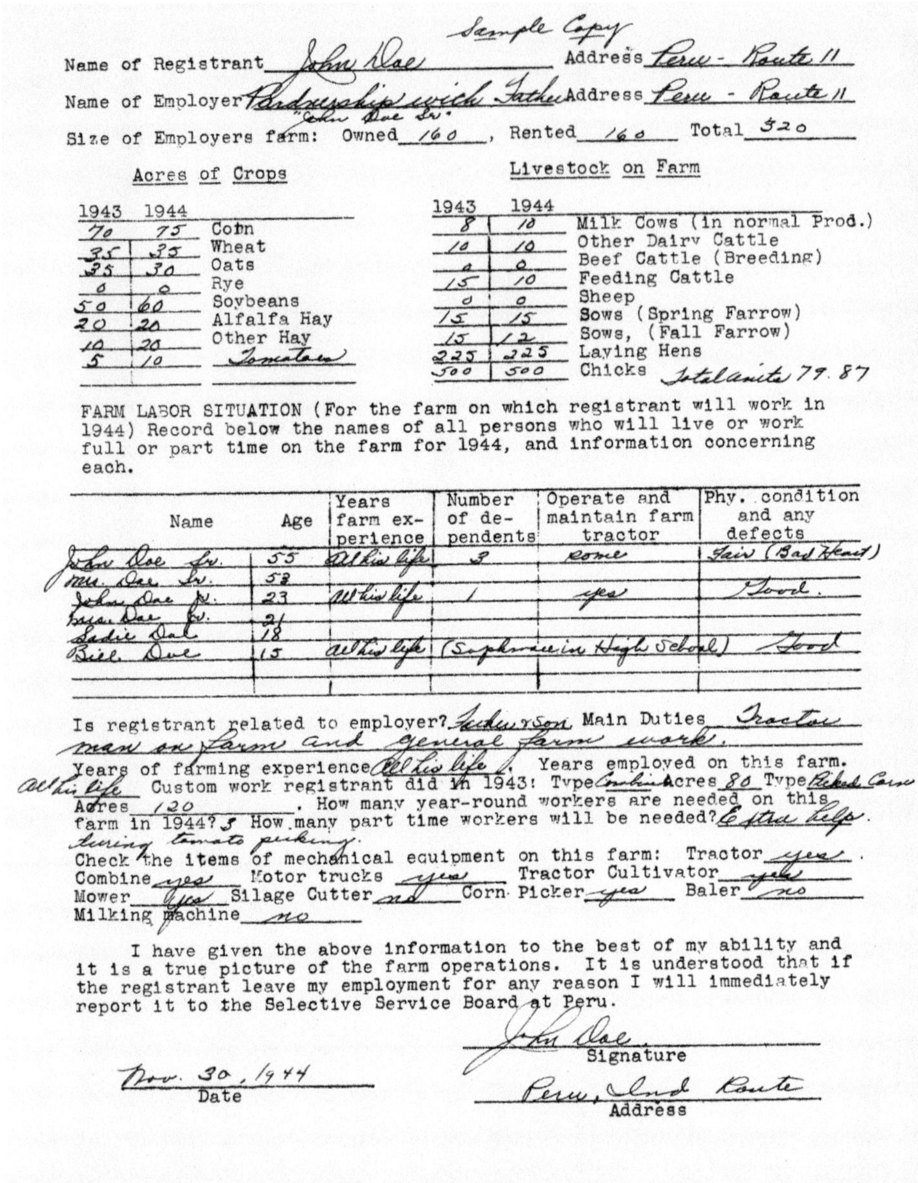

Sample Copy

Name of Registrant John Doe Address Peru - Route 11

Name of Employer Partnership with Father "John Doe Sr." Address Peru - Route 11

Size of Employers farm: Owned 160, Rented 160 Total 320

Acres of Crops

1943	1944	
70	75	Corn
35	35	Wheat
35	30	Oats
0	0	Rye
50	60	Soybeans
20	20	Alfalfa Hay
10	20	Other Hay
5	10	Tomatoes

Livestock on Farm

1943	1944	
8	10	Milk Cows (in normal Prod.)
10	10	Other Dairy Cattle
0	0	Beef Cattle (Breeding)
15	10	Feeding Cattle
0	0	Sheep
15	15	Sows (Spring Farrow)
15	12	Sows, (Fall Farrow)
225	225	Laying Hens
500	500	Chicks

Total units 79.87

FARM LABOR SITUATION (For the farm on which registrant will work in 1944) Record below the names of all persons who will live or work full or part time on the farm for 1944, and information concerning each.

Name	Age	Years farm experience	Number of dependents	Operate and maintain farm tractor	Phy. condition and any defects
John Doe Sr.	55	All his life	3	some	Fair (Bad Heart)
Mrs. Doe Sr.	53				
John Doe Jr.	23	All his life	1	yes	Good.
Mrs. Doe Jr.	21				
Sadie Doe	18				
Bill Doe	15	All his life	(Sophmore in High School)		Good

Is registrant related to employer? Father & Son Main Duties Tractor man on farm and general farm work.

Years of farming experience All his life. Years employed on this farm. All his life Custom work registrant did in 1943: Type Combine Acres 80 Type Picked Corn Acres 120. How many year-round workers are needed on this farm in 1944? 3 How many part time workers will be needed? Extra help during tomato picking.

Check the items of mechanical equipment on this farm: Tractor yes. Combine yes Motor trucks yes Tractor Cultivator yes Mower yes Silage Cutter no Corn Picker yes Baler no Milking machine no

I have given the above information to the best of my ability and it is a true picture of the farm operations. It is understood that if the registrant leave my employment for any reason I will immediately report it to the Selective Service Board at Peru.

John Doe
Signature

Nov. 30, 1944
Date

Peru, Ind Route
Address

This sample worksheet prepared by Miami County agent John Wylie in 1944 shows the type of information agents collected to decide how essential a draft-eligible individual was to a farm operation. The federal government set guidelines regarding how many workers could be retained based on the amount of essential agricultural products being produced for the war effort. Courtesy of Purdue University Libraries, Karnes Archives and Special Collections.

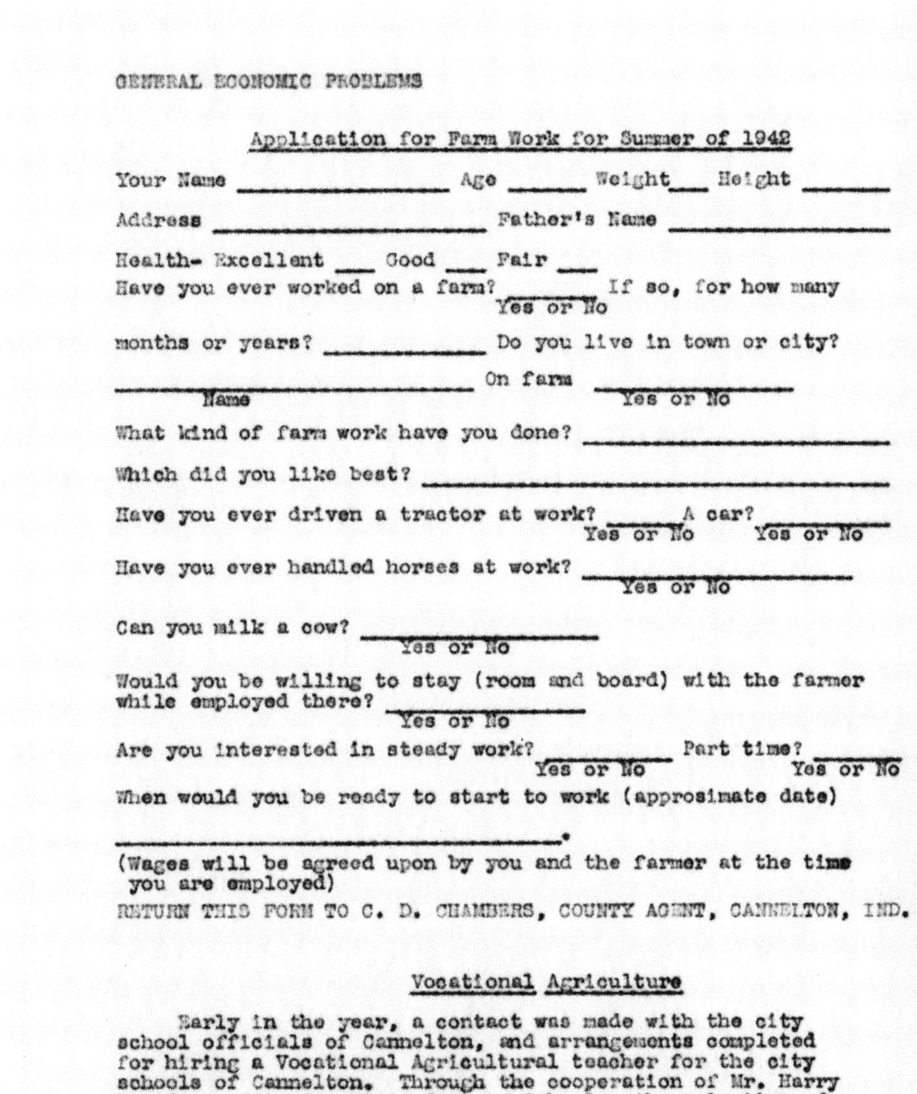
GENERAL ECONOMIC PROBLEMS

Application for Farm Work for Summer of 1942

Your Name ______________ Age _____ Weight___ Height ______

Address ________________ Father's Name ______________

Health- Excellent ___ Good ___ Fair ___

Have you ever worked on a farm? ______ If so, for how many
Yes or No

months or years? __________ Do you live in town or city?

_______________________ On farm ____________
Name Yes or No

What kind of farm work have you done? ______________

Which did you like best? ____________________

Have you ever driven a tractor at work? ______ A car? ______
Yes or No Yes or No

Have you ever handled horses at work? ______________
Yes or No

Can you milk a cow? ______________
Yes or No

Would you be willing to stay (room and board) with the farmer while employed there? __________
Yes or No

Are you interested in steady work? ______ Part time? ______
Yes or No Yes or No

When would you be ready to start to work (approsimate date)

______________________.

(Wages will be agreed upon by you and the farmer at the time you are employed)

RETURN THIS FORM TO C. D. CHAMBERS, COUNTY AGENT, CANNELTON, IND.

Vocational Agriculture

Early in the year, a contact was made with the city school officials of Cannelton, and arrangements completed for hiring a Vocational Agricultural teacher for the city schools of Cannelton. Through the cooperation of Mr. Harry Davidson, City Superintendent of Schools, the selection of Mr. Bruce Swinford was made, and he reported for work July 1.

As the agricultural labor pool shrank during the war, the federal government charged Extension with finding full- and part-time labor to help farmers meet federal production quotas. In 1942, Perry County agent Charles Chambers collected information from those seeking work, which included assessing any prior experience working on a farm. Courtesy of Purdue University Libraries, Karnes Archives and Special Collections.

FARM LABOR

PLACEMENT CENTER

"YOU CAN GET LABOR FROM THE TOWNSHIP POOL BY PUTTING LABOR INTO IT"

Here's a new war tool for your use

-- HOW TO DO IT --

Enlist at the most convenient placement center when you can spare a day or two to help some other farmer that's 'in-a-pinch' for help. The Center will record your name. The man 'in-a-pinch' will call the Center to learn who is available. He will then contact the most convenient farmer listed and arrange to exchange work with him. The volunteer will have a day or two's help coming to him later on in exchange, depending on the situation.

-- A PATRIOTIC CONTRIBUTION --

By enlisting at the placement center you are perfomring a patriotic act of first importance to food production. You are also building a labor supply for yourself which you can call upon when you need it. It is better than trying out an inexperienced green hand. The men exchanging work make their own deal.

-- LOCATION OF PLACEMENT CENTERS --

Your most convenient placement center is listed on the back of this sheet. Use it!

Build a labor account to draw upon later

United States Department of Agriculture
Purdue University Agricultural
Extension Department and Miami County
Extension Committee, Co-operating.

John E. Wylie
John E. Wylie
County Agricultural Agent
Miami County

In 1943, agent John Wylie advertised the Miami County Farm Labor Placement Center, noting that participation was "a patriotic act of first importance to food production." In this instance, Wylie encouraged area farmers to contribute time to help each other as one way to address the labor shortage. Courtesy of Purdue University Libraries, Karnes Archives and Special Collections.

Women work in a foundry for the Harrison Manufacturing Company in Fountain County in 1944. With the United States drafting so many men to serve during World War II, the industrial complex depended on labor provided in part by women to produce the equipment and supplies required by the military. Courtesy of J. C. Allen and Son.

An unidentified woman operates a tractor in Lake County in 1944. In addition to working in factories, women also provided a significant portion of the required farm labor during the war. Courtesy of J. C. Allen and Son.

It was often the case that women such as this one pictured in Marion County in 1942 had to bring their children to the field when working as agricultural laborers. Courtesy of Purdue University Libraries, Karnes Archives and Special Collections.

Workers from Gary were brought in to pick tomatoes in rural Lake County in 1936. Each morning and night, men and women were transported between the city and the fields, which amounted to approximately an 80-mile round trip each day. Courtesy of Purdue University Libraries, Karnes Archives and Special Collections.

Eleven Jamaicans arrive for work at the Glenwood Canning Company in Rush County in May 1945. Foreign and domestic migrant workers provided an important source of labor during the war along with others such as prisoners of war. Courtesy of Purdue University Libraries, Karnes Archives and Special Collections.

Memorandum of Understanding for Employment of Boys and Girls by Truck Growers in 1945.

The following rules were agreed upon by Lake County tfuck growers at their meeting in Munster Town Hall on May 16, 1945:

1. The minimum wage shall be 25¢ per hour.
2. There is to be a 15-minute rest period each morning a6 the expense of the employer.
3. There is to be an adequate lunch period of at least 30 minutes.
4. No boy or girl is to work more than 8 hours, exclusive of lunch period, unless he or she so desires.
5. Transportation is to be furnished at the expense of the farmer, providing the worker stays until the end of the day.
6. A weekly pay day is to be set by the employer and adhered to.

Truck growers are requested to post a copy of these rules where they can be seen. Violations should be reported to the County Agent's Office in Crown Point in writing.

Lloyd E. Cutler
County Agent

Older schoolchildren were hired to provide limited agricultural labor—mostly to plant or harvest crops—to supplement the limited number of adult workers who were available for completing the work. In 1945, Lake County agent Lloyd Cutler worked with local farmers to establish requirements for the working conditions and pay for these boys and girls. This was part of a statewide system to ensure that children were paid the prevailing wage rate for agricultural labor during the war. Courtesy of Purdue University Libraries, Karnes Archives and Special Collections.

In 1943, a farmer meets a busload of boys from a Marion County city school as they arrive to work in the farm fields. Courtesy of Purdue University Libraries, Karnes Archives and Special Collections.

These Marion County boys, who were too young for heavier farm jobs, harvested green beans in 1943. Courtesy of Purdue University Libraries, Karnes Archives and Special Collections.

Boys from Marion County Public School Number 8 and Number 41 harvested 380 hampers of beans in six hours in 1944. Courtesy of Purdue University Libraries, Karnes Archives and Special Collections.

This group of Clinton County boys, ages 10–14, pulled and packed 80,000 tomato plants in a single day in 1943. Courtesy of Purdue University Libraries, Karnes Archives and Special Collections.

Boys from Marion County learn how to properly detassel corn in 1943. With the wide adoption of hybrid varieties, companies began to plant more acres of corn that required detasseling. This was an effort to produce the seed that farmers wanted and needed to increase crop production for wartime demands. Courtesy of Purdue University Libraries, Karnes Archives and Special Collections.

Boys from a camp for student workers in Kentland pose for a photo prior to starting their work to detassel corn in 1944. Students were especially in demand by hybrid seed corn companies, which turned to Extension to find the needed detasselers along with bus drivers and field supervisors. One agent reported in 1943 that this camp operated by Edward J. Funk and Sons at the Newton County Fairgrounds was equipped with tents to house and feed 500 boys who came from as many as twenty-one different counties. Courtesy of Purdue University Libraries, Karnes Archives and Special Collections.

These workers came from Dubois County in southern Indiana to detassel corn at Kentland in Newton County in 1946. Courtesy of Purdue University Libraries, Karnes Archives and Special Collections.

This photo shows a 4-H club pin (left) and a 4-H Victory Club pin (right) that were awarded during World War II. Members of 4-H were encouraged to contribute to the war effort through a variety of methods, which included activities such as planting and tending a victory garden and participating in community drives to collect scrap metal and sell war bonds. Courtesy of Becky Holbert.

In Newton County, a crew of women poses for a picture in 1943 while detasseling corn for Edward J. Funk and Sons in Kentland. Courtesy of J. C. Allen and Son.

These members of the Frankfort Lions Club in Clinton County spent one morning detasseling 36 acres of corn for a commercial seed company in 1943. The ongoing agricultural labor shortage prompted county agents to recruit volunteers for short-term agricultural assignments from service clubs, labor unions, and local businesses. These men were paid $60 for their work, which they then donated back to their club. Courtesy of Purdue University Libraries, Karnes Archives and Special Collections.

CHAPTER 35

Explaining the Red Tape That Farmers Must Go Through to Get Electricity

During the 1930s, approximately 90 percent of urban populations in the United States had electricity, while only 10 percent of rural people had it. Utilities were unwilling to spend money providing electrical service to rural areas due to the high cost of building the infrastructure needed to deliver the power to a small customer base.

By the end of 1942, seven out of ten Indiana farms were powered with electricity. Many of those had only received access within the preceding few years.[1] Farmers appreciated the fact that, as one agent observed, electricity "provided light at both ends of the day, which at certain times of the year, meant longer hours for the farmer if he was to get his work done."[2]

In addition to allowing longer work days, electrical power also meant that farmers could use electrical equipment such as hog heaters, water pumps, and feed grinders. This allowed them to increase efficiency in the face of the wartime labor shortages and still meet the ever-increasing production quotas set by the federal government (see Chapter 27). For instance, with the demand for pork products at an all-time high in 1942, the use of electric pig hovers saved 20 percent more young pigs at farrowing.[3] Using electric milking machines and milking coolers allowed farmers to expand their dairy herds, which meant the military would have more of the milk powder it needed. And with the farmer's time, tires, and fuel in short supply, Orange County agent Ross Atkinson observed in his 1942 annual report that electric "grinders are paying for themselves two to four times a year in savings in hauling feed to commercial grinders and grinding bills."[4] Electrical power had clearly revolutionized how farmers operated in Indiana.

Farmers who had not previously signed up with the local rural electric membership corporation (REMC) were now envious of what their neighbors could get done with electrical equipment, especially as farmers were asked to produce more with less available labor to do so. As Pulaski County agent Carl Olson noted in his 1942 annual report, "The spread of these REMC lines in the county have been of great value to the farmers, especially to those farmers who because of this are able to purchase power equipment, which makes it invaluable under the present labor shortage."[5] By 1944, as the labor situation only continued to worsen, Knox County agent Henry Benson would write, "With most farms suffering from a serious labor shortage, electric power has suddenly become a necessity and it is extremely important that unserviced farms be connected to the rural power lines with the least possible delay."[6]

The reason agents such as Benson still needed to advocate to have farms connected to electrical power well into the war years is because the federal government had put a hold on expanding rural electric lines early in the war. In fact, many farmers who had been ready to sign up with an electric cooperative were surprised to learn that only those farmers who met specific criteria could qualify to receive electricity on their farms.[7] The main reason that construction of rural lines had been stopped was that the federal government had begun rationing the required copper conductors, transformers, meters, steel, and aluminum.[8] These items—especially copper—were being redirected to the military, which in turn created a public shortage.[9]

The percentage of farms that already had electricity prior to the rationing varied among counties, but one common denominator was how aggressive the county REMC and county Extension agents had been in getting people to agree to become members of the electric cooperative during the previous decade. In places where enough people had agreed to join, the electric poles and lines had been built into the rural parts of the county. So in counties such as Johnson, Scott, and Whitley—where the local REMCs and Extension agents had actively worked to sign up new members during the 1930s—most farms had already been hooked up to the power grid by World War II.[10] But in counties such as Pulaski and Vigo—where REMC recruitment efforts hadn't been as successful—less than half of the farms had electricity and now had little hope of receiving power until after the war ended.[11]

Farmers were demanding power, but many of these demands would go unanswered for the duration of the war.[12]

During the 1930s, Extension—and agents, in particular—had provided significant assistance in the creation of REMCs across the state (see Chapter 13). So agents were understandably perturbed that the federal government had imposed restrictions on constructing new power lines without notifying them in advance. Agents were also put into the difficult position of explaining the evolving requirements to farmers as the government continually changed the rules regarding who was eligible. In 1942, Knox County agent Henry Benson explained the problem he faced with the ever-changing policies and regulations, saying, "Restrictions were first limited to extensions not to exceed one-thousand feet and not over five-hundred dollars in value, then reduced successively to five-hundred feet, two-hundred fifty feet and finally prohibited altogether. I am happy to say that, with rare exceptions, members desiring service accept the situation with good grace and are content to await the end of the war for this service."[13]

The construction hold remained in place until January 1943, when the government informed the REMCs that they could build short extensions to reach agricultural operations, making a limited number of farmers eligible to receive electrical service. As one agent explained, if farms were located "near an existing line and if the farmers could get the materials to wire his premises, they have been given the service by the REMC, but where they are not located near an existing line, they will not be served until after the war."[14]

Before a power line to a farm could be established, there were a handful of stipulations that the farmer had to meet first, which in 1943 included:

1. No connection would be provided if the electricity was to be used only for convenience, such as lighting the house;
2. The farmer had to own or be able to purchase specific farm equipment, including water pumps for livestock, milking machines and coolers, incubators, brooders, feed grinders, etc.;
3. The county war board would have to verify that the electricity would substantially increase the farm's productivity or provide a savings in farm labor; and
4. The farm had to be large enough to meet the minimum animal units.[15]

According to government regulations, an animal unit consisted of one milk cow; 10 beef cattle; 30 breeding ewes; three brood sows; 75 laying hens; 40 turkeys or geese; 600 broilers; or an equivalent combination of the required number of animals.[16] In his 1943 annual report for Starke County, agent Arthur Shideler further explained how animal units were used to qualify a farm for an electrical extension, writing, "A number of new electric connections were approved by the War Board where the animal units were sufficient to qualify. One animal unit was necessary for each 100 feet of line required with a 10 unit minimum. Later in the year, 5 units was the minimum for a connection."[17]

A member of the county war board—often the agent—would visit the farm where additional electrical service was desired to determine how the added electricity would increase production.[18] Typically, it was also the agent who explained to farmers what paperwork was required and the criteria that the war board would use to approve or reject a request to extend electrical lines. Switzerland County agent Oran McNary reported on his efforts in 1943, saying, "Electricity is in big demand by several farmers who are operating dairy farms and who have not as yet been hooked on to a power line. Some time has been spent in visiting some of the people and helping explain the necessary red tape that they must go through with if they get electricity."[19]

However, counties still had to contend with copper being in short supply. By August 1943, the federal government set quotas for how much copper wiring a county could receive.[20] For example, that year the Pike County allotment was 455 pounds of copper, which was to be used exclusively for farm electrical extensions.[21]

As with most other rationed products during World War II, the demand for copper to extend electrical connections far exceeded the supply that the county received, so farmers wanting to hook up to the power grid or to expand electricity on their farms had to compete with others wanting to do the same. Each farmer had to submit an application to members of the county war board, who would then pass judgment on the request.[22] This meant that the war board would assign part of the county's copper quota to each successful applicant only until the rationed supply had been exhausted.

In 1943, Henry Benson talked about the difficulties the Knox County REMC was having even completing short extensions from existing lines onto farm properties, writing:

> For a period of time it was impossible to secure an adequate supply of materials and all construction work was "frozen." During the late spring, steel wire was authorized for primary use and the REMC [was authorized] to begin construction on a modest scale of a number of extension[s] authorized by [the] local County War Board, and where the service would be used for food production activities. About 200 of these approved applications have been received, and there still is about 35 miles of line to construct. This is quite a task when it is remembered that they are attempting to build these lines with a crew of men only large enough for their normal maintenance work. Heretofore, new construction was done by contractors, but these contractors are no longer available. There has been considerable delay in completing these extensions, even after the necessary materials [were] secured.
>
> The REMC is completely out of these materials now, and it may be another three to four months before they can secure sufficient new materials to permit them to proceed with the work.[23]

Once the supply of materials ran out, then even those with approved applications would have to wait until more materials were allocated and shipped to the county. By 1944, Benson noted that the shortage of materials was having an impact on even those farms that could show increased production with electricity. That year he reported:

> Until recently, consideration could only be given those applications where electricity would definitely be used for farm productive efforts, but even in the most urgent cases, there have been some applications for extensions

> which could not be completed because of the inability to secure the necessary materials. . . . [C]opper conductor[s], transformers and meters are more difficult to secure now than at any time since war started. At the present time, factory deliveries are promised on conductors for some time in February of 1945 and there is no assurance as to when the transformer situation will be eased.[24]

By the end of 1944, 30 percent of Indiana farmers still didn't have electricity.[25] But by 1945, the limitations on connecting to nearby electrical lines or expanding existing service on farms had largely been eliminated as the government began rescinding many of its mandates on electrical equipment and materials.[26] With the availability of copper wire, meters, and transformers no longer a limiting factor to expansion, the government provided REMCs with additional funds to add mile after mile of poles, stretching electrical lines further into rural areas in order to bring electricity to the majority of Hoosiers. By 1950, 91.5 percent of farms in Indiana had electrical service.[27]

World War II had shown how important electricity was to the productivity and profits of the farming community. While many farmers had initially been hesitant to sign up for electrical service due to concerns about their monthly electric bill, the time saved by using electricity more than offset what it cost them each month. As more farms received electrical service and added electrical equipment, county agents would continue to promote the educational programs offered by Extension on how to best take advantage of electricity, both on the farm and within the home.

While many connections to the electric grid had largely been put on hold during World War II, the pent-up demand for power created a golden era for the REMCs during the postwar years. But that rapid expansion belied how long rural electrification had actually taken in Indiana. In 1880, Wabash, Indiana, had become the first electrically lighted city when electrically powered lamps were installed in the Wabash County Courthouse. On July 22, 1935, the Boone County Rural Electric Membership Corporation had become the first federally sponsored electric project in Indiana. But it would be 1965 before almost all Hoosier farms would have electricity, leading historian Ethan Chitty to write, "[I]t took nearly eight decades of sustained effort for most rural Hoosiers to gain access to one of the utilities that we so often take for granted today."[28]

CHAPTER 36

Machinery Will Need to Be Replaced Before the Supply of Baling Wire Is Exhausted

Following World War I, most farmers continued to use horses to do the majority of their farm work because their operations were too small to justify purchasing tractors or other labor-saving devices.[1] In 1920, manufacturers sold only 20,000 tractors, which represented use on approximately 3.6 percent of U.S. farms.[2] From 1925 to 1930, the number of tractors on farms rose by nearly 750,000 units nationwide. In 1930, manufacturers sold 200,000 tractors, but just two years later, only 19,000 tractors were sold, showing the negative impact of the Great Depression on sales.

As the economy improved and sales rebounded, farmers renewed their interest in acquiring equipment, and manufacturers would build 160,000 new tractors by 1935.[3] By that year, there were more than 1 million tractors in use, representing approximately 20 percent of U.S. farms.[4] Indiana seemed to follow this national trend, with only 12 percent of Indiana farms having tractors in 1925, and 32 percent having them a decade later.[5]

World War II would continue the push toward a greater adoption and acceptance of mechanized equipment, primarily for two reasons. First, using horse-drawn implements required more labor, but those workers whom farmers had taken for granted were now hard to come by, having been absorbed by the military draft and jobs in manufacturing plants.[6] As one agent noted, the lack "of farm help makes it necessary to use more machinery"—an opinion shared by county Extension agents throughout Indiana.[7] One person using a tractor could now pull larger implements that outworked what it took many men using multiple horses to accomplish in a day.[8]

The second factor pushing the conversion to mechanized equipment was that farms were becoming larger, which meant that, with the better crop and livestock prices brought on by the war, purchasing equipment made sense economically.[9] The increase in mechanization would also allow farmers to operate more efficiently.[10]

Annual reports by agents documented the rapid adoption of mechanized equipment in Indiana. For example, Scott County agent John McCulloch reported in 1944 that mechanized pickers were used to harvest about a quarter of the corn acres as compared to 2 percent just four years earlier.[11] Early in the war, many farmers were ready to convert to tractors that could pull larger implements, especially on the more productive farms.[12] But those who now had the need and the economic means to obtain the equipment would soon encounter other problems.

In some cases, farmers had the equipment but couldn't use it. In the fall of 1942, Warren County agent Herman Madaus indicated that farmers were feeling "a shortage of trained help required on most farms for operating power machinery."[13] That same year, Lake County agent Lloyd Cutler reported:

> Some difficulty has been experienced in the fall of 1942 in getting crops harvested. Shortage of labor has usually been the main factor. In many cases, however, combines and corn pickers have not been available for harvesting the soy bean and corn crops. In some cases the owners of harvesting machinery needed within the community have let their machines stand in the shed because competent help was not available to run it, or because they were afraid they would not be able to get necessary repairs in case of breakage.[14]

But the bigger issue facing farmers was the reduced supply of the equipment they so desperately needed to increase their output. Converting factories to produce military hardware and other supplies had limited the availability of raw materials such as rubber, steel, and iron needed to build agricultural equipment.[15] Consequently, there just wasn't enough equipment to meet the ramped-up demand.

Agents bemoaned the shortage of farm equipment at a time when the U.S. Department of Agriculture (USDA) was pressuring farmers to increase production. To many in agriculture, the government's policies seemed to be counterproductive.[16] As Kosciusko County agent Joseph Clark pointed out in his 1942 annual report, "It seems foolish to waste time in driving two horses when the farmer and everybody knows he could produce from two to three times as much with power equipment. It also seems foolish to milk eight cows by hand when the farmer could and would milk twice that many if he could secure a milking machine. Power equipment needs some sincere consideration in our 'All Out For War' program."[17]

A 1943 Jay County survey confirmed to agent Chester Langston that the lack of machinery was a serious issue for local farmers. That year, he wrote, "Approximately one farmer out of 5 reported that crop production could not be maintained in 1943 with present machinery. The need for new or additional machinery is more evident on the larger farms where 1 operator out of 4 reported that their machinery was inadequate for maintaining production. New or additional machinery listed as necessary in order of needs were: combines, tractors, grain drills, corn planters, disc harrows, manure spreaders, and corn pickers."[18]

At the same time, farmers were also dealing with an aging fleet of farm trucks that couldn't be replaced since fewer were being built during the war. This posed challenges for delivering agricultural goods to market. In Marion County,

Horace Abbott observed in 1942 that "growers have reduced their number of trips to market, and are pooling their loads in order to conserve their tires and trucks."[19] That same year, Wabash County agent William Delaplane expressed concern that it was becoming increasingly difficult to move corn and livestock from the farm to the market because fewer trucks were available.[20] The trucks that were available were often driven to a much higher mileage, which made repairs an issue.[21] In 1944, one agent reported that two truckers indicated their vehicles had been out of operation for four weeks waiting on repair parts.[22]

Simply put, farmers needed equipment and wanted to buy it, but the supplies just weren't available.[23] Those who ordered new equipment in the winter of 1941 still hadn't received it by the following spring.[24] Of course, this often curtailed the ability of farmers to expand their operations and limited their production.

The federal government responded by enacting a rationing program for critical farm machinery and electrical equipment on September 17, 1942.[25] Rationed equipment was placed into two categories known as Group A and Group B. Group A included equipment such as combines, tractors, corn pickers, disc harrows, feed grinders, fertilizer spreaders, milk coolers, milking machines, pickup balers, potato diggers, and shredders. Equipment not listed in Group A was by default placed on the Group B list. Farmers would now need to apply to the county farm machinery rationing board to obtain a purchase certificate in order to buy any Group A equipment and any Group B equipment that cost more than twenty-five dollars.[26]

The USDA, which oversaw the program, set quotas at the federal level, which states then administered through the local war boards. As was often the case with rationing programs, the amount of equipment that was actually available for farmers to purchase ended up being much less than what rationing boards had been told they would receive. In Pike County, the quota for 1943 was 23 percent less than what had been expected, and agent Edward Bogue stated, "In the case of tractors, combines, and corn pickers, the quota for the county was less than 10% of the number sold under the 1940–1941 average."[27] This showed how limited new equipment had become, leading St. Joseph County agent Eugene Bird to observe the following year, "Production could have been even more abundant if the man-power and machinery curbs had not been so severely invoked."[28]

It's no wonder that agents such as Charles Chambers in Perry County were outspoken and a little more than perturbed. In his 1943 report, he wrote, "In looking back over the rules and regulations handed down, it seems that most of the rules didn't fit. There has been more complaining and dissatisfaction about the rationing of farm machinery among farmers, than about all the other programs combined."[29] Farmers didn't understand how one farmer was selected to receive equipment over the next. Without a clear understanding of the rationing procedures, farmers who had the money but were prevented from purchasing equipment directed their wrath toward the agents.

As one agent noted, the challenge was "in making a fair distribution of this small amount of machinery in face of the large demand."[30] For example, in 1943 Knox County only had nine corn pickers to distribute among the ninety applicants who desired them.[31] To address the situation, county war boards relied on the local farm machinery rationing boards to more fairly distribute the small allotments provided.[32]

Agents met with farm machinery dealers to discuss the process by which equipment in inventory and newly arriving equipment would be distributed and sold.[33] Each farmer submitted an application for machinery to the local rationing board and, in return, received a preference rating based on factors such as farm size and past production records.[34] Those who had served in the military received a Veteran's Preference Certificate, which put them at the head of the line.[35]

Typically, a portion of the requests had to be denied, with only a small portion of applicants approved for the more sought-after items. But those who were lucky enough to be approved would receive a purchase certificate to authorize the transaction.[36] In a 1943 annual report, agent James Guthrie provided the following details on the applications received for equipment in Union County:

> During the year, a total of 620 applications for rationed equipment were received with 476 purchase certificates being issued. Eight of the most important farm implements that were rationed in the county, the number of applications received, and purchase certificates issued were as follows:

Implement Type	Number of Applications Received	Number of Purchase Certificates Issued
Corn Pickers	36	6
Combines	10	9
Tractors	40	21
Tractor Discs	20	17
Tractor Cultivators	36	28
Hay Loaders	10	4
Mowers	20	10
Grain Elevators	10	5[37]

Other items for which purchase certificates were issued included hammer mills, electric pumps, side delivery rakes, milking machines, cream separators, manure spreaders, power shellers, breaking plows, corn shredders, grain drills, gas engines, hay balers, wagons, corn planters, rotary hoes, buck rakes, cultipackers, and pressure cookers.

In a 1944 report, agent Edward Bogue explained how the farm machinery rationing board in Pike County used the purchase certificates to get farmers to share the limited equipment, writing:

> The Farm Machinery Rationing Board of the County . . . held regular weekly meetings each Saturday afternoon during the spring and summer and every other week during the fall. The agent [Bogue] served in an advisory capacity with the board. The most difficult problem handled by the board was the rationing of seven corn pickers allocated to the county to 34 applicants, the majority of which were eligible from the standpoint of acreage. This problem was settled by the decision of the board and the Agent to call in all applicants for a meeting to pool the applications according to neighborhoods. Applicants were grouped to enable 2 to 5 farmers to either purchase a picker together, or with one applicant deciding to purchase one and agreeing to harvest the other neighbors' crops. This plan worked exceedingly well and could be recommended to other counties with a problem of similar nature.[38]

The limited supply of new equipment meant that selling used machinery brought much higher prices than normal. Equipment less than five years old could now fetch more money on the used market than what it had cost new.[39] In 1944, Randolph County agent Lewis Husted reported that he had tried to work out procedures with local dealers to redistribute used equipment to those with the greatest need rather than just making the easiest sale.[40]

Naturally, farm sales grew in popularity, drawing hundreds to thousands of farmers to bid on used equipment.[41] At a large sale held in Wayne County in 1943, attendees bid on one- and two-row corn cultivators, double discs, walking plows, sulky plows, one-horse wheat drills, combines, binders, mowers, and hand tools. The auction, which included 124 pieces of used equipment, netted $2,510.[42] Those who purchased used farm machinery and equipment often paid the sellers using war bonds and defense stamps.[43]

By the spring of 1944, the demand for farm equipment remained strong, but the need for producing more landing barges for the military was affecting the production. By April 1, machinery production nationwide included combines, 36,000; tractors, 121,000; corn pickers, 15,000; spreaders, 44,000; side racks, 23,000; and balers, 9,000.[44] These were larger inventories that provided some relief to the ration boards and the farmers who wanted the equipment.[45] With the exception of corn pickers, the rationing for farm machinery ended on September 1, 1944.[46]

While rationing was halted by the federal government in the fall of 1944, it would take many years for the supply to meet the pent-up demands by agriculture.[47] In 1945, Guthrie reported, "Machinery on most of the farms in Noble County is pretty badly in need of replacement. Farmers are using in many cases what they had during the 1932 depression, and as a result it is in very bad shape."[48] A year later, Sullivan County agent J. Howard Telfer reported, "Much of the machinery will need to be replaced as soon as new equipment is available and 'before the supply of baling wire is exhausted.'"[49]

During the years when new equipment was in short supply, Extension worked to encourage farmers to take care of their existing machinery and make necessary repairs sooner.[50] For example, at the start of the war, Tippecanoe County had $3 million worth of farming equipment, and agent Sheldon Pershing observed that "to keep it rolling without major replacements or repairs for the duration was the job which faced farmers in 1942."[51] That year in Jackson County, agent Willis Stall echoed Pershing's sentiments, writing, "[S]upplies for the Nation's heavily mechanized armed forces are taking huge quantities of metals and rubber that formerly were available for agricultural and industrial uses, and that it is most urgent for every farmer to prolong the life of his machinery by careful operation, adjustment and repair."[52]

Extension became focused on encouraging farmers to keep their existing equipment running through better maintenance and operation, but agents also knew there was the distinct possibility that repair parts would soon be harder to get. LaPorte County agent Everett Hartman reported, "One of the major problems faced by local farmers during 1942 was the scarcity of new farm machinery to be used as replacements. Anticipating the possible delays in securing repairs and replacements, every possible effort was made to bring the situation to the attention of both the farmers and the machinery dealers."[53] Hartman's expectation that parts would soon be in short supply would become a reality by 1943 and remained a problem for the duration of the war.[54]

In January 1942, agents brought together machinery dealers, blacksmiths, and repair shops to inventory the expected needs for repairs and parts for agricultural equipment.[55] Agents also began contacting farmers to assess their needs, with Pershing reporting the results from Tippecanoe County in his 1942 annual report:

> A repair list form was sent out with the notices of the meeting [a machinery repair school offered by Extension] to be returned at the time of the meeting or taken directly to the implement dealers. One out of every six farms in the county turned in a report. Of these, 10 per cent indicated they needed no important new parts. The remaining 90 per cent listed an average of six new parts that would be needed to put their machinery in operating condition for spring work. . . . These figures indicated the seriousness of the repair problem.[56]

Agents asked machinery dealers to increase inventory while parts were still available to help cover those farmers who might not have planned ahead.[57] In Marion County, dealers agreed to the agent's request to stock 50 percent more parts than were used the year before.[58] Dealers also worked out ordering schedules with farmers, with Madison County setting up the following: January for tractors, plows, discs, and peg harrows; February for planters, seeders, and cultivators; March for harvesting machinery, mowers, and combines; and April for corn harvesters, pickers, and binders.[59]

Extension agents used every possible tactic—including letters to farmers, machinery schools, farmer meetings, and radio broadcasts—to get their message about planning ahead for repairs across to farmers.[60] Large posters were placed around the county at implement dealers, stores, and other areas frequented by farmers.[61] And ads and articles in newspapers urged farmers to order early just in case repair parts became rationed at some future time.[62]

Scott County agent John McCulloch held a farm meeting in 1942 to discuss the importance of getting machinery repaired during the winter, when time and parts were not in short supply. McCulloch described the goals of the meeting as follows:

1. To direct more attention and effort toward the repair and adjustment of farm equipment.
2. To encourage the early repair and servicing of farm machinery.
3. To encourage the immediate ordering of needed repair parts for farm machines.
4. To explain systematic methods of checking, repairing, and adjusting farm machinery.[63]

Initially, the message from some agents didn't seem to resonate with farmers, with one indicating that "most people were doing nothing about it."[64] But with repeated reminders, the response gradually turned into an avalanche as farmers began placing orders with their dealers in record numbers.[65]

In the 1942 annual report for LaPorte County, agent Everett Hartman reported that mail-order companies indicated orders were 20 percent higher than normal.[66] And after Posey County agent Oral Riggs made his pitch to farmers, he was told by a local garage that twenty-four tractors were on a list to be repaired.[67]

In Porter County, machinery dealers had sold $5,000 worth of repair parts from January 1, 1941, to March 31, 1941. With Extension's emphasis on obtaining parts to make repairs, farmers increased their spending to $16,600 during the same three-month period in 1942.[68] In LaGrange County, orders for repair parts were up 48 percent in 1942, which the agent attributed to "publicity and urging to repair farm machinery early by machinery companies, dealers and the extension service."[69]

Since farmers couldn't count on replacing broken equipment with new machinery, there were national concerns that a lack of timely repairs might lead to a decline in agricultural output.[70] To make sure needed repairs were completed, Extension directed a statewide campaign by January 1942 to offer farm machinery repair schools.[71] Extension had already been offering these schools prior to the war, but they became even more important in light of the challenges brought on by the war.[72]

Extension counted on implement dealers to advertise these events, with the Huntington County agent noting that local dealers had distributed advertisements and issued personal invitations to customers to attend.[73] Purdue sent agricultural engineers Seigel A. Anderson and Russell Shipman to meetings around the state to lead discussions of equipment repairs and to show how to make the necessary adjustments so that equipment operated correctly.[74]

When Shipman spoke at a 1942 Jackson County farm repair school, agent Willis Stall provided an account of the presentation in his annual report that year. He indicated that Shipman had warned farmers to focus their full attention on repairing equipment or face the consequences of broken machinery that couldn't be fixed in the spring, noting that Shipman said:

> The problem of keeping farm machinery going this year can be met only by finding out what repairs are needed and ordering them now. If this is not done, a good many folks will be caught short when the busy season really gets under way. There is in prospect only half of the usual output of new machinery. This means that old machinery will have to be repaired, if this is possible, and be placed in order for the year's work. Because most of the machinery manufacturers must know in advance what the actual needs are, it is important that orders be placed at once.[75]

A 1943 farm machinery repair school in Vigo County shows the topics covered by the invited speakers, which included Anderson:

Morning Session

The Farm Machinery Situation	S. A. Anderson, Purdue
Machinery Rationing Problems	Emery Mack, County Machinery Rationing Board
Machinery Dealers' Problems	J. W. Torrens, I.H.C. [International Harvester Company] and Chairman [of] Farm Transportation Committee
Household Equipment Situation	Public Service Company Representative
Labor-Saving Practices and Devices	S. A. Anderson

Afternoon Session

War on Wear Film

Men:	repair, care and operation of tractor discs, harrows, plows, mowers, drills, and other farm implements
Women:	care and repair of electrical cords, fuses, plugs and motors, washing machines, refrigerators, and sweepers[76]

At many of the schools hosted by Extension, local implement dealers and farmers brought in small to large equipment for speakers such as Anderson, Shipman, agents, and vocational agriculture teachers to use in their presentations.

A 1943 repair school in Montgomery County featured three different makes of tractors, which were used to show the audience how to clean an oil filter, how to properly lubricate equipment, and how to adjust breaking plows.[77] In Jay County, farmers learned how to fine-tune a combine's adjustments to prevent the loss of wheat at harvest at a 1944 event.[78] And in Steuben County, a 1942 school focused on lubricating essential parts, caring for the cooling system, and maintaining tractor tire pressure.[79]

In 1943, Scott County agent John McCulloch went the extra mile, driving 200 miles looking for machinery that was left unprotected during the winter. Over the course of his visual survey, McCulloch found corn binders, combines, hay equipment, plows, threshers, tractors, and wagons not protected against the elements. The agent told what he had seen at meetings as he did his best to encourage farmers to do a better job protecting their farm equipment.[80]

As the machinery repair schools grew in popularity, the larger-than-expected attendance required a change in the program format.[81] In lieu of using real equipment, which became too difficult for everyone in the room to see, Extension staff began using films, colored lantern slides, and publications to support the speakers.[82]

Major manufacturers such as the Standard Oil Company and Goodyear Tire and Rubber Company produced multiple films that illustrated various aspects of machinery maintenance and repair.[83] One produced by Standard Oil was called *War on Wear*, which showed the importance of a lubrication program for reducing wear on farm equipment.[84] Agents indicated in their evaluations that using slides and films wasn't as effective as using real equipment, but they understood that the change was necessary given the large number of attendees.

While most farmers had the skills and experience to perform general maintenance and adjustments on their equipment, larger jobs were typically completed at local repair shops.[85] But experienced mechanics had skills that industry and the military highly coveted, which created a high demand for them.[86] Mechanics and blacksmiths were in such short supply that two garages in Scott County were forced to close their doors because workers to fill open positions were simply unavailable.[87]

The shortage of basic materials used by blacksmiths also stymied equipment repairs. In Posey County, Oral Riggs reported in 1943 that blacksmiths were having trouble "getting welding rods, rod material and flatiron for the repair of farm machinery. They seemed to be unable to find themselves classified in any rating high enough to get the equipment and material."[88] In some cases, agents were asked to intervene on behalf of the blacksmiths to get those in authority to provide welding rods for repair shops.[89]

During the last years of the 1930s, the agricultural community had begun to rebound from the deleterious impact of the Depression. As President Franklin Roosevelt began sending food aid to Great Britain early in 1941, the prices paid for grain, eggs, meat, and dairy reached historic highs. With money in their pockets, farmers now wanted to purchase supplies and materials to make their farming operations more efficient.

In 1946, Knox County agent Martin Huxley would observe that the "change over to tractors during the last ten years and particularly during the War years has been marked."[90] While horse numbers had been declining for years, World War II, with its shortage of farm labor, brought the era of using horses for farm work to an abrupt end.[91] Instead, with money to spend and enormous production challenges facing them, farmers were ready and willing to acquire more equipment. So it must have been very frustrating for those who wanted and needed to purchase tractors, combines, corn pickers, and many other pieces of farm equipment that these items had been severely curtailed by manufacturers, who themselves were consumed with meeting the demands of the military.

With a limited number of farmers able to obtain rationed equipment, the only recourse was to keep existing equipment running and repaired to the best of their abilities and to share limited supplies with neighbors in the short term. Shortly after the end of the war, farmers would purchase new equipment that would meet their pent-up demands and serve to modernize their operations for decades to come.

CHAPTER 37

Patriotic Duty to Get the Most Possible Mileage From Tires

The United States and its allies found themselves in a vulnerable strategic position when their sources of natural rubber in Southeast Asia fell to the Japanese early in 1942.[1] This was a significant blow to the military, which relied on equipment made with rubber. A gas mask contained one pound of rubber; a life raft, 100 pounds; a scout vehicle, 300 pounds; a U.S. Army transport truck, 370 pounds; a U.S. Navy bomber, 1,825 pounds; and a Sherman tank, more than 20,000 rubber parts.[2]

Rubber was essential to military operations, but by March 1942, with the Japanese controlling roughly 90 percent of the natural rubber supply, the U.S. faced a huge challenge to replace the lost resource to supply its demands.[3] The government developed several strategies to deal with the potential shortage. One of the first involved stockpiling and conserving what was already on hand for the military as the government pursued alternative sources.

U.S. officials also moved forward with an ambitious project to identify new botanical plants that could be used to produce rubber substitutes.[4] Botanists were dispatched across the globe, with some traveling to the Amazon Rain Forest to search for latex-producing plants that could be converted to rubber. In January 1942, the *Indianapolis Star* reported, "[R]ubber chemists are pushing through the wilderness of South America, re-exploring the 'original rubber country' where the first rubber trees were found and transplanted to the Malayan states for cultivation."[5] Another plant drawing wide interest for its latex substance was a shrub called guayule (*Parthenium argentatum*) growing in the southwestern U.S. and northern Mexico.[6]

Officials also fast-tracked a national effort to find improved and quicker methods for making synthetic rubber. With the government bankrolling the project, academicians as well as industry and government experts worked cooperatively in 1942 to standardize the formula for making synthetic rubber.[7] Once established, the government began releasing federal funds for building new plants to manufacture synthetic tires.[8] The government also provided $117 million to fund three manufacturing plants where grain alcohol was used to make synthetic rubber.[9]

Recycling and reclaiming rubber from existing products provided a more immediate avenue for increasing rubber supplies. A national campaign to collect tires, boots, raincoats, and thousands of other items eventually produced 450,000 tons of scrap rubber across the country that was used to manufacture new products.[10]

In Indiana, a network of volunteers known as neighborhood leaders (see Chapter 39) facilitated the local collection of rubber-containing items. Those spearheading these county drives often went house-to-house and used their personal trucks to bring the collected items to central locations.[11] Gasoline delivery companies, service stations, garages, and local businesses often agreed to be drop-off points where these leaders and members of the general public could bring the items.[12] Those disposing of rubber items at these sites were paid one cent per pound, but many refused the money. Businesses that hosted drop-off collection sites were paid 1.25 cents per pound, with their profits going toward the purchase of war bonds.[13]

County leaders were told how much salvaged rubber they were expected to gather as part of the national recycling campaign. In 1942, Kosciusko County collected 12 pounds of rubber per person, which exceeded the county's goal of 9.6 pounds per person.[14] That same year, the Morgan County quota was 236,612 pounds of rubber, with the collection falling short at 225,830 pounds.[15] Agents across the state reported that their counties collected from 64,000 pounds to 315,000 pounds in 1942.[16]

Collecting unwanted rubber helped in the near term as other solutions were being pursued by the federal officials. But the only viable answer to supply enough rubber for military use was to drastically cut the availability of new tires to the general public, so in January 1942, tires became the first item to be rationed by the government.[17] This seemed like a logical step to take since automobile tires consumed two-thirds of the rubber used in the United States. The rubber saved by not producing eight passenger tires could now be used to make a tire for a bomber.[18] By that fall, rubber boots and work shoes would be added to the list of rationed items.

Government estimates were that the U.S. had a year's supply of rubber on hand, but that was designated for military uses following the Pearl Harbor attack. Estimates in 1942 indicated that the nation had an inventory of 10 million new passenger tires. Given that Americans purchased approximately 4 million to 4.5 million tires each month, that total would last less than three months.[19] America was on the verge of a transportation meltdown and military crisis without the Roosevelt administration taking draconian measures.

For the month of January 1942, the public was told only 357,000 new tires would be available for purchase nationwide. This number represented only one-eleventh of the total number of new tires sold per month prior to the war. Whether more tires would be made available at some future time depended on the military situation and future demands by the Allied forces.[20]

With few choices at hand, the government put into place a frugal program that blocked most Americans from obtaining new tires for their personal vehicles. As the *Times* in Munster reported on January 1, 1942, "Outbreak of war in the Pacific has touched off a series of government orders affecting the automobile industry, all designed primarily to conserve rubber for military uses. They included: slashing passenger car production for the months of December and January from 60 to 75 per cent, respectively, of the industry's output a year ago. . . . Elimination of the spare tire on all new cars hereafter produced by the industry."[21] That same day, the *Lafayette Journal and Courier* reported, "A large portion of the American people meanwhile faces an immediate sacrifice—the sacrifice of new automobile tires. . . . In government business circles, the tire sacrifices were regarded as trivial. Government experts say the people must work longer, pay higher taxes and do without more and more non-defense articles in 1942—to help America on the way to victory."[22]

The federal government directed each county to initiate a tire rationing program by January 5, 1942, under the control of a local tire rationing board, which had the full authority to act on behalf of the government.[23] It's important to note that these tire rationing boards were not controlled by the county war boards. In fact, the local war boards only got involved when the actions of the tire rationing boards negatively impacted farm productivity, which at times pitted the responsibilities of one with the needs of the other.[24]

The tire rationing board consisted of three members who were hand-selected by the director of the Civilian Defense Council for the county.[25] For example, in Franklin County, the tire rationing board included a coal dealer, realtor, and farmer.[26] In a few cases, the county Extension agent was appointed as a board member.[27] The only persons excluded from serving on these rationing boards were tire dealers and fleet truck operators.[28]

Each board member served without compensation, accepting the position as a "patriotic duty and as a service important to the successful prosecution of the all-out war effort," according to one newspaper report.[29] Board members were summoned to Indianapolis on January 4, 1942, to learn how to administer the newly created program and to be officially sworn in as employees of the federal Office of Price Administration.[30] In Indiana, the boards were supervised by James Strickland, who had been appointed by the state Civilian Defense Council.[31] As one agent noted, the rationing boards were faced with a "tire shortage becoming more and more acute as time went on following Pearl Harbor."[32] Board members had rules to follow, but it was going to be difficult for them to deal with so many people whose requests had to be denied.

In 1942, Indiana received an allotment of 2,717 tires and 2,274 inner tubes for automobiles, motorcycles, and light trucks that were divided up among the county rationing boards. In addition, the state received 6,413 tires and 5,361 inner tubes for larger trucks and buses.[33] During the first month of the rationing program, Bartholomew County received 80 new tires and 67 new inner tubes while, by comparison, Brown County only received two tires.[34] According to a report by the *Indianapolis News*, tire rationing boards were "prohibited from rationing more than 25 per cent of the month's total each week."[35] Any tires remaining at the end of the month would be subtracted from the following month's allotment.[36]

Access to new tires and inner tubes would be limited to a very narrow range of professionals that excluded most drivers.[37] The *Indianapolis Star* reported, "Those who drive passenger cars for personal transportation will be permitted to buy no tires or tubes . . . at least so long as the Japanese control the Pacific in the region where the United States gets its crude rubber supply."[38] And a story in Richmond's *Palladium-Item* warned, "War will reach out and grab America's 32,000,000 motor car owners by the scruff of the neck this coming week. From now on most of them will be unable to buy tires because of rationing forced by the cutoff of crude rubber from the Far East."[39]

Federal regulations spelled out exactly who could and could not obtain new tires and inner tubes. According to the new regulations, these items could only be purchased by people in these professions or for these types of vehicles:

1. Physicians, surgeons, visiting nurses, and veterinarians;
2. Ambulances;
3. Firefighting equipment, police vehicles, garbage removal trucks, and mail delivery cars;
4. Public-service vehicles with a capacity of ten or more passengers on regular transportation routes and school buses or cars to carry workers to and from industrial plants;
5. Trucks for ice and fuel delivery, road maintenance, public utilities facilities, essential repair services, waste and scrap dealers, common carriers, transportation for raw materials, semi-finished, or finished goods that are not moving directly to the household for consumers' use;
6. Farm tractors or other implements, except trucks or pleasure cars; and
7. Industrial, mining, and construction equipment, except trucks or other automobiles.[40]

Requests that didn't fall into one of these classifications would not be approved under any condition. Of course, there were those who thought they could use their influence to bypass the regulations, but the tire rationing boards would not show favoritism because members had sworn an oath to the United States to fulfill the obligations of the position.

Those who qualified to request new tires or inner tubes would first have to fill out a document that some equated to filling out an income tax form.[41] Individuals making false claims on a tire application were subject to a $10,000 fine and ten years in a federal penitentiary.[42]

The second step in the drawn-out process required obtaining a certified tire inspection. The eligible person would bring the vehicle to one of the approved sites, where an inspector would examine the tires and indicate whether they were beyond repair.[43] The inspector could not charge a fee unless the tires had to be removed in order to properly assess their condition, in which case a fee of fifty cents applied.[44] The applicant would then deliver the inspector's report to the tire rationing board, obtain the official purchase certificate, and go to the tire store with the necessary paperwork to complete the purchase.[45] Tire dealers participating in the government program were expected to produce the redeemed purchase certificates when requested by federal officials and had to keep records that were subject to auditing by federal inspectors from the federal Office of Price Administration.[46]

Days before the rationing program was to become effective, the federal government tried backstopping the tire rationing boards. An *Indianapolis News* article stated, "Persons feeling they may not be qualified under the regulations to purchase tires were urged today by defense officials not to 'pester' the rationing boards with requests."[47]

Clarence Jackson, director of the Indiana Civilian Defense Council, made a direct appeal in the media on January 4, 1942, noting that people could provide a patriotic service by refraining from making unnecessary calls to the tire rationing boards. He said, "If you are not a physician, surgeon, visiting nurse or veterinarian, you, as an individual, will not be able to obtain tires for passenger cars. Individuals who do not fall in the above classifications must not expect or hope to obtain a certificate for the purchase of new tires or tubes. Likewise, not even those individuals in the above classifications may obtain new tires until such time as their present tires are beyond repair or reconditioning by retreading."[48] Jackson reminded the public that the rationing boards had no discretionary powers, and their primary role was to determine whether the person requesting a tire purchase certificate was eligible as designated by federal regulations.[49]

Given the shortage of tires, Extension staff and others immediately worked to focus the public's attention on preserving the rubber on the tires they already owned. Speaking to a national radio audience, Harvey S. Firestone Jr., the founder of Firestone Tire and Rubber Company, said it was the "patriotic duty of every American to get the most possible mileage from his tires."[50] Consequently, drivers were asked to save rubber and tires by:

- reducing the miles driven;
- slowing down to keep speeds below 35 miles per hour;
- preventing abrupt stopping and starting;
- keeping tires properly inflated;
- rotating tires every 3,000 to 5,000 miles;
- carpooling when possible;
- riding a bike; and
- taking public transportation.[51]

The American Automobile Association indicated that there were more than 7.5 million automobiles made in 1940 and 1941 with tires that were expected to last two years. Those with older cars who did not qualify for new tires would have to purchase recapped tires, which were those with rubber that had been replaced.[52] Finding rubber for the recapped tires was one of the goals for the scrap drives that had been initiated in 1942. The expectation was that the reclaimed rubber could be turned into "war tires" for consumers.[53]

Dealers whose new tire sales were slashed to the bone tried making up the lost business by advertising that their reconditioned and recapped tires were just as good as newly manufactured tires but cost much less. An advertisement placed in the *Lafayette Journal and Courier* stated, "Don't Let Smooth Tires Worry You . . . Bring Them In Today! We'll Quickly Change Them From Old To New By Our Factory Process Re-capping. You'll Save One-Half The Cost Of New Tires."[54] And a classified ad for a used car in the *Daily Clintonian* highlighted the condition of the tires: "A very clean 1941 Hudson Deluxe Four Door Sedan, low mileage, all good tires."[55]

As with other World War II rationed goods, the limited supply of tires gave rise to a lucrative black market for new and used tires, price gouging, falsification on tire ration board forms, tire dealer fraud, and tire thefts.[56] One newspaper published a notice admonishing its readers, "Watch out for automobile tire thieves!"[57]

Motorists were instructed to record the serial numbers of their tires, to place spare tires under lock and key, and to make sure garage doors were secured.[58] In Evansville, the police asked tire dealers to keep records of all tires and inner tubes sold or repaired, which was "believed to be the first effort by Indiana authorities to secure ownership facts for use in tire theft cases," according to a newspaper report.[59]

Those caught stealing tires didn't get much sympathy. One Indiana judge told a defendant, "Tires are important in the defense of our country and the stealing of them at this time amounts almost to treason." The judge who made this remark set the bond at $5,000, which was a significant sum of money in 1942.[60]

Established procedures encouraged the public to turn in individuals who they believed were gaming the tire rationing program.[61] The federal Office of Price Administration relied on field inspectors and 500 employees from the federal

Department of Labor to investigate fraudulent claims, bootlegging of tires, and price gouging.[62]

By 1943, farmers were told that they could convert their steel-wheeled tractors to rubber tires by obtaining permits from the county war board. Given the important work that farmers were doing, this change was made to support their contribution to the war effort and help them meet agricultural production quotas. But those retrofitted tires would still count against the county quota.[63] Agent Oral Riggs explained his experience with this effort in the Posey County annual report that year, writing:

> Because the tires used in this conversion process came off the county quota, there was considerable uncertainty as to the possibility of utilizing these new regulations. Although there was considerable delay in getting it done, there were several tractors converted from steel to rubber during the year. Many more wanted it done. It seems that our job was to explain the processes that they should go through and to discuss with them the possibilities or perhaps impossibilities so that we expended more effort with those who did not get it done than with those who did.[64]

Beyond the farm fields, tire rationing also caused a huge challenge for farmers who needed to deliver agricultural products to market. Having dependable tires for their farm trucks was essential to the process, so sometimes agents had to step in to facilitate the continued flow of goods.[65] Leo Geyer described his efforts along this line in his 1942 annual report for Jefferson County:

> When the tire board was up against the proposition of keeping Agriculture commodities rolling, they called on the agent for help. . . . Trucks hauling milk, eggs, [agricultural] lime, [and] tomatoes were the four most important trucks so far as farm produce was concerned. The [Land Use] Planning Committee further recommended that the ration board call these truckers in and have a tire man give them instructions on proper care of tires. This was done by the rationing board and a man from U. S. Rubber Co. gave these truckers an instructive talk on [the] care of tires. This meeting took about 90% of the heat off of the tire board and things began to run more smoothly.[66]

That same year in Knox County, Henry Benson petitioned members of the local tire rationing board for inspections in hopes of getting more replacement tires on farm trucks hauling apples, cantaloupes, peaches, sweet potatoes, and watermelons by showing the existing tires were beyond repair.[67]

Collectively, tomato growers faced serious tire shortages. Agents worked out emergency procedures to help truckers who found themselves stranded alongside the highway from a blown tire as they headed to the processors.[68] In Switzerland County, agent Oran McNary reported in 1942 that with tomatoes identified as a war crop, members of the local tire rationing board—with support from state officials—indicated "that they would do all they possibly could to move the tomatoes."[69] In the Ripley County annual report that year, agent Guy Harris said local farmers were also "assured at that time that trucks hauling tomatoes would be taken care of."[70] The agents never mentioned in their reports whether the ration boards had been able to do more than just make promises.

Agents found other ways to help, including arranging for competitors to temporarily swap customers as a way to reduce the miles traveled by individual trucks.[71] In 1942, McNary explained the process used in one situation in Switzerland County, writing:

> At the request of some of the milk truck drivers, a county meeting was called for all milk truck drivers of the county. A large map, showing the roads and houses which has been drawn by the REA [Rural Electrification Administration] Office to the scale of 2 [inches] to the mile, was used to mark out all routes and customers. This showed many duplicate miles being driven. We have five or six trucks taking milk to the Condensed Market at Madison. Two or three trucks in competition with this company are taking milk to Osgood. We have three trucking concerns taking milk out of the county to the whole milk market in Cincinnati.[72]

After comparing routes, milk haulers temporarily exchanged customers for the duration of the war in order to allow for more efficient collection and transportation of milk within the county. McNary went on to say that most "of the larger trucking men were glad to put their business on the map and compare notes. We found some cooperation going on the routes in the hauling of one another's cans and milk in a few local places rather than trying to transfer farmers from one milk company to another."[73] This unusual cooperative effort among producers, haulers, and processors not only preserved tires, but it also reduced gasoline consumption and saved wear and tear on the trucks.[74]

Similar efforts were made elsewhere in the state. Two companies in Jefferson County—the Thompson-Glass Dairy and the Ewing–Von Allmen Dairy—saved 10 percent of their miles when they worked together over the five routes where their customers overlapped.[75] And in Porter County, shifting eight farms between two milk haulers saved 171,000 tire miles each year.[76]

Even before the end of the war, the tire shortage began to ease for farmers. In 1941, there had been 2 million tires produced for farm tractors and implements.[77] Early in 1944, production had once again increased to the point that, by eighteen months later, enough farm tires had been produced to eliminate the deficits experienced during the first two years of the war. In 1945, production would surpass 1941 levels.

By the end of 1945, the *Indianapolis News* reported that the flow of "300,000 tons of natural rubber will reach the United States from liberated and other areas in the next year."[78] With war-related activities winding down, the military had also reduced its demand for tires and indicated it would release 400,000 new tires for public use by January 1946.[79]

Passenger and truck tires remained rationed until January 1, 1946. After riding for four years on thin treads and recapped tires, one December 1945 newspaper article claimed, "Motorists who have nursed old tires through years of wartime restrictions will ask nothing better for Christmas than a few new ones."[80] Even though tires would no longer be rationed, the public was repeatedly warned that the shortage was expected to persist until the end of 1946.[81]

Production schedules showed 28 million passenger tires had been manufactured in 1945, which was 64 percent higher than the number of tires produced the year before. The goal for 1946 was 66 million passenger tires, which was 13 million more than the highest number ever produced before.[82] Even with the increased production levels, it would still take some time before new cars coming off the assembly line would be equipped with a spare tire or with special-ordered whitewall tires.[83]

With a high demand anticipated for new tires, there were concerns that businesses would run up prices as high as the market would bear. To deal with those worries, the government kept price controls on tires and inner tubes even after the rationing program had ended. Depending on size and ply ratings, the maximum that dealers could charge for a tire ranged from $13.53 to $20.90, and prices for inner tubes ranged from a low of $3.05 to a high of $4.40. Each tire dealer was required to display a poster to show what they could legally charge for tires and inner tubes.[84]

Even as the tire markets entered the postwar phase, one notable difference remained. Prior to the end of the war, advances in the manufacturing of synthetic tires meant they had largely begun replacing tires made of natural rubber. The synthetic tires had been tested under military combat conditions and had held up to the rigors of wartime conditions. As early as January 1944, the *Star Press* in Muncie reported that the government had ordered the United States Rubber Company "to convert an ammunition plant back to tire building . . . and virtually all the products will be made of synthetic" at one of the government-owned plants.[85]

By August 1945, an article in the *Lafayette Journal and Courier* stated, "Synthetic rubber is regarded as one of the miracles of the war." It also indicated that "synthetic tires and tubes are here for a long stay, perhaps permanently."[86] This statement would ultimately prove to be true, with more than half of the rubber used by the tire industry today being synthetic.[87]

CHAPTER 38

Farmers Were Second Only to the Army in Needing Fuel

Early in World War II, German submarines in the Atlantic Ocean began sinking fuel tankers heading toward the East Coast that were carrying oil and fuel from the Gulf Coast.[1] It was common to read newspaper headlines such as this one from January 1942 published in the *Lafayette Journal and Courier*: "Two American Tankers Are Attacked Off East Coast."[2]

To deal with fuel shortages on the East Coast that were caused by these attacks, the government required trains to haul fuel out of midwestern states, where gas was plentiful at first but soon became in short supply.[3] By the spring of 1942, newspapers were reporting that rationing was becoming more likely for midwestern and southwestern states because of the diverted supplies.[4] To make matters worse, ships normally filled with fuel and oil for the East Coast were being diverted by the fall of 1942 to Great Britain and other allies as the escalating war increased demand.[5]

This combination of events meant the government had little recourse but to initiate a rationing program for fuel.[6] Gas rationing had started first in the eastern U.S. on May 15, 1942. This initial round affected 3 million drivers in seventeen states and the District of Columbia.[7] While many Americans did not completely support the government's reaction to the fuel shortage, officials used the press to remind the public what was at stake, such as with this statement published in April 1942 in the *Daily Clintonian*: "[The] United States no longer will risk the lives of sailors so that civilians can have gasoline to go to a bridge party or to a ball game."[8] In addition to rationing, the government placed controls on the maximum price of a gallon of gasoline to help prevent price gouging.[9]

A person's occupation largely determined the amount of rationed fuel available to them.[10] Those considered nonessential drivers were provided with four gallons per week, while those in designated occupations received additional fuel.[11] Fuel allotments for nonessential drivers were based on driving 2,880 miles a year, or about half of what a typical person normally drove, in a vehicle getting 15 miles to the gallon.[12]

A gas ration book was required in order to purchase the fuel under one of three categories:

- Class A gas ration book: Nonessential drivers.
- Class B gas ration book: Essential drivers, including defense workers, public service officials, and others.
- Class X gas ration book: Unlimited gasoline purchases for those such as doctors, nurses, messengers, taxi drivers, and truckers.[13]

Farmers would likely have registered for Class B ration books as essential drivers, which would have allowed them to have extra fuel for conducting their work.[14]

Those wanting to renew their gas rationing books were required to show proof of having their tires inspected and certify that they owned no more than five tires for each vehicle. They also were told that exceeding the 35-mile-per-hour speed limit could cause their ration privileges to be revoked.[15]

However, the gas rationing program in the east failed to get the anticipated fuel savings that were predicted by the government experts. With the ever-increasing demand for gasoline to fuel the war, only one option remained: apply the gas rationing program nationwide.[16] By December 1, 1942, drivers from across the country would be placed on a modified version of the rationing program in the east.[17] Days prior to the start, long lines formed at the service stations as everyone not only filled up their automobiles but also, according to one newspaper report, filled "drums, cans and containers of all kinds . . . to hold the precious liquid."[18]

The updated rationing policies included the following categories:

- Class A gas ration book: Gas coupons for those not meeting the requirements of the other coupon classes. Amount of gas limited to three gallons per week.
- Class B gas ration book: Gas coupons for those in car-sharing pools. Amount of gas limited to four gallons per week.
- Class C gas ration book: Gas coupons for those employed by defense plants and government officials. Amount of gas determined by the employer or government agency.
- Class T gas ration book: Gas coupons for commercial vehicles, including farm trucks. Required applying for a Certificate of War Necessity and receiving approval from the Office of Defense Transportation. Amount of gas determined on a case-by-case basis during the application process.[19]

Across the country, government inspectors were enforcing rules that prevented driving for recreational and entertainment purposes. According to a January 8, 1943, article in the *Muncie Evening Press,* motorists in Miami, Florida, were called in front of judges to answer questions about why they were attending sporting events. The same article also indicated that in Rochester, New York, "O.P.A. [Office of Price Administration] agents seized gasoline ration books of several motorists attending a philharmonic concert," while Massachusetts planned to "start a rigid ban tonight by checking autos found near theaters, dine and dance halls, and winter sports resorts."[20] In an August 24, 1943, article, the *Daily Clintonian* reported that driving for pleasure or going to "State Parks, theatres, picnics, outings, carnivals, fairs, golf courses, and beer taverns" were considered in many circles unpatriotic and subject to fines by local police.[21]

Farmers needed fuel to run their farming operations and move agricultural products to market. Without a steady supply, production would decline, and they would be unable to help meet quotas for crop and livestock production set by the federal government. With rumors running rampant within the agricultural community that farmers were not going to get the fuel they needed, government leaders turned to newspapers to reassure them. An August 1943 article in Richmond's *Palladium-Item* stated, "Farmers are getting and will continue to get full preference for gasoline used on the farm to operate tractors and machinery . . . [and to] move livestock, poultry and other commodities. . . . For highway use, the farmer now has preferred mileage status, much the same as that of war workers, allowing him sufficient gasoline for his passenger car if it is used to carry produce to market."[22] Regarding the need for fuel, farmers were second only to the army.

Farmers would get the gasoline they required, but not without first jumping over the hurdle of government regulations. The federal Office of Defense Transportation required all farmers and other commercial businesses with trucks to obtain an authorization called a Certificate of War Necessity to purchase extra fuel beyond what most drivers were receiving.[23] Applications for these certificates were then reviewed by government officials in Detroit to determine whether to allocate the extra fuel.

The fuel regulations for farm trucks were divided into two categories: farm use and personal use. In order to get a Class T ration book, a farm truck had to be used to transport farm products and supplies 51 percent of the time. Anything less than this meant a Class A ration book would be issued instead.[24]

With the nationwide gasoline rationing program set to begin on December 1, 1942, farmers were originally told to complete their certificate applications by November 15, which was later changed to December 1.[25] The certificate set the maximum mileage and minimum loads.[26] A farmer doing custom work for other farmers could ask for additional fuel as well.[27]

Counties established transportation committees to process the applications. These five-person organizations typically included the chairman of the county war board, two farmers, one trucker, and one farm supply dealer.[28] In

Noble County, the Extension agent was also named as a member of the local transportation committee.[29]

Unfortunately, the application for obtaining a certificate was so difficult for farmers to complete correctly that the government had to create a thirty-two-page instruction book, which was said to be even more difficult to understand than the four-page application itself.[30] So farmers received one-on-one assistance with their applications from Extension agents, members of the transportation committee, war board members, and volunteers recruited by these officials.[31] The goal of this approach was to make sure that the request for gasoline, off-road fuel, and kerosene for use on the farm would not be denied based on incomplete information or inadvertent mistakes.

On October 15, 1942, forty recruited volunteers from central Indiana counties met in Indianapolis to learn how to properly fill out an application for a certificate.[32] District meetings were also conducted around the state to train volunteers from other areas.[33]

Nationwide, the federal Office of Defense Transportation held a three-day farm truck registration event October 22–24, 1942, during which the local transportation committee members were expected to assist farmers in applying for the Certificate of War Necessity.[34] The Pulaski County Extension office sent farmers a letter telling them the specific times and places to receive help in filling out their applications.[35] Farmers were told to come prepared with the information needed to help the volunteers fill out the forms.[36]

In Pulaski County, agent Carl Olson seemed surprised that out of the 450 farm trucks in the county, only 240 of them were registered during the three-day period. In his 1942 annual report, he commented, "It was very difficult for farmers to give the information asked for very accurately as very few of them had kept any record."[37] Olson did not speculate as to how those not registering their trucks obtained a certificate.

Agent Henry Benson wrote the following account of the application event in Knox County:

> Thursday morning, October 22nd, will long be remembered in the county extension office as the day that farmers from all over the county flooded the office to obtain assistance in filling out applications for certificates of War Necessity for their trucks. It was very gratifying to see the confidence placed in the extension office as farm truck owners from every township in the county passed by their local center to drive several miles extra to have their application filled out in the county extension office. The mad rush for assistance continued for three days.[38]

In most counties, it took the full three days to deal with the large influx of farmers who participated. In Wayne County, 600 farm truck owners asked for assistance with their applications.[39] In Tipton County, 42 volunteers were required to help fill out the forms.[40] In Jackson County, transportation committee members and 50 volunteers helped 250 farm truck owners.[41]

The transportation committee typically asked agents to do a final review of the completed applications.[42] Once that was accomplished, the committee would make its recommendation on the number of miles to allocate to each farmer.[43] The applications were then sent to Detroit, where officials at the Office of Defense Transportation (ODT) could reject or accept the recommendations. According to an October 1942 article in Richmond's *Palladium-Item*, one county official indicated that "some worker at the Detroit ODT office that had never seen a cow or a horse would undoubtedly use the red pencil on some of the applications if they did not carry the proper information."[44]

Extension agents became frustrated when ODT officials ignored the recommendations regarding the miles assigned to a specific farmer.[45] In the 1942 annual report for Warrick County, agent Bryce Bond showed his displeasure with those working for ODT, writing, "I think this perhaps is due to the fact of the very deplorable errors that [have] been made in issuing certificates of War necessity for farm trucks where gasoline allowance had been reduced to less than ⅓ of what they [farmers] had normally used. Although we have been assured that farmers will secure all the necessary gasoline that they need, this is casting a gloom over the rural population."[46]

If an application passed the review in Detroit, the farmer received a signed copy of acceptance that listed the gallons of gas allotted for farm use. A truck owner could appeal to the county transportation committee when the outcome was too restrictive.[47]

Farmers with approved certificates obtained their Class T ration books on a quarterly basis from the gas rationing board, which was separate from the county war board.[48] Unless the farmer's situation changed, the certificates were typically valid for the duration of the war, which saved the farmer time by cutting down on bureaucratic red tape.[49]

Agent Jean Sliger reported in 1943 that he had been placed in charge of rationing gasoline for tractors in Sullivan County. According to his annual report that year, this was because War Price and Rationing program officials "believed that the county agent was in a better position to know the needs of farmers than his board."[50] Sliger went on to explain that the process "was a haphazard program for the first two or three months until a system was worked out whereby farmers' needs were estimated for a six months' period. This makes it necessary for farmers to file applications only twice each year, in March and April and again in September and October."[51]

While federal and state officials reassured farmers that there would be enough gas for their operations in 1943 and 1944, farmers were more or less leery of such remarks, knowing the paperwork and the limitations they had faced in 1942.[52] The challenges with getting enough gas for farm work would persist throughout the war years.

Agricultural production was a key factor in whether or not the United Stated and the Allied forces would win the war, so it was absolutely critical that farmers had access to the supplies they needed to meet their production quotas. At a time when farmers were already coping with a heavier workload and daily challenges to keeping their operations running, Extension agents and others in the county were instrumental in helping farmers break through the red tape and get the fuel needed to keep their equipment running and make sure their vehicles could deliver agricultural goods to market. Without this assistance, many farmers would have been hard-pressed to deal with the bureaucracy that surrounded the Certificate of War Necessity.

The gas rationing program ended with a government announcement on August 15, 1945, with gas supplies rebounding over the course of the following year.

CHAPTER 39

Extension Meetings Will Be Curtailed Due to Gas Rationing, Thin Tires, and Busier People

During World War II, farmers faced gasoline, tire, and equipment rationing in addition to an acute labor shortage, so they were less likely to attend traditional Extension meetings, programs, and tours.[1] When Lake County hog producers were asked if they wanted a summer meeting in 1942, the agent reported that "they thought we had better save tires and time this year."[2] In Starke County, the agent wrote in his 1943 annual report, "With everybody so busy, attendance at meetings was small."[3]

Jackson County agent Willis Stall explained the situation this way:

> Many of the agencies of extension work have been hampered during the year [1942]. Farm people have been busier with their own affairs than at any time during the past 20 years. Travel has been restricted all during the year, since most of this time most people realized that they will probably be unable to travel at all after their present tire supply is exhausted. . . .
>
> As the year advanced towards its end, with its restricting and rationing programs, it became more apparent that the forces of agricultural education will be increasingly difficult during the coming years, were it not for the splendid cooperation shown by our people in support of the biggest job ahead of all of us, which is the winning of our War, at as early a time as possible.[4]

Lawrence County agent John Armstrong shared Stall's assessment, noting that typical Extension programs would have to "be greatly curtailed due to gas rationing, thin tires and busier people."[5] In order to still provide critical outreach, agents had to become more efficient "so that one meeting may accomplish as much in 1943 as two or three did in 1942," as one agent noted.[6] In addition to holding meetings, agents also relied on alternative methods for reaching their constituents. Naturally, that included their old standbys, including telephone calls, circular letters, direct mail, and newspaper articles.[7]

Agents found circular letters, or unsolicited newsletters, especially useful for targeting specific agricultural groups, such as those raising hogs, beef or dairy cattle, or poultry or those growing specific crops, such as corn, mint, or potatoes.[8] Often written by Purdue campus specialists and countersigned by the local agent, these newsletters provided reminders and advice for implementing a specific set of farm management actions at a particular time. Newspapers gave agents the opportunity to reach a wide audience since many had allowed agents to write weekly columns since the inception of the Cooperative Extension Service in 1912.[9] But in addition to these communication channels, agents relied on two other methods to reach their intended audience during the war: radio broadcasts and a newly formed coalition of neighborhood leaders who would serve as proxy Extension agents.

Farmers had quickly adapted to and depended upon the radio for updates on the markets and other information.[10] So radio broadcasting, which had already been used by Extension for two decades, became critical in Extension's arsenal of agricultural education options throughout the war. Using radio, agents were able to broadcast agricultural, war-related, and family-oriented information to a wider audience than they could ever have attracted to meetings. Extension shared with listeners the latest on a diverse set of topics, including:

- soliciting participants for the Victory Garden Program and encouraging people to raise more homegrown vegetables;
- updating growers on the latest economic news;
- relaying information on fire prevention for the home and farm;
- providing information on temporarily storing corn when bins were filled to capacity;
- discussing the life cycle and control of termites infesting the home; and
- providing updates on the relative value of different woods for winter fuel.[11]

Speaking over the radio was as close to being in front of farmers as the agent could get without meeting them face-to-face. The agents could connect with farm families, who listened in the comfort of their living rooms or even while repairing equipment in the barn. Since most households had at least one radio at the start of World War II, it made absolute sense for Extension to rely more on this form of communication.[12]

Agents used radio broadcasts to relay important wartime requests, such as when the U.S. secretary of agriculture, Indiana's own Claude R. Wickard, asked in 1942 for more soybeans and tomatoes to be planted.[13] And they used it to provide information of interest to their audience, such as one broadcast described by Posey County agent Oral Riggs in 1944:

> A radio report was given summarizing the study of the use of air travel in transporting perishable products after the war as it was presented in Marketing and Transportation, a publication put out by the [U.S.] Bureau of Agricultural Economics . . . based on the [Douglas] C-47 plane. . . . This report was given over the air because we feel that it is necessary that farm leaders

and farm people be alert to change. Change has a way of creeping upon us unnoticed, and it is necessary to be alert to recognize change before it arrives.[14]

In southwest Indiana, agents from Posey, Spencer, Vanderburgh, and Warrick Counties pooled their resources to offer the Pocket Neighbor radio program, a fifteen-minute daily program that first aired April 28, 1941. The program was broadcast every Monday through Friday at 12:15 P.M., during the lunch hour when farmers were apt to listen as they ate their noontime meal.[15]

As interest in using the radio as an educational medium grew, Purdue sent communications professional Glenn W. Sample to meet with agents and vocational agriculture teachers from east-central Indiana to discuss taking full advantage of the popularity of radio broadcasting. The Howard County agent reported that, over the course of the five-hour program, "Mr. Sample gave a very interesting discussion as to the procedure for presenting radio programs and certain fundamentals that should be followed."[16] Two years later, Howard County was still transmitting five 10-minute broadcasts each week based on the instruction provided by Sample.[17]

As Extension staff gained experience and confidence using radio broadcasts to transmit their messages, the educational programs became more targeted toward particular audiences within agriculture. The agents teamed up with Purdue campus Extension specialists to present what became known as Schools of the Air. Offerings included the Poultry School of the Air, Hoosier Hog School of the Air, and Victory Garden School of the Air—all important topics because eggs and poultry, pork, and homegrown vegetables were vitally important to the war effort.

The U.S. Department of Agriculture (USDA) had pioneered this approach during the 1920s, when radio-based outreach to farmers was still in its infancy. People could sign up to attend the school, then tune in to hear lessons that were broadcast over a number of days or weeks on radio stations scattered throughout Indiana. At the end of the course, participants received a certificate after successfully completing a final examination by mail.[18]

In 1943, Extension offered two radio schools for poultry, with one in March and the other in October.[19] The schools were broadcast on WEOA and WGBF (Evansville), WOWO (Ft. Wayne), WIRE (Lebanon), WLBC (Muncie), WBOW (Terre Haute), WAOV (Vincennes), WBAA (West Lafayette), WIND (Chicago), WKMO (Lebanon Junction, Kentucky), and WHAS (Louisville, Kentucky).[20]

People who were interested in the Poultry School of the Air signed an enrollment card and listed the radio station they could receive at their farm. When they enrolled, they received in the mail the following items:

- six Extension bulletins and leaflets;
- two postcards for requesting additional bulletins, asking questions, or offering comments and suggestions;
- ten lesson plans that were timed to reach the participant two or three days ahead of the radio lesson; and
- a certificate that the agent described as "suitable for framing" when the lessons were completed and the participant passed a final examination over the materials covered.[21]

Miami County agent John Wylie advertised the March Poultry School of the Air as "Ten snappy, concise, 15-minute lessons over your radio. . . . The broadcasts can be heard in every county of the state. . . . Enrollment is limited to 10,000. . . . You will save precious gasoline and tires and at the same time gain knowledge which will help you to profitably produce more eggs and poultry in the Food for Freedom effort."[22] In his annual report that year, he also listed the following topics:

March Workshop Topics	October Workshop Topics
Chick fever—planning early for high egg prices	Figuring your 1944 egg crop
What kind of chicks shall I buy?	Housing the pullets
Furniture for the chick "nursery"	Ventilation and damp litter
Preparing the brooder house	Save labor in the hen house
The chicks arrive	How shall I feed the layers?
Feeding the Victory chicks	Feeding the hens during war times
Sanitation in the brooder house	Managing to get the most eggs
Watch the little (management) things	Keeping hens healthy
As the "youngsters" grow	Cannibalism, egg eating, and feather picking
Review and final examination	Review and final examination[23]

With the wartime emphasis on increasing pork production (see Chapter 28), Extension also took to the airwaves to share information about raising more hogs and getting them to market faster.[24] For the 1944 hog school, participants received a schedule for each broadcast before it aired along with written materials, which included the following:

- How to Grow Thrifty Fall Pigs
- Selection, Management, and Feeding of Breeding Hogs
- How to Handle the Brood Sow and Her Litter
- Fight Hog Cholera—Vaccinate
- Hog Diseases
- Economic Outlook for Hogs[25]

The programs were hosted by the Purdue Extension Department of Animal Husbandry during the week of December 6 on Monday, Wednesday, and

Saturday.[26] Agents were well pleased with the program, with one reporting, "These [hog] broadcasts are always well received and we know that the farmers are listening because of the questions asked regarding the various broadcasts."[27] At the conclusion of the hog school, agents and Purdue Extension specialists provided to all of the participants answers to the questions raised by farmers.[28]

Marion County agent Horace Abbott described the Victory Garden Program offered in 1945, which showed the close cooperation among the agents and campus Extension specialists:

> A series of radio broadcasts over station WIBC [Indianapolis] were started on Monday, March 5. These broadcasts were given each Monday, Tuesday and Wednesday at 7:45 to 8:00 A.M. . . . About one-half of these broadcasts were given from transcriptions which were recorded by the writer [Abbott] and Purdue specialists on different phases of gardening. The other broadcasts were live broadcasts given by the writer . . . [and entomologist] Glen Lehker on insects; [plant pathologist] C[harles] T. Gregory on Plant diseases; [agronomist] George Enfield on garden soils and fertilizer; [horticultural assistant] W[illard] B. Ward on general garden cultural practices; [horticulturist] E[arl] R. Honeywell on flowers in the garden.[29]

The use of radio as an educational medium proved so successful during World War II that Extension staff would continue to rely on it to get their messages out for decades to come.

In 1942, the U.S. secretary of agriculture asked the federal Cooperative Extension Service to involve county Extension agents nationwide in preparing the country for potential emergencies arising from the war, such as the unlikely event that German and Japanese planes bombed American industrial war facilities.[30] Near the end of May, agents received materials from the federal government explaining a program that would use volunteers as a link between the agents and farm families.[31] In Indiana, this program—known as the Neighborhood Leaders Program—was directed by Philip K. Connelly, assistant state Extension leader at Purdue.[32]

During the spring of 1942, agents organized individuals within their counties, which included creating an advisory committee made up of members representing the federal Agricultural Adjustment Administration and Office of Civil Defense, farm bureaus, 4-H clubs, home economics clubs, and land use planning committees.[33] By 1945, the statewide program would grow to 60,630 neighborhood leaders working on behalf of the agents and Extension.[34] According to Pulaski County agent Carl E. Olson, the program provided the means "to get necessary war information to each family in the county as quickly as possible."[35] From an Extension perspective, the neighborhood leaders would provide another way to facilitate personal communication between the agents and their constituents.[36]

According to one agent, the USDA originally indicated that the neighborhood system "would be made available to the Red Cross, U.S.O. [United Service Organizations], Civilian Defense, War Bonds, as well as to the Extension Service."[37] Indeed, county organizations and war committees from these organizations as well as others such as local salvage committees and rationing boards all wanted to use the volunteers to support their work as well. But agents and administrators at Purdue decided early on that since Extension had organized the program, the county Extension staff should decide how to utilize these volunteers.[38]

Limiting the scope of the volunteers' work created power struggles within some counties, as agent Gordon Sowers described in his 1942 annual report:

> From the beginning of the organization of the Neighborhood Leader System in Montgomery County, it was apparent to the county agricultural extension agent that numerous groups in the county were desirous of obtaining a list of the Neighborhood Leaders so that they might send them materials of various kinds and get them to do certain jobs. It was obvious that the Neighborhood Leaders' time must be protected and that they must not be exploited by outside agencies, so after consulting with the State Agricultural Extension Department at Purdue, the county agent refused to allow any group to have a list of Neighborhood Leaders. This resulted in a great deal of misunderstanding, both on the part of the Neighborhood Leaders and of leaders of other groups, and the misunderstanding was finally cleared up when the county agent sent a mimeographed letter to each Neighborhood Leader explaining the situation and assuring him that he [the agent] would keep faith with the Neighborhood Leaders and not allow them to be exploited by other groups. A number of Neighborhood Leaders stated that they appreciated this clarification of the situation.[39]

The agents strongly felt the need to protect the neighborhood leaders from excessive demands on their time since these volunteers were critical to helping Extension meet its increased wartime responsibilities.

It was left to the agent and members of the local advisory committee to decide how to break the county into a series of neighborhoods.[40] Quite often, the agent formed temporary township-based committees to aid in the process. In Pulaski County, Carl Olson used that approach, later reporting that "the group was asked to gather around a table and with a township map which had already been prepared, they were asked to write in the names of the people in their township. After all of the families had been listed, they were divided into neighborhood groups ranging from 4 to 22 families depending on distribution."[41]

The number of people assigned to the men and women who served as volunteer leaders depended on the county, the number of available volunteers, and the density of people within a given area, which affected driving distances.[42] For instance, Sullivan County created 200 neighborhoods that were divided up among 400 neighborhood leaders.[43] And Ripley County mapped out 168 neighborhoods with 335 leaders serving 2,500 families.[44]

The agents and members of the temporary township committee provided the names of potential neighborhood leaders, who were then contacted to ascertain if they would serve. Those accepting were officially notified of their appointment with a letter signed by Purdue Dean of Agriculture Harry Reed and the local agent.[45] Most leaders enthusiastically accepted and remained in that capacity until the end of the war.[46]

The leaders were then sent follow-up letters explaining the duties they might be asked to take on. In Parke County, Noah Hadley produced and distributed a newsletter titled "Neighborhood News" to keep volunteers informed on "current economic and technical information and governmental regulations."[47]

Neighborhood leaders were also called to township meetings to discuss the program in greater detail and answer their questions.[48] In Huntington County, Walter Rusk created the following agenda for the multiple neighborhood leader meetings he conducted in 1942:

A. Explain the neighborhood organization
  1. The purpose and need for the neighborhood organization
  2. Jobs that are to be done
  3. Duties of the leader
B. Presents economic information
  1. Discusses costs of living
  2. Price ceilings
  3. Uses economic charts that were furnished
C. Conducts a short discussion period
  1. Gives leaders a chance to ask questions
D. Elects township chairman
  1. This election is conducted by written ballot
E. Assigned each group of leaders the job of preparing a complete mailing list of all farmers in their neighborhood.
  1. This list to be sent to the county office
  2. After it is retyped, a copy is then kept at the county office and one copy to each township leader, and one copy to each of the representatives of the three neighborhood leaders.[49]

At the township meetings, the leaders received the names of families assigned to their neighborhood. Then each farm family received a letter from the agent telling them about the newly created program along with the name of their neighborhood leader.[50]

Across Indiana, these leaders were credited with assisting their communities with many programs. Their duties varied from county to county, but examples of activities—both Extension- and war-related—that they were often asked to take on included:

- drives for community and national organizations, such as the Red Cross, Salvation Army, YMCA, Navy Relief, United War Relief, United Service Organizations (USO), and other wartime charities
- tasks to support wartime needs, such as selecting rural air-raid wardens and participating in community collections, for example, to recycle rubber and scrap metal;
- tasks related to farm production, including helping to recruit farm laborers, distributing information about farm safety, and conducting surveys on corn borers, equipment availability, and farm labor needs;
- tasks related to government programs and wartime agricultural markets, such as providing explanations about dairy subsidies and federal price controls;
- tasks related to home food production, including promoting victory gardens, distributing information about food preservation, and conducting surveys on home canning equipment; and
- promotion of Extension-related activities, including publicizing 4-H club recruitments and information about Purdue Winter Short Courses in Agriculture.[51]

The neighborhood leaders in Tippecanoe County conducted seven programs for agent Sheldon Pershing in 1942, which he reported as:

1. The "Scrap Harvest"–with a seven-ton quota per neighborhood unit to meet the critical need for scrap iron and steel.
2. "Salute to Heroes"—September bond purchase program, including participation in a District War Rally; promotion of a definite purchase plan urging adoption of the following: Victory Pig, one-day egg receipts, or milk receipts per week or monthly, definite per cent of receipts, or some other specific plan.
3. "Save Waste Fats"—Pledge to keep a container handy in the kitchen; and also pledging a volunteer quota for delivery monthly.
4. For wheat growers. "Pledge to Not Sow Wheat in Standing Corn"—To make possible complete disposal of corn stalks in corn borer control; also information on fly-free date, September 28, etc.
5. For hog producers. "Feed some Wheat"—For faster gain and to save corn for future needs.
6. For hay makers. "Prevent Spontaneous Combustion Fires," being careful of moisture content, and if in doubt, get special hay mow thermometer at County Agent's office and check up on temperature.
7. "Perfect Neighborhood Rolls"—Check for completeness, use of first name or initial of head of family.[52]

Pershing gave every neighborhood leader a series of leaflets that described the objectives of each program to make sure the leaders understood what needed to be done.[53]

In annual reports for Scott County, agent John McCulloch listed the following leaflets and circular letters he sent out in 1942 and 1943 to neighborhood leaders:

1942

1. How Can Firewood Help Win The War?
2. Farm Fire Facts
3. Important facts on the life history, habits, and control of the European Corn Borer
4. Letter from [Purdue Extension] Director H[arry] J. Reed
5. Lists of families in each neighborhood
6. What can I do (booklet) and Home Storage of Fruits and Vegetables
7. Fire hazard check sheets
8. Letter urging people to get in scrap anytime
9. Details on scrap drive December 7 and Information on collecting tin cans

1943

10. War Food Communique No. 1, Proclamation by Governor H[enry] F. Schricker and Save your waste fats to make explosives
11. Letter with enclosures—Our good earth—keep it ours and Properly fed sows farrow large vigorous pigs
12. Scrap iron drive
13. Safety bulletin with check sheets—Purdue Poultry Pointers
14. Second war loan
15. Poultry feeding with limited feed supplies
16. Third war loan
17. War fund rally—October 16
18. War fund drive including solicitation cards[54]

Agents had their work cut out for them trying to maintain enthusiasm among the leaders. The agents also had to find substitutes for those leaders who had failed to perform their duties or just got tired of the work and resigned.[55] In 1943, Lawrence County agent John Armstrong reported that he had asked local leaders to do a half dozen tasks, but only 75 percent of the volunteers had completed the work.[56] Armstrong noted, "Old leaders that have been working must be retained by cajoling, bribery or other means."[57]

In spite of the negatives, agents seemed pleased that the program was reaching the agricultural community. In Jackson County, the agent reported that 821 leaders had contributed 3,500 hours of effort in relaying information about the war programs to their neighbors during 1942.[58] And in 1945, Orange County agent Ross Atkinson observed, "Since there is a neighborhood system in rather good working condition, we find it not too hard to present new practices to the county through these leaders."[59]

Over the course of the war, neighborhood leaders conducted many activities on behalf of the Extension office and others that the agent deemed essential, but three signature programs in particular stood out: rural fire prevention, scrap material recovery, and war bond drives.

Fire prevention in rural areas was one of the earliest programs that Extension asked the neighborhood leaders to conduct as part of a national fire prevention program.[60] With building materials in short supply (see Chapter 42), leaders were asked to talk to farmers about steps they could take to reduce the fire risk for homes, barns, and outbuildings.

One task that leaders reminded farmers to do was to periodically measure the temperature in the middle of any stored hay in a barn to detect the risk for spontaneous combustion.[61] To check the temperature, farmers could borrow a hay probe from the local Extension office, with Porter County agent Stewart Leaming describing the process: "[W]henever a farmer has trouble with hot hay in his mow, he borrows the probe, observes the temperature and compares it to the scale printed in the leaflet printed by Purdue—then taking necessary precautionary measures if necessary."[62]

Across the state, leaders were given printed materials to distribute to area farmers, which included a checklist to guide rural families through an inspection of their homes and barns for locating potential fire hazards, a did-you-know leaflet on fire prevention, a card to use in reporting any hazards to the agent, and fire prevention posters to place around their local communities.[63] For example, LaGrange County and Huntington County leaders distributed 1,200 and 2,500 fire prevention checklists, respectively.[64]

In addition to distributing publications, leaders discussed with farmers how to use a fire extinguisher, methods of storing tractor fuels, conditions to look for when inspecting the chimney, examining electrical wiring for defects, preventing lightning strikes by rodding and grounding buildings, eliminating smoking around barns, organizing volunteer firefighters, and asking farmers to bring two milk cans of water in the event of a nearby fire.[65]

Agents worked with county committees to coordinate the collection of scrap materials in 1942 and 1943. The agents called on the neighborhood leaders to contact their neighbors about the kinds of scrap the government was looking to collect and recycle, including iron, steel, tin, rubber, paper, rags, and waste fat from cooking.[66]

The federal government set dates for some collections, such as for scrap metal. Each state was given a quota for the expected amount of scrap to collect.[67] The 1943 collection quota for Pulaski County was 6,400 pounds for tin cans, 8 tons for other tin, 500 tons for iron, 100 tons for paper, and 8,050 pounds of waste fat from cooking.[68]

Neighborhood leaders and others across the state made sure that every household in the county received a pledge card to encourage participation in the scrap collections.[69] Agent Walter Rusk provided the following explanation for how the pledge cards were used in Huntington County in 1942:

> These pledge cards indicated the amount of scrap iron, paper, rubber and rags that farm people had available and how they were to be re-imbursed, whether they were to sell for cash, defense stamps, or give the money to [a] charitable organization which they indicated. All farmers were encouraged to deliver their scrap materials to the local junk dealers. . . . The local papers also carried pledge blanks and many of the schools assisted in distributing these pledge blanks so that everyone in the county was reached.[70]

Neighborhood leaders often took a prominent role in scrap collections, helping to distribute information, remind community members of the dates to encourage their involvement, and at times, picking up materials from donors. Depending on the material being called for by the government, various community groups got involved, such as 4-H and home demonstration clubs. Area businesses provided support as well by advertising the event in their establishments, sending out information to their customers to promote participation, and offering to be a collection site. For example, rural electric cooperatives placed flyers in their customers' monthly statements to remind them of collection dates and drop-off points in the county.[71]

In describing a 1942 collection, Jay County agent Chester Langston noted that neighborhood leaders "called on each farm family and encouraged those having scrap materials to deliver this to the depots [collection points] on designated days. A record was kept of those people who had scrap material and were unable to get it to the depots and this record was turned over to the township salvage committee who in turn directed trucks to pick it up."[72] Extension offices often sent reminder letters to rural homes before a scrap campaign was launched.[73]

The collection of scrap metal would prove to be a particular focus of the scrap collection drives. By the summer of 1942, manufacturers making hardware for the military were concerned that the demand for equipment would outstrip the supply of iron and steel.[74] Keeping the mills working at full capacity would require large amounts of scrap metal.[75]

Government-directed publicity campaigns reminded the public that scrap iron and steel lying unused around the farm had high value for manufacturing equipment for the men and women overseas. The Scott County salvage committees used the slogan "The Japs [Japanese] Bombed Us On December 7, 1941, Let's Bomb Them With Scott County Scrap on December 7, 1942" as one way of energizing citizens to donate any scrap metal they could collect on their farms.[76]

The public was repeatedly told that half of every gun, tank, and ship was made from scrap metal. Posters reminded people what could be manufactured from donated metals:

1 old plow equals 100 armor piercing shells
1 old disc equals 210 rifles
1 old flat iron equals 30 hand grenades
1 old stove equals 4 10-inch shells
1 bucket equals 3 bayonets[77]

Agents were especially interested in asking farmers to salvage dilapidated equipment, but as Marion County agent Horace Abbott explained, "Farmers have been rather hesitant to sacrifice used farm machinery due to the shortage of new machinery and the outlook for even more critical conditions in this respect. . . . Only those farm implements that are hopelessly beyond repair have been relegated to this scrap heap."[78]

It was commonplace for locations such as auto dealers, rural schools, filling stations, and junkyards to serve as central repositories for the collected scrap metal.[79] Abbott described one such arrangement in Marion County, writing, "On account of the distance of buyers from the average farmer, twelve weighing centers were established in the rural areas where farmers and other people having heavy metal might sell this metal into commercial channels."[80]

County highway departments and rural electric cooperatives donated trucks to pick up and deliver any scrap metal that owners could not haul to drop-off points.[81] When a truck was sent to a farm, the scrap metal was weighed and a receipt provided to the owner. In St. Joseph County, the agent used the receipts to calculate how much was owed. He then issued a check signed by him and the chairman of the scrap salvage committee.[82]

Local banks provided no-interest loans to allow for the purchase of scrap metals along with other salvaged materials. In the 1942 Jay County annual report, agent Langston described the process, saying, "The Citizens Bank of Portland financed the township salvage committees in this purchase of scrap materials. After the salvage materials were collected, the committee sold them to an authorized junk dealer and returned the money to the bank."[83]

Junk dealers paid 55–60 cents for every hundred pounds of scrap metal delivered to their facilities. The price was reduced to 40–50 cents when the scrap material had been picked up by a third party and brought to the recycling center.[84]

Contributors were often paid in war stamps, which could be saved toward the purchase of war bonds.[85] Others who received cash would often donate their earnings to organizations that supported military troops.[86] In 1942, agent Eugene Bird reported raising $1,500 from the sale of scrap metal in St. Joseph County, with the proceeds divided as follows: "one-fifth to Salvation Army, two-fifths to Red Cross and two-fifths to the USO."[87] Other organizations also benefited at times, with Marion County students raising $6,000 for student education through a 1943 scrap metal drive.[88]

With tremendous public support, most counties exceeded their quotas. In 1943, Allen County led statewide collections of scrap metal with 27,000 tons, while Marion County claimed second place with 12,170 tons.[89] The previous year, agent Scott Milligan reported that the Wayne County fall drive yielded 3,200 tons of scrap metal, which averaged 110 pounds per person and qualified the county for a War Production Board pennant in recognition of the achievement.[90]

The last step in the process was to haul the collected scrap metal to larger distribution points. In 1942, agent Guy Harris noted that Ripley County trucks picked up "the scrap iron in the daytime and hauled it to the central point, which was Cincinnati, during the night. The yards at Cincinnati became so full of scrap that it was impossible for the trucks to get unloaded and back into the county for the next day's work."[91] This showed that members of the public were fully supporting the various recycling drives as one way to show their patriotism.

The third area where neighborhood leaders would lend important support to the work of the agents was during drives to sell bonds for the war effort. On May 1, 1941, the United States had begun issuing defense bonds to raise cash to pay for the mounting debt incurred for the war overseas. These defense bonds were officially renamed war bonds when the U.S. entered the war later that year. From 1942 to 1945, the federal government conducted eight war bond drives, which generated $185.7 billion.[92] The dates for the nationwide drives were:

- First War Loan Drive: November 30 to December 23, 1942
- Second War Loan Drive: April 12 to May 1, 1943
- Third War Loan Drive: September 9 to October 1, 1943
- Fourth War Loan Drive: January 18 to February 15, 1944
- Fifth War Loan Drive: June 12 to July 8, 1944
- Sixth War Loan Drive: November 20 to December 16, 1944
- Seventh War Loan Drive: May 14 to June 30, 1945
- Victory Loan Drive: October 29 to December 8, 1945[93]

The face value of the war bonds ranged from $25.00 to $1,000.00, giving everyone the opportunity to pledge according to their ability to do so.[94] A person paid $18.75 for a $25.00 war bond that could be redeemed for its full value in ten years. War stamps could also be purchased for ten cents each, saved in an album, and redeemed for a war bond when the album was filled.[95]

Indiana received its quota from the federal government, then in turn, the State War Loan Office determined how much money individual counties were expected to collect.[96] Agents were put in charge of bond drives in the rural parts of the county, while others in the community were charged with coordinating efforts in urban areas, factories, and offices.[97] Ernest Brown described what he went through to get the first war bond drive in Rush County off the ground in 1942, writing:

> After receiving the county quota from the state organization, township quotas are established, based on the percentage of the total county assessed valuation within their respective townships, resulting in larger quotas in townships having a higher assessed valuation and vice-versa. Every week the issuing agents (post office, banks, etc.) made a report of the [war] bonds sold and to what township they are accredited, which is summarized and published in the paper to inform the townships of their standing. This publicity arouses township competitiveness, [and] publicizes and promotes the sale of war bonds in the county.[98]

As the war debt mounted, state and county quotas became increasingly larger with each new loan drive.[99] With counties asked to sell more bonds at each succeeding drive, the agents relied on neighborhood leaders to organize a door-to-door sales force to make sure that those living in rural areas were contacted personally and asked to buy bonds.[100] During one of the 1943 bond drives, LaPorte County neighborhood leaders visited 1,750 homes, which led to bond pledges totaling $100,000.[101]

There were many approaches and even a few gimmicks to encourage those from rural areas to invest in bonds. For the 1942 bond drive, agents asked each farm family to invest 10 percent of their net earnings each quarter or at least twice a year.[102] That same year, Union County agent James Guthrie described a promotion called the Victory Pig Club program, which "secure[d] pledges by farmers to put the proceeds of the sale of one or more pigs into War Bonds on December 7. . . . 210 members pledged a total of 250 pigs."[103]

Youth also actively participated in war bond drives alongside adults. In 1945, Miami County agent John Wylie said schools that sold at least $3,000 worth of bonds "would receive a plaque which would be placed on the bed of a wounded serviceman in one of the State Service Hospitals and a copy be placed in the school. Over half of the schools [in the county] have gone over the top and [at] the time of this writing, indications are that practical[ly] all will go over."[104] And in LaPorte County, the home demonstration agent and assistant county agricultural agent organized a 4-H war bond project in 1942, with agent Everett Hartman reporting, "This project was organized so that the boys and girls might save their spending money and money obtained from the collection of waste materials to buy stamps and bonds."[105]

Counties largely exceeded their dollar quotas because people believed it was their patriotic duty to support those serving by purchasing war bonds.[106] In Wayne County, agent Scott Milligan reported that in the summer of 1945 during the seventh drive, $7,692,221 worth of bonds were sold, with every township exceeding its quota.[107]

But by the end of 1945, fatigue among the neighborhood leaders was setting in. Huntington County agent Walter Rusk reported, "[P]eople are showing a let-down and the [neighborhood] leaders were not as enthusiastic in soliciting. The War Bond Drives are the most difficult drives to conduct and the Victory Bond Drive [eighth drive; held in November and December 1945] that is being organized indicates even more lack of interest."[108]

While the interest shown by the neighborhood leaders may have waned at the end of 1945, their work throughout the war years had provided essential support to the agents for carrying out the work of Extension at a critical time in the nation's history.

World War II brought many changes to the county Extension program. From the inception of the Cooperative Extension Service, farmers and their families had driven their horse-drawn wagons and trucks to Extension-sponsored events. With government-mandated tire and gas rationing firmly in place during the war years, agents adapted by delivering information to their constituents in ways that did not depend on face-to-face meetings.[109] As the war concluded, Extension returned to holding meetings and field days, opting to rely once again on in-person connections and hands-on demonstrations to deliver the latest agricultural information.

While radio broadcasts, newspaper columns, and circular letters continued to be used for communicating Extension information long after the war, the Neighborhood Leaders Program was discontinued. This program, which one agent described as "held together by the cement of patriotism," was disbanded once the overwhelming needs of the war had dissipated.[110]

CHAPTER 40

School Children of America! Help Save Your Fathers', Brothers', and Neighbors' Lives by Collecting Milkweed Pods

American sailors, soldiers, marines, and flight crews depended on life jackets, life preservers, rafts, and flight suits to stay afloat when enemy fire sank their ships or downed their planes over water.[1] Those using life-saving devices such as these had a better chance of surviving in water until search-and-rescue operations could find them.

Life jackets and similar equipment manufactured prior to World War II were filled with plant fibers processed from the fruit of the kapok tree, which primarily grew around the Dutch East Indies (today known as Indonesia). The waxy fibers of the kapok fruit were resistant to water and held thirty times their weight in water.[2]

The United States imported 20 million pounds, or roughly 90 percent, of the fiber incorporated as a filler in those life-saving devices.[3] From early in the war, Japan controlled the one location in the world that had previously supplied the U.S. with the fiber it needed for military use.[4] While kapok from South America was a potential source of fiber, there just wasn't enough production there to fill all of the devices that were expected to be manufactured for American and Allied military personnel.[5] And the cargo ships that would carry the South American fiber to the United States faced a serious threat from German submarines patrolling those shipping lanes.

Finding a substitute for kapok fiber became a high priority for Congress and the country's military brass.[6] As an article in the *Scott County Journal* would explain in 1944, "These life jackets are standard equipment in the navy and merchant marine. Fliers who range over the ocean, soldiers and marines engaged in amphibious warfare, and all other servicemen who face danger of death by drowning depend on these life jackets to keep them afloat until help comes."[7] If a suitable substitute could not be readily found in quantities needed to manufacture new equipment, military planners had no choice but to assume that more individuals would drown at sea before rescue teams could reach them in time.[8]

Scientists and inventors proposed a number of alternative fibers, but early on, the floss derived from milkweed pods topped the list due to what was already known about the physical properties of the plant's fibers.[9] Boris Berkman, a Russian-born physician and inventor in Chicago, had researched the commercial opportunities for milkweed and by 1939 had patented a process for separating the floss from the attached seed.[10]

Berkman appeared before a congressional agricultural committee in the spring of 1942 to discuss the suitability of using milkweed floss as a filler for life jackets, the lining of flight suits, the insulation of military aircraft, and countless other military and civilian uses.[11] He told the elected officials that individual milkweed fibers were essentially a hollow tube filled with air and coated in wax, making them lightweight, buoyant, and water-resistant.[12] An article in the *Muncie Evening Press* highlighted the benefits, saying, "[The] fiber of milkweed has been found equal to kapok and even superior in some respects. The waxy coating of the fiber makes it rot-resistant under most conditions and it is said to break less readily than kapok. The milkweed floss is also considered less likely to pack or lump after continued use."[13] The plant that had long been considered a troubling weed in farm fields and flower gardens, and along roadsides and ditches was now highly sought after. As one newspaper article stated, "After years of being considered a pest, milkweed finds itself suddenly on the nation's priority list as a 'war plant' because the floss its pods contain can be used to replace war-scarce kapok."[14]

To verify Berkman's claims, the U.S. Navy conducted its own experiments, which confirmed that milkweed fiber was perfectly suited as a substitute for kapok fiber.[15] The one significant glitch that could have derailed the optimism of using it was the immediate need for the replacement fiber as kapok supplies dwindled. Milkweed plants could be cultivated as a crop, but producing an economically viable crop would take three years.[16] That meant the only immediate source of fiber was from uncultivated milkweed plants growing as a weed on farms and undisturbed land, in ditch banks, and alongside fencerows and highways. This created a second predicament, which was, Who would harvest the pods from such scattered areas? Labor was already in short supply due to the war, and the remaining agricultural workers had their hands full. It seemed the only workable solution would be to organize the nation's elementary and secondary students to go out and handpick the milkweed pods.[17]

Based on the building and manufacturing designs Berkman provided, Congress appropriated $225,000 to build a single processing plant at Petoskey, Michigan, that would separate the milkweed floss from the seeds.[18] In August 1942, the government took over a building owned by the Preston Feather and Sons Lumber Company to build the plant. The facility was enlarged to 60,000 square feet to meet the anticipated influx of milkweed pods.[19]

The U.S. Department of Agriculture (USDA) supported the Petoskey site for two reasons. First, the lumberyard was easily accessible by road, rail, and water, which made it easier to receive the raw milkweed pods and ship out the collected fibers to manufacturers.[20] A second but equally important reason was that the plant was situated near where 60 percent of the harvested milkweed pods were expected to be collected.[21]

The USDA established a three-year contract to run from 1943 to 1945 with the goal of collecting enough milkweed fiber to fill 1 million life jackets.[22] Due to delays in obtaining equipment, the plant didn't run at full capacity until the fall of 1943, but eventually it would operate twenty-four hours a day for seven days a week with a staff of approximately eighty employees in order to meet the emergency needs of the military.[23]

Approximately every three days, a boxcar of baled milkweed floss left Petoskey heading to factories that used the material to pack life jackets.[24] All of the fiber produced at the plant went only to military contractors, with one newspaper article noting, "None of the floss will be available for civilian goods until after the war."[25]

The processing plant faced an immediate setback in October 1943, when the anticipated supply of milkweed pods from the surrounding areas was much less than what had been anticipated. That month, a newspaper article published by the *Times* in Munster, Indiana, noted that "northern Michigan is not yielding a sufficient supply of the floss to meet requirements and that assistance is being asked of other areas where sufficient quantities of milkweed grow to justify a collection program."[26] What started out as basically a single-state program now turned into a much larger regional collection effort.

In an attempt to get more states participating in the milkweed program, the federal War Food Administration and the Office of Education jointly called for a national collection effort.[27] In support, the Public Roads Administration sent letters to state and county highway commissioners asking them not to cut milkweed plants growing in rights-of-way until the pods could be hand-harvested by area students.[28]

By 1944, the USDA asked each state where milkweed grew to designate a full-time director to spearhead a pod-collection program. In July, Purdue assigned weed Extension specialist Oliver C. Lee as the director of milkweed collection for Indiana.[29] Lee was told that Indiana's assigned quota for 1944 was 10,000 bags of milkweed pods, with the size of the bags based on the mesh ones that were usually used to hold fifty pounds of onions.[30] Lee then set quotas for each county based on his knowledge of how milkweed plants were distributed across the state.[31] For example, the quota for Ripley County was 400 bags, while the Pike County quota was 1,000 bags.[32]

Each bag held from 600 to 800 pods, which meant that the Pike County students would have to pick roughly 600,000 to 800,000 pods to fill their quota of

1,000 bags.[33] The floss from one bag yielded roughly one bushel, or four-fifths of a pound, of floss. It took 1.5 pounds of floss to fill a single life jacket—or approximately two bags of pods—so the floss from those 1,000 bags would be enough to pack approximately 500 life jackets.[34]

In the spring of 1944, Clement T. Malan, Indiana superintendent of public instruction, wrote to county superintendents of education and school principals statewide, asking them to encourage students to harvest milkweed pods in September.[35] Lee often teamed up with the local county Extension agent to meet with the county superintendent and area principals and teachers in order to get their buy-in.[36] Lee and the agents encouraged local leaders to establish a county milkweed collection program.[37]

There were cases where the agent became the de facto chair of the collection program when superintendents or principals refused to participate or were largely incompetent in managing these duties.[38] In the 1944 Posey County annual report, agent Oral Riggs explained, "Milkweed pod collection, sandwiched in with all the other extension programs, almost became the straw that broke the camel's back. The County Superintendent of Schools said that he just could not possibly handle the program. The War Board chairman thought they had more to do than they could get done. No one else wanted to attack the program, so it just naturally fell into my lap."[39] Luckily for most agents across Indiana, the Posey County situation was more or less an outlier.

All participating schools received a pamphlet prepared by the federal government that explained how to select, pick, and store milkweed pods. The publication emphasized the importance of what the students were being asked to do by saying, "School children of *America!* Help save your fathers', brothers', and neighbors' lives by collecting milkweed pods."[40]

The success of the statewide collection program hinged on getting students to harvest the pods in a timely fashion. Through newspaper articles and word-of-mouth promotions, agents made patriotic appeals to students, their parents, and anyone who would listen that collecting milkweed pods could save lives. One article in the *Tipton Daily Tribune* noted, "Home front Hoosiers who want to help their fighting loved ones all they can are being offered another collecting task—the gathering of milkweed floss to make life jackets. . . . Those wishing to aid the life jacket project may contact their county agricultural extension agent."[41] Catchy headlines such as "Save the Milkweed and Save a Life" and "Two Bags Save One Life" easily conveyed the difference that a single person could make.[42] And life jackets were often displayed in storefront businesses as a way of keeping the importance of the project in front of people walking past the windows.[43]

In September 1944, agent Willis Stall published the following column in the *Tribune* in Seymour:

> By this time, we feel that every person in Indiana knows that milkweed floss is badly needed for the purpose of making life jackets for the Navy. . . . Fortunately milkweeds grow in every part of Indiana. There are probably some on every farm in Jackson County.
>
> We need the help of every person in seeing that not one milkweed pod goes unharvested this year, for this very necessary use. We appeal to parents and land owners to see that school boys and girls be urged to help get milkweed pods gathered. These pods, in most instances, are ready to gather now. Some have already been gathered. Gathering should be done before the pods burst and the floss starts to scatter. . . .
>
> Payment at the rate of twenty cents per bag will be made to those bringing in properly dried milkweed pods. We know of several spots in Jackson county where any school child can make good wages collecting pods at this rate. However, the price paid is a small consideration in proportion to the importance of saving every bit of this important material. Every bit of this material that remains unharvested may mean that some American service man may be forced to abandon ship without a life jacket.
>
> We respectfully ask the assistance of every person in this county in seeing that every milkweed pod in Jackson county is harvested this year. If any person knows of milkweed pods and is unable to secure them, either Mr. [James] Tatlock, [Jackson County school superintendent,] his assistant Lester Hanner, County School Attendance officer, or I will be glad to hear of it and we will arrange for some school group or some youth group, such as the Boy Scouts or the Girl Scouts, to see that none of this material goes unharvested.
>
> Jackson county should harvest at least 1,000 bags of this material which, if properly dried and handled, will be enough to make 500 life jackets. This job is small if we all help. Its importance is tremendous. Let's not fail in this.[44]

As Stall's column indicated, the government paid students for their efforts. Participants received twenty cents per bag, which included fifteen cents for handpicking the pods and an additional five cents for drying the pods before bringing them to a drop-off point, such as a local school.[45] The Michigan processing plant sent the school a payment for the total number of bags, which was then distributed to the students based on how many bags they had collected.[46]

In the fall of 1944, students from Wells County collectively earned nearly $1,000 dollars from their 4,994 bags.[47] However, the payment was a secondary consideration for most, as an article in the *Princeton Clarion-News* pointed out: "The payment is an added incentive but even without it we think the school pupils of Gibson county could have been counted on to help in this essential project.

Gathering milkweed pods may seem a small thing to do but it is important—it will save lives."[48] As such, students often donated their hard-earned proceeds to the United Service Organizations (USO) and to the Red Cross.[49] Other students used their earnings to purchase war stamps and war bonds.[50]

Before joining the effort, students were explicitly told that in no way could the collection program interfere with their school attendance and homework assignments, leading one newspaper to report, "No school time was used for the collection of pods, students volunteering to help with the war effort [collected pods] outside of school hours."[51]

As director of the Indiana effort, Oliver Lee, in cooperation with the agents, made sure that everything was in place and that the students had received training prior to sending them out to harvest the pods.[52] Students had to learn how to identify milkweed plants, how to know when the pods were ripe, and how to dry and store the pods. As one newspaper reported, those doing the work were "urged to watch for the purplish flowers and fat green pods of milkweed, making notes on locations where they are to be found and the areas where they are the most abundant."[53]

Collection efforts focused on two different milkweed species: the common milkweed (*Asclepias syriaca*), which grew in large numbers in the northern half of the state, and the bluevine milkweed (*Cynanchum laeve*), which was primarily found in the southern counties.[54] Typically, agents announced when the pods were ready to harvest, usually sometime in September.[55] The students only had a seven- to ten-day window during which they could gather unopened pods.[56] Their instructions were to harvest only ripe pods—verified by cracking open a few pods to check for brown seeds—and not to include stems, leaves, and immature pods.[57]

Some counties had a rather coordinated operation, with the agent assisting in "securing trucks to haul pods, driving school busses to transport pickers, supervising pickers and spotting fields to pick in," as one agent described it.[58] In Vanderburgh County, agent Albert Bishea got permission from farmers with property along the river bottoms to have students harvest the bluevine milkweed plants from their fields. Within a four-hour period, school buses transported 1,400 students to Knight and Union Townships, where they filled 2,990 bags with pods.[59]

It was commonplace to see bags drying in barn lots, hung over fences, tied up in trees, or set on wooden frames.[60] The students were instructed to hang bags at least one foot off the ground to allow air movement and speed drying time.[61] It took three to five weeks for the pods to dry completely.[62] Pods that crackled when squeezed were dry enough to be moved inside the house, shed, or barn until the specified collection date.[63]

In the late fall, students were directed to bring the dried pods to their school.[64] Then state and county highway trucks were used to collect the bags and transport them to a central location, where they would be combined with those from other schools.[65]

In Williamsport, a government grain bin was used to store 2,000 bags collected from Warren County schools in 1944.[66] Bags of the collected pods often remained in storage before being loaded onto railcars for shipment to the Michigan processing plant.[67] Delays in shipping were common, which an article in the *Kokomo Tribune* noted was "because that plant, which processes them and removes the floss, has a heavy supply of pods on hand."[68]

When it was time to ship the bags, Lee made the arrangements for the railroads to pick them up at centralized locations across the state.[69] He was often present on the appointed day to make sure things ran smoothly at the site and to help agents and volunteers load the bags of pods onto the railcars.[70]

To make better use of the railcars, Lee tried to have 5,000 to 6,000 bags on site for each shipment.[71] Trucks often brought bags from one county to another, combining smaller collections into larger ones to make shipping more efficient.[72] During the later months of 1944 and into the winter of 1945, the plant at Petoskey received approximately three carloads of pods from Indiana each week.[73]

In 1944, Indiana's collection of pods was by any measure successful, with 89 of the 92 counties participating.[74] In Warrick County, even students from the one-room schoolhouses took part.[75] The Indiana quota of 10,000 bags was surpassed many times over that year, with the schoolchildren turning in 200,000 bags of milkweed pods, which were estimated to be worth $30,000.[76] The students must have felt quite proud knowing that their efforts helped fill approximately 100,000 life jackets.[77]

Many counties collected more than 1,000 bags of pods.[78] For example, the final tally for Posey County was 15,000 bags of bluevine pods.[79] Knox County schools harvested 8,000 bags consisting of 6.5 million milkweed pods, while the 40 participating schools in Madison County collected 2,527 bags.[80]

In Wells County, the collection of 4,000 bags produced enough floss to make life jackets for every person on one battleship, which was quite an achievement that the students and people in the county could talk about for years to come.[81] And statewide, the 4-H clubs contributed 50,000 bags out of the overall total.[82]

County officials often recognized those students who topped the leader board for the most bags turned in.[83] James and John Peer from Newport were singled out as the champion pod pickers in Vermillion County when they turned in 50 bags.[84] John Manship Jr. from Pendleton collected 41 bags to lead the campaign in Madison County.[85]

A few counties presented prizes such as a $25 war bond or $5–$10 in war stamps to those students who picked the most pods.[86] In Tipton County, agent Walter Clary announced that the Tipton Kiwanis Club awarded three monetary prizes to the schools in the county that had collected the highest number of pods on a per-student basis.[87]

In Pike County, champion milkweed picker Thomas Johnson became a local celebrity when his accomplishments appeared in the *Indiana Prairie Farmer*. Agent Edward Bogue discussed the story in his 1944 annual report, writing:

> The Indiana edition of 'Prairie Farmer' carried a story concerning the county winner, Thomas Johnson of Madison township. Thomas, a 14 year old student at Burkhart School, assisted by his two nephews, Freddie McCrary, 8 years old, and Froman McCrary, age 11 years, picked a total of 287 bags filled to capacity at the last count for which they received $57.40. Thomas picked 156 bags to win the $25 War bond. The magazine reporter asked Thomas what he planned to do with the money and the following answer was received. "I'm going to put the money for the pods in war stamps," but he declared he wasn't picking them for the money. "I'm just picking them because those boys need the life jackets."[88]

The students' work to collect milkweed pods should not be underestimated. While it will never be known how many lives they may have helped saved, there is no doubt that gathering the pods produced the fiber needed to manufacture essential military equipment at a critical time. As a secondary benefit, this participation by these young people also allowed them to make a positive and very important contribution to the war effort.

Approximately two dozen states supplied milkweed pods to the processing plant at Petoskey during the war.[89] A February 1945 article in the *Lafayette Journal and Courier* reported, "Michigan was top producer with 700,000 bushel bags. Other leading states were Wisconsin, New York, Pennsylvania and Minnesota, with Ohio, Illinois, Indiana, Iowa and Missouri close behind."[90]

Indiana's contribution was part of a larger national effort where more than 2.5 million bags were harvested and processed, with the fiber used to fill approximately 1.2 million life jackets. This amount exceeded the goal of collecting enough bags to fill 1 million life jackets.[91] In fact, the program had been so successful that the USDA abandoned it after the 1944 pods had been harvested and processed.[92]

Milkweed expert Boris Berkman and others hoped that the milkweed plant would transition from a weed to a cultivated crop that could be turned into lubricants, plastics, fabrics, edible oils, and livestock feedstocks.[93] A June 1945 article in the *Hancock Democrat* stated that the oil from milkweed seeds had "potential value in protective coatings such as exterior paints, enamels and baked finishes. New uses for these seeds, until now a waste product in the production of milkweed floss, will make possible the introduction of milkweed cultivation in the permanent agricultural economy after the war."[94] University experiments also pointed to the fact that 300 to 400 pounds of floss per acre could be produced with better selection and cross-breeding.[95]

Berkman was quoted in 1943 as saying that "the milkweed crop is here to stay as one of the postwar period developments and that kapok, formerly imported from lands now held by the Japanese, will not return as a serious competition after the war."[96] That prediction by Berkman would soon be proven wrong, and all hopes of turning milkweed into a commercially viable crop evaporated as milkweed floss was quickly replaced with kapok once the Japanese had been defeated.[97] And within a few years after the end of World War II, synthetic materials would replace both kapok and milkweed as the fiber of choice.

Today, milkweed is known as an agricultural pest or, conversely, a host plant to the migratory monarch butterfly. But for the duration of World War II, milkweed became an essential war product as its fiber was successfully used to protect those serving in the military from drowning at sea. Young people representing large and small schools, 4-H clubs, Boy Scouts, Girl Scouts, Camp Fire Girls, and Junior Red Cross enthusiastically took on the challenge of collecting milkweed pods as part of their patriotic responsibilities.[98] In 1944, more than 120,000 students walked across their counties to fill bags with milkweed pods in hopes that their efforts would make an important contribution to the war effort.[99]

In the 1944 annual report for Montgomery County, agent Gordon A. Sowers wrote, "[E]very boy and girl who participated can justly take pride in their recollection of their part in the great war. . . . In this way the county's school children helped save lives of many men in the service."[100] It was quite a noble accomplishment for the students to fill their onion bags with one handpicked pod after another for the benefit of those who they didn't know personally.

CHAPTER 41

American Hemp Will Go On Duty Again

The United States Army, Army Air Forces, Marines, and Navy depended heavily on rope to conduct military operations on land, in the air, and at sea during World War II. To make its supply of rope, the U.S. used Manila hemp, or abaca (*Musa textilis*), nearly all of which was imported.[1] The Davao Province in the Philippines was the epicenter of Manila hemp production, with nearly a half million acres devoted to growing this fiber crop that was largely exported across the globe.[2]

When the Japanese military took control of the port at Davao City on the island of Mindanao in the southern Philippines in December 1941, the U.S. lost access to its supply of Manila hemp. Other natural fibers affected by the war in the South Pacific and Asia included jute (*Corchorus capsularis*) from India and sisal hemp (*Agave sisalana*) from Indonesia.[3]

The term *hemp* had become synonymous with describing strong fibers from many different and unrelated plant species, such as abaca, sisal, and others. While each has been referred to as hemp, none of them are true hemp species.[4] A January 1942 article in the *Indianapolis Star* explained that abaca is a "super-long vegetable fiber that has practically shouldered ordinary hemp out the rope industry and appropriated its name. The fiber comes from the stalk of a type of banana 'tree.' So essential is the fiber for ropes, cables and cord that the United States has placed it on the list of strictly rationed strategic materials."[5]

A second article in the *Star* a few days later noted, "Japan had no hemp, manganese, rubber, tin or chromium. Now. . . they have them all and we have none except for the supplies on hand."[6] The problem was that the stored abaca supplies were expected to be depleted within a year's time, considering that every battleship required 34,000 feet of rope and new ships were being built and launched from American shipyards.[7] While nautical and rigging rope took priority, the lack of abaca fibers would also disrupt the manufacturing of uniforms, fire hoses, thread for shoes, parachute cords, twine, canvas, and countless other products used by both the military and civilians.

Military planners understood that this emergency needed a quick solution, much like the situation where milkweed fibers had been identified as a replacement for kapok fibers in life jackets and other equipment (see Chapter 40). There was immediate interest in substituting alternative fibers from plants growing in

Central and South America, but with German submarines patrolling shipping lanes, getting those fibers to the United States was problematic at best.[8] One by one, proposed alternatives were erased from the list for various reasons, including problems with transportation, accessibility, inconsistent quality, and insufficient volume.

It quickly became apparent that the United States would have to focus attention on domestic production.[9] Industrial hemp (*Cannabis sativa*), which had once been grown on thousands of acres in the U.S., had been shown to be an excellent fiber for meeting the needs for the military.[10] One newspaper account named it as the "best known substitute which can be produced on a practicable basis."[11]

President Franklin Roosevelt gave the U.S. Department of Agriculture (USDA) significant regulatory oversight of the domestic hemp production program.[12] The USDA announced its hemp-to-rope strategy on January 7, 1942. Kentucky would be asked to plant 30,000–37,000 acres of hemp to harvest for seed, forming the basis of a "multi-million-dollar program for revival of the century old hemp industry in Kentucky, Minnesota, Wisconsin, Illinois, Iowa, and Indiana," according to one newspaper report.[13]

These midwestern states had been chosen to participate because hemp production required the same fertile, well-drained soils as used for corn production.[14] The USDA goal was to get farmers from these six states to use the seed from Kentucky to grow 300,000 acres of hemp strictly for use by the government.[15] This was quite an ambitious target, considering that hemp production had fallen to 7,100 acres nationally by 1938.[16] The steady decline of domestic hemp production was primarily because growers in the Far East could produce fiber for rope and other goods cheaper than U.S. growers could.

However, when Kentucky produced less seed than expected in 1942, the USDA announced in January 1943 that the goal had been revised downward to 185,000 acres of hemp.[17] An additional 50,000 acres would be grown in Kentucky for seed to plant the 1944 crop.[18]

Reports vary, but it appears that Indiana was originally assigned 32,000–42,000 acres out of the 300,000-acre goal.[19] But when Kentucky produced the smaller seed crop, Indiana's quota was reduced to 20,000 acres for 1943.[20]

The federal government originally committed to building 70 mills to turn the 300,000 acres of raw hemp into commercial-grade fiber.[21] By April 1943, with fewer acres expected, the number of hemp mills on the drawing board was reduced to 42 scattered across the six hemp-growing states.[22] The total construction and equipment cost for each plant was expected to average in the neighborhood of $330,000.[23]

Initially, the plan proposed in November 1942 included five processing mills in Indiana at an estimated total cost of $1.5 million, with six potential sites listed. The number of mills was based on the state's 1943 allocation of 20,000 acres as well as an expectation that enough hemp for processing would be grown close to each plant.[24] Just a month later, the list of Indiana sites for the USDA-financed hemp mills was further refined, with the cities under consideration including Attica in Fountain County, Flora in Carroll County, Frankfort in Clinton County, Nappanee in Elkhart and Kosciusko Counties, Remington in Jasper County, and Warren in Huntington County.[25]

The processing plants had the potential to bring an investment and jobs to the communities where they were located, but by December 7, 1942, U.S. Senator Raymond E. Willis, a Republican representing Indiana, reminded his constituents that the construction of the mills hinged on the farmers signing contracts to grow enough hemp for the government.[26] Indeed, the location of the mills would be determined by three factors. First and most important, farmers near the proposed plant had to agree to plant a minimum of 4,000 acres of hemp. Second, those fields needed to be within twelve to fifteen miles of the mill, although the USDA did not hold firmly to this requirement.[27] And third, the farmers had to sign contracts that stipulated how the crop was to be handled.

State and federal officials faced several serious roadblocks in getting farmers to grow hemp in Indiana: (1) the federal Marihuana Tax Act of 1937, often called the Marijuana Tax Act in modern-day references; (2) public backlash for eliminating long-standing marijuana eradication programs; and (3) uncertainty over whether Indiana farmers would violate state laws by growing hemp under the federal program.

President Roosevelt had signed the Marijuana Tax Act into law in August of 1937 as a measure of controlling—but not outright banning—the use of *Cannabis*. But the act made no distinction between marijuana as a drug and industrial hemp as a rope fiber.[28] Trying to discourage the planting of hemp as a crop without making it illegal seemed to be the goal of the sponsors of this legislation.

The U.S. Treasury was charged with enforcing the law, which included burdensome taxing, licensing, and reporting requirements that effectively brought the production of hemp to a standstill. As author Nick Johnson noted, "By coupling strict requirements with harsh consequences, the [Marijuana] Tax Act so discouraged legal cannabis activity that it did, indeed function like a ban."[29] Given the likely prospect that the military would run out of rope sooner rather than later, the federal government needed to work around the provisions of the law in order to allow farmers to grow *Cannabis* and to deal with those who protested government subsidies for something viewed only months earlier as being illegal.

In December 1941, Roosevelt signed an executive order that provided a legal pathway for farmers to grow hemp without fear of prosecution by law enforcement.[30] While switching the classification of hemp from an illegal crop to a legal one was controversial, the military simply needed rope, and the war effort trumped the concerns about the change in policy. The government's new position worked this way: When the farmer signed a contract to grow hemp, the government paid for the registration and tax stamp, making it legal for the registrant to grow, harvest, and deliver hemp to a mill for use only by the military.[31] Farmers were repeatedly warned that if they crossed the line between legal rope and the illegal drug, they would find themselves in court.

Still, when the federal government announced that it would subsidize hemp production, Indiana officials quickly heard from protesters who feared that reintroducing the crop would subvert programs designed to eradicate marijuana.[32] In a November 1942 article titled "Hemp Growing in State Opposed by Groups Fighting Marijuana," the *Indianapolis Star* reported:

> Church and women's groups who last year asked Governor Henry F. Schricker to continue enforcement of the marijuana eradication program by the prevention of hemp growing on farm lands are determined that Indiana must not share in the proposed 400,000 [acre] hemp crop planned by the government for 1943. . . . Last year [1941] Governor Schricker assigned the Indiana state police to co-operate with the Federal Narcotics Bureau in total elimination of the marijuana trade and weed. Rural organizations joined in achieving what was believed [to be] complete eradication of every hemp plant from all of Indiana's 92 counties.[33]

Even if the federal government had authorized growing hemp, for many there was still a question as to whether state laws prevented this. A September 1942 article in the *Indianapolis News* explained that in 1939 the Indiana General Assembly

> forbade the growing of *Cannabis sativa* . . . and required its destruction whenever found. The same session of the assembly approved a new act regulating narcotics and included *Cannabis sativa*, its seeds and resins, but the act does not class as a narcotic the stalks of the plant, the fiber produced from stalks or the oils made from the seeds. . . . The disputed point now is whether the Indiana law actually forbids hemp growing. If it does, Indiana farmers who wish to grow it probably will take a chance that they will not be prosecuted for complying with a war program which the WPB [War Production Board] has marked essential to the war effort.[34]

In Indiana, there were no acres of hemp grown in 1942, so meeting a state quota of 20,000 acres in 1943 would be a challenge.[35] Farmers and county Extension agents in Indiana generally were skeptical about turning corn and soybean fields over to hemp. Their less-than-exuberant response wasn't too surprising, given that the state's farmers had repeatedly been told to destroy hemp in support of the marijuana eradication effort.[36] For some agents, that initial skepticism would eventually give way as they were told that Extension would publicly support the growing of hemp in Indiana.

In an effort to convince farmers to grow the crop, the government had launched a public relations campaign in February 1942 called Hemp for Victory, so prior to the 1943 growing season, agents began showing farmers the USDA-produced film of the same name, which promoted raising hemp for the war effort.[37] The *Indianapolis News* noted in a November 1942 article that Shelby County agent Dwight Brinson urged farmers "to grow as much hemp as possible next year. He made the appeal after he was informed by the United States Department of Agriculture that several hemp mills would be erected in Indiana. The reason for the changed attitude, of course, is that the nation's hemp sources have been cut off by the war in the Pacific."[38]

In counties where the government proposed building hemp-processing plants, the local agents arranged for government officials to speak at grower meetings, where the hemp program was fully discussed and debated.[39] In December 1942, the *Kokomo Tribune* reported on a meeting held in Clinton County, stating, "The large courtroom was filled and much interest was shown in the discussion of the cultivation of hemp, and of contract prices for the 4,000 acres which these farmers are asked to raise next summer. . . . The signing of contracts for the county's acreage [in Clinton County] will start this week."[40]

While previous generations of the state's farmers would have known how to grow hemp, the vast majority of those working in 1943 lacked that knowledge. This meant that researchers at the Indiana Agricultural Experiment Station at Purdue, campus Extension specialists, and the county Extension agents would have to coordinate a statewide effort to train a new generation of hemp growers. In 1943, the statewide annual report for Extension described staff efforts to educate farmers as "explaining the need, helping contract acreage, advising in the preparation of the ground and the seed; and giving detailed instructions concerning the handling of the crop, both when it was growing and when it was mature."[41] Purdue assigned the responsibilities of Indiana's hemp program to Extension agronomist George Enfield, who was helped by the agents where contracted hemp acres were to be grown.[42]

An *Indianapolis News* article published in July 1943 noted that, at first, farmers "just shook their heads and lifted their eyebrows" when asked "in the late winter and early spring to come to Uncle Sam's rescue and devote a part of lands to the production of a new war crop."[43] In the 1943 Huntington County annual report, agent Walter Rusk explained the response he received as he tried to convince local farmers to participate in the federal program:

> Hemp, being a new crop to this area, made a very difficult problem to sell to the farmers as a war-time emergency crop. The County Agent [Rusk] conducted six educational meetings with the attendance of 530 farmers and was assisted by the State Extension office in discussing soil requirements, seed bed preparation, and harvesting. It was difficult to overcome the many rumors and false statements that were circulating in the community concerning the production of hemp. . . . Toward the last of February, when it seemed that it was going to be impossible to get sufficient acreage for the operation of a plant, a few farmers were signed up which influenced then spread rapidly so that the 4,000 acres were secured in a very short time and some farmers were even on the waiting list.[44]

Rusk may have had a difficult time, but his perseverance paid off. By securing the minimum 4,000-acre commitment within a fourteen-mile radius of Warren, he helped bring a new hemp-processing facility to the county.

While Rusk was eventually successful in getting enough area farmers to participate, his results weren't typical. In Newton County, agent Russell Zell found it was "a rather difficult educational campaign" to get enough farmers to agree to plant the crop in order to reach the 4,000-acre minimum.[45] Through his efforts, he eventually got forty-five farmers to sign contracts to grow 451 acres of hemp. That wasn't enough to secure a processing plant, so the resulting crop would have to be processed outside the county.[46]

In the end, Indiana farmers contracted to grow 8,500 acres of hemp in 1943, which justified building only two processing mills of the proposed five. One mill was built in Remington in Jasper County for $261,000 and completed in April 1944. The second mill, which was located in Warren in Huntington County, cost $245,000, but construction ran behind schedule, so it wasn't completed until July 1944.[47]

The Indiana mills were each equipped to process 1,000 tons of hemp per month.[48] For each 2–3 tons of fiber, which was the amount typically produced by an acre, the mills could manufacture approximately 1,200 pounds of rope.[49]

While Indiana farmers fell short of the 20,000 acres of hemp assigned to the state, farmers in the other states contracted with USDA to plant 240,000 acres for the 1943 growing season.[50]

Indiana growers signing contracts to grow hemp for the 1943 season had to agree to the following stipulations:

- grow the hemp for government use only;
- permit no hemp to be taken off the farm unless it's to the processing mill;
- plant the acres agreed to in the contract;
- purchase all hemp seed from the government;
- plant 1.25 bushels of hemp seed per acre;
- follow good cultural and management practices as outlined by Extension; and
- rent USDA-owned equipment to harvest the crop.[51]

In return, the government agreed to supply the federal registration permit and tax stamp to allow the grower to legally plant and harvest hemp. The government contract also set guaranteed prices for harvested hemp straw based on the quality of the fiber, which included: Class 1, $50.00 per ton; Class 2, $40.00 per ton; Class 3, $35.00 per ton; and Class 4, $30.00 per ton.[52]

While the program was in effect, the USDA would distribute and sell 400,000 pounds of hemp seeds to farmers under government contract.[53] Seed cost $11.00–$12.50 an acre, but farmers would not have to pay for the seed until the crop was delivered to the mills. According to a July 1943 article in the *Indianapolis News*, farmers would not be expected to pay for the seed at all if the crop failed.[54]

Seed was planted at approximately 1.25–1.50 bushels per acre using a wheat drill. Hemp required 300–400 pounds of commercial fertilizer per acre to produce a yield of 2–3 tons per acre.[55] The same *Indianapolis News* article noted that the material was graded based on "uniform length of fiber and the proper degree of retting—or rotting. The preferred commercial length is about four feet, although if left to itself, the hemp plant will grow much taller. The straw delivered to the mill consists wholly of the tough, fibrous stems or stalks. Hemp defoliates itself in the process of ripening or retting. Thickly grown under cultivation, the plant is said to lose much of its narcotic content."[56]

The main problem that growers faced was the lack of harvest equipment. The USDA agreed to pay for the construction of 3,500 units of specialty hemp-harvesting equipment, with John Deere making the harvesters and the International Harvester Company making the binders. This equipment was slow in coming, but as it became available, it was loaned to equipment dealers near where contracted hemp was grown. These dealers were paid to rent, maintain, and repair the equipment.

In 1943, Huntington County received twenty-nine harvesters to help with the September harvest.[57] In order to spread the equipment and labor among the hemp producers, Huntington County agent Walter Rusk noted, "All hemp growers were organized into hemp rings of eight to fifteen farmers who assisted each other in getting the crop harvested."[58] The farmers paid five dollars per acre to rent the machinery from the equipment dealers.[59]

Harvesting hemp was difficult work, with Rusk reporting, "Most of the hemp has been turned by hand but has been slow in retting so that the pick-up binders have been slow in binding the hemp into bundles for shocking or hauling to the hemp mill."[60]

Agent Mervin Smith in Wells County, which was located just to the east of Huntington County and the mill at Warren, provided a comprehensive overview of the hemp program in that area in his 1943 annual report, writing:

> Since this [hemp] crop requires a soil with a relatively high fertility level, such as is also required for good corn production, Wells county was selected as one of the counties where the crop could advantageously be produced.
>
> A site for the mill was selected at the edge of the town of Warren, in Huntington county. Average quotas were set up for the surrounding counties as follows: Huntington—2,000 acres; Wells—1,500 acres; Grant—400 acres; Blackford—100 acres. In promoting the project, the Extension Service had charge of the education part of the program and cooperated with the AAA [Agricultural Adjustment Administration] in scheduling township meetings in Jackson, Chester, Liberty, Rockcreek, Nottingham, and Harrison townships [in Wells County].
>
> The requirements of the crop, the national need for its production, and provisions of the proposed contract were explained in these meetings, and farmers urged to sign contracts. While the contract price offered was such that a fair profit could be expected from the crop, yet the primary motive of those signing contracts for its production was that of a patriotic desire to make their contribution to the War Effort.
>
> Wells County farmers were the first to go over the top in meeting the acreage quota assigned, with contracts totaling 2,300 acres. Some of this excessive acreage above our county quota was used to absorb a part of the quota for the other counties which failed to meet their assigned quotas.
>
> Meetings were held to instruct growers on the cultural practices to be followed in producing the crop. George Enfield, Extension Agronomist, Purdue, assisted in conducting these meetings.
>
> Five carloads of seed were shipped into the area from Kentucky, for the planting of the crop. Three of these cars were received at Liberty Center, in Wells County, and two at Warren, in Huntington county. The treating of the seed with one of the commercial mercury dust treatments, was supervised by the Wells County Agent [Smith]. . . .

> The management of the mill was placed in the hands of Von Crow, former Wells County Highway Superintendent, who began his duties on May 25 [1943], assisted by Robert Bricker, of Grant County. Mr. Crow and Mr. Bricker organized the growers into twenty-nine rings averaging 125 to 130 acres each, for purposes of harvesting the crop. Thirty-one hemp harvesting machines built by the John Deere Company, were shipped [to] the area and used in cutting the crop. This work was started in early September. Following a retting period of six to eight weeks, during which time approximately one-half of the hemp was turned, the crop was bound by use of thirty binding machines built by the International Harvester Company. . . .
>
> A total of 3,468 acres were seeded in the entire area. Of this number, 2,811 acres were harvested, and 657 acres abandoned due to flooding and other causes. . . . Early receipts of hemp at the mill indicated probable average yield of 2.5 tons per acre, approximately two-thirds of which was grading number one. Many farmers making early deliveries to the mill had net returns of over $100 per acre after expenses were deducted. Considering the handicaps of weather, and the inexperience of our growers with the crop, the enterprise has been considered on the whole, quite successful. While it is not contemplated that the crop will continue to be produced after the war in competition with our established crops, yet we have proven our ability to meet the challenges of this war emergency.[61]

Smith seems to reference deliveries to the Warren mill in this 1943 report, but with Huntington County agent Walter Rusk reporting that this mill wasn't operational until July 7, 1944, it's likely that the fiber grown in the area may have been processed by another mill.[62]

Prior to growing hemp for the government, farmers had asked their local agents whether the subsidized hemp program would be more profitable than planting corn, soybeans, and or oats.[63] Smith's 1944 report provided these details about the results obtained by local farmers: "Yields ranging from less than a ton to more than four tons per acre were secured, with an average of approximately two and one-half tons per acre. . . . Net returns reported ranged from a few dollars per acre, to a return of $187 per acre, after seed, fertilizer and machinery expense had been paid. The returns were generally very satisfactory to the growers. . . ."[64]

By January 1944, with the advantages going to the Allied forces in the Mediterranean and with German submarines posing less of a threat, military planners felt more certain that they could get the material they needed to manufacture rope from Central America, South America, India, and Italy as well as other areas in the Mediterranean at cheaper prices.[65] A *Muncie Evening Press* article published on April 7, 1944, reported, "All essential military and civilian needs for rope are being met with far fewer domestic hemp mills than previously were believed necessary, a War Production Board official said today. Ed. R. Metcalf, head of the W. P. B.'s cordage branch, told a Senate agriculture subcommittee that while the demand for rope now is two and a half times greater than the peace-time normal, production is up to requirements."[66]

This meant that the USDA hemp program would take a 180-degree turn that spring. Based on the tonnage of hemp produced in 1943, the government reduced the 1944 goal to 60,000 acres nationwide since officials believed that would be sufficient to meet the needs of the military for the duration of the war.[67]

Those employed in the mills and the farmers who had grown hemp in 1943 were soon told that the government would no longer be subsidizing the crop. In Wells County, Smith wrote, "It now appears unlikely that the crop will be grown again in this area [in 1944], since supplies are now becoming available from countries that can produce it more economically."[68] His projection proved correct, as no hemp would be grown for the military in Indiana that year.[69]

As the United States once again began importing rope-making fiber from foreign sources, farmers who had grown hemp for the military were upset that their crop was being pushed aside for the cheaper imports. To add insult to injury, farmers were told that the quality of the imported fiber was better than what was produced domestically.[70]

As soon as the need for domestic hemp started to wane, the USDA also began the formal process of decommissioning the hemp mills. By the end of 1944, the USDA had closed twenty-eight hemp-processing plants.[71] Approximately one-third of the mills were left to operate into 1945, but this was just to process the hemp remaining from the previous fall since the government had canceled all acreage as of 1945.[72]

As the mills closed, the government began advertising the shuttered facilities for sale, with a January 1945 article in the *Lafayette Journal and Courier* reporting, "Uncle Sam has tacked 'for sale' signs on federal hemp plants."[73]

Twelve of the hemp mills were declared surplus property and put up for sale or lease by July 30, 1945, including the two in Indiana.[74] In fact, most of the mills had only been used for one year, and some had just been completed when the decision was made to discontinue the domestic hemp program.[75] The federal government had poured a lot of money into building mills to process the domestic hemp crop, but with construction delays due to lack of workers and supplies, many of the hemp plants around the country weren't completed before the Allies were able to gain the upper hand and the domestic fiber was no longer needed.

In May 1946, the Warren mill was sold to the Hooker Glass and Paint Manufacturing Company in Chicago for $53,600. After investing an additional

$30,000, the company turned the facility into a factory that produced mirrors and flat glass.[76] The Remington hemp mill was sold in October 1946 to a Vermont company that would use the plant to process marble from its East Coast quarries.[77]

Domestic hemp production only lasted briefly, with U.S. farmers planting 178,000 acres in six states in 1943 and 40,000 acres in four states in 1944.[78] At the end of the war, the United States had surplus hemp supplies, which it subsequently distributed to Belgium, Canada, and France.[79]

With domestic hemp no longer needed by the government, it once again became strictly regulated under the original 1937 tax act.[80] Under those regulations, Indiana did not have a single hemp producer in 1945.[81]

While hemp for making rope was largely replaced with synthetic fibers such as nylon after World War II, that didn't take away from the fact that 20,000 American farmers supplied enough hemp to see the military through much of the war.[82] It was rather impressive that Indiana farmers had started 1943 with zero acres of hemp being grown but ended by planting 8,500 acres of hemp for the government.[83]

During the war, Purdue researchers, campus Extension specialists, and county Extension agents showed that they could be relied upon to provide instruction on how to plant, grow, and harvest hemp on Indiana soils. Those farmers who answered the production call on behalf of the U.S. military could always take pride in knowing that their efforts had helped the nation and its allies win the war against Germany, Italy, and Japan.

CHAPTER 42

Draining the Woodlots of the Larger Oaks, Walnuts, Maples, and Sycamores

Wood products were in short supply and subject to rationing during World War II for the same reasons that gasoline, rubber, and many other items were: the military needed these resources to meet wartime demands.[1] The military used lumber in planes and ships, but it was also needed for crates to ship materiel overseas.[2]

To ensure an adequate supply for the military, government regulations reduced domestic building activity to a minimum.[3] This meant farmers had to compete with one another for the lumber to rebuild a barn that had burned down, repair grain cribs, or put up new buildings such as machinery sheds and poultry or hog houses.[4] That's not to say that a farmer couldn't obtain these construction materials, but the federally rationed lumber was reserved for those who met government guidelines. Farmers who wanted to purchase lumber at a local store had to go through the same process as for other rationed goods, which was to fill out the appropriate paperwork, get approval and obtain a purchase certificate from the county board in charge of the item, then present that certificate to the local lumber store.[5]

County Extension agents helped farmers fill out the government forms to request lumber for use on their farms. Everett E. Fultz provided the following information in his 1942 annual report for Jasper County: "Applications for priority assistance in securing materials or to begin construction have been considered at regular U.S.D.A. War Board meetings each Saturday. The applications include construction of farm buildings essential to production, remodeling to make possible increased production, replacement of worn parts or machines to maintain production and securing supplies in order that repair services may be available."[6] That same year in the Howard County report, Roscoe Stangland further explained, "It was necessary in every case for the county agent to spend considerable time in rewriting these applications to get them in condition to be approved by the State and National Boards. These requests for permits were not just something that someone wanted—they were a matter of necessity to maintain agricultural production."[7]

The justification for approval was pretty much the same as for any rationed item during the war: Would issuing the product increase agricultural production by the recipient? For each farmer who received approval, there were countless others whose requests were denied.[8]

In Union County, fifty-three farmers received permission to purchase a total of 56,000 board feet of lumber.[9] With a board foot equal to 144 cubic inches of wood, it might have only amounted to enough for each individual to build a floor measuring approximately 32 feet square, such as in a hay loft, if they had each received an equal amount.[10]

In 1944, Randolph County agent Lewis Husted observed, "[T]hose farmers so unfortunate as to lose a building [to fire] have in most cases been unable to replace satisfactorily," indicating that extenuating circumstances didn't necessarily prioritize the request.[11] In 1945, agent Joseph Clark summed up the situation in Kosciusko County by saying, "Very little new building and major repairs have been made on the average farm."[12]

Some farmers were lucky enough to get timber by harvesting trees from their own woodlots, with Perry County agent Charles Chambers observing in 1943, "Farm buildings, generally in this area, are constructed with rough lumber, generally sawed from trees on the farm, and the only critical materials except the lumber itself are nails and roofing."[13] That same year in the Orange County annual report, agent Ross Atkinson described a similar resourcefulness by local farmers, writing, "To read the rules and consider the limitations on materials for new construction, is to think that there will be none for the duration. On the other hand, to travel the roads of Orange County, we find more new construction than usual. Legal? Yes! Because native lumber, skills, and nonrationed materials have enabled many farmers to better equip their farm properties to care for the greatly increased livestock and crop enterprises."[14]

Other agents reported a different type of resourcefulness by noting that local farmers had asked for more information on construction techniques using concrete and concrete blocks.[15]

The military demand for lumber created new opportunities for those owning woodlots to sell quality timber for much-higher-than-normal prices.[16] As early as 1942, agents were reporting significant increases in the sale of timber largely due to premium prices offered by the sawmills. Even smaller-diameter trees were now selling for more than what larger ones did just a few years earlier. With prices 400 percent higher than normal and the demand for timber nearly unlimited, Ross Atkinson indicated in his 1942 Orange County report, "Many owners are of the opinion that timber will again be cheap [following the war], and are, therefore, pushing their small trees on the market."[17]

As a result, timber from privately owned woodlots was quickly disappearing due to the rapid harvesting.[18] Agents were confronting the unusually heavy timber harvest that was "draining the countryside and the woodlots of the larger oaks, walnuts, maples, sycamores," as one agent put it.[19] This approach to harvesting had continued from Depression-era forestry practices, when woodlots had been overharvested as farmers sold off timber for an additional source of income. But to some extent, it was also an age-old problem that had persisted since the pioneers had first settled the land.

In a 1944 report, Atkinson drew on his own ancestry to describe how farmers had historically viewed trees first as an obstacle to farming and later as an undervalued asset, writing:

> Great-great-grandfather Atkinson was the first man to enter land and build a cabin on it in Paoli township, Orange County, Indiana. That was in February, 1811 and of that lineage there have been farm owners and operators in the township ever since. The point of parallel is that timber has always been a problem of clearing for the purpose of gaining the fertile virgin soil for crop land. This lasted for about two generations. Then the next two generations hoped to sell timber for something more than the cost of labor of getting it to market. This present generation has talked much of proper management for greater over-all profit and proper land use. Now that the world at war has created such unusual demand for all kinds of timber products, the U.S. Forest Service, State Conservation Service, and the Extension Service have been brought to a point of very close cooperation in the bringing out [of] the answer that has been needed though not demanded by every farmer with trees on his farm.[20]

Getting farmers to understand forestry principles had always been a stumbling block to improving the quality of trees coming from private woodlots. Perry County agent Elmo Chatham observed in his 1945 annual report:

> One of the major problems of the Extension Service is to get local farmers to appreciate the value of timber and consider it as a crop and teach them that timber properly managed will earn a good rate of interest. Far too much of our timber production possibilities are nullified by land owners selling off a tract of timber for a lump sum of money to local saw mill men who cut everything that will make a railroad tie. Many of the land owners have not realized 50 cents [per] a hundred [board feet] for timber at the stump.[21]

As more farmers wanted to take advantage of higher timber prices, agents strived to bring the science of forestry to landowners so they could make better management and financial decisions for their woodlots. They joined forces with representatives from the U.S. Forest Service and Indiana Department of Conservation, and began contacting owners of forest tracts to discuss marketing their timber for wartime use but in a way that allowed for sustainable growth.[22] As Atkinson reported in 1943:

> The new plan of the [federal] forestry service is to personally instruct them [landowners] in the present war time demand for timber products of the various hard woods as well as some of what we call soft woods, such as sycamore and willows; and methods of estimating the measurements and values. They [U.S. Forest Service representatives] are also advising that the farmer contact as many potential markets as are available for his particular kind and class of timber before he sells. Through this kind of work, farmers are getting the idea of selling certain marked trees rather than permitting the 'Timber Man' to buy the whole woods with privileges of cutting what ever he can sell.[23]

As prices continued to rise, owners sought to understand how to value their woodlots. Some questions commonly fielded by agents included:

- How is timber priced?
- What kinds of timber are worth more?
- What is the best way to handle a timber sale?
- How is the timber used by the lumber industry?[24]

This wasn't unexpected since, as one agent observed, "Timber is a product which most farmers have not been in the habit of selling and they are not very well informed upon the kinds of timber, the value of this timber or its use in the lumber industry."[25]

Agents and farmers relied heavily on advice provided by Purdue forestry Extension specialists Thomas Shaw and John L. Van Camp in addition to that from federal and state foresters.[26] Shaw provided technical advice to farmers interested in managing and selling timber at the best prices. His expertise was invaluable when a powerful spring storm toppled 1.5 million board feet of marketable timber in and around LaPorte County. Shaw and local agent Everett Hartman surveyed the aftermath to help farmers assess the value of the trees left standing and that of the downed trees that could be salvaged for lumber.[27]

Extension also held demonstrations at woodlots that were managed using the latest recommended practices. At these events, attendees would get hands-on opportunities to learn how to value trees according to species, size, and quality of the wood.[28] They also heard from forestry Extension specialists such as Shaw and Van Camp, who typically advised landowners to cut all trees that were badly deformed or had little value, sell those with the highest market value, and use the remainder for firewood.[29] Selectively cutting the less desirable tree species from the forest stand was a management approach that provided more space and sunlight for those trees deemed to be more valuable.[30]

Agent John McCulloch offered a description of a 1945 Extension event held at the Merrill C. Turley Farm in Scott County, writing, "The purpose of the demonstration was to mark trees, in this large wooded area, so that Mr. Turley would be an example to other farmers, in the county, in proper selection of trees that are ready to be marketed. It also showed the method of calculating the amount to be sold so that farmers could intelligently deal with timber buyers."[31] Extension used demonstrations such as these to encourage farmers to sell select trees ready for harvest at premium prices rather than selling a whole woodlot.[32] This strategy required more effort to manage the forestland but had the potential to provide short-term returns as well as sustained returns in the future.

On December 1, 1942, the federal government commissioned the construction of the Naval Ammunition Depot near Burns City in southern Indiana. When completed, it would primarily be located in Martin County but also in parts of Greene and Lawrence Counties. The property needed to be fenced off, but with only minimal timber available on the property, depot officials asked the Indiana Division of Forestry for help in securing the required fence posts. Jefferson, Ripley, and Switzerland Counties soon received orders for 10,000 posts each.[33]

Agent Guy Harris discussed the depot's order in the 1942 Ripley County annual report, writing:

> The Naval Ammunition Depot paid the farmers 40 cents per post not less than 4.5 inches in diameter and six feet long. At a later date they agreed to buy posts from 3 to 4.5 inches in diameter and all of the same length. They paid 30 cents [each] for these. A total of 10,000 fence posts were shipped from Ripley County to Burns City, netting these farmers a gross of $4,100. The net price for #1 [grade] posts was 30 cents and #2 [grade] posts was 20 cents. The 10 cent difference was paid [to] the truck driver for hauling the posts from Ripley County to Burns City.[34]

Agent Leo Geyer provided this information on how the Jefferson County group that he had helped organize worked to get its 10,000 posts to the depot:

> The farmers were to deliver the posts to the roadside and notify the trucker when he had his allotment ready. The trucker was then to get the posts at his convenience and do the gradings as he loaded and then give the farmer a receipt as to the number of posts he had accepted. The trucker then delivered the posts to [officials at] Burns City, who in return, after checking, sent a receipt to Mr. [Harold] Clashman, the manager. No one, including the trucker, was to get paid until the contract had been fulfilled, at which time the Navy Department paid Mr. Clashman and he in turn paid both the farmers and the trucker. The trucker was to get 10 cents per post for hauling and the farmer 30 cents at the Roadside for all No. 1 posts. For all culls or no. 2's he was to receive 20 cents.[35]

In his report, Geyer also described the specifications for the posts, noting that the No. 1 grade had to be "7 feet long, 4 to 6 inches at the minimum or top dimension with only 6 inches off axis allowed for crookedness. A round post was preferred but if a farmer could split his post so as to stay with minimum dimension, he was allowed to do so. . . . The No. 2 posts had to meet the same specifications, only they had to measure 3 inches or larger at the small end."[36]

In 1943, the naval depot would order a total of 14,000 more posts from these counties, paying 43 cents each for the higher graded posts.[37]

With lumber in short supply, Randolph County agent Lewis Husted reported that the county "should really 'knock on wood' for luck because we have had fewer farm buildings lost from fire than any year for a long time."[38] This may have been due to luck, but it's also possible that it was a result of the fire prevention efforts made by Extension and others in the community. To reduce the need for lumber, Extension staff worked to make sure that farmers did everything possible to prevent damage to their farm structures and woodlots from fires and lightning strikes (see Chapter 39). They also became involved with officials from other state and federal agencies in creating a network of volunteers trained to protect woodlots and larger tracts of forest from fires started by lightning, carelessness, and arson.[39]

For example, fifty volunteers were recruited in Jefferson County in 1942 under the leadership of the local Civilian Defense Council.[40] In Owen County, neighborhood leaders received firefighting training.[41]

Since many adults who typically would have been asked to volunteer were those who were already serving in the military or working in industry, high-school students were recruited to serve their communities in this capacity.[42] Student training included classroom instruction on methods used in getting a fire under control as well as hands-on practice outdoors provided by the state's Forest Fire Fighters Service.[43] In 1942, Jackson County agent Willis Stall provided this account of the effort:

> The schools of the county have cooperated by sending from 5 to 10 older boys to the training meetings held. The first two of these training meetings, which have been held in the service building of the Jackson County State Forest, were devoted to explanation of the use of forest fire fighting tools and equipment. . . . The third day of training was devoted to actual work with tools in putting out a woods fire. . . . The period of training continued with three more days devoted to outdoor work. As a result of this effort, Jackson County has 50 to 75 additional men with some experience and training in firefighting . . . [that can replace] the trained men who have gone into military service or are working in the defense industries.[44]

Forestry subjects had long been in the wheelhouse of the county Extension agents, who had experience working with landowners on reforestation projects to control soil erosion, getting woodlots classified for tax breaks, and planting trees as windbreaks for homes, barns, and fields, and as a future investment.[45] But it would take a second world war to bring forestry science to the forefront like nothing prior to the war had done. Agents and forestry Extension specialists at Purdue took advantage of the interest in timber to educate the state's landowners regarding the value of a properly managed woodlot. Extension was in a perfect position to deliver all of the science developed by Purdue agricultural researchers to woodlot owners, which allowed them to take advantage of the high prices that timber was selling for. For many years after the war, the management practices promoted by Extension would provide a significant source of income for the agricultural community.[46]

CHAPTER 43

Think More of the Soil as a Heritage to Be Conserved and Passed On to Those Who Follow

During the early 1940s, county Extension agents had little choice but to provide full-time attention to help farmers ramp up agricultural production to meet the needs brought about by World War II. Wabash County agent William K. Delaplane summed up the situation in his 1942 annual report, writing, "Under present conditions, it is quite apparent that in order to meet the national needs, our entire energies and resources must be directed to all-out production."[1] For Delaplane and his colleagues across Indiana, the food and fiber needs of the military would trump all other Extension projects.

From the earliest days of the war, agents were reporting that the all-out production had the real and potential negative consequences of increasing soil erosion and depleting the soil of nutrients. Sullivan County agent Jean Sliger noted this in his 1942 report, saying, "Production of food, of course, will be the primary objective and every effort will be bent toward this goal. There is another job too, that of soil conservation. The fact that some land will be mined of its fertility and other land neglected should not be overlooked."[2]

Despite wartime demands, some agents still managed to report progress on soil conservation projects. In the 1943 Huntington County annual report, Walter Rusk cited the Ivan Hardin Farm as an example to show how Extension had successfully helped restore worn-out farmland over a period of several years, writing, "Two or three older men living near this farm testified how this farm was badly eroded, had a low fertility level, and was practically abandoned five years ago, and how this rehabilitation program [by Extension] has established this farm to a fertility and land use basis that will make it a profitable farm again."[3]

But the truth was that most agents were forced to back off of soil conservation efforts as the war required farmers to push their acres to produce more in order to reach government quotas.[4] Still, whether it was before, during, or after the war, agents continued wherever possible to integrate the message of production and conservation into their educational outreach programs and practical field demonstrations, with one agent noting that "the greatest production of essential food and fiber without impairing our soils and depleting our agricultural resources will require carefully thought out extension programs."[5]

Agents only had to look back a few years to the Great Depression to know what they were up against when it came to raising crops on broken-down land. Marion County agent Horace Abbott wrote in 1945 that what agents had done the previous decade had allowed for substantial gains in farm productivity during the first half of the 1940s. They had rolled up their sleeves, heavily investing time and resources to host educational programs and field demonstrations that focused on reducing the loss of topsoil and reversing the decline of soil nutrients. The results spoke for themselves, with Abbott noting, "The value of a sound, long-time program of improved agricultural methods and soil-building practices can never be better demonstrated than during this period of emergency war food production. Vast stores of soil reserves accumulated during the last decade through the use of soil-building practices, have been converted to record yields of food through the use of good varieties, well fertilized and well cultivated."[6]

But those record yields were already putting a new strain on the soil during the early 1940s, with most agents reporting that the farm ground in their counties was showing signs of deterioration due to wartime production levels.[7] Steuben County agent Thomas Grabill offered his point of view by writing, "How long the bank account of soil fertility will be able to hold out is a question. The danger lies in the depletion of organic matter which in turn means lack of moisture holding capacity and soil erosion. A liberal use of commercial fertilizer has contributed to the bank account. . . . This is on the credit side of the ledger but it appears that the close of the war will see our lands in need of repair and restoration."[8]

If the conservation gains weren't maintained and new problems prevented, the land would quickly return to the same poor condition as during the Great Depression. Indeed, by 1945, agent Arthur K. Wissman would observe that repair efforts in Jennings County "include proper soil management such as contour tillage, standard and diversion terraces, winter cover crops, as well as reforestation of lands too badly eroded for cropping or permanent pasture."[9]

Agents had a hard time getting farmers to focus attention on protecting soils as long as agricultural prices remained above average.[10] For those growers who had somehow survived the lows of the Great Depression, making up for lost income and trying to pay down debt was as important as doing their patriotic duty by meeting the federal agricultural quotas.

A key problem that Extension needed to address was how to get farmers to put highly erodible lands and those not suited for row crops back into pastures and forestland. With the government demanding increased agricultural production and supporting higher prices, Jay County agent Chester Langston noted,

"[F]armers have had the tendency of placing into cultivation those areas which formerly were not profitable for corn and soybean production."[11] Agent Sumner J. Brown had warned Vermillion County farmers about doing this in 1942, noting that "[the] mistakes of the last war, such as plowing up land that should have been left to grass . . . must be avoided."[12] But comments such as this went largely unheeded by farmers and government officials. The United States needed production at all cost, and if erosion was one of the consequences, then that was a small price to pay toward keeping the troops fed and clothed, and winning the war.

By the end of the war, Jackson County agent Willis Stall estimated that "1,500,000 acres of agricultural land had been essentially ruined for agricultural production through man-made erosion."[13] Agents reported that county drainage ditches were filling in with sediment washed in from fields bordering the waterways.[14] In some cases, farms eroded so badly that, without any conservation measures taken to improve it, the only recourse was to abandon the land as farm ground.[15]

To agents and many others in the agricultural community, the evidence was clear that years of raising corn and wheat had not only depleted the soil nutrients but had also caused erosion issues that extended well beyond the farm.[16] Agents had their work cut out for them as they tried to address the issues of lower soil fertility and eroded topsoil, but they were in a better position to do so after working on the same problems during the Great Depression. And by the early 1940s, an effort to create soil conservation districts that had begun in the late 1930s would gain momentum and give them a partner to help focus attention on protecting Indiana's soil and water resources.

The origination of soil conservation districts occurred when President Franklin Roosevelt signed the Soil Conservation Act in 1935. This law established the Soil Conservation Service as an agency of the U.S. Department of Agriculture (USDA) under the Secretary of Agriculture Henry A. Wallace. In the summer of 1935, a USDA committee recommended that the most effective way to deal with the problems caused by erosion was to have the states create soil conservation districts to address problems unique to their geography. These districts would be managed by local volunteer leaders who primarily focused on agriculture-related conservation issues.

To foster cooperation between the USDA and the states, in 1937 Roosevelt sent each governor a Standard State Soil Conservation Districts Law to use as a model within their own legislative bodies. By following the federal model, states could obtain federal aid to help support projects.[17] That same year, the Indiana General Assembly passed the Soil Conservation District Law, which allowed for the creation of the districts.[18] The state elected to set up county-based districts.

Indiana created the State Soil and Water Conservation Committee under the umbrella of the Indiana Department of Conservation. This five-member group would oversee the creation and management of the local districts.[19] By 1944, the director and associate director of Extension would be appointed to serve on this committee.[20]

Each county that wanted a soil conservation district had specific protocols to follow. The first step was to submit a petition signed by at least twenty-five landowners who agreed that a soil conservation district was needed. This initiated a process that would eventually lead to a referendum to vote on the proposal.[21] As part of this process, agents led educational meetings throughout the community to inform farmers and other landowners about what a soil conservation district meant to them, how it would function within the county, and the pros and cons of setting up such an organization.[22] In Lake County, agent Lloyd E. Cutler in 1944 reported, "A series of five informational meetings was held during the week of March 27 with [Extension soil conservationist] R[uall] O. Cole of Purdue as one of the principal speakers. Every possible effort was made through meetings, newspapers, circular letters, and personal contacts to tell local people just what they might expect from the establishment of a Soil Conservation District."[23] When Ellis Christen reported on meetings he led in Benton County in 1939, he noted, "The local Extension Agent endeavors only to give the facts in the law and makes no recommendations as to whether a district should be established or not."[24]

In 1945, Purdue agricultural economist J. Carroll Bottum provided an excellent account of how the Jasper County Soil Conservation District had become a reality:

> During the months of December [1944] and January [1945], meetings were held to inform the people about the possible formation of a Soil Conservation District. Petitions were circulated in every township with the exception of Carpenter. To get the latest information regarding Soil Conservation Districts, either R[uall] O. Cole from Purdue, or Lawrence Bassitt, Soil Conservation Service, conducted these meetings and answered questions regarding the working of a Soil Conservation District. . . .
>
> On January 31 a hearing was held on the formation of a Soil Conservation District. This hearing was conducted by Dr. [William V.] Lambert [associate director of the Agricultural Experiment Station] of Purdue and Kenneth Welton, State Conservationist. Representatives from every township in the county were present and had an opportunity to express themselves regarding the formation of such a district. All commented very favorably for the formation of the district.
>
> The State Committee, after reviewing the testimony of the hearing, ruled favorably for the district, which is to include substantially the following: All of Jasper County except all towns, incorporated and unincorporated, and all platted areas. The date for the referendum was set for March 15 [1945].

> The County Agent held nineteen educational meetings with a total of seven hundred one people attending these meetings. In order to complete the voting on time, a township committee was formed and territories assigned to each individual to canvas. He was to see that everyone had an opportunity to vote either for or against the district. . . . More than 100 volunteer leaders assisted in carrying ballots and did a very thorough job of contacting each land owner. . . .
>
> The voting on the formation of a Soil Conservation District was completed on March 15 with the following results; Total votes cast, 1,676; number in favor, 1,566; number voting against, 110; percent participation of eligible voters, 73.2.[25]

Following a referendum, members of the state committee held a meeting in the county to validate the vote, then agreements were signed between the new county-level district, the USDA Soil Conservation Service, and Purdue.[26]

During the late 1930s, several counties attempted to form districts, but while there was often interest, these attempts were sometimes voted down due to a lack of information or when landowners became concerned that the government would dictate to them how to use their land.[27] In 1940, the Vanderburgh County Soil Conservation District became the first officially recognized district in the state, while the Tipton County district would be the last to organize, doing so in 1974. Most of the districts were formed between the years of 1945 and 1960.[28]

Once established, the local soil conservation district was led by a five-member board of supervisors.[29] Two of the supervisors were appointed by the state-level committee, and three at-large members were elected by landowners to serve two consecutive three-year terms.[30]

Upon signing an agreement with the state, a person trained in soil conservation would be assigned to the local district by Kenneth Welton, director of the Indiana Soil Conservation Service. Welton was housed at the Purdue campus in West Lafayette, which showed the close working relationship between the university and the state agency.[31] In 1946, Harold Creech of the U.S. Soil Conservation Service was sent to Martin County to host meetings, provide conservation tours to farmers, work with 4-H projects, and develop conservation plans for farms.[32] Not surprisingly, many county Extension advisory boards would set soil management as the focus of the local Extension program after a soil conservation district had been established.[33]

For the most part, a district included everything but towns and cities, but early on, many districts included only those townships that were agreeable to forming such an organization.[34] The lack of support for a district among some farmers didn't mean that soil conservation wasn't an important aspect of agriculture that needed to be addressed. Instead, many farmers opposed the district because they saw it as the government telling them what they could and couldn't do on their farms.[35]

The district was, in fact, a governmental subdivision of the state, but according to a 1940 article in the *Indianapolis News*, it was one with broad regulatory powers. The article stated that

> [district] supervisors have authority to adopt land-use programs, give financial aid, buy land and provide materials. Damages may be assessed against a farmer who lives in the district and fails to use his land as the supervisors order. District supervisors may act as agents of the federal government in carrying out any of its soil-conservation or erosion-control projects. Land-use programs are to be submitted at special elections and ordinances having legal force are not to be adopted unless a majority votes for them.[36]

These broad powers to regulate were precisely what concerned farmers in Sullivan County. In 1943, agent Jean Sliger reported that "an element of farmers protested strenuously to the land use clause in the laws governing the district. Since the meeting, the opposition has made known that it would oppose a soil conservation district again."[37] Two years later, the new agent, Howard Telfer, made what would be a fourth attempt in getting support for a soil conservation district in Sullivan County, with his report showing that he had made progress among some farmers in the county. He wrote, "[I]t appeared that objections were of such a form that it would not be desirable or advisable to include the entire county. . . . Consequently, the farmers of Fairbanks, Turman, and upper Gill Townships petitioned the State Committee for a hearing on the matter."[38] Sullivan County would successfully create its district that year with only a portion of the townships participating.

Sullivan County was not the only county to exclude certain townships in order to move forward with creating a district. In 1944, Wabash County only included Lagro, Liberty, Noble, and Waltz Townships in its district, with agent William Delaplane noting it was "with the idea that the other three townships or any of them might be added at later date. Our farm leaders all felt that the district should include the entire county but recognized [that] the difficulties in securing the required number of votes would call for a great deal of work on everyone's part."[39]

The previous year, Monroe County had only included 23,000 acres from Bean Blossom and Washington Townships when petitioning to create its district. But agent Walter Rogers reported that, in response, the state committee "suggested, however, that an area of this size was too small to work effectively and suggested that the territory be enlarged before the State Committee acted upon the proposition. . . . [T]he proposed district included an area of 44,800 acres in parts of Washington, Bean Blossom, Van Buren, and Bloomington Townships."[40]

Over time, all districts would expand to include the entire county—minus cities and towns—as the fears that some farmers had about the government taking over their farms never materialized.[41]

Agents generally attended and participated in the monthly meetings of the county soil conservation district.[42] Most of the initial meetings focused on what the local soil conservation program would entail, with the state committee having final approval over the plan of work.[43]

It was clear from the outset that the county Extension program would be an important asset to the soil conservation district and vice versa.[44] After all, the agents were well informed on installing grass waterways to improve drainage, converting land not suited for cultivation back into pastures and forests, and using different cultivation practices to reduce erosion.[45] But while the agents could be depended on for educational programming, they would have nothing to do with the management and operation of the district.[46]

It was typical for the soil conservation district, county Extension agent, and federal Agricultural Adjustment Administration to work together on programs and projects to install tiles, create ditches to divert water, dynamite ditches to increase their depth, construct ponds, build terraces, establish sod waterways, seed and fertilize pastures, and plant trees.[47] The 1943 and 1944 Knox County annual reports documented what agent Henry Benson and his partners had accomplished through focused conservation efforts. The local soil conservation district had rented equipment it owned, including scrapers, terracers, and a grader, to help farmers with establishing terraces, building dams, and cleaning ditches, with Benson estimating that the improvement in drainage had increased corn production by 80,000 bushels.[48] He also indicated that 350 farmers had received help in planning for or applying conservation measures and that new developments in the county included 62 stock ponds, 86 miles of terraces on 110 farms, and nearly 100,000 feet of drainage ditches.[49]

Of course, it was easier to sell soil conservation tactics when the results could easily be seen. As Whitley County agent Blanchard Widney explained, "The value of soil conservation is recognized in a general way by farmers but only a few are sufficiently convinced to try out the practices on their own fields."[50] Farmers had to see the practices being done by another farmer before using those practices on their own farms.

In his 1944 annual report, agent Noah Hadley described the response to contour cultivation plots by those seeing them for the first time:

> Parke County farmers who have used contour cultivation this year and those who have inspected its use on the several tours conducted are enthusiastic about the results obtained. Marked differences in yields and in soil movement

> are obvious between land farmed on the contour and that operated up and down slopes. Farmers who are using this method report that on most fields somewhat more time is required for contour cultivation but that there is a considerable saving in power costs.[51]

One of the more interesting projects showed the importance of soil depth as it related directly to yield in Knox County, with Benson offering these details in his 1942 report:

> A cooperative study in which the Extension Service and Soil Conservation Service participated showed that this year the average value of each inch of top soil in terms of production was 5 bushels of corn per acre. Areas on which no top soil remained averaged less than 20 bushels of corn per acre. A typical example of the yields in comparison to the depth of top soil on one of the better farms in the county was as follows:
>
> 11–12" top soil—95 bushels per acre
> 8"—78 bushels [per acre]
> 6"—46 bushels [per acre]
> 4"—40 bushels [per acre]
> 0"—8 bushels [per acre] . . .
>
> The seriousness of erosion losses in Washington Township of Knox County is relevant by an analysis of 621 acres of crop land which is representative of the township. The following table shows the depth of top soil remaining on this 621 acres of cultivated up-land:
>
> 12"—6.4% (represents normal depth of top soil under virgin conditions)
> 9"—6.8%
> 6"—44.6%
> 3"—16.2%
> 0"—22.3%
> Eroded Sub-Soil: 3.7%[52]

With such numbers linked directly to farm productivity, it was easier for agents to convince farmers of the importance of dealing with erosion on their land.

In October 1942, the Noble County Soil Conservation District worked with Extension to hold the first-ever contour plowing contest in Indiana. The purpose was to show farmers the benefits of and best practices for this type of plowing on hilly land. The event took place on the William Clouse Farm, located four miles southwest of Albion.[53] Agent Milo A. Nye provided this description:

> The program started at 1:30 P.M. and was finished by 4:30 P.M. After calling the crowd together, an explanation of the purposes of the contest was made by R[uall] O. Cole, Extension Soil Conservationist. It was pointed out that the rotation being followed on this field made it necessary to plow for the next crop since the field was in corn this year (planted on the contour) and corn

> borer control makes plowing in the fall, winter or spring (before May 15) desirable and practically necessary. It was also shown that if the field was to be bare for a while after plowing that erosion would be greatly reduced if the furrows were on the contour rather than up and down the slope. The connection between corn borer and erosion control and increased production of "Food for Freedom" was very definitely pointed out. The savings in labor, fuel, and wear and tear on machinery by contour operations were also emphasized.[54]

Russell Shipman, Extension agricultural engineer at Purdue, explained the rules to the 107 audience members—mostly farmers—and how the four contestants would be scored, with Nye noting, "Emphasis [was] placed upon adherence to the contour, leaving of waterways, leaving of land ends with irregular edges, and clean plowing for corn borer control."[55]

Plowing started at 2:00 P.M. and ended one hour later. Participants drew for their section of ground to cultivate and used their own tractor and two-bottom plow. The award for the best at contour plowing went to John Schermerhorn Jr., followed by William Gerver, Nelson Roscoe, and Alvin Wysong. Schermerhorn received $15, while the others got either $10 or $5. Each participant also received a bushel of hybrid seed corn.[56]

During World War II, agents continued to work with the Agricultural Adjustment Administration (AAA), just as they had during the Great Depression.[57] In his 1942 report, Monroe County agent Walter Rogers noted that the "extension office has cooperated to the fullest extent in all the educational activities connected with the A.A.A. program. It has not, however, participated in any of the office work as it is our understanding that the job of the County Agent is to be purely educational in matters which related to the A.A.A. program."[58] That same year, Howard County agent Roscoe Stangland reported, "The Extension Service and AAA continued [their] relationship as in the last few years, with the county agent providing the scientific information and advising them on many of their activities. They proceed but whenever they 'get out on a limb,' they come to the county agent for advice and help."[59]

Agent Harold K. Rippey reported on the educational activities that he undertook for the AAA in Jay County in 1946, writing that the "adoption of practices approved were, (1) Establishing sod water ways. (2) Contouring cultivated row crops. (3) Contouring drilled crops. (4) Dams for livestock water. (5) Establishing new pastures or reseeding noncrop open pastures. (6) Green manure crops. (7) Harvesting hay seed. (8) Liming material. (9) Fertilizer."[60]

Before World War II, the AAA program largely paid farmers for implementing contour cultivation, adding sodded waterways, and applying limestone and fertilizer.[61] During the war, many Indiana farmers still cooperated with AAA directives and received cash payments from the government for implementing selected practices.[62] But by that time, the intent of the federal program was to support increased food production by getting farmers to use more fertilizer, plant the crops and raise the livestock the government needed, put in cover crops, and build emergency bins to store excess grain.[63]

The AAA helped pay for lime to adjust the pH of the soil and for fertilizer applications that would return nutrients back to the soil because these additives would help farmers raise the crops needed to feed livestock and meet quotas.[64] But in order to meet AAA requirements, farmers first had to get their soil tested by the local county Extension agent. This, of course, led to a higher demand for testing and requests from farmers for Extension recommendations for applying lime, phosphorus, and potassium.

Agents seemed pleased to provide the soil-testing service. As Marion County agent Horace Abbott reported in 1944, he was analyzing 3,000–8,000 soil samples each year from 500–1,000 farms.[65] He noted that the tests "have been far reaching in their effects. They have probably contributed more to the constructive and economic use of fertilizer on Marion County farms than we realize. Many farmers who are using this service, depend entirely from year to year upon recommendation of the Extension Office as to [the] kind and amount of fertilizer used on the various farm crops."[66]

Unfortunately, wartime shortages meant there was a limited amount of lime allocated for farm use, and often the supply needed for a county was oversold, leaving some farmers short during spring planting.[67] Prior to the war, Switzerland County was getting upwards of 9,000 tons of limestone annually, but agent Oran McNary noted in 1945 that local farmers had hoped to get 3,000 tons.[68]

There were several reasons why limestone was in short supply during the war. Limited supplies from quarries had resulted from labor shortages to mine the material, the lack of railcars to deliver the limestone to central locations and of trucks to bring the material to farms, and a shortage of trucks equipped with spreaders to complete applications onto farm fields.[69] The government also needed limestone for use in ordnance plants, leading Parke County agent Noah Hadley to observe that an "ordnance plant in Vermillion County used the entire output of local quarries for a period of about ninety days and a large portion of the output for a much longer period."[70] Farmers would have to compete for limited supplies of both lime and fertilizer for the duration of the war.

County Extension agents understood that their primary role during World War II was to help farmers meet production quotas. However, as they did this, most also expressed concerns about how the production increases were depleting the soil's resources and creating problems with erosion. Their annual reports included many observations about gullies too deep to cross with farm equipment, drainage ditches filling up with soil, and decreasing levels of soil fertility.[71]

Extension agents would resort to Depression-era soil-conservation strategies that had worked to renew interest in conserving soil. These included instituting crop rotations, growing soil-building crops, increasing pasture acreage, planting trees on eroded land, incorporating cover crops, sodding waterways, using more commercial fertilizers, and raising livestock that required less feed corn.[72]

Soil conservation teams leaned on local cooperators to show that soil conservation practices were practical and well worth the effort for keeping farms productive.[73] Since many of the recommended techniques had been widely used prior to World War II, agricultural researchers and Extension staff had gained the experience needed to help farmers adopt them at this critical stage.

During the 1940s, agents and local soil conservation districts began to identify soil conservation priorities. This, in turn, would allow them to focus on educational programs that could eventually repair the damage caused to Indiana's soil resources by years of intense agricultural production demanded by World War II. As Knox County agent Henry Benson expressed in his 1942 report, the goal was for people to "think more of the soil as a heritage to be conserved and passed on to those who follow as well preserved as when it was received."[74]

A man uses a Ford 9N tractor to pull a disc in 1941. Starting in the late 1930s, more farmers began replacing their horses with tractors. During World War II, those who could obtain tractors readily did so to compensate for the lack of available labor and still increase production to take advantage of higher commodity prices. However, with wartime rationing limiting supplies of agricultural equipment, most operations would not become fully mechanized until rationing programs ended after the war. Courtesy of J. C. Allen and Son.

Workers load hay onto a horse-drawn flatbed trailer in 1943. During the war, farmers used powered equipment when possible, but shortages of this equipment, fuel, and labor meant that they also continued to rely on horses to meet wartime production quotas. Courtesy of J. C. Allen and Son.

This Rush County show featuring Belgian draft horses drew a large crowd in 1942. However, shows such as this became less common as the 1940s progressed. As equipment became available and farmers had the money to purchase it, they soon replaced their draft horses and mules with tractors and other powered agricultural machinery. Courtesy of Purdue University Libraries, Karnes Archives and Special Collections.

Inside the shop of Major Rafferty of Morocco in Newton County, Paul E. Merriman repairs a Farmall tractor in 1944. During World War II, steel and other materials were diverted to make wartime equipment such as tanks, planes, and guns. This forced farmers to put more effort into maintaining their existing equipment since few replacements were available. Courtesy of J. C. Allen and Son.

FARM MACHINERY

REPAIR SCHOOL

Peru H.S. Old Gym.
Wednesday, Jan.28th,1942

10 a.m.

to

3 p.m.

UNCLE SAM HAS ABIG JOB TO DO

He Is Depending on You to Do Your Part

Is Your Machinery Ready To Go To Work?

All Farmers of the County are urged to attend this

Farm Machinery School.

Tell your neighbors about this meeting and bring them with you.

John E. Wylie

County Agricultural Agent

Miami County

This flyer advertised a farm machinery school to be held in Miami County in January 1942. Extension had offered events such as this before the war, but they took on new importance with the need to keeping farm machinery running as long as possible during the war. Courtesy of Purdue University Libraries, Karnes Archives and Special Collections.

CO-OPERATIVE EXTENSION WORK
IN AGRICULTURE AND HOME ECONOMICS
STATE OF INDIANA

PURDUE UNIVERSITY AGRICULTURAL EXTENSION DEPARTMENT, U. S. DEPARTMENT OF AGRICULTURE COUNTY BOARD OF EDUCATION AND COUNTY EXTENSION COMMITTEES CO-OPERATING

January 27, 1943

EXTENSION SERVICE
COUNTY EXTENSION WORK

<u>F A R M</u> <u>M A C H I N E R Y</u> <u>S C H O O L</u>

February 5, 1943

PERU HIGH SCHOOL OLD GYM

10:00 a.m. - 3:00 p.m.

P R O G R A M

10:00 a.m.	The Machinery Situation- Russell Shipman, Agri. Eng. Dept.
10:45 a.m.	Machinery rationing problems- H. R. Wilson, Chrm. County Machinery Board
10:55 a.m.	Machinery dealers problems- James Pearcy
11:05 a.m.	Welders Problems- Wm. Volpert
11:15 a.m.	Labor saving practices and divices and training new operators. Russell Shipman
1:00 p.m.	Discussion on repairs, care and operation of farm machines
2:45 p.m.	Announcements and discussions.

COME AND BRING A CAR LOAD WITH YOU!

John E. Wylie

John E. Wylie
County Agricultural Agent
Miami County

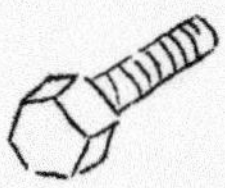

For want of a bolt
A machine was lost

For want of a machine
A crop was lost

For want of a crop
A soldier was lost

For want of a soldier
A battle was lost

For want of a battle
A victory was lost ____S O ____O R D E R R E P A I R S N O W

The agenda for the Miami County farm machinery school held in February 1943, just more than a year after the United States had entered World War II, focused on various equipment and service problems farmers were facing because of the war. The messaging beside and below the agenda reminded farmers to be proactive with their machinery repairs so production wouldn't be hampered and negatively affect the war effort. Courtesy of Purdue University Libraries, Karnes Archives and Special Collections.

(Top and bottom) Purdue University staff members swapped their automobiles for bicycles during World War II as a way of doing their part to save gas and tires. These photographs were taken in 1943 in front of the Agricultural Experiment Station Building, which today is called the Agricultural Administration Building. Courtesy of J. C. Allen and Son.

In Daviess County, a countywide drive collected 200,000 pounds of products containing rubber in June 1942. With new imports of natural rubber supplies cut off due to the war, collections such as these were aimed at stretching existing resources. The sign in front of the pile of rubber reminded everyone that these items would be recycled as part of the war effort. Courtesy of Purdue University Libraries, Karnes Archives and Special Collections.

It seems that everyone was involved in collecting products containing rubber, including these Westwood 4-H Club members. In all, their Henry County club collected 520 pounds of rubber products to be recycled in 1942. Courtesy of Purdue University Libraries, Karnes Archives and Special Collections.

In 1942, a Marion County grower loads melons to be sold at a market four miles away from his farm. During the war, farmers faced significant challenges in delivering produce, grain, and livestock to consumers and agricultural markets, given the difficulty of getting fuel, tires, and trucks due to rationing by the federal government. Courtesy of Purdue University Libraries, Karnes Archives and Special Collections.

In Tippecanoe County, Dale Brown places cans of milk into a truck for delivery to a creamery in 1941. In January 1942, tires would become the first item to be rationed by the federal government, which meant farmers and agricultural businesses would largely be unable to replace worn-out tires with new ones for the duration of the war. Some dairy processors that picked up milk and butterfat from dairy farms partially addressed this problem by also collecting supplies for their competitors or by temporarily swapping customers with them. This reduced the number of miles each company needed to cover and made the routes more efficient. Courtesy of J. C. Allen and Son.

COOPERATIVE EXTENSION WORK
IN
AGRICULTURE AND HOME ECONOMICS
STATE OF INDIANA

Purdue University Agricultural Extension Dept., U. S. Dept. of Agriculture and Fountain County Board of Education Cooperating.

Extension Service
County Agent Work
Covington, Indiana
July 7, 1942

To All 4-H Club Members and Parents:

Dear Friends:

As a loyal 4-H Club member we are urging you to be ready for all kinds of experiences, both pleasures and disappointments. 1941 was a successful year in 4-H Club work, and with a war going on in 1942 we must do our work as well.

The County War Board has suggested, and the Vocational Teachers approved, that due to the war we will not hold a County 4-H Club Show this year. There will be a local exhibit in each township in September, after school is under way. Each member will be given an opportunity then to exhibit after his project is completed. So steam into it and finish with a bang. Let's keep the colors flyin' by 4-H'ers keepin' tryin'.

We as loyal 4-H Club Members must feel that we are big enough not to be disappointed; but, loyal in contributing our part toward VICTORY. If our club show cannot be held we would not be patriotic to continue our plans for camp. This of course is a disappointment at first, but when we think seriously we can understand why this has been done.

You are old enough to understand when I tell you that two-thirds of all the rubber in the U. S. is now on the vehicles of the country.

A one day show shouldn't mean much on our tires, but your influence for good and for safety will help our great country. 4-H'ers will be setting an example to all groups to cut down activities involving the use of rubber. It must be done or we are sunk. Let's respect our government's wishes, whether all others do or not.

Sincerely,

Alfred J. Hesler

ALFRED J. HESLER,
County Agricultural Agent.

mep

Fountain County agent Alfred Hesler wrote this letter dated July 7, 1942, informing local 4-H members that the annual county show and camp would be canceled due to a need to conserve tires. Over the course of the war, county Extension agents would make many adjustments to their meeting and event schedules to conserve tires and fuel, and because many in the farming community were already busy trying to handle increased production levels with fewer workers. Courtesy of Debbie Moore.

On May 2, 1945, Vigo County Extension staff members celebrated their 4,000th radio broadcast held in cooperation with WBOW in Terre Haute. Radio programs were a popular option for Extension agents to use in reaching their audiences during the war, when everyone was concerned about saving fuel and tires. Courtesy of Purdue University Libraries, Karnes Archives and Special Collections.

COOPERATIVE EXTENSION WORK IN AGRICULTURE AND HOME ECONOMICS
U. S. DEPARTMENT OF AGRICULTURE AND STATE LAND-GRANT COLLEGES COOPERATING

Please enroll me for the Purdue Poultry School of the Air—Care and Management of Layers.

Name (Please print)__________

Address (Please print)__________

County__________ Size of Flock__________

I understand that I will receive:
1. 6 Purdue Bulletins.
2. Request cards (no postage necessary).
3. One Guide Book—Care and Management of Layers.
4. A certificate if I write and pass the final examination.

I am a: (Check one)
Farmer
Rural homemaker
Backyard Poultry Keeper
Teacher
High school student
4-H Club Member
Home Ec. Club Member

In 1943, farmers across Indiana could enroll in the Purdue Poultry School of the Air by completing this card. Upon registering for this educational course, participants received lesson plans, Purdue Extension bulletins, and other course materials. Then they listened to ten lessons that were broadcast over the radio. Those passing the final examination received a certificate of completion through the mail. Courtesy of Purdue University Libraries, Karnes Archives and Special Collections.

In 1942, members of the Neighborhood Leaders Program, a statewide network established by Extension to collect and distribute war-related information, meet in the Marion County home of Milo Mendenhall. Agents divided the rural areas of their counties into neighborhoods and assigned the leadership responsibilities for each to one woman and one man. Then the agents worked through those leaders to share information with the people living in those areas. Courtesy of Purdue University Libraries, Karnes Archives and Special Collections.

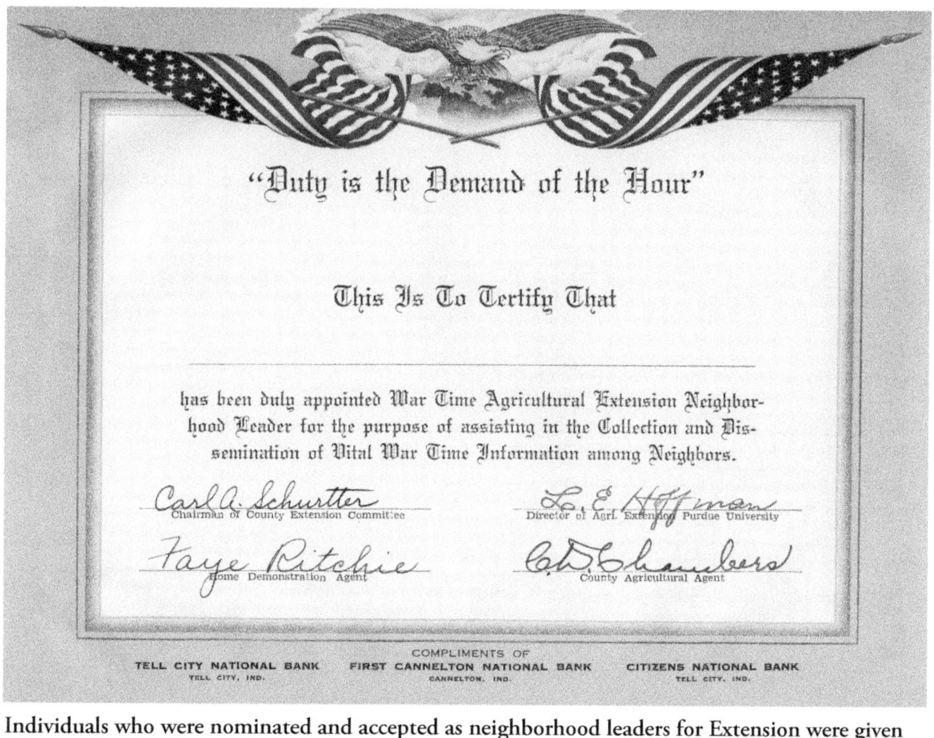

"Duty is the Demand of the Hour"

This Is To Certify That

has been duly appointed War Time Agricultural Extension Neighborhood Leader for the purpose of assisting in the Collection and Dissemination of Vital War Time Information among Neighbors.

Carl A. Schutter — Chairman of County Extension Committee

L. E. Hoffman — Director of Agri. Extension Purdue University

Faye Ritchie — Home Demonstration Agent

C. W. Chambers — County Agricultural Agent

COMPLIMENTS OF

TELL CITY NATIONAL BANK, TELL CITY, IND. — FIRST CANNELTON NATIONAL BANK, CANNELTON, IND. — CITIZENS NATIONAL BANK, TELL CITY, IND.

Individuals who were nominated and accepted as neighborhood leaders for Extension were given this official certificate from Purdue University. Most readily accepted the appointment and served for the duration of the war. Courtesy of Purdue University Libraries, Karnes Archives and Special Collections.

This 1942 Grant County Neighborhood Leader window sticker could be displayed to identify the location of the contact person for the people in that area. Courtesy of Purdue University Libraries, Karnes Archives and Special Collections.

In 1945, volunteers from Swayzee in Grant County plan a program to collect waste fats to be used by the military for making munitions. People saved up these fats when cooking, then delivered them to collection points located throughout the county. Swift and Company, a business that sold meat, picked up the collected fats and transferred them to the Civil Defense Office, which sold the material to renderers. Approximately 8,000–10,000 pounds of waste fats were collected in Grant County each month in 1945. Courtesy of Purdue University Libraries, Karnes Archives and Special Collections.

Two Grant County 4-H members turn in their collection of waste fats in 1945. Local butchers and retail meat dealers often served as collection sites. Courtesy of Purdue University Libraries, Karnes Archives and Special Collections.

A truck caravan carries scrap metal collected from Green Township in Wayne County in 1942. The military needed large quantities of metal to build tanks, planes, and jeeps for World War II, so as the war continued, the government promoted drives to recycle scrap iron, steel, and other metals. In many counties, the neighborhood leaders played a pivotal role in promoting the dates, gathering information about the amount of scrap metal available, and helping to coordinate the collection. Courtesy of Purdue University Libraries, Karnes Archives and Special Collections.

Benton County Highway Department trucks filled with scrap metal line the side of the road in 1941. It was common for county highway trucks to haul donated scrap metal to storage depots during collection drives. For this collection, Benton County received a statewide award for the top donation. Residents there collected about 3,760 pounds of scrap metal, which was three times more than any other county on a per capita basis. At the Indiana Corn Festival, Indiana Governor Henry Schricker presented Benton County officials with a certificate, which was framed and displayed in the courthouse. Courtesy of Purdue University Libraries, Karnes Archives and Special Collections.

This scrap metal collection in October 1942 was a community effort in Daviess County. Over six days, residents collected 2,850 tons of scrap metal. As with so many other government-sponsored programs during World War II, citizens were told that it was patriotic to participate. Courtesy of Purdue University Libraries, Karnes Archives and Special Collections.

A man sits in front of the scrap iron collected in Clinton County in 1943. The campaign had used the slogan "Nothing Too Big, Nothing Too Small" to encourage people to donate. Courtesy of Purdue University Libraries, Karnes Archives and Special Collections.

In 1944, Marion County assistant Extension agent Charles Murphy (center) explains to this group of children—(from left to right) Amos and Helen Mills, and Furnal and Chyron Wallen—when this patch of milkweed plants would be ready to harvest. When the war cut off supplies of the kapok fiber used in life-saving devices such as life jackets, the fiber from milkweed pods was identified as a substitute. With the military draft and wartime industries using much of the available adult labor, schoolchildren were called upon to pick the milkweed pods instead. Courtesy of Purdue University Libraries, Karnes Archives and Special Collections.

These milkweed pods were collected by members of the Junior Red Cross of Wayne County in 1944. That fall, the county's youth collected a total of 1,773 bags of milkweed pods. Since approximately two bags of pods provided enough fill for one life jacket, young people across Indiana understood that harvesting milkweed pods was equivalent to saving the lives of soldiers fighting overseas. Courtesy of Purdue University Libraries, Karnes Archives and Special Collections.

In 1944, Thomas Johnson (center) collected 287 bags of milkweed pods with the help of his nephews, Froman McCrary (left) and Freddie McCrary. Johnson was featured in an *Indiana Prairie Farmer* article and won a $25 war bond for the accomplishment. Courtesy of Purdue University Libraries, Karnes Archives and Special Collections.

Alice Rohlfing, a fourth grader from Napoleon, collected the most bags of milkweed pods in Ripley County during the fall of 1944, contributing 44 bags out of the 958 bags collected. Courtesy of Purdue University Libraries, Karnes Archives and Special Collections.

A portion of the 1,200 bags of milkweed pods collected in Pike County in 1944 hang outside to dry. Without proper drying, milkweed pods would become moldy. Onion bags that held 600–800 pods each were used to collect and store the pods. Courtesy of Purdue University Libraries, Karnes Archives and Special Collections.

Oliver Lee, a weed Extension specialist at Purdue, loads a railcar with dried milkweed pods from Knox County in 1945. Purdue assigned Lee to coordinate the milkweed pod collection effort with county Extension agents across Indiana. Lee would arrange for the agents to bring the collected pods to a rail station, where he would oversee the loading process and coordinate the delivery to Petoskey, Michigan, with the railroad company. Courtesy of Purdue University Libraries, Karnes Archives and Special Collections.

Joseph Eckstein and Sons in Jasper in Dubois County could saw 15,000 board feet of lumber per day in 1941, bringing a return of $80–$100 per thousand board feet. With the military needing wood for crates to pack up supplies being sent overseas, mills such as this one operated at full capacity during the war. The high demand prompted agents and Purdue forestry Extension specialists to intensify their efforts to teach landowners how to manage their woodlots and maximize profits when selling timber to a mill. Courtesy of Purdue University Libraries, Karnes Archives and Special Collections.

In 1942, a group of men load a truck with fence posts in Jefferson County. That year, the federal government contracted with woodlot owners in Jefferson, Ripley, and Switzerland Counties to supply posts for the new Naval Ammunition Depot near Burns City, providing yet another source of revenue for individuals with woodlots who could supply posts at the specified size. Courtesy of Purdue University Libraries, Karnes Archives and Special Collections.

PART 4

Beyond World War II (1945–1946)

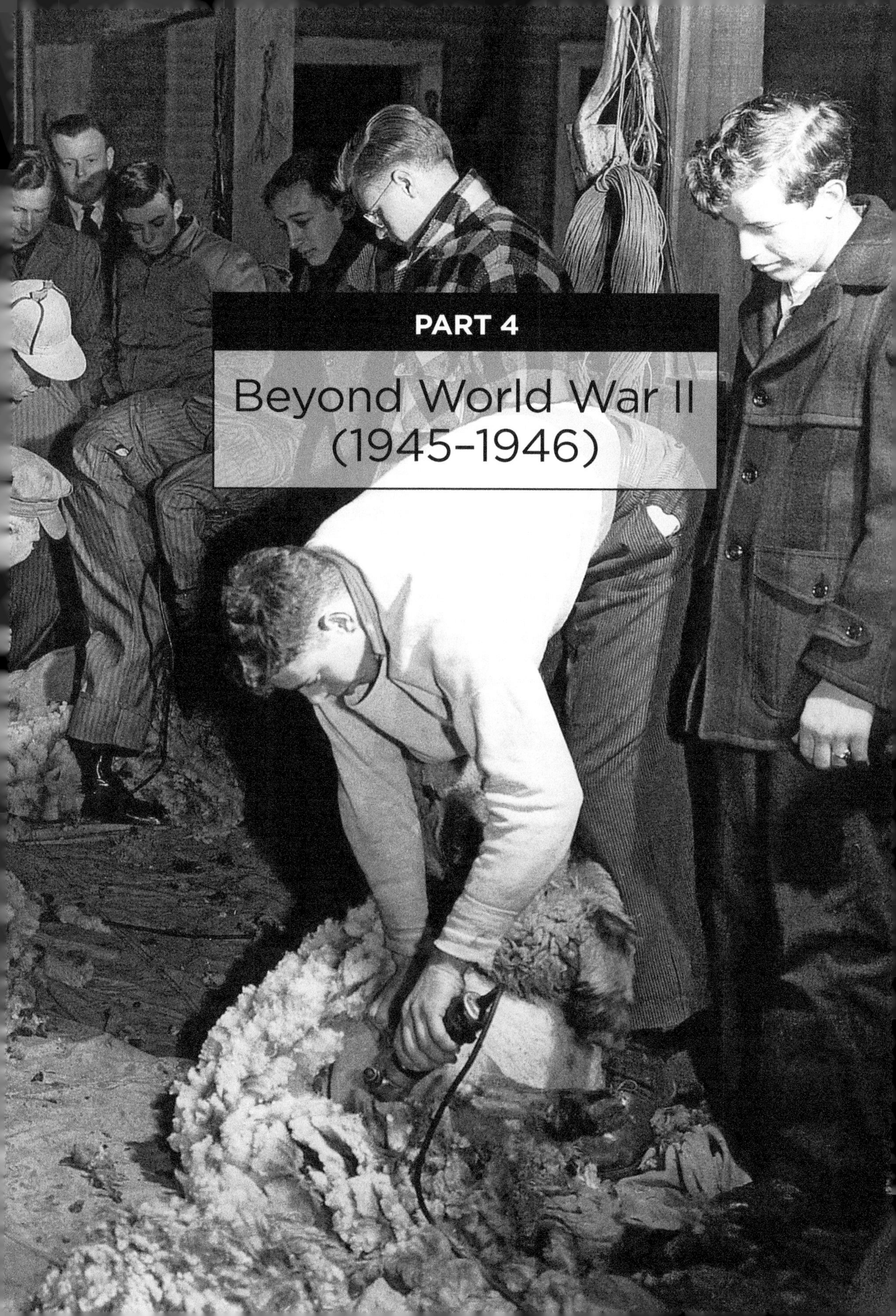

CHAPTER 44

Boys Are Having a Hard Time Making "Fox Hole" Dreams Come True

On May 7, 1945, Germany surrendered to the Allied forces, and May 8 was declared V-E Day, or Victory in Europe Day. A few months later, Japan surrendered on August 14, effectively ending World War II. The date of September 2 would later be named as V-J Day, or Victory over Japan Day, since that was when Japanese officials boarded the USS *Missouri*, moored in Tokyo Bay, to sign the surrender papers. Six years after the start of hostilities, World War II was finally over.

Even before the war ended, though, the U.S. government had started preparing resources that would be needed to help veterans assimilate back into civilian life. Those who had been drafted or enlisted voluntarily would now return to their homes, and when they did, they could draw upon the support provided in the Servicemen's Readjustment Act, better known historically as the GI Bill. This federal legislation, which had been passed by Congress and signed into law by President Franklin Roosevelt on June 22, 1944—less than a year before the war ended—helped discharged veterans obtain low-interest loans to purchase a home, business, or farm; tuition support and stipends to attend college or receive vocational training; or one year of unemployment benefits as they transitioned back into U.S. society. This last benefit was sometimes referred to as the 52-20 program because eligible veterans received $20 per week for up to 52 weeks while they were looking for work.

Many veterans who had farmed returned not knowing exactly what they wanted to do. At least one county Extension agent observed that "many workers employed on farms when they went into the service failed to return to the farms when they were discharged."[1] Orange County agent Ross Atkinson offered this explanation for why some veterans were looking beyond the farm:

> The returned "G.I.s" who entered the service from Orange County farms are not interested in returning to the old jobs. This is as should be expected for two reasons. A very large percentage left home soon after high school and had not yet decided upon a vocation. Therefore, many of them would have left the farm normally. The number two reason is that even present high farm

> prices do not compare with industrial wages, and G.I. privileges, such as subsidized education and others including the "52-20" program.[2]

As veterans began considering their options, the county Extension agents were well positioned to serve as advisors who could help explain the benefits available under the GI Bill. As the war was ending, federal law had established various committees to help returning veterans readjust to civilian life, with agriculture being just one of the committees formed in the counties.[3] By the fall of 1945, Purdue Extension had established veterans agricultural advisory committees in all ninety-two counties in Indiana.[4]

According to a newspaper report, the goal of these five-person advisory committees was "[t]o advise, inform and guide interested veterans and the general public on matters pertinent to agriculture in each individual county."[5] Specifically, Extension charged the committees with performing three pivotal roles: providing one-on-one consultations with veterans to explain GI Bill benefits, especially as they applied to agriculture; developing guidelines to show what it would take to farm after the war for those who planned to take government loans to purchase farmland; and helping design vocational agriculture classes that would be approved for tuition assistance under the GI Bill.[6]

In 1945, Orange County agent Ross Atkinson shared his perception of the veterans and how he viewed his role as a counselor to them, writing:

> The rural youth who entered military service went through the greatest change of his life at that time. There was army personnel on hand to receive him and direct him in every detail from there on throughout all his military relations. Now he is back home in ever increasing numbers. . . . However, there is no one to direct his thinking and actions. Neither is he looking for anyone to tell him what to do even if they could because it is most natural for him to lounge in this returned freedom of civilian life. We consider this situation one of our greatest opportunities to serve in the matter of supplying every possible request with the most exact information possible.[7]

In Scott County, agent John McCulloch reported that in 1945 he had received more than 150 calls from veterans asking about the GI Bill.[8] Miami County agent John Wylie wrote that he had "devoted considerable time with returning veterans discussing the farming possibilities in the County and the advantages of the G. I. Bill. It has been the policy of the County Agent to encourage the returning veterans to complete their education and also to discourage them from acquiring a heavy indebtedness."[9] Wayne County agent Scott Milligan indicated that most of the questions he fielded were from veterans inquiring about starting their own farms, while in Vermillion County, agent Maurice J. Peterson reported that veterans were seeking his assistance in finding suitable farms for rent and how they could get credit to purchase supplies.[10]

One agent expressed a measure of caution related to the counseling work, noting that the local committee "favored the assistance to ex-servicemen who have farming experience and ability, and other farm boys and war workers who show desirable qualifications. It was felt that agriculture should not be the dumping ground during the reconversion period of industry and the conclusion of the war."[11] And not all counties reported successes in helping veterans, with another agent noting that "veterans asked no advice from anyone before embarking into the taxi business or deciding to buy a truck and haul lime instead of returning to farm work."[12]

To help veterans set realistic expectations of what it took to become a successful farmer in Indiana, the veterans advisory committees held planning meetings and solicited public input that they then used to develop the written guidelines that would be distributed to the veterans.[13] Purdue agricultural economists J. Carroll Bottum and J. Edwin Losey as well as Hugh Heckard, the assistant state leader of county Extension agents, also worked with agents to prepare information specific to their counties.[14] Most county guidelines were similar to one another in that they provided recommendations regarding the minimum number of acres needed to farm in the area, the amount and type of livestock to purchase when starting out, and how much savings to have on hand.[15]

In preparing the guidelines, one concern shared by agents and other committee members was that the high market prices for agricultural products would be misleading to veterans who might decide to start farming. In 1945, Miami County agent John Wylie noted, "Present farm prices can be expected to remain fairly stable until the close of the war, but thereafter this same level of prices should not be a criterion to be followed in planning the farming program."[16] Owen County agent Fred Hoover cautioned, "After the exports of food to Europe for rehabilitation is discontinued, profits from farming will depend more upon efficiency of production than upon maximum production."[17]

In 1944, agent Chester Langston reported that the topics covered by the Jay County committee members as they prepared the local guidelines included: "1. Agricultural production and price situation that exists. 2. Agricultural price and production situation from now until the war ends. 3. What may we expect in the way of price, production and demand for agricultural products during the period of one to three years immediately following the war. 4. Problems which will face agriculture in the post-war period proper." Langston also noted, "Considerable discussion followed as to how agriculture will avert similar conditions which existed during the depression period following the last war."[18]

He was referring to the agricultural depression that had started just after World War I and persisted through the end of the 1920s. In that case, farmers had expected that the high return on agricultural commodities they had received during the war would continue, but when the war ended, they suddenly faced surpluses, low prices for commodities, and high debt in the postwar years that resulted in economic hardship.[19] The Jay County planning committee and, indeed, everyone in the agricultural community hoped that better planning would prevent a similar situation from developing following World War II.

The preamble to the guidelines that Howard County provided to veterans noted that the committee members "have thoroughly discussed many phases of the agriculture in this county that will be of information to any returning war veteran. These statements are for the purpose of giving information which we think a returning veteran will profit by considering carefully. We also believe that this information will be equally valuable to any man starting in agriculture in Howard County."[20] The committee members also explained that they provided the guidelines so that "the returning veteran does not start himself on a program that will only mean an existence in agriculture."[21] In other words, they wanted those who were new to farming to do more than just get by; they wanted them to thrive.

Agent Scott Milligan provided veterans who were interested in farming in Wayne County with the following recommendations:

> *Types of Farming:* In Wayne County general farming is usually practiced with hogs and dairy cattle as the principal enterprises. Most farms produce enough feed for their livestock needs with the exception of protein supplements. The opportunity for small acreages of intensive crops is limited because of markets and soil types. Specialized poultry production at the present time is limited.
>
> *Land Tenure:* At the present time ⅔ of the farm land is owner-operated. Older operators are continuing during the present war emergency. After the end of the war, opportunities may be available in each community for returning veterans to take over farm land. It would seem more desirable for returning veterans to attach themselves to successful operating units rather than attempting immediate purchase of land and equipment.
>
> *Farming Units:* Farming units at the present time are increasing in size and larger operations are necessary to justify modern equipment. Livestock operations require definite time and regular care on every farm having such enterprises. Small farming units unless specialized do not provide income for adequate living standards and to pay off indebtedness.
>
> *Land Values:* Land values are determined by the long time prospective income. Present inflated land prices and present returns for farm products are misleading.
>
> *Agricultural Credit:* Returning veterans interested in Agricultural Credit are advised to investigate private capital, local banks, insurance companies and federal farm loan agencies. Any purchaser of farm land should be prepared to

make substantial down payment. Safe margin of indebtedness depends upon many factors, e.g.—size and age of family, individual adaptability to farm life and willingness to work.

Production Methods: Future production in agriculture will require increasing attention to the efficient use of machinery, time, seed, livestock and conservation of soil resources. Efficient use of machinery represents care of equipment, proper housing and purchases limited to essential needs. Use good seed of proven adaptability. Breed and care for livestock that can make the best use of home grown grains and roughages. Practice conservation of soil resources by careful attention to rotation, fertilization and management to increase production and maintain the top soil from erosion and mechanical loss.[22]

Throughout the state, members of the advisory committees thought it was especially important to help veterans who expected to farm full time to understand the financial requirements. These committee members were often local farm leaders who were well versed in the economics of farming in the county, and as one agent noted, all remembered "the influences of World War No. I on American agriculture and the resulting disaster which came to the agricultural industry following the boom conditions which occurred during and after the war, and the acute [agricultural] depression which wiped out the equity of many ambitious farmers."[23] By providing guidance up front, they hoped to spare new farmers from suffering similar hardships.

Some of the information and advice offered by various committees around the state included:

1. Veterans previously employed on farms will be able to find work in agriculture upon their discharge from the military.
2. Part-time farmers will need to have a full-time, nonfarming job to add to the household income.
3. Land values are inflated, which means it will take longer to pay off any loans.
4. Farming over the next few years will continue to be profitable.
5. A person should have a farm of around 80–125 acres of land or more.
6. It will cost $10,000–$20,000 to purchase and begin operating a family-sized farm with good soil and containing average buildings.
7. Those purchasing a farm should have a down payment at least one half of the cost of the land and equipment.
8. It will take a minimum of $1,200–$3,500 to equip this type of farm, excluding productive livestock.
9. It will take another $800–$1,250 to purchase productive livestock, which should include eight dairy cows, four brood sows, and 200 hens.

10. There are a few farms available to rent, but it will take $3,000–$4,000 to provide farm equipment and livestock.
11. Credit is readily available through banks, insurance companies, and government agencies.
12. Excellent opportunities are available for part- or full-time work as blacksmiths, carpenters, electricians, mechanics, and welders.[24]

Most committee members believed a veteran needed to put down 50–60 percent of the farm's cost before borrowing the remainder.[25] As Montgomery County agent Gordon Sowers observed in 1945, "[T]he first two or three years following the European War will not be a good time to start farming or to buy a farm unless the farm can be bought without going into debt."[26] Indeed, committee members believed that, in particular, the inflated land prices should cause the prospective farmers to be cautious. In Marion County, land values were so high in 1945 that not a single veteran bought a farm in the county. Instead, veterans were told by agent Horace Abbott to purchase farms in surrounding counties, where land prices were more reasonable.[27]

Veterans in other areas—especially those with limited or no farm experience—were advised to work full-time as a hired hand until land prices returned to more normal levels.[28] As Owen County agent Fred Hoover explained, "Agriculture is a business that requires ability, training and experience. An individual that does not have farm training or experience would do better to hire out as a hand, rather than start as a tenant or owner. Then if he is unsuited to agriculture, he can leave without a financial loss. Any soldier with a non-farm reared wife should make sure she will be happy living on a farm before investing any money."[29]

More than anything, agents and committee members tried to get veterans to proceed carefully and with a clear understanding about what they were getting into in order to avoid an untenable situation, such as one described by Orange County agent Ross Atkinson. In his 1945 annual report, he wrote, "Every day boys are having a hard time making 'Fox hole' dreams come true. When a New York City soldier and his young bride move to an off-the-road shack and ask the county agent how to farm their thirty acres of rock and mud and make money after the $20 per week [unemployment benefit] runs out, we wish we knew."[30]

Under the GI Bill, veterans could earn credit toward vocational training, so members of county veterans advisory committees worked to set up vocational agriculture classes under the Institutional On-Farm Training Program for interested veterans.[31] A January 1946 article in the *Hancock Democrat* would describe the effort this way:

> To bring vocational agriculture training to veterans who for some reason cannot go to Purdue University, an approved program of on-the-job training for veterans under provisions of the G. I. Bill of Rights, has been devised by the three-way cooperation of the State Department of Education, the Veterans Administration and Purdue University. . . . [T]he courses have been devised in order to aid Indiana veterans interested in farming in quickly catching up on agricultural progress made during the war, and that the courses apply to the Indiana farm veteran whether he be a hired man, a renter or owner-operator, who cannot pursue a course of formal education in agriculture. . . . The programs are built around the vocational agricultural teacher and the high school employing him, as well as the county extension service administered by the county agent, in collaboration with Purdue extension specialists.[32]

To set up the program, federal and state officials met with advisory committee members along with county school superintendents, school principals, and local vocational agriculture teachers to explain what it would take for an agricultural course to qualify under the legislation.[33] Purdue sent education professor Kenneth W. Kiltz and agricultural economist Edwin Losey to participate in many of these informational meetings and explain how best to meet the requirements.[34] Those putting on an accredited program would be reimbursed by the federal government for expenses.[35]

Members of the advisory committees were responsible for setting a budget, filing the government paperwork for approval, securing the instructors and working with them to determine the course content, and registering veterans to participate.[36] The GI Bill required that the curriculum contain 200 hours of classroom and shop training, and 100 hours of on-the-farm instruction, which agent Eugene Bird viewed as overly cumbersome.[37] In the 1946 annual report for St. Joseph County, he wrote:

> The program handed down from above [GI Bill] is full of the usual swivel chair goat feathers, any of which should be and would be eliminated if local people had the authority. Such requirements as the 100 and 200 hours of farm visits and classroom work per pupil, can easily lend themselves to padding because they are impractical of accomplishment. Such visionary requirements, of course, limit the number of men that an individual instructor can handle. The teaching could be greatly expedited by a practical adjustment in this number of hours required.[38]

A common problem for the advisory committees was that the demand for the vocational agriculture classes was high but the lack of qualified teachers and available facilities limited the number of courses that could be offered.[39] The first training course at the Laketon School in Wabash County filled up, leaving 30–40 veterans on a waiting list.[40] And in Spencer County, 85 veterans were placed on a waiting list to attend a second course when one became available.[41]

Jay County agent Harold Rippey made the following comments in his 1946 annual report:

> Since the beginning of the first class with the veterans "On the Job Training," a sufficient number of additional requests have been made with the Extension Agents to start one or two more departments. The present situation is that Vocational teachers are not available nor are teachers who are qualified to teach the work. It is hoped that, within the very near future, additional vocational teachers, or teachers who can qualify, may be obtained to take care of the veterans requesting the G. I. training.[42]

Some counties were able to hire trained vocational agriculture teachers to lead these courses for veterans.[43] For example, Rushville High School in Rush County released its vocational agriculture teacher half time from his normal duties to focus on conducting a farm training for World War II veterans.[44] But from the onset, it was difficult for local vocational agriculture teachers to work with their usual students during the day and, in addition, teach the veterans on two or three nights during the week.

Most vocational agriculture programs instead relied on other instructors to teach the classes, but finding qualified individuals proved challenging, even with funding from the Veterans Administration. In Spencer County, which was one of the first three counties in Indiana to start its vocational agriculture program under the GI Bill, agent Lewis E. Cooper reported in 1946:

> William F. Richardson, Rockport, . . . was appointed Instructor of the Veterans-on-the-farm Training Program. William is a graduate of Luce Township High School, graduated from Purdue University in 1942 with B.S. [bachelor of science] degree in Agriculture. Three weeks following his graduation from the University, he enlisted in the Service. He was discharged in the spring of 1946. Spencer County is very fortunate in having a local boy who can meet the requirements necessary to be the instructor on such a program.[45]

While agents at times helped to teach specific topics, most provided instruction through hands-on demonstrations at field days on crop cultivation, dairy, fertilizers, forestry, hybrid corn, poultry, tobacco, wheat, and a host of other topics. Agent Scott Milligan provided this list of demonstrations that Wayne County veterans would participate in:

> This committee prepared a number of crops demonstrations to be carried out by farm veterans in their on-the-farm training programs. These outlines provided for two demonstrations on fertilizing corn, two on fertilizing wheat, one on fertilizing oats, one on fertilizing and inoculation of soybeans and two on permanent pasture renovation and fertilization. These suggested demonstrations are to be used in the course 'C' where one-half time will be devoted to projects offered by the extension service.[46]

Programs such as the Purdue Agricultural Outlook and Farmers' Institutes as well as visits to Purdue farms were also included as part of the approved training.[47]

Over time, this little-known program under the GI Bill was responsible for providing training to thousands of returning veterans following World War II.[48]

Before the war was over, county Extension agents were already preparing to help veterans assimilate back into local communities throughout the state under the auspices of the GI Bill, which offered important benefits to help the veterans make this transition. The agents were in a perfect position to harness those benefits on behalf of individuals who wanted to take on farming as a career. By working together with prominent local farmers and business leaders, the Extension agents and specialists developed and shared customized advice designed to help veterans understand exactly what they needed to know.

The agents and other members of the Extension veterans agricultural advisory committees also rose to the demand from veterans who wanted to further their education, setting up vocational agriculture classes and training that met the criteria of the GI Bill as well as the needs of the veterans. Extension supplemented the classroom instruction received by the veterans with hands-on demonstrations that provided science-based educational opportunities on a wide range of agricultural topics.

In many ways, the agents acted as counselors, providing the best advice and training that they could muster for the veterans, many of whom the agents must have known as they grew up in the community, participated in 4-H club work, or worked on area farms owned by their families.[49] In 1944, Tippecanoe County agent Sheldon Pershing wrote about the importance of training the veterans, observing, "These are serious responsibilities of the peace toward which everyone is looking, often anticipating only the happy return of those who have survived."[50] Helping veterans take steps toward becoming successful farmers or establishing a career in the agricultural industry must have been one of the more pleasant tasks that the county Extension agents got involved with.

CHAPTER 45

Effort Must Be Directed Toward Building for the Future

When World War II ended, Americans celebrated in the streets as newspaper headlines declared in large bold letters that the war was over. But the moment was also sobering because thousands of American lives had been lost and the United States had accumulated a $4 trillion debt.

The generation that had lived through the Second World War could never forget the enduring images of hundreds of thousands of soldiers fighting and dying on the battlefields, beaches, or oceans. For its part, Indiana had sent 338,000 soldiers to fight. In addition, 118,000 women from the state had served in the military as nurses, truck drivers, clerks, and pilots.[1]

The war would claim almost 12,000 Hoosiers, both men and women. Some of the fallen were returned to Indiana to be laid to rest in local cemeteries, while many were buried in the countries they died in or were lost at sea.[2] There was hardly anyone who did not personally know a family who had lost loved ones in the war. Thus, the celebration of victory was mixed with reflections on the sacrifices made by real heroes.

With the war finally in the rearview mirror, Extension would now be charged with helping the country and its people move from a wartime economy to a more normal peacetime economy.[3] Agents were being asked by community leaders and officials at all levels of government to direct their efforts toward moving beyond the war.[4]

Transitioning away from a wartime economy would be difficult because the agents had to reverse what they had told farmers and their local communities for four years.[5] What that meant exactly in terms of Extension activity and programming was still unclear to many agents as the soldiers returned home. As 1945 drew to a close, Tippecanoe County agent Sheldon Pershing summed up what many of his colleagues were thinking, writing, "Farmers are facing the first peace New Year in five years, and it is surely a responsibility of extension service to make the anticipated joys of peace develop with the wisest possible adjustments to the changing and uncertain times ahead, and to quickly re-establish the many facilities and condition which were changed by the war period."[6]

The uncertainty about what the future held was on everyone's minds. Campus Extension leaders and county agents alike were asking, What should Extension focus on now?[7] Lawrence County agent John Armstrong wrote in 1946, "The agent, without the war programs, campaigns and drives, is a little like the returned service man, wondering what to do and when to do it. Just what are the most important problems confronting farmers and just how can Agricultural Extension help farmers solve these problems?"[8]

Agents were looking at rebuilding an Extension program that had been critically disrupted by the wartime needs of the country.[9] Now they would have to make a 180-degree turn from maximum production at all cost. Instead, economic production would now take center stage, and Extension programs would resemble those of the Great Depression years, only now they would be updated with new research and novel technologies.[10]

It would take a number of years for Extension and the economy to wean itself "from the nationalistic war-time set up of war activities related to agriculture," as one agent put it.[11] The agents knew that a return to a peacetime economy meant that farmers would need to quickly adjust their production practices.[12] But as agent Martin Huxley observed in his annual report for Knox County, "The year 1946, being the first post-war production year for farmers, has seen production of major crops in the county carried forward at approximately the same high level as during the final year of the war. While farm prices were under government control at planting time and prevailed throughout the first six months of the year, prices were relatively high for most farm products, which encouraged farmers in general to produce to capacity."[13]

It was only natural that farmers had continued producing at wartime rates in order to capitalize on the higher market prices being supported by the government. Indeed, in the first year following the war, one agent reported that farmers had "received some of the highest prices on record for their farm products."[14] But some in agriculture feared that the situation was shaping up in much the same way as the years following World War I, when a severe postwar agricultural depression had put many people at risk of losing their farms and livelihoods.

In that case, the federal government had continued to push farmers to produce during the postwar period in order to provide food to countries where the war had obliterated agricultural production, but when the demand for exports fell as countries recovered, the U.S. agricultural economy suffered.[15] With the federal government encouraging the same approach following World War II, the fear was that the demand for U.S. agricultural products would soon suffer a sharp decline. Prices remained high for now, but for how long would they remain so? That was the hundred-dollar question.

In an effort to help farmers avoid a downward economic spiral, Rush County agent Clarence L. E. Spuller, named the following areas as the most important ones for Extension to focus on:

- conserving soil resources,
- having less tillable land and growing more soil-conserving crops,
- promoting the reduction of debt,
- helping to prevent inflation,
- keeping the supply of farm products in line with consumer demand,
- getting farm businesses ready for less profitable periods,
- adjusting farm labor and new farm machinery purchases,
- discouraging the purchase of farms at inflated prices unless most of the payment could be in cash, and
- assisting returning veterans who want to farm.[16]

The Extension staff believed that sooner rather than later the government would remove price supports and controls, leading to declining farm profits. As Steuben County agent Thomas Grabill observed, "The expected slump in farm prices as a result of smaller markets for farm products looms like a specter in the background."[17] As a result, Extension repeatedly told the agricultural community that the "farm markets should not be depended upon to remain stable for any length of time," as Putnam County agent Daniel J. Holmes stated it.[18]

What concerned agents and economists alike was the purchase of overpriced farm ground. In Johnson County, agent Sylvester Scott noted, "Many farms have sold too high for safe investments."[19] Extension's unifying message that it was prudent for farmers to pay down existing debts and not to incur further debt by purchasing high-priced land as a way of surviving the eventual drop in prices. LaPorte County agent Everett Hartman reported in 1946 that farmers in that area seemed to be heeding the warning. He observed that the "general feeling among farmers seems to be particularly sound from the standpoint of avoiding heavy financial commitments that will require a long period of years to liquidate, as well as the feeling that now is the time to make as much progress as possible in the liquidation of outstanding financial obligations."[20]

With available money to spend, it was natural for farmers to look at expanding their farm acreage by buying up smaller tracts of land, but they had competition from many sources for those purchases.[21] As more returning GIs and industrial workers who had previously worked as farmhands purchased farms or rented ground, land values increased.[22] In Jennings County, agent Arthur Wissman reported in 1946, "Sales to veterans, former tenants and farmers, and other land seekers have been numerous. Many are able to pay a large part of the purchase price. Even once-abandoned farms are being sold and are being brought into production."[23]

A major influence on the upward movement of land prices also included urban people who were in the market for land in the country upon which to build their homes.[24] For example, in Lake and LaPorte Counties, some farmland near cities had been purchased to build houses and businesses.[25] In 1942, LaPorte County agent Everett Hartman had reported that much of the farmland was "being purchased by 'city money.' Those tracts purchased can be divided into two classifications: one, large tracts being purchased by individuals looking for good investments; and two, wage earners purchasing small tracts as homes and a place to maintain their families."[26] Two years later, he reported continued "demand for farm property which has exchanged hands in increased numbers and at higher prices. Purchasers include both farmers and nonresident non-farmers."[27]

As for purchases by farmers, agents understood that the war years had been prosperous, but it was risky to buy land during a time of inflated prices and profits, with Howard County agent Everett Johnson cautioning in 1946 that farmers "should not expect prices and farm profits to be as great over a period of years as has been the case the past few years."[28] Purdue agricultural economists had predicted that 1947 would see sharp decreases in commodity prices, so Extension would continue to focus farmers' attention on anticipating that prices would eventually fall to prewar levels.[29]

Farmers planning next year's crops or making adjustments in what livestock they carried tended to draw heavily on their prior experiences. While relying on one's experiences might have worked well under normal conditions, it did not provide the information needed to make the best decisions in times of radical change.

The better farmers soon realized that economic trends delivered at Purdue-sponsored meetings and bulletins written by agricultural economists could put those prior experiences into context.[30] In the new world that would follow World War II, Extension provided farmers with the following information through the Purdue Agricultural Outlook meetings: what economists thought future prices would be, the overall demand for crops and livestock, the expected levels of exports, the cost of farm inputs, the availability of labor and machinery, changing government policies, and tariffs and trade barriers.[31]

In 1946, Tippecanoe County agent Sheldon Pershing shared a key message that farmers heard from the Purdue agricultural economists at meetings during the first postwar year, writing, "They pointed out that with the removal of government controls, prices would again go up and down according to the general influence made by the people who produce and consume, as well as by the general economic trends; in other words, the regular forces of supply and demand and general price level will affect prices, instead of government regulations as in recent years."[32]

Farmers were attending outlook meetings in larger-than-expected numbers. As Spencer County agent Lewis Cooper remarked, "Farmers are eager to have a general picture of conditions for 1947 due to the large attendance at the Outlook meeting and for the request of Outlook bulletins."[33] While farmers had still experienced high market prices for what they grew and raised in 1946, they were anticipating that the market could drop at any time, so before they planned their next growing season, they wanted to know, Was the market going to collapse in 1947?

During the uncertain times of the post–World War II years, farm records and marketing strategies became even more pivotal to a farmer's ability to plan for the future. As a result, Extension doubled down on getting farmers to better understand their net profits and losses by keeping detailed records for their operations. In 1943, agent Ross Atkinson had talked to Orange County farmers who kept records, then observed, "Those who appreciate records cannot understand why anyone should wait to be driven to keeping farm records. Be that as it may, we find that a very large majority of farm folk find it to be one of the harder chores to keep good accurate records for which there are many practical reasons."[34] While it seemed to many that records were a common-sense strategy, it was still a hard sell in the first year after the war as prices and demand remained high. In spite of this, Extension was relentless in broaching the benefits of keeping good records at every opportunity.[35]

Without detailed records for operations, agents believed wholeheartedly that farmers were more apt to feel the pain when commodity prices dropped, input costs increased, and inflation reduced the value of their incomes.[36] As Starke County agent Arthur Shideler remarked, "The rise in costs of what the farmer buys will catch up with the rise in farm commodity prices so that there will be less spread," meaning less net profitability.[37]

The shortage of labor, which was a big challenge for agricultural production during the war, was expected to continue after the war until the movement of the workforce from industry and the military back to the farm stabilized.[38] Returning veterans provided some short-term relief, but labor was still critically low to meet the demands of agriculture, a situation that was exacerbated by the fact that many workers were choosing not to seek out farm positions in the postwar economy.[39] Of the more than 1,000 Jackson County war veterans, only a few returned back to the farm after the war to take up positions that they had once held.[40]

Farmers across the state could sense that the labor supply they had grown accustomed to prior to the war would not return to prewar levels.[41] Former agricultural workers were no longer willing to put in the hours or effort for what they now considered to be substandard wages.[42] Many of those finding employment in the industrial plants decided to remain in those positions in the postwar years.[43]

In Lake County, agent Lloyd Cutler observed that agricultural labor was, at best, difficult to find, writing:

> Some young men who were anxious to have farm jobs in order to avoid service in the armed forces during the war were no longer interested in farm work. It also seemed that most of the veterans who had received their discharge were anxious to take advantage of the high industrial wages offered in order to build up reserve funds which they might use later to start farming when farm machinery became available at more reasonable prices. Truck growers reported that many young people who had been willing to work during the war were no longer very much interested because their fathers were making good wages and the children felt that it was not necessary for them to work unless they wanted to.[44]

With a negative prognosis for labor, farmers looked in earnest at investing in labor-saving equipment to compensate. Farm machinery remained in short supply after the war, but manufacturers were rapidly converting back to producing farm equipment, which farmers hoped to see at local implement dealers sooner rather than later.[45] Jasper County agent Frank DeLaCroix understood the need to modernize farm equipment, but like many other agents, he cautioned farmers about carrying debt with uncertainties ahead for agriculture. In his 1944 annual report, he wrote, "The trend is likely to be over-stocking in farm machinery as soon as farm machinery becomes available. This will be brought on by the fact that the farmers have uninvested capital which they have been forced to save through rationing. Then because their present machinery has been seriously depleted, overbuying will result. All farmers should be cautioned about over-stocking their farm with new machinery. Carrying this overhead will dissipate most of their profits."[46]

Still, many producers had money in the bank to make their purchases, so agents anticipated that programs on equipment would remain popular for years to come.[47]

As the postwar conversation centered around the economic future of farming, there was no doubt that Extension would once again focus attention on the importance of soil conservation and land rehabilitation. As Starke County agent Arthur Shideler pointed out in 1945, "The increase in grain and food crops to meet the demands of war has drawn heavily upon the plant food resources of our soils. The future welfare of our farms will require greater attention to soil conservation practices."[48] The following year, Lewis Cooper wrote frankly in his Spencer County report, "Soil conservation must be shoved by all known means."[49]

Everyone understood that farmers had to push production limits higher and higher throughout the war to meet the ever-increasing quotas set by the federal government. But with the war now past, Extension knew it was time to address the consequences of this practice. As Wabash County agent William Delaplane explained, "We have used our soil to the point where it has been depleted, for more than good business methods could justify. This has been necessary to win the war and we have done it with a full knowledge of the effects. These soil damages must be repaired, in order that agriculture may be in a position to accept any future responsibility. The bringing back of our soil, involves the use of all good management practices."[50]

Those counties that had established a soil conservation district (see Chapter 43) found a partner willing to promote practices that focused attention on the land.[51] By 1950, fifty Indiana counties had organized officially recognized soil conservation districts.[52]

One critical problem that Extension and its partners would have to address was encouraging farmers to restore pastures and forests that had been converted to cropland for wartime production.[53] As Posey County agent Ray A. Burger observed in 1945, "During the war, many acres of land were put under cultivation which should not have been put to any kind of row crops. Consequently, the end of the war has brought about the increased necessity of planning for soil conservation."[54] And in the 1946 annual report for Spencer County, agent Cooper noted, "Pasture improvement will be one of the 'musts' for grain crops have been on a larger acreage, and with [the] possibility of farm prices to decrease, cheap pasture will mean profit or loss. Pasture improvement is needed to rest some land and to keep land from becoming eroded."[55]

By 1946, it was clear that the fertility levels of soils across the state had largely been depleted.[56] But even before the war had ended, agents had already been expressing concern about loss of soil fertility and making plans to address it. In the 1944 annual report for Huntington County, Walter Rusk had reported:

> The Huntington County AAA [Agricultural Adjustment Administration] committeemen and the Post-war Planning committee are recognizing that soil and its fertility are the most valuable possession we have, and the basis of prosperity, and that something needs to be done about it. With Huntington County farmers committing themselves to an all-out food production program in the war effort, it means that larger amounts of plant food are being removed from the soils of the county. More and more row crops are being grown with a result that there is less in grass or that has the winter green cover protection; consequently much of the soil is at the point of being lost through erosion.[57]

In 1945, Blanchard Widney identified soil fertility as a top postwar priority in Whitley County, writing, "Restoration of fertility levels and conservation of the fertility the soil still possess[es], are problems which must receive attention in the 1946 Extension program."[58] His effort would be one of many made throughout the state as agents worked with soil conservation districts to offer farm demonstrations showcasing production practices that made money and improved the soil.

According to White County agent W. Johnson Emerson, events such as these would highlight "soil conservation practices, soil and crop improvement in varieties, rotations, fertilization and land use."[59]

Like many industries operating during World War II, agriculture was a very profitable enterprise due to the high demand for food by the international community and because farmers were receiving government subsidies for their production. Based on what had happened after World War I, those who understood the agricultural economy seemed certain that at some point in the near future demand for agricultural products would drop, and prices would follow suit.

It was natural for Extension agents to advise farmers to settle debts as quickly as possible and to really watch the prices they were paying for land, equipment, and other expensive goods. This conservative messaging about saving and limiting spending was supported by the agricultural economists at Purdue and by those who had been farming at the conclusion of the First World War.

What was not fully understood at the end of World War II was that overproduction from the modernization and mechanization of agriculture would dominate the economic front for many decades.[60] These permanent changes would influence agricultural research and Extension outreach in ways that no one could have predicted at the time.

President Franklin Roosevelt (seated) signs the Servicemen's Readjustment Act on June 22, 1944. The legislation, which became better known as the GI Bill, provided veterans returning from World War II with opportunities to obtain college and vocational training, low-interest loans for buying farms, and unemployment benefits as they reintegrated into their communities. Courtesy of U.S. Department of Defense.

INFORMATION FOR VETERANS Who Plan to FARM in DeKalb County

Around the state, county Extension agents worked with prominent area farmers to develop documents, such as this one for DeKalb County, that provided guidelines and advice to veterans who wanted to farm. Recommendations included how much to pay for farmland, the size of the farm needed to make a good living, what percentage of the farm's value should be paid for with cash, the crops and livestock that should be grown, and other conditions to help ensure that these new farmers would be successful. Courtesy of Purdue University Libraries, Karnes Archives and Special Collections.

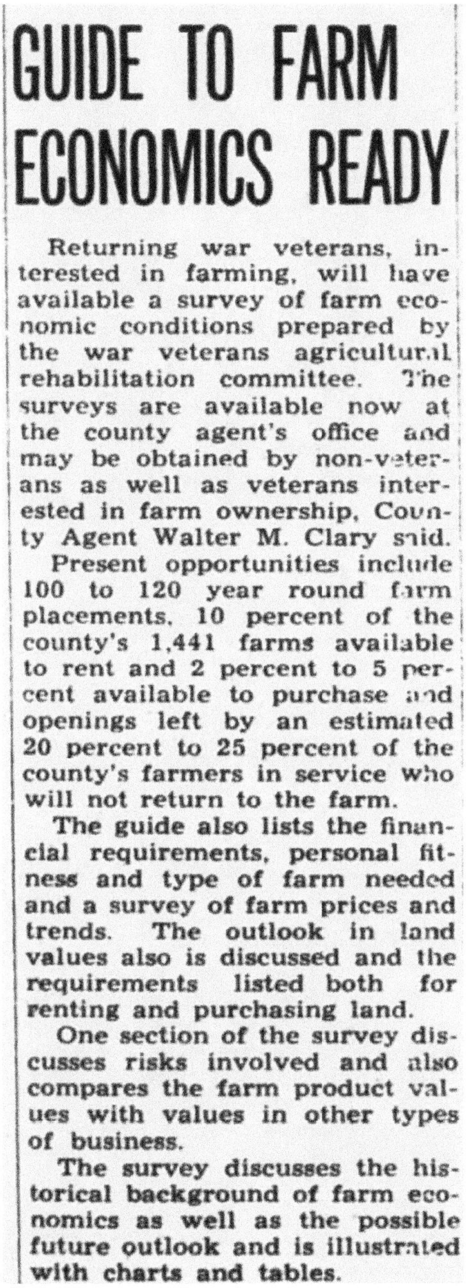

GUIDE TO FARM ECONOMICS READY

Returning war veterans, interested in farming, will have available a survey of farm economic conditions prepared by the war veterans agricultural rehabilitation committee. The surveys are available now at the county agent's office and may be obtained by non-veterans as well as veterans interested in farm ownership, County Agent Walter M. Clary said.

Present opportunities include 100 to 120 year round farm placements, 10 percent of the county's 1,441 farms available to rent and 2 percent to 5 percent available to purchase and openings left by an estimated 20 percent to 25 percent of the county's farmers in service who will not return to the farm.

The guide also lists the financial requirements, personal fitness and type of farm needed and a survey of farm prices and trends. The outlook in land values also is discussed and the requirements listed both for renting and purchasing land.

One section of the survey discusses risks involved and also compares the farm product values with values in other types of business.

The survey discusses the historical background of farm economics as well as the possible future outlook and is illustrated with charts and tables.

In January 1945, Tipton County agent Walter Clary publicized the availability of a farming guide for veterans in the *Tipton Daily Tribune*. One concern was that new farmers would not adequately prepare for an expected drop in market prices following the war, so county guides sought to help veterans set realistic expectations and avoid the financial mistakes made by many farmers after World War I. Courtesy of *Tipton Daily Tribune*.

The GI Bill paid for veterans to receive vocational agriculture training, which meant that hands-on programs such as this 1947 sheep-shearing demonstration at a Purdue farm counted toward the educational credits outlined by the federal government. With demand for agricultural training high, there often weren't enough teachers available, so county Extension agents would sometimes assist by teaching hands-on demonstrations such as this. Courtesy of J. C. Allen and Son.

At this 1947 demonstration, the discussion focused on raising Hereford and Angus livestock. Field events such as these allowed veterans who wanted to start farming the opportunity to ask questions of the presenters, who were typically experienced farmers, county Extension agents, and vocational agriculture teachers. Courtesy of J. C. Allen and Son.

From 1942 to 1945, Edgar Greene served in the U.S. Navy during World War II. Following his discharge, he received $100 per month under the GI Bill to continue his education. He attended the Purdue University Poultry Short Course in 1946 and 1947, then he and his father, Ray Greene, started a poultry dressing business in Sandborn, located in Knox County. In 1950, Edgar moved to Linton in Greene County, where the business remained operational until 1973. (Top) Edgar Greene (third row, third from right) poses in 1946 in front of Fowler Hall with other Purdue short course attendees; (bottom) Edgar and his wife, Dorothy, handle turkeys at their business, circa 1953. Courtesy of Danny Greene.

After serving in the U.S. Navy during World War II, James T. Cummings used the benefits provided by the GI Bill to attend Purdue, where he earned a bachelor's degree in agricultural education in 1949 and a master's degree in horticulture in 1951. He would go on to teach agriculture and horticulture for thirty-nine years at Cowan High School in Delaware County and at Whiteland High School and Central 9 Vocational School, both in Johnson County. (Top) James Cummings, June 1943; (bottom) Cummings milking a cow, circa 1950. Courtesy of Don Cummings.

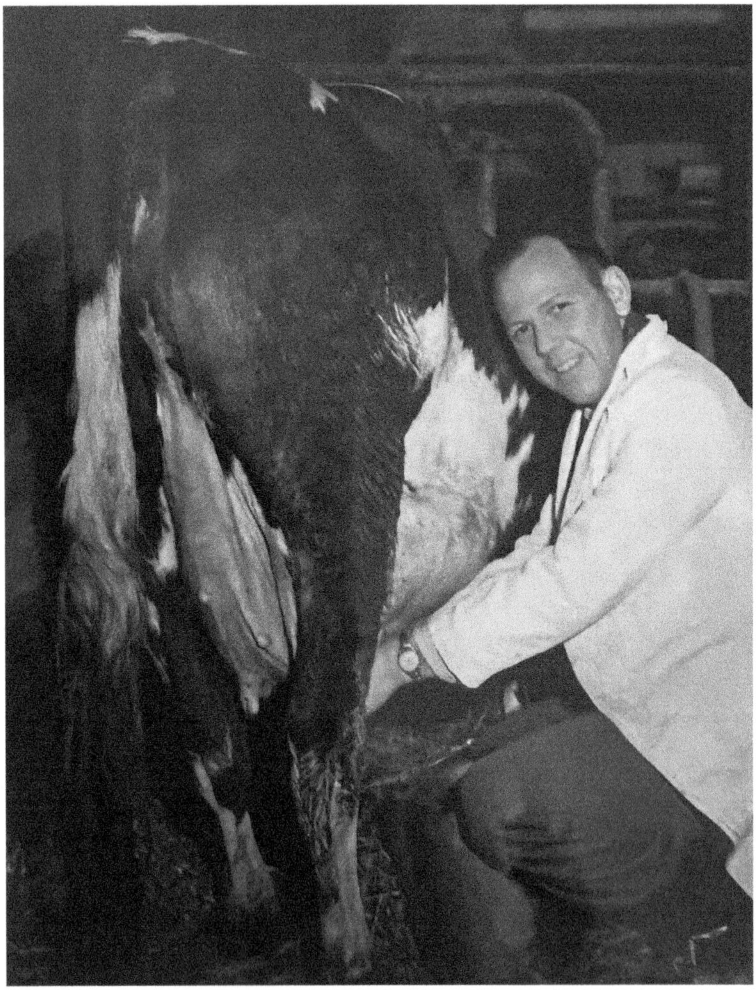

Epilogue

Now as Never Before Farmers Have Put Into Use Many Practices Advocated by the Extension Service

No one could ever have ever imagined at the inception of the Cooperative Extension Service that Extension staff in Indiana would be asked to help the agricultural community cope with four historic events within its first three decades. Extension had been established in 1914 to deliver the latest scientific findings from the Agricultural Experiment Station at Purdue University to the state's farmers, homemakers, and children in an effort to improve their lives and livelihoods. But Extension was still in its infancy as World War I began and then was followed in quick succession by an agricultural depression in the 1920s, the Great Depression in the 1930s, and World War II in the 1940s.

During the early years, campus specialists and county agents earned their stripes by working shoulder-to-shoulder with farm families, business owners, and community leaders to offer technical guidance and personal support through the ups and downs that the 1910s and 1920s brought. By the time the events of the Great Depression began unfolding, Extension had been in existence for less than two decades. However, the learning curve experienced during that short time by Extension staff on campus and in the field would serve as the foundation from which they could offer help and hope when prices for crops, livestock, and other commodities fell to record lows and the agricultural sector spiraled into chaos.

At the start of the Depression, farmers, who were still recovering from the 1920s agricultural depression, had to quickly figure out how to make enough money to feed their families, pay their farm mortgages, purchase coal to heat their homes, and buy seed, lime, and fertilizer to plant the spring crop. Extension agents were well positioned to draw on university resources to provide this much-needed help to their constituents, who were also their friends and neighbors. The agricultural research showed that strategic economic investments had the potential to bring in much higher revenues, but how could Extension get farmers to invest more when they were struggling to cover the basics?

Agents got the attention of farmers by standing in front of crowds at on-farm demonstrations to present the evidence firsthand: which practices were changed; what it cost in time, labor, and money; and the net returns for making those changes. With support from respected local farmers, the demonstrations became key to helping farmers squeeze more returns out of every dollar invested. During the Depression, the agricultural community leaned heavily on agents for sound, practical advice on becoming more efficient with their scant resources—producing more bushels of corn per acre, more pigs per sow, more gallons of milk and pounds of butterfat per cow, more eggs per chicken, and more canned produce from the home garden.

Becoming more efficient with the dollars invested in the farm became the centerpiece of Extension education. Agents not only presented the work through demonstrations, but they also used farm ledgers analyzed by the Purdue Department of Agricultural Extension's Division of Farm Management to show unequivocally how expenses related to revenues when implementing a new practice. Farm records became the method by which agents and campus specialists alike could prove beyond a reasonable doubt whether a practice supported by Extension paid for itself and how much it returned beyond what it cost to implement. The business side of farming had taken hold out of necessity during the troubled times of the Great Depression.

The Depression was pivotal for getting farmers to listen to the advice of Purdue researchers, campus Extension specialists, and especially the agents. With the one-two punch of practical demonstrations backed by farm ledgers, Extension continued to persuade farmers to change their practices when they had been unwilling or unlikely to do so before. What is largely lost to history is that it had taken the agents' presence in their communities and personal interactions with their clients to gain a much wider acceptance of the information that Purdue experts had to offer.

Even as farmers were returning to normal times after the Depression, there were already rumblings of war in Europe. As the American public learned about the brewing conflict in newspapers and on their radios, it seemed early on that the nation could face another world war. Prior to the official declaration of war, the federal government demanded that farmers significantly increase their production to help feed the troops and populations of allies overseas that had already felt the brutality of the war. By early 1941, cargo ships filled with food had already begun flowing from the United States to England.

When the U.S. entered the war as a combatant, federal decrees would require even more of American farmers. Instead of efficient production, World War II pushed agricultural production to ever-increasing levels encouraged by government subsidies and other market incentives. The mantra was now to do whatever it took to maximize bushels per acre, gallons of milk and pounds of butterfat per cow, eggs per hen, and quicker gains per hog.

This demand for more production was made even as the federal government implemented dozens of rationing programs that affected every American. Farmers would be asked to reach the higher production quotas at a time when the required farm inputs were in extremely short supply. In addition, farm labor became an acute problem as the Selective Service draft moved a large number of men from the workforce to the battlefields. At the same time, many of the men and women who remained left the farm in order to take lucrative jobs with industries that had converted their production to meet military needs.

Farmers who had seen the lowest prices for their production during the Depression were now receiving some of the highest returns they had even seen for their labors. Agricultural production quotas were continually raised as the war continued, and farmers worked hard to deliver the farm products in the quantities that the government asked for.

Throughout it all, the county Extension agent served as an intermediary between the U.S. Department of Agriculture, the state, and the agricultural community as farmers dealt with rationing programs and labor shortages. Agents were asked to explain what farmers needed to do to comply with the county production quotas while at the same time addressing how they could secure labor and obtain fuel, tires, machinery, and lumber under the government's rationing programs. Their success in navigating this crucial role was rooted in the relationship that they had established throughout the decades with the people and communities they served. People knew and trusted them. Agents provided that personal touch that others in government could never have hoped to develop with their constituents, which had allowed Extension to become an important resource and support system for the agricultural community.

During the time period covered by this book, Indiana agriculture and Extension underwent many radical changes as scientific technologies were more fully utilized. Some of those changes included:

- the electrification of rural homes and farms,
- the shift toward replacing horses with mechanical farm equipment,
- a reduced reliance on manual labor, even as farm sizes and production levels increased,
- more widespread use of hybrid corn in place of open-pollinated varieties,
- the acceptance of fertilizers as a needed input and more reliance on soil analyses to apply the proper kind and rates,
- better feeding of livestock to improve productivity and profit,
- the use of milking machines, and
- an increased reliance on genetics to improve dairy herds.

In general, these advancements were made possible during the 1930s and 1940s due to a greater acceptance of the education offered by Extension and the more widespread adoption of better management practices grounded in agricultural research.

While any number of these developments could show how traditional practices were growing out of favor as new scientific-based ideas emerged, the advancement of corn during this time period provides a clear example of how quickly the state's farmers were adapting to new ideas advocated by county Extension agents and the impact that this had during critical years for Indiana agriculture. The introduction of hybrid corn doubled and tripled county yields that had long averaged 40 bushels an acre, pushing them to 150 bushels per acre during World War II. These new record-breaking yields soon became commonplace as farmers planted hybrid corn and incorporated best management practices to maximize yields. Farmers began asking agents more questions about which hybrid corn to purchase and what it took to obtain and harvest these astronomically high yields. Farmers were taught about the need for fertilizers, incorporation of rotations that brought in sources of nitrogen, planting cover crops to protect the soil and add organic matter, selecting adaptable hybrid varieties that fit a specific locale, proper planting dates, and using higher planting rates and narrower rows. The importance of the five-acre corn competitions in demonstrating to farmers what was possible with good management of the corn crop cannot be understated.

Hybrid corn, which began to take hold among farmers during the late 1930s, achieved near universal adoption across Indiana during World War II. This was a critical development because corn served as a main ingredient for feeding the record number of hogs being raised and marketed for the Allied forces, among others. The doors to agricultural research that had been partially opened during the Great Depression were blown off their hinges with the rapid adoption of hybrid corn during the 1940s.

Similar advancements were taking place across the spectrum of all farming activities, including with poultry, beef and dairy cattle, soybeans, and tomatoes as well as other endeavors. Those farming throughout the 1930s and 1940s began to appreciate how science was revolutionizing and advancing agriculture for the better, leading Marion County agent Horace Abbott to observe in 1944, "Now as never before farmers have put into use many practices advocated by the Extension Service. Such practices have resulted in record crops of food. They include the extensive use of lime and fertilizers, not only more fertilizer, but more of the right analysis. They include the use of better and more highly productive varieties of crops, better feeding of dairy cows for higher production, the more extensive production of better type hogs, and the use of better marketing methods."[1]

All of this was made possible as agricultural research by Purdue scientists was moved to the campus Extension specialists and then distributed throughout the state by the agents, who proved to be essential to their communities during the ups and downs of those years. It was the scientific-based practices adopted by farmers that allowed Abbott to repeatedly use the word *better* in his summary, but it was the work done by Abbott and his Extension colleagues to promote that research that led the way to those improvements. Research and outreach now went hand-in-hand; one without the other would have resulted in fewer advances.

Extension rose to each new challenge as it helped the agricultural community cope with the economic lows of the Great Depression and the diverse demands and obstacles during World War II, but skill, resourcefulness, and commitment allowed its people to persevere. The downside for Extension was that the heavy demands eventually took a toll on the agents, resulting in nearly half leaving their positions for less stressful employment. By the start of 1946, there were 52 new county agents who were receiving training on Extension program planning, leadership training, teaching methods, and organizational skills. Working alongside these county agricultural agents were now 50 home demonstration agents, 15 assistant county agricultural agents, 60 campus specialists, and 13 administrators and supervisors, all trying to refocus Extension on the future needs of agriculture in the state.[2]

While Extension had lost experienced agents, the influx of new men and women into the ranks provided much-needed energy into a system that had been drained by the overwhelming demands of the previous decades. In 1945, the Department of Agricultural Extension annual report stated, "The size of the job can be appreciated more when we consider that there are nearly 185,000 farms in Indiana with 813,000 [individuals] or 24 per cent of the Indiana population living on them. An additional 21 per cent of the Indiana population live in towns of less than 2,500 and are assisted in many ways by the Department of Agricultural Extension."[3] Everyone understood that even as the Depression and war faded into the rearview mirror, a lot of work lay ahead.

True to form, Extension was already planning for the what-ifs and the must-dos. Tasks on the immediate docket included getting farmers to replace the all-out production of the war with efficient production as prices were certain to fall to more normal levels; warning farmers to pay down debt and prepare for land prices to eventually drop, as they had after World War I; helping returning soldiers take advantage of the GI Bill; and beginning the process of helping to repair lands that were damaged by wartime production practices.

While it is often programs and policies that historians analyze and write articles and books about, examining vintage photographs offers a different perspective, allowing us to see the worries of the day etched on the faces of those who were beaten down by economic calamity or burdened by the cares of war. One can only imagine what it was like to be an Extension agent during those unprecedented times—seeing friends and neighbors lose their homes, farms, and savings, or watching as many whom the agent had taught in the classroom didn't return from the war.

For these agents, the despair of the Depression and the wartime loss of life were personal, but their jobs demanded that they focus on helping others move forward. While no one could know what the future would bring, the county Extension agents who worked during the Great Depression and World War II had set the bar high for those who followed in their footsteps.

Notes

Much of the material for this book was drawn from the annual reports filed by Indiana county Extension agents and by administrators in the Purdue University Department of Agricultural Extension. The Extension reports cited here are at the Virginia Kelly Karnes Archives and Special Collections Research Center, Purdue University Libraries, West Lafayette, IN. The finding aid dated 15 June 2012 was used to locate the county reports, which are found in the College of Agriculture Extension Reports Collection, 1913–1969.

In notes that contain multiple sources, the first entry is the source for any direct quotes included in the text. In cases where more than one person served as the Extension agent in a county during a given year, the agent who filed the cited report was named for the attribution. The quoted material has been presented as written with the exception of some limited changes made to correct minor grammatical issues in instances that might have affected readability or understanding.

Unless otherwise indicated, newspaper articles were accessed at Newspapers.com, https://www.newspapers.com.

Abbreviations

CAERC	College of Agriculture Extension Reports Collection, 1913–1969. In the Virginia Kelly Karnes Archives and Special Collections Research Center, Purdue University Libraries, West Lafayette, IN.
CEAR	*County Extension Annual Report*

CHAPTER 1

Neither the Agent nor His Farmer-Constituents Knew Very Much About What to Expect of One Another

1. M. J. Huxley, *Fulton CEAR,* 1939 (CAERC box 85, folder 50).
2. E. A. Johnson, *Cass CEAR,* 1939 (CAERC box 69, folder 44).
3. Ibid.
4. *Purdue University Agricultural Experiment Station Annual Report*, 1909, 22.
5. H. E. Young, *25 Years of Extension Work in Indiana: Historical Narrative and Achievement Summary* (West Lafayette, IN: Purdue University, 1939), 7, https://babel.hathitrust.org/cgi/pt?id=coo.31924000899157&view=1up&seq=5.
6. M. J. Huxley, *Fulton CEAR,* 1939 (CAERC box 85, folder 50); F. E. Conder, *Grant CEAR,* 1939 (CAERC box 11, folder 18); C. A. Alcorn, *Dearborn CEAR,* 1939 (CAERC box 75, folder 34); M. M. Hamilton, *Decatur CEAR,* 1938 (CAERC box 76, folder 44).
7. E. M. Rowe, *Clinton CEAR,* 1939 (CAERC box 72, folder 45); C. M. East, *Floyd CEAR,* 1938 (CAERC box 82, folder 39).
8. M. M. Hamilton, *Decatur CEAR,* 1938 (CAERC box 76, folder 44); J. H. Brown, *Elkhart CEAR,* 1939 (CAERC box 80, folder 37).
9. F. E. Conder, *Grant CEAR,* 1939 (CAERC box 11, folder 18).
10. M. M. Hamilton, *Decatur CEAR,* 1938 (CAERC box 76, folder 44).
11. *Purdue University Department of Agricultural Extension Annual Report,* 1920, 20.
12. A. V. Keesling, *Bartholomew CEAR,* 1939 (CAERC box 63, folder 30).
13. D. S. Bishopp, *Fayette CEAR,* 1939 (CAERC box 81, folder 40).
14. H. L. Royce, *Clay CEAR,* 1932 (CAERC box 71, folder 44).

15. E. S. Miles, *Clark CEAR,* 1939 (CAERC box 70, folder 50).
16. A. V. Keesling, *Bartholomew CEAR,* 1939 (CAERC box 63, folder 30).
17. E. S. Miles, *Clark CEAR,* 1939 (CAERC box 70, folder 50).
18. A. Haseman, *Greene CEAR,* 1939 (CAERC box 89, folder 14).
19. C. V. Kimmell, *Allen CEAR,* 1939 (CAERC box 62, folder 25).
20. C. M. East, *Floyd CEAR,* 1939 (CAERC box 82, folder 40).
21. E. M. Christen, *Benton CEAR,* 1939 (CAERC box 65, folder 27).
22. F. E. Conder, *Grant CEAR,* 1939 (CAERC box 11, folder 18); E. M. Christen, *Benton CEAR,* 1939 (CAERC box 65, folder 27).
23. R. D. McHargue, *Clay CEAR,* 1939 (CAERC box 71, folder 51).
24. M. M. Hamilton, *Decatur CEAR,* 1939 (CAERC box 76, folder 45); R. D. McHargue, *Clay CEAR,* 1939 (CAERC box 71, folder 51).
25. F. E. Conder, *Grant CEAR,* 1939 (CAERC box 11, folder 18).
26. J. L. Morrow, *Crawford CEAR,* 1939 (CAERC box 73, folder 40).
27. E. M. Christen, *Benton CEAR,* 1939 (CAERC box 65, folder 27); M. E. Cromer, *Delaware CEAR,* 1930 (CAERC box 78, folder 32).
28. R. Gwaltney, *Gibson CEAR,* 1939 (CAERC box 87, folder 9).
29. E. M. Christen, *Benton CEAR,* 1939 (CAERC box 65, folder 27).
30. C. V. Kimmell, *Allen CEAR,* 1939 (CAERC box 62, folder 25); W. U. Rusk, *Blackford CEAR,* 1939 (CAERC box 66, folder 16).
31. L. E. Archbold, *Adams CEAR,* 1930 (CAERC box 61, folder 16).
32. C. A. Nicholson, *Dubois CEAR,* 1933 (CAERC box 79, folder 42).
33. David B. Danbom, "The Agricultural Extension System and the First World War," *The Historian* 41, no. 2 (Feb. 1979): 316, https://www.jstor.org/stable/24445110.
34. *Purdue University Department of Agricultural Extension Annual Report,* 1917, 12; *Purdue University Department of Agricultural Extension Annual Report,* 1919, 7; M. M. Hamilton, *Decatur CEAR,* 1938 (CAERC box 76, folder 44).
35. E. A. Johnson, *Cass CEAR,* 1939 (CAERC box 69, folder 44); J. B. Todd, *Hamilton CEAR,* 1939 (CAERC box 90, folder 5); A. V. Keesling, *Bartholomew CEAR,* 1939 (CAERC box 63, folder 30).
36. E. M. Christen, *Benton CEAR,* 1939 (CAERC box 65, folder 27).
37. Ibid.; A. V. Keesling, *Bartholomew CEAR,* 1939 (CAERC box 63, folder 30).
38. *Purdue University Department of Agricultural Extension Annual Report,* 1918, 11.
39. C. V. Kimmell, *Allen CEAR,* 1939 (CAERC box 62, folder 25).
40. R. D. McHargue, *Clay CEAR,* 1939 (CAERC box 71, folder 51); C. V. Kimmell, *Allen CEAR,* 1939 (CAERC box 62, folder 25).
41. E. S. Miles, *Clark CEAR,* 1939 (CAERC box 70, folder 50).
42. J. L. Morrow, *Crawford CEAR,* 1939 (CAERC box 73, folder 40); E. M. Christen, *Benton CEAR,* 1939 (CAERC box 65, folder 27).
43. R. D. McHargue, *Clay CEAR,* 1939 (CAERC box 71, folder 51).
44. A. V. Keesling, *Bartholomew CEAR,* 1938 (CAERC box 63, folder 28).
45. W. U. Rusk, *Blackford CEAR,* 1931 (CAERC box 66, folder 8).
46. R. Gwaltney, *Gibson CEAR,* 1939 (CAERC box 87, folder 9).

CHAPTER 2

The Shattering, Sledge Hammer Economic Blows of the Depression

1. J. L. Morrow, *Crawford CEAR,* 1939 (CAERC box 73, folder 40).
2. R. Gwaltney, *Gibson CEAR,* 1939 (CAERC box 87, folder 9).

3. Ibid.; E. M. Christen, *Benton CEAR,* 1939 (CAERC box 65, folder 27); L. E. Moore, *Clark CEAR,* 1931 (CAERC box 70, folder 38); C. A. Nicholson, *Dubois CEAR,* 1932 (CAERC box 79, folder 41); J. L. Morrow, *Crawford CEAR,* 1937 (CAERC box 73, folder 38).
4. L. E. Archbold, *Adams CEAR,* 1939 (CAERC box 1, folder 1).
5. H. S. Heckard, *Allen CEAR,* 1930 (CAERC box 62, folder 15).
6. W. G. Smith, *Henry CEAR,* 1933 (CAERC box 94, folder 2).
7. *Purdue University Department of Agricultural Extension Annual Report,* 1935, 7; A. T. Marvel, *Elkhart CEAR,* 1935 (CAERC box 80, folder 33).
8. *Purdue University Department of Agricultural Extension Annual Report,* 1935, 15.
9. M. E. Cromer, *Delaware CEAR,* 1933 (CAERC box 78, folder 35); M. F. Detrick, *Fayette CEAR,* 1933 (CAERC box 81, folder 34); C. A. Alcorn, *Floyd CEAR,* 1930 (CAERC box 82, folder 31).
10. *Purdue University Department of Agricultural Extension Annual Report,* 1935, 15.
11. A. M. Bishea, *Daviess CEAR,* 1937 (CAERC box 74, folder 31); H. L. Royce, *Clay CEAR,* 1930 (CAERC box 71, folder 42); H. S. Heckard, *Allen CEAR,* 1931 (CAERC box 62, folder 16); E. S. Miles, *Clark CEAR,* 1938 (CAERC box 70, folder 48).
12. W. G. Smith, *Henry CEAR,* 1934 (CAERC box 94, folder 3).
13. C. A. Alcorn, *Dearborn CEAR,* 1933 (CAERC box 75, folder 28); M. M. Hamilton, *Decatur CEAR,* 1938 (CAERC box 76, folder 44).
14. E. S. Miles, *Clark CEAR,* 1938 (CAERC box 70, folder 48); E. M. Rowe, *Clinton CEAR,* 1939 (CAERC box 72, folder 45).
15. N. S. Hadley, *Franklin CEAR,* 1933 (CAERC box 84, folder 33).
16. C. V. Kimmell, *Allen CEAR,* 1937 (CAERC box 62, folder 23).
17. R. L. Zell, *Franklin CEAR,* 1935 (CAERC box 84, folder 35); E. S. Miles, *Clark CEAR,* 1939 (CAERC box 70, folder 50).
18. R. L. Zell, *Franklin CEAR,* 1939 (CAERC box 84, folder 39); E. M. Christen, *Benton CEAR,* 1937 (CAERC box 65, folder 23).
19. M. E. Cromer, *Delaware CEAR,* 1933 (CAERC box 78, folder 35).
20. H. S. Heckard, *Allen CEAR,* 1930 (CAERC box 62, folder 15); H. L. Royce, *Clay CEAR,* 1930 (CAERC box 71, folder 42); H. S. Heckard, *Allen CEAR,* 1932 (CAERC box 62, folder 17).
21. *Purdue University Department of Agricultural Extension Annual Report,* 1934, 19.
22. A. J. Hesler, *Fountain CEAR,* 1939 (CAERC box 83, folder 35).
23. W. O'Hara, *Hancock CEAR,* 1931 (CAERC box 91, folder 14); R. E. Johnson, *Harrison CEAR,* 1932 (CAERC box 92, folder 27).
24. R. E. Johnson, *Harrison CEAR,* 1932 (CAERC box 92, folder 27); W. G. Smith, *Henry CEAR,* 1932 (CAERC box 94, folder 1); H. L. Royce, *Clay CEAR,* 1931 (CAERC box 71, folder 43).
25. V. J. Mann, *Clinton CEAR,* 1930 (CAERC box 72, folder 34); E. E. Stinson, *DeKalb CEAR,* 1930 (CAERC box 77, folder 25); I. L. Thurston, *Decatur CEAR,* 1931 (CAERC box 76, folder 37); J. L. Morrow, *Crawford CEAR,* 1936 (CAERC box 73, folder 37); C. A. Alcorn, *Dearborn CEAR,* 1936 (CAERC box 75, folder 31); N. S. Hadley, *Fulton CEAR,* 1934 (CAERC box 85, folder 42); A. Haseman, *Greene CEAR,* 1937 (CAERC box 89, folder 12); N. S. Hadley, *Franklin CEAR,* 1933 (CAERC box 84, folder 33); M. M. Hamilton, *Decatur CEAR,* 1937 (CAERC box 76, folder 43).
26. E. E. Stinson, *DeKalb CEAR,* 1930 (CAERC box 77, folder 25); W. U. Rusk, *Blackford CEAR,* 1936 (CAERC box 66, folder 13); C. A. Alcorn, *Dearborn CEAR,* 1936 (CAERC box 75, folder 31).

27. C. A. Alcorn, *Dearborn CEAR,* 1933 (CAERC box 75, folder 28); M. F. Goode, *Harrison CEAR,* 1936 (CAERC box 92, folder 31); R. E. Johnson, *Harrison CEAR,* 1931 (CAERC box 92, folder 26); H. J. Yarling, *Hancock CEAR,* 1930 (CAERC box 91, folder 13).
28. E. S. Miles, *Clark CEAR,* 1938 (CAERC box 70, folder 48).
29. C. A. Nicholson, *Dubois CEAR,* 1937 (CAERC box 79A, folder 1).
30. E. M. Christen, *Benton CEAR,* 1938 (CAERC box 65, folder 25).
31. C. K. Whistler, *Boone CEAR,* 1938 (CAERC box 67, folder 3).
32. A. Haseman, *Greene CEAR,* 1939 (CAERC box 89, folder 14); J. L. Morrow, *Crawford CEAR,* 1939 (CAERC box 73, folder 40).

CHAPTER 3

Building Support Through Advisory Boards

1. W. A. Douglass, *Bartholomew CEAR,* 1930 (CAERC box 63, folder 17); W. O'Hara, *Hancock CEAR,* 1936 (CAERC box 91, folder 19); E. E. Stinson, *DeKalb CEAR,* 1930 (CAERC box 77, folder 25); J. L. Morrow, *Crawford CEAR,* 1939 (CAERC box 73, folder 40); E. C. Parker, *Hamilton CEAR,* 1930 (CAERC box 89, folder 45); H. R. Rosenbury, *Fulton CEAR,* 1930 (CAERC box 85, folder 38).
2. E. E. Stinson, *DeKalb CEAR,* 1930 (CAERC box 77, folder 25); J. Jensen, *Daviess CEAR,* 1930 (CAERC box 74, folder 24).
3. M. E. Cromer, *Delaware CEAR,* 1933 (CAERC box 78, folder 35); T. E. Myers, *DeKalb CEAR,* 1931 (CAERC box 77, folder 26); C. A. Alcorn, *Dearborn CEAR,* 1937 (CAERC box 75, folder 32); C. C. Madison, *Dearborn CEAR,* 1930 (CAERC box 75, folder 25); R. Gwaltney, *Gibson CEAR,* 1935 (CAERC box 87, folder 5).
4. N. S. Hadley, *Franklin CEAR,* 1930 (CAERC box 84, folder 30).
5. R. Gwaltney, *Gibson CEAR,* 1936 (CAERC box 87, folder 6).
6. H. R. Rosenbury, *Fulton CEAR,* 1930 (CAERC box 85, folder 38).
7. C. A. Nicholson, *Dubois CEAR,* 1934 (CAERC box 79, folder 43).
8. R. D. McHargue, *Clay CEAR,* 1937 (CAERC box 71, folder 49).
9. L. E. Archbold, *Adams CEAR,* 1930 (CAERC box 61, folder 16).
10. V. D. Sexson, *Greene CEAR,* 1930 (CAERC box 89, folder 5).
11. E. M. Christen, *Benton CEAR,* 1938 (CAERC box 65, folder 25); A. V. Keesling, *Bartholomew CEAR,* 1937 (CAERC box 63, folder 26); E. M. Rowe, *Clinton CEAR,* 1938 (CAERC box 72, folder 44); W. O'Hara, *Hancock CEAR,* 1937 (CAERC box 91, folder 20); E. L. Hawk, *Hendricks CEAR,* 1933 (CAERC box 93, folder 20).
12. A. M. Bishea, *Daviess CEAR,* 1939 (CAERC box 74, folder 32); E. M. Rowe, *Clinton CEAR,* 1938 (CAERC box 72, folder 44).
13. E. S. Miles, *Clark CEARs,* 1938–39 (CAERC box 70, folders 48 and 50); E. A. Johnson, *Cass CEAR,* 1937 (CAERC box 69, folder 41); C. K. Whistler, *Boone CEAR,* 1937 (CAERC box 67, folder 2); W. U. Rusk, *Blackford CEAR,* 1932 (CAERC box 66, folder 9); J. B. Todd, *Hamilton CEAR,* 1939 (CAERC box 90, folder 5); W. O'Hara, *Hancock CEAR,* 1937 (CAERC box 91, folder 20).
14. E. A. Johnson, *Cass CEAR,* 1939 (CAERC box 69, folder 44); E. S. Miles, *Clark CEAR,* 1938 (CAERC box 70, folder 48).
15. E. M. Christen, *Benton CEAR,* 1938 (CAERC box 65, folder 25).
16. C. A. Alcorn, *Dearborn CEAR,* 1937 (CAERC box 75, folder 32).
17. A. M. Bishea, *Daviess CEAR,* 1937 (CAERC box 74, folder 31).
18. A. V. Keesling, *Bartholomew CEAR,* 1937 (CAERC box 63, folder 26); T. E. Myers, *DeKalb CEAR,* 1938 (CAERC box 77, folder 33).

19. E. M. Rowe, *Clinton CEAR,* 1938 (CAERC box 72, folder 44).
20. V. J. Mann, *Clinton CEAR,* 1930 (CAERC box 72, folder 34).
21. N. S. Hadley, *Franklin CEAR,* 1931 (CAERC box 84, folder 31); R. Gwaltney, *Gibson CEAR,* 1937 (CAERC box 87, folder 7); N. S. Hadley, *Fulton CEAR,* 1934 (CAERC box 85, folder 42).
22. E. M. Christen, *Benton CEAR,* 1939 (CAERC box 65, folder 27); M. E. Cromer, *Delaware CEAR,* 1937 (CAERC box 78, folder 39); J. L. Morrow, *Crawford CEAR,* 1939 (CAERC box 73, folder 40).
23. J. L. Morrow, *Crawford CEAR,* 1938 (CAERC box 73, folder 39); C. A. Alcorn, *Floyd CEAR,* 1931 (CAERC box 82, folder 32).
24. H. J. Yarling, *Hendricks CEAR,* 1938 (CAERC box 93, folder 21).
25. A. M. Bishea, *Daviess CEAR,* 1937 (CAERC box 74, folder 31).
26. E. M. Christen, *Benton CEAR,* 1939 (CAERC box 65, folder 27).
27. E. M. Christen, *Benton CEAR,* 1938 (CAERC box 65, folder 25).

CHAPTER 4

Does the County Agent Do Anybody or Group of Farmers Any Good or Justify His Expense?

1. J. L. Morrow, *Crawford CEAR,* 1939 (CAERC box 73, folder 40).
2. C. C. Madison, *Dearborn CEAR,* 1932 (CAERC box 75, folder 27).
3. J. Jensen, *Daviess CEAR,* 1933 (CAERC box 74, folder 27); J. E. Little, *Cass CEARs,* 1931–32 (CAERC box 69, folders 34–35).
4. G. L. Reed, *Benton CEAR,* 1931 (CAERC box 65, folder 16).
5. R. E. Johnson, *Harrison CEAR,* 1931 (CAERC box 92, folder 26).
6. R. E. Johnson, *Harrison CEARs,* 1931–32 (CAERC box 92, folders 26–27).
7. R. E. Johnson, *Harrison CEAR,* 1932 (CAERC box 92, folder 27); M. F. Goode, *Harrison CEAR,* 1933 (CAERC box 92, folder 28).
8. C. L. Dyer, *Gibson CEAR,* 1933 (CAERC box 87, folder 4).
9. R. E. Johnson, *Harrison CEAR,* 1932 (CAERC box 92, folder 27).
10. W. A. Douglass, *Bartholomew CEAR,* 1930 (CAERC box 63, folder 17).
11. C. A. Nicholson, *Dubois CEAR,* 1933 (CAERC box 79, folder 42).
12. R. E. Johnson, *Harrison CEAR,* 1932 (CAERC box 92, folder 27).
13. V. D. Sexson, *Greene CEAR,* 1930 (CAERC box 89, folder 5).
14. W. G. Smith, *Henry CEAR,* 1933 (CAERC box 94, folder 2).
15. C. M. East, *Floyd CEAR,* 1939 (CAERC box 82, folder 40).
16. Ibid.
17. V. D. Sexson, *Greene CEAR,* 1932 (CAERC box 89, folder 7).
18. L. E. Archbold, *Adams CEAR,* 1932 (CAERC box 61, folder 18).
19. W. A. Douglass, *Bartholomew CEAR,* 1931 (CAERC box 63, folder 18).
20. J. E. Little, *Cass CEAR,* 1931 (CAERC box 69, folder 34).
21. A. V. Keesling, *Bartholomew CEAR,* 1932 (CAERC box 63, folder 19); J. Jensen, *Daviess CEAR,* 1932 (CAERC box 74, folder 26).
22. R. E. Johnson, *Harrison CEAR,* 1932 (CAERC box 92, folder 27).
23. A. V. Keesling, *Bartholomew CEAR,* 1932 (CAERC box 63, folder 19); C. A. Nicholson, *Dubois CEAR,* 1931 (CAERC box 79, folder 40); L. E. Moore, *Clark CEAR,* 1932 (CAERC box 70, folder 39).
24. L. E. Archbold, *Adams CEAR,* 1932 (CAERC box 61, folder 18).
25. Ibid.

26. L. E. Archbold, *Adams CEAR,* 1933 (CAERC box 61, folder 19).
27. E. M. Christen, *Benton CEAR,* 1933 (CAERC box 65, folder 18).
28. J. L. Morrow, *Crawford CEAR,* 1936 (CAERC box 73, folder 37).
29. J. Jensen, *Daviess CEAR,* 1931 (CAERC box 74, folder 25).
30. E. L. Hawk, *Hendricks CEAR,* 1933 (CAERC box 93, folder 20).
31. Ibid.
32. Ibid.
33. H. J. Yarling, *Hendricks CEAR,* 1939 (CAERC box 93, folder 22).
34. H. J. Yarling, *Hendricks CEAR,* 1937 (CAERC box 93, folder 21).
35. Ibid.
36. J. Jensen, *Daviess CEAR,* 1931 (CAERC box 74, folder 25).
37. A. J. Hesler, *Fountain CEAR,* 1939 (CAERC box 83, folder 35).
38. W. T. Merrilees, *Crawford CEAR,* 1933 (CAERC box 73, folder 35); R. Gwaltney, *Gibson CEAR,* 1939 (CAERC box 87, folder 9).
39. W. T. Merrilees, *Crawford CEAR,* 1933 (CAERC box 73, folder 35).
40. Ibid.
41. C. L. Dyer, *Gibson CEAR,* 1933 (CAERC box 87, folder 4).
42. Ibid.
43. A. V. Keesling, *Bartholomew CEAR,* 1935 (CAERC box 63, folder 22); C. V. Kimmell, *Allen CEAR,* 1936 (CAERC box 62, folder 22).
44. E. M. Rowe, *Clinton CEAR,* 1935 (CAERC box 72, folder 39).
45. E. M. Rowe, *Clinton CEAR,* 1939 (CAERC box 72, folder 45).
46. E. M. Christen, *Benton CEAR,* 1937 (CAERC box 65, folder 23).

CHAPTER 5

Live Out of the Garden, the Smoke House, and Cellar

1. M. F. Detrick, *Fayette CEAR,* 1933 (CAERC box 81, folder 34); C. A. Nicholson, *Dubois CEAR,* 1933 (CAERC box 79, folder 42); V. J. Mann, *Clinton CEAR,* 1932 (CAERC box 72, folder 42); J. Jensen, *Daviess CEAR,* 1931 (CAERC box 74, folder 25).
2. C. A. Nicholson, *Dubois CEAR,* 1932 (CAERC box 79, folder 41); J. Jensen, *Daviess CEAR,* 1931 (CAERC box 74, folder 25).
3. C. C. Madison, *Dearborn CEAR,* 1931 (CAERC box 75, folder 26); J. Jensen, *Daviess CEAR,* 1933 (CAERC box 74, folder 27).
4. W. T. Merrilees, *Crawford CEAR,* 1933 (CAERC box 73, folder 35).
5. C. A. Nicholson, *Dubois CEAR,* 1932 (CAERC box 79, folder 41).
6. J. Jensen, *Daviess CEAR,* 1931 (CAERC box 74, folder 25); C. M. East, *Floyd CEAR,* 1933 (CAERC box 82, folder 34).
7. V. J. Mann, *Clinton CEAR,* 1937 (CAERC box 72, folder 36); A. V. Keesling, *Bartholomew CEAR,* 1933 (CAERC box 63, folder 20); W. T. Merrilees, *Crawford CEAR,* 1933 (CAERC box 73, folder 35).
8. A. V. Keesling, *Bartholomew CEAR,* 1933 (CAERC box 63, folder 20); E. M. Rowe, *Clinton CEAR,* 1933 (CAERC box 72, folder 37).
9. R. J. Maggart, *Carroll CEAR,* 1933 (CAERC box 68, folder 58).
10. Ibid.
11. E. S. Miles, *Clark CEAR,* 1939 (CAERC box 70, folder 50); M. D. Butler, *Grant CEAR,* 1933 (CAERC box 87, folder 50).
12. V. J. Mann, *Clinton CEAR,* 1931 (CAERC box 72, folder 35).
13. E. S. Miles, *Clark CEAR,* 1933 (CAERC box 70, folder 40).

14. H. S. Heckard, *Allen CEAR,* 1932 (CAERC box 62, folder 17).
15. H. S. Heckard, *Allen CEAR,* 1934 (CAERC box 62, folder 19).
16. L. E. Moore, *Clark CEAR,* 1931 (CAERC box 70, folder 38).
17. H. S. Heckard, *Allen CEAR,* 1934 (CAERC box 62, folder 19); R. E. Johnson, *Harrison CEAR,* 1931 (CAERC box 92, folder 26).
18. M. E. Cromer, *Delaware CEAR,* 1932 (CAERC box 78, folder 34).
19. *Purdue University Department of Agricultural Extension Annual Report,* 1939, 8.
20. J. L. Morrow, *Crawford CEAR,* 1937 (CAERC box 73, folder 38); E. S. Miles, *Clark CEAR,* 1938 (CAERC box 70, folder 48).
21. N. S. Hadley, *Franklin CEAR,* 1933 (CAERC box 84, folder 33).
22. *Purdue University Department of Agricultural Extension Annual Report,* 1936, 78.
23. C. M. East, *Floyd CEAR,* 1933 (CAERC box 82, folder 34); C. A. Alcorn, *Dearborn CEAR,* 1937 (CAERC box 75, folder 32).
24. *Purdue University Department of Agricultural Extension Annual Report,* 1936, 78.
25. A. J. Hesler, *Fountain CEAR,* 1935 (CAERC box 83, folder 31).
26. E. M. Christen, *Benton CEAR,* 1937 (CAERC box 65, folder 23).
27. E. C. Parker, *Hamilton CEAR,* 1932 (CAERC box 89, folder 47); M. D. Butler, *Grant CEAR,* 1933 (CAERC box 87, folder 50).
28. H. E. Young, "Community and Industrial Gardens in Indiana, Even in Face of Drought, Close the Growing Season with Returns Estimated at Three Million Dollars," *Indianapolis News,* 25 Nov. 1933, 9.
29. Ibid.
30. V. J. Mann, *Clinton CEAR,* 1932 (CAERC box 72, folder 36); M. D. Butler, *Grant CEAR,* 1933 (CAERC box 87, folder 50).
31. H. S. Heckard, *Allen CEAR,* 1932 (CAERC box 62, folder 17); M. F. Detrick, *Fayette CEAR,* 1933 (CAERC box 81, folder 34); W. U. Rusk, *Blackford CEAR,* 1933 (CAERC box 66, folder 10).
32. M. F. Detrick, *Fayette CEAR,* 1933 (CAERC box 81, folder 34); E. C. Parker, *Hamilton CEAR,* 1932 (CAERC box 89, folder 47).
33. Young, "Community and Industrial Gardens in Indiana," 9.
34. M. E. Cromer, *Delaware CEAR,* 1932 (CAERC box 78, folder 34).
35. M. E. Cromer, *Delaware CEAR,* 1933 (CAERC box 78, folder 35).
36. M. D. Butler, *Grant CEAR,* 1933 (CAERC box 87, folder 50).
37. J. E. Little, *Cass CEAR,* 1932 (CAERC box 69, folder 35).
38. C. A. Nicholson, *Dubois CEAR,* 1933 (CAERC box 79, folder 42).
39. A. V. Keesling, *Bartholomew CEAR,* 1932 (CAERC box 63, folder 19); M. F. Detrick, *Fayette CEAR,* 1933 (CAERC box 81, folder 34); L. E. Moore, *Clark CEAR,* 1931 (CAERC box 70, folder 38); C. A. Nicholson, *Dubois CEAR,* 1933 (CAERC box 79, folder 42); W. U. Rusk, *Blackford CEAR,* 1933 (CAERC box 66, folder 10); E. C. Parker, *Hamilton CEAR,* 1934 (CAERC box 89, folder 49).
40. C. M. East, *Floyd CEAR,* 1933 (CAERC box 82, folder 34).
41. C. A. Nicholson, *Dubois CEAR,* 1932 (CAERC box 79, folder 41).
42. V. J. Mann, *Clinton CEAR,* 1932 (CAERC box 72, folder 36).
43. E. M. Rowe, *Clinton CEAR,* 1933 (CAERC box 72, folder 37).
44. C. A. Nicholson, *Dubois CEAR,* 1933 (CAERC box 79, folder 42); E. C. Parker, *Hamilton CEAR,* 1933 (CAERC box 89, folder 48); W. U. Rusk, *Blackford CEAR,* 1933 (CAERC box 66, folder 10).
45. E. M. Rowe, *Clinton CEAR,* 1933 (CAERC box 72, folder 37).

46. V. J. Mann, *Clinton CEAR,* 1931 (CAERC box 72, folder 35); H. S. Heckard, *Allen CEAR,* 1933 (CAERC box 62, folder 18); R. E. Johnson, *Harrison CEAR,* 1931 (CAERC box 92, folder 26).
47. J. L. Morrow, *Crawford CEAR,* 1937 (CAERC box 73, folder 38); L. E. Archbold, *Adams CEAR,* 1930 (CAERC box 61, folder 16); E. S. Miles, *Clark CEARs,* 1938–39 (CAERC box 70, folders 48 and 50).

CHAPTER 6

Cash Is the One Article That Is Scarcest and Hardest to Get

1. C. A. Nicholson, *Dubois CEAR,* 1932 (CAERC box 79, folder 41); W. A. Douglass, *Bartholomew CEAR,* 1931 (CAERC box 63, folder 18).
2. Sam Moore, "U.S. Farmers during the Great Depression," Farm Life (website), Farm Collector, 19 Sept. 2011, https://www.farmcollector.com/farm-life/u-s-farmers-during-great-depression/.
3. W. A. Douglass, *Bartholomew CEAR,* 1931 (CAERC box 63, folder 18).
4. W. U. Rusk, *Blackford CEAR,* 1933 (CAERC box 66, folder 10); W. A. Douglass, *Bartholomew CEAR,* 1931 (CAERC box 63, folder 18).
5. W. A. Douglass, *Bartholomew CEAR,* 1931 (CAERC box 63, folder 18).
6. C. A. Nicholson, *Dubois CEAR,* 1932 (CAERC box 79, folder 41).
7. N. S. Hadley, *Franklin CEAR,* 1931 (CAERC box 84, folder 31).
8. "Banking 1929–1941," Historic Events for Students: The Great Depression (website), Encyclopedia.com, accessed 15 Apr. 2021, https://www.encyclopedia.com/education/news-and-education-magazines/banking-1929-1941.
9. W. U. Rusk, *Blackford CEAR,* 1934 (CAERC box 66, folder 11); W. A. Douglass, *Bartholomew CEAR,* 1931 (CAERC box 63, folder 18).
10. W. U. Rusk, *Blackford CEAR,* 1934 (CAERC box 66, folder 11).
11. W. U. Rusk, *Blackford CEAR,* 1933 (CAERC box 66, folder 10).
12. W. A. Douglass, *Bartholomew CEAR,* 1930 (CAERC box 63, folder 17).
13. M. E. Cromer, *Delaware CEAR,* 1933 (CAERC box 78, folder 35); E. S. Miles, *Clark CEAR,* 1933 (CAERC box 70, folder 40); W. U. Rusk, *Blackford CEAR,* 1933 (CAERC box 66, folder 10).
14. E. S. Miles, *Clark CEAR,* 1933 (CAERC box 70, folder 40); W. A. Douglass, *Bartholomew CEAR,* 1931 (CAERC box 63, folder 18).
15. E. S. Miles, *Clark CEAR,* 1933 (CAERC box 70, folder 40); C. M. East, *Floyd CEARs,* 1934 and 1936 (CAERC box 82, folders 35 and 37); A. J. Hesler, *Fountain CEAR,* 1938 (CAERC box 83, folder 34).
16. M. F. Goode, *Harrison CEAR,* 1933 (CAERC box 92, folder 28); C. A. Nicholson, *Dubois CEAR,* 1934 (CAERC box 79, folder 43).
17. W. T. Merrilees, *Crawford CEAR,* 1931 (CAERC box 73, folder 33); I. L. Thurston, *Decatur CEAR,* 1931 (CAERC box 76, folder 37); J. Jensen, *Daviess CEAR,* 1931 (CAERC box 74, folder 25).
18. A. V. Keesling, *Bartholomew CEAR,* 1934 (CAERC box 63, folder 21); H. S. Heckard, *Allen CEAR,* 1934 (CAERC box 62, folder 19).
19. A. J. Hesler, *Fountain CEAR,* 1933 (CAERC box 83, folder 29).
20. W. U. Rusk, *Blackford CEAR,* 1934 (CAERC box 66, folder 11).
21. C. M. East, *Floyd CEAR,* 1934 (CAERC box 82, folder 35); E. M. Rowe, *Clinton CEAR,* 1933 (CAERC box 72, folder 37); C. A. Nicholson, *Dubois CEAR,* 1934 (CAERC box 79, folder 43).

22. W. U. Rusk, *Blackford CEAR,* 1935 (CAERC box 66, folder 12); A. V. Keesling, *Bartholomew CEAR,* 1933 (CAERC box 63, folder 20); R. E. Johnson, *Harrison CEAR,* 1931 (CAERC box 92, folder 26).
23. R. E. Johnson, *Harrison CEAR,* 1931 (CAERC box 92, folder 26).
24. J. Jensen, *Daviess CEAR,* 1933 (CAERC box 74, folder 27); A. M. Bishea, *Daviess CEAR,* 1934 (CAERC box 74, folder 28); M. E. Cromer, *Delaware CEAR,* 1934 (CAERC box 78, folder 36); W. T. Merrilees, *Crawford CEAR,* 1933 (CAERC box 73, folder 35); R. E. Johnson, *Harrison CEAR,* 1931 (CAERC box 92, folder 26); J. E. Little, *Cass CEAR,* 1932 (CAERC box 69, folder 35).
25. A. T. Marvel, *Elkhart CEAR,* 1934 (CAERC box 80, folder 32); N. S. Hadley, *Fulton CEAR,* 1934 (CAERC box 85, folder 42); M. D. Butler, *Grant CEAR,* 1933 (CAERC box 87, folder 50).
26. C. M. East, *Floyd CEAR,* 1935 (CAERC box 82, folder 36).
27. A. J. Hesler, *Fountain CEAR,* 1933 (CAERC box 83, folder 29).
28. A. V. Keesling, *Bartholomew CEAR,* 1938 (CAERC box 63, folder 28).
29. C. A. Nicholson, *Dubois CEAR,* 1934 (CAERC box 79, folder 43).
30. J. L. Morrow, *Crawford CEAR,* 1937 (CAERC box 73, folder 38).
31. M. E. Cromer, *Delaware CEAR,* 1933 (CAERC box 78, folder 35).
32. I. L. Thurston, *Decatur CEAR,* 1933 (CAERC box 76, folder 39); A. M. Bishea, *Daviess CEAR,* 1939 (CAERC box 74, folder 32); W. T. Merrilees, *Crawford CEAR,* 1933 (CAERC box 73, folder 35); M. F. Goode, *Harrison CEAR,* 1933 (CAERC box 92, folder 28).
33. C. M. East, *Floyd CEAR,* 1937 (CAERC box 82, folder 38).
34. J. E. Little, *Cass CEAR,* 1932 (CAERC box 69, folder 35).
35. M. F. Detrick, *Fayette CEAR,* 1934 (CAERC box 81, folder 35).
36. T. E. Myers, *DeKalb CEAR,* 1933 (CAERC box 77, folder 28).
37. E. M. Christen, *Benton CEAR,* 1935 (CAERC box 65, folder 20); W. O'Hara, *Hancock CEARs,* 1936–37 (CAERC box 91, folders 19–20); A. V. Keesling, *Bartholomew CEAR,* 1936 (CAERC box 63, folder 24).
38. W. U. Rusk, *Blackford CEAR,* 1936 (CAERC box 66, folder 13); E. M. Christen, *Benton CEAR,* 1935 (CAERC box 65, folder 20).
39. W. U. Rusk, *Blackford CEAR,* 1936 (CAERC box 66, folder 13); E. M. Christen, *Benton CEAR,* 1935 (CAERC box 65, folder 20).
40. E. M. Christen, *Benton CEAR,* 1935 (CAERC box 65, folder 20).
41. "Corn Loans to be Available First of Week," *Evening Republican* (Columbus, IN), 12 Dec. 1935, 1; "Warehouse Board Elected in County," *Hancock Democrat* (Greenfield, IN), 12 Dec. 1935, 8; "Corn Loans Ready for Farmers Here," *Rushville Republican* (Rushville, IN), 16 Dec. 1935, 1; "Corn Sealers to Begin Work," *Star Press* (Muncie, IN), 21 Dec. 1935, 12; "Cass Corn Loan Unit Is Formed," *Logansport Pharos-Tribune* (Logansport, IN), 27 Nov. 1935, 13; "Corn Loan of 45 Cents on Bushel Announced by AAA," *Garrett Clipper* (Garrett, IN), 31 Oct. 1935, 1; E. M. Christen, *Benton CEAR,* 1934 (CAERC box 65, folder 19).
42. C. K. Whistler, *Boone CEAR,* 1938 (CAERC box 67, folder 3); E. A. Johnson, *Cass CEAR,* 1939 (CAERC box 69, folder 44); R. J. Maggart, *Carroll CEAR,* 1937 (CAERC box 68, folder 63); W. O'Hara, *Hancock CEAR,* 1938 (CAERC box 91, folder 21); R. Gwaltney, *Gibson CEAR,* 1939 (CAERC box 87, folder 9).
43. *Purdue University Department of Agricultural Extension Annual Report,* 1938, 17.
44. W. O'Hara, *Hancock CEAR,* 1934 (CAERC box 91, folder 17).

45. R. L. Zell, *Franklin CEAR,* 1938 (CAERC box 84, folder 38); M. F. Goode, *Harrison CEAR,* 1938 (CAERC box 92, folder 33); W. A. Douglass, *Bartholomew CEAR,* 1931 (CAERC box 63, folder 18).
46. A. J. Hesler, *Fountain CEAR,* 1934 (CAERC box 83, folder 30); T. E. Myers, *DeKalb CEAR,* 1934 (CAERC box 77, folder 29).
47. R. L. Zell, *Franklin CEAR,* 1938 (CAERC box 84, folder 38).
48. T. E. Myers, *DeKalb CEAR,* 1937 (CAERC box 77, folder 32).
49. Bryant Putney, "Farm Tenancy in the United States," *Editorial Research Reports* 1 (1935), "Extent and Causes of Increase in Tenancy Since 1880" section (Washington, DC: CQ Press, 1935), accessed through CQ Researcher, http://library.cqpress.com/cqresearcher/cqresrre1935032000.
50. Paul Wallace Gates, "Land Policy and Tenancy in the Prairie Counties of Indiana," *Indiana Magazine of History* 35, no. 1 (Mar. 1939): 1, https://scholarworks.iu.edu/journals/index.php/imh/article/view/7107; E. M. Christen, *Benton CEAR,* 1937 (CAERC box 65, folder 23).
51. Gates, "Land Policy and Tenancy," 1; E. S. Miles, *Clark CEAR,* 1939 (CAERC box 70, folder 50).
52. C. A. Nicholson, *Dubois CEAR,* 1937 (CAERC box 79A, folder 1).
53. W. U. Rusk, *Blackford CEAR,* 1933 (CAERC box 66, folder 10).
54. A. J. Hesler, *Fountain CEAR,* 1939 (CAERC box 83, folder 35).
55. A. M. Bishea, *Daviess CEAR,* 1939 (CAERC box 74, folder 32); E. M. Christen, *Benton CEAR,* 1939 (CAERC box 65, folder 27).
56. "Bankhead-Jones Farm Tenant Act of 1937," Encyclopedia of the Great Depression (website), Encyclopedia.com, 16 Apr. 2021, https://www.encyclopedia.com/economics/encyclopedias-almanacs-transcripts-and-maps/bankhead-jones-farm-tenant-act-1937.
57. A. V. Keesling, *Bartholomew CEAR,* 1938 (CAERC box 63, folder 28); J. H. Brown, *Elkhart CEAR,* 1939 (CAERC box 80, folder 37); E. M. Christen, *Benton CEAR,* 1938 (CAERC box 65, folder 25); A. M. Bishea, *Daviess CEAR,* 1939 (CAERC box 74, folder 32); R. Gwaltney, *Gibson CEAR,* 1938 (CAERC box 87, folder 8); A. J. Hesler, *Fountain CEAR,* 1938 (CAERC box 83, folder 34); W. O'Hara, *Hancock CEAR,* 1938 (CAERC box 91, folder 21).
58. E. M. Christen, *Benton CEAR,* 1939 (CAERC box 65, folder 27); A. V. Keesling, *Bartholomew CEAR,* 1938 (CAERC box 63, folder 28).
59. A. J. Hesler, *Fountain CEAR,* 1938 (CAERC box 83, folder 34).
60. D. S. Bishopp, *Fayette CEAR,* 1938 (CAERC box 81, folder 39).
61. E. M. Christen, *Benton CEAR,* 1938 (CAERC box 65, folder 25).
62. W. O'Hara, *Hancock CEAR,* 1938 (CAERC box 91, folder 24).
63. Ibid.; E. M. Christen, *Benton CEAR,* 1939 (CAERC box 65, folder 27).
64. D. S. Bishopp, *Fayette CEAR,* 1938 (CAERC box 81, folder 39).
65. Encyclopedia.com, "Bankhead-Jones Farm Tenant Act of 1937."
66. A. V. Keesling, *Bartholomew CEAR,* 1938 (CAERC box 63, folder 28); C. M. East, *Floyd CEAR,* 1936 (CAERC box 82, folder 37); E. S. Miles, *Clark CEAR,* 1933 (CAERC box 70, folder 40); C. A. Nicholson, *Dubois CEAR,* 1936 (CAERC box 79, folder 45).
67. A. J. Hesler, *Fountain CEAR,* 1939 (CAERC box 83, folder 35); W. G. Smith, *Henry CEAR,* 1933 (CAERC box 94, folder 2).
68. W. O'Hara, *Hancock CEAR,* 1939 (CAERC box 91, folder 26).

CHAPTER 7

The Man With the High-Producing Soil, Hen, Cow, and Sow That Kept Operating Expense Down Was Able to Return a Profit

1. E. M. Christen, *Benton CEAR,* 1939 (CAERC box 65, folder 27).
2. E. C. Parker, *Hamilton CEARs,* 1930–31 (CAERC box 89, folders 45–46); R. Gwaltney, *Gibson CEAR,* 1938 (CAERC box 87, folder 8).
3. Geoffrey Shepherd et al., *Economic Analysis of Trends in Beef Cattle and Hog Prices,* Iowa Agricultural Experiment Station Research Bulletin No. 405 (Ames, IA: Iowa State College of Agriculture and Mechanic Arts, Jan. 1954), 728, https://dr.lib.iastate.edu/handle/20.500.12876/62744; V. D. Sexson, *Greene CEAR,* 1931 (CAERC box 89, folder 6); C. A. Nicholson, *Dubois CEAR,* 1930 (CAERC box 79, folder 39); J. L. Morrow, *Crawford CEAR,* 1939 (CAERC box 73, folder 40).
4. M. F. Goode, *Harrison CEAR,* 1933 (CAERC box 92, folder 28).
5. W. G. Smith, *Henry CEAR,* 1933 (CAERC box 94, folder 2).
6. W. G. Smith, *Henry CEAR,* 1931 (CAERC box 93, folder 60).
7. C. M. East, *Floyd CEAR,* 1932 (CAERC box 82, folder 33).
8. C. M. East, *Floyd CEAR,* 1933 (CAERC box 82, folder 34).
9. A. T. Marvel, *Elkhart CEAR,* 1934 (CAERC box 80, folder 32).
10. A. J. Hesler, *Fountain CEAR,* 1933 (CAERC box 83, folder 29).
11. H. S. Heckard, *Allen CEAR,* 1930 (CAERC box 62, folder 15).
12. T. E. Myers, *DeKalb CEAR,* 1931 (CAERC box 77, folder 26).
13. W. G. Smith, *Henry CEAR,* 1935 (CAERC box 94, folder 4).
14. R. E. Johnson, *Harrison CEAR,* 1932 (CAERC box 92, folder 27).
15. V. D. Sexson, *Greene CEAR,* 1931 (CAERC box 89, folder 6).
16. R. Gwaltney, *Gibson CEAR,* 1939 (CAERC box 87, folder 9).
17. H. S. Heckard, *Allen CEARs,* 1934–35 (CAERC box 62, folders 19–20); A. V. Keesling, *Bartholomew CEAR,* 1938 (CAERC box 63, folder 28); E. M. Christen, *Benton CEARs,* 1935 and 1937 (CAERC box 65, folders 20 and 23).
18. M. E. Cromer, *Delaware CEAR,* 1934 (CAERC box 78, folder 36); E. M. Christen, *Benton CEAR,* 1936 (CAERC box 65, folder 21).
19. M. D. Butler, *Grant CEAR,* 1931 (CAERC box 87, folder 48).
20. H. S. Heckard, *Allen CEAR,* 1935 (CAERC box 62, folder 20).
21. A. J. Hesler, *Fountain CEAR,* 1938 (CAERC box 83, folder 34).
22. R. Gwaltney, *Gibson CEAR,* 1936 (CAERC box 87, folder 6).
23. R. Gwaltney, *Gibson CEAR,* 1937 (CAERC box 87, folder 7).
24. L. E. Archbold, *Adams CEAR,* 1930 (CAERC box 61, folder 16).
25. E. M. Christen, *Benton CEAR,* 1938 (CAERC box 65, folder 25).
26. C. A. Alcorn, *Floyd CEAR,* 1930 (CAERC box 82, folder 31).
27. E. M. Christen, *Benton CEAR,* 1936 (CAERC box 65, folder 21); A. J. Hesler, *Fountain CEAR,* 1938 (CAERC box 83, folder 34).
28. W. U. Rusk, *Blackford CEAR,* 1932 (CAERC box 66, folder 9).
29. H. S. Heckard, *Allen CEAR,* 1935 (CAERC box 62, folder 20); E. A. Johnson, *Cass CEAR,* 1938 (CAERC box 69, folder 43).
30. C. V. Kimmell, *Allen CEAR,* 1937 (CAERC box 62, folder 23); J. E. Little, *Cass CEAR,* 1930 (CAERC box 69, folder 33); M. D. Butler, *Grant CEAR,* 1930 (CAERC box 87, folder 47).
31. *Purdue University Department of Agricultural Extension Annual Report,* 1930, 45; *Purdue University Department of Agricultural Extension Annual Report,* 1931, 50.

32. H. S. Heckard, *Allen CEAR,* 1930 (CAERC box 62, folder 15); J. E. Little, *Cass CEAR,* 1930 (CAERC box 69, folder 33).
33. E. A. Johnson, *Cass CEAR,* 1938 (CAERC box 69, folder 43); W. U. Rusk, *Blackford CEAR,* 1933 (CAERC box 66, folder 10); G. L. Reed, *Benton CEAR,* 1932 (CAERC box 65, folder 17); J. E. Little, *Cass CEAR,* 1930 (CAERC box 69, folder 33); E. M. Christen, *Benton CEAR,* 1935 (CAERC box 65, folder 20).
34. H. S. Heckard, *Allen CEAR,* 1930 (CAERC box 62, folder 15); F. E. Conder, *Grant CEAR,* 1935 (CAERC box 87, folder 52).
35. E. M. Christen, *Benton CEAR,* 1936 (CAERC box 65, folder 21).
36. *Purdue University Department of Agricultural Extension Annual Report,* 1935, 62; C. M. East, *Floyd CEAR,* 1935 (CAERC box 82, folder 36).
37. J. E. Little, *Cass CEAR,* 1931 (CAERC box 69, folder 34); H. S. Heckard, *Allen CEAR,* 1930 (CAERC box 62, folder 15).
38. M. F. Detrick, *Fayette CEAR,* 1934 (CAERC box 81, folder 35).
39. W. G. Smith, *Henry CEAR,* 1932 (CAERC box 94, folder 1).
40. M. D. Butler, *Grant CEAR,* 1931 (CAERC box 87, folder 48).
41. C. A. Alcorn, *Floyd CEAR,* 1931 (CAERC box 82, folder 32).
42. W. U. Rusk, *Blackford CEAR,* 1933 (CAERC box 66, folder 10).
43. W. U. Rusk, *Blackford CEAR,* 1932 (CAERC box 66, folder 9).
44. F. E. Conder, *Grant CEAR,* 1938 (CAERC box 87, folder 55).
45. M. F. Detrick, *Fayette CEAR,* 1931 (CAERC box 81, folder 32).
46. L. E. Archbold, *Adams CEAR,* 1936 (CAERC box 61, folder 22); M. F. Detrick, *Fayette CEAR,* 1933 (CAERC box 81, folder 34); C. A. Alcorn, *Dearborn CEAR,* 1937 (CAERC box 75, folder 32).
47. C. A. Alcorn, *Floyd CEAR,* 1931 (CAERC box 82, folder 32); H. S. Heckard, *Allen CEAR,* 1935 (CAERC box 62, folder 20); V. J. Mann, *Clinton CEAR,* 1930 (CAERC box 72, folder 34); E. M. Christen, *Benton CEAR,* 1938 (CAERC box 65, folder 25); W. O'Hara, *Hancock CEAR,* 1937 (CAERC box 91, folder 20); H. R. Rosenbury, *Fulton CEAR,* 1932 (CAERC box 85, folder 40).
48. H. S. Heckard, *Allen CEAR,* 1933 (CAERC box 62, folder 18).
49. A. V. Keesling, *Bartholomew CEAR,* 1939 (CAERC box 63, folder 30); M. F. Detrick, *Fayette CEAR,* 1930 (CAERC box 81, folder 31); H. J. Yarling, *Hancock CEAR,* 1930 (CAERC box 91, folder 13).
50. C. K. Whistler, *Boone CEAR,* 1937 (CAERC box 67, folder 2); W. U. Rusk, *Blackford CEAR,* 1932 (CAERC box 66, folder 9); E. M. Christen, *Benton CEAR,* 1936 (CAERC box 65, folder 21).
51. H. S. Heckard, *Allen CEAR,* 1935 (CAERC box 62, folder 20); L. E. Archbold, *Adams CEAR,* 1937 (CAERC box 61, folder 23); E. A. Johnson, *Cass CEAR,* 1937 (CAERC box 69, folder 41); W. U. Rusk, *Blackford CEAR,* 1933 (CAERC box 66, folder 10); E. C. Parker, *Hamilton CEAR,* 1932 (CAERC box 89, folder 47); N. S. Hadley, *Fulton CEAR,* 1935 (CAERC box 85, folder 43); W. O'Hara, *Hancock CEAR,* 1932 (CAERC box 91, folder 15).
52. R. Gwaltney, *Gibson CEAR,* 1937 (CAERC box 87, folder 7).
53. J. E. Little, *Cass CEAR,* 1932 (CAERC box 69, folder 35).
54. J. Jensen, *Daviess CEAR,* 1932 (CAERC box 74, folder 26).
55. E. M. Christen, *Benton CEAR,* 1937 (CAERC box 65, folder 23); W. U. Rusk, *Blackford CEAR,* 1933 (CAERC box 66, folder 10); N. S. Hadley, *Franklin CEAR,* 1932 (CAERC box 84, folder 32).

56. A. J. Hesler, *Fountain CEAR,* 1939 (CAERC box 83, folder 35).
57. C. V. Kimmell, *Allen CEAR,* 1939 (CAERC box 62, folder 25).
58. L. E. Archbold, *Adams CEAR,* 1939 (CAERC box 1, folder 1).
59. Ibid.
60. A. J. Hesler, *Fountain CEAR,* 1939 (CAERC box 83, folder 35); R. O. Cole, *Clark CEAR,* 1930 (CAERC box 70, folder 37); V. J. Mann, *Clinton CEAR,* 1930 (CAERC box 72, folder 34).
61. M. F. Detrick, *Fayette CEAR,* 1932 (CAERC box 81, folder 33).
62. E. L. Hawk, *Hendricks CEAR,* 1931 (CAERC box 93, folder 17).
63. A. Haseman, *Greene CEAR,* 1939 (CAERC box 89, folder 14).
64. M. F. Detrick, *Fayette CEAR,* 1933 (CAERC box 81, folder 34).
65. E. L. Hawk, *Hendricks CEAR,* 1931 (CAERC box 93, folder 18).
66. N. S. Hadley, *Franklin CEAR,* 1931 (CAERC box 84, folder 31); E. M. Christen, *Benton CEAR,* 1936 (CAERC box 65, folder 21); H. S. Heckard, *Allen CEAR,* 1932 (CAERC box 62, folder 17); M. F. Detrick, *Fayette CEAR,* 1932 (CAERC box 81, folder 33); W. U. Rusk, *Blackford CEAR,* 1934 (CAERC box 66, folder 11).
67. H. S. Heckard, *Allen CEAR,* 1932 (CAERC box 62, folder 17); R. Gwaltney, *Gibson CEAR,* 1936 (CAERC box 87, folder 6).
68. M. F. Detrick, *Fayette CEAR,* 1931 (CAERC box 81, folder 32).
69. Ibid.
70. C. V. Kimmell, *Allen CEAR,* 1937 (CAERC box 62, folder 23).
71. M. D. Butler, *Grant CEAR,* 1930 (CAERC box 87, folder 47).
72. E. E. Stinson, *DeKalb CEAR,* 1930 (CAERC box 77, folder 25).
73. C. A. Alcorn, *Dearborn CEAR,* 1933 (CAERC box 82, folder 34).
74. *Purdue University Department of Agricultural Extension Annual Report,* 1935, 64; H. S. Heckard, *Allen CEAR,* 1935 (CAERC box 62, folder 20); W. U. Rusk, *Blackford CEAR,* 1934 (CAERC box 66, folder 11).
75. M. F. Detrick, *Fayette CEAR,* 1934 (CAERC box 81, folder 35).
76. W. U. Rusk, *Blackford CEAR,* 1933 (CAERC box 66, folder 10).
77. *Purdue University Department of Agricultural Extension Annual Report,* 1934, 57.
78. H. S. Heckard, *Allen CEAR,* 1931 (CAERC box 62, folder 16); M. E. Cromer, *Delaware CEAR,* 1933 (CAERC box 78, folder 35); N. S. Hadley, *Fulton CEAR,* 1934 (CAERC box 85, folder 42); C. M. East, *Floyd CEAR,* 1935 (CAERC box 82, folder 36).
79. E. A. Johnson, *Cass CEAR,* 1939 (CAERC box 69, folder 44); E. S. Miles, *Clark CEAR,* 1939 (CAERC box 70, folder 50); I. L. Thurston, *Decatur CEAR,* 1935 (CAERC box 76, folder 41); W. O'Hara, *Hancock CEAR,* 1933 (CAERC box 91, folder 16); E. L. Hawk, *Hendricks CEAR,* 1931 (CAERC box 93, folder 18).
80. E. M. Christen, *Benton CEAR,* 1936 (CAERC box 65, folder 21).
81. W. O'Hara, *Hancock CEAR,* 1933 (CAERC box 91, folder 16).
82. W. O'Hara, *Hancock CEAR,* 1934 (CAERC box 91, folder 17).
83. C. A. Alcorn, *Dearborn CEAR,* 1938 (CAERC box 75, folder 33).
84. T. E. Myers, *DeKalb CEAR,* 1934 (CAERC box 77, folder 29); W. U. Rusk, *Blackford CEAR,* 1933 (CAERC box 66, folder 10); C. M. East, *Floyd CEAR,* 1935 (CAERC box 82, folder 36); C. A. Alcorn, *Floyd CEAR,* 1930 (CAERC box 82, folder 31); W. O'Hara, *Hancock CEAR,* 1938 (CAERC box 91, folder 21).
85. W. U. Rusk, *Blackford CEAR,* 1931 (CAERC box 66, folder 8).
86. W. U. Rusk, *Blackford CEAR,* 1933 (CAERC box 66, folder 10).
87. C. M. East, *Floyd CEAR,* 1935 (CAERC box 82, folder 36).

CHAPTER 8

Farmers Hanging On by a Mere Thread Reached Out for Benefit Payments to Save Their Farms

1. Dewitt C. Wing, "Corn-Hog Situation Carefully Analyzed," *Richmond Item* (Richmond, IN), 12 Jan. 1934, 13.
2. W. U. Rusk, *Blackford CEAR,* 1939 (CAERC box 66, folder 16).
3. H. L. Royce, *Clay CEAR,* 1931 (CAERC box 71, folder 43); E. M. Christen, *Benton CEAR,* 1939 (CAERC box 65, folder 27).
4. G. L. Reed, *Benton CEAR,* 1932 (CAERC box 65, folder 17); R. J. Maggart, *Carroll CEAR,* 1931 (CAERC box 68, folder 56).
5. L. E. Moore, *Clark CEAR,* 1932 (CAERC box 70, folder 39); Samuel A. Jones et al., "Poultry Industry Fairly Resistant to the Depression," in "Poultry and Eggs," *United States Department of Agriculture Yearbook of Agriculture, 1933,* ed. Milton S. Eisenhower (Washington, DC: Government Printing Office, 1933), 303, https://archive.org/details/yoa1933/page/n1/mode/2up.
6. W. O'Hara, *Hancock CEARs,* 1933 and 1936–37 (CAERC box 91, folders 16 and 19–20).
7. A. J. Hesler, *Fountain CEAR,* 1934 (CAERC box 83, folder 30); W. U. Rusk, *Blackford CEAR,* 1939 (CAERC box 66, folder 16); John H. Skinner, "What 'New Deal' in Agriculture Means to Indiana," *Indianapolis Star,* 17 Sept. 1933, 55; A. J. Hesler, *Fountain CEAR,* 1933 (CAERC box 83, folder 29).
8. E. S. Miles, *Clark CEAR,* 1933 (CAERC box 70, folder 40); A. J. Hesler, *Fountain CEAR,* 1933 (CAERC box 83, folder 29).
9. H. S. Heckard, *Allen CEAR,* 1935 (CAERC box 62, folder 20); L. E. Archbold, *Adams CEAR,* 1935 (CAERC box 61, folder 21); C. C. Madison, *Dearborn CEAR,* 1930 (CAERC box 75, folder 25); J. L. Morrow, *Crawford CEAR,* 1937 (CAERC box 73, folder 38); R. L. Zell, *Franklin CEAR,* 1934 (CAERC box 84, folder 34); A. Haseman, *Greene CEAR,* 1934 (CAERC box 89, folder 9).
10. J. Jensen, *Daviess CEAR,* 1933 (CAERC box 74, folder 27); A. V. Keesling, *Bartholomew CEAR,* 1934 (CAERC box 63, folder 21).
11. "Wheat Plan Nets $52,000 in County," *South Bend Tribune* (South Bend, IN), 28 July 1933, 13; Skinner, "What 'New Deal' in Agriculture Means," 55; T. E. Myers, *DeKalb CEAR,* 1934 (CAERC box 77, folder 29).
12. Skinner, "What 'New Deal' in Agriculture Means," 56.
13. "County Agent Reviews Week's Rural Events," *Richmond Item* (Richmond, IN), 27 Oct. 1933, 5; Alfred Hall, "Friend of the Farmer," *Richmond Item* (Richmond, IN), 27 Oct. 1933, 5.
14. C. A. Nicholson, *Dubois CEAR,* 1934 (CAERC box 79, folder 43).
15. Ibid.
16. W. O'Hara, *Hancock CEAR,* 1933 (CAERC box 91, folder 16).
17. Skinner, "What 'New Deal' in Agriculture Means," 56.
18. E. M. Christen, *Benton CEAR,* 1934 (CAERC box 65, folder 19); W. O'Hara, *Hancock CEAR,* 1935 (CAERC box 91, folder 18); Skinner, "What 'New Deal' in Agriculture Means," 56.
19. A. J. Hesler, *Fountain CEAR,* 1934 (CAERC box 83, folder 30).
20. R. Gwaltney, *Gibson CEAR,* 1939 (CAERC box 87, folder 9); E. S. Miles, *Clark CEAR,* 1934 (CAERC box 70, folder 41).
21. A. J. Hesler, *Fountain CEAR,* 1934 (CAERC box 83, folder 30); L. E. Archbold, *Adams CEAR,* 1935 (CAERC box 61, folder 21).

22. W. U. Rusk, *Blackford CEAR,* 1939 (CAERC box 66, folder 16); H. S. Heckard, *Allen CEAR,* 1934 (CAERC box 62, folder 19); H. L. Royce, *Clay CEAR,* 1935 (CAERC box 71, folder 47); I. L. Thurston, *Decatur CEAR,* 1932 (CAERC box 76, folder 38).
23. C. A. Nicholson, *Dubois CEAR,* 1934 (CAERC box 79, folder 43); W. G. Smith, *Henry CEAR,* 1933 (CAERC box 94, folder 2); W. U. Rusk, *Blackford CEAR,* 1934 (CAERC box 66, folder 11); "Fifty Thousand Sign Contracts," *Lafayette Journal and Courier* (Lafayette, IN), 26 Feb. 1934, 3.
24. H. S. Heckard, *Allen CEAR,* 1934 (CAERC box 62, folder 19); W. U. Rusk, *Blackford CEAR,* 1934 (CAERC box 66, folder 11).
25. R. J. Maggart, *Carroll CEAR,* 1939 (CAERC box 68, folder 65); L. E. Moore, *Clark CEAR,* 1931 (CAERC box 70, folder 38); M. E. Cromer, *Delaware CEAR,* 1933 (CAERC box 78, folder 35).
26. W. G. Smith, *Henry CEAR,* 1934 (CAERC box 94, folder 3).
27. M. E. Cromer, *Delaware CEAR,* 1934 (CAERC box 78, folder 36).
28. E. M. Rowe, *Clinton CEAR,* 1933 (CAERC box 72, folder 37); M. F. Goode, *Harrison CEARs,* 1933 and 1935 (CAERC box 92, folders 28 and 30).
29. T. E. Myers, *DeKalb CEAR,* 1934 (CAERC box 77, folder 29).
30. W. G. Smith, *Henry CEAR,* 1933 (CAERC box 94, folder 2); A. J. Hesler, *Fountain CEAR,* 1933 (CAERC box 83, folder 29).
31. W. U. Rusk, *Blackford CEAR,* 1933 (CAERC box 66, folder 10); A. V. Keesling, *Bartholomew CEAR,* 1933 (CAERC box 63, folder 20); W. O'Hara, *Hancock CEAR,* 1933 (CAERC box 91, folder 16); J. Jensen, *Daviess CEAR,* 1933 (CAERC box 74, folder 27); E. S. Miles, *Clark CEAR,* 1933 (CAERC box 70, folder 40).
32. W. U. Rusk, *Blackford CEAR,* 1934 (CAERC box 66, folder 11); C. A. Nicholson, *Dubois CEAR,* 1934 (CAERC box 79, folder 43); M. F. Detrick, *Fayette CEAR,* 1934 (CAERC box 81, folder 35); E. M. Rowe, *Clinton CEAR,* 1934 (CAERC box 72, folder 38).
33. C. A. Nicholson, *Dubois CEAR,* 1933 (CAERC box 79, folder 42); E. M. Rowe, *Clinton CEAR,* 1934 (CAERC box 72, folder 38); "County Agent Explains Hog-Corn Reduction Program," *Argos Reflector* (Argos, IN), 18 Jan. 1934, 1; W. U. Rusk, *Blackford CEAR,* 1934 (CAERC box 66, folder 11).
34. W. U. Rusk, *Blackford CEAR,* 1933 (CAERC box 66, folder 10); C. A. Nicholson, *Dubois CEAR,* 1933 (CAERC box 79, folder 42).
35. A. V. Keesling, *Bartholomew CEAR,* 1933 (CAERC box 63, folder 20); A. J. Hesler, *Fountain CEAR,* 1933 (CAERC box 83, folder 29).
36. E. M. Rowe, *Clinton CEAR,* 1933 (CAERC box 72, folder 37).
37. A. J. Hesler, *Fountain CEAR,* 1933 (CAERC box 83, folder 29).
38. C. A. Nicholson, *Dubois CEAR,* 1933 (CAERC box 79, folder 42).
39. A. J. Hesler, *Fountain CEAR,* 1933 (CAERC box 83, folder 29); A. V. Keesling, *Bartholomew CEAR,* 1934 (CAERC box 63, folder 21).
40. M. E. Cromer, *Delaware CEAR,* 1933 (CAERC box 78, folder 35); A. V. Keesling, *Bartholomew CEAR,* 1934 (CAERC box 63, folder 21); C. A. Nicholson, *Dubois CEAR,* 1934 (CAERC box 79, folder 43).
41. C. A. Nicholson, *Dubois CEAR,* 1933 (CAERC box 79, folder 42).
42. Ibid.
43. E. M. Rowe, *Clinton CEAR,* 1933 (CAERC box 72, folder 37); T. E. Myers, *DeKalb CEAR,* 1934 (CAERC box 77, folder 29).

44. A. J. Hesler, *Fountain CEAR,* 1934 (CAERC box 83, folder 30); R. J. Maggart, *Carroll CEAR,* 1934 (CAERC box 68, folder 59).
45. M. E. Cromer, *Delaware CEAR,* 1933 (CAERC box 78, folder 35); A. J. Hesler, *Fountain CEAR,* 1933 (CAERC box 83, folder 29); A. V. Keesling, *Bartholomew CEAR,* 1934 (CAERC box 63, folder 21).
46. W. U. Rusk, *Blackford CEAR,* 1935 (CAERC box 66, folder 12); A. J. Hesler, *Fountain CEAR,* 1933 (CAERC box 83, folder 29); M. F. Detrick, *Fayette CEAR,* 1934 (CAERC box 81, folder 35); A. M. Bishea, *Daviess CEAR,* 1934 (CAERC box 74, folder 28).
47. A. J. Hesler, *Fountain CEAR,* 1933 (CAERC box 83, folder 29).
48. H. S. Heckard, *Allen CEAR,* 1934 (CAERC box 62, folder 19); W. U. Rusk, *Blackford CEAR,* 1934 (CAERC box 66, folder 11).
49. H. S. Heckard, *Allen CEAR,* 1934 (CAERC box 62, folder 19); E. M. Rowe, *Clinton CEAR,* 1934 (CAERC box 72, folder 38); A. V. Keesling, *Bartholomew CEAR,* 1934 (CAERC box 63, folder 21); W. U. Rusk, *Blackford CEAR,* 1934 (CAERC box 66, folder 11).
50. M. E. Cromer, *Delaware CEAR,* 1934 (CAERC box 78, folder 36); C. M. East, *Floyd CEAR,* 1934 (CAERC box 82, folder 35).
51. H. S. Heckard, *Allen CEAR,* 1934 (CAERC box 62, folder 19); R. J. Maggart, *Carroll CEAR,* 1934 (CAERC box 68, folder 59).
52. E. M. Rowe, *Clinton CEAR,* 1934 (CAERC box 72, folder 38); M. F. Detrick, *Fayette CEAR,* 1934 (CAERC box 81, folder 35); C. A. Nicholson, *Dubois CEAR,* 1934 (CAERC box 79, folder 43).
53. W. U. Rusk, *Blackford CEAR,* 1934 (CAERC box 66, folder 11).
54. *Purdue University Department of Agricultural Extension Annual Report,* 1934, 8.
55. "Fifty Thousand Sign Contracts," 3; *Purdue University Department of Agricultural Extension Annual Report,* 1934, 9.
56. A. J. Hesler, *Fountain CEAR,* 1934 (CAERC box 83, folder 30).
57. I. L. Thurston, *Decatur CEARs,* 1933–35 (CAERC box 76, folders 39–41).
58. "County Agent Reviews," 5; Hall, "Friend of the Farmer," 5; E. S. Miles, *Clark CEAR,* 1934 (CAERC box 70, folder 41); I. L. Thurston, *Decatur CEAR,* 1934 (CAERC box 76, folder 40).
59. I. L. Thurston, *Decatur CEAR,* 1934 (CAERC box 76, folder 40); R. Gwaltney, *Gibson CEAR,* 1935 (CAERC box 87, folder 5); A. J. Hesler, *Fountain CEAR,* 1934 (CAERC box 83, folder 30).
60. A. V. Keesling, *Bartholomew CEAR,* 1934 (CAERC box 63, folder 21); W. O'Hara, *Hancock CEAR,* 1935 (CAERC box 91, folder 18).
61. E. M. Rowe, *Clinton CEAR,* 1933 (CAERC box 72, folder 37).
62. R. J. Maggart, *Carroll CEAR,* 1934 (CAERC box 68, folder 59); A. V. Keesling, *Bartholomew CEAR,* 1934 (CAERC box 63, folder 21); C. M. East, *Floyd CEAR,* 1934 (CAERC box 82, folder 35); E. M. Rowe, *Clinton CEAR,* 1934 (CAERC box 72, folder 38).
63. E. M. Rowe, *Clinton CEAR,* 1933 (CAERC box 72, folder 37).
64. H. S. Heckard, *Allen CEAR,* 1934 (CAERC box 62, folder 19).
65. A. J. Hesler, *Fountain CEAR,* 1933 (CAERC box 83, folder 29); M. F. Detrick, *Fayette CEAR,* 1934 (CAERC box 81, folder 35).
66. A. Haseman, *Greene CEAR,* 1934 (CAERC box 89, folder 9).
67. H. S. Heckard, *Allen CEAR,* 1934 (CAERC box 62, folder 19); W. U. Rusk, *Blackford CEAR,* 1934 (CAERC box 66, folder 11); E. S. Miles, *Clark CEAR,* 1934 (CAERC box 70, folder 41).

68. W. U. Rusk, *Blackford CEAR,* 1934 (CAERC box 66, folder 11); C. V. Kimmell, *Allen CEAR,* 1936 (CAERC box 62, folder 21); A. J. Hesler, *Fountain CEAR,* 1934 (CAERC box 83, folder 30).
69. R. Gwaltney, *Gibson CEAR,* 1935 (CAERC box 87, folder 5).
70. W. O'Hara, *Hancock CEAR,* 1934 (CAERC box 91, folder 17).
71. W. O'Hara, *Hancock CEAR,* 1935 (CAERC box 91, folder 18); W. O'Hara, *Hancock CEAR,* 1934 (CAERC box 91, folder 17); R. J. Maggart, *Carroll CEAR,* 1934 (CAERC box 68, folder 59).
72. C. A. Nicholson, *Dubois CEAR,* 1934 (CAERC box 79, folder 43); N. S. Hadley, *Fulton CEAR,* 1934 (CAERC box 85, folder 42); A. V. Keesling, *Bartholomew CEAR,* 1935 (CAERC box 63, folder 22).
73. H. S. Heckard, *Allen CEAR,* 1934 (CAERC box 62, folder 19); W. G. Smith, *Henry CEAR,* 1934 (CAERC box 94, folder 3); W. U. Rusk, *Blackford CEAR,* 1934 (CAERC box 66, folder 11).
74. A. J. Hesler, *Fountain CEAR,* 1934 (CAERC box 83, folder 30); W. U. Rusk, *Blackford CEAR,* 1934 (CAERC box 66, folder 11).
75. A. J. Hesler, *Fountain CEAR,* 1934 (CAERC box 83, folder 30).
76. E. M. Rowe, *Clinton CEAR,* 1934 (CAERC box 72, folder 38); W. G. Smith, *Henry CEAR,* 1934 (CAERC box 94, folder 3).
77. H. S. Heckard, *Allen CEAR,* 1934 (CAERC box 62, folder 19); M. E. Cromer, *Delaware CEAR,* 1934 (CAERC box 78, folder 36); W. O'Hara, *Hancock CEAR,* 1934 (CAERC box 91, folder 17).
78. R. J. Maggart, *Carroll CEAR,* 1934 (CAERC box 68, folder 59); W. G. Smith, *Henry CEAR,* 1934 (CAERC box 94, folder 3); T. E. Myers, *DeKalb CEAR,* 1934 (CAERC box 77, folder 29).
79. E. S. Miles, *Clark CEAR,* 1934 (CAERC box 70, folder 41).
80. A. V. Keesling, *Bartholomew CEAR,* 1934 (CAERC box 63, folder 21).
81. R. J. Maggart, *Carroll CEAR,* 1935 (CAERC box 68, folder 60).
82. W. O'Hara, *Hancock CEAR,* 1934 (CAERC box 91, folder 17).
83. C. A. Nicholson, *Dubois CEAR,* 1934 (CAERC box 79, folder 43).
84. R. J. Maggart, *Carroll CEAR,* 1934 (CAERC box 68, folder 59).
85. "Disposal of Pigs in Excess of the Number Permitted Under the Contract," *Brook Reporter* (Brook, IN), 25 July 1934, 7.
86. W. O'Hara, *Hancock CEAR,* 1934 (CAERC box 91, folder 17).
87. H. S. Heckard, *Allen CEAR,* 1935 (CAERC box 62, folder 20); E. S. Miles, *Clark CEAR,* 1934 (CAERC box 70, folder 41); E. M. Rowe, *Clinton CEAR,* 1934 (CAERC box 72, folder 38).
88. C. A. Nicholson, *Dubois CEAR,* 1934 (CAERC box 79, folder 43); E. S. Miles, *Clark CEAR,* 1934 (CAERC box 70, folder 41); T. E. Myers, *DeKalb CEAR,* 1935 (CAERC box 77, folder 30).
89. E. M. Rowe, *Clinton CEAR,* 1934 (CAERC box 72, folder 38).
90. H. S. Heckard, *Allen CEAR,* 1935 (CAERC box 62, folder 20).
91. A. V. Keesling, *Bartholomew CEAR,* 1936 (CAERC box 63, folder 24); E. M. Rowe, *Clinton CEAR,* 1936 (CAERC box 72, folder 42); N. S. Hadley, *Fulton CEAR,* 1936 (CAERC box 85, folder 44); W. O'Hara, *Hancock CEAR,* 1936 (CAERC box 91, folder 19); Skinner, "What 'New Deal' in Agriculture Means," 55.
92. A. V. Keesling, *Bartholomew CEAR,* 1936 (CAERC box 63, folder 24).

93. A. J. Hesler, *Fountain CEAR,* 1937 (CAERC box 83, folder 33); A. J. Hesler, *Fountain CEAR,* 1934 (CAERC box 83, folder 30); M. F. Detrick, *Fayette CEAR,* 1934 (CAERC box 81, folder 35).
94. C. V. Kimmell, *Allen CEAR,* 1939 (CAERC box 62, folder 25); H. S. Heckard, *Allen CEAR,* 1935 (CAERC box 62, folder 20).
95. W. G. Smith, *Henry CEAR,* 1933 (CAERC box 94, folder 2).
96. C. A. Nicholson, *Dubois CEAR,* 1934 (CAERC box 79, folder 43).
97. E. S. Miles, *Clark CEAR,* 1937 (CAERC box 70, folder 46); E. M. Rowe, *Clinton CEAR,* 1934 (CAERC box 72, folder 38); M. E. Cromer, *Delaware CEAR,* 1934 (CAERC box 78, folder 36).
98. E. M. Rowe, *Clinton CEAR,* 1935 (CAERC box 72, folder 39).
99. H. S. Heckard, *Allen CEAR,* 1934 (CAERC box 62, folder 19).
100. C. A. Nicholson, *Dubois CEAR,* 1935 (CAERC box 79, folder 44).
101. A. T. Marvel, *Elkhart CEAR,* 1936 (CAERC box 80, folder 34).
102. C. A. Nicholson, *Dubois CEAR,* 1935 (CAERC box 79, folder 44).
103. C. A. Nicholson, *Dubois CEAR,* 1934 (CAERC box 79, folder 43).
104. C. A. Nicholson, *Dubois CEAR,* 1933 (CAERC box 79, folder 42).
105. H. S. Heckard, *Allen CEAR,* 1934 (CAERC box 62, folder 19); R. Gwaltney, *Gibson CEAR,* 1939 (CAERC box 87, folder 9); H. L. Royce, *Clay CEAR,* 1937 (CAERC box 71, folder 49).

CHAPTER 9

Conservation of Soil Is the Solution on Which Will Hang Future Extension Activities

1. C. K. Whistler, *Boone CEAR,* 1936 (CAERC box 67, folder 1); E. M. Rowe, *Clinton CEAR,* 1936 (CAERC box 72, folder 42); W. G. Smith, *Henry CEAR,* 1936 (CAERC box 94, folder 5).
2. C. A. Alcorn, *Dearborn CEAR,* 1936 (CAERC box 75, folder 31); J. B. Todd, *Hamilton CEAR,* 1939 (CAERC box 90, folder 5); C. M. East, *Floyd CEAR,* 1938 (CAERC box 82, folder 39); M. E. Cromer, *Delaware CEAR,* 1939 (CAERC box 78, folder 42).
3. C. K. Whistler, *Boone CEAR,* 1936 (CAERC box 67, folder 1); M. F. Goode, *Harrison CEAR,* 1937 (CAERC box 92, folder 32).
4. A. J. Hesler, *Fountain CEAR,* 1932 (CAERC box 83, folder 28); I. L. Thurston, *Decatur CEAR,* 1930 (CAERC box 76, folder 36); M. E. Cromer, *Delaware CEAR,* 1933 (CAERC box 78, folder 35); H. L. Royce, *Clay CEAR,* 1930 (CAERC box 71, folder 42).
5. J. L. Morrow, *Crawford CEAR,* 1939 (CAERC box 73, folder 40); S. B. Scott, *Crawford CEAR,* 1930 (CAERC box 73, folder 40).
6. G. L. Reed, *Benton CEAR,* 1930 (CAERC box 65, folder 15).
7. E. M. Christen, *Benton CEAR,* 1937 (CAERC box 65, folder 23).
8. J. L. Morrow, *Crawford CEAR,* 1936 (CAERC box 73, folder 37); C. M. East, *Floyd CEAR,* 1938 (CAERC box 82, folder 39); A. J. Hesler, *Fountain CEAR,* 1936 (CAERC box 83, folder 32).
9. E. M. Rowe, *Clinton CEAR,* 1936 (CAERC box 72, folder 42); A. J. Hesler, *Fountain CEAR,* 1936 (CAERC box 83, folder 32).
10. R. Gwaltney, *Gibson CEAR,* 1936 (CAERC box 87, folder 6); E. M. Christen, *Benton CEAR,* 1938 (CAERC box 65, folder 25).
11. C. M. East, *Floyd CEAR,* 1936 (CAERC box 82, folder 37); C. M. East, *Floyd CEAR,* 1937 (CAERC box 82, folder 38).
12. C. M. East, *Floyd CEAR,* 1936 (CAERC box 82, folder 37).

13. Ibid.
14. Ibid.
15. Ibid.
16. C. M. East, *Floyd CEAR,* 1937 (CAERC box 82, folder 38).
17. Ibid.
18. M. F. Goode, *Harrison CEAR,* 1937 (CAERC box 92, folder 32).
19. H. J. Yarling, *Hendricks CEAR,* 1939 (CAERC box 93, folder 23).
20. M. F. Goode, *Harrison CEAR,* 1936 (CAERC box 92, folder 31); J. L. Morrow, *Crawford CEARs,* 1936–37 (CAERC box 73, folders 37–38).
21. J. L. Morrow, *Crawford CEAR,* 1936 (CAERC box 73, folder 37).
22. E. M. Christen, *Benton CEAR,* 1937 (CAERC box 65, folder 23); E. A. Johnson, *Cass CEAR,* 1937 (CAERC box 69, folder 41); W. O'Hara, *Hancock CEAR,* 1938 (CAERC box 91, folder 21).
23. T. E. Myers, *DeKalb CEAR,* 1936 (CAERC box 77, folder 31); E. M. Christen, *Benton CEAR,* 1937 (CAERC box 65, folder 23).
24. M. F. Goode, *Harrison CEAR,* 1938 (CAERC box 92, folder 33).
25. C. V. Kimmell, *Allen CEAR,* 1936 (CAERC box 62, folder 22).
26. Ibid.
27. E. M. Christen, *Benton CEAR,* 1937 (CAERC box 65, folder 23).
28. Ibid.; T. E. Myers, *DeKalb CEAR,* 1936 (CAERC box 77, folder 31).
29. *Purdue University Department of Agricultural Extension Annual Report,* 1939, 19.
30. J. L. Morrow, *Crawford CEAR,* 1938 (CAERC box 73, folder 39); E. M. Christen, *Benton CEAR,* 1936 (CAERC box 65, folder 21); E. M. Rowe, *Clinton CEAR,* 1936 (CAERC box 72, folder 42); J. H. Brown, *Elkhart CEAR,* 1938 (CAERC box 80, folder 36).
31. A. J. Hesler, *Fountain CEAR,* 1936 (CAERC box 83, folder 32).
32. E. A. Johnson, *Cass CEAR,* 1938 (CAERC box 69, folder 43).
33. T. E. Myers, *DeKalb CEAR,* 1936 (CAERC box 77, folder 31); E. M. Rowe, *Clinton CEAR,* 1936 (CAERC box 72, folder 42); C. V. Kimmell, *Allen CEAR,* 1936 (CAERC box 62, folder 22).
34. C. V. Kimmell, *Allen CEAR,* 1936 (CAERC box 62, folder 22).
35. E. A. Johnson, *Cass CEAR,* 1936 (CAERC box 69, folder 39).
36. L. E. Archbold, *Adams CEAR,* 1936 (CAERC box 61, folder 22).
37. M. M. Hamilton, *Decatur CEAR,* 1936 (CAERC box 76, folder 42); C. A. Nicholson, *Dubois CEAR,* 1938 (CAERC box 79A, folder 2); A. V. Keesling, *Bartholomew CEAR,* 1938 (CAERC box 63, folder 28).
38. C. M. East, *Floyd CEAR,* 1936 (CAERC box 82, folder 37).
39. C. V. Kimmell, *Allen CEAR,* 1936 (CAERC box 62, folder 22).
40. A. V. Keesling, *Bartholomew CEAR,* 1937 (CAERC box 63, folder 26); W. O'Hara, *Hancock CEAR,* 1938 (CAERC box 91, folder 21); C. M. East, *Floyd CEAR,* 1936 (CAERC box 82, folder 37).
41. J. L. Morrow, *Crawford CEAR,* 1937 (CAERC box 73, folder 38).
42. W. O'Hara, *Hancock CEAR,* 1936 (CAERC box 91, folder 19).
43. E. A. Johnson, *Cass CEAR,* 1936 (CAERC box 69, folder 39); A. V. Keesling, *Bartholomew CEAR,* 1938 (CAERC box 63, folder 28); J. L. Morrow, *Crawford CEAR,* 1937 (CAERC box 73, folder 38).
44. C. A. Nicholson, *Dubois CEAR,* 1938 (CAERC box 79A, folder 2).
45. A. V. Keesling, *Bartholomew CEAR,* 1936 (CAERC box 63, folder 24).

46. C. K. Whistler, *Boone CEAR,* 1938 (CAERC box 67, folder 3); C. V. Kimmell, *Allen CEAR,* 1938 (CAERC box 62, folder 24); M. M. Hamilton, *Decatur CEAR,* 1936 (CAERC box 76, folder 42).
47. *Purdue University Department of Agricultural Extension Annual Report,* 1938, 17; R. Gwaltney, *Gibson CEAR,* 1938 (CAERC box 87, folder 8).
48. E. A. Johnson, *Cass CEAR,* 1937 (CAERC box 69, folder 41).
49. C. A. Alcorn, *Dearborn CEAR,* 1937 (CAERC box 75, folder 32).
50. E. M. Rowe, *Clinton CEAR,* 1936 (CAERC box 72, folder 42); E. M. Christen, *Benton CEAR,* 1937 (CAERC box 65, folder 23); E. A. Johnson, *Cass CEAR,* 1939 (CAERC box 69, folder 44); J. H. Brown, *Elkhart CEAR,* 1937 (CAERC box 80, folder 35).
51. C. E. Baker, *DeKalb CEAR,* 1939 (CAERC box 77, folder 34); E. M. Rowe, *Clinton CEAR,* 1938 (CAERC box 72, folder 44).
52. E. M. Rowe, *Clinton CEAR,* 1938 (CAERC box 72, folder 44).
53. R. L. Zell, *Franklin CEAR,* 1939 (CAERC box 84, folder 39); E. A. Johnson, *Cass CEAR,* 1938 (CAERC box 69, folder 43).
54. J. L. Morrow, *Crawford CEAR,* 1939 (CAERC box 73, folder 40); E. S. Miles, *Clark CEAR,* 1937 (CAERC box 70, folder 46); E. A. Johnson, *Cass CEAR,* 1938 (CAERC box 69, folder 43).
55. H. R. Rosenbury, *Fulton CEAR,* 1930 (CAERC box 85, folder 38).
56. A. V. Keesling, *Bartholomew CEAR,* 1937 (CAERC box 63, folder 26).
57. R. Gwaltney, *Gibson CEAR,* 1936 (CAERC box 87, folder 6); R. L. Zell, *Franklin CEAR,* 1939 (CAERC box 84, folder 39); E. S. Miles, *Clark CEAR,* 1936 (CAERC box 70, folder 44); J. Jensen, *Daviess CEAR,* 1930 (CAERC box 74, folder 24).
58. H. L. Royce, *Clay CEAR,* 1931 (CAERC box 71, folder 43).
59. E. A. Johnson, *Cass CEAR,* 1939 (CAERC box 69, folder 44).
60. A. J. Hesler, *Fountain CEAR,* 1937 (CAERC box 83, folder 33).
61. E. S. Miles, *Clark CEAR,* 1939 (CAERC box 70, folder 50); E. M. Christen, *Benton CEAR,* 1937 (CAERC box 65, folder 23); C. A. Alcorn, *Dearborn CEAR,* 1936 (CAERC box 75, folder 31); R. J. Maggart, *Carroll CEAR,* 1936 (CAERC box 68, folder 61).
62. R. L. Zell, *Franklin CEAR,* 1938 (CAERC box 84, folder 38); A. Haseman, *Greene CEAR,* 1939 (CAERC box 89, folder 14).
63. E. S. Miles, *Clark CEAR,* 1937 (CAERC box 70, folder 46).
64. A. J. Hesler, *Fountain CEAR,* 1936 (CAERC box 83, folder 32); J. L. Morrow, *Crawford CEAR,* 1939 (CAERC box 73, folder 40).
65. L. E. Archbold, *Adams CEAR,* 1938 (CAERC box 61, folder 24).
66. C. A. Alcorn, *Dearborn CEAR,* 1937 (CAERC box 75, folder 32); D. S. Bishopp, *Fayette CEAR,* 1939 (CAERC box 81, folder 40); A. J. Hesler, *Fountain CEAR,* 1936 (CAERC box 83, folder 32).

CHAPTER 10

Erosion Is One of the Major Problems Which Must Be Faced

1. Teena Ligman, "Civilian Conservation Corps Camps Helped Form Parts of the Hoosier National Forest," *Reporter-Times* (Martinsville, IN), 19 Sept. 2010, 51.
2. Ibid.
3. C. A. Nicholson, *Dubois CEAR,* 1933 (CAERC box 79, folder 42).
4. "Visitors Are Welcome at Conservation Corps Camps," *Garrett Clipper* (Garrett, IN), 2 Nov. 1933, 4; Ligman, "Civilian Conservation Corps Camps," 51; E. A. Johnson, *Cass CEAR,* 1939 (CAERC box 69, folder 44).

5. J. Jensen, *Daviess CEAR,* 1933 (CAERC box 74, folder 27); E. A. Johnson, *Cass CEAR,* 1939 (CAERC box 69, folder 44).
6. R. J. Maggart, *Carroll CEAR,* 1936 (CAERC box 68, folder 61); W. G. Smith, *Henry CEAR,* 1933 (CAERC box 94, folder 2).
7. E. A. Johnson, *Cass CEAR,* 1939 (CAERC box 69, folder 44).
8. A. V. Keesling, *Bartholomew CEAR,* 1936 (CAERC box 63, folder 24); C. A. Alcorn, *Dearborn CEAR,* 1937 (CAERC box 75, folder 32).
9. M. F. Detrick, *Fayette CEAR,* 1933 (CAERC box 81, folder 34); "CCC to Have 38 Camps in This State," *Hammond Times* (Hammond, IN), 7 July 1937, 5.
10. "CCC to Have 38 Camps," 5; "Days to Remember . . . Civilian Conservation Corps," *Hoosier Farmer* 70, no. 3 (Mar. 1985): 25.
11. "Seven of CCC Camps in Indiana Have Been Closed," *Garrett Clipper* (Garrett, IN), 30 Jan. 1936, 3; Frank A. White, "The Camera Fan," *Indianapolis Star,* 25 May 1941, 77.
12. E. M. Christen, *Benton CEAR,* 1938 (CAERC box 65, folder 25); A. Haseman, *Greene CEARs,* 1937 and 1939 (CAERC box 89, folders 12 and 14); R. J. Maggart, *Carroll CEAR,* 1936 (CAERC box 68, folder 61); C. A. Alcorn, *Dearborn CEAR,* 1937 (CAERC box 75, folder 32); R. L. Zell, *Franklin CEAR,* 1935 (CAERC box 84, folder 35); W. U. Rusk, *Blackford CEAR,* 1933 (CAERC box 66, folder 10); E. A. Johnson, *Cass CEAR,* 1939 (CAERC box 69, folder 44); E. S. Miles, *Clark CEAR,* 1937 (CAERC box 70, folder 46); Ligman, "Civilian Conservation Corps," 51.
13. A. V. Keesling, *Bartholomew CEAR,* 1933 (CAERC box 63, folder 20); E. M. Rowe, *Clinton CEAR,* 1938 (CAERC box 72, folder 44); R. Gwaltney, *Gibson CEAR,* 1936 (CAERC box 87, folder 6); R. J. Maggart, *Carroll CEAR,* 1936 (CAERC box 68, folder 61).
14. R. J. Maggart, *Carroll CEARs,* 1936–38 (CAERC box 68, folders 61 and 63–64).
15. W. U. Rusk, *Blackford CEAR,* 1936 (CAERC box 66, folder 13).
16. R. J. Maggart, *Carroll CEAR,* 1936 (CAERC box 68, folder 61); W. U. Rusk, *Blackford CEAR,* 1936 (CAERC box 66, folder 13).
17. C. V. Kimmell, *Allen CEAR,* 1937 (CAERC box 62, folder 23).
18. W. O'Hara, *Hancock CEAR,* 1937 (CAERC box 91, folder 22).
19. A. V. Keesling, *Bartholomew CEAR,* 1939 (CAERC box 63, folder 30).
20. C. V. Kimmell, *Allen CEAR,* 1936 (CAERC box 62, folder 22).
21. A. V. Keesling, *Bartholomew CEAR,* 1939 (CAERC box 63, folder 30); H. J. Yarling, *Hendricks CEAR,* 1939 (CAERC box 93, folder 22).
22. A. V. Keesling, *Bartholomew CEAR,* 1936 (CAERC box 63, folder 24); D. S. Bishopp, *Fayette CEAR,* 1938 (CAERC box 81, folder 39).
23. E. A. Johnson, *Cass CEAR,* 1939 (CAERC box 69, folder 44).
24. A. V. Keesling, *Bartholomew CEAR,* 1938 (CAERC box 63, folder 28); R. J. Maggart, *Carroll CEAR,* 1936 (CAERC box 68, folder 61); D. S. Bishopp, *Fayette CEAR,* 1938 (CAERC box 81, folder 39); E. M. Christen, *Benton CEAR,* 1939 (CAERC box 65, folder 27).
25. H. J. Yarling, *Hendricks CEAR,* 1938 (CAERC box 93, folder 21).
26. D. S. Bishopp, *Fayette CEAR,* 1938 (CAERC box 81, folder 39); C. A. Nicholson, *Dubois CEAR,* 1933 (CAERC box 79, folder 42); R. J. Maggart, *Carroll CEAR,* 1936 (CAERC box 68, folder 61).
27. W. U. Rusk, *Blackford CEAR,* 1939 (CAERC box 66, folder 16); "Classes to Tour Farms," *Star Press* (Muncie, IN), 8 May 1941, 2.
28. W. U. Rusk, *Blackford CEAR,* 1939 (CAERC box 66, folder 16).
29. A. Haseman, *Greene CEAR,* 1937 (CAERC box 89, folder 12).
30. A. Haseman, *Greene CEAR,* 1936 (CAERC box 89, folder 11).

31. "Effects of Erosion Explained at Purdue," *Brook Reporter* (Brook, IN), 13 Feb. 1936, 6.
32. E. M. Christen, *Benton CEAR,* 1938 (CAERC box 65, folder 25).
33. E. M. Christen, *Benton CEAR,* 1937 (CAERC box 65, folder 23).
34. Ibid.
35. E. M. Christen, *Benton CEAR,* 1938 (CAERC box 65, folder 25).
36. A. V. Keesling, *Bartholomew CEAR,* 1939 (CAERC box 63, folder 30); H. J. Yarling, *Hendricks CEAR,* 1938 (CAERC box 93, folder 21); E. M. Christen, *Benton CEAR,* 1938 (CAERC box 65, folder 25).
37. E. A. Johnson, *Cass CEARs,* 1938–39 (CAERC box 69, folders 43–44).
38. A. J. Hesler, *Fountain CEAR,* 1936 (CAERC box 83, folder 32); C. V. Kimmell, *Allen CEAR,* 1937 (CAERC box 62, folder 23); A. V. Keesling, *Bartholomew CEAR,* 1937 (CAERC box 63, folder 26).
39. E. S. Miles, *Clark CEAR,* 1938 (CAERC box 70, folder 48).
40. C. V. Kimmell, *Allen CEAR,* 1938 (CAERC box 62, folder 24).
41. E. M. Christen, *Benton CEAR,* 1934 (CAERC box 65, folder 19).
42. E. A. Johnson, *Cass CEAR,* 1938 (CAERC box 69, folder 43).
43. E. A. Johnson, *Cass CEARs,* 1937–38 (CAERC box 69, folders 41 and 43); E. S. Miles, *Clark CEAR,* 1937 (CAERC box 70, folder 46); N. S. Hadley, *Fulton CEAR,* 1937 (CAERC box 85, folder 45).
44. W. G. Smith, *Henry CEAR,* 1936 (CAERC box 94, folder 5); A. V. Keesling, *Bartholomew CEAR,* 1939 (CAERC box 63, folder 30).
45. C. V. Kimmell, *Allen CEAR,* 1938 (CAERC box 62, folder 24).
46. W. O'Hara, *Hancock CEAR,* 1938 (CAERC box 91, folder 21).
47. C. V. Kimmell, *Allen CEAR,* 1938 (CAERC box 62, folder 24).
48. J. B. Todd, *Hamilton CEAR,* 1938 (CAERC box 90, folder 4).

CHAPTER 11

Land Use Planning Not Altogether a New Idea

1. P. K. Connelly, *Huntington CEAR,* 1940 (CAERC box 96, folder 16).
2. M. J. Huxley, *Fulton CEAR,* 1939 (CAERC box 85, folder 50); R. L. Zell, *Franklin CEAR,* 1936 (CAERC box 84, folder 36).
3. M. M. Hamilton, *Decatur CEAR,* 1936 (CAERC box 76, folder 42).
4. E. S. Miles, *Clark CEAR,* 1938 (CAERC box 70, folder 48); A. V. Keesling, *Bartholomew CEAR,* 1939 (CAERC box 63, folder 30); R. L. Zell, *Franklin CEAR,* 1936 (CAERC box 84, folder 36); A. Haseman, *Greene CEAR,* 1938 (CAERC box 89, folder 13); J. B. Todd, *Hamilton CEAR,* 1939 (CAERC box 90, folder 5); W. G. Smith, *Henry CEAR,* 1936 (CAERC box 94, folder 5).
5. M. M. Hamilton, *Decatur CEAR,* 1936 (CAERC box 76, folder 42); A. J. Hesler, *Fountain CEAR,* 1936 (CAERC box 83, folder 32).
6. E. M. Christen, *Benton CEAR,* 1939 (CAERC box 65, folder 27).
7. T. E. Myers, *DeKalb CEAR,* 1936 (CAERC box 77, folder 31); M. F. Goode, *Harrison CEAR,* 1937 (CAERC box 92, folder 32); H. J. Yarling, *Hendricks CEAR,* 1939 (CAERC box 93, folder 22).
8. M. J. Huxley, *Fulton CEAR,* 1939 (CAERC box 85, folder 50).
9. Ibid.
10. A. V. Keesling, *Bartholomew CEAR,* 1938 (CAERC box 63, folder 28).

11. M. M. Hamilton, *Decatur CEAR,* 1936 (CAERC box 76, folder 42).
12. E. M. Christen, *Benton CEAR,* 1939 (CAERC box 65, folder 27).
13. N. S. Hadley, *Fulton CEAR,* 1936 (CAERC box 85, folder 44); W. G. Smith, *Henry CEAR,* 1936 (CAERC box 94, folder 5).
14. A. V. Keesling, *Bartholomew CEAR,* 1938 (CAERC box 63, folder 28); M. F. Goode, *Harrison CEAR,* 1938 (CAERC box 92, folder 33).
15. C. M. East, *Floyd CEAR,* 1938 (CAERC box 82, folder 39); E. S. Miles, *Clark CEAR,* 1938 (CAERC box 70, folder 48); M. F. Goode, *Harrison CEAR,* 1938 (CAERC box 92, folder 33).
16. E. S. Miles, *Clark CEAR,* 1938 (CAERC box 70, folder 48).
17. C. M. East, *Floyd CEAR,* 1939 (CAERC box 82, folder 40).
18. A. V. Keesling, *Bartholomew CEAR,* 1938 (CAERC box 63, folder 28).
19. C. A. Nicholson, *Dubois CEAR,* 1938 (CAERC box 79A, folder 2).
20. A. V. Keesling, *Bartholomew CEAR,* 1938 (CAERC box 63, folder 28).
21. W. G. Smith, *Henry CEAR,* 1936 (CAERC box 94, folder 5).
22. M. F. Detrick, *Fayette CEAR,* 1937 (CAERC box 81, folder 38).
23. C. M. East, *Floyd CEARs,* 1935 and 1938 (CAERC box 82, folders 36 and 39).
24. M. M. Hamilton, *Decatur CEAR,* 1936 (CAERC box 76, folder 42).
25. A. Haseman, *Greene CEAR,* 1938 (CAERC box 89, folder 13).
26. Ibid.
27. Ibid.
28. R. Gwaltney, *Gibson CEAR,* 1937 (CAERC box 87, folder 7).
29. M. M. Hamilton, *Decatur CEAR,* 1936 (CAERC box 76, folder 42); E. S. Miles, *Clark CEAR,* 1938 (CAERC box 70, folder 48); W. G. Smith, *Henry CEAR,* 1936 (CAERC box 94, folder 5); M. F. Goode, *Harrison CEAR,* 1937 (CAERC box 92, folder 32).
30. R. Gwaltney, *Gibson CEAR,* 1937 (CAERC box 87, folder 7).
31. A. M. Bishea, *Daviess CEAR,* 1939 (CAERC box 74, folder 32).
32. A. Haseman, *Greene CEAR,* 1938 (CAERC box 89, folder 13); A. J. Hesler, *Fountain CEAR,* 1936 (CAERC box 83, folder 32); M. F. Goode, *Harrison CEAR,* 1938 (CAERC box 92, folder 33).
33. R. L. Zell, *Newton CEAR,* 1942 (CAERC box 117, folder 43).
34. W. J. Record, *Morgan CEAR,* 1942 (CAERC box 117, folder 3).
35. *Purdue University Department of Agricultural Extension Annual Report,* 1945, 6; E. A. Johnson, *Howard CEAR,* 1942 (CAERC box 95, folder 10); J. E. Wylie, *Miami CEAR,* 1942 (CAERC box 114, folder 13); J. A. Wood, *Martin CEAR,* 1942 (CAERC box 113, folder 28); W. C. Haynes, *Madison CEAR,* 1943 (CAERC box 109, folder 39); J. A. Clark, *Kosciusko CEAR,* 1945 (CAERC box 104, folder 21).
36. E. A. Johnson, *Howard CEAR,* 1946 (CAERC box 95, folder 14); H. E. Abbott, *Marion CEARs,* 1943–45 (CAERC box 111, folders 10, 13, and 15); L. E. Cutler, *Lake CEARs,* 1945–46 (CAERC box 107, folders 38 and 40); J. F. Armstrong, *Lawrence CEAR,* 1945 (CAERC box 108, folder 44); A. M. Bishea, *Daviess CEAR,* 1939 (CAERC box 74, folder 32); A. Haseman, *Greene CEAR,* 1938 (CAERC box 89, folder 13); A. J. Hesler, *Fountain CEAR,* 1936 (CAERC box 83, folder 32); M. F. Goode, *Harrison CEAR,* 1938 (CAERC box 92, folder 33).

CHAPTER 12

Extension Work Interrupted by Extreme Droughts and Flood

1. T. E. Myers, *DeKalb CEAR,* 1933 (CAERC box 77, folder 28); C. A. Nicholson, *Dubois CEAR,* 1930 (CAERC box 79, folder 39); R. O. Cole, *Clark CEAR,* 1930 (CAERC box 70, folder 37).
2. W. A. Douglass, *Bartholomew CEAR,* 1930 (CAERC box 63, folder 17).
3. J. Jensen, *Daviess CEAR,* 1930 (CAERC box 74, folder 24); C. C. Madison, *Dearborn CEAR,* 1931 (CAERC box 75, folder 26); S. B. Scott, *Crawford CEAR,* 1930 (CAERC box 73, folder 32).
4. "Priority Urged for Funds Bill," *Star Press* (Muncie, IN), 12 Mar. 1930, 1.
5. "Local Drought Broken by Sunday Night's Rain," *Kokomo Tribune* (Kokomo, IN), 14 Apr. 1930, 1; "Rain Aids Crops," *Kokomo Tribune* (Kokomo, IN), 24 June 1930, 17.
6. "Quick Drought Relief in 35 Counties Urged," *Richmond Palladium* (Richmond, IN), 15 Aug. 1930, 1.
7. R. E. Johnson, *Harrison CEAR,* 1931 (CAERC box 92, folder 26); J. Jensen, *Daviess CEAR,* 1930 (CAERC box 74, folder 24); W. A. Douglass, *Bartholomew CEAR,* 1930 (CAERC box 63, folder 17); A. T. Marvel, *Elkhart CEAR,* 1936 (CAERC box 80, folder 34); C. A. Nicholson, *Dubois CEAR,* 1930 (CAERC box 79, folder 39).
8. V. D. Sexson, *Greene CEAR,* 1930 (CAERC box 89, folder 5).
9. "Grain Markets," *Star Press* (Muncie, IN), 29 Mar. 1930, 12; "Wheat Prices Turn Upward," *Kokomo Tribune* (Kokomo, IN), 29 Mar. 1930, 15; "Crops Needing Rain," *Kokomo Tribune* (Kokomo, IN), 11 June 1930, 19.
10. "Leslie Plans Road Jobs as Drought Aid," *Indianapolis Star,* 15 Aug. 1930, 1.
11. R. E. Johnson, *Harrison CEAR,* 1931 (CAERC box 92, folder 26).
12. "Conservation of Water Is Urged at New Paris," *Richmond Item* (Richmond, IN), 18 July 1930, 6.
13. "Hoover Outlines Drought Relief with Governors," *Indianapolis Star,* 15 Aug. 1930, 1; "Take Steps to Help Indiana Farmers in Drought Zone," *Lafayette Journal and Courier* (Lafayette, IN), 12 Aug. 1930, 1; Spencer Howard, "Herbert Hoover and the 1930 Drought," *Hoover Heads* (blog), National Archives, 16 Sept. 2020, https://hoover.blogs.archives.gov/2020/09/16/herbert-hoover-and-the-1930-drought/.
14. R. E. Johnson, *Harrison CEAR,* 1930 (CAERC box 92, folder 25).
15. A. J. Hesler, *Fountain CEAR,* 1930 (CAERC box 83, folder 26).
16. Ibid.
17. "Leslie to Show Need in Southern Indiana," *Indianapolis News,* 12 Aug. 1930, 13.
18. "Quick Drought Relief in 35 Counties Urged," 1.
19. "Hoover Outlines Drought Relief," 1.
20. "State Drought Aid Given Blow," *Star Press* (Muncie, IN), 29 Aug. 1930, 2; "Drought Relief Legislation Useless if Security Is Required," *Lafayette Journal and Courier* (Lafayette, IN), 6 Dec. 1930, 7.
21. "Hoover Outlines Drought Relief," 1; "Session on Drought Will Open Monday," *Richmond Palladium* (Richmond, IN), 16 Aug. 1930, 1; "Three-Point Drought Relief Program Is Mapped," *Indianapolis Star,* 17 Sept. 1930, 11.
22. S. B. Scott, *Crawford CEAR,* 1930 (CAERC box 73, folder 32); C. C. Madison, *Dearborn CEAR,* 1930 (CAERC box 75, folder 25); I. L. Thurston, *Decatur CEAR,* 1930 (CAERC box 76, folder 36); R. E. Johnson, *Harrison CEAR,* 1931 (CAERC box 92, folder 26).
23. E. S. Miles, *Clark CEAR,* 1938 (CAERC box 70, folder 48); E. M. Christen, *Benton CEAR,* 1934 (CAERC box 65, folder 19); "County Survey First Move in Drought Relief," *Star Press* (Muncie, IN), 8 Aug. 1930, 2; "Take Steps to Help Indiana Farmers," 1.

24. "Hoover Calls on Governors for Drought Relief Parley," *Lafayette Journal and Courier* (Lafayette, IN), 9 Aug. 1930, 1.
25. "Hoover Outlines Drought Relief," 1; R. L. Zell, *Franklin CEAR,* 1939 (CAERC box 84, folder 39).
26. I. L. Thurston, *Decatur CEAR,* 1930 (CAERC box 76, folder 36).
27. W. A. Douglass, *Bartholomew CEAR,* 1930 (CAERC box 63, folder 17).
28. R. H. Aishton, "Railroads' Plan for Drought Relief Reported Progressing Smoothly," *Star Press* (Muncie, IN), 19 Oct. 1930, 22.
29. "Hoover Outlines Drought Relief," 1.
30. S. B. Scott, *Crawford CEAR,* 1930 (CAERC box 73, folder 32).
31. R. O. Cole, *Clark CEAR,* 1930 (CAERC box 70, folder 37).
32. R. E. Johnson, *Harrison CEAR,* 1930 (CAERC box 92, folder 25).
33. I. L. Thurston, *Decatur CEAR,* 1930 (CAERC box 76, folder 36).
34. S. B. Scott, *Crawford CEAR,* 1930 (CAERC box 73, folder 32); C. C. Madison, *Dearborn CEAR,* 1930 (CAERC box 75, folder 25); C. A. Alcorn, *Floyd CEAR,* 1930 (CAERC box 82, folder 31).
35. C. A. Alcorn, *Floyd CEAR,* 1930 (CAERC box 82, folder 31).
36. R. E. Johnson, *Harrison CEAR,* 1930 (CAERC box 92, folder 25).
37. Ibid.; R. E. Johnson, *Harrison CEAR,* 1931 (CAERC box 92, folder 26).
38. S. B. Scott, *Crawford CEAR,* 1930 (CAERC box 73, folder 32).
39. W. O'Hara, *Hancock CEAR,* 1936 (CAERC box 91, folder 19).
40. A. J. Hesler, *Fountain CEAR,* 1930 (CAERC box 83, folder 26).
41. C. C. Madison, *Dearborn CEAR,* 1931 (CAERC box 75, folder 26); "Priority Urged for Funds Bill," 1.
42. C. C. Madison, *Dearborn CEAR,* 1930 (CAERC box 75, folder 25).
43. R. E. Johnson, *Harrison CEAR,* 1931 (CAERC box 92, folder 26).
44. C. C. Madison, *Dearborn CEAR,* 1931 (CAERC box 75, folder 26).
45. W. O'Hara, *Hancock CEAR,* 1931 (CAERC box 91, folder 14).
46. "Farming in Indiana," *Rushville Republican* (Rushville, IN), 16 Aug. 1930, 6.
47. "Northern Indiana Farmers Will Donate Seed Corn," *Garrett Clipper* (Garrett, IN), 18 Sept. 1930, 6.
48. R. E. Johnson, *Harrison CEAR,* 1931 (CAERC box 92, folder 26).
49. A. V. Keesling, *Bartholomew CEAR,* 1934 (CAERC box 63, folder 21); E. M. Christen, *Benton CEAR,* 1934 (CAERC box 65, folder 19).
50. "Mercury 103 Here; U. S. Drought Intensifies," *Indianapolis News,* 8 July 1936, 1; E. M. Christen, *Benton CEAR,* 1937 (CAERC box 65, folder 23); J. L. Morrow, *Crawford CEAR,* 1936 (CAERC box 73, folder 37).
51. J. L. Morrow, *Crawford CEAR,* 1937 (CAERC box 73, folder 38).
52. A. V. Keesling, *Bartholomew CEAR,* 1936 (CAERC box 63, folder 24).
53. E. S. Miles, *Clark CEAR,* 1936 (CAERC box 70, folder 44).
54. Ibid.
55. Ibid.
56. C. M. East, *Floyd CEAR,* 1936 (CAERC box 82, folder 37).
57. Ibid.
58. R. L. Zell, *Franklin CEAR,* 1936 (CAERC box 84, folder 36).
59. Ibid.
60. Ibid.
61. C. M. East, *Floyd CEAR,* 1936 (CAERC box 82, folder 37); C. A. Alcorn, *Dearborn CEAR,* 1936 (CAERC box 75, folder 31).

62. C. A. Alcorn, *Dearborn CEAR,* 1936 (CAERC box 75, folder 31).
63. Ibid.
64. J. L. Morrow, *Crawford CEAR,* 1937 (CAERC box 73, folder 38).
65. J. L. Morrow, *Crawford CEAR,* 1936 (CAERC box 73, folder 37).
66. "Ohio River Flood Comes Rushing Down," *Tipton Daily Tribune* (Tipton, IN), 20 Jan. 1937, 1; "Fresh Rains Swell Floods in Southern Area of State," *Lafayette Journal and Courier* (Lafayette, IN), 15 Jan. 1937, 1.
67. "Hazleton New Danger Spot in Indiana Flood," *Lafayette Journal and Courier* (Lafayette, IN), 20 Jan. 1937, 1.
68. "Congress Joins in Rush to Help Flood Victims," *Richmond Palladium* (Richmond, IN), 26 Jan. 1937, 9; "Ohio Floods Hammer Weakening Levees," *Richmond Palladium* (Richmond, IN), 28 Jan. 1937, 9.
69. "Ohio Floods Hammer Weakening Levees," 9.
70. "Railroads Appoint Liddy Co-ordinator for Flood Zone," *Indianapolis Star,* 31 Jan. 1937, 4; "Jeffersonville—300 Stay in Plant—No Deaths," *Indianapolis Star,* 30 Jan. 1937, 10.
71. "Jeffersonville, New Albany Flood Picture 'Incredible,'" *Indianapolis Star,* 25 Jan. 1937, 10.
72. "4,958 Troops Rush Flood Zone Cleanup," *Indianapolis Star,* 28 Jan. 1937, 1.
73. Ibid.
74. Ibid.; "Latest Reports from Flood-Stricken Towns and Cities," *Indianapolis News,* 23 Jan. 1937, 2.
75. "Latest Reports from Flood-Stricken Towns and Cities," 2; "4,958 Troops Rush Flood Zone Cleanup," 1.
76. "Latest Reports from Flood-Stricken Towns and Cities," 2.
77. Ibid.
78. "Congress Joins in Rush to Help Flood Victims," 9.
79. "4,958 Troops Rush Flood Zone Cleanup," 1.
80. "Railroads Appoint Liddy Co-ordinator," 4.
81. "Kokomo May Have 500 Flood Victims to Care for Within 24 Hours," *Kokomo Tribune* (Kokomo, IN), 25 Jan. 1937, 1.
82. "4,958 Troops Rush Flood Zone Cleanup," 1; "Kokomo May Have 500 Flood Victims to Care for," 1.
83. "4,958 Troops Rush Flood Zone Cleanup," 1.
84. C. A. Alcorn, *Dearborn CEAR,* 1937 (CAERC box 75, folder 32).
85. Ibid.
86. C. A. Nicholson, *Dubois CEAR,* 1937 (CAERC box 79A, folder 1).
87. Ibid.
88. Ibid.
89. C. M. East, *Floyd CEAR,* 1937 (CAERC box 82, folder 38); R. E. Johnson, *Harrison CEAR,* 1931 (CAERC box 92, folder 26).
90. S. B. Scott, *Crawford CEAR,* 1930 (CAERC box 73, folder 32).

CHAPTER 13

There Is Convenience and Satisfaction of Flipping a Switch and Getting Light

1. A. J. Hesler, *Fountain CEAR,* 1929 (CAERC box 83, folder 24).
2. A. J. Hesler, *Fountain CEAR,* 1935 (CAERC box 83, folder 31).
3. A. J. Hesler, *Fountain CEARs,* 1930 and 1935 (CAERC box 83, folders 26 and 31).
4. A. J. Hesler, *Fountain CEAR,* 1935 (CAERC box 83, folder 31).

5. "History," Indiana Electric Cooperatives (website), accessed 16 Apr. 2022, https://www.indianaec.org/who-we-are/history/; "History: The Story Behind America's Electric Cooperatives and NRECA," National Rural Electric Cooperative Association (website), accessed 16 Apr. 2022, https://www.electric.coop/our-organization/history; A. J. Hesler, *Fountain CEAR,* 1930 (CAERC box 83, folder 26); R. Douglas Hurt, "REA: A New Deal for Farmers," *Timeline* 2, no. 6 (Dec. 1985/Jan. 1986): 33.
6. R. L. Zell, *Franklin CEAR,* 1937 (CAERC box 84, folder 37); A. J. Hesler, *Fountain CEAR,* 1930 (CAERC box 83, folder 26); N. S. Hadley, *Fulton CEAR,* 1938 (CAERC box 85, folder 48).
7. C. A. Nicholson, *Dubois CEAR,* 1937 (CAERC box 79A, folder 1); Hurt, "REA," 36.
8. A. J. Hesler, *Fountain CEAR,* 1932 (CAERC box 83, folder 28).
9. A. M. Bishea, *Daviess CEAR,* 1935 (CAERC box 74, folder 29).
10. "Marshall County REMC: A Look Back at Where It All Began," *Indiana Connection* (digital magazine), 4 Oct. 2020, https://www.indianaconnection.org/a-look-back-at-where-it-all-began/.
11. A. M. Bishea, *Daviess CEAR,* 1939 (CAERC box 74, folder 32); A. V. Keesling, *Bartholomew CEAR,* 1936 (CAERC box 63, folder 24); R. J. Maggart, *Carroll CEAR,* 1936 (CAERC box 68, folder 61); N. S. Hadley, *Fulton CEAR,* 1937 (CAERC box 85, folder 45).
12. R. J. Maggart, *Carroll CEAR,* 1939 (CAERC box 68, folder 65); W. O'Hara, *Hancock CEAR,* 1935 (CAERC box 91, folder 18); R. J. Maggart, *Carroll CEAR,* 1936 (CAERC box 68, folder 61).
13. R. J. Maggart, *Carroll CEAR,* 1936 and 1938 (CAERC box 68, folders 61 and 64); W. O'Hara, *Hancock CEAR,* 1935 (CAERC box 91, folder 18).
14. A. V. Keesling, *Bartholomew CEAR,* 1936 (CAERC box 63, folder 24).
15. J. L. Morrow, *Crawford CEAR,* 1939 (CAERC box 73, folder 40).
16. W. G. Smith, *Henry CEAR,* 1936 (CAERC box 94, folder 5).
17. A. M. Bishea, *Daviess CEAR,* 1939 (CAERC box 74, folder 32); R. J. Maggart, *Carroll CEAR,* 1937 (CAERC box 68, folder 63).
18. C. A. Nicholson, *Dubois CEAR,* 1939 (CAERC box 79A, folder 3); W. O'Hara, *Hancock CEAR,* 1936 (CAERC box 91, folder 19); R. L. Zell, *Franklin CEAR,* 1936 (CAERC box 84, folder 36).
19. A. V. Keesling, *Bartholomew CEAR,* 1936 (CAERC box 63, folder 24).
20. C. A. Nicholson, *Dubois CEAR,* 1938 (CAERC box 79A, folder 2); W. G. Smith, *Henry CEAR,* 1936 (CAERC box 94, folder 5); W. O'Hara, *Hancock CEAR,* 1935 (CAERC box 91, folder 18).
21. A. V. Keesling, *Bartholomew CEAR,* 1936 (CAERC box 63, folder 24); R. J. Maggart, *Carroll CEAR,* 1936 (CAERC box 68, folder 61); W. G. Smith, *Henry CEAR,* 1936 (CAERC box 94, folder 5); W. O'Hara, *Hancock CEAR,* 1935 (CAERC box 91, folder 18).
22. W. G. Smith, *Henry CEAR,* 1936 (CAERC box 94, folder 5).
23. R. L. Zell, *Franklin CEAR,* 1936 (CAERC box 84, folder 36); M. M. Hamilton, *Decatur CEAR,* 1938 (CAERC box 76, folder 44).
24. M. M. Hamilton, *Decatur CEAR,* 1936 (CAERC box 76, folder 42); W. G. Smith, *Henry CEAR,* 1936 (CAERC box 94, folder 5).
25. A. Haseman, *Greene CEAR,* 1938 (CAERC box 89, folder 13).
26. E. M. Christen, *Benton CEAR,* 1939 (CAERC box 65, folder 27).
27. E. M. Christen, *Benton CEAR,* 1938 (CAERC box 65, folder 25).
28. A. V. Keesling, *Bartholomew CEAR,* 1936 (CAERC box 63, folder 24); W. G. Smith, *Henry CEAR,* 1936 (CAERC box 94, folder 5).

29. W. O'Hara, *Hancock CEAR,* 1935 (CAERC box 91, folder 18).
30. R. J. Maggart, *Carroll CEAR,* 1936 (CAERC box 68, folder 61); W. G. Smith, *Henry CEAR,* 1936 (CAERC box 94, folder 5); W. O'Hara, *Hancock CEAR,* 1936 (CAERC box 91, folder 19).
31. A. V. Keesling, *Bartholomew CEARs,* 1936–37 (CAERC box 63, folders 24 and 26); J. L. Morrow, *Crawford CEAR,* 1939 (CAERC box 73, folder 40); N. S. Hadley, *Fulton CEAR,* 1936 (CAERC box 85, folder 44); W. O'Hara, *Hancock CEAR,* 1936 (CAERC box 91, folder 19).
32. A. V. Keesling, *Bartholomew CEAR,* 1936 (CAERC box 63, folder 24).
33. R. L. Zell, *Franklin CEAR,* 1937 (CAERC box 84, folder 37).
34. A. V. Keesling, *Bartholomew CEARs,* 1936–37 (CAERC box 63, folders 24 and 26); J. L. Morrow, *Crawford CEAR,* 1939 (CAERC box 73, folder 40); N. S. Hadley, *Fulton CEAR,* 1936 (CAERC box 85, folder 44); W. O'Hara, *Hancock CEAR,* 1936 (CAERC box 91, folder 19).
35. W. O'Hara, *Hancock CEAR,* 1936 (CAERC box 91, folder 19).
36. W. G. Smith, *Henry CEAR,* 1936 (CAERC box 94, folder 5); R. J. Maggart, *Carroll CEAR,* 1936 (CAERC box 68, folder 61); A. V. Keesling, *Bartholomew CEAR,* 1938 (CAERC box 63, folder 28).
37. National Academy of Engineering, "Electrification History 2: Rural Electrification," Greatest Engineering Achievements of the 20th Century (website), National Academy of Sciences, accessed 13 May 2022, http://www.greatachievements.org/?id=2990; W. G. Smith, *Henry CEAR,* 1936 (CAERC box 94, folder 5); Hurt, "REA," 36.
38. Ethan Chitty, "The Decades-Long Struggle to Electrify Rural Hoosierdom," *Indiana History Blog,* Indiana Historical Bureau of the Indiana State Library, 4 Apr. 2017, https://blog.history.in.gov/the-decades-long-struggle-to-electrify-rural-hoosierdom/.
39. William Oatis, "Rea and Utility Firm Cooperate," *Star Press* (Muncie, IN), 23 Jan. 1939, 7.
40. W. W. Whitehead, *Boone CEAR,* 1939 (CAERC box 67, folder 4).
41. "REA Funds Granted Two Indiana Co-ops," *Indianapolis Star,* 30 Sept. 1936, 9; "$1,040,000 Is Allotted REA in Indiana," *Times* (Munster, IN), 15 Oct. 1936, 9.
42. R. J. Maggart, *Carroll CEARs,* 1936–37 (CAERC box 68, folders 61 and 63).
43. A. M. Bishea, *Daviess CEAR,* 1939 (CAERC box 74, folder 32).
44. R. Gwaltney, *Gibson CEAR,* 1939 (CAERC box 87, folder 9).
45. A. V. Keesling, *Bartholomew CEAR,* 1939 (CAERC box 63, folder 30); W. G. Smith, *Henry CEAR,* 1936 (CAERC box 94, folder 5).
46. C. K. Whistler, *Boone CEAR,* 1936 (CAERC box 67, folder 1); L. E. Archbold, *Adams CEAR,* 1939 (CAERC box 1, folder 1); A. V. Keesling, *Bartholomew CEAR,* 1938 (CAERC box 63, folder 28).
47. W. G. Smith, *Henry CEAR,* 1936 (CAERC box 94, folder 5).
48. E. A. Johnson, *Cass CEAR,* 1939 (CAERC box 69, folder 44); A. V. Keesling, *Bartholomew CEAR,* 1938 (CAERC box 63, folder 28).
49. B. V. Widney, *Noble CEAR,* 1920 (CAERC box 118, folder 14).
50. W. U. Rusk, *Blackford CEARs,* 1932–33 (CAERC box 66, folders 9–10); J. H. Brown, *Elkhart CEAR,* 1937 (CAERC box 80, folder 35); W. A. Douglass, *Bartholomew CEAR,* 1931 (CAERC box 63, folder 18); W. G. Smith, *Henry CEAR,* 1931 (CAERC box 93, folder 60); W. O'Hara, *Hancock CEAR,* 1938 (CAERC box 91, folder 21); H. J. Yarling, *Hendricks CEAR,* 1939 (CAERC box 93, folder 22).
51. W. U. Rusk, *Blackford CEAR,* 1937 (CAERC box 66, folder 14).

52. H. J. Yarling, *Hendricks CEAR,* 1939 (CAERC box 93, folder 22).
53. W. U. Rusk, *Blackford CEAR,* 1933 (CAERC box 66, folder 10); H. E. Abbott, *Vigo CEARs,* 1928–29 (CAERC box 50, folders 41–42).
54. R. J. Maggart, *Carroll CEAR,* 1939 (CAERC box 68, folder 65).
55. H. S. Heckard, *Allen CEAR,* 1930 (CAERC box 62, folder 15).
56. A. J. Hesler, *Fountain CEAR,* 1930 (CAERC box 83, folder 26); W. O'Hara, *Hancock CEAR,* 1937 (CAERC box 91, folder 20).
57. W. U. Rusk, *Blackford CEAR,* 1936 (CAERC box 66, folder 13).
58. A. J. Hesler, *Fountain CEAR,* 1932 (CAERC box 83, folder 28).
59. A. J. Hesler, *Fountain CEARs,* 1935–36 (CAERC box 83, folders 31–32); C. M. East, *Floyd CEAR,* 1936 (CAERC box 82, folder 37); A. Haseman, *Greene CEAR,* 1938 (CAERC box 89, folder 13).
60. W. O'Hara, *Hancock CEAR,* 1936 (CAERC box 91, folder 19); W. W. Whitehead, *Boone CEAR,* 1939 (CAERC box 67, folder 4); A. Haseman, *Greene CEAR,* 1939 (CAERC box 89, folder 14).
61. A. M. Bishea, *Daviess CEAR,* 1939 (CAERC box 74, folder 32); M. F. Detrick, *Fayette CEAR,* 1937 (CAERC box 81, folder 38); E. A. Johnson, *Cass CEAR,* 1939 (CAERC box 69, folder 44); W. O'Hara, *Hancock CEAR,* 1937 (CAERC box 91, folder 20); W. U. Rusk, *Blackford CEAR,* 1937 (CAERC box 66, folder 14).
62. *Purdue University Department of Agricultural Extension Annual Report,* 1936, 27–28.
63. *Purdue University Department of Agricultural Extension Annual Report,* 1937, 26.
64. A. J. Hesler, *Fountain CEAR,* 1930 (CAERC box 83, folder 26).
65. M. E. Cromer, *Delaware CEAR,* 1936 (CAERC box 78, folder 38).
66. H. S. Heckard, *Allen CEAR,* 1930 (CAERC box 62, folder 15).
67. M. E. Cromer, *Delaware CEAR,* 1936 (CAERC box 78, folder 38).
68. A. J. Hesler, *Fountain CEAR,* 1930 (CAERC box 83, folder 26).
69. C. A. Alcorn, *Dearborn CEAR,* 1939 (CAERC box 75, folder 34); W. U. Rusk, *Blackford CEAR,* 1936 (CAERC box 66, folder 13); A. Haseman, *Greene CEAR,* 1938 (CAERC box 89, folder 13).
70. *Purdue University Department of Agricultural Extension Annual Report,* 1936, 78.
71. C. K. Whistler, *Boone CEAR,* 1936 (CAERC box 67, folder 1).
72. R. J. Maggart, *Carroll CEAR,* 1937 (CAERC box 68, folder 63).
73. C. K. Whistler, *Boone CEAR,* 1937 (CAERC box 67, folder 2).
74. R. J. Maggart, *Carroll CEAR,* 1939 (CAERC box 68, folder 65).
75. Ibid.
76. W. W. Whitehead, *Boone CEAR,* 1939 (CAERC box 67, folder 4).
77. Ibid.
78. R. J. Maggart, *Carroll CEAR,* 1938 (CAERC box 68, folder 64); C. K. Whistler, *Boone CEAR,* 1937 (CAERC box 67, folder 2); A. Haseman, *Greene CEAR,* 1939 (CAERC box 89, folder 14).
79. A. V. Keesling, *Bartholomew CEAR,* 1938 (CAERC box 63, folder 28).
80. *Purdue University Department of Agricultural Extension Annual Report,* 1937, 25.
81. C. K. Whistler, *Boone CEARs,* 1936–37 (CAERC box 67, folders 1–2).
82. A. M. Bishea, *Daviess CEAR,* 1935 (CAERC box 74, folder 29); R. J. Maggart, *Carroll CEAR,* 1939 (CAERC box 68, folder 65).
83. A. J. Hesler, *Fountain CEAR,* 1930 (CAERC box 83, folder 26).

84. Chitty, "Decades-Long Struggle to Electrify Rural Hoosierdom."
85. E. M. Rowe, *Clinton CEAR,* 1939 (CAERC box 72, folder 45); R. J. Maggart, *Carroll CEAR,* 1937 (CAERC box 68, folder 63).
86. A. Haseman, *Greene CEAR,* 1938 (CAERC box 89, folder 13).
87. R. J. Maggart, *Carroll CEAR,* 1938 (CAERC box 68, folder 64).
88. A. J. Hesler, *Fountain CEAR,* 1939 (CAERC box 83, folder 35).
89. Ibid.; A. J. Hesler, *Fountain CEAR,* 1932 (CAERC box 83, folder 28).

CHAPTER 14

Shall I Sell One Team on a Four-Horse Farm and Buy a Tractor?

1. M. E. Cromer, *Delaware CEAR,* 1939 (CAERC box 78, folder 41).
2. A. J. Hesler, *Fountain CEAR,* 1936 (CAERC box 83, folder 32).
3. E. C. Parker, *Hamilton CEARs,* 1930–31 (CAERC box 89, folders 45–46); V. J. Mann, *Clinton CEAR,* 1931 (CAERC box 72, folder 35); C. B. Blosser, *Elkhart CEARs,* 1931–32 (CAERC box 80, folders 30–31); E. S. Miles, *Clark CEAR,* 1936 (CAERC box 70, folder 44); N. S. Hadley, *Fulton CEAR,* 1934 (CAERC box 85, folder 42).
4. M. E. Cromer, *Delaware CEAR,* 1933 (CAERC box 78, folder 35).
5. R. D. McHargue, *Clay CEAR,* 1938 (CAERC box 71, folder 50); C. L. Dyer, *Gibson CEAR,* 1932 (CAERC box 87, folder 3); R. Gwaltney, *Gibson CEAR,* 1939 (CAERC box 87, folder 9); M. E. Cromer, *Delaware CEAR,* 1939 (CAERC box 78, folder 41); G. L. Reed, *Benton CEAR,* 1931 (CAERC box 65, folder 16).
6. A. J. Hesler, *Fountain CEAR,* 1936 (CAERC box 83, folder 32).
7. A. J. Hesler, *Fountain CEAR,* 1939 (CAERC box 83, folder 35).
8. M. F. Goode, *Harrison CEAR,* 1933 (CAERC box 92, folder 28); H. L. Royce, *Clay CEAR,* 1932 (CAERC box 71, folder 44); E. S. Miles, *Clark CEAR,* 1935 (CAERC box 70, folder 42).
9. C. B. Blosser, *Elkhart CEAR,* 1931 (CAERC box 80, folder 30).
10. M. F. Detrick, *Fayette CEAR,* 1931 (CAERC box 81, folder 32); G. L. Reed, *Benton CEAR,* 1932 (CAERC box 65, folder 17).
11. M. D. Butler, *Grant CEAR,* 1932 (CAERC box 87, folder 49).
12. W. G. Smith, *Henry CEAR,* 1934 (CAERC box 94, folder 3).
13. L. E. Archbold, *Adams CEAR,* 1934 (CAERC box 61, folder 20); H. L. Royce, *Clay CEAR,* 1936 (CAERC box 71, folder 48).
14. C. B. Blosser, *Elkhart CEAR,* 1931 (CAERC box 80, folder 30).
15. W. G. Smith, *Henry CEAR,* 1932 (CAERC box 94, folder 1).
16. J. B. Todd, *Hamilton CEAR,* 1937 (CAERC box 90, folder 3); C. A. Nicholson, *Dubois CEAR,* 1934 (CAERC box 79, folder 43).
17. J. B. Todd, *Hamilton CEAR,* 1936 (CAERC box 90, folder 2).
18. Ibid.
19. M. E. Cromer, *Delaware CEAR,* 1933 (CAERC box 78, folder 35); W. G. Smith, *Henry CEAR,* 1931 (CAERC box 93, folder 60).
20. N. S. Hadley, *Franklin CEAR,* 1933 (CAERC box 84, folder 33); H. L. Royce, *Clay CEAR,* 1932 (CAERC box 71, folder 44).
21. J. Jensen, *Daviess CEAR,* 1933 (CAERC box 74, folder 27); J. E. Little, *Cass CEAR,* 1930 (CAERC box 69, folder 33); M. F. Detrick, *Fayette CEAR,* 1931 (CAERC box 81, folder 32); T. E. Myers, *DeKalb CEAR,* 1933 (CAERC box 77, folder 28).
22. W. G. Smith, *Henry CEAR,* 1932 (CAERC box 94, folder 1).

23. V. J. Mann, *Clinton CEAR,* 1932 (CAERC box 72, folder 36).
24. W. G. Smith, *Henry CEAR,* 1931 (CAERC box 93, folder 60).
25. E. L. Hawk, *Hendricks CEAR,* 1932 (CAERC box 93, folder 19).
26. V. J. Mann, *Clinton CEAR,* 1930 (CAERC box 72, folder 34).
27. Ibid.
28. V. J. Mann, *Clinton CEAR,* 1931 (CAERC box 72, folder 35).
29. M. E. Cromer, *Delaware CEAR,* 1930 (CAERC box 78, folder 32); H. J. Yarling, *Hendricks CEAR,* 1937 (CAERC box 93, folder 21); E. S. Miles, *Clark CEAR,* 1937 (CAERC box 70, folder 46).
30. John D. Thompson, "County Agricultural News," *Brook Reporter* (Brook, IN), 22 July 1937, 2.
31. M. E. Cromer, *Delaware CEAR,* 1933 (CAERC box 78, folder 35); H. L. Royce, *Clay CEAR,* 1932 (CAERC box 71, folder 44).
32. M. E. Cromer, *Delaware CEAR,* 1931 (CAERC box 78, folder 33).
33. E. A. Johnson, *Cass CEAR,* 1938 (CAERC box 69, folder 43); M. E. Cromer, *Delaware CEAR,* 1935 (CAERC box 78, folder 37).
34. M. F. Goode, *Harrison CEAR,* 1935 (CAERC box 92, folder 30).
35. Thompson, "County Agricultural News," 2.
36. "Seek Enrollments in Colt Club for Balance of Month," *Rushville Republican* (Rushville, IN), 15 July 1937, 1.
37. Ibid.
38. Mara Carter, "Farm News," *Muncie Evening Press* (Muncie, IN), 6 July 1936, 3.
39. W. G. Smith, *Henry CEAR,* 1931 (CAERC box 93, folder 60).
40. N. S. Hadley, *Fulton CEAR,* 1935 (CAERC box 85, folder 43).
41. A. Haseman, *Greene CEAR,* 1939 (CAERC box 89, folder 14).
42. J. E. Little, *Cass CEAR,* 1932 (CAERC box 69, folder 35).
43. Oliver Knight, "Machines Begin Displacing of Hands on American Farms," *Palladium-Item* (Richmond, IN), 1 June 1939, 16.
44. Ibid.
45. J. B. Todd, *Hamilton CEAR,* 1939 (CAERC box 90, folder 5).
46. J. B. Todd, *Hamilton CEAR,* 1936 (CAERC box 90, folder 2).
47. Ibid.
48. M. F. Detrick, *Fayette CEAR,* 1937 (CAERC box 81, folder 38).
49. E. M. Christen, *Benton CEAR,* 1939 (CAERC box 65, folder 27); H. S. Heckard, *Allen CEAR,* 1933 (CAERC box 62, folder 18).
50. E. M. Christen, *Benton CEAR,* 1939 (CAERC box 65, folder 27).

CHAPTER 15

The Average Farmer Has Not Learned the Principles of Economic Uses of Wood Lots

1. J. L. Morrow, *Crawford CEAR,* 1939 (CAERC box 73, folder 40).
2. A. Haseman, *Greene CEAR,* 1933 (CAERC box 89, folder 8).
3. R. E. Johnson, *Harrison CEAR,* 1930 (CAERC box 92, folder 25).
4. S. B. Scott, *Crawford CEAR,* 1930 (CAERC box 73, folder 32); A. M. Bishea, *Daviess CEAR,* 1939 (CAERC box 74, folder 32); C. A. Nicholson, *Dubois CEAR,* 1933 (CAERC box 79, folder 42).
5. A. V. Keesling, *Bartholomew CEAR,* 1933 (CAERC box 63, folder 20).
6. W. T. Merrilees, *Crawford CEAR,* 1933 (CAERC box 73, folder 35).

7. James H. Madison and Lee Ann Sandweiss, "The Great Depression and World War II," chap. 9 in *Hoosiers and the American Story* (Indianapolis: Indiana Historical Society, 2014), 243, https://indianahistory.org/education/educator-resources/curriculum/hoosiers-and-the-american-story/.
8. "State of Indiana to Have 24 CCC Camps," *Edinburg Daily Courier* (Edinburg, IN), 6 Apr. 1934, 1.
9. A. J. Hesler, *Fountain CEAR,* 1933 (CAERC box 83, folder 29).
10. N. S. Hadley, *Franklin CEAR,* 1933 (CAERC box 84, folder 33); A. J. Hesler, *Fountain CEAR,* 1933 (CAERC box 83, folder 29).
11. A. J. Hesler, *Fountain CEAR,* 1933 (CAERC box 83, folder 29); A. V. Keesling, *Bartholomew CEAR,* 1933 (CAERC box 63, folder 20); Matt Waggener, "Eleven C.C.C. Camps Are Stationed in Forests of Indiana," *Franklin Evening Star* (Franklin, IN), 7 Sept. 1933, 7.
12. Matt Waggener, "Change in Location Is Announced for Three C.C.C. Camps," *Franklin Evening Star* (Franklin, IN), 21 Oct. 1933, 7.
13. Ibid.
14. David E. Conrad, *The Land We Cared For: A History of the Forest Service's Eastern Region* (Carbondale, IL: American Resources Group Ltd., 1997), 105, https://foresthistory.org/wp-content/uploads/2017/01/The-Land-We-Cared-for-A-History-of-the-Forest-Service-Eastern-Region.pdf.
15. Ibid., 106.
16. L. E. Archbold, *Adams CEAR,* 1931 (CAERC box 61, folder 17).
17. E. M. Christen, *Benton CEAR,* 1937 (CAERC box 65, folder 23).
18. C. A. Nicholson, *Dubois CEAR,* 1935 (CAERC box 79, folder 44); W. A. Douglass, *Bartholomew CEAR,* 1931 (CAERC box 63, folder 18); A. V. Keesling, *Bartholomew CEAR,* 1939 (CAERC box 63, folder 30); M. F. Detrick, *Fayette CEAR,* 1931 (CAERC box 81, folder 32); L. E. Archbold, *Adams CEARs,* 1933–34 (CAERC box 61, folders 19–20); E. M. Christen, *Benton CEAR,* 1935 (CAERC box 65, folder 20); W. O'Hara, *Hancock CEAR,* 1933 (CAERC box 91, folder 16); W. U. Rusk, *Blackford CEAR,* 1935 (CAERC box 66, folder 12).
19. A. V. Keesling, *Bartholomew CEAR,* 1937 (CAERC box 63, folder 26).
20. C. A. Alcorn, *Dearborn CEAR,* 1937 (CAERC box 75, folder 32).
21. R. Gwaltney, *Gibson CEAR,* 1936 (CAERC box 87, folder 6).
22. C. M. East, *Floyd CEAR,* 1932 (CAERC box 82, folder 33).
23. C. A. Nicholson, *Dubois CEAR,* 1936 (CAERC box 79, folder 45).
24. C. A. Nicholson, *Dubois CEAR,* 1937 (CAERC box 79A, folder 1).
25. J. H. Brown, *Elkhart CEAR,* 1938 (CAERC box 80, folder 36).
26. C. A. Nicholson, *Dubois CEAR,* 1937 (CAERC box 79A, folder 1).
27. W. U. Rusk, *Blackford CEAR,* 1935 (CAERC box 66, folder 12); L. E. Archbold, *Adams CEAR,* 1933 (CAERC box 61, folder 19); E. M. Christen, *Benton CEAR,* 1937 (CAERC box 65, folder 23); N. S. Hadley, *Franklin CEAR,* 1931 (CAERC box 84, folder 31).
28. C. A. Nicholson, *Dubois CEAR,* 1937 (CAERC box 79A, folder 1); R. E. Johnson, *Harrison CEARs,* 1931–32 (CAERC box 92, folders 26–27); C. M. East, *Floyd CEAR,* 1934 (CAERC box 82, folder 35); C. A. Alcorn, *Dearborn CEAR,* 1936 (CAERC box 75, folder 31).
29. R. E. Johnson, *Harrison CEAR,* 1932 (CAERC box 92, folder 27); M. F. Goode, *Harrison CEAR,* 1936 (CAERC box 92, folder 31).
30. M. F. Goode, *Harrison CEAR,* 1938 (CAERC box 92, folder 33); M. F. Goode, *Harrison CEAR,* 1936 (CAERC box 92, folder 31).
31. C. M. East, *Floyd CEAR,* 1932 (CAERC box 82, folder 33).

32. C. A. Nicholson, *Dubois CEAR,* 1936 (CAERC box 79, folder 45); M. F. Goode, *Harrison CEAR,* 1937 (CAERC box 92, folder 32).
33. J. H. Brown, *Elkhart CEAR,* 1938 (CAERC box 80, folder 36).
34. *Purdue University Department of Agricultural Extension Annual Report,* 1936, 75.
35. R. E. Johnson, *Harrison CEAR,* 1932 (CAERC box 92, folder 27); C. A. Nicholson, *Dubois CEAR,* 1936 (CAERC box 79, folder 45).
36. C. A. Nicholson, *Dubois CEAR,* 1937 (CAERC box 79A, folder 1); C. M. East, *Floyd CEAR,* 1934 (CAERC box 82, folder 35).
37. R. E. Johnson, *Harrison CEAR,* 1930 (CAERC box 92, folder 25); R. Gwaltney, *Gibson CEAR,* 1937 (CAERC box 87, folder 7).
38. A. J. Hesler, *Fountain CEAR,* 1939 (CAERC box 83, folder 35).
39. J. L. Morrow, *Crawford CEAR,* 1939 (CAERC box 73, folder 40).
40. W. W. Whitehead, *Boone CEAR,* 1939 (CAERC box 67, folder 4).
41. C. V. Kimmell, *Allen CEAR,* 1939 (CAERC box 62, folder 25).
42. Ibid.
43. R. L. Zell, *Franklin CEAR,* 1939 (CAERC box 84, folder 39).
44. J. H. Brown, *Elkhart CEAR,* 1938 (CAERC box 80, folder 36).
45. *Purdue University Department of Agricultural Extension Annual Report,* 1936, 75.
46. *Purdue University Department of Agricultural Extension Annual Report,* 1938, 83.
47. A. M. Bishea, *Daviess CEAR,* 1939 (CAERC box 74, folder 32).

CHAPTER 16

Hybrid Corn Is With Us to Stay Until Something Is Found to Take Its Place

1. H. L. Royce, *Clay CEAR,* 1930 (CAERC box 71, folder 42).
2. E. M. Christen, *Benton CEAR,* 1936 (CAERC box 65, folder 21); E. A. Johnson, *Cass CEAR,* 1938 (CAERC box 69, folder 43).
3. R. D. McHargue, *Clay CEAR,* 1937 (CAERC box 71, folder 49).
4. M. M. Hamilton, *Decatur CEAR,* 1938 (CAERC box 76, folder 44).
5. R. J. Maggart, *Carroll CEAR,* 1931 (CAERC box 68, folder 56); H. L. Royce, *Clay CEAR,* 1930 (CAERC box 71, folder 42).
6. W. U. Rusk, *Blackford CEAR,* 1939 (CAERC box 66, folder 16); C. A. Nicholson, *Dubois CEAR,* 1930 (CAERC box 79, folder 39); M. E. Cromer, *Delaware CEARs,* 1933 and 1937 (CAERC box 78, folders 35 and 39).
7. C. K. Whistler, *Boone CEAR,* 1938 (CAERC box 67, folder 3); A. V. Keesling, *Bartholomew CEAR,* 1939 (CAERC box 63, folder 30).
8. M. D. Butler, *Grant CEAR,* 1932 (CAERC box 87, folder 49).
9. H. L. Royce, *Clay CEAR,* 1930 (CAERC box 71, folder 42); W. A. Douglass, *Bartholomew CEAR,* 1930 (CAERC box 63, folder 17); M. E. Cromer, *Delaware CEAR,* 1937 (CAERC box 78, folder 39); J. L. Morrow, *Crawford CEAR,* 1939 (CAERC box 73, folder 40); R. Gwaltney, *Gibson CEAR,* 1937 (CAERC box 87, folder 7); F. E. Conder, *Grant CEAR,* 1937 (CAERC box 87, folder 54); H. R. Rosenbury, *Fulton CEAR,* 1931 (CAERC box 85, folder 39).
10. W. W. Whitehead, *Boone CEAR,* 1939 (CAERC box 67, folder 4).
11. R. J. Maggart, *Carroll CEAR,* 1939 (CAERC box 68, folder 65).
12. N. S. Hadley, *Franklin CEAR,* 1931 (CAERC box 84, folder 31); E. C. Parker, *Hamilton CEAR,* 1932 (CAERC box 89, folder 47); R. Gwaltney, *Gibson CEAR,* 1938 (CAERC box 87, folder 8).
13. A. Haseman, *Greene CEAR,* 1938 (CAERC box 89, folder 13); M. D. Butler, *Grant CEAR,* 1931 (CAERC box 87, folder 48).

14. C. B. Blosser, *Elkhart CEAR,* 1932 (CAERC box 80, folder 31); M. E. Cromer, *Delaware CEAR,* 1932 (CAERC box 78, folder 34); H. R. Rosenbury, *Fulton CEAR,* 1931 (CAERC box 85, folder 39); C. C. Madison, *Dearborn CEAR,* 1930 (CAERC box 75, folder 25).
15. I. L. Thurston, *Decatur CEAR,* 1931 (CAERC box 76, folder 37); J. E. Little, *Cass CEAR,* 1931 (CAERC box 69, folder 34).
16. J. E. Little, *Cass CEAR,* 1931 (CAERC box 69, folder 34).
17. I. L. Thurston, *Decatur CEAR,* 1931 (CAERC box 76, folder 37).
18. H. S. Heckard, *Allen CEAR,* 1933 (CAERC box 62, folder 18); H. R. Rosenbury, *Fulton CEARs,* 1930–31 (CAERC box 85, folders 38–39); E. C. Parker, *Hamilton CEAR,* 1931 (CAERC box 89, folder 46).
19. M. F. Detrick, *Fayette CEAR,* 1930 (CAERC box 81, folder 31).
20. C. C. Madison, *Dearborn CEAR,* 1931 (CAERC box 75, folder 26).
21. J. Jensen, *Daviess CEAR,* 1930 (CAERC box 74, folder 24).
22. A. V. Keesling, *Bartholomew CEAR,* 1932 (CAERC box 63, folder 19).
23. C. M. East, *Floyd CEAR,* 1936 (CAERC box 82, folder 37).
24. A. V. Keesling, *Bartholomew CEAR,* 1932 (CAERC box 63, folder 19).
25. J. E. Little, *Cass CEAR,* 1930 (CAERC box 69, folder 33); C. M. East, *Floyd CEAR,* 1936 (CAERC box 82, folder 37); C. C. Madison, *Dearborn CEAR,* 1930 (CAERC box 75, folder 25); W. G. Smith, *Henry CEAR,* 1932 (CAERC box 94, folder 1).
26. E. L. Hawk, *Hendricks CEAR,* 1931 (CAERC box 93, folder 18).
27. C. A. Nicholson, *Dubois CEAR,* 1930 (CAERC box 79, folder 39); T. E. Myers, *DeKalb CEAR,* 1936 (CAERC box 77, folder 31).
28. H. J. Yarling, *Hancock CEAR,* 1930 (CAERC box 91, folder 13).
29. E. M. Rowe, *Clinton CEAR,* 1936 (CAERC box 72, folder 42).
30. E. A. Johnson, *Cass CEAR,* 1936 (CAERC box 69, folder 39).
31. F. E. Conder, *Grant CEAR,* 1936 (CAERC box 87, folder 53).
32. H. J. Yarling, *Hancock CEAR,* 1930 (CAERC box 91, folder 13).
33. A. Haseman, *Greene CEAR,* 1936 (CAERC box 89, folder 11).
34. E. C. Parker, *Hamilton CEAR,* 1931 (CAERC box 89, folder 46); R. E. Johnson, *Harrison CEAR,* 1932 (CAERC box 92, folder 27).
35. E. C. Parker, *Hamilton CEAR,* 1930 (CAERC box 89, folder 45); H. R. Rosenbury, *Fulton CEAR,* 1931 (CAERC box 85, folder 39).
36. E. M. Christen, *Benton CEAR,* 1936 (CAERC box 65, folder 21); H. L. Royce, *Clay CEAR,* 1935 (CAERC box 71, folder 47); W. A. Douglass, *Bartholomew CEAR,* 1931 (CAERC box 63, folder 18); C. K. Whistler, *Boone CEAR,* 1937 (CAERC box 67, folder 2).
37. E. M. Christen, *Benton CEAR,* 1934 (CAERC box 65, folder 19).
38. A. V. Keesling, *Bartholomew CEAR,* 1935 (CAERC box 63, folder 22).
39. L. E. Archbold, *Adams CEAR,* 1935 (CAERC box 61, folder 21).
40. C. C. Madison, *Dearborn CEAR,* 1930 (CAERC box 75, folder 25).
41. M. E. Cromer, *Delaware CEAR,* 1928 (CAERC box 78, folder 30).
42. A. Haseman, *Greene CEAR,* 1936 (CAERC box 89, folder 11); E. M. Christen, *Benton CEAR,* 1937 (CAERC box 65, folder 23); A. J. Hesler, *Fountain CEAR,* 1938 (CAERC box 83, folder 34); E. E. Stinson, *DeKalb CEAR,* 1930 (CAERC box 77, folder 25); A. M. Bishea, *Daviess CEAR,* 1937 (CAERC box 74, folder 31).
43. J. L. Morrow, *Crawford CEAR,* 1938 (CAERC box 73, folder 39).
44. R. J. Maggart, *Carroll CEAR,* 1939 (CAERC box 68, folder 65).

45. W. U. Rusk, *Blackford CEAR,* 1939 (CAERC box 66, folder 16); E. M. Christen, *Benton CEAR,* 1937 (CAERC box 65, folder 23); N. S. Hadley, *Franklin CEAR,* 1931 (CAERC box 84, folder 31); M. M. Hamilton, *Decatur CEAR,* 1937 (CAERC box 76, folder 43); M. F. Detrick, *Fayette CEAR,* 1935 (CAERC box 81, folder 36); C. B. Blosser, *Elkhart CEAR,* 1930 (CAERC box 80, folder 29).
46. J. B. Todd, *Hamilton CEAR,* 1938 (CAERC box 90, folder 4).
47. J. E. Little, *Cass CEAR,* 1932 (CAERC box 69, folder 35).
48. H. S. Heckard, *Allen CEAR,* 1935 (CAERC box 62, folder 20); T. E. Myers, *DeKalb CEAR,* 1936 (CAERC box 77, folder 31).
49. C. K. Whistler, *Boone CEAR,* 1937 (CAERC box 67, folder 2).
50. A. V. Keesling, *Bartholomew CEAR,* 1937 (CAERC box 63, folder 25).
51. A. V. Keesling, *Bartholomew CEAR,* 1939 (CAERC box 63, folder 30).
52. E. S. Miles, *Clark CEAR,* 1936 (CAERC box 70, folder 44).
53. J. L. Morrow, *Crawford CEAR,* 1938 (CAERC box 73, folder 39); E. S. Miles, *Clark CEAR,* 1938 (CAERC box 70, folder 48); M. F. Goode, *Harrison CEAR,* 1938 (CAERC box 92, folder 33); N. S. Hadley, *Fulton CEAR,* 1935 (CAERC box 85, folder 43).
54. N. S. Hadley, *Franklin CEAR,* 1934 (CAERC box 84, folder 34).
55. E. M. Rowe, *Clinton CEAR,* 1936 (CAERC box 72, folder 42).
56. H. L. Royce, *Clay CEAR,* 1936 (CAERC box 71, folder 48); E. S. Miles, *Clark CEAR,* 1938 (CAERC box 70, folder 48); A. V. Keesling, *Bartholomew CEAR,* 1932 (CAERC box 63, folder 19); J. B. Todd, *Hamilton CEAR,* 1936 (CAERC box 90, folder 2); N. S. Hadley, *Fulton CEAR,* 1935 (CAERC box 85, folder 43); J. L. Morrow, *Crawford CEAR,* 1938 (CAERC box 73, folder 39); A. M. Bishea, *Daviess CEAR,* 1939 (CAERC box 74, folder 32); T. E. Myers, *DeKalb CEAR,* 1936 (CAERC box 77, folder 31).
57. F. E. Conder, *Grant CEAR,* 1938 (CAERC box 87, folder 55).
58. N. S. Hadley, *Fulton CEAR,* 1938 (CAERC box 85, folder 48).
59. D. S. Bishopp, *Fayette CEAR,* 1938 (CAERC box 81, folder 39).
60. A. J. Hesler, *Fountain CEAR,* 1936 (CAERC box 83, folder 32).
61. W. G. Smith, *Henry CEAR,* 1934 (CAERC box 94, folder 3); F. E. Conder, *Grant CEAR,* 1937 (CAERC box 87, folder 54).
62. A. J. Hesler, *Fountain CEAR,* 1936 (CAERC box 83, folder 32).
63. R. D. McHargue, *Clay CEAR,* 1938 (CAERC box 71, folder 50).
64. E. S. Miles, *Clark CEAR,* 1938 (CAERC box 70, folder 48).
65. E. C. Parker, *Hamilton CEAR,* 1930 (CAERC box 89, folder 45); A. Haseman, *Greene CEAR,* 1939 (CAERC box 89, folder 14); F. E. Conder, *Grant CEAR,* 1937 (CAERC box 87, folder 54); R. Gwaltney, *Gibson CEAR,* 1937 (CAERC box 87, folder 7); R. J. Maggart, *Carroll CEAR,* 1939 (CAERC box 68, folder 65); I. L. Thurston, *Decatur CEAR,* 1931 (CAERC box 76, folder 37); M. F. Detrick, *Fayette CEAR,* 1935 (CAERC box 81, folder 36).
66. H. L. Royce, *Clay CEAR,* 1936 (CAERC box 71, folder 48).
67. C. K. Whistler, *Boone CEARs,* 1937–38 (CAERC box 67, folders 2–3).
68. C. V. Kimmell, *Allen CEAR,* 1936 (CAERC box 62, folder 21).
69. J. L. Morrow, *Crawford CEAR,* 1939 (CAERC box 73, folder 40).
70. R. L. Zell, *Franklin CEAR,* 1938 (CAERC box 84, folder 38); A. J. Hesler, *Fountain CEAR,* 1939 (CAERC box 83, folder 35); C. E. Baker, *DeKalb CEAR,* 1939 (CAERC box 77, folder 34); E. M. Rowe, *Clinton CEAR,* 1938 (CAERC box 72, folder 44).
71. A. J. Hesler, *Fountain CEAR,* 1936 (CAERC box 83, folder 32).

72. A. J. Hesler, *Fountain CEAR,* 1937 (CAERC box 83, folder 33).
73. W. O'Hara, *Hancock CEAR,* 1935 (CAERC box 91, folder 18).
74. R. Gwaltney, *Gibson CEAR,* 1939 (CAERC box 87, folder 9); A. Haseman, *Greene CEAR,* 1939 (CAERC box 89, folder 14).
75. J. B. Todd, *Hamilton CEAR,* 1936 (CAERC box 90, folder 2).
76. N. S. Hadley, *Fulton CEAR,* 1937 (CAERC box 85, folder 45).
77. R. Gwaltney, *Gibson CEAR,* 1939 (CAERC box 87, folder 9).
78. E. A. Johnson, *Cass CEAR,* 1939 (CAERC box 69, folder 44).
79. R. L. Zell, *Franklin CEAR,* 1935 (CAERC box 84, folder 35).
80. A. Haseman, *Greene CEAR,* 1936 (CAERC box 89, folder 11).
81. F. E. Conder, *Grant CEAR,* 1937 (CAERC box 87, folder 54).
82. E. A. Johnson, *Cass CEAR,* 1935 (CAERC box 69, folder 38).
83. R. J. Maggart, *Carroll CEAR,* 1931 (CAERC box 68, folder 56); R. J. Maggart, *Carroll CEAR,* 1932 (CAERC box 68, folder 57).
84. A. M. Bishea, *Daviess CEAR,* 1936 (CAERC box 74, folder 30); E. M. Christen, *Benton CEAR,* 1936 (CAERC box 65, folder 21); W. A. Douglass, *Bartholomew CEAR,* 1930 (CAERC box 63, folder 17).
85. R. J. Maggart, *Carroll CEAR,* 1939 (CAERC box 68, folder 65).
86. E. M. Christen, *Benton CEAR,* 1935 (CAERC box 65, folder 20).
87. Ibid.
88. E. M. Christen, *Benton CEAR,* 1936 (CAERC box 65, folder 21).
89. Ibid.; E. M. Christen, *Benton CEAR,* 1937 (CAERC box 65, folder 23).
90. E. M. Christen, *Benton CEAR,* 1937 (CAERC box 65, folder 23).
91. W. G. Smith, *Henry CEAR,* 1935 (CAERC box 94, folder 4); M. E. Cromer, *Delaware CEAR,* 1939 (CAERC box 78, folder 41).
92. R. J. Maggart, *Carroll CEAR,* 1933 (CAERC box 68, folder 58).
93. A. J. Hesler, *Fountain CEAR,* 1937 (CAERC box 83, folder 33).
94. R. L. Zell, *Franklin CEAR,* 1938 (CAERC box 84, folder 38).
95. M. F. Detrick, *Fayette CEAR,* 1935 (CAERC box 81, folder 36); L. E. Archbold, *Adams CEAR,* 1938 (CAERC box 61, folder 24); C. A. Nicholson, *Dubois CEAR,* 1937 (CAERC box 79A, folder 1); A. M. Bishea, *Daviess CEAR,* 1936 (CAERC box 74, folder 30).
96. W. U. Rusk, *Blackford CEAR,* 1936 (CAERC box 66, folder 13); W. O'Hara, *Hancock CEAR,* 1937 (CAERC box 91, folder 20).
97. A. V. Keesling, *Bartholomew CEAR,* 1937 (CAERC box 63, folder 25).
98. W. U. Rusk, *Blackford CEARs,* 1936 and 1939 (CAERC box 66, folders 13 and 16); C. B. Blosser, *Elkhart CEAR,* 1930 (CAERC box 80, folder 29); C. A. Nicholson, *Dubois CEAR,* 1937 (CAERC box 79A, folder 1); F. E. Conder, *Grant CEAR,* 1935 (CAERC box 87, folder 52).
99. E. M. Rowe, *Clinton CEAR,* 1938 (CAERC box 72, folder 44).
100. G. L. Reed, *Benton CEAR,* 1930 (CAERC box 65, folder 15); A. Haseman, *Greene CEAR,* 1938 (CAERC box 89, folder 13); R. Gwaltney, *Gibson CEAR,* 1938 (CAERC box 87, folder 8).
101. R. Gwaltney, *Gibson CEAR,* 1939 (CAERC box 87, folder 9).
102. Stewart Seeds, *Stewart's 100th Anniversary* (video), 3:07, 26 July 2018, https://www.stewartseeds.com/en-us/history.html.
103. M. M. Hamilton, *Decatur CEAR,* 1936 (CAERC box 76, folder 42).

104. R. J. Maggart, *Carroll CEARs,* 1936 and 1938 (CAERC box 68, folders 61 and 63); E. M. Christen, *Benton CEAR,* 1937 (CAERC box 65, folder 23); A. V. Keesling, *Bartholomew CEAR,* 1936 (CAERC box 63, folder 24); W. G. Smith, *Henry CEAR,* 1936 (CAERC box 94, folder 5).

105. T. E. Myers, *DeKalb CEAR,* 1931 (CAERC box 77, folder 26); C. C. Madison, *Dearborn CEAR,* 1930 (CAERC box 75, folder 25).

106. W. G. Smith, *Henry CEAR,* 1932 (CAERC box 94, folder 1).

107. W. G. Smith, *Henry CEAR,* 1931 (CAERC box 93, folder 60); W. G. Smith, *Henry CEAR,* 1932 (CAERC box 94, folder 1).

108. L. E. Archbold, *Adams CEAR,* 1938 (CAERC box 61, folder 24); M. E. Cromer, *Delaware CEAR,* 1937 (CAERC box 78, folder 39); F. E. Conder, *Grant CEAR,* 1937 (CAERC box 87, folder 54).

109. E. S. Miles, *Clark CEAR,* 1935 (CAERC box 70, folder 42).

110. W. O'Hara, *Hancock CEAR,* 1939 (CAERC box 91, folder 22).

111. L. E. Archbold, *Adams CEAR,* 1939 (CAERC box 1, folder 1).

112. J. B. Todd, *Hamilton CEAR,* 1939 (CAERC box 90, folder 5).

113. A. A. Dowell and O. B. Jesness, "Economic Aspects of Hybrid Corn," *Journal of Farm Economics* 21, no. 2 (May 1939): 479, https://www.jstor.org/stable/1230816.

114. E. M. Christen, *Benton CEAR,* 1938 (CAERC box 65, folder 25).

115. W. U. Rusk, *Blackford CEAR,* 1938 (CAERC box 66, folder 115).

116. A. V. Keesling, *Bartholomew CEAR,* 1936 (CAERC box 63, folder 24); R. J. Maggart, *Carroll CEAR,* 1936 (CAERC box 68, folder 61).

117. A. Haseman, *Greene CEAR,* 1936 (CAERC box 89, folder 11).

118. A. Haseman, *Greene CEAR,* 1938 (CAERC box 89, folder 13).

CHAPTER 17

The Necessity of Knowing the Soil Before a Good Crop Can Be Produced

1. A. M. Bishea, *Daviess CEAR,* 1936 (CAERC box 74, folder 30); E. C. Parker, *Hamilton CEAR,* 1930 (CAERC box 89, folder 45); R. Gwaltney, *Gibson CEAR,* 1939 (CAERC box 87, folder 9); A. Haseman, *Greene CEAR,* 1935 (CAERC box 89, folder 10).

2. C. A. Alcorn, *Dearborn CEAR,* 1937 (CAERC box 75, folder 32).

3. A. Haseman, *Greene CEAR,* 1937 (CAERC box 89, folder 12).

4. W. O'Hara, *Hancock CEAR,* 1936 (CAERC box 91, folder 19); H. J. Yarling, *Hendricks CEAR,* 1939 (CAERC box 93, folder 22); E. S. Miles, *Clark CEAR,* 1938 (CAERC box 70, folder 48).

5. I. L. Thurston, *Decatur CEAR,* 1935 (CAERC box 76, folder 41).

6. V. J. Mann, *Clinton CEAR,* 1930 (CAERC box 72, folder 34); A. J. Hesler, *Fountain CEAR,* 1938 (CAERC box 83, folder 34).

7. W. A. Douglass, *Bartholomew CEAR,* 1930 (CAERC box 63, folder 17).

8. W. U. Rusk, *Blackford CEAR,* 1936 (CAERC box 66, folder 13); W. A. Douglass, *Bartholomew CEAR,* 1931 (CAERC box 63, folder 18); M. E. Cromer, *Delaware CEAR,* 1935 (CAERC box 78, folder 37).

9. W. A. Douglass, *Bartholomew CEAR,* 1931 (CAERC box 63, folder 18); M. D. Butler, *Grant CEAR,* 1930 (CAERC box 87, folder 47).

10. E. C. Parker, *Hamilton CEAR,* 1930 (CAERC box 89, folder 45).

11. A. V. Keesling, *Bartholomew CEAR,* 1937 (CAERC box 63, folder 25).
12. H. R. Rosenbury, *Fulton CEAR,* 1932 (CAERC box 85, folder 40); R. Gwaltney, *Gibson CEAR,* 1939 (CAERC box 87, folder 9).
13. R. D. McHargue, *Clay CEAR,* 1937 (CAERC box 71, folder 49); R. E. Johnson, *Harrison CEAR,* 1932 (CAERC box 92, folder 27); C. A. Alcorn, *Floyd CEAR,* 1931 (CAERC box 82, folder 32).
14. Frederick Whitford, *Scattering the Seeds of Knowledge: The Words and Works of Indiana's Pioneer County Extension Agents* (West Lafayette, IN: Purdue University Press, 2017), 420; H. J. Yarling, *Hancock CEAR,* 1930 (CAERC box 91, folder 13); C. A. Nicholson, *Dubois CEAR,* 1930 (CAERC box 79, folder 39); E. A. Johnson, *Cass CEAR,* 1935 (CAERC box 69, folder 38); M. E. Cromer, *Delaware CEAR,* 1938 (CAERC box 78, folder 40); T. E. Myers, *DeKalb CEAR,* 1933 (CAERC box 77, folder 28).
15. J. Jensen, *Daviess CEAR,* 1930 (CAERC box 74, folder 24); H. J. Yarling, *Hendricks CEAR,* 1938 (CAERC box 93, folder 21); W. O'Hara, *Hancock CEAR,* 1937 (CAERC box 91, folder 20).
16. J. E. Little, *Cass CEAR,* 1931 (CAERC box 69, folder 34); I. L. Thurston, *Decatur CEAR,* 1935 (CAERC box 76, folder 41).
17. A. M. Bishea, *Daviess CEAR,* 1939 (CAERC box 74, folder 32).
18. C. A. Nicholson, *Dubois CEAR,* 1930 (CAERC box 79, folder 39).
19. M. D. Butler, *Grant CEAR,* 1931 (CAERC box 87, folder 48).
20. C. A. Alcorn, *Dearborn CEAR,* 1936 (CAERC box 75, folder 31); "Purdue Soil Test Train on B. & O. August 3 to 15," *Lafayette Journal and Courier* (Lafayette, IN), 27 June 1936, 5; *Purdue University Department of Agricultural Extension Annual Report,* 1937, 37.
21. "Purdue Soil Test Train on B. & O.," 5; "Purdue Train Is at Fountaintown," *Hancock Democrat* (Greenfield, IN), 20 Aug. 1936, 1; "B. & O. Soil Train Has Busy Day Here," *Mitchell Tribune* (Mitchell, IN), 13 Aug. 1936, 1.
22. "Was Surprised Interest Shown: Purdue Specialist with Soil Improvement Train Speaks of Hancock Farmers," *Greenfield Daily Reporter* (Greenfield, IN), 18 Aug. 1936, 1.
23. *Purdue University Department of Agricultural Extension Annual Report,* 1937, 37.
24. V. J. Mann, *Clinton CEAR,* 1932 (CAERC box 72, folder 36); C. A. Alcorn, *Dearborn CEAR,* 1937 (CAERC box 75, folder 32); E. M. Rowe, *Clinton CEAR,* 1934 (CAERC box 72, folder 38).
25. M. E. Cromer, *Delaware CEAR,* 1929 (CAERC box 78, folder 31); I. L. Thurston, *Decatur CEAR,* 1931 (CAERC box 76, folder 37).
26. H. L. Royce, *Clay CEAR,* 1935 (CAERC box 71, folder 47); R. J. Maggart, *Carroll CEAR,* 1938 (CAERC box 68, folder 63); J. E. Little, *Cass CEAR,* 1930 (CAERC box 69, folder 33).
27. J. L. Morrow, *Crawford CEAR,* 1939 (CAERC box 73, folder 40).
28. V. J. Mann, *Clinton CEAR,* 1932 (CAERC box 72, folder 36).
29. R. J. Maggart, *Carroll CEARs,* 1932 and 1935 (CAERC box 68, folders 57 and 60); R. E. Johnson, *Harrison CEAR,* 1932 (CAERC box 92, folder 27); M. E. Cromer, *Delaware CEAR,* 1935 (CAERC box 78, folder 37); L. E. Archbold, *Adams CEAR,* 1938 (CAERC box 61, folder 24).
30. M. D. Butler, *Grant CEAR,* 1930 (CAERC box 87, folder 47).
31. C. V. Kimmell, *Allen CEAR,* 1937 (CAERC box 62, folder 23).
32. E. A. Johnson, *Cass CEAR,* 1936 (CAERC box 69, folder 39).
33. C. V. Kimmell, *Allen CEAR,* 1937 (CAERC box 62, folder 23).

34. A. V. Keesling, *Bartholomew CEAR,* 1938 (CAERC box 63, folder 28); V. J. Mann, *Clinton CEAR,* 1930 (CAERC box 72, folder 34).
35. R. J. Maggart, *Carroll CEAR,* 1931 (CAERC box 68, folder 56).
36. A. M. Bishea, *Daviess CEAR,* 1939 (CAERC box 74, folder 32); H. J. Yarling, *Hendricks CEAR,* 1938 (CAERC box 93, folder 21).
37. M. E. Cromer, *Delaware CEAR,* 1935 (CAERC box 78, folder 37).
38. H. R. Rosenbury, *Fulton CEAR,* 1932 (CAERC box 85, folder 40).
39. V. D. Sexson, *Greene CEAR,* 1930 (CAERC box 89, folder 5).
40. C. A. Nicholson, *Dubois CEAR,* 1932 (CAERC box 79, folder 41); E. C. Parker, *Hamilton CEAR,* 1930 (CAERC box 89, folder 45); E. S. Miles, *Clark CEAR,* 1933 (CAERC box 70, folder 40); M. E. Cromer, *Delaware CEAR,* 1932 (CAERC box 78, folder 34).
41. C. L. Dyer, *Gibson CEAR,* 1932 (CAERC box 87, folder 3); M. F. Detrick, *Fayette CEAR,* 1931 (CAERC box 81, folder 32); I. L. Thurston, *Decatur CEAR,* 1931 (CAERC box 76, folder 37).
42. A. Haseman, *Greene CEAR,* 1937 (CAERC box 89, folder 12).
43. A. J. Hesler, *Fountain CEAR,* 1938 (CAERC box 83, folder 34).
44. A. J. Hesler, *Fountain CEAR,* 1939 (CAERC box 83, folder 35).
45. R. L. Zell, *Franklin CEAR,* 1938 (CAERC box 84, folder 38).
46. D. S. Bishopp, *Fayette CEAR,* 1938 (CAERC box 81, folder 39).
47. R. D. McHargue, *Clay CEAR,* 1937 (CAERC box 71, folder 49).

CHAPTER 18

Growing Wheat Is One Thing and Growing Quality Wheat Is Another

1. H. L. Royce, *Clay CEAR,* 1930 (CAERC box 71, folder 42); C. A. Nicholson, *Dubois CEAR,* 1930 (CAERC box 79, folder 39); C. A. Alcorn, *Dearborn CEAR,* 1933 (CAERC box 75, folder 28); F. E. Conder, *Grant CEARs,* 1937–38 (CAERC box 87, folders 54–55).
2. R. J. Maggart, *Carroll CEAR,* 1931 (CAERC box 68, folder 56).
3. M. F. Detrick, *Fayette CEAR,* 1933 (CAERC box 81, folder 34).
4. W. O'Hara, *Hancock CEAR,* 1935 (CAERC box 91, folder 18).
5. H. R. Rosenbury, *Fulton CEAR,* 1930 (CAERC box 85, folder 38); F. E. Conder, *Grant CEAR,* 1937 (CAERC box 87, folder 54).
6. W. O'Hara, *Hancock CEAR,* 1932 (CAERC box 91, folder 15); R. E. Johnson, *Harrison CEARs,* 1931–32 (CAERC box 92, folders 26–27).
7. A. J. Hesler, *Fountain CEAR,* 1938 (CAERC box 83, folder 34).
8. F. E. Conder, *Grant CEAR,* 1938 (CAERC box 87, folder 55).
9. C. L. Dyer, *Gibson CEAR,* 1932 (CAERC box 87, folder 3).
10. F. E. Conder, *Grant CEAR,* 1938 (CAERC box 87, folder 55); E. S. Miles, *Clark CEARs,* 1938–39 (CAERC box 70, folders 48 and 50); R. J. Maggart, *Carroll CEAR,* 1939 (CAERC box 68, folder 65).
11. E. C. Parker, *Hamilton CEAR,* 1930 (CAERC box 89, folder 45).
12. J. Jensen, *Daviess CEAR,* 1931 (CAERC box 74, folder 25).
13. A. M. Bishea, *Daviess CEAR,* 1937 (CAERC box 74, folder 31); E. S. Miles, *Clark CEAR,* 1938 (CAERC box 70, folder 48).
14. E. S. Miles, *Clark CEARs,* 1934–35 (CAERC box 70, folders 41–42).
15. C. A. Nicholson, *Dubois CEAR,* 1931 (CAERC box 79, folder 40); H. L. Royce, *Clay CEAR,* 1930 (CAERC box 71, folder 42); R. J. Maggart, *Carroll CEAR,* 1931 (CAERC box 68, folder 56).

16. R. Gwaltney, *Gibson CEAR,* 1936 (CAERC box 87, folder 6).
17. V. J. Mann, *Clinton CEAR,* 1930 (CAERC box 72, folder 34); R. Gwaltney, *Gibson CEAR,* 1936 (CAERC box 87, folder 6); J. H. Brown, *Elkhart CEAR,* 1938 (CAERC box 80, folder 36); E. A. Johnson, *Cass CEAR,* 1938 (CAERC box 69, folder 43); N. S. Hadley, *Franklin CEAR,* 1931 (CAERC box 84, folder 31); R. J. Maggart, *Carroll CEAR,* 1931 (CAERC box 68, folder 56).
18. J. E. Little, *Cass CEAR,* 1931 (CAERC box 69, folder 34).
19. A. V. Keesling, *Bartholomew CEAR,* 1938 (CAERC box 63, folder 28); W. A. Douglass, *Bartholomew CEAR,* 1930 (CAERC box 63, folder 17); R. J. Maggart, *Carroll CEAR,* 1931 (CAERC box 68, folder 56).
20. H. R. Rosenbury, *Fulton CEAR,* 1930 (CAERC box 85, folder 38).
21. C. A. Alcorn, *Floyd CEAR,* 1930 (CAERC box 82, folder 31).
22. J. Jensen, *Daviess CEAR,* 1930 (CAERC box 74, folder 24).
23. J. E. Little, *Cass CEAR,* 1930 (CAERC box 69, folder 33).
24. M. F. Detrick, *Fayette CEAR,* 1931 (CAERC box 81, folder 32).
25. Ibid.
26. R. J. Maggart, *Carroll CEAR,* 1931 (CAERC box 68, folder 56).
27. Ibid.
28. J. L. Morrow, *Crawford CEAR,* 1938 (CAERC box 73, folder 39).
29. M. F. Detrick, *Fayette CEAR,* 1931 (CAERC box 81, folder 32).
30. J. Jensen, *Daviess CEAR,* 1931 (CAERC box 74, folder 25); J. E. Little, *Cass CEAR,* 1931 (CAERC box 69, folder 34).
31. R. Gwaltney, *Gibson CEAR,* 1936 (CAERC box 87, folder 6); R. L. Zell, *Franklin CEAR,* 1936 (CAERC box 84, folder 36).
32. R. Gwaltney, *Gibson CEAR,* 1937 (CAERC box 87, folder 7).
33. A. M. Bishea, *Daviess CEAR,* 1935 (CAERC box 74, folder 29); J. Jensen, *Daviess CEAR,* 1932 (CAERC box 74, folder 26).
34. C. L. Dyer, *Gibson CEAR,* 1932 (CAERC box 87, folder 3).
35. A. M. Bishea, *Daviess CEAR,* 1935 (CAERC box 74, folder 29); C. L. Dyer, *Gibson CEAR,* 1932 (CAERC box 87, folder 3).
36. A. Haseman, *Greene CEAR,* 1937 (CAERC box 89, folder 12); W. O'Hara, *Hancock CEAR,* 1937 (CAERC box 91, folder 20); C. E. Skiver, "Wheat and Chaff," *Boonville Enquirer* (Boonville, IN), 27 Mar. 1936, 5; "Farmers, Grain Dealers to Buy Grain Cleaner," *Jasper Herald* (Jasper, IN), 12 June 1936, 1.
37. "New Approach to Seed Wheat Cleaning," *Extension Service Review* 5, no. 11 (Nov. 1934): 173, https://archive.org/details/sim_extension-review_1934-11_5_11/page/n16/mode/1up.
38. M. E. Cromer, *Delaware CEAR,* 1939 (CAERC box 78, folder 41); A. Haseman, *Greene CEAR,* 1937 (CAERC box 89, folder 12).
39. *Extension Service Review*, "New Approach to Seed Wheat Cleaning," 173.
40. C. A. Nicholson, *Dubois CEAR,* 1936 (CAERC box 79, folder 45); "Farmers, Grain Dealers to Buy Grain Cleaner," 1.
41. "Portable Seed Cleaner to Be Seen at Purdue," *Culver Citizen* (Culver, IN), 8 Jan. 1936, 11.
42. "Cleaning Seed Wheat," *Boonville Enquirer* (Boonville, IN), 21 Aug. 1936, 7; C. A. Nicholson, *Dubois CEAR,* 1936 (CAERC box 79, folder 45); R. L. Zell, *Franklin CEAR,* 1939 (CAERC box 84, folder 39); H. J. Yarling, *Hendricks CEAR,* 1939 (CAERC box 93, folder 22); W. O'Hara, *Hancock CEAR,* 1932 (CAERC box 91, folder 15); R. L. Zell, *Franklin CEAR,* 1937 (CAERC box 84, folder 37).

43. W. O'Hara, *Hancock CEAR,* 1937 (CAERC box 91, folder 20).
44. *Purdue University Department of Agricultural Extension Annual Report,* 1938, 39.
45. *Purdue University Department of Agricultural Extension Annual Report,* 1939, 45.
46. "From Defeat to Victory, 1930–1949," Growing a Nation (website), Utah State University National Center for Agricultural Literacy, accessed 13 May 2022, https://growinganation.org/content/show-content/from_defeat_to_victory/.
47. A. J. Hesler, *Fountain CEAR,* 1939 (CAERC box 83, folder 35).
48. "Great Wheat Loan Program Prepared to Boost Market," *Lafayette Journal and Courier* (Lafayette, IN), 9 May 1938, 5.
49. "Crop Insurance Plan Ready to Go into Effect," *Evening Republican* (Columbus, IN), 29 July 1938, 1; "Wheat Insurance Premiums Due Soon," *Argos Reflector* (Argos, IN), 9 Nov. 1938, 1; "Great Wheat Loan Program Prepared to Boost Market," 5.
50. "Crop Insurance Plan Ready to Go into Effect," 1; "Wheat Insurance Premiums Due Soon," 1; "Over 200 Taking Wheat Insurance," *Star Press* (Muncie, IN), 12 Oct. 1938, 7.
51. "Crop Insurance Plan Ready to Go into Effect," 1; "Great Wheat Loan Program Prepared to Boost Market," 5; "Over 200 Taking Wheat Insurance," 7.
52. "Crop Insurance Plan Ready to Go into Effect," 1.
53. W. O'Hara, *Hancock CEAR,* 1938 (CAERC box 91, folder 21).
54. Ibid.
55. M. E. Cromer, *Delaware CEAR,* 1939 (CAERC box 78, folder 41).
56. C. A. Nicholson, *Dubois CEAR,* 1938 (CAERC box 79A, folder 2); A. V. Keesling, *Bartholomew CEAR,* 1938 (CAERC box 63, folder 28); M. M. Hamilton, *Decatur CEAR,* 1938 (CAERC box 76, folder 44); R. L. Zell, *Franklin CEAR,* 1939 (CAERC box 84, folder 39).
57. W. W. Whitehead, *Boone CEAR,* 1939 (CAERC box 67, folder 4).
58. C. A. Alcorn, *Dearborn CEAR,* 1939 (CAERC box 75, folder 34).
59. A. V. Keesling, *Bartholomew CEAR,* 1939 (CAERC box 63, folder 30).
60. "Over 200 Taking Wheat Insurance," 7.

CHAPTER 19

Farmers on the Lookout for Some New or Different Crop That Offers More Promise for Fair Returns

1. C. L. Dyer, *Gibson CEAR,* 1932 (CAERC box 87, folder 3).
2. L. E. Archbold, *Adams CEAR,* 1931 (CAERC box 61, folder 17); C. B. Blosser, *Elkhart CEAR,* 1931 (CAERC box 80, folder 30); L. E. Archbold, *Adams CEAR,* 1933 (CAERC box 61, folder 19).
3. C. A. Alcorn, *Floyd CEAR,* 1930 (CAERC box 82, folder 31); E. S. Miles, *Clark CEAR,* 1938 (CAERC box 70, folder 48).
4. C. C. Madison, *Dearborn CEAR,* 1932 (CAERC box 75, folder 27).
5. C. M. East, *Floyd CEAR,* 1934 (CAERC box 82, folder 35).
6. A. V. Keesling, *Bartholomew CEAR,* 1937 (CAERC box 63, folder 26); G. L. Reed, *Benton CEAR,* 1931 (CAERC box 65, folder 16).
7. G. L. Reed, *Benton CEAR,* 1931 (CAERC box 65, folder 16).
8. C. L. Dyer, *Gibson CEAR,* 1932 (CAERC box 87, folder 3).
9. R. Gwaltney, *Gibson CEAR,* 1937 (CAERC box 87, folder 7).
10. A. M. Bishea, *Daviess CEAR,* 1939 (CAERC box 74, folder 32).
11. G. L. Reed, *Benton CEAR,* 1930 (CAERC box 65, folder 15); G. L. Reed, *Benton CEAR,* 1931 (CAERC box 65, folder 16).

12. C. B. Blosser, *Elkhart CEAR,* 1930 (CAERC box 80, folder 29).
13. H. S. Heckard, *Allen CEAR,* 1931 (CAERC box 62, folder 16).
14. L. E. Archbold, *Adams CEAR,* 1933 (CAERC box 61, folder 19).
15. Ibid.; W. A. Douglass, *Bartholomew CEAR,* 1930 (CAERC box 63, folder 17); G. L. Reed, *Benton CEAR,* 1931 (CAERC box 65, folder 16).
16. A. V. Keesling, *Bartholomew CEAR,* 1938 (CAERC box 63, folder 28).
17. E. C. Parker, *Hamilton CEAR,* 1930 (CAERC box 89, folder 45).
18. W. A. Douglass, *Bartholomew CEAR,* 1930 (CAERC box 63, folder 17).
19. "Much Smaller Tomato Yield Here This Year," *Evening Republican* (Columbus, IN), 5 June 1931, 1.
20. T. E. Myers, *DeKalb CEAR,* 1932 (CAERC box 77, folder 27).
21. R. J. Maggart, *Carroll CEAR,* 1937 (CAERC box 68, folder 63).
22. W. G. Smith, *Henry CEAR,* 1930 (CAERC box 93, folder 59).
23. W. U. Rusk, *Blackford CEAR,* 1937 (CAERC box 66, folder 14).
24. W. G. Smith, *Henry CEAR,* 1935 (CAERC box 94, folder 4).
25. R. J. Maggart, *Carroll CEAR,* 1937 (CAERC box 68, folder 63).
26. U.S. Department of Agriculture and Purdue University Agricultural Experiment Station, *Indiana Crops and Livestock: Annual Crop Summary, 1932,* Bulletin No. 87 (West Lafayette, IN: U.S. Department of Agriculture and Purdue University, Dec. 1932), 5, https://www.nass.usda.gov/Statistics_by_State/Indiana/Publications/Annual_Statistical_Bulletin/hist_pdf/as_1932_ebook.pdf.
27. W. A. Douglass, *Bartholomew CEAR,* 1930 (CAERC box 63, folder 17).
28. C. A. Nicholson, *Dubois CEAR,* 1930 (CAERC box 79, folder 39).
29. A. J. Hesler, *Fountain CEAR,* 1932 (CAERC box 83, folder 28).
30. W. O'Hara, *Hancock CEAR,* 1937 (CAERC box 91, folder 20); W. U. Rusk, *Blackford CEAR,* 1935 (CAERC box 66, folder 12); A. M. Bishea, *Daviess CEAR,* 1935 (CAERC box 74, folder 29); E. C. Parker, *Hamilton CEAR,* 1931 (CAERC box 89, folder 46).
31. R. O. Cole, *Clark CEAR,* 1930 (CAERC box 70, folder 37).
32. G. L. Reed, *Benton CEAR,* 1931 (CAERC box 65, folder 16).
33. R. J. Maggart, *Carroll CEAR,* 1938 (CAERC box 68, folder 63).
34. W. U. Rusk, *Blackford CEAR,* 1935 (CAERC box 66, folder 12).
35. W. G. Smith, *Henry CEAR,* 1935 (CAERC box 94, folder 4).
36. C. K. Whistler, *Boone CEAR,* 1938 (CAERC box 67, folder 3).
37. Ibid.
38. L. E. Archbold, *Adams CEAR,* 1938 (CAERC box 61, folder 24).
39. W. U. Rusk, *Blackford CEAR,* 1935 (CAERC box 66, folder 12).
40. W. U. Rusk, *Blackford CEAR,* 1939 (CAERC box 66, folder 16).
41. W. U. Rusk, *Blackford CEAR,* 1937 (CAERC box 66, folder 14).
42. E. S. Miles, *Clark CEAR,* 1938 (CAERC box 70, folder 48); M. E. Cromer, *Delaware CEAR,* 1933 (CAERC box 78, folder 35); C. A. Nicholson, *Dubois CEAR,* 1932 (CAERC box 79, folder 41); L. E. Archbold, *Adams CEAR,* 1933 (CAERC box 61, folder 19); V. J. Mann, *Clinton CEAR,* 1930 (CAERC box 72, folder 34).
43. C. L. Dyer, *Gibson CEAR,* 1932 (CAERC box 87, folder 3).
44. H. R. Rosenbury, *Fulton CEAR,* 1933 (CAERC box 85, folder 41); E. E. Stinson, *DeKalb CEAR,* 1930 (CAERC box 77, folder 25).
45. C. B. Blosser, *Elkhart CEAR,* 1931 (CAERC box 80, folder 30).

46. T. E. Myers, *DeKalb CEAR,* 1934 (CAERC box 77, folder 29); Fay C. Gaylord and Harry M. Cleaver, *Marketing Indiana Potatoes: Local Supply and Imports,* Purdue University Agricultural Experiment Station Bulletin No. 361 (West Lafayette, IN: Purdue University, Mar. 1932), 2, https://babel.hathitrust.org/cgi/pt?id=uiug.30112019548558&view=1up&seq=6.
47. M. E. Cromer, *Delaware CEAR,* 1935 (CAERC box 78, folder 37); H. S. Heckard, *Allen CEAR,* 1933 (CAERC box 62, folder 18).
48. L. E. Archbold, *Adams CEAR,* 1933 (CAERC box 61, folder 19).
49. H. S. Heckard, *Allen CEARs,* 1930 and 1933 (CAERC box 62, folders 15 and 18).
50. Gaylord and Cleaver, *Marketing Indiana Potatoes,* 7.
51. H. S. Heckard, *Allen CEAR,* 1933 (CAERC box 62, folder 18).
52. W. G. Smith, *Henry CEAR,* 1930 (CAERC box 93, folder 59); H. S. Heckard, *Allen CEAR,* 1930 (CAERC box 62, folder 15); H. R. Rosenbury, *Fulton CEAR,* 1933 (CAERC box 85, folder 41).
53. H. S. Heckard, *Allen CEAR,* 1930 (CAERC box 62, folder 15).
54. N. S. Hadley, *Fulton CEAR,* 1934 (CAERC box 85, folder 42).
55. Gaylord and Cleaver, *Marketing Indiana Potatoes,* 5.

CHAPTER 20

Not More Cows but Fewer and Better Dairy Cows Is the Imperative Need

1. J. B. Todd, *Hamilton CEAR,* 1937 (CAERC box 90, folder 3).
2. U.S. Department of Agriculture, "Specified Livestock and Livestock Products on Farms and Ranches" (PDF), in Indiana Census of Agriculture, 1935, p. 148, at Census of Agriculture Historical Archive (website), accessed 18 May 2022, https://agcensus.library.cornell.edu/census_year/1935-census/.
3. A. T. Marvel, *Elkhart CEARs,* 1934 and 1936 (CAERC box 80, folders 32 and 34); C. K. Whistler, *Boone CEAR,* 1937 (CAERC box 67, folder 2).
4. A. T. Marvel, *Elkhart CEAR,* 1936 (CAERC box 80, folder 34); A. T. Marvel, *Elkhart CEAR,* 1934 (CAERC box 80, folder 32).
5. C. K. Whistler, *Boone CEAR,* 1937 (CAERC box 67, folder 2); J. Jensen, *Daviess CEAR,* 1930 (CAERC box 74, folder 24); J. L. Morrow, *Crawford CEAR,* 1936 (CAERC box 73, folder 37).
6. N. S. Hadley, *Franklin CEAR,* 1932 (CAERC box 84, folder 32); J. Jensen, *Daviess CEAR,* 1932 (CAERC box 74, folder 26); W. G. Smith, *Henry CEAR,* 1932 (CAERC box 94, folder 1).
7. E. C. Parker, *Hamilton CEAR,* 1932 (CAERC box 89, folder 47).
8. R. E. Johnson, *Harrison CEAR,* 1932 (CAERC box 92, folder 27); R. E. Johnson, *Harrison CEAR,* 1931 (CAERC box 92, folder 26).
9. C. M. East, *Floyd CEAR,* 1932 (CAERC box 82, folder 33).
10. Ibid.
11. W. G. Smith, *Henry CEAR,* 1935 (CAERC box 94, folder 4).
12. R. D. McHargue, *Clay CEAR,* 1937 (CAERC box 71, folder 49); A. T. Marvel, *Elkhart CEAR,* 1936 (CAERC box 80, folder 34).
13. E. A. Johnson, *Cass CEAR,* 1936 (CAERC box 69, folder 39); W. U. Rusk, *Blackford CEAR,* 1937 (CAERC box 66, folder 14); R. E. Johnson, *Harrison CEAR,* 1931 (CAERC box 92, folder 26).
14. C. C. Madison, *Dearborn CEAR,* 1930 (CAERC box 75, folder 25).
15. E. M. Rowe, *Clinton CEAR,* 1935 (CAERC box 72, folder 39).

16. A. Haseman, *Greene CEAR,* 1938 (CAERC box 89, folder 13); C. A. Nicholson, *Dubois CEAR,* 1935 (CAERC box 79, folder 44).
17. J. Jensen, *Daviess CEAR,* 1930 (CAERC box 74, folder 24); W. U. Rusk, *Blackford CEAR,* 1937 (CAERC box 66, folder 14).
18. J. Jensen, *Daviess CEAR,* 1930 (CAERC box 74, folder 24); A. J. Hesler, *Fountain CEAR,* 1930 (CAERC box 83, folder 26); E. A. Johnson, *Cass CEAR,* 1936 (CAERC box 69, folder 39).
19. E. C. Parker, *Hamilton CEAR,* 1930 (CAERC box 89, folder 45).
20. M. F. Detrick, *Fayette CEAR,* 1933 (CAERC box 81, folder 34).
21. A. J. Hesler, *Fountain CEAR,* 1930 (CAERC box 83, folder 26).
22. Ibid.
23. C. C. Madison, *Dearborn CEAR,* 1930 (CAERC box 75, folder 25).
24. A. J. Hesler, *Fountain CEAR,* 1930 (CAERC box 83, folder 26).
25. Mitchell V. Palmer and W. Ray Waters, "Bovine Tuberculosis and the Establishment of an Eradication Program in the United States: Role of Veterinarians," *Veterinary Medicine International* 2011, article ID 816345 (17 May 2011): section 10, "Conception of a National Eradication Program," https://doi.org/10.4061/2011/816345.
26. E. A. Johnson, *Cass CEAR,* 1938 (CAERC box 69, folder 43).
27. W. A. Douglass, *Bartholomew CEAR,* 1930 (CAERC box 63, folder 17).
28. E. C. Parker, *Hamilton CEAR,* 1931 (CAERC box 89, folder 46); E. A. Johnson, *Cass CEAR,* 1938 (CAERC box 69, folder 43).
29. R. E. Johnson, *Harrison CEAR,* 1931 (CAERC box 92, folder 26).
30. L. E. Archbold, *Adams CEAR,* 1932 (CAERC box 61, folder 18).
31. C. A. Nicholson, *Dubois CEAR,* 1935 (CAERC box 79, folder 44); E. M. Rowe, *Clinton CEAR,* 1935 (CAERC box 72, folder 39); E. M. Christen, *Benton CEAR,* 1935 (CAERC box 65, folder 20); A. M. Bishea, *Daviess CEAR,* 1935 (CAERC box 74, folder 29).
32. R. L. Zell, *Franklin CEAR,* 1935 (CAERC box 84, folder 35); W. U. Rusk, *Blackford CEAR,* 1934 (CAERC box 66, folder 11); W. U. Rusk, *Blackford CEAR,* 1935 (CAERC box 66, folder 12).
33. F. E. Conder, *Grant CEAR,* 1935 (CAERC box 87, folder 52); R. L. Zell, *Franklin CEAR,* 1936 (CAERC box 84, folder 36); E. M. Rowe, *Clinton CEAR,* 1935 (CAERC box 72, folder 39); R. J. Maggart, *Carroll CEAR,* 1939 (CAERC box 68, folder 65).
34. E. A. Johnson, *Cass CEAR,* 1938 (CAERC box 69, folder 43).
35. C. M. East, *Floyd CEAR,* 1934 (CAERC box 82, folder 35); W. O'Hara, *Hancock CEAR,* 1935 (CAERC box 91, folder 18); L. E. Archbold, *Adams CEAR,* 1935 (CAERC box 61, folder 21).
36. C. A. Nicholson, *Dubois CEAR,* 1935 (CAERC box 79, folder 44); A. Haseman, *Greene CEAR,* 1938 (CAERC box 89, folder 13); H. L. Royce, *Clay CEAR,* 1936 (CAERC box 71, folder 48); N. S. Hadley, *Fulton CEAR,* 1937 (CAERC box 85, folder 45); E. A. Johnson, *Cass CEAR,* 1938 (CAERC box 69, folder 43).
37. E. M. Christen, *Benton CEAR,* 1935 (CAERC box 65, folder 20).
38. E. A. Johnson, *Cass CEAR,* 1938 (CAERC box 69, folder 43).
39. A. M. Bishea, *Daviess CEAR,* 1939 (CAERC box 74, folder 32); E. A. Johnson, *Cass CEAR,* 1939 (CAERC box 69, folder 44).
40. E. A. Johnson, *Cass CEAR,* 1939 (CAERC box 69, folder 44).
41. A. M. Bishea, *Daviess CEAR,* 1939 (CAERC box 74, folder 32).
42. J. B. Todd, *Hamilton CEAR,* 1939 (CAERC box 90, folder 5).
43. J. Jensen, *Daviess CEAR,* 1932 (CAERC box 74, folder 26); E. A. Johnson, *Cass CEAR,* 1935 (CAERC box 69, folder 38).

44. J. E. Little, *Cass CEAR,* 1932 (CAERC box 69, folder 35); D. S. Bishopp, *Fayette CEAR,* 1938 (CAERC box 81, folder 39).
45. H. L. Royce, *Clay CEAR,* 1930 (CAERC box 71, folder 42).
46. J. E. Little, *Cass CEAR,* 1932 (CAERC box 69, folder 35).
47. W. A. Douglass, *Bartholomew CEAR,* 1931 (CAERC box 63, folder 18).
48. E. L. Hawk, *Hendricks CEAR,* 1931 (CAERC box 93, folder 18).
49. M. E. Cromer, *Delaware CEAR,* 1933 (CAERC box 78, folder 35); E. L. Hawk, *Hendricks CEAR,* 1931 (CAERC box 93, folder 18).
50. M. E. Cromer, *Delaware CEAR,* 1933 (CAERC box 78, folder 35).
51. W. G. Smith, *Henry CEAR,* 1930 (CAERC box 93, folder 59).
52. I. L. Thurston, *Decatur CEAR,* 1931 (CAERC box 76, folder 37).
53. Ibid.
54. E. S. Miles, *Clark CEAR,* 1938 (CAERC box 70, folder 48).
55. L. E. Archbold, *Adams CEAR,* 1933 (CAERC box 61, folder 19).
56. L. E. Archbold, *Adams CEAR,* 1937 (CAERC box 61, folder 23).
57. R. J. Maggart, *Carroll CEAR,* 1933 (CAERC box 68, folder 58).
58. H. L. Royce, *Clay CEAR,* 1931 (CAERC box 71, folder 43); G. L. Reed, *Benton CEAR,* 1931 (CAERC box 65, folder 16); J. E. Little, *Cass CEAR,* 1930 (CAERC box 69, folder 33).
59. E. L. Hawk, *Hendricks CEAR,* 1932 (CAERC box 93, folder 19).
60. "125 Dairymen Gain Club Membership," *Indianapolis Star,* 12 Jan. 1929, 16.
61. E. C. Parker, *Hamilton CEAR,* 1933 (CAERC box 89, folder 48); C. K. Whistler, *Boone CEAR,* 1938 (CAERC box 67, folder 3); H. J. Yarling, *Hendricks CEAR,* 1939 (CAERC box 93, folder 22).
62. Indiana State Dairy Association, "Executive Committee Meeting," in *Thirty-Eighth Annual Report of the Indiana State Dairy Association, 1928* (Indianapolis: Wm. B. Burford, 1928), 45, https://babel.hathitrust.org/cgi/pt?id=wu.89044276178&view=1up&seq=671&skin=2021; "23 Proved Sires Awarded Medals for First Time," *Lafayette Journal and Courier* (Lafayette, IN), 26 Jan. 1929, 10.
63. "23 Proved Sires Awarded Medals for First Time," 10; "Livestock Men Meet at Purdue," *Indianapolis Star,* 18 Jan. 1929, 11.
64. H. J. Yarling, *Hendricks CEAR,* 1939 (CAERC box 93, folder 22).
65. L. E. Archbold, *Adams CEAR,* 1930 (CAERC box 61, folder 16).
66. R. J. Maggart, *Carroll CEAR,* 1932 (CAERC box 68, folder 57).
67. L. E. Archbold, *Adams CEAR,* 1935 (CAERC box 61, folder 21).
68. R. R. Graves and M. H. Fohrman,"Superior Germ Plasm in Dairy Herds," in *United States Department of Agriculture Yearbook of Agriculture, 1936* (Washington, DC: Government Printing Office, 1936), 997, https://archive.org/details/yoa1936/page/n1005/mode/2up; L. E. Archbold, *Adams CEAR,* 1936 (CAERC box 61, folder 22).
69. L. E. Archbold, *Adams CEAR,* 1936 (CAERC box 61, folder 22); H. S. Heckard, *Allen CEAR,* 1930 (CAERC box 62, folder 15).
70. R. J. Maggart, *Carroll CEAR,* 1930 (CAERC box 68, folder 55).
71. E. M. Christen, *Benton CEAR,* 1936 (CAERC box 65, folder 21).
72. H. S. Heckard, *Allen CEAR,* 1930 (CAERC box 62, folder 15).
73. R. J. Maggart, *Carroll CEAR,* 1934 (CAERC box 68, folder 59).
74. Indiana State Board of Health, "Report of the Traveling Milk Laboratory," in *Year Book of the State of Indiana for 1923* (Indianapolis: Wm. B. Burford, 1924), 143, https://indianamemory.contentdm.oclc.org/digital/collection/p15078coll21/id/87691.

75. Ibid.; M. D. Butler, *Grant CEAR,* 1931 (CAERC box 87, folder 48); E. C. Parker, *Hamilton CEAR,* 1931 (CAERC box 89, folder 46).
76. John Taylor, "Report of the Bureau of Dairy Products," in *Year Book of the State of Indiana for 1933* (Indianapolis: Wm. B. Burford, 1934), 667, https://indianamemory.contentdm.oclc.org/digital/collection/p15078coll21/id/101361; Indiana State Board of Health, "Report of the Traveling Milk Laboratory," 143.
77. Frank C. Wilson, "Report of Dairy Products Bureau," in *Year Book of the State of Indiana for 1931* (Indianapolis: Wm. B. Burford, 1932), 232, https://indianamemory.contentdm.oclc.org/digital/collection/p15078coll21/id/98082.
78. Wilson, "Report of Dairy Products Bureau," 234.
79. Wilson, "Report of Dairy Products Bureau," 233.
80. Ibid.; "Photography Used by Milk Association to Diagnose the Source of Infection," *Vidette-Messenger of Porter County* (Valparaiso, IN), 24 June 1931, 7; "Weekly Dairy Letter," *Kokomo Tribune* (Kokomo, IN), 6 July 1931, 5.
81. *Purdue University Department of Agricultural Extension Annual Report,* 1932, 39; E. C. Parker, *Hamilton CEAR,* 1930 (CAERC box 89, folder 45); A. M. Bishea, *Daviess CEAR,* 1934 (CAERC box 74, folder 28).
82. L. E. Moore, *Clark CEAR,* 1932 (CAERC box 70, folder 39); A. V. Keesling, *Bartholomew CEARs,* 1937 and 1939 (CAERC box 63, folders 26 and 30); E. C. Parker, *Hamilton CEAR,* 1931 (CAERC box 89, folder 46).
83. W. G. Smith, *Henry CEAR,* 1931 (CAERC box 93, folder 60); C. A. Alcorn, *Dearborn CEAR,* 1938 (CAERC box 75, folder 33).
84. E. C. Parker, *Hamilton CEAR,* 1931 (CAERC box 89, folder 46).
85. A. V. Keesling, *Bartholomew CEAR,* 1935 (CAERC box 63, folder 22).
86. M. D. Butler, *Grant CEAR,* 1931 (CAERC box 87, folder 48).
87. R. Gwaltney, *Gibson CEAR,* 1939 (CAERC box 87, folder 9); G. L. Reed, *Benton CEAR,* 1930 (CAERC box 65, folder 15).

CHAPTER 21

Sheep Have a Place on Most Every Farm

1. H. L. Royce, *Clay CEAR,* 1930 (CAERC box 71, folder 42); E. C. Parker, *Hamilton CEAR,* 1935 (CAERC box 90, folder 1).
2. H. L. Royce, *Clay CEAR,* 1932 (CAERC box 71, folder 44); A. V. Keesling, *Bartholomew CEAR,* 1938 (CAERC box 63, folder 28).
3. C. B. Blosser, *Elkhart CEAR,* 1931 (CAERC box 80, folder 30).
4. R. J. Maggart, *Carroll CEAR,* 1930 (CAERC box 68, folder 55); M. E. Cromer, *Delaware CEAR,* 1931 (CAERC box 78, folder 33).
5. H. L. Royce, *Clay CEAR,* 1932 (CAERC box 71, folder 44).
6. M. E. Cromer, *Delaware CEAR,* 1933 (CAERC box 78, folder 35); T. E. Myers, *DeKalb CEAR,* 1935 (CAERC box 77, folder 30).
7. M. E. Cromer, *Delaware CEAR,* 1933 (CAERC box 78, folder 35); E. C. Parker, *Hamilton CEAR,* 1935 (CAERC box 90, folder 1).
8. W. U. Rusk, *Blackford CEAR,* 1935 (CAERC box 66, folder 12).
9. T. E. Myers, *DeKalb CEAR,* 1935 (CAERC box 77, folder 30).
10. C. V. Kimmell, *Allen CEAR,* 1938 (CAERC box 62, folder 24); C. A. Nicholson, *Dubois CEAR,* 1930 (CAERC box 79, folder 39); E. S. Miles, *Clark CEAR,* 1937 (CAERC box 70, folder 46).

11. E. C. Parker, *Hamilton CEAR,* 1933 (CAERC box 89, folder 48); R. L. Zell, *Franklin CEAR,* 1935 (CAERC box 84, folder 35); C. A. Nicholson, *Dubois CEAR,* 1930 (CAERC box 79, folder 39).
12. E. S. Miles, *Clark CEAR,* 1933 (CAERC box 70, folder 40); M. E. Cromer, *Delaware CEAR,* 1933 (CAERC box 78, folder 35).
13. C. V. Kimmell, *Allen CEAR,* 1937 (CAERC box 62, folder 23); W. U. Rusk, *Blackford CEAR,* 1934 (CAERC box 66, folder 11).
14. N. S. Hadley, *Franklin CEAR,* 1931 (CAERC box 84, folder 31); W. G. Smith, *Henry CEAR,* 1932 (CAERC box 94, folder 1).
15. N. S. Hadley, *Franklin CEAR,* 1933 (CAERC box 84, folder 33).
16. W. W. Whitehead, *Boone CEAR,* 1939 (CAERC box 67, folder 4); W. U. Rusk, *Blackford CEAR,* 1937 (CAERC box 66, folder 14).
17. W. O'Hara, *Hancock CEAR,* 1936 (CAERC box 91, folder 19).
18. C. V. Kimmell, *Allen CEAR,* 1938 (CAERC box 62, folder 24).
19. A. T. Marvel, *Elkhart CEAR,* 1936 (CAERC box 80, folder 34).
20. C. V. Kimmell, *Allen CEAR,* 1937 (CAERC box 62, folder 23).
21. Ibid.
22. R. L. Zell, *Franklin CEAR,* 1935 (CAERC box 84, folder 35); W. U. Rusk, *Blackford CEAR,* 1935 (CAERC box 66, folder 12).
23. H. L. Royce, *Clay CEAR,* 1930 (CAERC box 71, folder 42).
24. Ibid.
25. W. U. Rusk, *Blackford CEAR,* 1934 (CAERC box 66, folder 11).
26. M. F. Detrick, *Fayette CEAR,* 1936 (CAERC box 81, folder 37); N. S. Hadley, *Franklin CEAR,* 1933 (CAERC box 84, folder 33).
27. H. L. Royce, *Clay CEAR,* 1930 (CAERC box 71, folder 42).
28. C. V. Kimmell, *Allen CEAR,* 1937 (CAERC box 62, folder 23); M. M. Hamilton, *Decatur CEAR,* 1937 (CAERC box 76, folder 43); R. L. Zell, *Franklin CEAR,* 1938 (CAERC box 84, folder 38); W. U. Rusk, *Blackford CEAR,* 1937 (CAERC box 66, folder 14); M. J. Huxley, *Fulton CEAR,* 1939 (CAERC box 85, folder 50); L. E. Archbold, *Adams CEAR,* 1938 (CAERC box 61, folder 24).
29. C. V. Kimmell, *Allen CEAR,* 1937 (CAERC box 62, folder 23).
30. W. U. Rusk, *Blackford CEAR,* 1937 (CAERC box 66, folder 14); R. J. Maggart, *Carroll CEAR,* 1938 (CAERC box 68, folder 64).
31. R. D. McHargue, *Clay CEAR,* 1939 (CAERC box 71, folder 51); Alfred Hall, "Rural Notes Picked Up at Many Points," *Richmond Item* (Richmond, IN), 29 May 1935, 13.
32. L. E. Archbold, *Adams CEAR,* 1936 (CAERC box 61, folder 22).
33. R. L. Zell, *Franklin CEAR,* 1938 (CAERC box 84, folder 38).
34. E. A. Johnson, *Cass CEARs,* 1937 and 1939 (CAERC box 69, folders 41 and 44).
35. E. A. Johnson, *Cass CEARs,* 1938–39 (CAERC box 69, folders 43–44).
36. R. D. McHargue, *Clay CEAR,* 1939 (CAERC box 71, folder 51).
37. W. U. Rusk, *Blackford CEAR,* 1937 (CAERC box 66, folder 14); E. A. Johnson, *Cass CEAR,* 1937 (CAERC box 69, folder 41); R. J. Maggart, *Carroll CEAR,* 1938 (CAERC box 68, folder 64); L. E. Archbold, *Adams CEAR,* 1936 (CAERC box 61, folder 22); C. K. Whistler, *Boone CEAR,* 1938 (CAERC box 67, folder 3); C. E. Baker, *DeKalb CEAR,* 1939 (CAERC box 77, folder 34).
38. C. V. Kimmell, *Allen CEAR,* 1938 (CAERC box 62, folder 24).
39. L. E. Archbold, *Adams CEAR,* 1937 (CAERC box 61, folder 23).

40. R. L. Zell, *Franklin CEAR,* 1939 (CAERC box 84, folder 39).
41. W. B. Stout, *Sheep and Lamb Price Facts,* Purdue University Agricultural Extension Leaflet No. 166 (West Lafayette, IN: Purdue University, Aug. 1932), 4, https://earchives.lib.purdue.edu/digital/collection/AESLeaflet/id/1214/rec/186.
42. H. L. Royce, *Clay CEAR,* 1932 (CAERC box 71, folder 44); W. U. Rusk, *Blackford CEAR,* 1932 (CAERC box 66, folder 9).
43. H. L. Royce, *Clay CEAR,* 1930 (CAERC box 71, folder 42).
44. E. M. Christen, *Benton CEAR,* 1936 (CAERC box 65, folder 21).
45. L. E. Archbold, *Adams CEAR,* 1931 (CAERC box 61, folder 17).
46. M. F. Detrick, *Fayette CEAR,* 1930 (CAERC box 81, folder 31).
47. J. E. Little, *Cass CEAR,* 1930 (CAERC box 69, folder 33).
48. H. L. Royce, *Clay CEAR,* 1932 (CAERC box 71, folder 44); C. K. Whistler, *Boone CEAR,* 1938 (CAERC box 67, folder 3).
49. C. V. Kimmell, *Allen CEAR,* 1937 (CAERC box 62, folder 23); J. E. Little, *Cass CEAR,* 1930 (CAERC box 69, folder 33).
50. W. W. Whitehead, *Boone CEAR,* 1939 (CAERC box 67, folder 4); W. G. Smith, *Henry CEAR,* 1932 (CAERC box 94, folder 1).
51. R. Gwaltney, *Gibson CEAR,* 1937 (CAERC box 87, folder 7); E. C. Parker, *Hamilton CEAR,* 1935 (CAERC box 90, folder 1); A. J. Hesler, *Fountain CEAR,* 1939 (CAERC box 83, folder 35).
52. C. G. Randell, *Cooperative Marketing of Livestock in the United States by Terminal Associations,* U.S. Department of Agriculture Technical Bulletin No. 57 (Washington, DC: Government Printing Office, Feb. 1928), 66, https://naldc.nal.usda.gov/download/CAT86200054/PDF.
53. C. K. Whistler, *Boone CEAR,* 1937 (CAERC box 67, folder 2); A. J. Hesler, *Fountain CEAR,* 1936 (CAERC box 83, folder 32).
54. A. J. Hesler, *Fountain CEAR,* 1939 (CAERC box 83, folder 35).
55. W. U. Rusk, *Blackford CEAR,* 1938 (CAERC box 66, folder 15).
56. W. U. Rusk, *Blackford CEAR,* 1936 (CAERC box 66, folder 13); M. F. Detrick, *Fayette CEAR,* 1936 (CAERC box 81, folder 37).
57. M. E. Cromer, *Delaware CEAR,* 1930 (CAERC box 78, folder 32); W. U. Rusk, *Blackford CEAR,* 1938 (CAERC box 66, folder 15); C. V. Kimmell, *Allen CEAR,* 1938 (CAERC box 62, folder 24); N. S. Hadley, *Franklin CEAR,* 1931 (CAERC box 84, folder 31).
58. W. U. Rusk, *Blackford CEAR,* 1937 (CAERC box 66, folder 14); C. V. Kimmell, *Allen CEAR,* 1936 (CAERC box 62, folder 21).
59. M. E. Cromer, *Delaware CEAR,* 1937 (CAERC box 78, folder 39).
60. W. U. Rusk, *Blackford CEAR,* 1937 (CAERC box 66, folder 14).
61. Alfred Hall, "Harvey Whitton Awarded Gold Medal for Lambs," *Richmond Item* (Richmond, IN), 21 July 1937, 14.
62. W. U. Rusk, *Blackford CEAR,* 1936 (CAERC box 66, folder 13); N. S. Hadley, *Franklin CEAR,* 1933 (CAERC box 84, folder 33).
63. M. E. Cromer, *Delaware CEAR,* 1939 (CAERC box 78, folder 41).
64. C. V. Kimmell, *Allen CEAR,* 1938 (CAERC box 62, folder 24); M. E. Cromer, *Delaware CEAR,* 1939 (CAERC box 78, folder 41).
65. M. E. Cromer, *Delaware CEAR,* 1939 (CAERC box 78, folder 41); M. J. Huxley, *Fulton CEAR,* 1939 (CAERC box 85, folder 50).
66. M. E. Cromer, *Delaware CEAR,* 1933 (CAERC box 78, folder 35).

67. Untitled (article on scoring for sheep-shearing contest), *Noblesville Ledger* (Noblesville, IN), 19 Sept. 1939, 2; "Shearing Contest at Royerton," *Muncie Evening Press* (Muncie, IN), 25 Feb. 1933, 13.
68. "Sheep Shearing Contest March 15," *Richmond Palladium* (Richmond, IN), 25 Jan. 1933, 10.
69. "Eastern Indiana Sheep Schools Well Attended," *Star Press* (Muncie, IN), 18 Mar. 1933, 12.
70. "Indiana Will Have Prominent Place at International Show," *Richmond Item* (Richmond, IN), 22 Nov. 1938, 3; "Junior Farmers Convert 4-H Livestock into Cash," *Richmond Item* (Richmond, IN), 3 Dec. 1938, 1.
71. M. E. Cromer, *Delaware CEAR,* 1931 (CAERC box 78, folder 33).

CHAPTER 22

Runts and Diseased Pigs Seldom Lift the Mortgage

1. W. O'Hara, *Hancock CEAR,* 1937 (CAERC box 91, folder 20).
2. F. E. Conder, *Grant CEAR,* 1937 (CAERC box 87, folder 54).
3. H. S. Heckard, *Allen CEAR,* 1930 (CAERC box 62, folder 15).
4. C. A. Nicholson, *Dubois CEAR,* 1938 (CAERC box 79A, folder 2); J. H. Brown, *Elkhart CEAR,* 1937 (CAERC box 80, folder 35); H. S. Heckard, *Allen CEAR,* 1931 (CAERC box 62, folder 16); I. L. Thurston, *Decatur CEAR,* 1930 (CAERC box 76, folder 36); M. E. Cromer, *Delaware CEAR,* 1933 (CAERC box 78, folder 35); R. Gwaltney, *Gibson CEAR,* 1939 (CAERC box 87, folder 9).
5. U.S. Department of Agriculture and Purdue University Agricultural Experiment Station, *Indiana Crops and Livestock: Annual Livestock Summary, 1935,* Bulletin No. 113 (West Lafayette, IN: U.S. Department of Agriculture and Purdue University, Feb. 1935), 6, https://www.nass.usda.gov/Statistics_by_State/Indiana/Publications/Annual_Statistical_Bulletin/hist_pdf/as_1935_ebook.pdf.
6. U.S. Department of Agriculture and Purdue University Agricultural Experiment Station, *Indiana Crops and Livestock: Annual Livestock Summary, 1935,* 7; W. G. Smith, *Henry CEAR,* 1933 (CAERC box 94, folder 2).
7. M. E. Cromer, *Delaware CEAR,* 1936 (CAERC box 78, folder 38).
8. F. E. Conder, *Grant CEAR,* 1937 (CAERC box 87, folder 54).
9. R. Gwaltney, *Gibson CEAR,* 1939 (CAERC box 87, folder 9).
10. H. R. Rosenbury, *Fulton CEAR,* 1932 (CAERC box 85, folder 40); M. E. Cromer, *Delaware CEAR,* 1933 (CAERC box 78, folder 35).
11. R. Gwaltney, *Gibson CEAR,* 1939 (CAERC box 87, folder 9).
12. M. F. Detrick, *Fayette CEAR,* 1930 (CAERC box 81, folder 31).
13. J. E. Little, *Cass CEAR,* 1932 (CAERC box 69, folder 35).
14. M. E. Cromer, *Delaware CEAR,* 1932 (CAERC box 78, folder 34).
15. A. J. Hesler, *Fountain CEAR,* 1939 (CAERC box 83, folder 35).
16. C. A. Nicholson, *Dubois CEAR,* 1938 (CAERC box 79A, folder 2).
17. J. Jensen, *Daviess CEAR,* 1931 (CAERC box 74, folder 25); H. S. Heckard, *Allen CEAR,* 1932 (CAERC box 62, folder 17).
18. E. C. Parker, *Hamilton CEAR,* 1930 (CAERC box 89, folder 45).
19. A. J. Hesler, *Fountain CEAR,* 1937 (CAERC box 83, folder 33).
20. E. C. Parker, *Hamilton CEAR,* 1930 (CAERC box 89, folder 45).
21. M. F. Detrick, *Fayette CEAR,* 1934 (CAERC box 81, folder 35).
22. M. F. Detrick, *Fayette CEARs,* 1931 and 1934 (CAERC box 81, folders 32 and 35).

23. E. L. Hawk, *Hendricks CEAR,* 1931 (CAERC box 93, folder 18).
24. R. Gwaltney, *Gibson CEAR,* 1938 (CAERC box 87, folder 8).
25. W. O'Hara, *Hancock CEAR,* 1938 (CAERC box 91, folder 21).
26. W. O'Hara, *Hancock CEAR,* 1939 (CAERC box 91, folder 22).
27. I. L. Thurston, *Decatur CEAR,* 1933 (CAERC box 76, folder 39).
28. W. U. Rusk, *Blackford CEAR,* 1934 (CAERC box 66, folder 11).
29. V. J. Mann, *Clinton CEAR,* 1930 (CAERC box 72, folder 34); M. M. Hamilton, *Decatur CEAR,* 1936 (CAERC box 76, folder 42); J. E. Little, *Cass CEAR,* 1932 (CAERC box 69, folder 35).
30. A. V. Keesling, *Bartholomew CEAR,* 1936 (CAERC box 63, folder 24); E. L. Hawk, *Hendricks CEAR,* 1932 (CAERC box 93, folder 19).
31. H. J. Yarling, *Hancock CEAR,* 1930 (CAERC box 91, folder 13).
32. I. L. Thurston, *Decatur CEARs,* 1930–31 (CAERC box 76, folders 36–37); E. A. Johnson, *Cass CEAR,* 1938 (CAERC box 69, folder 43).
33. *Purdue University Department of Agricultural Extension Annual Report,* 1934, 30–31.
34. H. S. Heckard, *Allen CEAR,* 1933 (CAERC box 62, folder 18).
35. *Purdue University Department of Agricultural Extension Annual Report,* 1934, 32.
36. A. M. Bishea, *Daviess CEAR,* 1934 (CAERC box 74, folder 28).
37. J. E. Little, *Cass CEAR,* 1931 (CAERC box 69, folder 34).
38. A. M. Bishea, *Daviess CEAR,* 1934 (CAERC box 74, folder 28); E. A. Johnson, *Cass CEAR,* 1937 (CAERC box 69, folder 41); J. E. Little, *Cass CEAR,* 1932 (CAERC box 69, folder 35); E. C. Parker, *Hamilton CEAR,* 1935 (CAERC box 90, folder 1).
39. I. L. Thurston, *Decatur CEAR,* 1931 (CAERC box 76, folder 37).
40. M. F. Detrick, *Fayette CEAR,* 1934 (CAERC box 81, folder 35).
41. D. S. Bishopp, *Fayette CEAR,* 1938 (CAERC box 81, folder 39).
42. A. V. Keesling, *Bartholomew CEAR,* 1938 (CAERC box 63, folder 28).
43. Ibid.
44. Ibid.
45. A. V. Keesling, *Bartholomew CEAR,* 1939 (CAERC box 63, folder 30).
46. E. A. Johnson, *Cass CEAR,* 1938 (CAERC box 69, folder 43).

CHAPTER 23

A Bushel Basket of Eggs Brings In as Much Money as 100 Bushels of Corn

1. H. L. Royce, *Wabash CEAR,* 1923 (CAERC box 52, folder 8); W. E. Shrode, *Posey CEAR,* 1922 (CAERC box 125, folder 8).
2. C. A. Alcorn, *Dearborn CEAR,* 1937 (CAERC box 75, folder 32); H. S. Heckard, *Allen CEAR,* 1931 (CAERC box 62, folder 16).
3. R. J. Maggart, *Carroll CEAR,* 1931 (CAERC box 68, folder 56).
4. C. C. Madison, *Dearborn CEAR,* 1930 (CAERC box 75, folder 25).
5. E. E. Stinson, *DeKalb CEAR,* 1930 (CAERC box 77, folder 25).
6. J. E. Little, *Cass CEAR,* 1931 (CAERC box 69, folder 34); M. E. Cromer, *Delaware CEAR,* 1932 (CAERC box 78, folder 34); E. S. Miles, *Clark CEAR,* 1937 (CAERC box 70, folder 46); C. A. Alcorn, *Dearborn CEAR,* 1937 (CAERC box 75, folder 32).
7. C. L. Dyer, *Gibson CEAR,* 1932 (CAERC box 87, folder 3).
8. E. L. Hawk, *Hendricks CEAR,* 1930 (CAERC box 93, folder 18).

9. C. A. Nicholson, *Dubois CEAR,* 1935 (CAERC box 79, folder 44); I. L. Thurston, *Decatur CEAR,* 1930 (CAERC box 76, folder 36); J. Jensen, *Daviess CEAR,* 1932 (CAERC box 74, folder 26); W. G. Smith, *Henry CEAR,* 1930 (CAERC box 93, folder 59); R. E. Johnson, *Harrison CEAR,* 1930 (CAERC box 92, folder 25).
10. W. O'Hara, *Hancock CEAR,* 1931 (CAERC box 91, folder 14).
11. E. S. Miles, *Clark CEAR,* 1936 (CAERC box 70, folder 44); C. A. Alcorn, *Floyd CEAR,* 1930 (CAERC box 82, folder 31).
12. W. O'Hara, *Hancock CEAR,* 1931 (CAERC box 91, folder 14); R. E. Johnson, *Harrison CEAR,* 1930 (CAERC box 92, folder 25).
13. V. D. Sexson, *Greene CEAR,* 1931 (CAERC box 89, folder 6).
14. J. E. Little, *Cass CEAR,* 1930 (CAERC box 69, folder 33).
15. J. Jensen, *Daviess CEAR,* 1931 (CAERC box 74, folder 25); J. E. Little, *Cass CEAR,* 1930 (CAERC box 69, folder 33); R. E. Johnson, *Harrison CEAR,* 1930 (CAERC box 92, folder 25).
16. J. E. Little, *Cass CEAR,* 1930 (CAERC box 69, folder 33).
17. M. F. Detrick, *Fayette CEAR,* 1930 (CAERC box 81, folder 31); J. E. Little, *Cass CEAR,* 1930 (CAERC box 69, folder 33).
18. E. M. Rowe, *Clinton CEAR,* 1933 (CAERC box 72, folder 37).
19. J. Jensen, *Daviess CEAR,* 1931 (CAERC box 74, folder 25).
20. E. S. Miles, *Clark CEAR,* 1937 (CAERC box 70, folder 46).
21. A. V. Keesling, *Bartholomew CEAR,* 1935 (CAERC box 63, folder 22).
22. I. L. Thurston, *Decatur CEAR,* 1930 (CAERC box 76, folder 36); H. L. Royce, *Clay CEAR,* 1932 (CAERC box 71, folder 44); W. A. Douglass, *Bartholomew CEAR,* 1930 (CAERC box 63, folder 17); J. L. Morrow, *Crawford CEAR,* 1939 (CAERC box 73, folder 40).
23. R. J. Maggart, *Carroll CEAR,* 1931 (CAERC box 68, folder 56).
24. H. J. Yarling, *Hancock CEAR,* 1930 (CAERC box 91, folder 13); H. L. Royce, *Clay CEAR,* 1932 (CAERC box 71, folder 44); R. Gwaltney, *Gibson CEAR,* 1939 (CAERC box 87, folder 9).
25. A. V. Keesling, *Bartholomew CEAR,* 1935 (CAERC box 63, folder 22).
26. C. A. Nicholson, *Dubois CEAR,* 1930 (CAERC box 79, folder 39).
27. E. M. Rowe, *Clinton CEAR,* 1934 (CAERC box 72, folder 38).
28. J. Jensen, *Daviess CEAR,* 1930 (CAERC box 74, folder 24); R. J. Maggart, *Carroll CEAR,* 1931 (CAERC box 68, folder 56).
29. J. Jensen, *Daviess CEAR,* 1932 (CAERC box 74, folder 26).
30. C. A. Alcorn, *Floyd CEAR,* 1930 (CAERC box 82, folder 31); J. B. Todd, *Hamilton CEAR,* 1936 (CAERC box 90, folder 2); J. E. Little, *Cass CEAR,* 1931 (CAERC box 69, folder 34); W. O'Hara, *Hancock CEAR,* 1931 (CAERC box 91, folder 14).
31. R. E. Johnson, *Harrison CEAR,* 1930 (CAERC box 92, folder 25).
32. J. Jensen, *Daviess CEAR,* 1931 (CAERC box 74, folder 25).
33. H. L. Royce, *Clay CEAR,* 1931 (CAERC box 71, folder 43).
34. J. E. Little, *Cass CEAR,* 1930 (CAERC box 69, folder 33).
35. A. V. Keesling, *Bartholomew CEAR,* 1935 (CAERC box 63, folder 22).
36. A. J. Hesler, *Fountain CEAR,* 1934 (CAERC box 83, folder 30).
37. Ibid.; J. E. Little, *Cass CEAR,* 1931 (CAERC box 69, folder 34).
38. A. Haseman, *Greene CEAR,* 1933 (CAERC box 89, folder 8); J. E. Little, *Cass CEAR,* 1931 (CAERC box 69, folder 34).

39. J. E. Little, *Cass CEAR,* 1931 (CAERC box 69, folder 34); C. A. Nicholson, *Dubois CEAR,* 1935 (CAERC box 79, folder 44); J. L. Morrow, *Crawford CEAR,* 1937 (CAERC box 73, folder 38); W. O'Hara, *Hancock CEAR,* 1931 (CAERC box 91, folder 14).
40. A. V. Keesling, *Bartholomew CEAR,* 1938 (CAERC box 63, folder 28).
41. R. J. Maggart, *Carroll CEAR,* 1931 (CAERC box 68, folder 56).
42. R. J. Maggart, *Carroll CEAR,* 1932 (CAERC box 68, folder 57).
43. L. E. Archbold, *Adams CEAR,* 1938 (CAERC box 61, folder 24).
44. R. J. Maggart, *Carroll CEAR,* 1931 (CAERC box 68, folder 56).
45. M. D. Butler, *Grant CEARs,* 1930–31 (CAERC box 87, folders 47–48); L. E. Archbold, *Adams CEAR,* 1933 (CAERC box 61, folder 19); E. C. Parker, *Hamilton CEAR,* 1933 (CAERC box 89, folder 48).
46. E. C. Parker, *Hamilton CEAR,* 1933 (CAERC box 89, folder 48).
47. V. D. Sexson, *Greene CEAR,* 1932 (CAERC box 89, folder 7); J. H. Brown, *Elkhart CEAR,* 1937 (CAERC box 80, folder 35).
48. A. J. Hesler, *Fountain CEAR,* 1935 (CAERC box 83, folder 31); R. J. Maggart, *Carroll CEAR,* 1931 (CAERC box 68, folder 56).
49. E. M. Christen, *Benton CEAR,* 1936 (CAERC box 65, folder 21); E. M. Christen, *Benton CEAR,* 1933 (CAERC box 65, folder 18); E. M. Christen, *Benton CEAR,* 1935 (CAERC box 65, folder 20).
50. E. M. Christen, *Benton CEAR,* 1939 (CAERC box 65, folder 27).
51. J. Jensen, *Daviess CEAR,* 1931 (CAERC box 74, folder 25); H. S. Heckard, *Allen CEAR,* 1931 (CAERC box 62, folder 16).
52. G. L. Reed, *Benton CEAR,* 1932 (CAERC box 65, folder 17).
53. L. E. Archbold, *Adams CEAR,* 1937 (CAERC box 61, folder 23).
54. H. S. Heckard, *Allen CEAR,* 1930 (CAERC box 62, folder 15).
55. R. D. McHargue, *Clay CEAR,* 1937 (CAERC box 71, folder 49).
56. J. E. Little, *Cass CEAR,* 1932 (CAERC box 69, folder 35).
57. Ibid.; E. M. Christen, *Benton CEAR,* 1935 (CAERC box 65, folder 20).
58. S. B. Scott, *Crawford CEAR,* 1930 (CAERC box 73, folder 32); R. E. Johnson, *Harrison CEAR,* 1932 (CAERC box 92, folder 27).
59. E. M. Christen, *Benton CEAR,* 1933 (CAERC box 65, folder 18).
60. M. E. Cromer, *Delaware CEAR,* 1933 (CAERC box 78, folder 35).
61. J. Jensen, *Daviess CEAR,* 1931 (CAERC box 74, folder 25); E. S. Miles, *Clark CEAR,* 1937 (CAERC box 70, folder 46); C. M. East, *Floyd CEARs,* 1932 and 1934 (CAERC box 82, folders 33 and 35).
62. R. D. McHargue, *Clay CEAR,* 1937 (CAERC box 71, folder 49).
63. E. S. Miles, *Clark CEAR,* 1938 (CAERC box 70, folder 48).
64. H. S. Heckard, *Allen CEARs,* 1931 and 1933 (CAERC box 62, folders 16 and 18).
65. L. E. Archbold, *Adams CEAR,* 1932 (CAERC box 61, folder 18).
66. H. S. Heckard, *Allen CEARs,* 1932–33 (CAERC box 62, folders 17–18); C. V. Kimmell, *Allen CEAR,* 1936 (CAERC box 62, folder 21); E. A. Johnson, *Cass CEAR,* 1938 (CAERC box 69, folder 43).
67. Untitled (article on 1931 Corydon poultry school), *Kokomo Tribune* (Kokomo, IN), 29 Jan. 1931, 15.
68. R. E. Johnson, *Harrison CEAR,* 1931 (CAERC box 92, folder 26).
69. H. S. Heckard, *Allen CEAR,* 1935 (CAERC box 62, folder 20).

CHAPTER 24

The Life of an Extension Worker Is an Honorable Occupation and an Interesting One

1. J. H. Brown, *Elkhart CEAR,* 1939 (CAERC box 80, folder 37).
2. R. J. Maggart, *Carroll CEAR,* 1939 (CAERC box 68, folder 65).
3. C. V. Kimmell, *Allen CEAR,* 1939 (CAERC box 62, folder 25).
4. A. V. Keesling, *Bartholomew CEAR,* 1939 (CAERC box 63, folder 30); R. L. Zell, *Franklin CEAR,* 1939 (CAERC box 84, folder 39).
5. D. S. Bishopp, *Fayette CEAR,* 1939 (CAERC box 81, folder 40).
6. E. M. Christen, *Benton CEAR,* 1939 (CAERC box 65, folder 27).
7. Ibid.; A. V. Keesling, *Bartholomew CEAR,* 1939 (CAERC box 63, folder 30).
8. C. A. Nicholson, *Dubois CEAR,* 1934 (CAERC box 79, folder 43).
9. A. V. Keesling, *Bartholomew CEAR,* 1939 (CAERC box 63, folder 30).
10. E. M. Christen, *Benton CEAR,* 1939 (CAERC box 65, folder 27).
11. J. H. Brown, *Elkhart CEAR,* 1939 (CAERC box 80, folder 37); E. M. Christen, *Benton CEAR,* 1939 (CAERC box 65, folder 27); A. V. Keesling, *Bartholomew CEAR,* 1939 (CAERC box 63, folder 30).
12. E. M. Christen, *Benton CEAR,* 1939 (CAERC box 65, folder 27).
13. Ibid.
14. E. M. Rowe, *Clinton CEAR,* 1935 (CAERC box 72, folder 39).
15. E. M. Christen, *Benton CEAR,* 1939 (CAERC box 65, folder 27).
16. R. Gwaltney, *Gibson CEAR,* 1937 (CAERC box 87, folder 7).
17. E. M. Christen, *Benton CEAR,* 1939 (CAERC box 65, folder 27).
18. R. D. McHargue, *Clay CEAR,* 1939 (CAERC box 71, folder 51).
19. A. V. Keesling, *Bartholomew CEAR,* 1939 (CAERC box 63, folder 30).
20. W. U. Rusk, *Blackford CEAR,* 1939 (CAERC box 66, folder 16).
21. L. E. Archbold, *Adams CEAR,* 1939 (CAERC box 1, folder 1).
22. D. S. Bishopp, *Fayette CEAR,* 1939 (CAERC box 81, folder 40).
23. Ibid.
24. R. L. Zell, *Franklin CEAR,* 1939 (CAERC box 84, folder 39).
25. R. D. McHargue, *Clay CEAR,* 1939 (CAERC box 71, folder 51).
26. E. S. Miles, *Clark CEAR,* 1939 (CAERC box 70, folder 50).
27. M. J. Huxley, *Fulton CEAR,* 1939 (CAERC box 85, folder 50).
28. R. L. Zell, *Franklin CEAR,* 1939 (CAERC box 84, folder 39).
29. R. D. McHargue, *Clay CEAR,* 1939 (CAERC box 71, folder 51); J. H. Brown, *Elkhart CEAR,* 1939 (CAERC box 80, folder 37); L. E. Archbold, *Adams CEAR,* 1939 (CAERC box 1, folder 1); M. J. Huxley, *Fulton CEAR,* 1939 (CAERC box 85, folder 50).
30. R. L. Zell, *Franklin CEAR,* 1939 (CAERC box 84, folder 39).
31. R. Gwaltney, *Gibson CEAR,* 1939 (CAERC box 87, folder 9).
32. M. J. Huxley, *Fulton CEAR,* 1939 (CAERC box 85, folder 50).
33. E. M. Christen, *Benton CEAR,* 1939 (CAERC box 65, folder 27).
34. C. V. Kimmell, *Allen CEAR,* 1939 (CAERC box 62, folder 25).
35. R. D. McHargue, *Clay CEAR,* 1939 (CAERC box 71, folder 51).
36. A. V. Keesling, *Bartholomew CEAR,* 1939 (CAERC box 63, folder 30).
37. W. U. Rusk, *Blackford CEAR,* 1939 (CAERC box 66, folder 16).

CHAPTER 25

Fitting the Extension Program to Wartime Conditions Has Required Some "Give and Take"

1. S. B. Pershing, *Tippecanoe CEAR,* 1942 (CAERC box 44, folder 12); H. E. Abbott, *Marion CEAR,* 1942 (CAERC box 111, folder 7).
2. A. C. Shideler, *Starke CEAR,* 1943 (CAERC box 40, folder 1).
3. "Lend-Lease and Military Aid to the Allies in the Early Years of World War II," U.S. Department of State—Office of the Historian (website), mid-2016, https://history.state.gov/milestones/1937-1945/lend-lease.
4. Frederick Simpich, "Farmers Keep Them Eating," *National Geographic Magazine* 83, no. 4 (Apr. 1943): 435, https://www.google.com/books/edition/The_National_Geographic_Magazine/4U6AAAAAMAAJ?hl=en&gbpv=1&pg=PA435&printsec=frontcover.
5. "U.S.S. Reuben James Sunk by Torpedo West of Iceland; First U. S. Warship to Go Down," *Boston Globe,* 1 Nov. 1941, 1.
6. Warren F. Kimball, "Franklin D. Roosevelt and World War II," *Presidential Studies Quarterly* 34, no. 1 (Mar. 2004): 83, https://www.jstor.org/stable/27552565.
7. Franklin D. Roosevelt, "Day of Infamy" (speech), 8 Dec. 1941, U.S. Library of Congress transcript, http://hdl.loc.gov/loc.afc/afc1986022.ms2201.
8. O. B. Riggs, *Posey CEAR,* 1942 (CAERC box 125, folder 27).
9. Simpich, "Farmers Keep Them Eating," 435.
10. F. E. DeLaCroix, *Jasper CEAR,* 1944 (CAERC box 98, folder 14).
11. R. L. Case, *LaGrange CEAR,* 1943 (CAERC box 106, folder 4); E. A. Johnson, *Howard CEAR,* 1942 (CAERC box 95, folder 10); C. D. Chambers, *Perry CEAR,* 1942 (CAERC box 122, folder 38).
12. C. A. Langston, *Jay CEAR,* 1942 (CAERC box 98, folder 51).
13. Simpich, "Farmers Keep Them Eating," 435.
14. S. Leaming, *Porter CEAR,* 1943 (CAERC box 124, folder 27).
15. R. N. Atkinson, *Orange CEAR,* 1942 (CAERC box 120, folder 25).
16. E. C. Bird, *St. Joseph CEAR,* 1942 (CAERC box 35, folder 28).
17. C. D. Chambers, *Perry CEAR,* 1942 (CAERC box 122, folder 38).
18. E. S. Bogue, *Pike CEAR,* 1942 (CAERC box 123, folder 34).
19. *Purdue University Department of Agricultural Extension Annual Report,* 1941, 3; *Purdue University Department of Agricultural Extension Annual Report,* 1941–43, 3.
20. R. N. Atkinson, *Orange CEAR,* 1942 (CAERC box 120, folder 25).
21. J. A. Clark, *Kosciusko CEAR,* 1942 (CAERC box 104, folder 18); F. E. DeLaCroix, *Jasper CEAR,* 1943 (CAERC box 98, folder 13); J. C. Sliger, *Sullivan CEAR,* 1943 (CAERC box 41, folder 54); E. A. Johnson, *Howard CEAR,* 1944 (CAERC box 95, folder 12); W. P. Stall, *Jackson CEAR,* 1943 (CAERC box 97, folder 8).
22. B. T. Bond, *Warrick CEAR,* 1943 (CAERC box 54, folder 37); H. E. Abbott, *Marion CEAR,* 1944 (CAERC box 111, folder 13); E. A. Johnson, *Howard CEAR,* 1942 (CAERC box 95, folder 10); N. S. Hadley, *Parke CEAR,* 1942 (CAERC box 121, folder 51); L. J. Fultz, *Washington CEAR,* 1943 (CAERC box 55, folder 45).
23. T. M. Grabill, *Steuben CEAR,* 1944 (CAERC box 40, folder 52).
24. L. E. Cutler, *Lake CEAR,* 1942 (CAERC box 107, folder 34); M. A. Nye, *Noble CEAR,* 1943 (CAERC box 118, folder 36); L. E. Husted, *Randolph CEAR,* 1943 (CAERC box 30, folder 42); J. F. Armstrong, *Lawrence CEAR,* 1942 (CAERC box 108, folder 41); R. L. Case, *LaGrange CEAR,* 1942 (CAERC box 106, folder 3); H. E. Abbott, *Marion CEAR,* 1944 (CAERC box 111, folder 13); W. C. Haynes, *Madison CEAR,* 1942 (CAERC box 109, folder 38); N. S. Hadley, *Parke CEAR,* 1942 (CAERC box 121, folder 51).

25. R. J. Maggart, *Carroll CEAR,* 1939 (CAERC box 68, folder 65); G. T. Harris, *Ripley CEAR,* 1943 (CAERC box 31, folder 45).

CHAPTER 26

The County Agent Is Expected to Be a Walking Encyclopedia on Government Programs

1. Gladys L. Baker and Wayne D. Rasmussen, "A Chronology of the War Food Administration, Including Predecessor and Successor Agencies, August 1939 to December 1946," U.S. Department of Agriculture Bureau of Agricultural Economics, Nov. 1950, 5, https://naldc.nal.usda.gov/catalog/6538554; W. J. Record, *Morgan CEAR,* 1942 (CAERC box 117, folder 3); S. Leaming, *Porter CEAR,* 1942 (CAERC box 124, folder 26); T. M. Grabill, *Steuben CEAR,* 1942 (CAERC box 40, folder 50); H. S. Benson, *Knox CEAR,* 1945 (CAERC box 103, folder 3); B. V. Widney, *Whitley CEAR,* 1946 (CAERC box 60, folder 7).
2. O. B. Riggs, *Posey CEAR,* 1942 (CAERC box 125, folder 27).
3. Baker and Rasmussen, "A Chronology of the War Food Administration," 6.
4. J. McCulloch, *Scott CEAR,* 1942 (CAERC box 36, folder 23); J. E. Wylie, *Miami CEAR,* 1943 (CAERC box 114, folder 14); H. E. Abbott, *Marion CEAR,* 1942 (CAERC box 111, folder 7); J. F. Armstrong, *Lawrence CEAR,* 1942 (CAERC box 108, folder 41).
5. G. A. Sowers, *Montgomery CEAR,* 1943 (CAERC box 115, folder 55); C. E. Olson, *Pulaski CEAR,* 1942 (CAERC box 28, folder 10); L. M. Butler, *Marshall CEAR,* 1942 (CAERC box 112, folder 32); E. L. Hartman, *LaPorte CEAR,* 1943 (CAERC box 105, folder 19); R. L. Case, *LaGrange CEAR,* 1942 (CAERC box 106, folder 3); W. P. Stall, *Jackson CEAR,* 1943 (CAERC box 97, folder 8).
6. S. Leaming, *Porter CEAR,* 1942 (CAERC box 124, folder 26).
7. F. I. Hoover, *Owen CEAR,* 1942 (CAERC box 121, folder 3).
8. B. T. Bond, *Warrick CEAR,* 1942 (CAERC box 54, folder 36); E. F. Downen, *Spencer CEAR,* 1942 (CAERC box 38, folder 48); G. T. Harris, *Ripley CEAR,* 1943 (CAERC box 31, folder 45); C. E. Olson, *Pulaski CEAR,* 1943 (CAERC box 28, folder 11); S. Leaming, *Porter CEAR,* 1942 (CAERC box 124, folder 26).
9. W. Rogers, *Monroe CEAR,* 1943 (CAERC box 115, folder 26).
10. J. E. Wylie, *Miami CEAR,* 1943 (CAERC box 114, folder 14); L. M. Butler, *Marshall CEAR,* 1943 (CAERC box 112, folder 33); G. A. Sowers, *Montgomery CEAR,* 1943 (CAERC box 115, folder 55); O. B. Riggs, *Posey CEAR,* 1942 (CAERC box 125, folder 27); E. S. Bogue, *Pike CEAR,* 1942 (CAERC box 123, folder 34).
11. W. K. Delaplane, *Wabash CEAR,* 1942 (CAERC box 52, folder 27); W. J. Emerson, *White CEAR,* 1944 (CAERC box 59, folder 7); G. T. Harris, *Ripley CEAR,* 1943 (CAERC box 31, folder 45).
12. N. S. Hadley, *Parke CEAR,* 1943 (CAERC box 121, folder 53).
13. Ibid.
14. E. A. Johnson, *Howard CEAR,* 1942 (CAERC box 95, folder 10).
15. R. C. Stangland, *Howard CEAR,* 1942 (CAERC box 95, folder 10); M. A. Nye, *Porter CEAR,* 1944 (CAERC box 124, folder 28); E. C. Bird, *St. Joseph CEAR,* 1942 (CAERC box 35, folder 28).
16. N. S. Hadley, *Parke CEARs,* 1943–44 (CAERC box 121, folders 53 and 55).
17. O. B. Riggs, *Posey CEAR,* 1943 (CAERC box 125, folder 29).
18. W. C. Haynes, *Madison CEAR,* 1942 (CAERC box 109, folder 38); E. E. Fultz, *Jasper CEAR,* 1942 (CAERC box 98, folder 12).
19. G. A. Sowers, *Montgomery CEAR,* 1942 (CAERC box 115, folder 54).
20. E. A. Johnson, *Howard CEAR,* 1944 (CAERC box 95, folder 12); J. C. Sliger, *Sullivan CEAR,* 1942 (CAERC box 41, folder 53).

21. E. F. Downen, *Spencer CEAR,* 1942 (CAERC box 38, folder 48); R. N. Atkinson, *Orange CEAR,* 1942 (CAERC box 120, folder 25).
22. R. C. Stangland, *Howard CEAR,* 1942 (CAERC box 95, folder 10).
23. J. C. Sliger, *Sullivan CEAR,* 1943 (CAERC box 41, folder 54).
24. R. L. Case, *LaGrange CEAR,* 1942 (CAERC box 106, folder 3).
25. J. C. Sliger, *Sullivan CEAR,* 1942 (CAERC box 41, folder 53); O. B. Riggs, *Posey CEAR,* 1942 (CAERC box 125, folder 27); R. L. Case, *LaGrange CEAR,* 1942 (CAERC box 106, folder 3); E. A. Johnson, *Howard CEAR,* 1944 (CAERC box 95, folder 12); S. Leaming, *Porter CEAR,* 1943 (CAERC box 124, folder 27); W. P. Stall, *Jackson CEAR,* 1943 (CAERC box 97, folder 8).
26. E. F. Downen, *Spencer CEAR,* 1942 (CAERC box 38, folder 48); E. C. Bird, *St. Joseph CEAR,* 1943 (CAERC box 35, folder 29); C. D. Chambers, *Perry CEAR,* 1943 (CAERC box 122, folder 39).
27. O. B. Riggs, *Posey CEAR,* 1942 (CAERC box 125, folder 27).

CHAPTER 27

Production Goals That Looked Impossible Were Reached

1. R. L. Case, LaGrange *CEAR,* 1942 (CAERC box 106, folder 3); *Purdue University Department of Agricultural Extension Annual Report,* 1941–43, 3.
2. T. M. Grabill, *Steuben CEAR,* 1942 (CAERC box 40, folder 50).
3. S. J. Brown, *Vermillion CEAR,* 1942 (CAERC box 49, folder 37).
4. C. D. Chambers, *Perry CEAR,* 1943 (CAERC box 122, folder 39); N. S. Hadley, *Parke CEAR,* 1942 (CAERC box 121, folder 51); J. E. Wylie, *Miami CEAR,* 1942 (CAERC box 114, folder 13).
5. N. S. Hadley, *Parke CEAR,* 1942 (CAERC box 121, folder 51); W. C. Haynes, *Madison CEAR,* 1944 (CAERC box 109, folder 40).
6. W. C. Haynes, *Madison CEAR,* 1944 (CAERC box 109, folder 40).
7. T. M. Grabill, *Steuben CEAR,* 1942 (CAERC box 40, folder 50).
8. W. J. Emerson, *White CEAR,* 1942 (CAERC box 59, folder 5); B. V. Widney, *Whitley CEAR,* 1942 (CAERC box 60, folder 3); C. A. Langston, *Jay CEAR,* 1943 (CAERC box 99, folder 1); L. J. Fultz, *Washington CEAR,* 1942 (CAERC box 55, folder 44); T. M. Grabill, *Steuben CEAR,* 1942 (CAERC box 40, folder 50).
9. "Food Production: Quotas for Twelve States Were Set in Chicago," *Tipton Daily Tribune* (Tipton, IN), 18 Sept. 1941, 6.
10. Ibid.; *Purdue University Department of Agricultural Extension Annual Report*, 1941–43, 12; "Poultry Chief Says Indiana Is Sure to Exceed Egg Quota," *Lafayette Journal and Courier* (Lafayette, IN), 3 Aug. 1942, 1.
11. S. Leaming, *Porter CEAR,* 1942 (CAERC box 124, folder 26); G. T. Harris, *Ripley CEAR,* 1942 (CAERC box 31, folder 44); S. B. Pershing, *Tippecanoe CEAR,* 1942 (CAERC box 44, folder 12); G. I. John, *Ohio CEAR,* 1942 (CAERC box 119, folder 30).
12. W. M. Clary, *Tipton CEAR,* 1942 (CAERC box 45, folder 23); H. H. Madaus, *Warren CEAR,* 1942 (CAERC box 53, folder 24).
13. W. C. Haynes, *Madison CEAR,* 1942 (CAERC box 109, folder 38); H. E. Abbott, *Marion CEAR,* 1942 (CAERC box 111, folder 7).
14. H. H. Madaus, *Warren CEAR,* 1942 (CAERC box 53, folder 24); W. M. Clary, *Tipton CEAR,* 1942 (CAERC box 45, folder 23).
15. H. H. Madaus, *Warren CEAR,* 1942 (CAERC box 53, folder 24).

16. *Purdue University Department of Agricultural Extension Annual Report,* 1941–43, 7.
17. J. E. Wylie, *Miami CEAR,* 1943 (CAERC box 114, folder 14).
18. W. C. Haynes, *Madison CEAR,* 1942 (CAERC box 109, folder 38).
19. H. E. Abbott, *Marion CEAR,* 1943 (CAERC box 111, folder 10).
20. J. E. Wylie, *Miami CEAR,* 1943 (CAERC box 114, folder 14).
21. E. A. Johnson, *Howard CEAR,* 1942 (CAERC box 95, folder 10).
22. L. M. Butler, *Marshall CEAR,* 1943 (CAERC box 112, folder 33).
23. H. E. Abbott, *Marion CEAR,* 1943 (CAERC box 111, folder 10).
24. S. Leaming, *Porter CEAR,* 1942 (CAERC box 124, folder 26).
25. J. E. Wylie, *Miami CEAR,* 1943 (CAERC box 114, folder 14); C. D. Chambers, *Perry CEAR,* 1943 (CAERC box 122, folder 39).
26. S. Leaming, *Porter CEAR,* 1943 (CAERC box 124, folder 27); S. B. Pershing, *Tippecanoe CEAR,* 1943 (CAERC box 44, folder 14); J. C. Sliger, *Sullivan CEAR,* 1943 (CAERC box 41, folder 54); E. S. Bogue, *Pike CEAR,* 1943 (CAERC box 123, folder 35).
27. M. J. Peterson, *Vermillion CEAR,* 1943 (CAERC box 49, folder 38); J. E. Wylie, *Miami CEAR,* 1943 (CAERC box 114, folder 14).
28. E. S. Bogue, *Pike CEAR,* 1943 (CAERC box 123, folder 35).
29. Ibid.; W. M. Clary, *Tipton CEAR,* 1943 (CAERC box 45, folder 24); J. E. Wylie, *Miami CEAR,* 1943 (CAERC box 114, folder 14).
30. J. E. Wylie, *Miami CEAR,* 1943 (CAERC box 114, folder 14); H. S. Benson, *Knox CEAR,* 1943 (CAERC box 102, folder 42).
31. Annette Scherber, "Corn, Tomatoes, and POWs: Hoosier Agriculture during World War II," *Indiana History Blog,* Indiana Historical Bureau of the Indiana State Library, 17 Nov. 2016, https://blog.history.in.gov/corn-tomatoes-pows-hoosier-agriculture-during-world-war-ii/.
32. C. A. Langston, *Jay CEAR,* 1943 (CAERC box 99, folder 1); B. T. Bond, *Warrick CEAR,* 1942 (CAERC box 54, folder 36).
33. "Agricultural Mobilization," American Home Front in World War II (website), Encyclopedia.com, 23 May 2018, https://www.encyclopedia.com/plants-and-animals/agriculture-and-horticulture/agriculture-general/agricultural-revolution; C. E. Olson, *Pulaski CEAR,* 1942 (CAERC box 28, folder 10).
34. H. H. Madaus, *Warren CEAR,* 1944 (CAERC box 53, folder 26).
35. A. C. Shideler, *Starke CEAR,* 1944 (CAERC box 40, folder 2); L. E. Cooper, *Spencer CEAR,* 1946 (CAERC box 38, folder 52); O. B. Riggs, *Posey CEAR,* 1942 (CAERC box 125, folder 27); L. E. Cutler, *Lake CEAR,* 1943 (CAERC box 107, folder 36).
36. L. M. Butler, *Marshall CEAR,* 1942 (CAERC box 112, folder 32); J. F. Armstrong, *Lawrence CEAR,* 1946 (CAERC box 109, folder 1).
37. A. C. Shideler, *Starke CEAR,* 1944 (CAERC box 40, folder 2).
38. H. K. Rippey, *Jay CEAR,* 1945 (CAERC box 99, folder 3).
39. G. I. John, *Ohio CEAR,* 1942 (CAERC box 119, folder 30).
40. D. L. Grimes, *Putnam CEAR,* 1943 (CAERC box 29, folder 27).
41. H. E. Abbott, *Marion CEAR,* 1943 (CAERC box 111, folder 10).
42. O. B. Riggs, *Posey CEAR,* 1944 (CAERC box 125, folder 31).
43. C. A. Langston, *Jay CEAR,* 1943 (CAERC box 99, folder 1); J. E. Wylie, *Miami CEAR,* 1945 (CAERC box 114, folder 16).
44. L. E. Cooper, *Spencer CEAR,* 1946 (CAERC box 38, folder 52).
45. A. C. Shideler, *Starke CEAR,* 1944 (CAERC box 40, folder 2).
46. H. H. Madaus, *Warren CEAR,* 1944 (CAERC box 53, folder 26).

47. H. K. Rippey, *Jay CEAR,* 1945 (CAERC box 99, folder 3).
48. L. E. Cutler, *Lake CEAR,* 1946 (CAERC box 107, folder 40).
49. S. B. Pershing, *Tippecanoe CEAR,* 1942 (CAERC box 44, folder 12).
50. R. N. Atkinson, *Orange CEAR,* 1942 (CAERC box 120, folder 25).
51. W. K. Delaplane, *Wabash CEAR,* 1942 (CAERC box 52, folder 27); L. E. Husted, *Randolph CEAR,* 1944 (CAERC box 30, folder 43).
52. L. M. Butler, *Marshall CEAR,* 1944 (CAERC box 112, folder 34).
53. C. A. Langston, *Jay CEAR,* 1943 (CAERC box 99, folder 1); C. E. Baker, *DeKalb CEAR,* 1939 (CAERC box 77, folder 34).
54. H. H. Madaus, *Warren CEAR,* 1942 (CAERC box 53, folder 24).
55. N. S. Hadley, *Parke CEAR,* 1942 (CAERC box 121, folder 51).
56. J. C. Sliger, *Sullivan CEAR,* 1942 (CAERC box 41, folder 53).
57. B. T. Bond, *Warrick CEAR,* 1942 (CAERC box 54, folder 36).
58. O. H. McNary, *Switzerland CEAR,* 1942 (CAERC box 42, folder 56); C. E. Olson, *Pulaski CEAR,* 1942 (CAERC box 28, folder 10).
59. S. B. Pershing, *Tippecanoe CEAR,* 1943 (CAERC box 44, folder 14).
60. J. McCulloch, *Scott CEAR,* 1942 (CAERC box 36, folder 23); E. F. Downen, *Spencer CEAR,* 1942 (CAERC box 38, folder 48).
61. O. B. Riggs, *Posey CEAR,* 1944 (CAERC box 125, folder 31).
62. S. Leaming, *Porter CEAR,* 1942 (CAERC box 124, folder 26).
63. S. B. Pershing, *Tippecanoe CEAR,* 1942 (CAERC box 44, folder 12).
64. O. B. Riggs, *Posey CEAR,* 1943 (CAERC box 125, folder 29).
65. S. B. Pershing, *Tippecanoe CEAR,* 1942 (CAERC box 44, folder 12); W. K. Delaplane, *Wabash CEAR,* 1943 (CAERC box 52, folder 28).
66. W. P. Stall, *Jackson CEAR,* 1942 (CAERC box 97, folder 7).
67. B. V. Widney, *Whitley CEAR,* 1943 (CAERC box 60, folder 4); S. Leaming, *Porter CEAR,* 1942 (CAERC box 124, folder 26); A. M. Bishea, *Vanderburgh CEAR,* 1942 (CAERC box 48, folder 4); O. B. Riggs, *Posey CEAR,* 1943 (CAERC box 125, folder 29).
68. B. V. Widney, *Whitley CEAR,* 1943 (CAERC box 60, folder 4).
69. *Purdue University Department of Agricultural Extension Annual Report,* 1945, 3.

CHAPTER 28

Higher Hog, Dairy, and Poultry Prices Created an Interest Like Never Before

1. S. B. Pershing, *Tippecanoe CEAR,* 1945 (CAERC box 44, folder 18); G. T. Harris, *Ripley CEAR,* 1942 (CAERC box 31, folder 44).
2. "Meat Outlook Good for 1944," *Lafayette Journal and Courier* (Lafayette, IN), 1 Jan. 1944, 1.
3. W. M. Clary, *Tipton CEAR,* 1942 (CAERC box 45, folder 23); C. D. Chambers, *Perry CEAR,* 1942 (CAERC box 122, folder 38); J. A. Clark, *Kosciusko CEAR,* 1942 (CAERC box 104, folder 18); S. J. Brown, *Vermillion CEAR,* 1942 (CAERC box 49, folder 37).
4. W. U. Rusk, *Huntington CEAR,* 1941 (CAERC box 96, folder 17); W. J. Emerson, *White CEAR,* 1942 (CAERC box 59, folder 5).
5. J. C. Sliger, *Sullivan CEAR,* 1942 (CAERC box 41, folder 53).
6. A. C. Shideler, *Starke CEAR,* 1943 (CAERC box 40, folder 1); G. A. Sowers, *Montgomery CEAR,* 1942 (CAERC box 115, folder 54); B. T. Bond, *Warrick CEAR,* 1942 (CAERC box 54, folder 36); C. E. Olson, *Pulaski CEAR,* 1942 (CAERC box 28, folder 10); E. A. Chatham, *Perry CEAR,* 1944 (CAERC box 122, folder 40).
7. W. U. Rusk, *Huntington CEAR,* 1941 (CAERC box 96, folder 17).

8. L. M. Butler, *Marshall CEAR,* 1942 (CAERC box 112, folder 32).
9. C. D. Chambers, *Perry CEAR,* 1942 (CAERC box 122, folder 38).
10. D. L. Grimes, *Putnam CEAR,* 1942 (CAERC box 29, folder 26).
11. A. C. Shideler, *Starke CEAR,* 1943 (CAERC box 40, folder 1); A. M. Bishea, *Vanderburgh CEAR,* 1944 (CAERC box 48, folder 9); H. H. Madaus, *Warren CEAR,* 1943 (CAERC box 53, folder 25); E. F. Brown, *Rush CEAR,* 1943 (CAERC box 33, folder 10).
12. L. E. Husted, *Randolph CEAR,* 1943 (CAERC box 30, folder 42).
13. S. B. Pershing, *Tippecanoe CEAR,* 1944 (CAERC box 44, folder 16).
14. N. S. Hadley, *Parke CEAR,* 1942 (CAERC box 121, folder 51); J. McCulloch, *Scott CEAR,* 1944 (CAERC box 36, folder 25).
15. C. E. Olson, *Pulaski CEAR,* 1942 (CAERC box 28, folder 10).
16. H. H. Madaus, *Warren CEAR,* 1943 (CAERC box 53, folder 25); B. T. Bond, *Warrick CEAR,* 1944 (CAERC box 54, folder 38); S. B. Pershing, *Tippecanoe CEAR,* 1945 (CAERC box 44, folder 18).
17. J. C. Sliger, *Sullivan CEAR,* 1943 (CAERC box 41, folder 54); B. T. Bond, *Warrick CEAR,* 1944 (CAERC box 54, folder 38); E. S. Bogue, *Pike CEAR,* 1944 (CAERC box 123, folder 36); A. C. Shideler, *Starke CEAR,* 1944 (CAERC box 40, folder 2).
18. R. C. Garrison, *Union CEAR,* 1944 (CAERC box 46, folder 26); C. E. Olson, *Pulaski CEAR,* 1942 (CAERC box 28, folder 10); S. Leaming, *Porter CEAR,* 1942 (CAERC box 124, folder 26); O. B. Riggs, *Posey CEAR,* 1943 (CAERC box 125, folder 29).
19. C. E. Olson, *Pulaski CEAR,* 1942 (CAERC box 28, folder 10); H. E. Abbott, *Marion CEAR,* 1942 (CAERC box 111, folder 7).
20. J. A. Wood, *Martin CEAR,* 1942 (CAERC box 113, folder 28); J. F. Armstrong, *Lawrence CEAR,* 1943 (CAERC box 108, folder 42).
21. M. S. Smith, *Wells CEAR,* 1944 (CAERC box 58, folder 30); C. E. Olson, *Pulaski CEAR,* 1942 (CAERC box 28, folder 10); W. J. Record, *Morgan CEAR,* 1942 (CAERC box 117, folder 3).
22. W. J. Record, *Morgan CEAR,* 1942 (CAERC box 117, folder 3).
23. H. E. Abbott, *Marion CEAR,* 1942 (CAERC box 111, folder 7).
24. W. C. Haynes, *Madison CEAR,* 1942 (CAERC box 109, folder 38); J. F. Armstrong, *Lawrence CEAR,* 1943 (CAERC box 108, folder 42); W. J. Record, *Morgan CEAR,* 1942 (CAERC box 117, folder 3).
25. W. J. Record, *Morgan CEAR,* 1942 (CAERC box 117, folder 3); C. E. Olson, *Pulaski CEAR,* 1942 (CAERC box 28, folder 10).
26. C. D. Chambers, *Perry CEAR,* 1943 (CAERC box 122, folder 39); B. T. Bond, *Warrick CEAR,* 1944 (CAERC box 54, folder 38).
27. M. J. Peterson, *Vermillion CEAR,* 1944 (CAERC box 49, folder 39).
28. W. J. Record, *Morgan CEAR,* 1942 (CAERC box 117, folder 3).
29. H. E. Abbott, *Marion CEAR,* 1942 (CAERC box 111, folder 7).
30. E. S. Bogue, *Pike CEAR,* 1943 (CAERC box 123, folder 35).
31. W. C. Haynes, *Madison CEAR,* 1943 (CAERC box 109, folder 39).
32. E. S. Bogue, *Pike CEAR,* 1943 (CAERC box 123, folder 35).
33. J. E. Wylie, *Miami CEAR,* 1944 (CAERC box 114, folder 15); E. F. Downen, *Spencer CEAR,* 1943 (CAERC box 38, folder 49).
34. M. S. Smith, *Wells CEAR,* 1944 (CAERC box 58, folder 30); B. T. Bond, *Warrick CEAR,* 1942 (CAERC box 54, folder 36).

35. S. B. Pershing, *Tippecanoe CEAR,* 1944 (CAERC box 44, folder 16).
36. H. H. Madaus, *Warren CEAR,* 1944 (CAERC box 53, folder 26).
37. B. V. Widney, *Whitley CEAR,* 1944 (CAERC box 60, folder 5).
38. J. McCulloch, *Scott CEAR,* 1943 (CAERC box 36, folder 24); E. A. Chatham, *Perry CEAR,* 1944 (CAERC box 122, folder 40); H. H. Madaus, *Warren CEAR,* 1944 (CAERC box 53, folder 26).
39. B. V. Widney, *Whitley CEAR,* 1944 (CAERC box 60, folder 5).
40. T. M. Grabill, *Steuben CEAR,* 1943 (CAERC box 40, folder 51); E. A. Chatham, *Perry CEAR,* 1944 (CAERC box 122, folder 40); B. V. Widney, *Whitley CEAR,* 1943 (CAERC box 60, folder 4).
41. J. A. Clark, *Kosciusko CEAR,* 1943 (CAERC box 104, folder 19).
42. J. A. Clark, *Kosciusko CEAR,* 1944 (CAERC box 104, folder 20); E. A. Chatham, *Perry CEAR,* 1944 (CAERC box 122, folder 40); J. L. Guthrie, *Noble CEAR,* 1944 (CAERC box 118, folder 37).
43. E. S. Bogue, *Pike CEAR,* 1944 (CAERC box 123, folder 36); A. C. Shideler, *Starke CEAR,* 1944 (CAERC box 40, folder 2).
44. J. McCulloch, *Scott CEAR,* 1944 (CAERC box 36, folder 25); J. E. Wylie, *Miami CEAR,* 1944 (CAERC box 114, folder 15).
45. L. E. Husted, *Randolph CEAR,* 1944 (CAERC box 30, folder 43).
46. E. A. Chatham, *Perry CEAR,* 1944 (CAERC box 122, folder 40).
47. R. N. Atkinson, *Orange CEAR,* 1942 (CAERC box 120, folder 25); G. I. John, *Ohio CEAR,* 1946 (CAERC box 119, folder 34).
48. B. V. Widney, *Whitley CEAR,* 1942 (CAERC box 60, folder 3).
49. G. I. John, *Ohio CEAR,* 1943 (CAERC box 119, folder 31).
50. O. B. Riggs, *Posey CEAR,* 1944 (CAERC box 125, folder 31); A. M. Bishea, *Vanderburgh CEAR,* 1942 (CAERC box 48, folder 4); J. L. Guthrie, *Noble CEAR,* 1944 (CAERC box 118, folder 37); M. A. Nye, *Noble CEAR,* 1943 (CAERC box 118, folder 36).
51. J. L. Guthrie, *Noble CEAR,* 1944 (CAERC box 118, folder 37); G. I. John, *Ohio CEAR,* 1943 (CAERC box 119, folder 31).
52. M. A. Nye, *Porter CEAR,* 1944 (CAERC box 124, folder 28); W. M. Clary, *Tipton CEAR,* 1942 (CAERC box 45, folder 23).
53. L. E. Husted, *Randolph CEAR,* 1942 (CAERC box 30, folder 41).
54. J. C. Sliger, *Sullivan CEAR,* 1943 (CAERC box 41, folder 54).
55. R. N. Atkinson, *Orange CEAR,* 1942 (CAERC box 120, folder 25).
56. L. E. Cutler, *Lake CEAR,* 1943 (CAERC box 107, folder 36).
57. J. E. Wylie, *Miami CEAR,* 1943 (CAERC box 114, folder 14).
58. G. I. John, *Ohio CEAR,* 1943 (CAERC box 119, folder 31); J. E. Wylie, *Miami CEAR,* 1942 (CAERC box 114, folder 13); L. M. Butler, *Marshall CEAR,* 1942 (CAERC box 112, folder 32).
59. J. E. Wylie, *Miami CEAR,* 1943 (CAERC box 114, folder 14).
60. S. Leaming, *Porter CEAR,* 1943 (CAERC box 124, folder 27).
61. J. E. Wylie, *Miami CEAR,* 1943 (CAERC box 114, folder 14).
62. Ibid.
63. J. E. Wylie, *Miami CEAR,* 1942 (CAERC box 114, folder 13).
64. J. E. Wylie, *Miami CEAR,* 1944 (CAERC box 114, folder 15).
65. G. A. Sowers, *Montgomery CEAR,* 1942 (CAERC box 115, folder 54).
66. E. L. Hartman, *LaPorte CEAR,* 1942 (CAERC box 105, folder 17); E. S. Bogue, *Pike CEAR,* 1943 (CAERC box 123, folder 35); W. C. Haynes, *Madison CEAR,* 1944 (CAERC box 109, folder 40); R. N. Atkinson, *Orange CEAR,* 1942 (CAERC box 120, folder 25).

67. G. A. Sowers, *Montgomery CEAR,* 1942 (CAERC box 115, folder 54); M. A. Nye, *Noble CEAR,* 1942 (CAERC box 118, folder 35); L. M. Butler, *Marshall CEAR,* 1944 (CAERC box 112, folder 34); R. N. Atkinson, *Orange CEAR,* 1942 (CAERC box 120, folder 25).
68. H. S. Benson, *Knox CEAR,* 1942 (CAERC box 102, folder 40).
69. G. I. John, *Ohio CEAR,* 1942 (CAERC box 119, folder 30).
70. R. N. Atkinson, *Orange CEAR,* 1942 (CAERC box 120, folder 25).
71. M. A. Nye, *Porter CEAR,* 1946 (CAERC box 124, folder 31).
72. L. J. Fultz, *Washington CEAR,* 1944 (CAERC box 55, folder 46).
73. H. E. Abbott, *Marion CEAR,* 1942 (CAERC box 111, folder 7).
74. O. B. Riggs, *Posey CEAR,* 1942 (CAERC box 125, folder 27); L. P. Geyer, *Jefferson CEAR,* 1944 (CAERC box 100, folder 11); S. Leaming, *Porter CEAR,* 1942 (CAERC box 124, folder 26).
75. O. H. McNary, *Switzerland CEAR,* 1943 (CAERC box 42, folder 57).
76. E. S. Bogue, *Pike CEAR,* 1942 (CAERC box 123, folder 34); R. N. Atkinson, *Orange CEAR,* 1942 (CAERC box 120, folder 25).
77. J. L. Guthrie, *Noble CEAR,* 1944 (CAERC box 118, folder 37); H. S. Benson, *Knox CEAR,* 1944 (CAERC box 103, folder 2); L. P. Geyer, *Jefferson CEAR,* 1944 (CAERC box 100, folder 11); W. C. Haynes, *Madison CEAR,* 1944 (CAERC box 109, folder 40); E. L. Hartman, *LaPorte CEAR,* 1944 (CAERC box 105, folder 21).
78. A. C. Shideler, *Starke CEAR,* 1944 (CAERC box 40, folder 2); W. U. Rusk, *Huntington CEAR,* 1944 (CAERC box 96, folder 20); W. P. Stall, *Jackson CEAR,* 1945 (CAERC box 97, folder 10).
79. W. U. Rusk, *Huntington CEAR,* 1944 (CAERC box 96, folder 20).
80. H. E. Abbott, *Marion CEAR,* 1944 (CAERC box 111, folder 13).
81. Ibid.; J. McCulloch, *Scott CEAR,* 1944 (CAERC box 36, folder 25).
82. W. Rogers, *Monroe CEAR,* 1944 (CAERC box 115, folder 27); J. C. Sliger, *Sullivan CEAR,* 1942 (CAERC box 41, folder 53).
83. N. S. Hadley, *Parke CEAR,* 1944 (CAERC box 121, folder 55); W. J. Record, *Morgan CEAR,* 1942 (CAERC box 117, folder 3).
84. N. S. Hadley, *Parke CEAR,* 1944 (CAERC box 121, folder 55).
85. T. M. Grabill, *Steuben CEAR,* 1943 (CAERC box 40, folder 51); S. W. Milligan, *Wayne CEAR,* 1942 (CAERC box 56, folder 29); C. E. Olson, *Pulaski CEAR,* 1944 (CAERC box 28, folder 12).
86. B. T. Bond, *Warrick CEAR,* 1942 (CAERC box 54, folder 36).
87. W. J. Record, *Morgan CEAR,* 1942 (CAERC box 117, folder 3); J. C. Sliger, *Sullivan CEAR,* 1942 (CAERC box 41, folder 53).
88. N. S. Hadley, *Parke CEAR,* 1942 (CAERC box 121, folder 51); E. C. Bird, *St. Joseph CEAR,* 1943 (CAERC box 35, folder 29); C. E. Olson, *Pulaski CEAR,* 1942 (CAERC box 28, folder 10).
89. *Purdue University Department of Agricultural Extension Annual Report,* 1941–43, 12.
90. Frederick Simpich, "Farmers Keep Them Eating," *National Geographic Magazine* 83, no. 4 (Apr. 1943): 435, https://www.google.com/books/edition/The_National_Geographic_Magazine/4U6AAAAAMAAJ?hl=en&gbpv=1&pg=PA435&printsec=frontcover.
91. J. C. Sliger, *Sullivan CEAR,* 1942 (CAERC box 41, folder 53).
92. W. C. Haynes, *Madison CEAR,* 1942 (CAERC box 109, folder 38).
93. J. E. Wylie, *Miami CEAR,* 1942 (CAERC box 114, folder 13).
94. R. N. Atkinson, *Orange CEAR,* 1944 (CAERC box 120, folder 27).

95. E. A. Chatham, *Perry CEAR,* 1944 (CAERC box 122, folder 40); E. F. Downen, *Spencer CEAR,* 1942 (CAERC box 38, folder 48); B. Jewell, *Pike CEAR,* 1945 (CAERC box 123, folder 37); E. S. Bogue, *Pike CEAR,* 1943 (CAERC box 123, folder 35); F. I. Hoover, *Owen CEAR,* 1942 (CAERC box 121, folder 3).

96. G. T. Harris, *Ripley CEARs,* 1942–43 (CAERC box 31, folders 44–45); E. S. Bogue, *Pike CEAR,* 1942 (CAERC box 123, folder 34); R. L. Case, *LaGrange CEAR,* 1943 (CAERC box 106, folder 4).

97. R. N. Atkinson, *Orange CEAR,* 1942 (CAERC box 120, folder 25); C. D. Chambers, *Perry CEAR,* 1942 (CAERC box 122, folder 38).

98. J. A. Wood, *Martin CEAR,* 1945 (CAERC box 113, folder 31); W. J. Record, *Morgan CEAR,* 1940 (CAERC box 117, folder 1); W. C. Haynes, *Madison CEAR,* 1945 (CAERC box 109, folder 41).

99. J. E. Wylie, *Miami CEARs,* 1943 and 1945 (CAERC box 114, folders 14 and 16).

100. G. T. Harris, *Ripley CEAR,* 1943 (CAERC box 31, folder 45).

101. W. C. Haynes, *Madison CEAR,* 1944 (CAERC box 109, folder 40); A. C. Shideler, *Starke CEAR,* 1944 (CAERC box 40, folder 2).

102. J. McCulloch, *Scott CEAR,* 1944 (CAERC box 36, folder 25); E. S. Bogue, *Pike CEAR,* 1944 (CAERC box 123, folder 36).

103. E. S. Bogue, *Pike CEAR,* 1944 (CAERC box 123, folder 36).

104. A. C. Shideler, *Starke CEAR,* 1944 (CAERC box 40, folder 2).

105. J. E. Wylie, *Miami CEAR,* 1942 (CAERC box 114, folder 13); C. A. Langston, *Jay CEAR,* 1943 (CAERC box 99, folder 1); R. N. Atkinson, *Orange CEAR,* 1942 (CAERC box 120, folder 25); A. M. Bishea, *Vanderburgh CEAR,* 1942 (CAERC box 48, folder 4); W. M. Clary, *Tipton CEAR,* 1942 (CAERC box 45, folder 23).

106. R. N. Atkinson, *Orange CEAR,* 1944 (CAERC box 120, folder 27); L. M. Butler, *Marshall CEAR,* 1942 (CAERC box 112, folder 32); S. W. Milligan, *Wayne CEAR,* 1942 (CAERC box 56, folder 29); T. M. Grabill, *Steuben CEAR,* 1943 (CAERC box 40, folder 51).

CHAPTER 29

For Patriotic Reasons as well as for Profit, Acreage Has Been Expanded

1. P. K. Connelly, *Huntington CEAR,* 1940 (CAERC box 96, folder 16).

2. J. L. Guthrie, *Union CEAR,* 1942 (CAERC box 46, folder 24); G. T. Harris, *Ripley CEAR,* 1942 (CAERC box 31, folder 44).

3. E. A. Johnson, *Howard CEAR,* 1945 (CAERC box 95, folder 13); A. K. Wissman, *Jennings CEAR,* 1946 (CAERC box 101, folder 2); J. F. Armstrong, *Lawrence CEAR,* 1942 (CAERC box 108, folder 41); H. E. Abbott, *Marion CEAR,* 1944 (CAERC box 111, folder 13).

4. B. V. Widney, *Whitley CEAR,* 1942 (CAERC box 60, folder 3).

5. O. B. Riggs, *Posey CEAR,* 1943 (CAERC box 125, folder 29); C. D. Chambers, *Perry CEAR,* 1942 (CAERC box 122, folder 38); H. E. Abbott, *Marion CEAR,* 1944 (CAERC box 111, folder 13).

6. W. P. Stall, *Jackson CEAR,* 1942 (CAERC box 97, folder 7).

7. L. J. Fultz, *Washington CEAR,* 1942 (CAERC box 55, folder 44); O. B. Riggs, *Posey CEAR,* 1943 (CAERC box 125, folder 29); J. L. Guthrie, *Noble CEAR,* 1945 (CAERC box 118, folder 38).

8. B. V. Widney, *Whitley CEAR,* 1942 (CAERC box 60, folder 3); B. T. Bond, *Warrick CEAR,* 1942 (CAERC box 54, folder 36); H. H. Madaus, *Warren CEAR,* 1944 (CAERC box 53, folder 26); A. M. Bishea, *Vanderburgh CEAR,* 1944 (CAERC box 48, folder 9); J. L. Guthrie, *Union CEAR,* 1942 (CAERC box 46, folder 24); J. McCulloch, *Scott CEAR,* 1942 (CAERC box 36, folder 23); L. E. Husted, *Randolph CEAR,* 1943 (CAERC box 30,

folder 42); C. E. Olson, *Pulaski CEAR,* 1943 (CAERC box 28, folder 11); S. B. Pershing, *Tippecanoe CEAR,* 1942 (CAERC box 44, folder 12).

9. L. M. Butler, *Marshall CEAR,* 1944 (CAERC box 112, folder 34); J. A. Clark, *Kosciusko CEAR,* 1942 (CAERC box 104, folder 18).

10. H. E. Abbott, *Marion CEAR,* 1944 (CAERC box 111, folder 13).

11. B. T. Bond, *Warrick CEAR,* 1942 (CAERC box 54, folder 36); C. L. Brown, *Vigo CEAR,* 1942 (CAERC box 51, folder 1); J. McCulloch, *Scott CEAR,* 1942 (CAERC box 36, folder 23); E. C. Bird, *St. Joseph CEAR,* 1942 (CAERC box 35, folder 28).

12. H. E. Abbott, *Marion CEAR,* 1942 (CAERC box 111, folder 7).

13. C. A. Langston, *Jay CEAR,* 1944 (CAERC box 99, folder 2).

14. H. S. Benson, *Knox CEAR,* 1942 (CAERC box 102, folder 40).

15. R. N. Atkinson, *Orange CEAR,* 1944 (CAERC box 120, folder 27); R. N. Atkinson, *Orange CEAR,* 1942 (CAERC box 120, folder 25).

16. W. P. Stall, *Jackson CEAR,* 1942 (CAERC box 97, folder 7); W. C. Haynes, *Madison CEAR,* 1942 (CAERC box 109, folder 38); E. L. Hartman, *LaPorte CEAR,* 1945 (CAERC box 105, folder 23).

17. W. P. Stall, *Jackson CEAR,* 1942 (CAERC box 97, folder 7).

18. H. E. Abbott, *Marion CEAR,* 1944 (CAERC box 111, folder 13); L. J. Fultz, *Washington CEAR,* 1943 (CAERC box 55, folder 45).

19. W. P. Stall, *Jackson CEAR,* 1942 (CAERC box 97, folder 7).

20. H. H. Madaus, *Warren CEAR,* 1943 (CAERC box 53, folder 25).

21. R. L. Case, *LaGrange CEAR,* 1945 (CAERC box 106, folder 6).

22. H. H. Madaus, *Warren CEAR,* 1943 (CAERC box 53, folder 25); P. K. Connelly, *Huntington CEAR,* 1940 (CAERC box 96, folder 16); M. J. Huxley, *Knox CEAR,* 1946 (CAERC box 103, folder 6).

23. G. T. Harris, *Ripley CEAR,* 1942 (CAERC box 31, folder 44); F. I. Hoover, *Owen CEAR,* 1942 (CAERC box 121, folder 3); E. S. Bogue, *Pike CEAR,* 1944 (CAERC box 123, folder 36); J. L. Guthrie, *Noble CEAR,* 1945 (CAERC box 118, folder 38).

24. W. P. Stall, *Jackson CEAR,* 1942 (CAERC box 97, folder 7); M. S. Smith, *Wells CEAR,* 1943 (CAERC box 58, folder 29); A. C. Shideler, *Starke CEAR,* 1945 (CAERC box 40, folder 3); J. McCulloch, *Scott CEARs,* 1944–45 (CAERC box 36, folders 25–26).

25. H. E. Abbott, *Marion CEAR,* 1942 (CAERC box 111, folder 7); R. L. Case, *LaGrange CEAR,* 1945 (CAERC box 106, folder 6).

26. R. L. Case, *LaGrange CEAR,* 1945 (CAERC box 106, folder 6); E. A. Johnson, *Howard CEAR,* 1945 (CAERC box 95, folder 13); D. L. Grimes, *Putnam CEAR,* 1942 (CAERC box 29, folder 26); W. C. Haynes, *Madison CEAR,* 1945 (CAERC box 109, folder 41); H. H. Madaus, *Warren CEAR,* 1943 (CAERC box 53, folder 25).

27. R. L. Case, *LaGrange CEAR,* 1945 (CAERC box 106, folder 6).

28. R. L. Case, *LaGrange CEARs,* 1942–43 (CAERC box 106, folders 3–4).

29. W. C. Haynes, *Madison CEAR,* 1945 (CAERC box 109, folder 41); R. L. Case, *LaGrange CEAR,* 1945 (CAERC box 106, folder 6); E. A. Johnson, *Howard CEAR,* 1945 (CAERC box 95, folder 13).

30. W. U. Rusk, *Huntington CEAR,* 1942 (CAERC box 96, folder 18); J. A. Wood, *Martin CEAR,* 1944 (CAERC box 113, folder 30); W. C. Haynes, *Madison CEAR,* 1945 (CAERC box 109, folder 41).

31. R. L. Case, *LaGrange CEAR,* 1943 (CAERC box 106, folder 4); J. A. Clark, *Kosciusko CEARs,* 1942–43 (CAERC box 104, folders 18–19); M. J. Huxley, *Knox CEAR,* 1946 (CAERC box 103, folder 6); W. U. Rusk, *Huntington CEAR,* 1942 (CAERC box 96, folder 18).

32. W. U. Rusk, *Huntington CEAR,* 1942 (CAERC box 96, folder 18).
33. N. S. Hadley, *Franklin CEAR,* 1934 (CAERC box 84, folder 34); C. L. Dyer, *Gibson CEAR,* 1932 (CAERC box 87, folder 3); R. Gwaltney, *Gibson CEAR,* 1937 (CAERC box 87, folder 7).
34. H. R. Rosenbury, *Fulton CEAR,* 1931 (CAERC box 85, folder 39).
35. R. E. Johnson, *Harrison CEAR,* 1932 (CAERC box 92, folder 27).
36. C. A. Nicholson, *Dubois CEAR,* 1935 (CAERC box 79, folder 44).
37. A. Haseman, *Greene CEAR,* 1934 (CAERC box 89, folder 9); F. E. Conder, *Grant CEAR,* 1938 (CAERC box 87, folder 55).
38. S. B. Pershing, *Tippecanoe CEAR,* 1942 (CAERC box 44, folder 12); E. S. Bogue, *Pike CEAR,* 1942 (CAERC box 123, folder 34); C. D. Chambers, *Perry CEAR,* 1942 (CAERC box 122, folder 38).
39. W. P. Stall, *Jackson CEAR,* 1942 (CAERC box 97, folder 7); J. L. Guthrie, *Union CEAR,* 1942 (CAERC box 46, folder 24).
40. L. M. Butler, *Marshall CEAR,* 1944 (CAERC box 112, folder 34); E. F. Downen, *Spencer CEAR,* 1942 (CAERC box 38, folder 48).
41. W. C. Haynes, *Madison CEAR,* 1942 (CAERC box 109, folder 38).
42. E. A. Johnson, *Howard CEARs,* 1944–45 (CAERC box 95, folders 12–13).
43. B. V. Widney, *Whitley CEAR,* 1943 (CAERC box 60, folder 4); B. V. Widney, *Whitley CEAR,* 1942 (CAERC box 60, folder 3).
44. J. McCulloch, *Scott CEAR,* 1942 (CAERC box 36, folder 23); H. R. Rosenbury, *Fulton CEAR,* 1931 (CAERC box 85, folder 39); A. J. Hesler, *Fountain CEAR,* 1939 (CAERC box 83, folder 35).
45. A. M. Bishea, *Vanderburgh CEAR,* 1943 (CAERC box 48, folder 7).
46. W. J. Emerson, *White CEAR,* 1942 (CAERC box 59, folder 5); H. H. Madaus, *Warren CEAR,* 1942 (CAERC box 53, folder 24); J. McCulloch, *Scott CEAR,* 1942 (CAERC box 36, folder 23); E. M. Christen, *Benton CEAR,* 1939 (CAERC box 65, folder 27); C. A. Nicholson, *Dubois CEAR,* 1933 (CAERC box 79, folder 42); J. Jensen, *Daviess CEAR,* 1930 (CAERC box 74, folder 24); M. E. Cromer, *Delaware CEAR,* 1933 (CAERC box 78, folder 35); C. V. Kimmell, *Allen CEAR,* 1936 (CAERC box 62, folder 21).
47. E. F. Downen, *Spencer CEAR,* 1942 (CAERC box 38, folder 48); J. E. Wylie, *Miami CEAR,* 1942 (CAERC box 114, folder 13); W. K. Delaplane, *Wabash CEAR,* 1943 (CAERC box 52, folder 28); T. M. Grabill, *Steuben CEAR,* 1942 (CAERC box 40, folder 50); F. E. DeLaCroix, *Jasper CEAR,* 1943 (CAERC box 98, folder 13).
48. W. J. Record, *Morgan CEAR,* 1943 (CAERC box 117, folder 4).
49. A. M. Bishea, *Vanderburgh CEAR,* 1943 (CAERC box 48, folder 7); J. H. Telfer, *Sullivan CEAR,* 1945 (CAERC box 41, folder 56); L. E. Husted, *Randolph CEAR,* 1943 (CAERC box 30, folder 42).
50. J. F. Armstrong, *Lawrence CEAR,* 1942 (CAERC box 108, folder 41); E. A. Johnson, *Howard CEAR,* 1944 (CAERC box 95, folder 12).
51. J. H. Telfer, *Sullivan CEAR,* 1946 (CAERC box 42, folder 1); S. B. Pershing, *Tippecanoe CEAR,* 1946 (CAERC box 44, folder 20); M. S. Smith, *Wells CEAR,* 1945 (CAERC box 58, folder 31); L. E. Husted, *Randolph CEAR,* 1942 (CAERC box 30, folder 41).
52. R. R. Zell, *Newton CEARs,* 1942 and 1944 (CAERC box 117, folders 43 and 45); W. J. Record, *Morgan CEAR,* 1942 (CAERC box 117, folder 3); L. M. Butler, *Marshall CEAR,* 1942 (CAERC box 112, folder 32).

53. W. J. Record, *Morgan CEAR,* 1942 (CAERC box 117, folder 3).
54. S. W. Milligan, *Wayne CEAR,* 1942 (CAERC box 56, folder 29).
55. H. E. Abbott, *Marion CEAR,* 1943 (CAERC box 111, folder 10).
56. J. C. Sliger, *Sullivan CEAR,* 1942 (CAERC box 41, folder 53).
57. N. S. Hadley, *Parke CEAR,* 1942 (CAERC box 121, folder 51); *Purdue University Department of Agricultural Extension Annual Report,* 1941–43, 18.
58. H. H. Madaus, *Warren CEAR,* 1942 (CAERC box 53, folder 24).
59. Ibid.
60. H. S. Benson, *Knox CEAR,* 1942 (CAERC box 102, folder 40).
61. T. M. Grabill, *Steuben CEAR,* 1942 (CAERC box 40, folder 50); B. T. Bond, *Warrick CEAR,* 1942 (CAERC box 54, folder 36); S. W. Milligan, *Wayne CEAR,* 1942 (CAERC box 56, folder 29).

CHAPTER 30

Tomatoes Have Become a Major Crop

1. H. E. Abbott, *Marion CEAR,* 1942 (CAERC box 111, folder 7); L. E. Cutler, *Lake CEAR,* 1942 (CAERC box 107, folder 34); W. J. Record, *Morgan CEAR,* 1940 (CAERC box 117, folder 1).
2. H. E. Abbott, *Marion CEAR,* 1942 (CAERC box 111, folder 7).
3. Ibid.
4. H. E. Abbott, *Marion CEAR,* 1943 (CAERC box 111, folder 10).
5. Ibid.; H. E. Abbott, *Marion CEAR,* 1942 (CAERC box 111, folder 7).
6. L. E. Cutler, *Lake CEAR,* 1942 (CAERC box 107, folder 34); L. P. Geyer, *Jefferson CEAR,* 1942 (CAERC box 100, folder 9); *Purdue University Department of Agricultural Extension Annual Report,* 1941–43, 8.
7. W. P. Stall, *Jackson CEAR,* 1942 (CAERC box 97, folder 7).
8. E. S. Bogue, *Pike CEAR,* 1942 (CAERC box 123, folder 34); R. N. Atkinson, *Orange CEAR,* 1942 (CAERC box 120, folder 25); C. D. Chambers, *Perry CEAR,* 1942 (CAERC box 122, folder 38).
9. C. L. Brown, *Vigo CEAR,* 1942 (CAERC box 51, folder 1).
10. H. E. Abbott, *Marion CEAR,* 1942 (CAERC box 111, folder 7); C. D. Chambers, *Perry CEAR,* 1942 (CAERC box 122, folder 38); W. P. Stall, *Jackson CEAR,* 1943 (CAERC box 97, folder 8).
11. E. A. Johnson, *Howard CEAR,* 1946 (CAERC box 95, folder 14).
12. H. E. Abbott, *Marion CEAR,* 1943 (CAERC box 111, folder 10).
13. Ibid.; G. T. Harris, *Ripley CEAR,* 1943 (CAERC box 31, folder 45); S. B. Scott, *Johnson CEAR,* 1946 (CAERC box 101, folder 45).
14. L. P. Geyer, *Jefferson CEAR,* 1942 (CAERC box 100, folder 9); M. S. Smith, *Wells CEAR,* 1943 (CAERC box 58, folder 29).
15. E. S. Bogue, *Pike CEAR,* 1942 (CAERC box 123, folder 34); H. E. Abbott, *Marion CEAR,* 1942 (CAERC box 111, folder 7); W. P. Stall, *Jackson CEAR,* 1942 (CAERC box 97, folder 7).
16. G. A. Sowers, *Montgomery CEAR,* 1944 (CAERC box 115, folder 56); C. D. Chambers, *Perry CEAR,* 1942 (CAERC box 122, folder 38).
17. W. U. Rusk, *Huntington CEAR,* 1942 (CAERC box 96, folder 18); G. A. Sowers, *Montgomery CEAR,* 1944 (CAERC box 115, folder 56); E. A. Johnson, *Howard CEAR,* 1945 (CAERC box 95, folder 13).

18. W. U. Rusk, *Huntington CEAR,* 1942 (CAERC box 96, folder 18); W. P. Stall, *Jackson CEAR,* 1942 (CAERC box 97, folder 7); J. E. Wylie, *Miami CEAR,* 1942 (CAERC box 114, folder 13).
19. E. S. Bogue, *Pike CEAR,* 1944 (CAERC box 123, folder 36).
20. W. P. Stall, *Jackson CEAR,* 1944 (CAERC box 97, folder 9); L. P. Geyer, *Jefferson CEAR,* 1944 (CAERC box 100, folder 11).
21. L. P. Geyer, *Jefferson CEAR,* 1943 (CAERC box 100, folder 10).
22. G. T. Harris, *Ripley CEAR,* 1942 (CAERC box 31, folder 44); L. P. Geyer, *Jefferson CEAR,* 1942 (CAERC box 100, folder 9); W. U. Rusk, *Huntington CEAR,* 1941 (CAERC box 96, folder 17).
23. *Purdue University Department of Agricultural Extension Annual Report,* 1941–43, 8.
24. L. P. Geyer, *Jefferson CEAR,* 1942 (CAERC box 100, folder 9).
25. E. S. Bogue, *Pike CEAR,* 1943 (CAERC box 123, folder 35); J. C. Sliger, *Sullivan CEAR,* 1943 (CAERC box 41, folder 54); W. M. Clary, *Tipton CEAR,* 1943 (CAERC box 45, folder 24).
26. L. E. Cutler, *Lake CEAR,* 1944 (CAERC box 107, folder 37).
27. C. D. Chambers, *Perry CEAR,* 1942 (CAERC box 122, folder 38); G. I. John, *Ohio CEAR,* 1943 (CAERC box 119, folder 31).
28. E. A. Johnson, *Howard CEAR,* 1946 (CAERC box 95, folder 14).
29. S. B. Pershing, *Tippecanoe CEAR,* 1942 (CAERC box 44, folder 12); *Purdue University Department of Agricultural Extension Annual Report,* 1941–43, 8.
30. L. P. Geyer, *Jefferson CEARs,* 1942–43 (CAERC box 100, folders 9–10); C. A. Langston, *Jay CEARs,* 1943–44 (CAERC box 99, folders 1–2).
31. W. P. Stall, *Jackson CEAR,* 1943 (CAERC box 97, folder 8).
32. C. A. Langston, *Jay CEAR,* 1943 (CAERC box 99, folder 1).
33. B. Jewell, *Pike CEAR,* 1945 (CAERC box 123, folder 37); W. M. Clary, *Tipton CEAR,* 1943 (CAERC box 45, folder 24).
34. W. M. Clary, *Tipton CEAR,* 1943 (CAERC box 45, folder 24).
35. H. E. Abbott, *Marion CEAR,* 1943 (CAERC box 111, folder 10).
36. S. B. Scott, *Johnson CEAR,* 1942 (CAERC box 101, folder 41); W. M. Clary, *Tipton CEAR,* 1943 (CAERC box 45, folder 24).
37. L. P. Geyer, *Jefferson CEAR,* 1943 (CAERC box 100, folder 10).
38. W. M. Clary, *Tipton CEAR,* 1943 (CAERC box 45, folder 24).
39. *Purdue University Department of Agricultural Extension Annual Report,* 1945, 13; *Purdue University Department of Agricultural Extension Annual Report,* 1944, 11.

CHAPTER 31

The Total Increase in Home Production and Consumption Would Be a Staggering Amount of Food

1. M. F. Detrick, *Fayette CEAR,* 1936 (CAERC box 81, folder 37); A. V. Keesling, *Bartholomew CEAR,* 1937 (CAERC box 63, folder 26); T. E. Myers, *DeKalb CEAR,* 1938 (CAERC box 77, folder 33).
2. T. E. Myers, *DeKalb CEAR,* 1938 (CAERC box 77, folder 33); A. M. Bishea, *Daviess CEAR,* 1939 (CAERC box 74, folder 32).
3. *Purdue University Department of Agricultural Extension Annual Report,* 1944, 4.
4. J. H. Brown, *Elkhart CEAR,* 1937 (CAERC box 80, folder 35); C. V. Kimmell, *Allen CEAR,* 1938 (CAERC box 62, folder 24); J. E. Little, *Cass CEAR,* 1930 (CAERC box 69, folder 33); J. B. Todd, *Hamilton CEAR,* 1939 (CAERC box 90, folder 5).

5. *Purdue University Department of Agricultural Extension Annual Report,* 1944, 10.
6. R. L. Zell, *Franklin CEAR,* 1939 (CAERC box 84, folder 39); W. O'Hara, *Hancock CEAR,* 1937 (CAERC box 91, folder 20).
7. M. A. Nye, *Noble CEAR,* 1942 (CAERC box 118, folder 35); G. T. Harris, *Ripley CEAR,* 1942 (CAERC box 31, folder 44).
8. Wilbur Bush, "Meat Rationing during World War II," *Daviess County of Northwest Missouri* (blog), Daviess County Historical Society, 14 Dec. 2009, https://daviesscountyhistoricalsociety.com/2009/12/14/meat-rationing-during-world-war-ii/; E. A. Johnson, *Howard CEAR,* 1945 (CAERC box 95, folder 13).
9. C. D. Chambers, *Perry CEAR,* 1943 (CAERC box 122, folder 39); G. A. Sowers, *Montgomery CEAR,* 1943 (CAERC box 115, folder 55); J. E. Wylie, *Miami CEAR,* 1943 (CAERC box 114, folder 14); E. L. Hartman, *LaPorte CEAR,* 1943 (CAERC box 105, folder 19); Bush, "Meat Rationing during World War II."
10. J. F. Armstrong, *Lawrence CEAR,* 1943 (CAERC box 108, folder 42); J. A. Clark, *Kosciusko CEAR,* 1943 (CAERC box 104, folder 19); M. S. Smith, *Wells CEAR,* 1943 (CAERC box 58, folder 29); L. E. Cutler, *Lake CEAR,* 1943 (CAERC box 107, folder 36).
11. A. C. Shideler, *Starke CEAR,* 1944 (CAERC box 40, folder 2); C. E. Olson, *Pulaski CEAR,* 1943 (CAERC box 28, folder 11).
12. J. L. Guthrie, *Union CEAR,* 1943 (CAERC box 46, folder 25).
13. B. Jewell, *Pike CEAR,* 1945 (CAERC box 123, folder 37).
14. W. A. Douglass, *Bartholomew CEAR,* 1930 (CAERC box 63, folder 17); C. C. Madison, *Dearborn CEAR,* 1930 (CAERC box 75, folder 25).
15. J. L. Guthrie, *Union CEAR,* 1943 (CAERC box 46, folder 25); J. F. Armstrong, *Lawrence CEAR,* 1943 (CAERC box 108, folder 42).
16. *Purdue University Department of Agricultural Extension Annual Report,* 1939, 92; *Purdue University Department of Agricultural Extension Annual Report,* 1941–43, 20.
17. Adee Braun, "Turning Bacon into Bombs: The American Fat Salvage Committee," *The Atlantic* (website), 18 Apr. 2014, https://www.theatlantic.com/health/archive/2014/04/reluctantly-turning-bacon-into-bombs-during-world-war-ii/360298/.
18. J. E. Wylie, *Miami CEAR,* 1945 (CAERC box 114, folder 16).
19. E. A. Chatham, *Perry CEAR,* 1945 (CAERC box 122, folder 41).
20. B. Jewell, *Pike CEAR,* 1945 (CAERC box 123, folder 37); R. L. Case, *LaGrange CEAR,* 1943 (CAERC box 106, folder 4).
21. W. U. Rusk, *Huntington CEAR,* 1942 (CAERC box 96, folder 18).
22. R. L. Case, *LaGrange CEARs,* 1942–43 (CAERC box 106, folders 3–4).
23. R. R. Zell, *Newton CEAR,* 1943 (CAERC box 117, folder 44).
24. H. E. Abbott, *Marion CEAR,* 1942 (CAERC box 111, folder 7); R. L. Case, *LaGrange CEAR,* 1942 (CAERC box 106, folder 3); A. C. Shideler, *Starke CEAR,* 1943 (CAERC box 40, folder 1); E. L. Hartman, *LaPorte CEAR,* 1943 (CAERC box 105, folder 19); H. H. Madaus, *Warren CEAR,* 1942 (CAERC box 53, folder 24).
25. B. Jewell, *Pike CEAR,* 1945 (CAERC box 123, folder 37); H. H. Madaus, *Warren CEAR,* 1942 (CAERC box 53, folder 24); D. L. Grimes, *Putnam CEAR,* 1943 (CAERC box 29, folder 27).
26. H. E. Abbott, *Marion CEAR,* 1943 (CAERC box 111, folder 10); E. L. Hartman, *LaPorte CEAR,* 1942 (CAERC box 105, folder 17); S. B. Pershing, *Tippecanoe CEAR,* 1943 (CAERC box 44, folder 14).

27. D. L. Grimes, *Putnam CEAR,* 1942 (CAERC box 29, folder 26); R. N. Atkinson, *Orange CEAR,* 1942 (CAERC box 120, folder 25).
28. N. S. Hadley, *Parke CEAR,* 1942 (CAERC box 121, folder 51); W. J. Record, *Morgan CEAR,* 1942 (CAERC box 117, folder 3).
29. L. E. Cutler, *Lake CEAR,* 1943 (CAERC box 107, folder 36).
30. H. E. Abbott, *Marion CEAR,* 1943 (CAERC box 111, folder 10).
31. Ibid.
32. W. P. Stall, *Jackson CEAR,* 1943 (CAERC box 97, folder 8); H. E. Abbott, *Marion CEAR,* 1942 (CAERC box 111, folder 7); D. L. Grimes, *Putnam CEAR,* 1942 (CAERC box 29, folder 26); L. M. Butler, *Marshall CEAR,* 1942 (CAERC box 112, folder 32); L. E. Cutler, *Lake CEARs,* 1944–45 (CAERC box 107, folders 37–38).
33. W. J. Emerson, *White CEAR,* 1942 (CAERC box 59, folder 5); H. E. Abbott, *Marion CEAR,* 1942 (CAERC box 111, folder 7).
34. J. McCulloch, *Scott CEAR,* 1943 (CAERC box 36, folder 24).
35. W. P. Stall, *Jackson CEAR,* 1944 (CAERC box 97, folder 9); H. S. Benson, *Knox CEAR,* 1944 (CAERC box 103, folder 2); W. C. Haynes, *Madison CEAR,* 1944 (CAERC box 109, folder 40); J. L. Guthrie, *Noble CEAR,* 1944 (CAERC box 118, folder 37); E. A. Johnson, *Howard CEAR,* 1944 (CAERC box 95, folder 12).
36. L. P. Geyer, *Jefferson CEAR,* 1944 (CAERC box 100, folder 11).
37. W. P. Stall, *Jackson CEAR,* 1943 (CAERC box 97, folder 8).
38. W. U. Rusk, *Huntington CEAR,* 1943 (CAERC box 96, folder 19); C. D. Chambers, *Perry CEAR,* 1942 (CAERC box 122, folder 38); J. E. Wylie, *Miami CEAR,* 1942 (CAERC box 114, folder 13); B. T. Bond, *Warrick CEAR,* 1942 (CAERC box 54, folder 36).
39. H. E. Abbott, *Marion CEAR,* 1942 (CAERC box 111, folder 7); L. J. Fultz, *Washington CEAR,* 1942 (CAERC box 55, folder 44).
40. W. P. Stall, *Jackson CEAR,* 1942 (CAERC box 97, folder 7).
41. W. U. Rusk, *Huntington CEAR,* 1942 (CAERC box 96, folder 18).
42. B. V. Widney, *Whitley CEAR,* 1942 (CAERC box 60, folder 3); W. P. Stall, *Jackson CEAR,* 1942 (CAERC box 97, folder 7); J. McCulloch, *Scott CEAR,* 1943 (CAERC box 36, folder 24).
43. M. S. Smith, *Wells CEAR,* 1943 (CAERC box 58, folder 29); J. F. Armstrong, *Lawrence CEAR,* 1943 (CAERC box 108, folder 42); C. A. Langston, *Jay CEAR,* 1943 (CAERC box 99, folder 1); W. U. Rusk, *Huntington CEAR,* 1942 (CAERC box 96, folder 18); L. E. Cutler, *Lake CEAR,* 1943 (CAERC box 107, folder 36).
44. W. U. Rusk, *Huntington CEAR,* 1943 (CAERC box 96, folder 19).
45. W. C. Haynes, *Madison CEAR,* 1944 (CAERC box 109, folder 40).
46. M. S. Smith, *Wells CEAR,* 1943 (CAERC box 58, folder 29); J. F. Armstrong, *Lawrence CEAR,* 1943 (CAERC box 108, folder 42); L. E. Cutler, *Lake CEAR,* 1943 (CAERC box 107, folder 36); W. U. Rusk, *Huntington CEAR,* 1943 (CAERC box 96, folder 19).
47. J. F. Armstrong, *Lawrence CEAR,* 1943 (CAERC box 108, folder 42).
48. M. S. Smith, *Wells CEAR,* 1943 (CAERC box 58, folder 29).
49. J. McCulloch, *Scott CEAR,* 1943 (CAERC box 36, folder 24).
50. S. B. Pershing, *Tippecanoe CEAR,* 1943 (CAERC box 44, folder 14).
51. W. U. Rusk, *Huntington CEAR,* 1943 (CAERC box 96, folder 19); C. D. Chambers, *Perry CEAR,* 1942 (CAERC box 122, folder 38); C. D. Chambers, *Perry CEAR,* 1942 (CAERC box 122, folder 38); W. M. Clary, *Tipton CEAR,* 1942 (CAERC box 45, folder 23).
52. J. McCulloch, *Scott CEAR,* 1943 (CAERC box 36, folder 24); W. M. Clary, *Tipton CEAR,* 1942 (CAERC box 45, folder 23).

53. J. McCulloch, *Scott CEAR,* 1943 (CAERC box 36, folder 24).
54. E. A. Chatham, *Perry CEAR,* 1945 (CAERC box 122, folder 41).
55. R. R. Zell, *Newton CEAR,* 1944 (CAERC box 117, folder 45).
56. F. I. Hoover, *Owen CEAR,* 1944 (CAERC box 121, folder 5).
57. R. N. Atkinson, *Orange CEAR,* 1945 (CAERC box 120, folder 28).
58. M. S. Smith, *Wells CEAR,* 1943 (CAERC box 58, folder 29); R. C. Stangland, *Howard CEAR,* 1943 (CAERC box 14, folder 1).
59. W. M. Clary, *Tipton CEAR,* 1942 (CAERC box 45, folder 23); R. L. Campbell, *Vermillion CEAR,* 1945 (CAERC box 49, folder 40); H. E. Abbott, *Marion CEAR,* 1943 (CAERC box 111, folder 10); M. J. Peterson, *Vermillion CEAR,* 1943 (CAERC box 49, folder 38).
60. W. U. Rusk, *Huntington CEAR,* 1942 (CAERC box 96, folder 18).
61. R. R. Zell, *Newton CEAR,* 1943 (CAERC box 117, folder 44).
62. M. J. Peterson, *Vermillion CEAR,* 1943 (CAERC box 49, folder 38); R. R. Zell, *Newton CEAR,* 1943 (CAERC box 117, folder 44).
63. E. S. Bogue, *Pike CEAR,* 1944 (CAERC box 123, folder 36).
64. W. U. Rusk, *Huntington CEAR,* 1944 (CAERC box 96, folder 20).
65. L. P. Geyer, *Jefferson CEAR,* 1944 (CAERC box 100, folder 11).
66. J. McCulloch, *Scott CEAR,* 1942 (CAERC box 36, folder 23); F. I. Hoover, *Owen CEAR,* 1942 (CAERC box 121, folder 3); W. J. Emerson, *White CEAR,* 1943 (CAERC box 59, folder 6); C. D. Chambers, *Perry CEAR,* 1942 (CAERC box 122, folder 38); L. J. Fultz, *Washington CEAR,* 1942 (CAERC box 55, folder 44).
67. J. F. Armstrong, *Lawrence CEAR,* 1943 (CAERC box 108, folder 42); R. L. Case, *LaGrange CEAR,* 1942 (CAERC box 106, folder 3); F. I. Hoover, *Owen CEAR,* 1942 (CAERC box 121, folder 3).
68. C. E. Olson, *Pulaski CEAR,* 1943 (CAERC box 28, folder 11).
69. M. J. Peterson, *Vermillion CEAR,* 1943 (CAERC box 49, folder 38); J. McCulloch, *Scott CEAR,* 1945 (CAERC box 36, folder 26); R. L. Case, *LaGrange CEAR,* 1942 (CAERC box 106, folder 3).
70. A. C. Shideler, *Starke CEAR,* 1943 (CAERC box 40, folder 1); C. E. Olson, *Pulaski CEAR,* 1943 (CAERC box 28, folder 11).
71. A. C. Shideler, *Starke CEAR,* 1944 (CAERC box 40, folder 2).
72. G. T. Harris, *Ripley CEAR,* 1942 (CAERC box 31, folder 44); T. M. Grabill, *Steuben CEAR,* 1942 (CAERC box 40, folder 50); M. A. Nye, *Noble CEAR,* 1942 (CAERC box 118, folder 35).
73. E. E. Fultz, *Jasper CEAR,* 1942 (CAERC box 98, folder 12).
74. J. F. Armstrong, *Lawrence CEAR,* 1942 (CAERC box 108, folder 41); S. B. Scott, *Johnson CEAR,* 1942 (CAERC box 101, folder 41).
75. G. T. Harris, *Ripley CEAR,* 1942 (CAERC box 31, folder 44).
76. W. P. Stall, *Jackson CEAR,* 1942 (CAERC box 97, folder 7).
77. W. U. Rusk, *Huntington CEAR,* 1944 (CAERC box 96, folder 20); T. M. Grabill, *Steuben CEAR,* 1943 (CAERC box 40, folder 51); E. F. Brown, *Rush CEAR,* 1942 (CAERC box 33, folder 8); L. E. Husted, *Randolph CEAR,* 1944 (CAERC box 30, folder 43).
78. B. T. Bond, *Warrick CEAR,* 1943 (CAERC box 54, folder 37); A. C. Shideler, *Starke CEAR,* 1943 (CAERC box 40, folder 1); W. C. Haynes, *Madison CEAR,* 1943 (CAERC box 109, folder 39); J. McCulloch, *Scott CEAR,* 1944 (CAERC box 36, folder 25).
79. L. M. Butler, *Marshall CEAR,* 1942 (CAERC box 112, folder 32).
80. A. C. Shideler, *Starke CEAR,* 1943 (CAERC box 40, folder 1).

81. L. E. Cooper, *Spencer CEAR,* 1946 (CAERC box 38, folder 52); J. McCulloch, *Scott CEAR,* 1945 (CAERC box 36, folder 26).
82. R. R. Zell, *Newton CEAR,* 1944 (CAERC box 117, folder 45).
83. R. L. Campbell, *Vermillion CEAR,* 1945 (CAERC box 49, folder 40); W. K. Delaplane, *Wabash CEAR,* 1946 (CAERC box 52, folder 31).
84. W. P. Stall, *Jackson CEAR,* 1944 (CAERC box 97, folder 9).
85. A. V. Keesling, *Bartholomew CEAR,* 1939 (CAERC box 63, folder 30).
86. M. S. Smith, *Wells CEAR,* 1946 (CAERC box 58, folder 32); F. I. Hoover, *Owen CEAR,* 1945 (CAERC box 121, folder 6).
87. O. H. McNary, *Switzerland CEAR,* 1944 (CAERC box 42, folder 58); E. A. Chatham, *Perry CEAR,* 1945 (CAERC box 122, folder 41).
88. J. L. Guthrie, *Union CEAR,* 1943 (CAERC box 46, folder 25).
89. Ibid.
90. R. C. Garrison, *Union CEAR,* 1944 (CAERC box 46, folder 26).
91. H. H. Madaus, *Warren CEAR,* 1942 (CAERC box 53, folder 24).
92. M. S. Smith, *Wells CEAR,* 1946 (CAERC box 58, folder 32); L. E. Cooper, *Spencer CEAR,* 1946 (CAERC box 38, folder 52).
93. G. T. Harris, *Ripley CEAR,* 1943 (CAERC box 31, folder 45); W. J. Emerson, *White CEAR,* 1943 (CAERC box 59, folder 6); W. Rogers, *Monroe CEAR,* 1945 (CAERC box 115, folder 28); H. S. Benson, *Knox CEAR,* 1942 (CAERC box 102, folder 40); R. C. Stangland, *Howard CEAR,* 1943 (CAERC box 14, folder 1); E. S. Bogue, *Pike CEAR,* 1942 (CAERC box 123, folder 34).
94. L. P. Geyer, *Jefferson CEAR,* 1943 (CAERC box 100, folder 10).
95. C. D. Chambers, *Perry CEAR,* 1943 (CAERC box 122, folder 39).
96. S. W. Milligan, *Wayne CEAR,* 1944 (CAERC box 56, folder 32); E. L. Hartman, *LaPorte CEAR,* 1943 (CAERC box 105, folder 19).
97. R. N. Atkinson, *Orange CEAR,* 1943 (CAERC box 120, folder 26).
98. H. E. Abbott, *Marion CEAR,* 1942 (CAERC box 111, folder 7).
99. Ibid.
100. W. P. Stall, *Jackson CEAR,* 1944 (CAERC box 97, folder 9).
101. W. P. Stall, *Jackson CEAR,* 1942 (CAERC box 97, folder 7); R. L. Case, *LaGrange CEAR,* 1942 (CAERC box 106, folder 3).
102. F. I. Hoover, *Owen CEAR,* 1942 (CAERC box 121, folder 3).
103. R. L. Case, *LaGrange CEAR,* 1943 (CAERC box 106, folder 4).
104. Ibid.
105. C. E. Olson, *Pulaski CEAR,* 1942 (CAERC box 28, folder 10).
106. E. L. Hartman, *LaPorte CEAR,* 1943 (CAERC box 105, folder 19).
107. *Purdue University Department of Agricultural Extension Annual Report,* 1941–43, 108; W. K. Delaplane, *Wabash CEAR,* 1946 (CAERC box 52, folder 31); H. E. Abbott, *Marion CEAR,* 1942 (CAERC box 111, folder 7); W. P. Stall, *Jackson CEAR,* 1942 (CAERC box 97, folder 7).
108. *Purdue University Department of Agricultural Extension Annual Report,* 1945, 13.
109. J. F. Armstrong, *Lawrence CEAR,* 1945 (CAERC box 108, folder 44).
110. E. S. Bogue, *Pike CEAR,* 1944 (CAERC box 123, folder 36); L. E. Husted, *Randolph CEAR,* 1944 (CAERC box 30, folder 43); W. P. Stall, *Jackson CEAR,* 1944 (CAERC box 97, folder 9); L. P. Geyer, *Jefferson CEAR,* 1944 (CAERC box 100, folder 11).
111. H. E. Abbott, *Marion CEAR,* 1944 (CAERC box 111, folder 13).

112. L. E. Cutler, *Lake CEARs,* 1943–44 (CAERC box 107, folders 36–37); E. S. Bogue, *Pike CEAR,* 1944 (CAERC box 123, folder 36); M. S. Smith, *Wells CEAR,* 1944 (CAERC box 58, folder 30); S. B. Scott, *Johnson CEAR,* 1944 (CAERC box 101, folder 43).
113. H. E. Abbott, *Marion CEAR,* 1946 (CAERC box 111, folder 15); A. K. Wissman, *Jennings CEAR,* 1946 (CAERC box 101, folder 2); G. A. Sowers, *Montgomery CEAR,* 1946 (CAERC box 115, folder 59).

CHAPTER 32

The Armed Forces Have Taken 1,500 Men, Including Farmers. Why Wouldn't It Create Many Problems?

1. E. C. Bird, *St. Joseph CEAR,* 1943 (CAERC box 35, folder 29); H. E. Abbott, *Marion CEAR,* 1942 (CAERC box 111, folder 7); "New Rules for Draft Change Old Policies," *Times* (Munster, IN), 15 Jan. 1942, 2.
2. "Draft Board Studying New Regulations," *Noblesville Ledger* (Noblesville, IN), 5 Jan. 1942, 1.
3. "Manpower: Sweeping Changes Halt Enlistments, Cut Top Draft Age to 38, Give McNutt Selective Service Control," *Life* 13, no. 25 (21 Dec. 1942): 27, https://books.google.com/books?id=NVEEAAAAMBAJ&pg=PA27#v=onepage&q&f=false; "Farm Deferments May Be Revoked, Asserts Hershey," *Indianapolis News,* 11 Jan. 1944, 4; "Farm Draft 'Units' Scrapped but Maximum Call Is Urged," *South Bend Tribune* (South Bend, IN), 2 Apr. 1944, 1.
4. "Induction Statistics: World War II," U.S. Selective Service System (website), accessed 20 Apr. 2022, https://www.sss.gov/history-and-records/induction-statistics/.
5. "Draft Board Studying New Regulations," 1.
6. Ibid.; "Feb. 16 Is Set as Draft Date for Men 20–45," *Franklin Evening Star* (Franklin, IN), 5 Jan. 1942, 1.
7. Editors of Encyclopaedia Britannica, "Selective Service System," *Encyclopedia Britannica* (website), 24 May 2016, https://www.britannica.com/topic/Selective-Service-System#ref1193841.
8. "Draft Regulations Are Streamlined," *Indianapolis Star,* 1 Jan. 1942, 26.
9. "Draft Board Rechecks 1H's," *Republic* (Columbus, IN), 6 Jan. 1942, 2.
10. "Draft Boards Will Call Men Over 28 Soon," *Times* (Munster, IN), 7 Jan. 1942, 1.
11. "28 Year Old Men to Be Classified Again for Draft," *Garrett Clipper* (Garrett, IN), 5 Jan. 1942, 4; "Army Physical Standards Lowered; Draft Boards Ordered to Make Reclassification," *Palladium-Item* (Richmond, IN), 17 Feb. 1942, 5.
12. C. E. Olson, *Pulaski CEAR,* 1942 (CAERC box 28, folder 10); L. P. Geyer, *Jefferson CEAR,* 1944 (CAERC box 100, folder 11); "New Rules for Draft Change Old Policies," 2.
13. O. B. Riggs, *Posey CEAR,* 1943 (CAERC box 125, folder 29).
14. M. A. Nye, *Noble CEAR,* 1942 (CAERC box 118, folder 35).
15. E. C. Bird, *St. Joseph CEAR,* 1942 (CAERC box 35, folder 28).
16. M. S. Smith, *Wells CEAR,* 1943 (CAERC box 58, folder 29); M. S. Smith, *Wells CEAR,* 1944 (CAERC box 58, folder 30).
17. O. B. Riggs, *Posey CEAR,* 1943 (CAERC box 125, folder 29).
18. L. M. Butler, *Marshall CEAR,* 1942 (CAERC box 112, folder 32); H. E. Abbott, *Marion CEAR,* 1943 (CAERC box 111, folder 10).
19. L. M. Butler, *Marshall CEAR,* 1943 (CAERC box 112, folder 33); G. T. Harris, *Ripley CEAR,* 1942 (CAERC box 31, folder 44).
20. S. B. Pershing, *Tippecanoe CEAR,* 1944 (CAERC box 44, folder 16).

21. William A. DeHart and Mapheus Smith, *Agricultural Deferment* (U.S. Selective Service System) (Washington, DC: Government Printing Office, 1947), 50, https://www.google.com/books/edition/_/gcVfAAAAIAAJ?hl=en.
22. J. A. Clark, *Kosciusko CEAR,* 1943 (CAERC box 104, folder 19).
23. O. B. Riggs, *Posey CEAR,* 1943 (CAERC box 125, folder 29).
24. L. P. Geyer, *Jefferson CEAR,* 1942 (CAERC box 100, folder 9).
25. C. E. Olson, *Pulaski CEAR,* 1943 (CAERC box 28, folder 11); J. F. Armstrong, *Lawrence CEAR,* 1943 (CAERC box 108, folder 42).
26. C. A. Langston, *Jay CEAR,* 1943 (CAERC box 99, folder 1).
27. W. J. Record, *Morgan CEAR,* 1943 (CAERC box 117, folder 4); L. P. Geyer, *Jefferson CEAR,* 1942 (CAERC box 100, folder 9).
28. "Farmers to Keep Data for Draft," *Indianapolis News,* 5 Jan. 1943, 15.
29. N. S. Hadley, *Parke CEAR,* 1944 (CAERC box 121, folder 55); R. N. Atkinson, *Orange CEAR,* 1945 (CAERC box 120, folder 28).
30. "Draft Boards Given a Pattern," *Times* (Munster, IN), 5 Jan. 1943, 2; "Point System for Farm Deferments Is Explained," *Evening Republican* (Columbus, IN), 5 Dec. 1942, 2.
31. R. N. Atkinson, *Orange CEAR,* 1943 (CAERC box 120, folder 26).
32. W. C. Haynes, *Madison CEARs,* 1943–44 (CAERC box 109, folders 39–40).
33. E. S. Bogue, *Pike CEAR,* 1943 (CAERC box 123, folder 35).
34. I. O. Haug, *Ripley CEAR,* 1944 (CAERC box 31, folder 46).
35. N. S. Hadley, *Parke CEAR,* 1944 (CAERC box 121, folder 55).
36. F. E. DeLaCroix, *Jasper CEAR,* 1944 (CAERC box 98, folder 14).
37. E. A. Johnson, *Howard CEAR,* 1944 (CAERC box 95, folder 12); W. U. Rusk, *Huntington CEAR,* 1944 (CAERC box 96, folder 20); G. A. Sowers, *Montgomery CEAR,* 1942 (CAERC box 115, folder 54); J. A. Clark, *Kosciusko CEAR,* 1945 (CAERC box 104, folder 21).
38. I. O. Haug, *Ripley CEAR,* 1944 (CAERC box 31, folder 46).
39. E. C. Bird, *St. Joseph CEAR,* 1944 (CAERC box 35, folder 30); A. C. Shideler, *Starke CEAR,* 1945 (CAERC box 40, folder 3); R. N. Atkinson, *Orange CEAR,* 1944 (CAERC box 120, folder 27).
40. J. C. Sliger, *Sullivan CEAR,* 1943 (CAERC box 41, folder 54).
41. O. B. Riggs, *Posey CEAR,* 1944 (CAERC box 125, folder 31).
42. Ibid.
43. Ibid.; E. A. Johnson, *Howard CEAR,* 1944 (CAERC box 95, folder 12); M. S. Smith, *Wells CEAR,* 1943 (CAERC box 58, folder 29); C. A. Langston, *Jay CEAR,* 1942 (CAERC box 98, folder 51).
44. O. B. Riggs, *Posey CEAR,* 1944 (CAERC box 125, folder 31).
45. E. C. Bird, *St. Joseph CEAR,* 1945 (CAERC box 35, folder 31).
46. W. P. Stall, *Jackson CEAR,* 1944 (CAERC box 97, folder 9); G. A. Sowers, *Montgomery CEAR,* 1942 (CAERC box 115, folder 54); W. U. Rusk, *Huntington CEAR,* 1943 (CAERC box 96, folder 19).
47. G. A. Sowers, *Montgomery CEAR,* 1942 (CAERC box 115, folder 54); J. F. Armstrong, *Lawrence CEAR,* 1944 (CAERC box 108, folder 43).
48. J. L. Guthrie, *Union CEAR,* 1942 (CAERC box 46, folder 24).
49. E. C. Bird, *St. Joseph CEAR,* 1944 (CAERC box 35, folder 30).
50. W. P. Stall, *Jackson CEAR,* 1944 (CAERC box 97, folder 9).
51. O. H. McNary, *Switzerland CEAR,* 1943 (CAERC box 42, folder 57).
52. W. U. Rusk, *Huntington CEAR,* 1944 (CAERC box 96, folder 20); L. E. Husted, *Randolph CEAR,* 1943 (CAERC box 30, folder 42).

53. J. F. Armstrong, *Lawrence CEAR,* 1943 (CAERC box 108, folder 42).
54. C. E. Olson, *Pulaski CEAR,* 1943 (CAERC box 28, folder 11).
55. A. K. Wissman, *Jennings CEAR,* 1945 (CAERC box 101, folder 1).
56. L. P. Geyer, *Jefferson CEAR,* 1944 (CAERC box 100, folder 11); "Returned Yank Says Front-Line G.I.'s Favor Compulsory Military Training After the War," *Palladium-Item* (Richmond, IN), 18 Jan. 1945, 10; "Farm Draft 'Units' Scrapped," 1.
57. "Many under 26 Hold Farm Deferments," *Palladium-Item* (Richmond, IN), 3 May 1944, 5.
58. "Farm Deferments May Be Revoked," 4.
59. "Farmers Face Tighter Draft," *South Bend Tribune* (South Bend, IN), 11 Feb. 1944, 4; "Farm Draft 'Units' Scrapped," 1.
60. "Farm Draft 'Units' Scrapped," 10.
61. L. P. Geyer, *Jefferson CEAR,* 1944 (CAERC box 100, folder 11).
62. J. E. Wylie, *Miami CEAR,* 1944 (CAERC box 114, folder 15).
63. Ibid.
64. H. K. Rippey, *Jay CEAR,* 1945 (CAERC box 99, folder 3); C. E. Olson, *Pulaski CEAR,* 1945 (CAERC box 28, folder 13); J. F. Armstrong, *Lawrence CEAR,* 1944 (CAERC box 108, folder 43); J. E. Wylie, *Miami CEAR,* 1945 (CAERC box 114, folder 16); M. A. Nye, *Porter CEAR,* 1945 (CAERC box 124, folder 29); A. C. Shideler, *Starke CEAR,* 1945 (CAERC box 40, folder 3).
65. H. K. Rippey, *Jay CEAR,* 1945 (CAERC box 99, folder 3).
66. F. E. DeLaCroix, *Jasper CEAR,* 1944 (CAERC box 98, folder 14); N. S. Hadley, *Parke CEAR,* 1944 (CAERC box 121, folder 55); J. E. Wylie, *Miami CEAR,* 1944 (CAERC box 114, folder 15); W. C. Haynes, *Madison CEAR,* 1944 (CAERC box 109, folder 40).
67. J. E. Wylie, *Miami CEAR,* 1945 (CAERC box 114, folder 16); B. V. Widney, *Whitley CEAR,* 1945 (CAERC box 60, folder 6); R. N. Atkinson, *Orange CEAR,* 1945 (CAERC box 120, folder 28).
68. J. L. Guthrie, *Noble CEAR,* 1945 (CAERC box 118, folder 38).
69. "No Loosening of Farm Deferments," *Lafayette Journal and Courier* (Lafayette, IN), 20 Mar. 1945, 1.
70. J. F. Armstrong, *Lawrence CEAR,* 1945 (CAERC box 108, folder 44); J. A. Clark, *Kosciusko CEAR,* 1945 (CAERC box 104, folder 21); A. K. Wissman, *Jennings CEAR,* 1945 (CAERC box 101, folder 1); C. L. Spuller, *Rush CEAR,* 1945 (CAERC box 33, folder 13).
71. W. P. Stall, *Jackson CEAR,* 1944 (CAERC box 97, folder 9).
72. R. N. Atkinson, *Orange CEAR,* 1944 (CAERC box 120, folder 27).
73. B. V. Widney, *Whitley CEAR,* 1945 (CAERC box 60, folder 6).
74. A. C. Shideler, *Starke CEAR,* 1945 (CAERC box 40, folder 3).
75. "Indiana and Another World War," The History Museum (of South Bend, IN) (website), accessed 13 May 2022, https://www.historymuseumsb.org/indiana-and-another-world-war/.

CHAPTER 33

All Agricultural Workers Seeking Employment in Industrial Factories Must Have a Statement of Transfer From the County Agent

1. L. E. Cutler, *Lake CEAR,* 1942 (CAERC box 107, folder 34); W. K. Delaplane, *Wabash CEAR,* 1942 (CAERC box 52, folder 27); J. E. Wylie, *Miami CEAR,* 1944 (CAERC box 114, folder 15).
2. W. K. Delaplane, *Wabash CEAR,* 1944 (CAERC box 52, folder 29); L. E. Cutler, *Lake CEAR,* 1942 (CAERC box 107, folder 34).
3. R. N. Atkinson, *Orange CEAR,* 1946 (CAERC box 120, folder 29).
4. L. M. Butler, *Marshall CEARs,* 1942–43 (CAERC box 112, folders 32–33).

5. E. S. Bogue, *Pike CEAR,* 1942 (CAERC box 123, folder 34); B. V. Widney, *Whitley CEAR,* 1943 (CAERC box 60, folder 4); J. E. Wylie, *Miami CEAR,* 1942 (CAERC box 114, folder 13); A. M. Bishea, *Vanderburgh CEAR,* 1942 (CAERC box 48, folder 4).
6. T. M. Grabill, *Steuben CEAR,* 1942 (CAERC box 40, folder 50); O. B. Riggs, *Posey CEAR,* 1942 (CAERC box 125, folder 27).
7. J. F. Armstrong, *Lawrence CEAR,* 1942 (CAERC box 108, folder 41).
8. J. F. Armstrong, *Lawrence CEARs,* 1943 and 1945 (CAERC box 108, folders 42 and 44).
9. J. F. Armstrong, *Lawrence CEAR,* 1942 (CAERC box 108, folder 41).
10. A. M. Bishea, *Vanderburgh CEAR,* 1943 (CAERC box 48, folder 7); C. E. Olson, *Pulaski CEAR,* 1942 (CAERC box 28, folder 10); O. B. Riggs, *Posey CEAR,* 1944 (CAERC box 125, folder 31); J. E. Wylie, *Miami CEAR,* 1944 (CAERC box 114, folder 15); G. A. Sowers, *Montgomery CEAR,* 1944 (CAERC box 115, folder 56).
11. O. B. Riggs, *Posey CEAR,* 1943 (CAERC box 125, folder 29).
12. C. E. Olson, *Pulaski CEAR,* 1945 (CAERC box 28, folder 13).
13. J. McCulloch, *Scott CEAR,* 1942 (CAERC box 36, folder 23); W. P. Stall, *Jackson CEAR,* 1943 (CAERC box 97, folder 8); L. P. Geyer, *Jefferson CEAR,* 1943 (CAERC box 100, folder 10).
14. E. F. Brown, *Rush CEAR,* 1943 (CAERC box 33, folder 10).
15. H. E. Abbott, *Marion CEAR,* 1942 (CAERC box 111, folder 7).
16. E. C. Bird, *St. Joseph CEAR,* 1942 (CAERC box 35, folder 28).
17. H. E. Abbott, *Marion CEAR,* 1942 (CAERC box 111, folder 7); H. E. Abbott, *Marion CEAR,* 1943 (CAERC box 111, folder 10); O. B. Riggs, *Posey CEAR,* 1944 (CAERC box 125, folder 31).
18. J. F. Armstrong, *Lawrence CEAR,* 1942 (CAERC box 108, folder 41).
19. C. E. Olson, *Pulaski CEAR,* 1942 (CAERC box 28, folder 10).
20. M. A. Nye, *Porter CEAR,* 1944 (CAERC box 124, folder 28); J. McCulloch, *Scott CEAR,* 1943 (CAERC box 36, folder 24).
21. W. C. Haynes, *Madison CEAR,* 1943 (CAERC box 109, folder 39).
22. L. E. Cutler, *Lake CEAR,* 1942 (CAERC box 107, folder 34); H. E. Abbott, *Marion CEAR,* 1946 (CAERC box 111, folder 17).
23. A. M. Bishea, *Vanderburgh CEAR,* 1944 (CAERC box 48, folder 9); C. A. Langston, *Jay CEAR,* 1943 (CAERC box 99, folder 1); R. N. Atkinson, *Orange CEAR,* 1943 (CAERC box 120, folder 26); B. T. Bond, *Warrick CEAR,* 1943 (CAERC box 54, folder 37).
24. C. L. Spuller, *Rush CEAR,* 1945 (CAERC box 33, folder 13); E. F. Brown, *Rush CEAR,* 1943 (CAERC box 33, folder 10); C. A. Langston, *Jay CEAR,* 1943 (CAERC box 99, folder 1); R. N. Atkinson, *Orange CEAR,* 1944 (CAERC box 120, folder 27); B. T. Bond, *Warrick CEAR,* 1943 (CAERC box 54, folder 37).
25. W. C. Haynes, *Madison CEAR,* 1945 (CAERC box 109, folder 41); D. L. Grimes, *Putnam CEAR,* 1942 (CAERC box 29, folder 26); J. C. Sliger, *Sullivan CEAR,* 1943 (CAERC box 41, folder 54).
26. J. McCulloch, *Scott CEAR,* 1942 (CAERC box 36, folder 23).
27. O. B. Riggs, *Posey CEAR,* 1944 (CAERC box 125, folder 31).
28. A. M. Bishea, *Vanderburgh CEAR,* 1943 (CAERC box 48, folder 7).
29. W. C. Haynes, *Madison CEAR,* 1945 (CAERC box 109, folder 41).
30. J. C. Sliger, *Sullivan CEAR,* 1944 (CAERC box 41, folder 55).
31. Ibid.
32. A. M. Bishea, *Vanderburgh CEAR,* 1943 (CAERC box 48, folder 7); E. A. Johnson, *Howard CEAR,* 1944 (CAERC box 95, folder 12).

33. H. E. Abbott, *Marion CEAR,* 1944 (CAERC box 111, folder 13); D. L. Grimes, *Putnam CEAR,* 1942 (CAERC box 29, folder 26).
34. M. A. Nye, *Noble CEAR,* 1943 (CAERC box 118, folder 36); B. V. Widney, *Whitley CEAR,* 1944 (CAERC box 60, folder 5).
35. O. B. Riggs, *Posey CEAR,* 1943 (CAERC box 125, folder 29); J. McCulloch, *Scott CEAR,* 1944 (CAERC box 36, folder 25).
36. L. M. Butler, *Marshall CEAR,* 1942 (CAERC box 112, folder 32).
37. L. E. Cutler, *Lake CEAR,* 1946 (CAERC box 107, folder 40).
38. C. E. Olson, *Pulaski CEAR,* 1943 (CAERC box 28, folder 11).
39. W. C. Haynes, *Madison CEAR,* 1943 (CAERC box 109, folder 39); L. M. Butler, *Marshall CEAR,* 1943 (CAERC box 112, folder 33); M. S. Smith, *Wells CEAR,* 1944 (CAERC box 58, folder 30).
40. M. S. Smith, *Wells CEAR,* 1944 (CAERC box 58, folder 30); L. M. Butler, *Marshall CEAR,* 1943 (CAERC box 112, folder 33); A. M. Bishea, *Vanderburgh CEAR,* 1943 (CAERC box 48, folder 7).
41. C. E. Olson, *Pulaski CEAR,* 1943 (CAERC box 28, folder 11); W. C. Haynes, *Madison CEAR,* 1944 (CAERC box 109, folder 40); J. E. Wylie, *Miami CEAR,* 1945 (CAERC box 114, folder 16); E. L. Hartman, *LaPorte CEAR,* 1944 (CAERC box 105, folder 21); N. S. Hadley, *Parke CEAR,* 1944 (CAERC box 121, folder 55).
42. W. C. Haynes, *Madison CEAR,* 1944 (CAERC box 109, folder 40).
43. Ibid.
44. M. S. Smith, *Wells CEAR,* 1944 (CAERC box 58, folder 30); L. M. Butler, *Marshall CEAR,* 1943 (CAERC box 112, folder 33); J. E. Wylie, *Miami CEAR,* 1944 (CAERC box 114, folder 15); C. A. Langston, *Jay CEAR,* 1944 (CAERC box 99, folder 2); A. M. Bishea, *Vanderburgh CEAR,* 1943 (CAERC box 48, folder 7); J. L. Guthrie, *Noble CEAR,* 1944 (CAERC box 118, folder 37).
45. H. S. Benson, *Knox CEAR,* 1945 (CAERC box 103, folder 3); J. F. Armstrong, *Lawrence CEAR,* 1943 (CAERC box 108, folder 42).
46. L. E. Husted, *Randolph CEAR,* 1945 (CAERC box 30, folder 44).
47. T. M. Grabill, *Steuben CEAR,* 1944 (CAERC box 40, folder 52); L. P. Geyer, *Jefferson CEAR,* 1944 (CAERC box 100, folder 11).
48. W. C. Haynes, *Madison CEAR,* 1945 (CAERC box 109, folder 41); L. M. Butler, *Marshall CEAR,* 1943 (CAERC box 112, folder 33); T. M. Grabill, *Steuben CEAR,* 1944 (CAERC box 40, folder 52); C. A. Langston, *Jay CEAR,* 1943 (CAERC box 99, folder 1).
49. T. M. Grabill, *Steuben CEAR,* 1944 (CAERC box 40, folder 52).
50. L. P. Geyer, *Jefferson CEAR,* 1944 (CAERC box 100, folder 11).
51. O. H. McNary, *Switzerland CEAR,* 1944 (CAERC box 42, folder 58); L. P. Geyer, *Jefferson CEAR,* 1945 (CAERC box 100, folder 12).
52. C. E. Olson, *Pulaski CEAR,* 1944 (CAERC box 28, folder 12).
53. L. M. Butler, *Marshall CEAR,* 1944 (CAERC box 112, folder 34).
54. T. M. Grabill, *Steuben CEAR,* 1943 (CAERC box 40, folder 51).
55. T. M. Grabill, *Steuben CEAR,* 1944 (CAERC box 40, folder 52).
56. E. A. Johnson, *Howard CEAR,* 1944 (CAERC box 95, folder 12); H. H. Madaus, *Warren CEAR,* 1942 (CAERC box 53, folder 24).
57. L. E. Cutler, *Lake CEAR,* 1943 (CAERC box 107, folder 36); E. A. Johnson, *Howard CEAR,* 1944 (CAERC box 95, folder 12).

58. W. P. Stall, *Jackson CEAR,* 1943 (CAERC box 97, folder 8).
59. E. S. Bogue, *Pike CEAR,* 1943 (CAERC box 123, folder 35).
60. S. B. Scott, *Johnson CEAR,* 1942 (CAERC box 101, folder 41).
61. F. I. Hoover, *Owen CEAR,* 1942 (CAERC box 121, folder 3).
62. T. M. Grabill, *Steuben CEAR,* 1942 (CAERC box 40, folder 50).
63. W. P. Stall, *Jackson CEAR,* 1942 (CAERC box 97, folder 7); E. S. Bogue, *Pike CEAR,* 1942 (CAERC box 123, folder 34).
64. E. E. Fultz, *Jasper CEAR,* 1942 (CAERC box 98, folder 12); J. C. Sliger, *Sullivan CEAR,* 1943 (CAERC box 41, folder 54).
65. H. H. Madaus, *Warren CEAR,* 1942 (CAERC box 53, folder 24).
66. C. E. Olson, *Pulaski CEAR,* 1942 (CAERC box 28, folder 10); G. T. Harris, *Ripley CEAR,* 1943 (CAERC box 31, folder 45); L. J. Fultz, *Washington CEAR,* 1942 (CAERC box 55, folder 44); W. K. Delaplane, *Wabash CEAR,* 1944 (CAERC box 52, folder 29).
67. T. M. Grabill, *Steuben CEAR,* 1942 (CAERC box 40, folder 50); B. V. Widney, *Whitley CEAR,* 1945 (CAERC box 60, folder 6).
68. L. E. Cutler, *Lake CEAR,* 1942 (CAERC box 107, folder 34); S. Leaming, *Porter CEAR,* 1943 (CAERC box 124, folder 27); D. L. Grimes, *Putnam CEAR,* 1943 (CAERC box 29, folder 27); J. C. Sliger, *Sullivan CEAR,* 1943 (CAERC box 41, folder 54); W. J. Emerson, *White CEAR,* 1942 (CAERC box 59, folder 5).
69. N. S. Hadley, *Parke CEAR,* 1942 (CAERC box 121, folder 51).
70. H. S. Benson, *Knox CEAR,* 1943 (CAERC box 102, folder 42).
71. W. P. Stall, *Jackson CEAR,* 1943 (CAERC box 97, folder 8).
72. J. E. Wylie, *Miami CEAR,* 1944 (CAERC box 114, folder 15).

CHAPTER 34

Farm Women and Children Will Ride the Machinery to the Desired 10 Percent Increase in Production of Meat, Milk, and Eggs

1. W. P. Stall, *Jackson CEAR,* 1943 (CAERC box 97, folder 8); E. A. Johnson, *Howard CEAR,* 1942 (CAERC box 95, folder 10); C. E. Olson, *Pulaski CEAR,* 1943 (CAERC box 28, folder 11); C. A. Langston, *Jay CEAR,* 1942 (CAERC box 98, folder 51).
2. S. B. Pershing, *Tippecanoe CEAR,* 1943 (CAERC box 44, folder 14).
3. C. A. Langston, *Jay CEAR,* 1943 (CAERC box 99, folder 1).
4. W. P. Stall, *Jackson CEAR,* 1943 (CAERC box 97, folder 8); C. E. Olson, *Pulaski CEAR,* 1943 (CAERC box 28, folder 11); A. C. Shideler, *Starke CEAR,* 1943 (CAERC box 40, folder 1); S. Leaming, *Porter CEAR,* 1943 (CAERC box 124, folder 27); O. B. Riggs, *Posey CEAR,* 1943 (CAERC box 125, folder 29).
5. C. A. Langston, *Jay CEAR,* 1943 (CAERC box 99, folder 1); W. P. Stall, *Jackson CEAR,* 1943 (CAERC box 97, folder 8); *Purdue University Department of Agricultural Extension Annual Report*, 1941–43, 4.
6. *Purdue University Department of Agricultural Extension Annual Report,* 1941–43, 5.
7. T. M. Grabill, *Steuben CEAR,* 1942 (CAERC box 40, folder 50).
8. C. A. Langston, *Jay CEAR,* 1943 (CAERC box 99, folder 1); R. R. Zell, *Newton CEAR,* 1943 (CAERC box 117, folder 44); E. C. Bird, *St. Joseph CEAR,* 1943 (CAERC box 35, folder 29); L. J. Fultz, *Washington CEAR,* 1943 (CAERC box 55, folder 45); E. S. Bogue, *Pike CEAR,* 1943 (CAERC box 123, folder 35); G. T. Harris, *Ripley CEAR,* 1943 (CAERC box 31, folder 45).
9. G. T. Harris, *Ripley CEAR,* 1943 (CAERC box 31, folder 45).

10. M. A. Nye, *Noble CEAR,* 1942 (CAERC box 118, folder 35); S. W. Milligan, *Wayne CEAR,* 1942 (CAERC box 56, folder 29).
11. F. E. DeLaCroix, *Jasper CEAR,* 1943 (CAERC box 98, folder 13).
12. J. F. Armstrong, *Lawrence CEAR,* 1943 (CAERC box 108, folder 42).
13. W. P. Stall, *Jackson CEAR,* 1943 (CAERC box 97, folder 8).
14. Ibid.
15. W. U. Rusk, *Huntington CEAR,* 1943 (CAERC box 96, folder 19).
16. W. J. Record, *Morgan CEAR,* 1943 (CAERC box 117, folder 4); W. P. Stall, *Jackson CEAR,* 1943 (CAERC box 97, folder 8); S. B. Scott, *Johnson CEAR,* 1943 (CAERC box 101, folder 42); W. C. Haynes, *Madison CEAR,* 1945 (CAERC box 109, folder 41); S. W. Milligan, *Wayne CEAR,* 1943 (CAERC box 56, folder 30).
17. W. P. Stall, *Jackson CEAR,* 1943 (CAERC box 97, folder 8).
18. S. B. Scott, *Johnson CEAR,* 1943 (CAERC box 101, folder 42).
19. J. F. Armstrong, *Lawrence CEAR,* 1944 (CAERC box 108, folder 43).
20. C. A. Langston, *Jay CEAR,* 1943 (CAERC box 99, folder 1); E. A. Johnson, *Howard CEAR,* 1944 (CAERC box 95, folder 12); W. J. Emerson, *White CEAR,* 1943 (CAERC box 59, folder 6).
21. C. A. Langston, *Jay CEAR,* 1943 (CAERC box 99, folder 1).
22. L. M. Butler, *Marshall CEAR,* 1943 (CAERC box 112, folder 33).
23. D. L. Grimes, *Putnam CEAR,* 1943 (CAERC box 29, folder 27).
24. C. A. Langston, *Jay CEAR,* 1943 (CAERC box 99, folder 1).
25. E. S. Bogue, *Pike CEAR,* 1943 (CAERC box 123, folder 35); S. Leaming, *Porter CEAR,* 1942 (CAERC box 124, folder 26); S. B. Pershing, *Tippecanoe CEAR,* 1943 (CAERC box 44, folder 14); W. J. Record, *Morgan CEAR,* 1943 (CAERC box 117, folder 4).
26. E. A. Johnson, *Howard CEARs,* 1944–45 (CAERC box 95, folders 12–13); F. E. DeLaCroix, *Jasper CEAR,* 1944 (CAERC box 98, folder 14).
27. E. C. Bird, *St. Joseph CEAR,* 1945 (CAERC box 35, folder 31).
28. M. A. Nye, *Porter CEAR,* 1944 (CAERC box 124, folder 28).
29. S. Leaming, *Porter CEAR,* 1943 (CAERC box 124, folder 27).
30. C. E. Olson, *Pulaski CEAR,* 1943 (CAERC box 28, folder 11).
31. L. P. Geyer, *Jefferson CEAR,* 1943 (CAERC box 100, folder 10); J. F. Armstrong, *Lawrence CEAR,* 1943 (CAERC box 108, folder 42); S. Leaming, *Porter CEAR,* 1943 (CAERC box 124, folder 27); H. H. Madaus, *Warren CEAR,* 1943 (CAERC box 53, folder 25); *Purdue University Department of Agricultural Extension Annual Report,* 1941–43, 6.
32. F. E. DeLaCroix, *Jasper CEAR,* 1943 (CAERC box 98, folder 13); W. U. Rusk, *Huntington CEAR,* 1943 (CAERC box 96, folder 19); B. V. Widney, *Whitley CEAR,* 1943 (CAERC box 60, folder 4).
33. H. H. Madaus, *Warren CEAR,* 1943 (CAERC box 53, folder 25); J. A. Clark, *Kosciusko CEAR,* 1944 (CAERC box 104, folder 20).
34. R. L. Case, *LaGrange CEAR,* 1943 (CAERC box 106, folder 4).
35. S. Leaming, *Porter CEAR,* 1943 (CAERC box 124, folder 27).
36. R. C. Stangland, *Howard CEAR,* 1943 (CAERC box 14, folder 1); E. A. Johnson, *Howard CEAR,* 1945 (CAERC box 95, folder 13).
37. E. A. Johnson, *Howard CEAR,* 1944 (CAERC box 95, folder 12).
38. B. V. Widney, *Whitley CEAR,* 1943 (CAERC box 60, folder 4).
39. L. E. Cutler, *Lake CEAR,* 1943 (CAERC box 107, folder 36).
40. J. C. Sliger, *Sullivan CEAR,* 1943 (CAERC box 41, folder 54).

41. J. L. Guthrie, *Union CEAR,* 1943 (CAERC box 46, folder 25).
42. J. C. Sliger, *Sullivan CEAR,* 1943 (CAERC box 41, folder 54).
43. J. F. Armstrong, *Lawrence CEAR,* 1944 (CAERC box 108, folder 43).
44. Nancy Nakano Conner, "From Internment to Indiana: Japanese Americans, the War Relocation Authority, the Disciples of Christ, and Citizen Committees in Indianapolis," *Indiana Magazine of History* 102, no. 2 (June 2006): 100, https://scholarworks.iu.edu/journals/index.php/imh/article/view/12193; A. C. Shideler, *Starke CEAR,* 1944 (CAERC box 40, folder 2); M. S. Smith, *Wells CEAR,* 1944 (CAERC box 58, folder 30).
45. D. L. Grimes, *Putnam CEAR,* 1942 (CAERC box 29, folder 26); L. E. Cutler, *Lake CEAR,* 1942 (CAERC box 107, folder 34).
46. M. S. Smith, *Wells CEAR,* 1943 (CAERC box 58, folder 29); L. E. Cutler, *Lake CEAR,* 1942 (CAERC box 107, folder 34).
47. E. C. Bird, *St. Joseph CEAR,* 1942 (CAERC box 35, folder 28); E. C. Bird, *St. Joseph CEAR,* 1945 (CAERC box 35, folder 31); E. L. Hartman, *LaPorte CEAR,* 1944 (CAERC box 105, folder 21); W. C. Haynes, *Madison CEAR,* 1942 (CAERC box 109, folder 38); C. A. Langston, *Jay CEAR,* 1943 (CAERC box 99, folder 1).
48. J. A. Clark, *Kosciusko CEAR,* 1945 (CAERC box 104, folder 21); L. E. Cutler, *Lake CEAR,* 1942 (CAERC box 107, folder 34); C. E. Olson, *Pulaski CEAR,* 1944 (CAERC box 28, folder 12).
49. W. U. Rusk, *Huntington CEAR,* 1943 (CAERC box 96, folder 19).
50. R. N. Atkinson, *Orange CEAR,* 1942 (CAERC box 120, folder 25); M. A. Nye, *Noble CEAR,* 1942 (CAERC box 118, folder 35).
51. J. A. Wood, *Martin CEAR,* 1945 (CAERC box 113, folder 31).
52. B. T. Bond, *Warrick CEAR,* 1944 (CAERC box 54, folder 38).
53. S. W. Milligan, *Wayne CEAR,* 1944 (CAERC box 56, folder 32).
54. R. N. Atkinson, *Orange CEAR,* 1942 (CAERC box 120, folder 25).
55. Ibid.; F. E. DeLaCroix, *Jasper CEAR,* 1943 (CAERC box 98, folder 13); H. K. Rippey, *Jay CEAR,* 1946 (CAERC box 99, folder 5); H. E. Abbott, *Marion CEAR,* 1943 (CAERC box 111, folder 10); C. A. Langston, *Jay CEAR,* 1943 (CAERC box 99, folder 1).
56. C. A. Langston, *Jay CEAR,* 1942 (CAERC box 98, folder 51).
57. S. B. Scott, *Johnson CEAR,* 1943 (CAERC box 101, folder 42); O. H. McNary, *Switzerland CEAR,* 1942 (CAERC box 42, folder 56).
58. F. I. Hoover, *Owen CEAR,* 1945 (CAERC box 121, folder 6).
59. L. M. Butler, *Marshall CEAR,* 1943 (CAERC box 112, folder 33); F. I. Hoover, *Owen CEARs,* 1943–44 (CAERC box 121, folders 4–5); L. P. Geyer, *Jefferson CEAR,* 1945 (CAERC box 100, folder 12).
60. H. S. Benson, *Knox CEAR,* 1943 (CAERC box 102, folder 42).
61. H. H. Madaus, *Warren CEAR,* 1943 (CAERC box 53, folder 25); O. B. Riggs, *Posey CEAR,* 1943 (CAERC box 125, folder 29); E. S. Bogue, *Pike CEAR,* 1943 (CAERC box 123, folder 35).
62. E. F. Downen, *Spencer CEAR,* 1943 (CAERC box 38, folder 49); B. V. Widney, *Whitley CEAR,* 1943 (CAERC box 60, folder 4).
63. "Agricultural Mobilization," American Home Front in World War II (website), Encyclopedia.com, 23 May 2018, https://www.encyclopedia.com/plants-and-animals/agriculture-and-horticulture/agriculture-general/agricultural-revolution.
64. H. K. Rippey, *Jay CEAR,* 1945 (CAERC box 99, folder 3); E. A. Johnson, *Howard CEAR,* 1944 (CAERC box 95, folder 12); W. U. Rusk, *Huntington CEAR,* 1943 (CAERC box 96, folder 19); M. J. Huxley, *Knox CEAR,* 1946 (CAERC box 103, folder 6); J. E. Wylie, *Miami CEAR,* 1945 (CAERC box 114, folder 16).

65. W. U. Rusk, *Huntington CEAR,* 1943 (CAERC box 96, folder 19).
66. A. C. Shideler, *Starke CEAR,* 1943 (CAERC box 40, folder 1); H. K. Rippey, *Jay CEAR,* 1945 (CAERC box 99, folder 3); E. C. Bird, *St. Joseph CEAR,* 1943 (CAERC box 35, folder 29); W. U. Rusk, *Huntington CEAR,* 1943 (CAERC box 96, folder 19); M. A. Nye, *Noble CEAR,* 1943 (CAERC box 118, folder 36).
67. R. L. Campbell, *Vermillion CEAR,* 1945 (CAERC box 49, folder 40).
68. L. M. Butler, *Marshall CEAR,* 1944 (CAERC box 112, folder 34); J. L. Guthrie, *Noble CEAR,* 1944 (CAERC box 118, folder 37).
69. E. C. Bird, *St. Joseph CEARs,* 1943–44 (CAERC box 35, folders 29–30); A. C. Shideler, *Starke CEAR,* 1943 (CAERC box 40, folder 1); M. S. Smith, *Wells CEAR,* 1944 (CAERC box 58, folder 30).
70. A. C. Shideler, *Starke CEAR,* 1943 (CAERC box 40, folder 1).
71. W. U. Rusk, *Huntington CEAR,* 1942 (CAERC box 96, folder 18).
72. M. S. Smith, *Wells CEAR,* 1943 (CAERC box 58, folder 29).
73. M. S. Smith, *Wells CEAR,* 1945 (CAERC box 58, folder 31).
74. H. K. Rippey, *Jay CEAR,* 1945 (CAERC box 99, folder 3); C. L. Brown, *Vigo CEAR,* 1945 (CAERC box 51, folder 7); B. V. Widney, *Whitley CEAR,* 1945 (CAERC box 60, folder 6).
75. H. K. Rippey, *Jay CEAR,* 1945 (CAERC box 99, folder 3).
76. M. S. Smith, *Wells CEAR,* 1945 (CAERC box 58, folder 31).
77. F. E. DeLaCroix, *Jasper CEAR,* 1944 (CAERC box 98, folder 14).
78. T. M. Grabill, *Steuben CEAR,* 1944 (CAERC box 40, folder 52).
79. C. E. Olson, *Pulaski CEAR,* 1944 (CAERC box 28, folder 12).
80. M. A. Nye, *Noble CEAR,* 1943 (CAERC box 118, folder 36); W. J. Record, *Morgan CEAR,* 1944 (CAERC box 117, folder 5); E. A. Johnson, *Howard CEAR,* 1945 (CAERC box 95, folder 13); M. J. Huxley, *Knox CEAR,* 1946 (CAERC box 103, folder 6); J. E. Wylie, *Miami CEAR,* 1943 (CAERC box 114, folder 14).
81. F. E. DeLaCroix, *Jasper CEAR,* 1943 (CAERC box 98, folder 13); J. E. Wylie, *Miami CEAR,* 1943 (CAERC box 114, folder 14).
82. L. E. Husted, *Randolph CEAR,* 1945 (CAERC box 30, folder 44); E. L. Hartman, *LaPorte CEAR,* 1944 (CAERC box 105, folder 21).
83. Brian Spangle, "German POWs Were Placed in Vincennes Camp," *Vincennes Sun-Commercial* (Vincennes, IN), 26 Aug. 2012, 7.
84. Annette Scherber, "Corn, Tomatoes, and POWs: Hoosier Agriculture during World War II," *Indiana History Blog,* Indiana Historical Bureau of the Indiana State Library, 17 Nov. 2016, https://blog.history.in.gov/corn-tomatoes-pows-hoosier-agriculture-during-world-war-ii/.
85. Encyclopedia.com, "Agricultural Mobilization"; Scherber, "Corn, Tomatoes, and POWs."
86. Scherber, "Corn, Tomatoes, and POWs."
87. S. B. Scott, *Johnson CEAR,* 1944 (CAERC box 101, folder 43); W. C. Haynes, *Madison CEARs,* 1944–45 (CAERC box 109, folders 40–41); C. L. Spuller, *Rush CEAR,* 1945 (CAERC box 33, folder 13); Scherber, "Corn, Tomatoes, and POWs"; "Camp Thomas A. Scott, Fort Wayne, Indiana," Waymarking.com, Groundspeak Inc., 5 Oct. 2007, https://www.waymarking.com/waymarks/WM2B13_Camp_Thomas_A_Scott_Fort_Wayne_Indiana%20for%20information%20on%20Camp%20Scott; *Purdue University Department of Agricultural Extension Annual Report,* 1945, 18.
88. L. E. Cutler, *Lake CEAR,* 1945 (CAERC box 107, folder 38).
89. B. Jewell, *Pike CEAR,* 1945 (CAERC box 123, folder 37); *Purdue University Department of Agricultural Extension Annual Report,* 1945, 18.

90. Scherber, "Corn, Tomatoes, and POWs."
91. H. E. Abbott, *Marion CEAR,* 1944 (CAERC box 111, folder 13); J. A. Wood, *Martin CEAR,* 1944 (CAERC box 113, folder 30); J. E. Wylie, *Miami CEAR,* 1945 (CAERC box 114, folder 16); Scherber, "Corn, Tomatoes, and POWs."
92. H. E. Abbott, *Marion CEAR,* 1944 (CAERC box 111, folder 13).
93. W. U. Rusk, *Huntington CEAR,* 1945 (CAERC box 96, folder 21).
94. J. E. Wylie, *Miami CEAR,* 1945 (CAERC box 114, folder 16); Scherber, "Corn, Tomatoes, and POWs."
95. L. E. Cutler, *Lake CEAR,* 1945 (CAERC box 107, folder 38).
96. J. E. Wylie, *Miami CEAR,* 1945 (CAERC box 114, folder 16).
97. S. B. Scott, *Johnson CEAR,* 1943 (CAERC box 101, folder 42).
98. W. C. Haynes, *Madison CEAR,* 1944 (CAERC box 109, folder 40).
99. Ken de la Bastide, "Windfall Once Home to a WWII POW Camp," *Kokomo Tribune* (Kokomo, IN), 4 Sept. 2006, https://www.kokomotribune.com/news/local_news/windfall-once-home-to-a-wwii-pow-camp/article_bc39a268-8d04-5aec-94cc-a0711768694a.html.
100. E. A. Johnson, *Howard CEAR,* 1945 (CAERC box 95, folder 13).
101. E. A. Johnson, *Howard CEAR,* 1944 (CAERC box 95, folder 12); W. C. Haynes, *Madison CEAR,* 1945 (CAERC box 109, folder 41); J. E. Wylie, *Miami CEAR,* 1945 (CAERC box 114, folder 16); J. McCulloch, *Scott CEAR,* 1943 (CAERC box 36, folder 24); W. U. Rusk, *Huntington CEAR,* 1945 (CAERC box 96, folder 21).
102. W. J. Record, *Morgan CEAR,* 1943 (CAERC box 117, folder 4).
103. W. C. Haynes, *Madison CEAR,* 1943 (CAERC box 109, folder 39).
104. E. A. Johnson, *Howard CEAR,* 1945 (CAERC box 95, folder 13).
105. Scherber, "Corn, Tomatoes, and POWs."
106. H. S. Benson, *Knox CEAR,* 1945 (CAERC box 103, folder 3).
107. W. J. Record, *Morgan CEAR,* 1945 (CAERC box 117, folder 6).
108. H. E. Abbott, *Marion CEAR,* 1944 (CAERC box 111, folder 13).
109. Waymarking.com, "Camp Thomas A. Scott, Fort Wayne, Indiana."
110. Scherber, "Corn, Tomatoes, and POWs."
111. Waymarking.com, "Camp Thomas A. Scott, Fort Wayne, Indiana."
112. *Purdue University Department of Agricultural Extension Annual Report,* 1945, 17.
113. C. D. Chambers, *Perry CEAR,* 1942 (CAERC box 122, folder 38); J. A. Carroll, *Jasper CEAR,* 1945 (CAERC box 98, folder 15); C. A. Langston, *Jay CEAR,* 1944 (CAERC box 99, folder 2); J. A. Clark, *Kosciusko CEAR,* 1943 (CAERC box 104, folder 19); E. F. Brown, *Rush CEAR,* 1943 (CAERC box 33, folder 10).
114. Encyclopedia.com, "Agricultural Mobilization"; Scherber, "Corn, Tomatoes, and POWs."
115. Encyclopedia.com, "Agricultural Mobilization."
116. H. E. Abbott, *Marion CEAR,* 1944 (CAERC box 111, folder 13).
117. L. E. Cutler, *Lake CEAR,* 1944 (CAERC box 107, folder 37); W. K. Delaplane, *Wabash CEAR,* 1943 (CAERC box 52, folder 28); H. E. Abbott, *Marion CEAR,* 1944 (CAERC box 111, folder 13).
118. E. F. Brown, *Rush CEAR,* 1943 (CAERC box 33, folder 10).
119. R. L. Case, *LaGrange CEAR,* 1943 (CAERC box 106, folder 4).
120. S. Leaming, *Porter CEAR,* 1943 (CAERC box 124, folder 27).
121. D. L. Grimes, *Putnam CEAR,* 1942 (CAERC box 29, folder 26); C. A. Langston, *Jay CEAR,* 1943 (CAERC box 99, folder 1); L. E. Cutler, *Lake CEAR,* 1942 (CAERC box 107, folder 34).

122. G. A. Sowers, *Montgomery CEAR,* 1944 (CAERC box 115, folder 56).
123. A. C. Shideler, *Starke CEAR,* 1943 (CAERC box 40, folder 1).
124. L. E. Cutler, *Lake CEAR,* 1942 (CAERC box 107, folder 34).
125. I. O. Haug, *Ripley CEAR,* 1944 (CAERC box 31, folder 46).
126. C. A. Langston, *Jay CEAR,* 1942 (CAERC box 98, folder 51); H. E. Abbott, *Marion CEAR,* 1942 (CAERC box 111, folder 7); N. S. Hadley, *Parke CEAR,* 1944 (CAERC box 121, folder 55).
127. G. A. Sowers, *Montgomery CEAR,* 1944 (CAERC box 115, folder 56); B. T. Bond, *Warrick CEAR,* 1942 (CAERC box 54, folder 36); C. E. Olson, *Pulaski CEAR,* 1944 (CAERC box 28, folder 12).
128. L. P. Geyer, *Jefferson CEAR,* 1943 (CAERC box 100, folder 10).
129. B. T. Bond, *Warrick CEAR,* 1943 (CAERC box 54, folder 37).
130. L. M. Butler, *Marshall CEAR,* 1943 (CAERC box 112, folder 33).
131. J. F. Armstrong, *Lawrence CEAR,* 1943 (CAERC box 108, folder 42); G. A. Sowers, *Montgomery CEAR,* 1943 (CAERC box 115, folder 55); L. P. Geyer, *Jefferson CEAR,* 1943 (CAERC box 100, folder 10).
132. A. M. Bishea, *Vanderburgh CEAR,* 1943 (CAERC box 48, folder 7).
133. H. E. Abbott, *Marion CEAR,* 1943 (CAERC box 111, folder 10).
134. E. F. Brown, *Rush CEAR,* 1943 (CAERC box 33, folder 10).
135. C. A. Langston, *Jay CEAR,* 1944 (CAERC box 99, folder 2).
136. R. N. Atkinson, *Orange CEAR,* 1943 (CAERC box 120, folder 26).
137. H. S. Benson, *Knox CEAR,* 1944 (CAERC box 103, folder 2).
138. E. S. Bogue, *Pike CEAR,* 1943 (CAERC box 123, folder 35).
139. G. A. Sowers, *Montgomery CEAR,* 1944 (CAERC box 115, folder 56).
140. H. E. Abbott, *Marion CEAR,* 1944 (CAERC box 111, folder 13); G. A. Sowers, *Montgomery CEAR,* 1944 (CAERC box 115, folder 56).
141. G. A. Sowers, *Montgomery CEAR,* 1945 (CAERC box 115, folder 57); E. A. Chatham, *Perry CEAR,* 1946 (CAERC box 123, folder 1); O. B. Riggs, *Posey CEAR,* 1944 (CAERC box 125, folder 31); S. Leaming, *Porter CEAR,* 1943 (CAERC box 124, folder 27); C. L. Brown, *Vigo CEAR,* 1944 (CAERC box 51, folder 5).
142. G. A. Sowers, *Montgomery CEAR,* 1944 (CAERC box 115, folder 56).
143. Ibid.
144. R. R. Zell, *Newton CEAR,* 1944 (CAERC box 117, folder 45).
145. R. R. Zell, *Newton CEAR,* 1943 (CAERC box 117, folder 44); H. E. Abbott, *Marion CEAR,* 1946 (CAERC box 111, folder 17); N. S. Hadley, *Parke CEAR,* 1944 (CAERC box 121, folder 55).
146. R. R. Zell, *Newton CEAR,* 1943 (CAERC box 117, folder 44).
147. R. R. Zell, *Newton CEAR,* 1944 (CAERC box 117, folder 45).
148. E. F. Downen, *Spencer CEAR,* 1943 (CAERC box 38, folder 49).
149. B. Jewell, *Pike CEAR,* 1945 (CAERC box 123, folder 37).
150. J. A. Wood, *Martin CEAR,* 1946 (CAERC box 113, folder 32); W. Rogers, *Monroe CEAR,* 1943 (CAERC box 115, folder 26).
151. B. Jewell, *Pike CEAR,* 1946 (CAERC box 123, folder 38).
152. L. J. Fultz, *Washington CEAR,* 1944 (CAERC box 55, folder 46); E. F. Brown, *Rush CEAR,* 1943 (CAERC box 33, folder 10).
153. C. J. Murphy, *Marion CEAR,* 1942 (CAERC box 111, folder 6).
154. E. F. Brown, *Rush CEAR,* 1943 (CAERC box 33, folder 10).

155. N. S. Hadley, *Parke CEAR,* 1943 (CAERC box 121, folder 53).
156. B. T. Bond, *Warrick CEAR,* 1944 (CAERC box 54, folder 38).
157. C. L. Brown, *Vigo CEAR,* 1943 (CAERC box 51, folder 3).
158. Encyclopedia.com, "Agricultural Mobilization"; Scherber, "Corn, Tomatoes, and POWs."
159. Editors of Encyclopaedia Britannica, "Women's Land Army," *Encyclopedia Britannica* (website), 11 Sept. 2018, https://www.britannica.com/topic/Womens-Land-Army.
160. W. U. Rusk, *Huntington CEAR,* 1942 (CAERC box 96, folder 18).
161. H. E. Abbott, *Marion CEAR,* 1942 (CAERC box 111, folder 7).
162. J. A. Clark, *Kosciusko CEAR,* 1943 (CAERC box 104, folder 19); R. N. Atkinson, *Orange CEAR,* 1943 (CAERC box 120, folder 26).
163. C. A. Langston, *Jay CEAR,* 1942 (CAERC box 98, folder 51); R. R. Zell, *Newton CEAR,* 1943 (CAERC box 117, folder 44).
164. F. E. DeLaCroix, *Jasper CEAR,* 1943 (CAERC box 98, folder 13); H. E. Abbott, *Marion CEAR,* 1943 (CAERC box 111, folder 10).
165. L. E. Husted, *Randolph CEAR,* 1943 (CAERC box 30, folder 42).
166. H. S. Benson, *Knox CEAR,* 1943 (CAERC box 102, folder 42).
167. R. R. Zell, *Newton CEAR,* 1943 (CAERC box 117, folder 44).
168. W. K. Delaplane, *Wabash CEAR,* 1943 (CAERC box 52, folder 28).
169. F. E. DeLaCroix, *Jasper CEAR,* 1944 (CAERC box 98, folder 14); Charles B. Hirsch, "Conscientious Objectors in Indiana during World War II," *Indiana Magazine of History* 46, no. 1 (Mar. 1950): 61, https://www.jstor.org/stable/27787823?seq=1.
170. Hirsch, "Conscientious Objectors in Indiana during World War II," 67.
171. Gary Weimberg and Catherine Ryan, *Soldiers of Conscience* (film), *POV,* 16 Oct. 2008, http://archive.pov.org/soldiersofconscience/background/.
172. Hirsch, "Conscientious Objectors in Indiana during World War II," 69.
173. J. C. Sliger, *Sullivan CEAR,* 1942 (CAERC box 41, folder 53).
174. Ibid.
175. J. C. Sliger, *Sullivan CEAR,* 1943 (CAERC box 41, folder 54); Hirsch, "Conscientious Objectors in Indiana during World War II," 71.
176. C. E. Olson, *Pulaski CEAR,* 1946 (CAERC box 28, folder 14).
177. C. E. Olson, *Pulaski CEAR,* 1944 (CAERC box 28, folder 12).
178. E. A. Johnson, *Howard CEAR,* 1945 (CAERC box 95, folder 13).
179. *Purdue University Department of Agricultural Extension Annual Report,* 1945, 17.

CHAPTER 35

Explaining the Red Tape That Farmers Must Go Through to Get Electricity

1. *Purdue University Department of Agricultural Extension Annual Report,* 1941–43, 18.
2. J. A. Clark, *Kosciusko CEAR,* 1943 (CAERC box 104, folder 19); B. T. Bond, *Warrick CEAR,* 1943 (CAERC box 54, folder 37).
3. G. A. Sowers, *Montgomery CEAR,* 1942 (CAERC box 115, folder 54).
4. R. N. Atkinson, *Orange CEAR,* 1942 (CAERC box 120, folder 25).
5. C. E. Olson, *Pulaski CEAR,* 1942 (CAERC box 28, folder 10); G. A. Sowers, *Montgomery CEAR,* 1946 (CAERC box 115, folder 59); B. T. Bond, *Warrick CEAR,* 1943 (CAERC box 54, folder 37); O. B. Riggs, *Posey CEAR,* 1942 (CAERC box 125, folder 27).
6. H. S. Benson, *Knox CEAR,* 1944 (CAERC box 103, folder 2).
7. J. H. Telfer, *Sullivan CEAR,* 1946 (CAERC box 42, folder 1).

8. C. L. Brown, *Vigo CEAR,* 1942 (CAERC box 51, folder 1); C. E. Olson, *Pulaski CEAR,* 1942 (CAERC box 28, folder 10); "Morgan Co. REMC Meeting Feb. 24," *Brown County Democrat* (Nashville, IN), 18 Feb. 1943, 4.
9. F. E. DeLaCroix, *Jasper CEAR,* 1943 (CAERC box 98, folder 13); C. E. Olson, *Pulaski CEAR,* 1943 (CAERC box 28, folder 11); E. S. Bogue, *Pike CEAR,* 1942 (CAERC box 123, folder 34); C. D. Chambers, *Perry CEAR,* 1942 (CAERC box 122, folder 38); F. I. Hoover, *Owen CEAR,* 1942 (CAERC box 121, folder 3); J. F. Armstrong, *Lawrence CEAR,* 1943 (CAERC box 108, folder 42); H. S. Benson, *Knox CEAR,* 1944 (CAERC box 103, folder 2).
10. J. McCulloch, *Scott CEAR,* 1944 (CAERC box 36, folder 25); G. A. Sowers, *Montgomery CEAR,* 1946 (CAERC box 115, folder 59); S. B. Scott, *Johnson CEAR,* 1945 (CAERC box 101, folder 44); B. V. Widney, *Whitley CEAR,* 1943 (CAERC box 60, folder 4).
11. C. L. Brown, *Vigo CEAR,* 1942 (CAERC box 51, folder 1); C. E. Olson, *Pulaski CEAR,* 1942 (CAERC box 28, folder 10).
12. F. E. DeLaCroix, *Jasper CEAR,* 1943 (CAERC box 98, folder 13).
13. H. S. Benson, *Knox CEAR,* 1942 (CAERC box 102, folder 40).
14. J. McCulloch, *Scott CEAR,* 1943 (CAERC box 36, folder 24); M. A. Nye, *Noble CEAR,* 1942 (CAERC box 118, folder 35).
15. "Some Farmers to Obtain New Electric Lines," *Evening Republican* (Columbus, IN), 29 Jan. 1943, 1; F. E. DeLaCroix, *Jasper CEAR,* 1943 (CAERC box 98, folder 13).
16. "Some Farmers to Obtain New Electric Lines," 1.
17. A. C. Shideler, *Starke CEAR,* 1943 (CAERC box 40, folder 1).
18. E. S. Bogue, *Pike CEAR,* 1943 (CAERC box 123, folder 35); J. A. Clark, *Kosciusko CEAR,* 1943 (CAERC box 104, folder 19); W. U. Rusk, *Huntington CEAR,* 1943 (CAERC box 96, folder 19).
19. O. H. McNary, *Switzerland CEAR,* 1943 (CAERC box 42, folder 57).
20. C. E. Olson, *Pulaski CEAR,* 1943 (CAERC box 28, folder 11); H. E. Abbott, *Marion CEAR,* 1943 (CAERC box 111, folder 10).
21. E. S. Bogue, *Pike CEAR,* 1943 (CAERC box 123, folder 35); E. F. Downen, *Spencer CEAR,* 1943 (CAERC box 38, folder 49); J. L. Guthrie, *Union CEAR,* 1943 (CAERC box 46, folder 25).
22. C. E. Olson, *Pulaski CEAR,* 1943 (CAERC box 28, folder 11); L. J. Fultz, *Washington CEAR,* 1943 (CAERC box 55, folder 45); C. L. Brown, *Vigo CEAR,* 1943 (CAERC box 51, folder 3).
23. H. S. Benson, *Knox CEAR,* 1943 (CAERC box 102, folder 42).
24. H. S. Benson, *Knox CEAR,* 1944 (CAERC box 103, folder 2).
25. "Wickard Urges REMC Expansion," *Indianapolis Star,* 3 Nov. 1944, 8.
26. A. C. Shideler, *Starke CEAR,* 1945 (CAERC box 40, folder 3).
27. *Department of Agriculture Appropriations for 1955: Hearings before the Subcommittee of the Committee on Appropriations, House of Representatives,* 83rd Congress, second session (1954) (part 3, statement of the secretary, the budget for the Department of Agriculture) (Washington, DC: Government Printing Office, 1954), 1220–21, https://www.google.com/books/edition/Department_of_Agriculture_Appropriations/B0KZvwEACAAJ?hl=en&gbpv=1&pg=PA852-IA7&printsec=frontcover.
28. Ethan Chitty, "The Decades-Long Struggle to Electrify Rural Hoosierdom," *Indiana History Blog,* Indiana Historical Bureau of the Indiana State Library, 4 Apr. 2017, https://blog.history.in.gov/the-decades-long-struggle-to-electrify-rural-hoosierdom/.

CHAPTER 36

Machinery Will Need to Be Replaced Before the Supply of Baling Wire Is Exhausted

1. N. S. Hadley, *Parke CEAR,* 1942 (CAERC box 121, folder 51); J. A. Wood, *Martin CEAR,* 1943 (CAERC box 113, folder 29); W. J. Emerson, *White CEAR,* 1942 (CAERC box 59, folder 5).
2. Amelia Fogarty, "Five Ways the Tractor Changed American Farming," *Smithsonian Sparks* (blog), Smithsonian Institution, 1 Mar. 2018, https://www.si.edu/stories/five-ways-tractor-changed-american-farming.
3. Bill Ganzel, "The March of Machines," Farming in the 1930s: Machines, Wessels Living History Farm (York, NE) (website), 2003, https://livinghistoryfarm.org/farminginthe30s/machines/.
4. Fogarty, "Five Ways the Tractor Changed American Farming"; Alan L. Olmstead and Paul W. Rhode, "Reshaping the Landscape: The Impact and Diffusion of the Tractor in American Agriculture, 1910–1960," *Journal of Economic History* 61, no. 3 (Sept. 2001): 663, https://www.jstor.org/stable/2698132; Farm Equipment Institute Research Department, "Tractors Increase Despite Depression," *Better Farm Equipment and Methods* 8, no. 2 (Oct. 1935): 5, https://www.google.com/books/edition/Big_Farmer/fSlPAAAAIAAJ?hl=en&gbpv=1&pg=RA7-PA4&printsec=frontcover.
5. Farm Equipment Institute Research Department, "Tractors Increase Despite Depression," 5.
6. H. E. Abbott, *Marion CEAR,* 1946 (CAERC box 111, folder 17); E. S. Bogue, *Pike CEAR,* 1942 (CAERC box 123, folder 34).
7. L. E. Cutler, *Lake CEAR,* 1942 (CAERC box 107, folder 34); W. C. Haynes, *Madison CEAR,* 1943 (CAERC box 109, folder 39).
8. E. C. Bird, *St. Joseph CEAR,* 1943 (CAERC box 35, folder 29); M. A. Nye, *Noble CEAR,* 1942 (CAERC box 118, folder 35); H. E. Abbott, *Marion CEAR,* 1946 (CAERC box 111, folder 17).
9. *Purdue University Department of Agricultural Extension Annual Report,* 1941, 34; M. A. Nye, *Noble CEAR,* 1942 (CAERC box 118, folder 35); H. E. Abbott, *Marion CEAR,* 1946 (CAERC box 111, folder 17); L. E. Cutler, *Lake CEAR,* 1945 (CAERC box 107, folder 38); E. S. Bogue, *Pike CEAR,* 1942 (CAERC box 123, folder 34).
10. C. A. Langston, *Jay CEAR,* 1943 (CAERC box 99, folder 1).
11. J. McCulloch, *Scott CEAR,* 1944 (CAERC box 36, folder 25).
12. E. F. Brown, *Rush CEAR,* 1942 (CAERC box 33, folder 8); D. J. Holmes, *Putnam CEAR,* 1945 (CAERC box 29, folder 29); O. B. Riggs, *Posey CEAR,* 1942 (CAERC box 125, folder 27); E. L. Hartman, *LaPorte CEAR,* 1945 (CAERC box 105, folder 23).
13. H. H. Madaus, *Warren CEAR,* 1942 (CAERC box 53, folder 24).
14. L. E. Cutler, *Lake CEAR,* 1942 (CAERC box 107, folder 34).
15. C. A. Langston, *Jay CEAR,* 1942 (CAERC box 98, folder 51).
16. T. M. Grabill, *Steuben CEAR,* 1942 (CAERC box 40, folder 50); J. C. Sliger, *Sullivan CEAR,* 1943 (CAERC box 41, folder 54); O. B. Riggs, *Posey CEAR,* 1944 (CAERC box 125, folder 31).
17. J. A. Clark, *Kosciusko CEAR,* 1942 (CAERC box 104, folder 18).
18. C. A. Langston, *Jay CEAR,* 1943 (CAERC box 99, folder 1).
19. H. E. Abbott, *Marion CEAR,* 1942 (CAERC box 111, folder 7).
20. W. K. Delaplane, *Wabash CEAR,* 1942 (CAERC box 52, folder 27).
21. S. W. Milligan, *Wayne CEAR,* 1943 (CAERC box 56, folder 30).
22. L. E. Husted, *Randolph CEAR,* 1944 (CAERC box 30, folder 43).
23. J. L. Guthrie, *Union CEAR,* 1942 (CAERC box 46, folder 24).
24. T. M. Grabill, *Steuben CEAR,* 1942 (CAERC box 40, folder 50).

25. W. P. Stall, *Jackson CEAR,* 1942 (CAERC box 97, folder 7); Gladys L. Baker and Wayne D. Rasmussen, "A Chronology of the War Food Administration, Including Predecessor and Successor Agencies, August 1939 to December 1946," U.S. Department of Agriculture Bureau of Agricultural Economics, Nov. 1950, 10, https://naldc.nal.usda.gov/catalog/6538554.
26. "Farmers! You Can Still Get the Farm Equipment Necessary to Produce Your Crops," *Vidette-Messenger of Porter County* (Valparaiso, IN), 7 Oct. 1942, 10.
27. E. S. Bogue, *Pike CEAR,* 1943 (CAERC box 123, folder 35).
28. E. C. Bird, *St. Joseph CEAR,* 1944 (CAERC box 35, folder 30).
29. C. D. Chambers, *Perry CEAR,* 1943 (CAERC box 122, folder 39).
30. T. M. Grabill, *Steuben CEAR,* 1943 (CAERC box 40, folder 51); " '43 Farm Tool Quota Listed by Ration Head," *Times-Mail* (Bedford, IN), 29 Dec. 1942, 1.
31. H. S. Benson, *Knox CEAR,* 1943 (CAERC box 102, folder 42).
32. J. L. Guthrie, *Union CEAR,* 1942 (CAERC box 46, folder 24); J. E. Wylie, *Miami CEAR,* 1943 (CAERC box 114, folder 14).
33. C. E. Olson, *Pulaski CEAR,* 1942 (CAERC box 28, folder 10).
34. J. L. Guthrie, *Union CEAR,* 1942 (CAERC box 46, folder 24); C. A. Langston, *Jay CEAR,* 1942 (CAERC box 98, folder 51); J. McCulloch, *Scott CEAR,* 1943 (CAERC box 36, folder 24).
35. H. S. Benson, *Knox CEAR,* 1945 (CAERC box 103, folder 3); M. J. Huxley, *Knox CEAR,* 1946 (CAERC box 103, folder 6).
36. C. E. Olson, *Pulaski CEAR,* 1942 (CAERC box 28, folder 10); J. L. Guthrie, *Union CEAR,* 1943 (CAERC box 46, folder 25).
37. J. L. Guthrie, *Union CEAR,* 1943 (CAERC box 46, folder 25); J. L. Guthrie, *Union CEAR,* 1942 (CAERC box 46, folder 24).
38. E. S. Bogue, *Pike CEAR,* 1944 (CAERC box 123, folder 36).
39. S. W. Milligan, *Wayne CEAR,* 1943 (CAERC box 56, folder 30); J. L. Guthrie, *Union CEAR,* 1942 (CAERC box 46, folder 24).
40. L. E. Husted, *Randolph CEAR,* 1944 (CAERC box 30, folder 43).
41. S. J. Brown, *Vermillion CEAR,* 1942 (CAERC box 49, folder 37); J. L. Guthrie, *Union CEAR,* 1943 (CAERC box 46, folder 25); A. M. Bishea, *Vanderburgh CEAR,* 1942 (CAERC box 48, folder 4).
42. S. W. Milligan, *Wayne CEAR,* 1943 (CAERC box 56, folder 30).
43. J. L. Guthrie, *Union CEAR,* 1943 (CAERC box 46, folder 25).
44. "Agriculture: Machinery," *Bremen Enquirer* (Bremen, IN), 27 Apr. 1944, 1.
45. J. McCulloch, *Scott CEAR,* 1944 (CAERC box 36, folder 25).
46. H. S. Benson, *Knox CEAR,* 1944 (CAERC box 103, folder 2).
47. J. L. Guthrie, *Noble CEARs,* 1945–46 (CAERC box 118, folders 38–39).
48. J. L. Guthrie, *Noble CEAR,* 1945 (CAERC box 118, folder 38).
49. J. H. Telfer, *Sullivan CEAR,* 1946 (CAERC box 42, folder 1).
50. L. M. Butler, *Marshall CEAR,* 1942 (CAERC box 112, folder 32); L. E. Cooper, *Spencer CEAR,* 1946 (CAERC box 38, folder 52); W. C. Haynes, *Madison CEAR,* 1942 (CAERC box 109, folder 38).
51. S. B. Pershing, *Tippecanoe CEAR,* 1942 (CAERC box 44, folder 12).
52. W. P. Stall, *Jackson CEAR,* 1942 (CAERC box 97, folder 7).
53. E. L. Hartman, *LaPorte CEAR,* 1942 (CAERC box 105, folder 17).
54. J. C. Sliger, *Sullivan CEAR,* 1943 (CAERC box 41, folder 54); N. S. Hadley, *Parke CEAR,* 1942 (CAERC box 121, folder 51); W. P. Stall, *Jackson CEAR,* 1945 (CAERC box 97, folder 10); O. H. McNary, *Switzerland CEAR,* 1943 (CAERC box 42, folder 57).

55. S. W. Milligan, *Wayne CEAR,* 1942 (CAERC box 56, folder 29).
56. S. B. Pershing, *Tippecanoe CEAR,* 1942 (CAERC box 44, folder 12).
57. C. D. Chambers, *Perry CEAR,* 1942 (CAERC box 122, folder 38).
58. H. E. Abbott, *Marion CEAR,* 1942 (CAERC box 111, folder 7); W. C. Haynes, *Madison CEAR,* 1942 (CAERC box 109, folder 38).
59. W. C. Haynes, *Madison CEAR,* 1942 (CAERC box 109, folder 38).
60. W. C. Haynes, *Madison CEAR,* 1943 (CAERC box 109, folder 39).
61. L. E. Cutler, *Lake CEAR,* 1942 (CAERC box 107, folder 34); W. C. Haynes, *Madison CEAR,* 1942 (CAERC box 109, folder 38).
62. E. L. Hartman, *LaPorte CEAR,* 1942 (CAERC box 105, folder 17).
63. J. McCulloch, *Scott CEAR,* 1942 (CAERC box 36, folder 23).
64. W. P. Stall, *Jackson CEAR,* 1942 (CAERC box 97, folder 7).
65. S. Leaming, *Porter CEAR,* 1942 (CAERC box 124, folder 26); L. J. Fultz, *Washington CEAR,* 1943 (CAERC box 55, folder 45).
66. E. L. Hartman, *LaPorte CEAR,* 1942 (CAERC box 105, folder 17); H. E. Abbott, *Marion CEAR,* 1942 (CAERC box 111, folder 7); J. C. Sliger, *Sullivan CEAR,* 1942 (CAERC box 41, folder 53).
67. O. B. Riggs, *Posey CEAR,* 1943 (CAERC box 125, folder 29).
68. S. Leaming, *Porter CEAR,* 1942 (CAERC box 124, folder 26).
69. R. L. Case, *LaGrange CEAR,* 1942 (CAERC box 106, folder 3); R. R. Zell, *Newton CEAR,* 1943 (CAERC box 117, folder 44).
70. W. U. Rusk, *Huntington CEAR,* 1942 (CAERC box 96, folder 18).
71. H. E. Abbott, *Marion CEAR,* 1942 (CAERC box 111, folder 7); C. A. Langston, *Jay CEAR,* 1942 (CAERC box 98, folder 51); F. I. Hoover, *Owen CEAR,* 1942 (CAERC box 121, folder 3).
72. M. F. Detrick, *Fayette CEARs,* 1931 and 1936 (CAERC box 81, folders 32 and 37); H. S. Heckard, *Allen CEAR,* 1933 (CAERC box 62, folder 18); J. Jensen, *Daviess CEAR,* 1933 (CAERC box 74, folder 27); C. A. Alcorn, *Dearborn CEAR,* 1938 (CAERC box 75, folder 33).
73. W. U. Rusk, *Huntington CEAR,* 1945 (CAERC box 96, folder 21).
74. W. U. Rusk, *Huntington CEAR,* 1942 (CAERC box 96, folder 18); W. P. Stall, *Jackson CEAR,* 1942 (CAERC box 97, folder 7).
75. W. P. Stall, *Jackson CEAR,* 1942 (CAERC box 97, folder 7).
76. C. L. Brown, *Vigo CEAR,* 1943 (CAERC box 51, folder 3).
77. G. A. Sowers, *Montgomery CEAR,* 1943 (CAERC box 115, folder 55).
78. C. A. Langston, *Jay CEAR,* 1944 (CAERC box 99, folder 2).
79. T. M. Grabill, *Steuben CEAR,* 1942 (CAERC box 40, folder 50).
80. J. McCulloch, *Scott CEAR,* 1943 (CAERC box 36, folder 24); G. A. Sowers, *Montgomery CEAR,* 1942 (CAERC box 115, folder 54).
81. G. T. Harris, *Ripley CEAR,* 1942 (CAERC box 31, folder 44).
82. E. L. Hartman, *LaPorte CEAR,* 1943 (CAERC box 105, folder 19); W. U. Rusk, *Huntington CEARs,* 1942 and 1945 (CAERC box 96, folders 18 and 21); L. E. Husted, *Randolph CEAR,* 1942 (CAERC box 30, folder 41); J. McCulloch, *Scott CEAR,* 1942 (CAERC box 36, folder 23).
83. S. W. Milligan, *Wayne CEAR,* 1943 (CAERC box 56, folder 30); T. M. Grabill, *Steuben CEAR,* 1942 (CAERC box 40, folder 50).
84. W. M. Clary, *Tipton CEAR,* 1943 (CAERC box 45, folder 24); S. B. Scott, *Johnson CEAR,* 1943 (CAERC box 101, folder 42).

85. C. E. Olson, *Pulaski CEAR,* 1943 (CAERC box 28, folder 11).
86. B. T. Bond, *Warrick CEAR,* 1942 (CAERC box 54, folder 36); R. R. Zell, *Newton CEAR,* 1943 (CAERC box 117, folder 44); L. E. Husted, *Randolph CEAR,* 1944 (CAERC box 30, folder 43); R. C. Garrison, *Union CEAR,* 1945 (CAERC box 46, folder 27); C. A. Langston, *Jay CEAR,* 1943 (CAERC box 99, folder 1).
87. J. McCulloch, *Scott CEAR,* 1943 (CAERC box 36, folder 24); L. E. Husted, *Randolph CEAR,* 1943 (CAERC box 30, folder 42).
88. E. S. Bogue, *Pike CEAR,* 1943 (CAERC box 123, folder 35).
89. O. B. Riggs, *Posey CEAR,* 1943 (CAERC box 125, folder 29); C. E. Olson, *Pulaski CEAR,* 1943 (CAERC box 28, folder 11).
90. M. J. Huxley, *Knox CEAR,* 1946 (CAERC box 103, folder 6); D. J. Holmes, *Putnam CEAR,* 1945 (CAERC box 29, folder 29); C. E. Olson, *Pulaski CEAR,* 1942 (CAERC box 28, folder 10).
91. J. L. Guthrie, *Union CEAR,* 1942 (CAERC box 46, folder 24).

CHAPTER 37

Patriotic Duty to Get the Most Possible Mileage From Tires

1. American Chemical Society, "United States Synthetic Rubber Program, 1939–1945" (commemorative booklet [PDF]), 3, at National Historic Chemical Landmarks (website), https://www.acs.org/content/acs/en/education/whatischemistry/landmarks/syntheticrubber.html.
2. U.S. War Production Board Bureau of Industrial Conservation, "America Needs Your Scrap Rubber" (poster) (Washington, DC: U.S. Government Printing Office, 1942), https://www.mnhs.org/mgg/artifact/rubber.
3. American Chemical Society, "United States Synthetic Rubber Program, 1939–1945," 3; R. L. Case, *LaGrange CEAR,* 1942 (CAERC box 106, folder 3); "Begin Strict Rationing of Gas for U.S.," *Vidette-Messenger of Porter County* (Valparaiso, IN), 1 Dec. 1942, 1.
4. "Patriotic Duty to Save Tires," *Lafayette Journal and Courier* (Lafayette, IN), 1 Jan. 1942, 3.
5. "1,311 New Casings for January—County Has 141,000 Vehicles," *Indianapolis Star,* 2 Jan. 1942, 1.
6. "Patriotic Duty to Save Tires," 3.
7. American Chemical Society, "United States Synthetic Rubber Program, 1939–1945," 4.
8. "1,311 New Casings for January," 1.
9. "400,000 New Tires Released by Military," *Indianapolis News,* 5 Dec. 1945, 1; J. C. Sliger, *Sullivan CEAR,* 1942 (CAERC box 41, folder 53).
10. Sarah Sundin, "Make It Do: Tire Rationing in World War II," *Sarah Sundin* (blog), 5 Jan. 2022, http://www.sarahsundin.com/make-it-do-tire-rationing-in-world-war-ii/.
11. G. T. Harris, *Ripley CEAR,* 1942 (CAERC box 31, folder 44).
12. Ibid.; G. I. John, *Ohio CEAR,* 1942 (CAERC box 119, folder 30).
13. G. T. Harris, *Ripley CEAR,* 1942 (CAERC box 31, folder 44); G. I. John, *Ohio CEAR,* 1942 (CAERC box 119, folder 30).
14. J. A. Clark, *Kosciusko CEAR,* 1942 (CAERC box 104, folder 18).
15. W. J. Record, *Morgan CEAR,* 1942 (CAERC box 117, folder 3).
16. B. T. Bond, *Warrick CEAR,* 1942 (CAERC box 54, folder 36); G. T. Harris, *Ripley CEAR,* 1942 (CAERC box 31, folder 44); J. A. Wood, *Martin CEAR,* 1942 (CAERC box 113, folder 28); S. Leaming, *Porter CEAR,* 1942 (CAERC box 124, folder 26); J. L. Guthrie, *Union CEAR,* 1942 (CAERC box 46, folder 24).

17. R. L. Case, *LaGrange CEAR,* 1942 (CAERC box 106, folder 3); "Cut in Rubber to Affect All," *South Bend Tribune* (South Bend, IN), 1 Jan. 1942, 17; "OPA Fixes State Quota on Tires," *Indianapolis Star,* 1 Jan. 1942, 1.
18. "Begin Strict Rationing of Gas for U.S.," 1.
19. "Rubber Vital Commodity for Auto Industry," *Times* (Munster, IN), 2 Jan. 1942, 30; "Tire, Tube Quotas Issued for State," *Indianapolis News,* 1 Jan. 1942, 25; "Fix Quotas on Tires, Tubes," *Republic* (Columbus, IN), 1 Jan. 1942, 2.
20. "Tire, Tube Quotas Issued for State," 25.
21. "Rubber Vital Commodity for Auto Industry," 30.
22. "Roosevelt Leads U.S. in New Year Prayer," *Lafayette Journal and Courier* (Lafayette, IN), 1 Jan. 1942, 9.
23. E. F. Downen, *Spencer CEAR,* 1942 (CAERC box 38, folder 48); C. E. Olson, *Pulaski CEAR,* 1942 (CAERC box 28, folder 10); S. B. Pershing, *Tippecanoe CEAR,* 1942 (CAERC box 44, folder 12); "Calumet Tire Ration Board Starts Work," *Times* (Munster, IN), 5 Jan. 1942, 2.
24. C. E. Olson, *Pulaski CEAR,* 1943 (CAERC box 28, folder 11).
25. "Members of Tire Rationing Board of County to Attend Meeting," *Princeton Clarion-News* (Princeton, IN), 1 Jan. 1942, 1; "Local Rationing Board Members to Be Sworn in Sunday," *Star Press* (Muncie, IN), 1 Jan. 1942, 1.
26. "County Tire Ration Board to Get Full Instructions," *Franklin Evening Star* (Franklin, IN), 1 Jan. 1942, 1.
27. G. T. Harris, *Ripley CEAR,* 1942 (CAERC box 31, folder 44).
28. "Local Rationing Board Members to Be Sworn in Sunday," 1.
29. Ibid.
30. G. T. Harris, *Ripley CEAR,* 1942 (CAERC box 31, folder 44); "County Gets 80 Tires as First Ration Quota," *Republic* (Columbus, IN), 1 Jan. 1942, 1; "Members of Tire Rationing Board of County to Attend Meeting," 1.
31. "Local Rationing Board Members to Be Sworn in Sunday," 1; "Tire Rationing Setup Described," *Palladium-Item* (Richmond, IN), 1 Jan. 1942, 2.
32. W. J. Record, *Morgan CEAR,* 1942 (CAERC box 117, folder 3).
33. "Tire, Tube Quotas Issued for State," 25.
34. "County Gets 80 Tires as First Ration Quota," 1.
35. "Tire, Tube Quotas Issued for State," 25.
36. "Tire Inspectors Are Selected for DeKalb County," *Garrett Clipper* (Garrett, IN), 8 Jan. 1942, 1.
37. "Tire, Tube Quotas Issued for State," 25.
38. "OPA Fixes State Quota on Tires," 1.
39. "Tire Famine," *Palladium-Item* (Richmond, IN), 4 Jan. 1942, 1.
40. "Tire, Tube Quotas Issued for State," 25; "Drastic January Tire Quotas," *Star Press* (Muncie, IN), 1 Jan. 1942, 1.
41. "Many to Lose Jobs, but Traffic Deaths Will Drop under Tire-Rationing Plan," *Indianapolis Star,* 2 Jan. 1942, 1.
42. "Tire Rationing Violators Face Stiff Penalties," *Palladium-Item* (Richmond, IN), 8 Jan. 1942, 1.
43. "Tire, Tube Quotas Issued for State," 25; G. T. Harris, *Ripley CEAR,* 1942 (CAERC box 31, folder 44); "Tire Inspectors Are Selected for DeKalb County," 1.
44. "Tire Inspectors Are Selected for DeKalb County," 1.
45. "Tire, Tube Quotas Issued for State," 25; G. T. Harris, *Ripley CEAR,* 1942 (CAERC box 31, folder 44); "Tire Inspectors Are Selected for DeKalb County," 1.

46. "Tire Dealers to Keep Records," *Muncie Evening Press* (Muncie, IN), 9 Jan. 1942, 7.
47. "Tire, Tube Quotas Issued for State," 25.
48. "Public Asked Not to Make Tire Requests," *Palladium-Item* (Richmond, IN), 4 Jan. 1942, 2.
49. Ibid.
50. "Patriotic Duty to Save Tires," 3.
51. "Rubber Vital Commodity for Auto Industry," 30; F. I. Hoover, *Owen CEAR,* 1942 (CAERC box 121, folder 3).
52. "Retreaded Tires May Soon Be on Rationing List," *Princeton Clarion-News* (Princeton, IN), 1 Jan. 1942, 1; "1,311 New Casings for January," 1.
53. Maurice Early, "The Day in Indiana," *Indianapolis Star,* 2 Jan. 1942, 1.
54. "Don't Let Smooth Tires Worry You," *Lafayette Journal and Courier* (Lafayette, IN), 1 Jan. 1942, 16.
55. Advertisement, *Daily Clintonian* (Clinton, IN), 12 Jan. 1945, 8.
56. "Motorists Warned to Watch Tires from Theft," *Princeton Clarion-News* (Princeton, IN), 1 Jan. 1942, 1.
57. "Miscellany," *Rushville Republican* (Rushville, IN), 1 Jan. 1942, 4.
58. "Motorists Warned to Watch Tires from Theft," 1; "Calumet Tire Ration Board Starts Work," 2.
59. "Warn against Tire Thievery," *Lafayette Journal and Courier* (Lafayette, IN), 1 Jan. 1942, 17.
60. "Tire Thief Is Held under $10,000 Bond," *Lafayette Journal and Courier* (Lafayette, IN), 1 Jan. 1942, 1.
61. "Tire Complaint Blanks Arrive," *Star Press* (Muncie, IN), 26 Jan. 1942, 10.
62. "Tire Rationing Violators Face Heavy Penalty," *Vidette-Messenger of Porter County* (Valparaiso, IN), 2 Feb. 1942, 5.
63. O. B. Riggs, *Posey CEAR,* 1943 (CAERC box 125, folder 29); O. H. McNary, *Switzerland CEAR,* 1943 (CAERC box 42, folder 57); W. C. Haynes, *Madison CEAR,* 1943 (CAERC box 109, folder 39); C. L. Brown, *Vigo CEAR,* 1943 (CAERC box 51, folder 3).
64. O. B. Riggs, *Posey CEAR,* 1943 (CAERC box 125, folder 29).
65. O. H. McNary, *Switzerland CEAR,* 1942 (CAERC box 42, folder 56); M. A. Nye, *Noble CEAR,* 1943 (CAERC box 118, folder 36); C. E. Olson, *Pulaski CEAR,* 1943 (CAERC box 28, folder 11).
66. L. P. Geyer, *Jefferson CEAR,* 1942 (CAERC box 100, folder 9).
67. H. S. Benson, *Knox CEAR,* 1942 (CAERC box 102, folder 40).
68. W. M. Clary, *Tipton CEAR,* 1943 (CAERC box 45, folder 24).
69. O. H. McNary, *Switzerland CEAR,* 1942 (CAERC box 42, folder 56).
70. G. T. Harris, *Ripley CEAR,* 1942 (CAERC box 31, folder 44).
71. S. Leaming, *Porter CEAR,* 1943 (CAERC box 124, folder 27); R. R. Zell, *Newton CEAR,* 1944 (CAERC box 117, folder 45).
72. O. H. McNary, *Switzerland CEAR,* 1942 (CAERC box 42, folder 56).
73. Ibid.
74. S. Leaming, *Porter CEAR,* 1943 (CAERC box 124, folder 27); C. E. Olson, *Pulaski CEAR,* 1943 (CAERC box 28, folder 11); L. P. Geyer, *Jefferson CEAR,* 1943 (CAERC box 100, folder 10); J. A. Clark, *Kosciusko CEAR,* 1942 (CAERC box 104, folder 18).
75. L. P. Geyer, *Jefferson CEAR,* 1943 (CAERC box 100, folder 10).
76. M. A. Nye, *Porter CEAR,* 1944 (CAERC box 124, folder 28).
77. "Farm Tires Output Tops Peace Totals," *Times* (Munster, IN), 9 Aug. 1945, 11.
78. "400,000 New Tires Released by Military," 1.
79. Ibid.

80. "Christmas Gift," *Rushville Republican* (Rushville, IN), 3 Dec. 1945, 2.
81. "Rationing Shifts to Tire Dealers," *Princeton Clarion-News* (Princeton, IN), 1 Jan. 1946, 6; "Farm and Industrial Type Tires Released from Rationing List," *Muncie Evening Press* (Muncie, IN), 20 Aug. 1945, 7.
82. "Tire Rationing," *Indianapolis News,* 21 Dec. 1945, 16.
83. Ibid.
84. "Tire Ceiling Prices Listed by OPA District Director," *Princeton Clarion-News* (Princeton, IN), 7 Jan. 1946, 6.
85. "Critical Tire Shortage Is Forecast for Year," *Star Press* (Muncie, IN), 2 Jan. 1944, 19.
86. "Tire Rationing May End Soon," *Lafayette Journal and Courier* (Lafayette, IN), 21 Aug. 1945, 1.
87. Michelin, "An Unknown Object: The Tire—Materials," The Tire Digest (website), accessed 20 Apr. 2022, https://thetiredigest.michelin.com/an-unknown-object-the-tire-materials.

CHAPTER 38

Farmers Were Second Only to the Army in Needing Fuel

1. "Gas Rationing Shenanigans," *Palladium-Item* (Richmond, IN), 9 June 1942, 6.
2. "Two American Tankers Are Attacked off East Coast," *Lafayette Journal and Courier* (Lafayette, IN), 28 Jan. 1942, 1.
3. "Gas Rationing Shenanigans," 6.
4. Drew Pearson and Robert Allen, "The Daily Washington Merry-Go-Round," *Star Press* (Muncie, IN), 14 May 1942, 6; "Gas Rationing for East First," *Muncie Evening Press* (Muncie, IN), 12 Mar. 1942, 23.
5. "Dash Hopes for East Coast Gas Ration Increase," *Hammond Times* (Hammond, IN), 30 Dec. 1942, 18.
6. "Gas Rationing on East Coast Starts May 15," *Call-Leader* (Elwood, IN), 9 May 1942, 1.
7. "Gas Rationing: East Coast First," *Scott County Journal* (Scottsburg, IN), 14 May 1942, 3; "Gas Rationing on East Coast Starts May 17," *Daily Clintonian* (Clinton, IN), 23 Apr. 1942, 1; "Gas Rationing on East Coast Starts May 15," *Call-Leader* (Elwood, IN), 9 May 1942, 1.
8. "Gas Rationing on East Coast Starts May 17," *Daily Clintonian* (Clinton, IN), 23 Apr. 1942, 1.
9. Ibid.
10. "Gasoline Shortage," *Tipton Daily Tribune* (Tipton, IN), 9 May 1942, 4.
11. Ibid.
12. "Gas Rationing Is Decreed for Entire Nation," *Muncie Evening Press* (Muncie, IN), 26 Sept. 1942, 1; "Gas Rationing Cuts Driving to 5,000 Mile-Annual-Average OPA Reveals," *Greenfield Daily Reporter* (Greenfield, IN), 12 Oct. 1942, 1.
13. "Gas Rationing: East Coast First," 3; "Gasoline Shortage," 4.
14. "B Classification Leads All Registrations for Gasoline," *Scranton Times* (Scranton, PA), 13 May 1942, 3.
15. "Gas Rationing Registration Starts Nov. 9," *Tribune* (Seymour, IN), 13 Oct. 1942, 1.
16. "Dash Hopes for East Coast Gas Ration Increase," 18.
17. "Gas Rationing Is Decreed for Entire Nation," 1; "Begin Strict Rationing of Gas for U.S.," *Vidette-Messenger of Porter County* (Valparaiso, IN), 1 Dec. 1942, 1.
18. "Gas Rationed for Nation's Drivers," *Indianapolis News,* 1 Dec. 1942, 1; "Begin Strict Rationing of Gas for U.S.," 1.
19. O. B. Riggs, *Posey CEAR,* 1943 (CAERC box 125, folder 29); "Gas Rationed for Nation's Drivers," 8; "Gas Rationing Is Decreed for Entire Nation," 1; H. S. Benson, *Knox CEAR,* 1943 (CAERC box 102, folder 42).

20. "Shivery Coast Walks to Work," *Muncie Evening Press* (Muncie, IN), 8 Jan. 1943, 9.
21. "ODT Head Raps Improper Use of Commercial Trucks," *Daily Clintonian* (Clinton, IN), 24 Aug. 1943, 5; S. B. Scott, *Johnson CEAR,* 1942 (CAERC box 101, folder 41).
22. "Farmers Will Continue to Get 'Liberal Priority' for Gas," *Palladium-Item* (Richmond, IN), 5 Aug. 1943, 22.
23. W. P. Stall, *Jackson CEAR,* 1942 (CAERC box 97, folder 7); "Truck Owners Must Get New Permits Soon," *Rushville Republican* (Rushville, IN), 16 Oct. 1942, 1.
24. "Registration Instructions Are Presented to Wayne Farm Transportation Committee," *Palladium-Item* (Richmond, IN), 21 Oct. 1942, 3.
25. Ibid.; "Truck Owners Warned about Gas Rationing," *Vidette-Messenger of Porter County* (Valparaiso, IN), 25 Nov. 1942, 1.
26. "U.S.D.A. Board Names Members of Committee," *Reporter-Times* (Martinsville, IN), 15 Oct. 1942, 1.
27. B. V. Widney, *Whitley CEAR,* 1943 (CAERC box 60, folder 4).
28. "Gas Rationing Cuts Driving to 5,000 Mile," 1; "U.S.D.A. Board Names Members of Committee," 1.
29. M. A. Nye, *Noble CEAR,* 1942 (CAERC box 118, folder 35).
30. "Probers Charge Government Complicates Gas Rationing," *Franklin Evening Star* (Franklin, IN), 11 Dec. 1942, 8.
31. "Registration Instructions Are Presented," 3.
32. "Gas Rationing Cuts Driving to 5,000 Mile," 1; "Truck Owners Must Get New Permits Soon," 1.
33. C. E. Olson, *Pulaski CEAR,* 1942 (CAERC box 28, folder 10).
34. G. T. Harris, *Ripley CEAR,* 1942 (CAERC box 31, folder 44); W. M. Clary, *Tipton CEAR,* 1942 (CAERC box 45, folder 23); "Gas Rationing Cuts Driving to 5,000 Mile," 1; W. P. Stall, *Jackson CEAR,* 1942 (CAERC box 97, folder 7); F. I. Hoover, *Owen CEAR,* 1942 (CAERC box 121, folder 3).
35. C. E. Olson, *Pulaski CEAR,* 1942 (CAERC box 28, folder 10); "Registration Instructions Are Presented," 3; "Gas Rationing Cuts Driving to 5,000 Mile," 1.
36. "Registration Instructions Are Presented," 3.
37. C. E. Olson, *Pulaski CEAR,* 1942 (CAERC box 28, folder 10).
38. H. S. Benson, *Knox CEAR,* 1942 (CAERC box 102, folder 40).
39. S. W. Milligan, *Wayne CEAR,* 1942 (CAERC box 56, folder 29).
40. W. M. Clary, *Tipton CEAR,* 1942 (CAERC box 45, folder 23).
41. W. P. Stall, *Jackson CEAR,* 1942 (CAERC box 97, folder 7); S. Leaming, *Porter CEAR,* 1943 (CAERC box 124, folder 27); E. S. Bogue, *Pike CEAR,* 1942 (CAERC box 123, folder 34).
42. W. C. Haynes, *Madison CEAR,* 1942 (CAERC box 109, folder 38).
43. G. T. Harris, *Ripley CEAR,* 1942 (CAERC box 31, folder 44); E. S. Bogue, *Pike CEAR,* 1942 (CAERC box 123, folder 34); H. E. Abbott, *Marion CEARs,* 1943 and 1945 (CAERC box 111, folders 10 and 15).
44. "Registration Instructions Are Presented," 3.
45. E. S. Bogue, *Pike CEAR,* 1942 (CAERC box 123, folder 34); J. A. Wood, *Martin CEAR,* 1943 (CAERC box 113, folder 29).
46. B. T. Bond, *Warrick CEAR,* 1942 (CAERC box 54, folder 36).
47. "Errors Noted on Truck War Certificates," *Palladium-Item* (Richmond, IN), 27 Nov. 1942, 15.
48. G. T. Harris, *Ripley CEAR,* 1942 (CAERC box 31, folder 44); C. E. Olson, *Pulaski CEAR,* 1943 (CAERC box 28, folder 11).

49. "Truck Certificates Good for Duration," *Star Press* (Muncie, IN), 23 Dec. 1943, 7.
50. J. C. Sliger, *Sullivan CEAR,* 1943 (CAERC box 41, folder 54).
51. Ibid.
52. "ODT Assures Farmers Ample Gas Allotments for All Truck Needs," *Argos Reflector* (Argos, IN), 8 Apr. 1943, 1; "Truck Certificates Good for Duration," 7.

CHAPTER 39

Extension Meetings Will Be Curtailed Due to Gas Rationing, Thin Tires, and Busier People

1. E. A. Johnson, *Howard CEAR,* 1944 (CAERC box 95, folder 12); C. D. Chambers, *Perry CEAR,* 1943 (CAERC box 122, folder 39); L. E. Cutler, *Lake CEAR,* 1942 (CAERC box 107, folder 34); G. A. Sowers, *Montgomery CEAR,* 1943 (CAERC box 115, folder 55); W. P. Stall, *Jackson CEAR,* 1943 (CAERC box 97, folder 8).
2. L. E. Cutler, *Lake CEAR,* 1942 (CAERC box 107, folder 34); E. A. Johnson, *Howard CEAR,* 1944 (CAERC box 95, folder 12).
3. A. C. Shideler, *Starke CEAR,* 1943 (CAERC box 40, folder 1).
4. W. P. Stall, *Jackson CEAR,* 1942 (CAERC box 97, folder 7).
5. J. F. Armstrong, *Lawrence CEAR,* 1942 (CAERC box 108, folder 41).
6. G. A. Sowers, *Montgomery CEAR,* 1942 (CAERC box 115, folder 54); E. F. Brown, *Rush CEAR,* 1943 (CAERC box 33, folder 10); W. J. Record, *Morgan CEAR,* 1942 (CAERC box 117, folder 3); E. S. Bogue, *Pike CEAR,* 1944 (CAERC box 123, folder 36).
7. E. A. Johnson, *Howard CEARs,* 1942 and 1944 (CAERC box 95, folders 10 and 12); W. J. Record, *Morgan CEAR,* 1942 (CAERC box 117, folder 3); S. B. Pershing, *Tippecanoe CEAR,* 1942 (CAERC box 44, folder 12); S. Leaming, *Porter CEAR,* 1942 (CAERC box 124, folder 26).
8. T. M. Grabill, *Steuben CEAR,* 1943 (CAERC box 40, folder 51); F. E. DeLaCroix, *Jasper CEAR,* 1943 (CAERC box 98, folder 13).
9. A. C. Shideler, *Starke CEAR,* 1943 (CAERC box 40, folder 1); E. A. Johnson, *Howard CEAR,* 1942 (CAERC box 95, folder 10); N. S. Hadley, *Parke CEAR,* 1942 (CAERC box 121, folder 51).
10. H. S. Heckard, *Allen CEAR,* 1935 (CAERC box 62, folder 20).
11. O. B. Riggs, *Posey CEAR,* 1942 (CAERC box 125, folder 27); L. E. Cutler, *Lake CEAR,* 1945 (CAERC box 107, folder 38); E. A. Johnson, *Howard CEAR,* 1942 (CAERC box 95, folder 10); B. V. Widney, *Whitley CEAR,* 1942 (CAERC box 60, folder 3).
12. F. I. Hoover, *Owen CEAR,* 1942 (CAERC box 121, folder 3).
13. E. A. Johnson, *Howard CEAR,* 1942 (CAERC box 95, folder 10); H. E. Abbott, *Marion CEAR,* 1943 (CAERC box 111, folder 10).
14. O. B. Riggs, *Posey CEAR,* 1944 (CAERC box 125, folder 31).
15. A. M. Bishea, *Vanderburgh CEAR,* 1942 (CAERC box 48, folder 4); L. E. Cooper, *Spencer CEAR,* 1946 (CAERC box 38, folder 52).
16. R. C. Stangland, *Howard CEAR,* 1943 (CAERC box 14, folder 1).
17. E. A. Johnson, *Howard CEAR,* 1945 (CAERC box 95, folder 13).
18. N. S. Hadley, *Parke CEAR,* 1943 (CAERC box 121, folder 53); I. O. Haug, *Parke CEAR,* 1946 (CAERC box 122, folder 2); F. I. Hoover, *Owen CEAR,* 1943 (CAERC box 121, folder 4).
19. R. L. Case, *LaGrange CEAR,* 1943 (CAERC box 106, folder 4); S. Leaming, *Porter CEAR,* 1943 (CAERC box 124, folder 27); J. C. Sliger, *Sullivan CEAR,* 1943 (CAERC box 41, folder 54); A. C. Shideler, *Starke CEAR,* 1943 (CAERC box 40, folder 1); E. S. Bogue, *Pike CEAR,* 1943 (CAERC box 123, folder 35).

20. J. E. Wylie, *Miami CEAR,* 1943 (CAERC box 114, folder 14); E. L. Hartman, *LaPorte CEAR,* 1943 (CAERC box 105, folder 19).
21. J. E. Wylie, *Miami CEAR,* 1943 (CAERC box 114, folder 14).
22. Ibid.
23. A. C. Shideler, *Starke CEAR,* 1943 (CAERC box 40, folder 1); J. E. Wylie, *Miami CEAR,* 1943 (CAERC box 114, folder 14).
24. C. E. Olson, *Pulaski CEAR,* 1943 (CAERC box 28, folder 11); H. S. Benson, *Knox CEAR,* 1944 (CAERC box 103, folder 2).
25. A. M. Bishea, *Vanderburgh CEAR,* 1944 (CAERC box 48, folder 9).
26. H. S. Benson, *Knox CEAR,* 1944 (CAERC box 103, folder 2).
27. Ibid.
28. G. A. Sowers, *Montgomery CEAR,* 1944 (CAERC box 115, folder 56).
29. H. E. Abbott, *Marion CEAR,* 1945 (CAERC box 111, folder 15).
30. W. C. Haynes, *Madison CEAR,* 1942 (CAERC box 109, folder 38); R. N. Atkinson, *Orange CEAR,* 1942 (CAERC box 120, folder 25).
31. C. E. Olson, *Pulaski CEAR,* 1942 (CAERC box 28, folder 10).
32. G. A. Sowers, *Montgomery CEAR,* 1942 (CAERC box 115, folder 54); S. B. Pershing, *Tippecanoe CEAR,* 1942 (CAERC box 44, folder 12); J. C. Sliger, *Sullivan CEAR,* 1942 (CAERC box 41, folder 53); E. A. Chatham, *Perry CEAR,* 1944 (CAERC box 122, folder 40).
33. W. M. Clary, *Tipton CEAR,* 1942 (CAERC box 45, folder 23); H. E. Abbott, *Marion CEAR,* 1944 (CAERC box 111, folder 13); J. A. Wood, *Martin CEAR,* 1942 (CAERC box 113, folder 28); L. M. Butler, *Marshall CEAR,* 1942 (CAERC box 112, folder 32); E. A. Johnson, *Howard CEAR,* 1942 (CAERC box 95, folder 10).
34. *Purdue University Department of Agricultural Extension Annual Report*, 1945, 3.
35. C. E. Olson, *Pulaski CEAR,* 1942 (CAERC box 28, folder 10).
36. W. C. Haynes, *Madison CEAR,* 1942 (CAERC box 109, folder 38).
37. S. B. Scott, *Johnson CEAR,* 1942 (CAERC box 101, folder 41).
38. Ibid.
39. G. A. Sowers, *Montgomery CEAR,* 1942 (CAERC box 115, folder 54).
40. W. M. Clary, *Tipton CEAR,* 1942 (CAERC box 45, folder 23); W. C. Haynes, *Madison CEAR,* 1942 (CAERC box 109, folder 38); W. U. Rusk, *Huntington CEAR,* 1943 (CAERC box 96, folder 19); G. I. John, *Ohio CEAR,* 1942 (CAERC box 119, folder 30).
41. C. E. Olson, *Pulaski CEAR,* 1942 (CAERC box 28, folder 10); C. A. Langston, *Jay CEAR,* 1942 (CAERC box 98, folder 51); E. S. Bogue, *Pike CEAR,* 1942 (CAERC box 123, folder 34); G. I. John, *Ohio CEAR,* 1942 (CAERC box 119, folder 30); R. R. Zell, *Newton CEAR,* 1942 (CAERC box 117, folder 43); W. J. Record, *Morgan CEAR,* 1942 (CAERC box 117, folder 3).
42. S. B. Pershing, *Tippecanoe CEAR,* 1942 (CAERC box 44, folder 12); A. M. Bishea, *Vanderburgh CEAR,* 1942 (CAERC box 48, folder 4); M. A. Nye, *Noble CEAR,* 1942 (CAERC box 118, folder 35); W. C. Haynes, *Madison CEAR,* 1942 (CAERC box 109, folder 38); J. F. Armstrong, *Lawrence CEAR,* 1943 (CAERC box 108, folder 42).
43. J. C. Sliger, *Sullivan CEAR,* 1942 (CAERC box 41, folder 53).
44. I. O. Haug, *Ripley CEAR,* 1944 (CAERC box 31, folder 46).
45. J. F. Armstrong, *Lawrence CEAR,* 1943 (CAERC box 108, folder 42).
46. E. F. Brown, *Rush CEAR,* 1942 (CAERC box 33, folder 8); C. L. Brown, *Vigo CEAR,* 1942 (CAERC box 51, folder 1); C. E. Olson, *Pulaski CEAR,* 1942 (CAERC box 28, folder 10); S. B. Scott, *Johnson CEAR,* 1942 (CAERC box 101, folder 41); W. C. Haynes, *Madison CEAR,* 1942 (CAERC box 109, folder 38).

47. N. S. Hadley, *Parke CEAR,* 1943 (CAERC box 121, folder 53).
48. W. J. Record, *Morgan CEAR,* 1942 (CAERC box 117, folder 3).
49. W. U. Rusk, *Huntington CEAR,* 1942 (CAERC box 96, folder 18).
50. G. I. John, *Ohio CEAR,* 1942 (CAERC box 119, folder 30).
51. G. T. Harris, *Ripley CEAR,* 1943 (CAERC box 31, folder 45); I. O. Haug, *Ripley CEAR,* 1944 (CAERC box 31, folder 46); D. L. Grimes, *Putnam CEAR,* 1942 (CAERC box 29, folder 26); C. E. Olson, *Pulaski CEAR,* 1942 (CAERC box 28, folder 10); C. A. Langston, *Jay CEAR,* 1943 (CAERC box 99, folder 1); W. U. Rusk, *Huntington CEAR,* 1943 (CAERC box 96, folder 19).
52. S. B. Pershing, *Tippecanoe CEAR,* 1942 (CAERC box 44, folder 12).
53. Ibid.; W. U. Rusk, *Huntington CEAR,* 1943 (CAERC box 96, folder 19).
54. J. McCulloch, *Scott CEAR,* 1943 (CAERC box 36, folder 24).
55. S. B. Scott, *Johnson CEAR,* 1943 (CAERC box 101, folder 42).
56. J. F. Armstrong, *Lawrence CEAR,* 1943 (CAERC box 108, folder 42).
57. Ibid.
58. W. P. Stall, *Jackson CEAR,* 1942 (CAERC box 97, folder 7).
59. R. N. Atkinson, *Orange CEAR,* 1945 (CAERC box 120, folder 28).
60. S. Leaming, *Porter CEAR,* 1942 (CAERC box 124, folder 26); W. P. Stall, *Jackson CEAR,* 1942 (CAERC box 97, folder 7).
61. S. Leaming, *Porter CEAR,* 1942 (CAERC box 124, folder 26).
62. S. Leaming, *Porter CEAR,* 1943 (CAERC box 124, folder 27).
63. S. B. Scott, *Johnson CEAR,* 1942 (CAERC box 101, folder 41); C. A. Langston, *Jay CEAR,* 1942 (CAERC box 98, folder 51); S. Leaming, *Porter CEAR,* 1942 (CAERC box 124, folder 26).
64. R. L. Case, *LaGrange CEAR,* 1942 (CAERC box 106, folder 3).
65. W. U. Rusk, *Huntington CEAR,* 1942 (CAERC box 96, folder 18); R. L. Case, *LaGrange CEAR,* 1942 (CAERC box 106, folder 3).
66. R. L. Case, *LaGrange CEAR,* 1943 (CAERC box 106, folder 4); W. P. Stall, *Jackson CEAR,* 1942 (CAERC box 97, folder 7); H. E. Abbott, *Marion CEAR,* 1942 (CAERC box 111, folder 7); G. A. Sowers, *Montgomery CEAR,* 1942 (CAERC box 115, folder 54); E. F. Brown, *Rush CEAR,* 1942 (CAERC box 33, folder 8); L. E. Cutler, *Lake CEAR,* 1942 (CAERC box 107, folder 34).
67. S. B. Pershing, *Tippecanoe CEAR,* 1942 (CAERC box 44, folder 12); G. T. Harris, *Ripley CEAR,* 1942 (CAERC box 31, folder 44).
68. C. E. Olson, *Pulaski CEAR,* 1943 (CAERC box 28, folder 11).
69. C. A. Langston, *Jay CEAR,* 1942 (CAERC box 98, folder 51).
70. W. U. Rusk, *Huntington CEAR,* 1942 (CAERC box 96, folder 18).
71. Ibid.
72. C. A. Langston, *Jay CEAR,* 1942 (CAERC box 98, folder 51); E. A. Johnson, *Howard CEAR,* 1942 (CAERC box 95, folder 10); R. L. Case, *LaGrange CEAR,* 1943 (CAERC box 106, folder 4).
73. G. A. Sowers, *Montgomery CEAR,* 1943 (CAERC box 115, folder 55).
74. G. I. John, *Ohio CEAR,* 1942 (CAERC box 119, folder 30); T. M. Grabill, *Steuben CEAR,* 1942 (CAERC box 40, folder 50).
75. J. E. Wylie, *Miami CEAR,* 1942 (CAERC box 114, folder 13).
76. J. McCulloch, *Scott CEAR,* 1943 (CAERC box 36, folder 24).

77. G. A. Sowers, *Montgomery CEAR,* 1942 (CAERC box 115, folder 54).
78. H. E. Abbott, *Marion CEAR,* 1942 (CAERC box 111, folder 7).
79. W. U. Rusk, *Huntington CEAR,* 1943 (CAERC box 96, folder 19).
80. H. E. Abbott, *Marion CEAR,* 1942 (CAERC box 111, folder 7); C. A. Langston, *Jay CEAR,* 1942 (CAERC box 98, folder 51).
81. L. P. Geyer, *Jefferson CEAR,* 1942 (CAERC box 100, folder 9); W. U. Rusk, *Huntington CEAR,* 1943 (CAERC box 96, folder 19); G. T. Harris, *Ripley CEAR,* 1942 (CAERC box 31, folder 44).
82. E. C. Bird, *St. Joseph CEAR,* 1942 (CAERC box 35, folder 28).
83. C. A. Langston, *Jay CEAR,* 1943 (CAERC box 99, folder 1).
84. E. A. Johnson, *Howard CEAR,* 1942 (CAERC box 95, folder 10).
85. L. M. Butler, *Marshall CEAR,* 1942 (CAERC box 112, folder 32).
86. C. L. Brown, *Vigo CEAR,* 1942 (CAERC box 51, folder 1); S. Leaming, *Porter CEAR,* 1942 (CAERC box 124, folder 26); E. F. Downen, *Spencer CEAR,* 1942 (CAERC box 38, folder 48); R. L. Case, *LaGrange CEAR,* 1943 (CAERC box 106, folder 4).
87. E. C. Bird, *St. Joseph CEAR,* 1942 (CAERC box 35, folder 28).
88. H. E. Abbott, *Marion CEARs,* 1942–43 (CAERC box 111, folders 7 and 10); G. I. John, *Ohio CEAR,* 1942 (CAERC box 119, folder 30).
89. H. E. Abbott, *Marion CEAR,* 1943 (CAERC box 111, folder 10).
90. S. W. Milligan, *Wayne CEAR,* 1942 (CAERC box 56, folder 29); B. T. Bond, *Warrick CEAR,* 1942 (CAERC box 54, folder 36); J. L. Guthrie, *Union CEAR,* 1942 (CAERC box 46, folder 24); E. S. Bogue, *Pike CEAR,* 1942 (CAERC box 123, folder 34).
91. G. T. Harris, *Ripley CEAR,* 1942 (CAERC box 31, folder 44).
92. E. C. Bird, *St. Joseph CEAR,* 1945 (CAERC box 35, folder 31); M. S. Smith, *Wells CEAR,* 1945 (CAERC box 58, folder 31); R. L. Case, *LaGrange CEAR,* 1945 (CAERC box 106, folder 6); S. W. Milligan, *Wayne CEAR,* 1943 (CAERC box 56, folder 30); L. J. Fultz, *Washington CEAR,* 1944 (CAERC box 55, folder 46).
93. L. J. Fultz, *Washington CEAR,* 1944 (CAERC box 55, folder 46); J. C. Sliger, *Sullivan CEAR,* 1943 (CAERC box 41, folder 54); E. F. Brown, *Rush CEAR,* 1942 (CAERC box 33, folder 8).
94. G. A. Sowers, *Montgomery CEAR,* 1942 (CAERC box 115, folder 54).
95. Sarah Sundin, "World War II War Bonds," *Sarah Sundin* (blog), 29 Nov. 2017, https://www.sarahsundin.com/world-war-ii-war-bonds/; E. F. Brown, *Rush CEAR,* 1942 (CAERC box 33, folder 8).
96. E. F. Downen, *Spencer CEAR,* 1943 (CAERC box 38, folder 49); G. I. John, *Ohio CEAR,* 1943 (CAERC box 119, folder 31).
97. L. J. Fultz, *Washington CEAR,* 1944 (CAERC box 55, folder 46); E. A. Chatham, *Perry CEAR,* 1944 (CAERC box 122, folder 40); C. D. Chambers, *Perry CEAR,* 1942 (CAERC box 122, folder 38); G. A. Sowers, *Montgomery CEAR,* 1942 (CAERC box 115, folder 54); J. E. Wylie, *Miami CEAR,* 1945 (CAERC box 114, folder 16).
98. E. F. Brown, *Rush CEAR,* 1942 (CAERC box 33, folder 8); W. J. Record, *Morgan CEAR,* 1944 (CAERC box 117, folder 5).
99. E. F. Downen, *Spencer CEAR,* 1943 (CAERC box 38, folder 49); J. McCulloch, *Scott CEAR,* 1943 (CAERC box 36, folder 24).
100. R. L. Case, *LaGrange CEAR,* 1945 (CAERC box 106, folder 6); J. L. Guthrie, *Union CEAR,* 1943 (CAERC box 46, folder 25).

101. E. L. Hartman, *LaPorte CEAR,* 1943 (CAERC box 105, folder 19).
102. E. F. Brown, *Rush CEAR,* 1942 (CAERC box 33, folder 8); O. B. Riggs, *Posey CEAR,* 1942 (CAERC box 125, folder 27); G. A. Sowers, *Montgomery CEAR,* 1942 (CAERC box 115, folder 54); C. A. Langston, *Jay CEAR,* 1942 (CAERC box 98, folder 51).
103. J. L. Guthrie, *Union CEAR,* 1942 (CAERC box 46, folder 24); S. B. Pershing, *Tippecanoe CEAR,* 1942 (CAERC box 44, folder 12).
104. J. E. Wylie, *Miami CEAR,* 1945 (CAERC box 114, folder 16).
105. E. L. Hartman, *LaPorte CEAR,* 1942 (CAERC box 105, folder 17).
106. L. P. Geyer, *Jefferson CEAR,* 1943 (CAERC box 100, folder 10); S. W. Milligan, *Wayne CEAR,* 1943 (CAERC box 56, folder 30); J. L. Guthrie, *Union CEAR,* 1943 (CAERC box 46, folder 25).
107. S. W. Milligan, *Wayne CEAR,* 1945 (CAERC box 56, folder 34).
108. W. U. Rusk, *Huntington CEAR,* 1945 (CAERC box 96, folder 21).
109. S. B. Pershing, *Tippecanoe CEAR,* 1942 (CAERC box 44, folder 12); J. F. Armstrong, *Lawrence CEAR,* 1942 (CAERC box 108, folder 41).
110. B. V. Widney, *Whitley CEAR,* 1945 (CAERC box 60, folder 6).

CHAPTER 40

School Children of America! Help Save Your Fathers', Brothers', and Neighbors' Lives by Collecting Milkweed Pods

1. "The Lowly Milkweed," *Bremen Enquirer* (Bremen, IN), 25 May 1944, 6.
2. Editors of Encyclopaedia Britannica, "Kapok," *Encyclopedia Britannica* (website), 23 Feb. 2017, https://www.britannica.com/topic/kapok.
3. "Lowly Milkweed to Play Important Part in War Effort and Children Will Be Asked to Collect Pods," *Rushville Republican* (Rushville, IN), 28 Apr. 1944, 1; "The Lowly Milkweed," *Bremen Enquirer* (Bremen, IN), 25 May 1944, 6; "Government Seeks Milkweed Pods as Kapok Substitute," *Indianapolis Star,* 23 Aug. 1943, 4.
4. "Pupils to Be Asked to Help Pick Pods for Milkweed Floss," *Palladium-Item* (Richmond, IN), 3 Sept. 1944, 7; "Milkweed Floss Will Be Ready about Sept. 1," *Rushville Republican* (Rushville, IN), 15 Aug. 1944, 2; "Prizes Offered to Collectors of Milkweed Pods," *Tipton Daily Tribune* (Tipton, IN), 8 Sept. 1944, 1.
5. "Government Seeks Milkweed Pods as Kapok Substitute," 4.
6. G. A. Sowers, *Montgomery CEAR,* 1945 (CAERC box 115, folder 57); M. S. Smith, *Wells CEAR,* 1944 (CAERC box 58, folder 30); I. O. Haug, *Ripley CEAR,* 1944 (CAERC box 31, folder 46).
7. "Milkweed Pickers Wanted: Silk Is Needed for War," *Scott County Journal* (Scottsburg, IN), 28 Dec. 1944, 7.
8. "Pupils to Pick Milkweed Pods," *South Bend Tribune* (South Bend, IN), 28 Apr. 1944, 36.
9. "Lowly Milkweed to Play Important Part in War Effort," 1.
10. "Navy Okays Milkweed," *Greenfield Daily Reporter* (Greenfield, IN), 29 July 1943, 5.
11. "See Place for Milkweed Plant in Postwar Era," *Hancock Democrat* (Greenfield, IN), 28 June 1945, 7; "Milkweed Pods to Be Collected by School Children," *Muncie Evening Press* (Muncie, IN), 12 May 1944, 8.
12. "Milkweed Earns a Job Filling Life Preservers," *Greenfield Daily Reporter* (Greenfield, IN), 21 Oct. 1943, 7; "The Milkweed Harvest," *Fairmount News* (Fairmount, IN), 7 Dec. 1944, 2.
13. "Milkweed Pods to Be Collected by School Children," 8.
14. "Milkweed—A War Plant," *Garrett Clipper* (Garrett, IN), 20 July 1944, 8.

15. Ibid.; "Navy Okays Milkweed," 5; "Investigators Find That Milkweed Is Defense Need," *Greenfield Daily Reporter* (Greenfield, IN), 13 July 1942, 3.
16. "Don't Chop Down the Milk Weed," *Call-Leader* (Elwood, IN), 21 June 1944, 4; "Milkweed—A War Plant," 8; "Investigators Find That Milkweed Is Defense Need," 3.
17. M. S. Smith, *Wells CEAR,* 1944 (CAERC box 58, folder 30); C. E. Olson, *Pulaski CEAR,* 1944 (CAERC box 28, folder 12); H. E. Abbott, *Marion CEAR,* 1944 (CAERC box 111, folder 13).
18. "Lowly Milkweed to Play Important Part in War Effort," 1.
19. "Michigan Mill to Process Milkweed Floss for Use in Life Belts and Flying Suits," *Tipton Daily Tribune* (Tipton, IN), 26 June 1942, 1.
20. "Government Seeks Milkweed Pods as Kapok Substitute," 4; "It's Who We Are," Preston Feather (website), accessed 20 Apr. 2022, https://prestonfeather.com/about-us/.
21. "Government Seeks Milkweed Pods as Kapok Substitute," 4.
22. "To Seek 10,000,000 Lbs. Milkweed Floss," *Republic* (Columbus, IN), 6 May 1944, 3; "Lowly Milkweed to Play Important Part in War Effort," 1.
23. "Michigan Mill to Process Milkweed Floss," 1.
24. "Flossy Harvest," *Lafayette Journal and Courier* (Lafayette, IN), 27 Feb. 1945, 6.
25. "Government Seeks Milkweed Pods as Kapok Substitute," 4; Preston Feather, "It's Who We Are."
26. "Milkweed Floss Salvage Sought," *Times* (Munster, IN), 3 Oct. 1943, 3.
27. W. C. Haynes, *Madison CEAR,* 1944 (CAERC box 109, folder 40); "Government Seeks Milkweed Pods as Kapok Substitute," 4.
28. "Milkweed—A War Plant," 8; "Milkweed Floss to Replace Kapok for Life Jackets," *Palladium-Item* (Richmond, IN), 16 July 1944, 2; "Don't Chop Down the Milk Weed," 4.
29. "Children's Milkweed Collection Will Make 100,000 Life Preservers," *Indianapolis Star,* 22 Nov. 1944, 24; A. M. Bishea, *Vanderburgh CEAR,* 1944 (CAERC box 48, folder 9); J. A. Clark, *Kosciusko CEAR,* 1945 (CAERC box 104, folder 21); L. E. Cutler, *Lake CEAR,* 1944 (CAERC box 107, folder 37); F. E. DeLaCroix, *Jasper CEAR,* 1944 (CAERC box 98, folder 14); *Purdue University Department of Agricultural Extension Annual Report,* 1945, 8.
30. C. E. Olson, *Pulaski CEAR,* 1944 (CAERC box 28, folder 12); W. C. Haynes, *Madison CEAR,* 1944 (CAERC box 109, folder 40).
31. E. S. Bogue, *Pike CEAR,* 1944 (CAERC box 123, folder 36).
32. Ibid.; I. O. Haug, *Ripley CEAR,* 1944 (CAERC box 31, folder 46).
33. Patterson Clark, "Milkweed Fruits: Pods of Plenty," *Urban Jungle, Washington Post* (website), 25 Sept. 2012, https://www.washingtonpost.com/wp-srv/special/metro/urban-jungle/pages/120925.html; "Children's Milkweed Collection," 24.
34. "Children's Milkweed Collection," 24.
35. "Lowly Milkweed to Play Important Part in War Effort," 1; "Milkweed Floss Collection Due during September," *Rushville Republican* (Rushville, IN), 3 Aug. 1944, 8.
36. "School Children in County Asked to Collect Pods," *Rushville Republican* (Rushville, IN), 9 Aug. 1944, 1; F. E. DeLaCroix, *Jasper CEAR,* 1944 (CAERC box 98, folder 14); O. B. Riggs, *Posey CEAR,* 1944 (CAERC box 125, folder 31).
37. "Use Milkweed in Life Jackets," *Star Press* (Muncie, IN), 19 Aug. 1944, 12; E. A. Chatham, *Perry CEAR,* 1944 (CAERC box 122, folder 40); S. B. Scott, *Johnson CEAR,* 1944 (CAERC box 101, folder 43); "Get Sacks for Milkweed Pod," *Franklin Evening Star* (Franklin, IN), 15 Sept. 1944, 1.

38. H. S. Benson, *Knox CEAR,* 1944 (CAERC box 103, folder 2); E. S. Bogue, *Pike CEAR,* 1944 (CAERC box 123, folder 36); O. B. Riggs, *Posey CEAR,* 1944 (CAERC box 125, folder 31).
39. O. B. Riggs, *Posey CEAR,* 1944 (CAERC box 125, folder 31).
40. Katherine Hauswirth, "The Heroic Milkweed," *Christian Science Monitor* (website), 26 Oct. 2008, https://www.csmonitor.com/The-Culture/Gardening/2008/1026/the-heroic-milkweed.
41. "Hoosiers Urged to Gather Floss for Life Jackets," *Tipton Daily Tribune* (Tipton, IN), 8 July 1944, 4; "Get Sacks for Milkweed Pod," 1.
42. "Government Seeks Milkweed Pods as Kapok Substitute," 4; "Save Milkweed and Save Life Is Campaign Slogan," *Tipton Daily Tribune* (Tipton, IN), 27 July 1944, 5.
43. H. S. Benson, *Knox CEAR,* 1945 (CAERC box 103, folder 3).
44. "Local Students Collect 120 Bags of Milkweed Pods to Fill Urgent Need," *Tribune* (Seymour, IN), 23 Sept. 1944, 1.
45. "Milkweed Floss Salvage Sought," 3; M. S. Smith, *Wells CEAR,* 1944 (CAERC box 58, folder 30); A. C. Shideler, *Starke CEAR,* 1944 (CAERC box 40, folder 2); R. R. Zell, *Newton CEAR,* 1944 (CAERC box 117, folder 45).
46. W. C. Haynes, *Madison CEAR,* 1944 (CAERC box 109, folder 40); E. A. Johnson, *Howard CEAR,* 1944 (CAERC box 95, folder 12).
47. M. S. Smith, *Wells CEAR,* 1944 (CAERC box 58, folder 30).
48. "Children to Assist in Milkweed Pod Collection," *Princeton Clarion-News* (Princeton, IN), 23 Sept. 1944, 2.
49. "Pupils to Be Asked to Help Pick Pods for Milkweed Floss," 7.
50. E. S. Bogue, *Pike CEAR,* 1944 (CAERC box 123, folder 36).
51. "School Students Gather Milkweed Floss for 15 Life Preservers," *Culver Citizen* (Culver, IN), 11 Oct. 1944, 1; "Milkweed Floss Collection Due during September," 8; "The Milkweed Harvest," 2.
52. "Hartford City," *Muncie Evening Press* (Muncie, IN), 21 July 1944, 7; "Milkweed Floss Has Gone to War," *Lafayette Journal and Courier* (Lafayette, IN), 8 July 1944, 5.
53. "Milkweed Pods to Be Collected by School Children," 8; M. S. Smith, *Wells CEAR,* 1944 (CAERC box 58, folder 30).
54. O. B. Riggs, *Posey CEAR,* 1944 (CAERC box 125, folder 31); E. A. Chatham, *Perry CEAR,* 1944 (CAERC box 122, folder 40); H. S. Benson, *Knox CEAR,* 1944 (CAERC box 103, folder 2).
55. "Pupils to Be Asked to Help Pick Pods for Milkweed Floss," 7; "Milkweed Floss Drive Scheduled," *Tipton Daily Tribune* (Tipton, IN), 11 July 1944, 6.
56. "Now Is Time to Collect Milkweed Pods for Life Jackets," *Jackson County Banner* (Brownstown, IN), 27 Sept. 1944, 1; "Milkweed Floss Replaces Kapok for Life Jackets," *Hancock Democrat* (Greenfield, IN), 13 July 1944, 2; "Local Students Collect 120 Bags of Milkweed Pods," 1; "Investigators Find That Milkweed Is Defense Need," 3.
57. "Milkweed Floss Collection Due during September," 8; "Local Students Collect 120 Bags of Milkweed Pods," 1.
58. E. A. Chatham, *Perry CEAR,* 1944 (CAERC box 122, folder 40).
59. A. M. Bishea, *Vanderburgh CEAR,* 1944 (CAERC box 48, folder 9).
60. H. S. Benson, *Knox CEAR,* 1944 (CAERC box 103, folder 2); C. E. Olson, *Pulaski CEAR,* 1944 (CAERC box 28, folder 12); "Milkweed Floss Drive Scheduled," 6.

61. "Local Students Collect 120 Bags of Milkweed Pods," 1; "Milkweed Pod Collection Is Planned Here," *Franklin Evening Star* (Franklin, IN), 3 Aug. 1944, 8.
62. E. A. Johnson, *Howard CEAR,* 1944 (CAERC box 95, folder 12); C. E. Olson, *Pulaski CEAR,* 1944 (CAERC box 28, folder 12).
63. "Local Students Collect 120 Bags of Milkweed Pods," 1; "Milkweed Pod Collection Is Planned Here," 8.
64. "Milkweed Pod Collection," *Brook Reporter* (Brook, IN), 2 Nov. 1944, 2.
65. "1,575 Bags of Pods Collected," *Rushville Republican* (Rushville, IN), 13 Dec. 1944, 1; "Plymouth," *Bremen Enquirer* (Bremen, IN), 8 Feb. 1945, 4.
66. H. H. Madaus, *Warren CEAR,* 1944 (CAERC box 53, folder 26); C. L. Spuller, *Rush CEAR,* 1945 (CAERC box 33, folder 13).
67. "Milkweed Pod Collection," *Brook Reporter* (Brook, IN), 2 Nov. 1944, 2; C. E. Olson, *Pulaski CEAR,* 1944 (CAERC box 28, folder 12); J. A. Clark, *Kosciusko CEAR,* 1944 (CAERC box 104, folder 20); E. A. Johnson, *Howard CEAR,* 1944 (CAERC box 95, folder 12).
68. "Milkweed Pods to Be Shipped to Hemp Plant," *Kokomo Tribune* (Kokomo, IN), 25 Jan. 1945, 25.
69. "1,575 Bags of Pods Collected," 1; "Milkweed Floss Salvage Sought," 3; W. P. Stall, *Jackson CEAR,* 1945 (CAERC box 97, folder 10).
70. "Jackson County Makes Largest Floss Contribution as Car Is Loaded Here," *Tribune* (Seymour, IN), 19 Jan. 1945, 1; W. P. Stall, *Jackson CEAR,* 1945 (CAERC box 97, folder 10).
71. C. E. Olson, *Pulaski CEAR,* 1945 (CAERC box 28, folder 13); "Milkweed Floss Salvage Sought," 3.
72. "Plymouth," 4; "Jackson County Makes Largest Floss Contribution," 1; W. P. Stall, *Jackson CEAR,* 1945 (CAERC box 97, folder 10).
73. "Plymouth," 4.
74. Ibid.
75. B. T. Bond, *Warrick CEAR,* 1944 (CAERC box 54, folder 38).
76. "No Inauguration Ceremony," *Indianapolis Star,* 26 Nov. 1944, 40; "War and Farming in Jackson County," *Jackson County Banner* (Brownstown, IN), 19 July 1944, 7.
77. G. A. Sowers, *Montgomery CEAR,* 1945 (CAERC box 115, folder 57); *Purdue University Department of Agricultural Extension Annual Report*, 1945, 8.
78. E. A. Chatham, *Perry CEAR,* 1944 (CAERC box 122, folder 40); L. E. Cutler, *Lake CEAR,* 1944 (CAERC box 107, folder 37); W. P. Stall, *Jackson CEAR,* 1945 (CAERC box 97, folder 10).
79. O. B. Riggs, *Posey CEAR,* 1944 (CAERC box 125, folder 31).
80. H. S. Benson, *Knox CEAR,* 1944 (CAERC box 103, folder 2); W. C. Haynes, *Madison CEAR,* 1945 (CAERC box 109, folder 41); E. S. Bogue, *Pike CEAR,* 1944 (CAERC box 123, folder 36); E. A. Johnson, *Howard CEAR,* 1944 (CAERC box 95, folder 12).
81. M. S. Smith, *Wells CEAR,* 1944 (CAERC box 58, folder 30); S. B. Pershing, *Tippecanoe CEAR,* 1944 (CAERC box 44, folder 16).
82. "4-H Club Achievement," *Indianapolis News,* 7 Nov. 1944, 6.
83. "Sullivan County Youth Collect Milkweed Pods," *Call-Leader* (Elwood, IN), 18 Oct. 1944, 6.
84. M. J. Peterson, *Vermillion CEAR,* 1944 (CAERC box 49, folder 39).
85. W. C. Haynes, *Madison CEAR,* 1944 (CAERC box 109, folder 40).
86. E. S. Bogue, *Pike CEAR,* 1944 (CAERC box 123, folder 36).
87. "Prizes Offered to Collectors of Milkweed Pods," 1.
88. E. S. Bogue, *Pike CEAR,* 1944 (CAERC box 123, folder 36).

89. "Prizes Offered to Collectors of Milkweed Pods," 1; "Milkweed—A War Plant," 8; "To Seek 10,000,000 Lbs. Milkweed Floss," 3.
90. "Flossy Harvest," 6.
91. Clark, "Milkweed Fruits."
92. "War and Farming in Jackson County," 7.
93. "To Seek 10,000,000 Lbs. Milkweed Floss," 3; "Milkweed Plant Provides War Use Fibers," *Indianapolis News,* 25 Mar. 1943, 10; "Milkweed Clothes," *Chronicle* (Scottsburg, IN), 4 Jan. 1945, 4.
94. "See Place for Milkweed Plant in Postwar Era," 7; "Investigators Find that Milkweed Is Defense Need," 3.
95. "Milkweed Yield," *Daily Clintonian* (Clinton, IN), 12 Nov. 1945, 4.
96. "Milkweed Plant Provides War Use Fibers," 10.
97. "Milkweed Floss Total," *Princeton Clarion-News* (Princeton, IN), 10 Aug. 1945, 5.
98. E. S. Bogue, *Pike CEAR,* 1944 (CAERC box 123, folder 36); "Michigan Mill to Process Milkweed Floss," 1; "Milkweed Floss Collection Due during September," 8.
99. "Children's Milkweed Collection," 24.
100. G. A. Sowers, *Montgomery CEAR,* 1944 (CAERC box 115, folder 56); G. A. Sowers, *Montgomery CEAR,* 1945 (CAERC box 115, folder 57).

CHAPTER 41

American Hemp Will Go On Duty Again

1. "Philippine Commonwealth Has 114,400 Square Miles," *Tribune* (Seymour, IN), 9 Jan. 1942, 4.
2. Ibid.
3. U.S. Department of Agriculture, *Hemp for Victory* (film), 1942, https://www.youtube.com/watch?v=d3rolyiTPr0; "Northern Indiana to Produce Hemp," *Argos Reflector* (Argos, IN), 26 Nov. 1942, 1.
4. Daryl T. Ehrensing, *Feasibility of Industrial Hemp Production in the United States Pacific Northwest,* Oregon State University Agricultural Experiment Station Bulletin 681 (Corvallis, OR: Oregon State University, May 1998), 2, https://catalog.extension.oregonstate.edu/sites/catalog/files/project/pdf/sb681.pdf; Jack Stinnett, "Washington Daybook," *Rushville Republican* (Rushville, IN), 4 Feb. 1942, 2.
5. "Davao 'Japanese Capital' of Isles," *Indianapolis Star,* 11 Jan. 1942, 16.
6. "Declares Russia Holds War Key," *Indianapolis Star,* 15 Jan. 1942, 5.
7. "Plans to Revive Hemp Industry to Be Studied," *Indianapolis News,* 16 Sept. 1942, 22; "Hemp Growing in State Opposed by Groups Fighting Marijuana," *Indianapolis Star,* 26 July 1942, 31.
8. Paul Gesner and John Beckley, "Hemispheric Economy Seen as Slow Task," *South Bend Tribune* (South Bend, IN), 30 Jan. 1942, 27.
9. M. S. Smith, *Wells CEAR,* 1943 (CAERC box 58, folder 29).
10. Nick Johnson, "Hemp for Victory: A Brief History—Part 1," *Hempirical Evidence* (blog), 11 Mar. 2019, https://hempiricalevidence.com/2019/03/11/hemp-for-victory-a-brief-overdue-history-part-1/; Anne L. Ash, "Hemp—Production and Utilization," *Economic Botany* 2 (Apr. 1948): 158, https://www.jstor.org/stable/4251894?seq=1.
11. "Northern Indiana to Produce Hemp," 1.
12. "WPB Approves 300,000-Acre Hemp Planting," *Kokomo Tribune* (Kokomo, IN), 24 Sept. 1942, 18.
13. "CCC Requests Hemp Planting Appropriations," *Palladium-Item* (Richmond, IN), 9 Oct. 1942, 5; Ash, "Hemp—Production and Utilization," 158; "Plan Increase in Production of U. S. Hemp," *Times* (Munster, IN), 19 May 1942, 8.

14. Ash, "Hemp—Production and Utilization," 158; C. P. Wilsie et al., *Hemp: A War Crop for Iowa,* Iowa Agricultural Experiment Station and Agricultural Extension Service Bulletin No. P49 (Ames, IA: Iowa State College of Agriculture and Mechanic Arts, Dec. 1942), 588, https://dr.lib.iastate.edu/handle/20.500.12876/11999.
15. "Plans to Revive Hemp Industry to Be Studied," 22.
16. Ibid.
17. "Hemp Industry to Become Permanent," *Muncie Evening Press* (Muncie, IN), 13 Apr. 1943, 3; "Curtailed Hemp Crop Acreage Is Predicted," *Indianapolis News,* 9 Jan. 1943, 18.
18. "Hemp Industry to Become Permanent," 3.
19. "5 Hemp Plants to Be Built in State," *Indianapolis News,* 21 Nov. 1942, 10; Bill Stahl, "Hoosier Farmers Find New Crop Interest as Season for Hemp Harvest Approaches," *Indianapolis News,* 2 July 1943, 25.
20. "5 Hemp Plants to Be Built in State," 10; Maurice Early, "The Day in Indiana," *Star Press* (Muncie, IN), 20 Nov. 1942, 1.
21. "CCC Requests Hemp Planting Appropriations," 5; Ash, "Hemp—Production and Utilization," 169.
22. Ash, "Hemp—Production and Utilization," 169; "Hemp Industry to Become Permanent," 3; "Curtailed Hemp Crop Acreage Is Predicted," 18.
23. Ash, "Hemp—Production and Utilization," 169; "Essential Needs for Rope Are Being Met," *Muncie Evening Press* (Muncie, IN), 7 Apr. 1944, 13.
24. "5 Hemp Plants to Be Built in State," 10; Stahl, "Hoosier Farmers Find New Crop Interest," 25; Maurice Early, "The Day in Indiana," *Star Press* (Muncie, IN), 20 Nov. 1942, 1.
25. "Six Indiana Sites for Hemp Mills Approved," *Muncie Evening Press* (Muncie, IN), 7 Dec. 1942, 1; "5 Hemp Plants to Be Built in State," 10; "Northern Indiana to Produce Hemp," 1.
26. "Six Indiana Sites for Hemp Mills Approved," 1; Maurice Early, "The Day in Indiana," *Star Press* (Muncie, IN), 20 Nov. 1942, 1.
27. "WPB Approves 300,000-Acre Hemp Planting," 18.
28. Trey Malone and Kevin Gomez, "Hemp in the United States: A Case Study of Regulatory Path Dependence," *Applied Economic Perspectives and Policy* 41, no. 2 (June 2019): 201, https://doi.org/10.1093/aepp/ppz001; Maurice Early, "The Day in Indiana," *Star Press* (Muncie, IN), 20 Nov. 1942, 1.
29. Johnson, "Hemp for Victory: A Brief History—Part 1"; Dennis Rens, "America's Hemp King," Newhead News (website), Feb. 1995, http://newheadnews.com/hemp/Rens.hempstory.Wis/.
30. Albert Chapman, email message to author, 22 Sept. 2021. Chapman wrote, "This was included in Executive Order 9280 signed by FDR [President Franklin D. Roosevelt] on December 5, 1942, and published in the December 8, 1942, *Federal Register.* . . .," https://www.govinfo.gov/content/pkg/FR-1942-12-08/pdf/FR-1942-12-08.pdf.
31. "Tax Financed New Industry to Be Shelved," *Chicago Daily Tribune,* 20 Jan. 1944, 23.
32. Maurice Early, "The Day in Indiana," *Star Press* (Muncie, IN), 20 Nov. 1942, 1; Stahl, "Hoosier Farmers Find New Crop Interest," 25; Tom Maddox, "Hemp Being Grown by Businessmen from Remington," *Palladium-Item* (Richmond, IN), 10 July 1943, 3.
33. "Hemp Growing in State Opposed by Groups Fighting Marijuana," 31.
34. "Hemp Growing," *Indianapolis News,* 28 Sept. 1942, 6.
35. U.S. Department of Agriculture, *Agricultural Statistics, 1944* (Washington, DC: Government Printing Office, 1944), 280, https://downloads.usda.library.cornell.edu/usda-esmis/files/j3860694x/sb397c286/9z903270s/Agstat-04-23-1944.pdf.

36. "Marijuana Planting Is Urged But It's for Hemp," *Edinburg Daily Courier* (Edinburg, IN), 20 Nov. 1942, 3.
37. Malone and Gomez, "Hemp in the United States," 201; U.S. Department of Agriculture, *Hemp for Victory* (film).
38. "Marijuana Planting Is Urged But It's for Hemp," *Indianapolis News*, 19 Nov. 1942, 42.
39. "Five Hemp Mills Recommended for Indiana Sections," *Lafayette Journal and Courier* (Lafayette, IN), 26 Nov. 1942, 1.
40. "Clinton County Farmers Discuss Hemp Growing," *Kokomo Tribune* (Kokomo, IN), 21 Dec. 1942, 9.
41. *Purdue University Department of Agricultural Extension Annual Report*, 1941–43, 14; Maddox, "Hemp Being Grown by Businessmen," 3; Stahl, "Hoosier Farmers Find New Crop Interest," 25.
42. Maddox, "Hemp Being Grown by Businessmen," 3; Stahl, "Hoosier Farmers Find New Crop Interest," 25.
43. Stahl, "Hoosier Farmers Find New Crop Interest," 25.
44. W. U. Rusk, *Huntington CEAR*, 1943 (CAERC box 96, folder 19).
45. R. R. Zell, *Newton CEAR*, 1943 (CAERC box 117, folder 44).
46. Ibid.
47. "Select 2 Indiana Hemp Plant Sites," *Indianapolis Star*, 9 Mar. 1943, 19; "Contract Let on Hemp Mills," *South Bend Tribune* (South Bend, IN), 28 Apr. 1943, 11; *Purdue University Department of Agricultural Extension Annual Report*, 1941–43, 14.
48. Stahl, "Hoosier Farmers Find New Crop Interest," 25.
49. Wilsie et al., *Hemp: A War Crop for Iowa*, 587; T. Maddox, "Merchants Close Stores Half Day to Produce Hemp," *Indianapolis Star*, 11 July 1943, 35; Maddox, "Hemp Being Grown by Businessmen," 3.
50. Stahl, "Hoosier Farmers Find New Crop Interest," 25.
51. Ibid.
52. Maddox, "Hemp Being Grown by Businessmen," 3; Stahl, "Hoosier Farmers Find New Crop Interest," 25.
53. Jack Herer, *The Emperor Wears No Clothes: Hemp and the Marijuana Conspiracy*, 12th ed. (Highlands, TX: AH HA Publishing, 2010), 2.
54. Stahl, "Hoosier Farmers Find New Crop Interest," 25.
55. Ibid.
56. Ibid.
57. "CCC Requests Hemp Planting Appropriations," 5; Stahl, "Hoosier Farmers Find New Crop Interest," 25.
58. W. U. Rusk, *Huntington CEAR*, 1943 (CAERC box 96, folder 19).
59. Herer, *The Emperor Wears No Clothes*, 45; Ash, "Hemp—Production and Utilization," 166.
60. W. U. Rusk, *Huntington CEAR*, 1943 (CAERC box 96, folder 19); Stahl, "Hoosier Farmers Find New Crop Interest," 25.
61. M. S. Smith, *Wells CEAR*, 1943 (CAERC box 58, folder 29); "Third of Warren Hemp Acreage Is Contracted," *Indianapolis News*, 29 Jan. 1943, 26.
62. W. U. Rusk, *Huntington CEARs*, 1943–44 (CAERC box 96, folders 19–20); M. S. Smith, *Wells CEAR*, 1944 (CAERC box 58, folder 30).
63. Maddox, "Hemp Being Grown by Businessmen," 3; Stahl, "Hoosier Farmers Find New Crop Interest," 25.
64. M. S. Smith, *Wells CEAR*, 1944 (CAERC box 58, folder 30).
65. "State Hemp Mills Will Close After Year," *Edinburg Daily Courier* (Edinburg, IN), 19 Jan. 1944, 2.

66. "Essential Needs for Rope Are Being Met," 13.
67. "Hemp Growing Program Will Be Cut a Third," *Evansville Press* (Evansville, IN), 23 Mar. 1944, 7; "U. S. Hemp: Cut Production," *Bremen Enquirer* (Bremen, IN), 27 Jan. 1944, 1.
68. M. S. Smith, *Wells CEAR,* 1944 (CAERC box 58, folder 30).
69. "State Hemp Mills Will Close After Year," 2.
70. "Hemp Plants for Sale," *Lafayette Journal and Courier* (Lafayette, IN), 1 Jan. 1945, 6.
71. "Essential Needs for Rope Are Being Met," 13; "Will Close Mid-West Hemp Plants This Year," *Star Press* (Muncie, IN), 19 Jan. 1944, 3.
72. "Hemp Plants for Sale," 6; "Essential Needs for Rope Are Being Met," 13.
73. "Hemp Plants for Sale," 6; "42 Government Hemp Mills for Sale or Lease," *South Bend Tribune* (South Bend, IN), 23 Aug. 1945, 23.
74. "Surplus Hemp Mills," *Lafayette Journal and Courier* (Lafayette, IN), 30 July 1945, 1; "State Hemp Mills Will Close After Year," 2.
75. Rens, "America's Hemp King."
76. "Hemp Mill Sold," *Lafayette Journal and Courier* (Lafayette, IN), 2 May 1946, 1.
77. "Abandoned U. S. Hemp Factory Sold," *Lafayette Journal and Courier* (Lafayette, IN), 12 Oct. 1946, 2.
78. "Hemp Plants for Sale," 6; "Indiana Hemp Production to Be Halted," *Union City Times-Gazette* (Union City, IN), 25 Jan. 1944, 4.
79. Ash, "Hemp—Production and Utilization," 168.
80. Ibid.; "State Hemp Mills Will Close After Year," 2; "Will Close Mid-West Hemp Plants This Year," 3.
81. U.S. Department of Agriculture, *Agricultural Statistics, 1944,* 280.
82. Johnson, "Hemp for Victory: A Brief History—Part 1."
83. *Purdue University Department of Agricultural Extension Annual Report,* 1941–43, 14.

CHAPTER 42

Draining the Woodlots of the Larger Oaks, Walnuts, Maples, and Sycamores

1. C. D. Chambers, *Perry CEAR,* 1942 (CAERC box 122, folder 38).
2. *Purdue University Department of Agricultural Extension Annual Report,* 1941–43, 17.
3. M. A. Nye, *Noble CEAR,* 1942 (CAERC box 118, folder 35).
4. L. E. Husted, *Randolph CEARs,* 1942–43 (CAERC box 30, folders 41–42); B. T. Bond, *Warrick CEAR,* 1943 (CAERC box 54, folder 37); O. B. Riggs, *Posey CEAR,* 1944 (CAERC box 125, folder 31); E. S. Bogue, *Pike CEAR,* 1943 (CAERC box 123, folder 35).
5. E. A. Johnson, *Howard CEAR,* 1942 (CAERC box 95, folder 10); J. L. Guthrie, *Union CEAR,* 1943 (CAERC box 46, folder 25); C. D. Chambers, *Perry CEAR,* 1942 (CAERC box 122, folder 38); W. M. Clary, *Tipton CEAR,* 1943 (CAERC box 45, folder 24); C. E. Olson, *Pulaski CEAR,* 1943 (CAERC box 28, folder 11); E. S. Bogue, *Pike CEAR,* 1943 (CAERC box 123, folder 35).
6. E. E. Fultz, *Jasper CEAR,* 1942 (CAERC box 98, folder 12).
7. R. C. Stangland, *Howard CEAR,* 1942 (CAERC box 95, folder 10).
8. C. D. Chambers, *Perry CEAR,* 1942 (CAERC box 122, folder 38); M. J. Huxley, *Knox CEAR,* 1946 (CAERC box 103, folder 6); O. B. Riggs, *Posey CEAR,* 1944 (CAERC box 125, folder 31).
9. J. L. Guthrie, *Union CEAR,* 1943 (CAERC box 46, folder 25).
10. Lenny Farlee, email message to author, 4 Jan. 2022.
11. L. E. Husted, *Randolph CEAR,* 1944 (CAERC box 30, folder 43).
12. J. A. Clark, *Kosciusko CEAR,* 1945 (CAERC box 104, folder 21).

13. C. D. Chambers, *Perry CEAR,* 1943 (CAERC box 122, folder 39); B. T. Bond, *Warrick CEAR,* 1943 (CAERC box 54, folder 37).
14. R. N. Atkinson, *Orange CEAR,* 1943 (CAERC box 120, folder 26).
15. L. E. Husted, *Randolph CEAR,* 1942 (CAERC box 30, folder 41); D. J. Holmes, *Putnam CEAR,* 1945 (CAERC box 29, folder 29); C. D. Chambers, *Perry CEAR,* 1943 (CAERC box 122, folder 39); M. J. Huxley, *Knox CEAR,* 1946 (CAERC box 103, folder 6).
16. I. O. Haug, *Ripley CEAR,* 1944 (CAERC box 31, folder 46); H. E. Abbott, *Marion CEAR,* 1942 (CAERC box 111, folder 7); H. S. Benson, *Knox CEAR,* 1943 (CAERC box 102, folder 42).
17. R. N. Atkinson, *Orange CEAR,* 1946 (CAERC box 120, folder 29); C. D. Chambers, *Perry CEAR,* 1942 (CAERC box 122, folder 38).
18. R. N. Atkinson, *Orange CEAR,* 1946 (CAERC box 120, folder 29); J. F. Armstrong, *Lawrence CEAR,* 1942 (CAERC box 108, folder 41).
19. H. E. Abbott, *Marion CEAR,* 1942 (CAERC box 111, folder 7); M. A. Nye, *Noble CEAR,* 1942 (CAERC box 118, folder 35).
20. R. N. Atkinson, *Orange CEAR,* 1944 (CAERC box 120, folder 27).
21. E. A. Chatham, *Perry CEAR,* 1945 (CAERC box 122, folder 41).
22. E. F. Downen, *Spencer CEAR,* 1943 (CAERC box 38, folder 49).
23. R. N. Atkinson, *Orange CEAR,* 1943 (CAERC box 120, folder 26).
24. T. M. Grabill, *Steuben CEAR,* 1944 (CAERC box 40, folder 52); R. N. Atkinson, *Orange CEAR,* 1944 (CAERC box 120, folder 27); W. Rogers, *Monroe CEAR,* 1946 (CAERC box 115, folder 29).
25. T. M. Grabill, *Steuben CEAR,* 1944 (CAERC box 40, folder 52); R. N. Atkinson, *Orange CEAR,* 1944 (CAERC box 120, folder 27).
26. J. McCulloch, *Scott CEAR,* 1945 (CAERC box 36, folder 26); E. L. Hartman, *LaPorte CEAR,* 1943 (CAERC box 105, folder 19); W. Rogers, *Monroe CEAR,* 1946 (CAERC box 115, folder 29).
27. E. L. Hartman, *LaPorte CEAR,* 1943 (CAERC box 105, folder 19); I. O. Haug, *Parke CEAR,* 1946 (CAERC box 122, folder 2); L. E. Cutler, *Lake CEAR,* 1944 (CAERC box 107, folder 37).
28. W. C. Haynes, *Madison CEAR,* 1946 (CAERC box 109, folder 42); G. T. Harris, *Ripley CEAR,* 1943 (CAERC box 31, folder 45).
29. L. P. Geyer, *Jefferson CEAR,* 1942 (CAERC box 100, folder 9).
30. M. S. Smith, *Wells CEAR,* 1944 (CAERC box 58, folder 30); N. S. Hadley, *Parke CEAR,* 1945 (CAERC box 122, folder 1); *Purdue University Department of Agricultural Extension Annual Report*, 1945, 11.
31. J. McCulloch, *Scott CEAR,* 1945 (CAERC box 36, folder 26).
32. J. F. Armstrong, *Lawrence CEAR,* 1942 (CAERC box 108, folder 41); F. I. Hoover, *Owen CEAR,* 1946 (CAERC box 121, folder 7).
33. L. P. Geyer, *Jefferson CEAR,* 1942 (CAERC box 100, folder 9); G. T. Harris, *Ripley CEAR,* 1942 (CAERC box 31, folder 44); O. H. McNary, *Switzerland CEAR,* 1942 (CAERC box 42, folder 56).
34. G. T. Harris, *Ripley CEAR,* 1942 (CAERC box 31, folder 44).
35. L. P. Geyer, *Jefferson CEAR,* 1942 (CAERC box 100, folder 9).
36. Ibid.
37. L. P. Geyer, *Jefferson CEAR,* 1943 (CAERC box 100, folder 10).
38. L. E. Husted, *Randolph CEAR,* 1943 (CAERC box 30, folder 42).
39. L. P. Geyer, *Jefferson CEAR,* 1942 (CAERC box 100, folder 9).

40. L. P. Geyer, *Jefferson CEAR,* 1944 (CAERC box 100, folder 11).
41. F. I. Hoover, *Owen CEAR,* 1942 (CAERC box 121, folder 3); N. S. Hadley, *Parke CEAR,* 1942 (CAERC box 121, folder 51).
42. W. P. Stall, *Jackson CEAR,* 1942 (CAERC box 97, folder 7); J. McCulloch, *Scott CEAR,* 1943 (CAERC box 36, folder 24); N. S. Hadley, *Parke CEAR,* 1945 (CAERC box 122, folder 1).
43. N. S. Hadley, *Parke CEAR,* 1945 (CAERC box 122, folder 1); W. P. Stall, *Jackson CEAR,* 1942 (CAERC box 97, folder 7).
44. W. P. Stall, *Jackson CEAR,* 1942 (CAERC box 97, folder 7).
45. L. E. Cutler, *Lake CEAR,* 1946 (CAERC box 107, folder 40); A. K. Wissman, *Jennings CEAR,* 1946 (CAERC box 101, folder 2); E. E. Fultz, *Jasper CEAR,* 1942 (CAERC box 98, folder 12); M. A. Nye, *Noble CEAR,* 1942 (CAERC box 118, folder 35).
46. A. K. Wissman, *Jennings CEARs,* 1945–46 (CAERC box 101, folders 1–2).

CHAPTER 43

Think More of the Soil as a Heritage to Be Conserved and Passed On to Those Who Follow

1. W. K. Delaplane, *Wabash CEAR,* 1942 (CAERC box 52, folder 27); J. C. Sliger, *Sullivan CEAR,* 1942 (CAERC box 41, folder 53).
2. J. C. Sliger, *Sullivan CEAR,* 1942 (CAERC box 41, folder 53).
3. W. U. Rusk, *Huntington CEAR,* 1943 (CAERC box 96, folder 19).
4. W. K. Delaplane, *Wabash CEAR,* 1942 (CAERC box 52, folder 27); C. A. Langston, *Jay CEAR,* 1943 (CAERC box 99, folder 1); H. E. Abbott, *Marion CEAR,* 1945 (CAERC box 111, folder 15).
5. O. B. Riggs, *Posey CEAR,* 1942 (CAERC box 125, folder 27); C. A. Langston, *Jay CEAR,* 1942 (CAERC box 98, folder 51).
6. H. E. Abbott, *Marion CEAR,* 1945 (CAERC box 111, folder 15).
7. T. M. Grabill, *Steuben CEAR,* 1944 (CAERC box 40, folder 52); H. K. Rippey, *Jay CEAR,* 1945 (CAERC box 99, folder 3); C. A. Langston, *Jay CEAR,* 1943 (CAERC box 99, folder 1).
8. T. M. Grabill, *Steuben CEAR,* 1944 (CAERC box 40, folder 52); J. F. Armstrong, *Lawrence CEAR,* 1942 (CAERC box 108, folder 41); S. B. Pershing, *Tippecanoe CEAR,* 1942 (CAERC box 44, folder 12).
9. A. K. Wissman, *Jennings CEAR,* 1945 (CAERC box 101, folder 1).
10. H. K. Rippey, *Jay CEAR,* 1946 (CAERC box 99, folder 5); L. E. Cutler, *Lake CEAR,* 1945 (CAERC box 107, folder 38); J. C. Sliger, *Sullivan CEAR,* 1944 (CAERC box 41, folder 55).
11. C. A. Langston, *Jay CEAR,* 1944 (CAERC box 99, folder 2).
12. S. J. Brown, *Vermillion CEAR,* 1942 (CAERC box 49, folder 37).
13. W. P. Stall, *Jackson CEAR,* 1946 (CAERC box 97, folder 11).
14. L. E. Cutler, *Lake CEAR,* 1943 (CAERC box 107, folder 36).
15. A. K. Wissman, *Jennings CEAR,* 1945 (CAERC box 101, folder 1).
16. R. Gwaltney, *Gibson CEAR,* 1939 (CAERC box 87, folder 9).
17. Huong N. Tran et al., *Natural Resources Conservation Laws: A Report on 17 States and Their Selected Counties and Townships,* U.S. Department of Agriculture—Natural Resources Conservation Service (n.p., July 1999), 7, https://www.google.com/books/edition/_/d8vuAAAAMAAJ?hl=en.
18. Indiana State Department of Agriculture, "Introduction to Indiana Soil and Water Conservation Districts (SWCDs)," chap. 1 in *District Operations Manual,* 1, at Indiana State Department of Agriculture (website), accessed 20 Apr. 2022, https://www.in.gov/isda/divisions/soil-conservation/districts-tools/district-operations-manual/.

19. "Soil Conservation," *Indianapolis News,* 11 Jan. 1940, 6.
20. *Purdue University Department of Agricultural Extension Annual Report*, 1945, 16.
21. J. A. Wood, *Martin CEARs,* 1944–45 (CAERC box 113, folders 30–31); L. E. Cutler, *Lake CEAR,* 1944 (CAERC box 107, folder 37).
22. C. A. Alcorn, *Dearborn CEAR,* 1938 (CAERC box 75, folder 33); E. M. Christen, *Benton CEAR,* 1938 (CAERC box 65, folder 25).
23. L. E. Cutler, *Lake CEAR,* 1944 (CAERC box 107, folder 37); H. S. Benson, *Knox CEAR,* 1945 (CAERC box 103, folder 23); E. L. Hartman, *LaPorte CEAR,* 1944 (CAERC box 105, folder 21); W. J. Record, *Morgan CEAR,* 1946 (CAERC box 117, folder 8).
24. E. M. Christen, *Benton CEAR,* 1939 (CAERC box 65, folder 27).
25. F. E. DeLaCroix, *Jasper CEAR,* 1945 (CAERC box 98, folder 15).
26. W. J. Record, *Morgan CEAR,* 1946 (CAERC box 117, folder 8).
27. E. M. Christen, *Benton CEAR,* 1938 (CAERC box 65, folder 25); C. A. Alcorn, *Dearborn CEAR,* 1938 (CAERC box 75, folder 33).
28. "Indiana's Soil and Water Conservation Districts," *Hoosier Farmer* 70, no. 3 (Mar. 1985): 14.
29. H. S. Benson, *Knox CEAR,* 1943 (CAERC box 102, folder 42); J. A. Wood, *Martin CEAR,* 1946 (CAERC box 113, folder 32); W. Rogers, *Monroe CEAR,* 1943 (CAERC box 115, folder 26).
30. W. Rogers, *Monroe CEAR,* 1943 (CAERC box 115, folder 26); B. Jewell, *Pike CEAR,* 1946 (CAERC box 123, folder 38); "Soil Conservation," 6; H. S. Benson, *Knox CEAR,* 1943 (CAERC box 102, folder 42); J. A. Wood, *Martin CEAR,* 1946 (CAERC box 113, folder 32).
31. *Purdue University Department of Agricultural Extension Annual Report*, 1941, 16; J. C. Sliger, *Sullivan CEAR,* 1943 (CAERC box 41, folder 54); L. E. Cutler, *Lake CEAR,* 1944 (CAERC box 107, folder 37); J. A. Wood, *Martin CEAR,* 1946 (CAERC box 113, folder 32); H. S. Benson, *Knox CEAR,* 1944 (CAERC box 103, folder 2).
32. J. A. Wood, *Martin CEAR,* 1946 (CAERC box 113, folder 32).
33. Ibid.
34. J. McCulloch, *Scott CEAR,* 1945 (CAERC box 36, folder 26); D. L. Grimes, *Putnam CEAR,* 1943 (CAERC box 29, folder 27); B. Jewell, *Pike CEAR,* 1945 (CAERC box 123, folder 37); S. B. Pershing, *Tippecanoe CEARs,* 1943 and 1946 (CAERC box 44, folders 14 and 20).
35. R. N. Atkinson, *Orange CEAR,* 1944 (CAERC box 120, folder 27); N. S. Hadley, *Parke CEAR,* 1942 (CAERC box 121, folder 51); E. M. Christen, *Benton CEAR,* 1939 (CAERC box 65, folder 27).
36. "Soil Conservation," 6.
37. J. C. Sliger, *Sullivan CEAR,* 1943 (CAERC box 41, folder 54).
38. J. H. Telfer, *Sullivan CEAR,* 1945 (CAERC box 41, folder 56).
39. W. K. Delaplane, *Wabash CEAR,* 1944 (CAERC box 52, folder 29).
40. W. Rogers, *Monroe CEAR,* 1943 (CAERC box 115, folder 26).
41. J. L. Guthrie, *Noble CEAR,* 1944 (CAERC box 118, folder 37).
42. J. A. Carroll, *Jasper CEAR,* 1946 (CAERC box 98, folder 16).
43. F. E. DeLaCroix, *Jasper CEAR,* 1944 (CAERC box 98, folder 14); J. A. Wood, *Martin CEAR,* 1945 (CAERC box 113, folder 31).
44. J. A. Wood, *Martin CEAR,* 1945 (CAERC box 113, folder 31); J. L. Guthrie, *Noble CEAR,* 1945 (CAERC box 118, folder 38); J. McCulloch, *Scott CEAR,* 1946 (CAERC box 36, folder 27); B. V. Widney, *Whitley CEAR,* 1945 (CAERC box 60, folder 6).
45. W. U. Rusk, *Huntington CEAR,* 1946 (CAERC box 96, folder 22).

46. Ibid.
47. E. F. Downen, *Spencer CEAR,* 1943 (CAERC box 38, folder 49); B. V. Widney, *Whitley CEAR,* 1945 (CAERC box 60, folder 6); N. S. Hadley, *Parke CEAR,* 1944 (CAERC box 121, folder 55); W. Rogers, *Monroe CEAR,* 1945 (CAERC box 115, folder 28); L. E. Cutler, *Lake CEAR,* 1942 (CAERC box 107, folder 34).
48. H. S. Benson, *Knox CEAR,* 1943 (CAERC box 102, folder 42).
49. H. S. Benson, *Knox CEAR,* 1944 (CAERC box 103, folder 2).
50. B. V. Widney, *Whitley CEAR,* 1945 (CAERC box 60, folder 6).
51. N. S. Hadley, *Parke CEAR,* 1944 (CAERC box 121, folder 55).
52. H. S. Benson, *Knox CEAR,* 1942 (CAERC box 102, folder 40).
53. M. A. Nye, *Noble CEAR,* 1942 (CAERC box 118, folder 35).
54. Ibid.
55. Ibid.
56. Ibid.
57. J. McCulloch, *Scott CEAR,* 1942 (CAERC box 36, folder 23).
58. W. Rogers, *Monroe CEAR,* 1942 (CAERC box 115, folder 25); C. A. Langston, *Jay CEAR,* 1942 (CAERC box 98, folder 51); W. U. Rusk, *Huntington CEAR,* 1941 (CAERC box 96, folder 17).
59. R. C. Stangland, *Howard CEAR,* 1942 (CAERC box 95, folder 10).
60. H. K. Rippey, *Jay CEAR,* 1946 (CAERC box 99, folder 5).
61. G. T. Harris, *Ripley CEAR,* 1943 (CAERC box 31, folder 45).
62. W. K. Delaplane, *Wabash CEAR,* 1942 (CAERC box 52, folder 27); S. B. Pershing, *Tippecanoe CEAR,* 1942 (CAERC box 44, folder 12); G. I. John, *Ohio CEAR,* 1942 (CAERC box 119, folder 30); G. A. Sowers, *Montgomery CEAR,* 1946 (CAERC box 115, folder 59); W. U. Rusk, *Huntington CEAR,* 1943 (CAERC box 96, folder 19).
63. S. B. Pershing, *Tippecanoe CEAR,* 1942 (CAERC box 44, folder 12); E. S. Bogue, *Pike CEAR,* 1943 (CAERC box 123, folder 35); E. C. Bird, *St. Joseph CEAR,* 1942 (CAERC box 35, folder 28); E. F. Downen, *Spencer CEAR,* 1943 (CAERC box 38, folder 49).
64. A. K. Wissman, *Jennings CEAR,* 1945 (CAERC box 101, folder 1); J. L. Guthrie, *Union CEAR,* 1943 (CAERC box 46, folder 25); S. W. Milligan, *Wayne CEAR,* 1943 (CAERC box 56, folder 30); L. J. Fultz, *Washington CEAR,* 1943 (CAERC box 55, folder 45); G. T. Harris, *Ripley CEAR,* 1943 (CAERC box 31, folder 45).
65. H. E. Abbott, *Marion CEAR,* 1944 (CAERC box 111, folder 13).
66. H. E. Abbott, *Marion CEAR,* 1942 (CAERC box 111, folder 7).
67. H. H. Madaus, *Warren CEAR,* 1946 (CAERC box 53, folder 28); J. C. Sliger, *Sullivan CEAR,* 1942 (CAERC box 41, folder 53); L. E. Husted, *Randolph CEAR,* 1946 (CAERC box 30, folder 45); E. S. Bogue, *Pike CEAR,* 1943 (CAERC box 123, folder 35); C. D. Chambers, *Perry CEAR,* 1943 (CAERC box 122, folder 39).
68. O. H. McNary, *Switzerland CEAR,* 1945 (CAERC box 43, folder 1).
69. S. B. Scott, *Johnson CEAR,* 1944 (CAERC box 101, folder 43); H. E. Abbott, *Marion CEAR,* 1945 (CAERC box 111, folder 15).
70. N. S. Hadley, *Parke CEAR,* 1942 (CAERC box 121, folder 51); E. F. Downen, *Spencer CEAR,* 1943 (CAERC box 38, folder 49).
71. L. E. Cutler, *Lake CEAR,* 1944 (CAERC box 107, folder 37).
72. E. L. Hartman, *LaPorte CEAR,* 1946 (CAERC box 105, folder 25); E. A. Chatham, *Perry CEAR,* 1945 (CAERC box 122, folder 41); M. J. Peterson, *Vermillion CEAR,* 1944 (CAERC box 49, folder 39); J. F. Armstrong, *Lawrence CEAR,* 1942 (CAERC box 108, folder 41).

73. C. D. Chambers, *Perry CEAR,* 1943 (CAERC box 122, folder 39).
74. H. S. Benson, *Knox CEAR,* 1942 (CAERC box 102, folder 40).

CHAPTER 44

Boys Are Having a Hard Time Making "Fox Hole" Dreams Come True

1. B. V. Widney, *Whitley CEAR,* 1946 (CAERC box 60, folder 7); J. E. Wylie, *Miami CEAR,* 1945 (CAERC box 114, folder 16); H. E. Abbott, *Marion CEAR,* 1946 (CAERC box 111, folder 16); W. P. Stall, *Jackson CEAR,* 1946 (CAERC box 97, folder 11).
2. R. N. Atkinson, *Orange CEAR,* 1946 (CAERC box 120, folder 29).
3. "Veterans' Bureau Is Opened in City," *Lafayette Journal and Courier* (Lafayette, IN), 2 Mar. 1945, 14; A. M. Bishea, *Vanderburgh CEAR,* 1943 (CAERC box 48, folder 7); G. A. Sowers, *Montgomery CEAR,* 1945 (CAERC box 115, folder 57); N. S. Hadley, *Parke CEAR,* 1944 (CAERC box 121, folder 55).
4. "Advice from County Agent," *Daily Clintonian* (Clinton, IN), 30 Apr. 1946, 3; "Farm Periscope," *Rushville Republican* (Rushville, IN), 12 July 1946, 5; "Veterans to Get Farmers' Advice," *Lafayette Journal and Courier* (Lafayette, IN), 15 Mar. 1945, 4; *Purdue University Department of Agricultural Extension Annual Report*, 1945, 6.
5. "Veterans to Get Farmers' Advice," 4; "Advice from County Agent," *Daily Clintonian* (Clinton, IN), 14 June 1946, 4.
6. N. S. Hadley, *Parke CEAR,* 1945 (CAERC box 122, folder 1); J. E. Wylie, *Miami CEAR,* 1945 (CAERC box 114, folder 16); O. B. Riggs, *Posey CEAR,* 1944 (CAERC box 125, folder 31); E. L. Hartman, *LaPorte CEAR,* 1945 (CAERC box 105, folder 23); J. A. Carroll, *Jasper CEAR,* 1945 (CAERC box 98, folder 15).
7. R. N. Atkinson, *Orange CEAR,* 1945 (CAERC box 120, folder 28).
8. J. McCulloch, *Scott CEAR,* 1945 (CAERC box 36, folder 26).
9. J. E. Wylie, *Miami CEAR,* 1945 (CAERC box 114, folder 16).
10. S. W. Milligan, *Wayne CEAR,* 1945 (CAERC box 56, folder 34); M. J. Peterson, *Vermillion CEAR,* 1944 (CAERC box 49, folder 39).
11. W. U. Rusk, *Huntington CEAR,* 1944 (CAERC box 96, folder 20).
12. W. P. Stall, *Jackson CEAR,* 1946 (CAERC box 97, folder 11).
13. N. S. Hadley, *Parke CEAR,* 1944 (CAERC box 121, folder 55); F. I. Hoover, *Owen CEAR,* 1944 (CAERC box 121, folder 5); *Purdue University Department of Agricultural Extension Annual Report*, 1945, 6.
14. N. S. Hadley, *Parke CEAR,* 1944 (CAERC box 121, folder 55); C. A. Langston, *Jay CEAR,* 1944 (CAERC box 99, folder 2).
15. N. S. Hadley, *Parke CEAR,* 1945 (CAERC box 122, folder 1).
16. J. E. Wylie, *Miami CEAR,* 1945 (CAERC box 114, folder 16).
17. F. I. Hoover, *Owen CEAR,* 1944 (CAERC box 121, folder 5).
18. C. A. Langston, *Jay CEAR,* 1944 (CAERC box 99, folder 2).
19. L. E. Cutler, *Lake CEAR,* 1945 (CAERC box 107, folder 38); C. A. Langston, *Jay CEAR,* 1942 (CAERC box 98, folder 51).
20. E. A. Johnson, *Howard CEAR,* 1945 (CAERC box 95, folder 13).
21. Ibid.
22. S. W. Milligan, *Wayne CEAR,* 1945 (CAERC box 56, folder 34).
23. W. U. Rusk, *Huntington CEAR,* 1944 (CAERC box 96, folder 20); H. E. Abbott, *Marion CEAR,* 1942 (CAERC box 111, folder 7).

24. L. J. Fultz, *Washington CEAR,* 1945 (CAERC box 55, folder 47); O. B. Riggs, *Posey CEAR,* 1944 (CAERC box 125, folder 31); F. I. Hoover, *Owen CEAR,* 1944 (CAERC box 121, folder 5); J. L. Guthrie, *Noble CEAR,* 1945 (CAERC box 118, folder 3); J. E. Wylie, *Miami CEAR,* 1945 (CAERC box 114, folder 16); W. U. Rusk, *Huntington CEAR,* 1944 (CAERC box 96, folder 20); E. A. Johnson, *Howard CEAR,* 1945 (CAERC box 95, folder 13); E. S. Bogue, *Pike CEAR,* 1944 (CAERC box 123, folder 36); A. K. Wissman, *Jennings CEAR,* 1946 (CAERC box 101, folder 2).
25. F. I. Hoover, *Owen CEAR,* 1944 (CAERC box 121, folder 5); J. E. Wylie, *Miami CEAR,* 1945 (CAERC box 114, folder 16); W. U. Rusk, *Huntington CEAR,* 1944 (CAERC box 96, folder 20).
26. G. A. Sowers, *Montgomery CEAR,* 1945 (CAERC box 115, folder 57); H. E. Abbott, *Marion CEAR,* 1945 (CAERC box 111, folder 15).
27. H. E. Abbott, *Marion CEAR,* 1945 (CAERC box 111, folder 15).
28. W. U. Rusk, *Huntington CEAR,* 1945 (CAERC box 96, folder 21).
29. F. I. Hoover, *Owen CEAR,* 1944 (CAERC box 121, folder 5); J. E. Wylie, *Miami CEAR,* 1945 (CAERC box 114, folder 16).
30. R. N. Atkinson, *Orange CEAR,* 1945 (CAERC box 120, folder 28).
31. S. B. Pershing, *Tippecanoe CEAR,* 1946 (CAERC box 44, folder 20); M. J. Huxley, *Knox CEAR,* 1946 (CAERC box 103, folder 6); Gary Moore, "When Johnny Comes Marching Home," *The Friday Footnote* (blog), North Carolina State University, 12 July 2019, https://footnote.wordpress.ncsu.edu/2019/07/11/when-johnny-comes-marching-home-7-12-2019/.
32. "Vocational Agriculture Training Planned for G. I.," *Hancock Democrat* (Greenfield, IN), 17 Jan. 1946, 2.
33. O. H. McNary, *Switzerland CEAR,* 1946 (CAERC box 43, folder 2); T. M. Grabill, *Steuben CEAR,* 1942 (CAERC box 40, folder 50); G. A. Sowers, *Montgomery CEAR,* 1946 (CAERC box 115, folder 59); W. J. Record, *Morgan CEAR,* 1946 (CAERC box 117, folder 8); E. L. Hartman, *LaPorte CEAR,* 1946 (CAERC box 105, folder 25); H. K. Rippey, *Jay CEAR,* 1946 (CAERC box 99, folder 5).
34. E. A. Chatham, *Perry CEAR,* 1946 (CAERC box 123, folder 1); J. McCulloch, *Scott CEAR,* 1945 (CAERC box 36, folder 26); E. A. Johnson, *Howard CEAR,* 1945 (CAERC box 95, folder 13); L. E. Cooper, *Spencer CEAR,* 1946 (CAERC box 38, folder 52).
35. L. M. Butler, *Marshall CEAR,* 1946 (CAERC box 112, folder 36).
36. L. E. Cooper, *Spencer CEAR,* 1946 (CAERC box 38, folder 52); A. K. Wissman, *Jennings CEAR,* 1946 (CAERC box 101, folder 2); S. W. Milligan, *Wayne CEAR,* 1946 (CAERC box 56, folder 35); S. B. Pershing, *Tippecanoe CEAR,* 1946 (CAERC box 44, folder 20).
37. Moore, "When Johnny Comes Marching Home."
38. E. C. Bird, *St. Joseph CEAR,* 1946 (CAERC box 35, folder 32).
39. M. A. Nye, *Porter CEAR,* 1946 (CAERC box 124, folder 32); W. K. Delaplane, *Wabash CEAR,* 1946 (CAERC box 52, folder 31); M. J. Huxley, *Knox CEAR,* 1946 (CAERC box 103, folder 6); H. K. Rippey, *Jay CEAR,* 1946 (CAERC box 99, folder 5).
40. W. K. Delaplane, *Wabash CEAR,* 1946 (CAERC box 52, folder 31); S. W. Milligan, *Wayne CEAR,* 1946 (CAERC box 56, folder 35).
41. L. E. Cooper, *Spencer CEAR,* 1946 (CAERC box 38, folder 52).
42. H. K. Rippey, *Jay CEAR,* 1946 (CAERC box 99, folder 5).
43. J. H. Telfer, *Sullivan CEAR,* 1946 (CAERC box 42, folder 1); S. W. Milligan, *Wayne CEAR,* 1946 (CAERC box 56, folder 35).
44. C. L. Spuller, *Rush CEAR,* 1946 (CAERC box 33, folder 15).

45. L. E. Cooper, *Spencer CEAR,* 1946 (CAERC box 38, folder 52).
46. S. W. Milligan, *Wayne CEAR,* 1946 (CAERC box 56, folder 35).
47. L. E. Cooper, *Spencer CEAR,* 1946 (CAERC box 38, folder 52).
48. Moore, "When Johnny Comes Marching Home."
49. S. B. Pershing, *Tippecanoe CEAR,* 1944 (CAERC box 44, folder 16).
50. Ibid.

CHAPTER 45

Effort Must Be Directed Toward Building for the Future

1. U.S. Department of Veterans Affairs, "World War II Veterans by the Numbers," VA fact sheet, circa Sept. 2003, http://dig.abclocal.go.com/ktrk/ktrk_120710_WWIIvetsfactsheet.pdf.
2. "World War II Veterans Information Center," Indiana War Memorials Home (website), accessed 20 Apr. 2022, https://www.in.gov/iwm/veterans-memorial-plaza/world-war-ii-veterans-information-center/#:~:text=The%20Second%20World%20War%20consumed,another%20half%20million%20were%20wounded.
3. E. A. Johnson, *Howard CEAR,* 1945 (CAERC box 95, folder 13); R. A. Burger, *Posey CEAR,* 1945 (CAERC box 126, folder 1).
4. R. L. Case, *LaGrange CEAR,* 1945 (CAERC box 106, folder 6).
5. R. A. Burger, *Posey CEAR,* 1945 (CAERC box 126, folder 1).
6. S. B. Pershing, *Tippecanoe CEAR,* 1945 (CAERC box 44, folder 18).
7. R. L. Case, *LaGrange CEAR,* 1946 (CAERC box 106, folder 8); W. J. Record, *Morgan CEAR,* 1946 (CAERC box 117, folder 8).
8. J. F. Armstrong, *Lawrence CEAR,* 1946 (CAERC box 109, folder 1).
9. J. E. Wylie, *Miami CEAR,* 1945 (CAERC box 114, folder 16); G. A. Sowers, *Montgomery CEAR,* 1945 (CAERC box 115, folder 57).
10. E. C. Bird, *St. Joseph CEAR,* 1946 (CAERC box 35, folder 32).
11. Ibid.
12. H. K. Rippey, *Jay CEAR,* 1946 (CAERC box 99, folder 5).
13. M. J. Huxley, *Knox CEAR,* 1946 (CAERC box 103, folder 6).
14. J. McCulloch, *Scott CEAR,* 1946 (CAERC box 36, folder 27).
15. S. B. Pershing, *Tippecanoe CEAR,* 1945 (CAERC box 44, folder 18); J. E. Wylie, *Miami CEAR,* 1945 (CAERC box 114, folder 16); E. C. Bird, *St. Joseph CEAR,* 1946 (CAERC box 35, folder 32).
16. C. L. Spuller, *Rush CEAR,* 1945 (CAERC box 33, folder 13).
17. T. M. Grabill, *Steuben CEAR,* 1944 (CAERC box 40, folder 52).
18. D. J. Holmes, *Putnam CEAR,* 1945 (CAERC box 29, folder 29); A. C. Shideler, *Starke CEAR,* 1944 (CAERC box 40, folder 2).
19. S. B. Scott, *Johnson CEAR,* 1942 (CAERC box 101, folder 41).
20. E. L. Hartman, *LaPorte CEAR,* 1946 (CAERC box 105, folder 25); O. B. Riggs, *Posey CEAR,* 1942 (CAERC box 125, folder 27); G. A. Sowers, *Montgomery CEAR,* 1946 (CAERC box 115, folder 59); S. W. Milligan, *Wayne CEAR,* 1945 (CAERC box 56, folder 34); J. E. Wylie, *Miami CEAR,* 1945 (CAERC box 114, folder 16).
21. R. N. Atkinson, *Orange CEAR,* 1946 (CAERC box 120, folder 29); F. I. Hoover, *Owen CEAR,* 1944 (CAERC box 121, folder 5).
22. A. K. Wissman, *Jennings CEAR,* 1945 (CAERC box 101, folder 1); G. A. Sowers, *Montgomery CEAR,* 1946 (CAERC box 115, folder 59); S. B. Scott, *Johnson CEAR,* 1945 (CAERC box 101, folder 44).

23. A. K. Wissman, *Jennings CEAR,* 1946 (CAERC box 101, folder 2); C. E. Olson, *Pulaski CEAR,* 1945 (CAERC box 28, folder 13).
24. E. L. Hartman, *LaPorte CEAR,* 1943 (CAERC box 105, folder 19); A. M. Bishea, *Vanderburgh CEAR,* 1944 (CAERC box 48, folder 9).
25. L. E. Cutler, *Lake CEARs,* 1942 and 1945 (CAERC box 107, folders 34 and 38); E. L. Hartman, *LaPorte CEAR,* 1942 (CAERC box 105, folder 17).
26. E. L. Hartman, *LaPorte CEAR,* 1942 (CAERC box 105, folder 17); L. E. Cutler, *Lake CEAR,* 1942 (CAERC box 107, folder 34).
27. E. L. Hartman, *LaPorte CEAR,* 1944 (CAERC box 105, folder 21).
28. E. A. Johnson, *Howard CEAR,* 1946 (CAERC box 95, folder 14).
29. Ibid.; T. M. Grabill, *Steuben CEAR,* 1942 (CAERC box 40, folder 50); M. J. Huxley, *Knox CEAR,* 1946 (CAERC box 103, folder 6).
30. M. S. Smith, *Wells CEAR,* 1945 (CAERC box 58, folder 31); J. McCulloch, *Scott CEAR,* 1942 (CAERC box 36, folder 23).
31. R. L. Campbell, *Vermillion CEAR,* 1945 (CAERC box 49, folder 40).
32. S. B. Pershing, *Tippecanoe CEAR,* 1946 (CAERC box 44, folder 20).
33. L. E. Cooper, *Spencer CEAR,* 1946 (CAERC box 38, folder 52).
34. R. N. Atkinson, *Orange CEAR,* 1943 (CAERC box 120, folder 26).
35. A. K. Wissman, *Jennings CEAR,* 1946 (CAERC box 101, folder 2).
36. S. W. Milligan, *Wayne CEAR,* 1945 (CAERC box 56, folder 34).
37. A. C. Shideler, *Starke CEAR,* 1944 (CAERC box 40, folder 2).
38. H. E. Abbott, *Marion CEAR,* 1946 (CAERC box 111, folder 16).
39. J. H. Telfer, *Sullivan CEAR,* 1946 (CAERC box 42, folder 1); B. V. Widney, *Whitley CEAR,* 1946 (CAERC box 60, folder 7); A. K. Wissman, *Jennings CEAR,* 1946 (CAERC box 101, folder 2).
40. W. P. Stall, *Jackson CEAR,* 1945 (CAERC box 97, folder 10).
41. A. M. Bishea, *Vanderburgh CEAR,* 1943 (CAERC box 48, folder 7).
42. A. C. Shideler, *Starke CEAR,* 1945 (CAERC box 40, folder 3); J. E. Wylie, *Miami CEAR,* 1945 (CAERC box 114, folder 16).
43. A. M. Bishea, *Vanderburgh CEAR,* 1943 (CAERC box 48, folder 7).
44. L. E. Cutler, *Lake CEAR,* 1946 (CAERC box 107, folder 40).
45. A. K. Wissman, *Jennings CEARs,* 1945–46 (CAERC box 101, folders 1–2); W. P. Stall, *Jackson CEAR,* 1946 (CAERC box 97, folder 11); J. McCulloch, *Scott CEAR,* 1946 (CAERC box 36, folder 27).
46. F. E. DeLaCroix, *Jasper CEAR,* 1944 (CAERC box 98, folder 14).
47. L. E. Husted, *Randolph CEAR,* 1946 (CAERC box 30, folder 45); E. A. Chatham, *Perry CEAR,* 1944 (CAERC box 122, folder 40); B. Jewell, *Pike CEAR,* 1946 (CAERC box 123, folder 38).
48. A. C. Shideler, *Starke CEAR,* 1945 (CAERC box 40, folder 3).
49. L. E. Cooper, *Spencer CEAR,* 1946 (CAERC box 38, folder 52).
50. W. K. Delaplane, *Wabash CEAR,* 1946 (CAERC box 52, folder 31); E. F. Downen, *Spencer CEAR,* 1943 (CAERC box 38, folder 49).
51. G. A. Sowers, *Montgomery CEAR,* 1946 (CAERC box 115, folder 59).
52. Peter Myers, "Soil Conservation Service at 50," *Hoosier Farmer* 70, no. 3 (Mar. 1985): 9.
53. R. L. Case, *LaGrange CEAR,* 1945 (CAERC box 106, folder 6); J. A. Carroll, *Jasper CEAR,* 1945 (CAERC box 98, folder 15); S. W. Milligan, *Wayne CEAR,* 1945 (CAERC box 56, folder 34).

54. R. A. Burger, *Posey CEAR,* 1945 (CAERC box 126, folder 1).
55. L. E. Cooper, *Spencer CEAR,* 1946 (CAERC box 38, folder 52).
56. R. N. Atkinson, *Orange CEAR,* 1946 (CAERC box 120, folder 29); J. H. Telfer, *Sullivan CEAR,* 1946 (CAERC box 42, folder 1); B. V. Widney, *Whitley CEAR,* 1944 (CAERC box 60, folder 5).
57. W. U. Rusk, *Huntington CEAR,* 1944 (CAERC box 96, folder 20).
58. B. V. Widney, *Whitley CEAR,* 1945 (CAERC box 60, folder 6).
59. W. J. Emerson, *White CEAR,* 1946 (CAERC box 59, folder 9).
60. Mike Moffatt, "Farming Post World-War II," ThoughtCo. (website), Dotdash Meredith Publishing, 27 Jan. 2020, https://www.thoughtco.com/farming-post-world-war-ii-1146852.

EPILOGUE

Now as Never Before Farmers Have Put Into Use Many Practices Advocated by the Extension Service

1. H. E. Abbott, *Marion CEAR,* 1944 (CAERC box 111, folder 13).
2. W. C. Haynes, *Madison CEAR,* 1946 (CAERC box 109, folder 42); *Purdue University Department of Agricultural Extension Annual Report,* 1945, 3.
3. *Purdue University Department of Agricultural Extension Annual Report,* 1945, 3.

Works Cited

American Chemical Society. "United States Synthetic Rubber Program, 1939–1945" (1998 commemorative booklet [PDF]). At National Historic Chemical Landmarks (website), accessed 16 May 2022. https://www.acs.org/content/acs/en/education/whatischemistry/landmarks/syntheticrubber.html.

Argos Reflector (Argos, IN), 9 Nov. 1938–8 Apr. 1943.

Ash, Anne L. "Hemp—Production and Utilization." *Economic Botany* 2 (Apr. 1948): 158–69. https://doi.org/10.1007/BF02858999.

Baker, Gladys L., and Wayne D. Rasmussen. "A Chronology of the War Food Administration, Including Predecessor and Successor Agencies, August 1939 to December 1946." U.S. Department of Agriculture Bureau of Agricultural Economics, Nov. 1950. https://naldc.nal.usda.gov/catalog/6538554.

Boonville Enquirer (Boonville, IN), 27 Mar. 1936–21 Aug. 1936.

Boston Globe (Boston, MA), 1 Nov. 1941.

Braun, Adee. "Turning Bacon into Bombs: The American Fat Salvage Committee." *The Atlantic* (website), 18 Apr. 2014. https://www.theatlantic.com/health/archive/2014/04/reluctantly-turning-bacon-into-bombs-during-world-war-ii/360298/.

Bremen Enquirer (Bremen, IN), 27 Jan. 1944–8 Feb. 1945.

Brook Reporter (Brook, IN), 25 July 1934–2 Nov. 1944.

Brown County Democrat (Nashville, IN), 7 Mar. 1940–18 Feb. 1943.

Bush, Wilbur. "Meat Rationing during World War II." *Daviess County of Northwest Missouri* (blog). Daviess County Historical Society, 14 Dec. 2009. https://daviesscountyhistoricalsociety.com/2009/12/14/meat-rationing-during-world-war-ii/.

Call-Leader (Elwood, IN), 9 May 1942–18 Oct. 1944.

Chicago Daily Tribune (Chicago, IL), 20 Jan. 1944.

Chitty, Ethan. "The Decades-Long Struggle to Electrify Rural Hoosierdom." *Indiana History Blog.* Indiana Historical Bureau of the Indiana State Library, 4 Apr. 2017. https://blog.history.in.gov/the-decades-long-struggle-to-electrify-rural-hoosierdom/.

Chronicle (Scottsburg, IN), 4 Jan. 1945.

Clark, Patterson. "Milkweed Fruits: Pods of Plenty." *Urban Jungle. Washington Post* (website), 25 Sept. 2012. https://www.washingtonpost.com/wp-srv/special/metro/urban-jungle/pages/120925.html.

Conrad, David E. *The Land We Cared For: A History of the Forest Service's Eastern Region.* Carbondale, IL: American Resources Group Ltd., 1997. https://foresthistory.org/wp-content/uploads/2017/01/The-Land-We-Cared-for-A-History-of-the-Forest-Service-Eastern-Region.pdf.

Culver Citizen (Culver, IN), 8 Jan. 1936–11 Oct. 1944.

Daily Clintonian (Clinton, IN), 23 Apr. 1942–14 June 1946.

Danbom, David B. "The Agricultural Extension System and the First World War." *The Historian* 41, no. 2 (Feb. 1979): 315–31. https://www.jstor.org/stable/24445110.

DeHart, William A., and Mapheus Smith. *Agricultural Deferment* (U.S. Selective Service System). Washington, DC: Government Printing Office, 1947. https://www.google.com/books/edition/_/gcVfAAAAIAAJ?hl=en.

Dowell, A. A., and O. B. Jesness. "Economic Aspects of Hybrid Corn." *Journal of Farm Economics* 21, no. 2 (May 1939): 479–88. https://www.jstor.org/stable/1230816.

Edinburg Daily Courier (Edinburg, IN), 6 Apr. 1934–19 Jan. 1944.

Ehrensing, Daryl T. *Feasibility of Industrial Hemp Production in the United States Pacific Northwest.* Oregon State University Agricultural Experiment Station Bulletin 681. Corvallis, OR: Oregon State University, May 1998. https://catalog.extension.oregonstate.edu/sites/catalog/files/project/pdf/sb681.pdf.

Encyclopaedia Britannica, Editors of. "Kapok." *Encyclopedia Britannica* (website), 23 Feb. 2017. https://www.britannica.com/topic/kapok.

———. "Selective Service System." *Encyclopedia Britannica* (website), 24 May 2016. https://www.britannica.com/topic/Selective-Service-System#ref1193841.

———. "Women's Land Army." *Encyclopedia Britannica* (website), 11 Sept. 2018. https://www.britannica.com/topic/Womens-Land-Army.

Encyclopedia.com. "Agricultural Mobilization." American Home Front in World War II (website), 23 May 2018. https://www.encyclopedia.com/plants-and-animals/agriculture-and-horticulture/agriculture-general/agricultural-revolution.

———. "Bankhead-Jones Farm Tenant Act of 1937." Encyclopedia of the Great Depression (website), accessed 16 Apr. 2021. https://www.encyclopedia.com/economics/encyclopedias-almanacs-transcripts-and-maps/bankhead-jones-farm-tenant-act-1937.

———. "Banking 1929–1941." Historic Events for Students: The Great Depression (website), accessed 15 Apr. 2021. https://www.encyclopedia.com/education/news-and-education-magazines/banking-1929-1941.

Evansville Press (Evansville, IN), 23 Mar. 1944.

Evening Republican (Columbus, IN), 5 June 1931–29 Jan. 1943.

Extension Service Review. "New Approach to Seed Wheat Cleaning." *Extension Service Review* 5, no. 11 (Nov. 1934): 173–74. https://archive.org/details/sim_extension-review_1934-11_5_11/page/n16/mode/1up.

Fairmount News (Fairmount, IN), 7 Dec. 1944.

Farm Equipment Institute Research Department. "Tractors Increase Despite Depression." *Better Farm Equipment and Methods* 8, no. 2 (Oct. 1935): 4–5. https://www.google.com/books/edition/Big_Farmer/fSlPAAAAIAAJ?hl=en&gbpv=1&pg=RA7-PA4&printsec=frontcover.

Fogarty, Amelia. "Five Ways the Tractor Changed American Farming." *Smithsonian Sparks* (blog). Smithsonian Institution, 1 Mar. 2018. https://www.si.edu/stories/five-ways-tractor-changed-american-farming.

Franklin Evening Star (Franklin, IN), 7 Sept. 1933–15 Sept. 1944.

Ganzel, Bill. "The March of Machines." Farming in the 1930s: Machines. Wessels Living History Farm (York, NE) (website), 2003. https://livinghistoryfarm.org/farminginthe30s/machines/.

Garrett Clipper (Garrett, IN), 18 Sept. 1930–20 July 1944.

Gates, Paul Wallace. "Land Policy and Tenancy in the Prairie Counties of Indiana." *Indiana Magazine of History* 35, no. 1 (Mar. 1939): 1–26. https://scholarworks.iu.edu/journals/index.php/imh/article/view/7107.

Gaylord, Fay C., and Harry M. Cleaver. *Marketing Indiana Potatoes: Local Supply and Imports.* Purdue University Agricultural Experiment Station Bulletin No. 361. West Lafayette, IN: Purdue University, Mar. 1932. https://babel.hathitrust.org/cgi/pt?id=uiug.30112019548558&view=1up&seq=6.

Graves, R. R., and M. H. Fohrman. "Superior Germ Plasm in Dairy Herds." In *United States Department of Agriculture Yearbook of Agriculture, 1936,* 997–1141. Washington, DC: Government Printing Office, 1936. https://archive.org/details/yoa1936/page/n1005/mode/2up.

Greenfield Daily Reporter (Greenfield, IN), 18 Aug. 1936–21 Oct. 1943.

Hammond Times (Hammond, IN), 7 July 1937–30 Dec. 1942.

Hancock Democrat (Greenfield, IN), 12 Dec. 1935–17 Jan. 1946.

Hauswirth, Katherine. "The Heroic Milkweed." *Christian Science Monitor* (website), 26 Oct. 2008. https://www.csmonitor.com/The-Culture/Gardening/2008/1026/the-heroic-milkweed.

Herer, Jack. *The Emperor Wears No Clothes: Hemp and the Marijuana Conspiracy.* 12th ed. Highlands, TX: AH HA Publishing, 2010.

Hirsch, Charles B. "Conscientious Objectors in Indiana during World War II." *Indiana Magazine of History* 46, no. 1 (Mar. 1950): 61–72. https://www.jstor.org/stable/27787823?seq=1.

History Museum (of South Bend, IN) (website), The. "Indiana and Another World War," accessed 13 May 2022. https://www.historymuseumsb.org/indiana-and-another-world-war/.

Hoosier Farmer. "Days to Remember . . . Civilian Conservation Corps." *Hoosier Farmer* 70, no. 3 (Mar. 1985): 25. Hicks Undergraduate Library, Purdue University Libraries, West Lafayette, IN.

———. "Indiana's Soil and Water Conservation Districts." *Hoosier Farmer* 70, no. 3 (Mar. 1985): 14. Hicks Undergraduate Library, Purdue University Libraries, West Lafayette, IN.

Howard, Spencer. "Herbert Hoover and the 1930 Drought." *Hoover Heads* (blog). National Archives, 16 Sept. 2020. https://hoover.blogs.archives.gov/2020/09/16/herbert-hoover-and-the-1930-drought/.

Hurt, R. Douglas. "REA: A New Deal for Farmers." *Timeline* 2, no. 6 (Dec. 1985/Jan. 1986): 32–47. Author's personal collection.

Indiana Connection (digital magazine). "Marshall County REMC: A Look Back at Where It All Began," 4 Oct. 2020. https://www.indianaconnection.org/a-look-back-at-where-it-all-began/.

Indiana Electric Cooperatives (website). "History," accessed 16 Apr. 2022. https://www.indianaec.org/who-we-are/history/.

Indianapolis News (Indianapolis, IN), 12 Aug. 1930–5 Dec. 1945.

Indianapolis Star (Indianapolis, IN), 12 Jan. 1929–26 Nov. 1944.

Indiana State Board of Health. "Report of the Traveling Milk Laboratory." In *Year Book of the State of Indiana for 1923,* 143–52. Indianapolis: Wm. B. Burford, 1924. https://indianamemory.contentdm.oclc.org/digital/collection/p15078coll21/id/87691.

Indiana State Dairy Association. "Executive Committee Meeting." In *Thirty-Eighth Annual Report of the Indiana State Dairy Association, 1928,* 45–46. Indianapolis: Wm. B. Burford, 1928. https://babel.hathitrust.org/cgi/pt?id=wu.89044276178&view=1up&seq=671&skin=2021.

Indiana State Department of Agriculture. "Introduction to Indiana Soil and Water Conservation Districts." Chap. 1 in *District Operations Manual,* 1–6. At Indiana State Department of Agriculture (website), accessed 20 Apr. 2022. https://www.in.gov/isda/divisions/soil-conservation/districts-tools/district-operations-manual/.

Indiana War Memorials Home (website). "World War II Veterans Information Center," accessed 20 Apr. 2022. https://www.in.gov/iwm/veterans-memorial-plaza/world-war-ii-veterans-information-center/#:~:text=The%20Second%20World%20War%2consumed,another%20half%20million%20were%20wounded.

Jackson County Banner (Brownstown, IN), 19 July 1944–27 Sept. 1944.

Jasper Herald (Jasper, IN), 12 June 1936.

Johnson, Nick. "Hemp for Victory: A Brief History—Part 1." *Hempirical Evidence* (blog), 11 Mar. 2019. https://hempiricalevidence.com/2019/03/11/hemp-for-victory-a-brief-overdue-history-part-1/.

Jones, Samuel A., Gordon W. Sprague, and Alexander Sturges. "Poultry Industry Fairly Resistant to the Depression." In "Poultry and Eggs," *United States Department of Agriculture Yearbook of Agriculture, 1933,* ed. Milton S. Eisenhower, 303–8. Washington, DC: Government Printing Office, 1933. https://archive.org/details/yoa1933/page/n1/mode/2up.

Kimball, Warren F. "Franklin D. Roosevelt and World War II." *Presidential Studies Quarterly* 34, no. 1 (Mar. 2004): 83–99. https://www.jstor.org/stable/27552565.

Kokomo Tribune (Kokomo, IN), 29 Mar. 1930–4 Sept. 2006.

Lafayette Journal and Courier (Lafayette, IN), 26 Jan. 1929–12 Oct. 1946.

Life (magazine). "Manpower: Sweeping Changes Halt Enlistments, Cut Top Draft Age to 38, Give McNutt Selective Service Control." *Life* 13, no. 25 (21 Dec. 1942): 27. https://books.google.com/books?id=NVEEAAAAMBAJ&pg=PA27#v=onepage&q&f=false.

Logansport Pharos-Tribune (Logansport, IN), 27 Nov. 1935.

Madison, James H., and Lee Ann Sandweiss. "The Great Depression and World War II." Chap. 9 in *Hoosiers and the American Story,* 229–59. Indianapolis: Indiana Historical Society, 2014. https://indianahistory.org/education/educator-resources/curriculum/hoosiers-and-the-american-story/.

Malone, Trey, and Kevin Gomez. "Hemp in the United States: A Case Study of Regulatory Path Dependence." *Applied Economic Perspectives and Policy* 41, no. 2 (June 2019): 199–214. https://doi.org/10.1093/aepp/ppz001.

Michelin. "An Unknown Object: The Tire—Materials." The Tire Digest (website), accessed 20 Apr. 2022. https://thetiredigest.michelin.com/an-unknown-object-the-tire-materials.

Mitchell Tribune (Mitchell, IN), 13 Aug. 1936.

Moffatt, Mike. "Farming Post World-War II." ThoughtCo. (website). Dotdash Meredith Publishing, 27 Jan. 2020. https://www.thoughtco.com/farming-post-world-war-ii-1146852.

Moore, Gary. "When Johnny Comes Marching Home." *The Friday Footnote* (blog). North Carolina State University, 12 July 2019. https://footnote.wordpress.ncsu.edu/2019/07/11/when-johnny-comes-marching-home-7-12-2019/.

Moore, Sam. "U.S. Farmers during the Great Depression." Farm Life (website). Farm Collector, 19 Sept. 2011. https://www.farmcollector.com/farm-life/u-s-farmers-during-great-depression/.

Muncie Evening Press (Muncie, IN), 25 Feb. 1933–20 Aug. 1945.

Myers, Peter. "Soil Conservation Service at 50." *Hoosier Farmer* 70, no. 3 (Mar. 1985): 9. Hicks Undergraduate Library, Purdue University Libraries, West Lafayette, IN.

Nakano Conner, Nancy. "From Internment to Indiana: Japanese Americans, the War Relocation Authority, the Disciples of Christ, and Citizen Committees in Indianapolis." *Indiana Magazine of History* 102, no. 2 (June 2006): 89–116. https://scholarworks.iu.edu/journals/index.php/imh/article/view/12193.

National Academy of Engineering. "Electrification History 2: Rural Electrification." Greatest Engineering Achievements of the 20th Century (website). National Academy of Sciences, accessed 13 May 2022. http://www.greatachievements.org/?id=2990.

National Rural Electric Cooperative Association (website). "History: The Story Behind America's Electric Cooperatives and NRECA," accessed 16 Apr. 2022. https://www.electric.coop/our-organization/history.

Noblesville Ledger (Noblesville, IN), 19 Sept. 1939–5 Jan. 1942.

Olmstead, Alan L., and Paul W. Rhode. "Reshaping the Landscape: The Impact and Diffusion of the Tractor in American Agriculture, 1910–1960." *Journal of Economic History* 61, no. 3 (Sept. 2001): 663–98. https://www.jstor.org/stable/2698132.

Palladium-Item (Richmond, IN), 1 June 1939–18 Jan. 1945.

Palmer, Mitchell V., and W. Ray Waters. "Bovine Tuberculosis and the Establishment of an Eradication Program in the United States: Role of Veterinarians." *Veterinary Medicine International* 2011, article ID 816345 (17 May 2011). https://doi.org/10.4061/2011/816345.

Preston Feather (website). "It's Who We Are," accessed 20 Apr. 2022. https://prestonfeather.com/about-us/.

Princeton Clarion-News (Princeton, IN), 1 Jan. 1942–7 Jan. 1946.

Purdue University. *Agricultural Experiment Station Annual Report,* 1909. Virginia Kelly Karnes Archives and Special Collections Research Center, Purdue University Libraries, West Lafayette, IN.

———. *County Extension Annual Reports,* 1920–1946. In College of Agriculture Extension Reports Collection, 1913–1969. Virginia Kelly Karnes Archives and Special Collections Research Center, Purdue University Libraries, West Lafayette, IN.

———. *Department of Agricultural Extension Annual Reports,* 1917–45. Virginia Kelly Karnes Archives and Special Collections Research Center, Purdue University Libraries, West Lafayette, IN.

Putney, Bryant. "Farm Tenancy in the United States." *Editorial Research Reports* 1 (1935): 201–19. Washington, DC: CQ Press, 1935. Accessed through CQ Researcher. http://library.cqpress.com/cqresearcher/cqresrre1935032000.

Randell, C. G. *Cooperative Marketing of Livestock in the United States by Terminal Associations.* U.S. Department of Agriculture Technical Bulletin No. 57. Washington, DC: Government Printing Office, Feb. 1928. https://naldc.nal.usda.gov/download/CAT86200054/PDF.

Rens, Dennis. "America's Hemp King." Newhead News (website), Feb. 1995. http://newheadnews.com/hemp/Rens.hempstory.Wis/.

Reporter-Times (Martinsville, IN), 15 Oct. 1942–19 Sept. 2010.

Republic (Columbus, IN), 1 Jan. 1942–6 May 1944.

Richmond Item (Richmond, IN), 18 July 1930–3 Dec. 1938.

Richmond Palladium (Richmond, IN), 15 Aug. 1930–28 Jan. 1937.

Roosevelt, Franklin D. "Day of Infamy" (speech), 8 Dec. 1941. U.S. Library of Congress transcript. http://hdl.loc.gov/loc.afc/afc1986022.ms2201.

Rushville Republican (Rushville, IN), 16 Aug. 1930–12 July 1946.

Scherber, Annette. "Corn, Tomatoes, and POWs: Hoosier Agriculture during World War II." *Indiana History Blog.* Indiana Historical Bureau of the Indiana State Library, 17 Nov. 2016. https://blog.history.in.gov/corn-tomatoes-pows-hoosier-agriculture-during-world-war-ii/.

Scott County Journal (Scottsburg, IN), 14 May 1942–28 Dec. 1944.

Scranton Times (Scranton, PA), 13 May 1942.

Shepherd, Geoffrey, J. C. Purcell, and L. V. Manderscheid. *Economic Analysis of Trends in Beef Cattle and Hog Prices.* Iowa Agricultural Experiment Station Research Bulletin No. 405. Ames, IA: Iowa State College of Agriculture and Mechanic Arts, Jan. 1954. https://dr.lib.iastate.edu/handle/20.500.12876/62744.

Simpich, Frederick. "Farmers Keep Them Eating." *National Geographic Magazine* 83, no. 4 (Apr. 1943): 435–58. https://www.google.com/books/edition/The_National_Geographic_Magazine/4U6AAAAAMAAJ?hl=en&gbpv=1&pg=PA435&printsec=frontcover.

South Bend Tribune (South Bend, IN), 28 July 1933–23 Aug. 1945.

Star Press (Muncie, IN), 12 Mar. 1930–19 Aug. 1944.

Stewart Seeds. *Stewart's 100th Anniversary.* Video, 3:07. 26 July 2018. https://www.stewartseeds.com/en-us/history.html.

Stout, W. B. *Sheep and Lamb Price Facts.* Purdue University Agricultural Extension Leaflet No. 166. West Lafayette, IN: Purdue University, Aug. 1932. https://earchives.lib.purdue.edu/digital/collection/AESLeaflet/id/1214/rec/186.

Sundin, Sarah. "Make It Do: Tire Rationing in World War II." *Sarah Sundin* (blog), 5 Jan. 2022. http://www.sarahsundin.com/make-it-do-tire-rationing-in-world-war-ii/.

———. "World War II War Bonds." *Sarah Sundin* (blog), 29 Nov. 2017. https://www.sarahsundin.com/world-war-ii-war-bonds/.

Taylor, John. "Report of the Bureau of Dairy Products." In *Year Book of the State of Indiana for 1933,* 666–73. Indianapolis: Wm. B. Burford, 1934. https://indianamemory.contentdm.oclc.org/digital/collection/p15078coll21/id/101361.

Times (Munster, IN), 15 Oct. 1936–9 Aug. 1945.

Times-Mail (Bedford, IN), 29 Dec. 1942.

Tipton Daily Tribune (Tipton, IN), 20 Jan. 1937–19 Jan. 1945.

Tran, Huong N., Liu-hsiung Chuang, and Carolyne L. Guss. *Natural Resources Conservation Laws: A Report on 17 States and Their Selected Counties and Townships.* U.S. Department of Agriculture—Natural Resources Conservation Service. N.p., July 1999. https://www.google.com/books/edition/_/d8vuAAAAMAAJ?hl=enhttps://www.google.com/books/edition/Natural_Resources_Conservation_Laws/d8vuAAAAMAAJ?hl=en&gbpv=1&dq=huong,+N+Tran+and+Liu+Chuang&pg=PR3&printsec=frontcover - v=onepage&q=huong%2C%20N%20Tran%20and%20Liu%20Chuang&f=false.

Tribune (Seymour, IN), 9 Jan. 1942–19 Jan. 1945.

Union City Times-Gazette (Union City, IN), 25 Jan. 1944.

U.S. Congress. *Department of Agriculture Appropriations for 1955: Hearings before the Subcommittee of the Committee on Appropriations, House of Representatives.* 83rd Congress, second session, 1954. Part 3, statement of the secretary, the budget for the Department of Agriculture. Washington, DC: Government Printing Office, 1954. https://www.google.com/books/edition/Department_of_Agriculture_Appropriations/B0KZvwEACAAJ?hl=en&gbpv=1&pg=PA852-IA7&printsec=frontcover.

U.S. Department of Agriculture. *Agricultural Statistics, 1944.* Washington, DC: Government Printing Office, 1944. https://downloads.usda.library.cornell.edu/usda-esmis/files/j3860694x/sb397c286/9z903270s/Agstat-04-23-1944.pdf.

———. *Hemp for Victory.* Film. 1942. https://www.youtube.com/watch?v=d3rolyiTPr0.

———. "Specified Livestock and Livestock Products on Farms and Ranches" (PDF). In Indiana Census of Agriculture, 1935, pp. 148–51. At Census of Agriculture Historical Archive (website), accessed 18 May 2022. https://agcensus.library.cornell.edu/census_year/1935-census/.

U.S. Department of Agriculture and Purdue University Agricultural Experiment Station. *Indiana Crops and Livestock: Annual Crop Summary, 1932.* Bulletin No. 87. West Lafayette, IN: U.S. Department of Agriculture and Purdue University, Dec. 1932. https://www.nass.usda.gov/Statistics_by_State/Indiana/Publications/Annual_Statistical_Bulletin/hist_pdf/as_1932_ebook.pdf.

———. *Indiana Crops and Livestock: Annual Livestock Summary, 1935.* Bulletin No. 113. West Lafayette, IN: U.S. Department of Agriculture and Purdue University, Feb. 1935. https://www.nass.usda.gov/Statistics_by_State/Indiana/Publications/Annual_Statistical_Bulletin/hist_pdf/as_1935_ebook.pdf.

U.S. Department of State. "Lend-Lease and Military Aid to the Allies in the Early Years of World War II." Office of the Historian (website), mid-2016. https://history.state.gov/milestones/1937-1945/lend-lease.

U.S. Department of Veterans Affairs. "World War II Veterans by the Numbers." VA fact sheet, circa Sept. 2003. http://dig.abclocal.go.com/ktrk/ktrk_120710_WWIIvetsfactsheet.pdf.

U.S. Selective Service System (website). "Induction Statistics: World War II," accessed 20 Apr. 2022. https://www.sss.gov/history-and-records/induction-statistics/.

U.S. War Production Board Bureau of Industrial Conservation. "America Needs Your Scrap Rubber." Poster. Washington, DC: U.S. Government Printing Office, 1942. https://www.mnhs.org/mgg/artifact/rubber.

Utah State University National Center for Agricultural Literacy. "From Defeat to Victory, 1930–1949." Growing a Nation (website), accessed 13 May 2022. https://growinganation.org.

Vidette-Messenger of Porter County (Valparaiso, IN), 24 June 1931–1 Dec. 1942.

Vincennes Sun-Commercial (Vincennes, IN), 26 Aug. 2012.

Waymarking.com. "Camp Thomas A. Scott, Fort Wayne, Indiana." Groundspeak Inc., 5 Oct. 2007. https://www.waymarking.com/waymarks/WM2B13_Camp_Thomas_A_Scott_Fort_Wayne_Indiana%20for%20information%20on%20Camp%20Scott.

Weimberg, Gary, and Catherine Ryan. *Soldiers of Conscience.* Film. *POV.* 16 Oct. 2008. http://archive.pov.org/soldiersofconscience/background/.
Whitford, Frederick. *Scattering the Seeds of Knowledge: The Words and Works of Indiana's Pioneer County Extension Agents.* West Lafayette, IN: Purdue University Press, 2017.
Wilsie, C. P., E. S. Dyas, and A. G. Norman. *Hemp: A War Crop for Iowa.* Iowa Agricultural Experiment Station and Agricultural Extension Service Bulletin No. P49. Ames, IA: Iowa State College of Agriculture and Mechanic Arts, Dec. 1942. https://dr.lib.iastate.edu/handle/20.500.12876/11999.
Wilson, Frank C. "Report of Dairy Products Bureau." In *Year Book of the State of Indiana for 1931,* 232–47. Indianapolis: Wm. B. Burford, 1932. https://indianamemory.contentdm.oclc.org/digital/collection/p15078coll21/id/98082.
Young, H. E. *25 Years of Extension Work in Indiana: Historical Narrative and Achievement Summary.* West Lafayette, IN: Purdue University, 1939. https://babel.hathitrust.org/cgi/pt?id=coo.31924000899157&view=1up&seq=5.

Index

Page numbers in bold refer to photos or captions.
Schools, townships, cities, and towns can be found under county entries.
Business names can be found under Businesses (specific), Canneries (specific companies), Dairies, *and* Seed corn dealers

Abbreviations

WWI World War I
WWII World War II